ECOLOGICAL EXPLOSIONS

ECOLOGICAL EXPLOSIONS

The History of Biological Invasions and Invasion Science

DANIEL SIMBERLOFF

The University of Chicago Press
Chicago and London

The University of Chicago Press, Chicago 60637
The University of Chicago Press, Ltd., London

Published 2025
Printed in the United States of America

34 33 32 31 30 29 28 27 26 25   1 2 3 4 5

ISBN-13: 978-0-226-84256-1 (cloth)
ISBN-13: 978-0-226-84257-8 (paper)
ISBN-13: 978-0-226-84258-5 (ebook)
DOI: https://doi.org/10.7208/chicago/9780226842585.001.0001

Library of Congress Control Number: 2025015475

♾ This paper meets the requirements of ANSI/NISO Z39.48-1992 (Permanence of Paper).

Authorized Representative for EU General Product Safety Regulation (GPSR) queries:
Easy Access System Europe—Mustamäe tee 50, 10621 Tallinn, Estonia,
gpsr.requests@easproject.com
Any other queries: https://press.uchicago.edu/press/contact.html

Contents

Preface

Invasion science—the study of organisms that people moved deliberately or inadvertently to new locations—is now a well-established, burgeoning discipline. But the modern field began recently (the 1980s) and the discipline from which it primarily arose (ecology) is itself a relatively young science, dating only to the turn of the twentieth century. Many people wrote earlier about non-native species and their impacts on agriculture or forests, and before the advent of a science of ecology some even wrote about what we would now call "ecological impacts." But with few exceptions, notably Charles Elton in the 1930s through the 1950s, they focused on particular species in particular locations, not on biological invasions as a general phenomenon with varied manifestations worldwide.

Not until the 1980s did scientific study of biological invasions finally began to coalesce into a modern science. Throughout that decade, many students of non-native species and their impacts assembled in a series of regional and international workshops—the Scientific Committee on Problems of the Environment (SCOPE) Programme—focused tightly on invasions, resulting in widely noted books and journal articles. Since then, invasion science has evolved rapidly, with a greatly expanded array of questions and technologies; new journals and textbooks wholly devoted to invasions; ecology and conservation biology textbooks assigning increasing attention to invasions; and high-profile reports of many specific invasions in popular media. During the years I have been writing this book, in North America alone, a lionfish, the emerald ash borer, bighead and silver carp, the Burmese python, and the spotted lanternfly have all been widely publicized. By the time this book is published, others will surely be similarly publicly reviled. The devastating fires on Maui in 2023 brought wide public attention to the alarming flammability of buffel grass and Guinea grass, both recognized as long-standing invasions in Hawaiian agricultural fields and natural areas. I am certain that, in the next few years, other older invasions will feature in popular media because of new or newly detected impacts.

As invasion science crystallized into a scientific field at the turn of the twenty-first century, efforts to manage invasions and limit their impacts also developed rapidly. Chemical, biological, and physical approaches were employed in the nineteenth century to poison, eat, infect, or uproot invasive species. In the mid-twentieth century, researchers occasionally proposed genetic approaches. But the rapid advent of molecular genetic technologies, starting in the late 1950s as modern invasion science exploded onto the scene, led to suggestions for new methods to control or even to eradicate invasive populations. These ideas galvanized researchers to rush to develop and implement genetic methods. Simultaneously, research on traditional methods—especially biological and chemical control—ramped up. Several proposed genetic approaches have triggered controversies in the broader arena of public opinion as well as in academic circles, just as controversies have roiled invasion science even before the modern era. Most of these controversies have arisen from outside the ranks of biologists who compose the core of invasion science, and some critics argue that invasion impacts are overblown and question the very notion that invasions should be managed even if their effects are recognized.

I first became intrigued by biological invasions during my graduate years in the 1960s when I noticed Elton's *The Ecology of Invasions by Animals and Plants* in the library of my doctoral advisor, Edward O. Wilson. He suggested it was a really interesting book and lent it to me to read (I am in good company—I later found a 1958 letter to Ed from Rachel Carson thanking him for directing her to the book). My own research in the 1970s and 1980s, largely in the areas of community ecology and biogeography, often led me to consider not only the frequency with which humans have moved species around but also how non-native species would seem to be useful in testing various ecological theories, especially at the community level.

However, two activities in the 1980s riveted my attention directly on the invasion phenomenon. In 1986, I began serving on the Board of Governors of The Nature Conservancy (TNC), and at each of our board meetings one or more land stewards would give presentations on the challenges managing their respective reserves. It quickly became apparent that most of the challenges involved impacts of non-native species, usually plants, and the heavy, ongoing commitment of resources devoted to managing them. Additionally, TNC land stewards and science staff sought me out during board meetings simply to discuss the ecology and problems in their preserves.

The prominence of invasions in management problems and the frequent complexity of their ecological role again focused my attention even more sharply on invasions. I came to realize just how major a part they play in the issues that have fostered and driven the conservation and environmental movements.

In addition, I was fortunate to be invited to two key meetings in the SCOPE Programme on the Ecology of Biological Invasions, which served as the springboard for modern invasion science. The first, in 1984 in Asilomar, California, focused on North America and Hawaii, and the second, in 1986 in Honolulu, engaged scientists from around the world who had participated in earlier regional workshops as part of the SCOPE Programme. These meetings afforded me the opportunity to meet scientists who had been studying invasions, including leading ecologists of the time. I was able to learn details of invasions of many types from many parts of the world and to participate in discussions that led to a synthesis of sorts, a common framework and set of questions that were the initial basis for modern invasion science.

My original plan for this book was simply to trace the scientific study of invasions, from the nineteenth century through Elton's 1958 monograph and ending with the SCOPE Programme. However, as I gathered materials and started writing, I realized that a better treatment would expand the ambit in several ways. First, what counts as "science" as opposed to simple observation and recording is not so clear, and several authors before the nineteenth century reported on invasions and their impacts without a recognized "scientific" context in which to embed their observations and conclusions. By the nineteenth century, when scientists clearly writing in a scientific tradition discussed invasions, they sometimes cited these earlier "prescientific" authors. It thus seemed that a proper rendering of the history of invasion science could not omit these early reports. Second, observations on invasions and their impacts could not have been presented unless invasions had occurred in the first place. An account of when and where invasions occurred, how and when non-native species first came to be recognized as non-native, and when and where non-native species were initially recognized as causing impacts therefore also seemed to be necessary parts of my account of the history of the field. Finally, the expansion of modern invasion science in the four decades since its inception has been rapid and dramatic. Because my research has largely explored this field during this remarkable evolution, I could not help but think that I would have either to write a second, post-SCOPE book or to include the entire history of the field, up to the present, in a single volume. I have chosen the second course, but that has entailed the frustrating problem that developments in invasion science arise so quickly that no sooner had I completed a chapter than new findings made it seem incomplete. I had to draw a line somewhere and have done so at the end of 2023, approximately eight years after I began. For readability, I have used species common names throughout the book. Scientific names are listed in the species index.

Many persons aided me by discussing parts of the book, reading sections, and providing facts and references. Among them are my students Sara Kuebbing, Martín Nuñez, Rafael Zenni, and Wieteke Holthuijzen, my postdoctoral fellow Sara Campbell, and my University of Tennessee colleagues Derek Alderman, Joe Bailey, Jim Fordyce, Laura Russo, and Jen Schweitzer. Several other persons provided great assistance on sections of the book: Leigh Boardman, Lauren Buckley, Jim Carlton, Angela Chuang, Hall Cushman, Rocky Gutierrez, Sandy Liebhold, Monica McCard, Laura Meyerson, Fritz Mulhauser, Tony Ricciardi, Louise Robbins, Don Schmitz, and Dick Tracy. Others answered questions about details of events and developments: David E. Brown, Rook Cleary, Michael Day, Jim Drake, Piero Genovesi, Roger Kitching, Ingolf Kühn, Mark Lonsdale, Richard Mack, Hal Mooney, Bill Murdoch, Bruce Schulte, Julianne Lutz Warren, and the late David Pimentel. Jennifer Dodd kindly constructed a figure from her research materials, and Joseph Antoine of Newgen rendered my suggestions and scribblings into elegant figures. Joseph Calamia (executive editor) and Matt Lang (editorial associate) at the University of Chicago Press provided extensive and prompt guidance on all stages of bringing this manuscript to completion.

I owe a special debt to the staff of the John C. Hodges Library of the University of Tennessee. Reference librarian Donna Braquet time and again aided me in literature searches involving obscure references as well as databases not readily accessible or sometimes not even known to me. The Resource Sharing (formerly Interlibrary Services) librarians have rapidly and successfully processed dozens of requests each year, including several arcane and ancient works held by few collections anywhere in the world. I could not have completed this project without this assistance. At the University of Tennessee, I hold an endowed chair—the Nancy Gore Hunger Professorship in Environmental Studies—that has supported all aspects of the research underlying this book and has afforded me the time to write it.

I am particularly grateful to Janice Moore and Nathan Sanders, my former postdoctoral fellows and longtime colleagues, for reading and rereading large portions of this book, and to my cousin Carol Bodian for reading and copiously criticizing all the chapters. They all know that I have not accepted some of their advice, but they also know I have accepted much of it, and they have continued to encourage me despite my obstinacy. Finally, my daughters, Ruth and Tander Simberloff, and my wife, Mary Tebo, have all read, edited, commented on, and reacted to this manuscript in various ways over many years, and I appreciate their unflagging support.

1

Introduction

Invasion science encompasses research on species that human actions have brought to new regions distant from their original geographic ranges. Nowadays biological invasions are recognized as one of the great ongoing global changes, along with climate warming, land use changes, and modified biogeochemical cycles, affecting both nature and human societies in myriad ways. Biological invasions have caused many species to go extinct and are one of the main threats to currently imperiled species.[1] They exact an enormous economic toll, tens of billions of dollars annually, and have even spawned a discipline of invasion economics.[2] The cost to human health is staggering, as witness the recent global spread of Zika virus and COVID-19.[3]

Humans have deliberately or inadvertently moved species around for millennia. Most people who study biological invasions consider the phenomenon and associated impacts to have arisen with the discovery of the New World by Europeans and the subsequent spread of European mercantilism and colonialism, depicted as the "Columbian exchange" and "ecological imperialism" in masterful works by environmental historian Alfred Crosby (more on him later, in chap. 8, sec. 8.3).[4] But early transport of various species into new regions consisted of far more than the establishment of "neo-Europes" in the last thousand years.[5] Inhabitants of the Indian subcontinent and eastern Africa exchanged many food plants between 2500 BC and 100 CE, and both animals and plants were spread around the Mediterranean at least as far back as the earliest years of the Christian era.[6]

However, the rate at which humans move species has increased enormously over the last few centuries, first as long-distance trade and travel by sailing ships increased, then as steam power in the 1800s and then petroleum fuels greatly increased the speed with which ships traversed oceans, and still later when airplanes reduced the travel time between any two points on the globe to a matter of hours. Not only have the volume of cargo and the number of passengers moving intercontinentally increased manyfold, especially in the twentieth century, but the great decrease in transit time means that a living organism is vastly more likely to survive than if the voyage had entailed a week or a month at sea.[7] Furthermore, the routes traversed by potential invaders have proliferated dramatically. Even occasional autonomous intercontinental dispersal events, such as that of the cattle egret from Africa to South America in the late nineteenth and early twentieth centuries, are aided by natural wind and water currents. Nowadays air and sea routes link distant continents and islands last accessible to one another more than one hundred million years ago before continents drifted apart.

Though humans moved species around thousands of years ago, and this movement increased in the last few centuries, concern with how species introduced by humans affect both nature and human welfare is much more recent. A few eighteenth-century travelers and scientists noted that particular non-native species damaged human enterprises (especially agriculture), but only during the nineteenth century did such observations became commonplace, as increasing intercontinental travel was greatly increasing the deliberate and inadvertent spread of species. At the same time, naturalist-explorers established an increasingly mature field of biogeography, depicting which species were found where. They often also noted species in places where they knew or suspected humans had brought them. Research on how invasions came about and what impacts they have had could not have been conducted without recognition that certain populations of some species are, in fact, non-native, and this recognition was not well developed until the nineteenth century for much of the world.

These mounting reports of harmful impacts of non-native species did not coalesce into a science of invasions in the nineteenth century for two main reasons. First, there was no discipline of ecology in the nineteenth century. Although both Carolus Linnaeus in the nineteenth century and Charles Darwin in the nineteenth century discussed many topics that we recognize today as ecological, no formal science of ecology existed until the end of the nineteenth century, so the increasingly noted ecological impacts wrought by non-native species could not be incorporated into a discipline with a clearly identified scientific literature. The German scientist Ernst Haeckel coined the term "ecology" in 1866, essentially bringing Darwinian evolutionary thinking to Linnaeus's static conception of species interacting with one another and with the physical environment.[8] It was not until the late nineteenth century that monographs in ecology first appeared, and ecology journals were first published in the early twentieth century. A second reason

Table 1.1 Numbers of Articles About Biological Invasions in Selected Leading Newspapers in the United States, Canada, and the United Kingdom, 1980–2019

	Decade			
Newspaper	*1980–1989*	*1990–1999*	*2000–2009*	*2010–2019*
New York Times	36	130	326	595
Wall Street Journal	2	8	43	132
Chicago Tribune	13	167	253	696
Los Angeles Times	30	144	228	234
Globe and Mail (Canada)	10	48	159	177
Times (UK)	3	67	177	446
Guardian (UK)	7	51	174	567

that a science of invasions did not arise in the nineteenth century is that observations of impacts at that time were, with a few exceptions, restricted to particular species in particular places affecting particular human activities or native species, without apparent recognition that such impacts were widespread and might be fruitfully examined in a general way.

However, reports of invasions increased in the early twentieth century, and the science of ecology matured rapidly. The early twentieth century saw several striking precursors to a science of invasions (as discussed in chap. 6), but only with Charles Elton's now famous 1958 book, *The Ecology of Invasions by Animals and Plants*, was a fully developed invasion science envisioned (as detailed in chap. 7, sec. 7.1). Elton described biological invasions as "ecological explosions," because they consist of "the bursting out from control of forces that were previously held in restraint by other forces."[9] However, although Elton's book was later cited as the founding document of the modern field, for twenty-five years after publication it really did not influence researchers to study invasions more systematically, nor did it catalyze formation of a full-fledged science.[10] The field did not crystallize into a modern science until a program in the 1980s of the international Scientific Committee on Problems of the Environment (SCOPE) assembled many scientists worldwide to focus on a wide range of invasions and their impacts (detailed in chap. 9).[11] Today entire textbooks and journals are devoted to invasion science, indicating the explosive growth of a burgeoning discipline (chronicled in chap. 10). Public interest in biological invasions has also increased greatly, beginning with a dramatic rise from the 1980s to the 1990s (see table 1.1), the same period during which scientific interest suddenly accelerated.[12]

This book describes the development of invasion science from its precursors through Elton and SCOPE to the present. First, however, the short definition of the field—"research on species that have arrived, by virtue of human activity, in new regions distant from their original geographical ranges"—must be unpacked, particularly as regards what "by virtue of human activity" means. Humans deliberately move species around for many reasons—as foodstuffs, as pets and ornamental plants, as game animals and sport fish. There have even been charges that various crop pests, such as the Colorado potato beetle (in fact, native to Mexico), have been dropped on non-native enemy territory during both hot and cold wars. Although evidence of such martial releases is ambiguous, governments did develop programs for this purpose.[13]

Humans also inadvertently introduce species to new regions. Insects hitchhike on non-native plants or in associated soil; parasites accompany non-native pets and game animals. Seeds, insects, and other organisms find their way into shipping containers; snails ride on ceramics and paving stones. But humans can less directly facilitate arrival of a species in a new and distant place. Tree plantings across the North American Great Plains formed stepping stones and corridors used by several eastern birds that reached the West.[14] The barred owl, currently threatening the endangered northern spotted owl in the Pacific Northwest through competition and possibly hybridization, likely crossed the Great Plains by virtue of activities by European settlers: tree plantings, devastation of native ungulates and beavers that suppressed trees, and fire suppression.[15] Humans also deliberately introduce species to stepping stones from which they disperse autonomously to nearby regions. The collared dove, brought from Eurasia to the Bahamas, escaped from captivity, easily reached Florida on its own, and from Florida spread to much of the United States.[16] An even more indirect influence of human activity is anthropogenic global warming, which facilitates the arrival and establishment of species in new locations (described in chap. 14, sec. 14.3).

Occasionally a species arrives in a distant region with no human assistance. Perhaps the best-known example is that of the cattle egret, a strong flier from Africa occasionally sighted in South America and finally establishing a population there in the early twentieth century, then spreading rapidly to much of the Americas (fig. 1.1).[17] Molecular genetic methods have recently revealed several such surprising intercontinental dispersal events in the distant geological past, before humans could have played a role.[18] Such cases have led some critics of invasion science to argue that the entire discipline is of little importance and overblown (as detailed in chap. 18), as new populations of species establishing in distant places constitute a "natural" phenomenon.[19] However, over the past few centuries, in the overwhelming majority of cases of long-distance dispersal that result in newly established populations, dispersal was aided by humans.[20] Although a broad understanding of impacts (or lack thereof) of non-native species can be enlightened by

Fig. 1.1 Spread of the cattle egret in the Americas from the 1940s by three routes. Credit: Adapted from Trinidad-Domínguez et al., "Geographic and Ecological Invasion of the Cattle Egret."

study of cases such as that of the cattle egret, they are not a major feature of invasion science, and this book will emphasize species whose arrival in new, distant sites was facilitated by humans.

The argument that species distributions have always changed, so biological invasions are not a new phenomenon deserving of scientific attention and public concern, ignores the staggering increase in the rates of change.[21] The flora of the Hawaiian Islands constitutes an excellent example.[22] Before Polynesians arrived between one and two thousand years ago, about one plant species from elsewhere established a population in the archipelago every one hundred thousand years.[23] Between the colonization by Polynesians and the arrival of Captain James Cook in 1778, the rate rose to about one newly established introduced plant species every fifty years. Subsequently, the rate rose to about twenty-two species each year![24] Similar massive increases in invasion rates have been estimated for groups of freshwater and marine species.[25] It is disingenuous to claim that modern biological invasions are just "the same old same old."

Part 1 of this book focuses on events and activities leading to the establishment of modern invasion science in the 1980s. Part 2 describes the evolution and influence of this new science. To do so requires brief discussion of the histories of several other sciences (e.g., evolution, climate science) to explain how they came to interact with invasion science. I then indulge in speculation on how invasion science will develop in the near future.

1.1 Definitional Matters: Invasive, Invasion, Naturalized, Introduced, Native, Non-Native, Alien

The term "invasive" has been used in various ways by scientists and even more loosely by lay media.[26] For scientists, most of the variation derives from whether an established non-native population has spread from its initial point of establishment, what sorts of habitats it has spread into, and whether it affects species or ecological processes in the sites into which it has spread. Virtually all workers in the field require that a species has spread in order to be termed "invasive." Some, but not all, also require that the sites into which a species has spread include "natural" sites and not simply human-modified systems such as lawns or reservoirs. And some of the latter scientists would also require that the spreading species have some sort of an impact on the species or ecological processes in a natural habitat in order to qualify as "invasive." In the scientific literature, it is almost always clear from context what authors mean by "invasive." In this book, "invasive" will mean spreading into quite natural habitats but not necessarily having a documented impact (yet!) on the resident species or ecosystem.

It is also important to distinguish between "non-native" and "invasive." Invasion scientists have traditionally restricted the term "invasive" to non-native species that reach areas distant from the introduction site with human assistance, reproduce there, then spread on their own, as previously noted. To define "non-native," we must define "native," and biologists have done so in a variety of ways.[27] Invasion scientists generally invoke two criteria for a site to be part of species' native range.[28] First, the species must not have arrived at the site with human assistance, either directly, by deliberate transport, or indirectly, by some human activity that removes a previously impassable barrier. Constructing a canal to link two oceans or river systems would be such an activity, as would the actions described earlier that allowed eastern bird species to reach the western United States. Second, there must be some date in the past such that all sites occupied by that time without human assistance are considered parts of the native range. By convention, many invasion scientists view 1492, the beginning of the Columbian exchange, as that date. However, the important point is not that specific date so much as that an author be specific when discussing a human-assisted introduction earlier than this. Thus, for instance, many European scientists refer to species recognized as introduced by humans but before 1492 as archeobiota and those introduced after 1492 as neobiota.[29] With these criteria, in a few cases it will still not be straightforward to classify an occupied region as part of the native range or not.[30] An example is a site from which a species has been extirpated by human activity, or has long been absent for unknown reasons, but to which humans have reintroduced the species. However, in the great majority of cases, in principle a classification is possible, although absence of data may make it impossible in practice, at least temporarily.

In fact, it is populations and not species that may be non-native or invasive, or both.[31] Any species is native somewhere. And a species is not generally invasive in its native range, although some species in their native ranges may do things under certain circumstances that might be termed "invasive" if the population exhibiting such characteristics were non-native (see chap. 18, sec. 18.4). However, the terms "non-native species" and "invasive species" are so well-entrenched in the scientific and popular literatures that I will use them, asking readers to bear in mind this technical detail.

Under this definition, not all non-native species are invasive. A non-native ornamental plant species maintained by humans in backyards but that has not spread autonomously would not be termed invasive. Often biologists have described a tripartite hierarchy for non-native species—introduced, naturalized, invasive—in which an introduced population of a species is simply a persistent population of any non-native species whose arrival at a site was facilitated by humans, whether or not its persistence requires human assistance. According to David Richardson and colleagues, a "naturalized" population is an introduced population that survives and reproduces without human assistance but need not have spread into a "natural" setting.[32] Others have long termed "naturalized" any population of a well-established species known to have been introduced to a country, no matter the habitats it occupies.[33]

Especially in the popular press, non-native populations are often termed "invasive" simply because they were introduced, but invasion scientists generally require that "invasive" populations, once introduced, spread widely and autonomously. Richardson and his collaborators argue that any introduced plant population that meets these criteria be termed "invasive" no matter the type of habitat to which it has spread and whether or not it impacts the species or ecosystem in that habitat.[34] This is how most invasion scientists have used the term, but President Bill Clinton's Executive Order

13112 of February 3, 1999, on Invasive Species sowed confusion in policy circles and popular accounts by requiring an impact. The executive order defined "invasive species" as "an alien species whose introduction does or is likely to cause economic or environmental harm or harm to human health," with "alien" defined as I have previously defined "introduced."[35]

It is probably irrelevant, at least from a historical standpoint, whether or not "invasive" status requires a demonstrable impact. It has long been a commonplace in invasion science that the great majority of non-native species do not have an impact, codified in the early days of modern invasion science by the "tens rule" (see chap. 18, sec. 18.3.1) for non-native plants and animals of Great Britain that has been frequently and loosely applied to invasions in general: one in ten arriving non-native species establish a population, and one in ten of these become pests (i.e., have an impact perceived as noxious).[36] However, three decades of intensive research on non-native populations have revealed many more impacts than had been originally suspected (whether or not anyone views them as noxious) and whole classes of impacts that are subtle and often take decades or longer to be manifested. Furthermore, it is difficult to imagine a newly established population having absolutely no impact on any ecosystem process or any population of a resident native species, if only on decomposers. The European Union definition of "invasive alien species" requires a negative impact to the natural environment or the economy.[37] As will be seen later in this book, much of the history of invasion science and an understanding of invasions deals with the developing recognition of impacts, and it is unimportant in describing that history whether a non-native species is known to have an impact.

One may also ask how distant a newly introduced population must be from the original range to qualify as invasive. No hard-and-fast rule exists. Although intercontinental introductions have received the most attention in research and publicity, humans have moved many species much shorter distances. Even without human facilitation, species' native ranges are dynamic, and innumerable short-distance range changes occur with no obvious link to human activity. However, some small range changes are known or strongly believed to have resulted from deliberate movement by humans. For example, many sport fish species native to the United States now occupy larger ranges than previously because people have carried them into drainages near their original ranges, then subsequently to sites increasingly distant from their original homes.[38] Thus, even though the spread is by small steps, such populations are every bit as non-native as would be an Asian fish in North America. Except by relatively rare geological events such as stream capture, it is unlikely that fish could disperse on their own to this extent.

Many legal instruments and policy documents define "non-native" or "introduced" in terms of a political boundary, especially a national boundary. However, from a biological standpoint, this criterion makes no sense.

Especially in a large nation with biotic communities with distinct evolutionary histories, an introduction within a nation can have ecological consequences as major as those of an international introduction. The introduction of smooth cordgrass from the Atlantic coast of the United States to the Pacific coast had enormous ecological and evolutionary consequences for native species.[39] Many fishes introduced for sport fishing and aquaculture between distinct Brazilian drainages, especially predators, have great potential ecological consequences.[40] With respect to the history of how introductions came to be recognized and studied, the matter of whether an introduction has crossed a political boundary would be of little importance except for the fact that some government agency programs relevant to invasions restrict their focus to species introduced from outside their own nation.

Once a non-native population is established, one may ask how rapidly and how far the species must spread to be termed "invasive." Many non-native populations have remained for decades in the vicinity of initial establishment or spread so slowly as not to be seen as invasive. For plants that spread by seeds, one suggested rule of thumb is that, for a species to qualify as "invasive," it must move of its own volition farther than one hundred meters in less than fifty years, and for plants that spread by roots, rhizomes, or stolons, farther than six meters in three years.[41] Observation of this degree of dispersal would certainly lead a manager to be concerned about impact whether or not the plant had reached more or less "natural" habitats. However, even aside from the arbitrariness of the numbers, the issue is beclouded by the fact that some non-native populations that remain highly localized for decades eventually, and often quite abruptly, begin to spread much more quickly and widely (a phenomenon explored in chap. 18, sec. 18.3.1).[42] But the precise definition of "invasive" is not relevant to the history of how non-native species came to be recognized and studied.

In the context of the spread of non-native species, the term "invasion" is of course a metaphor, and a somewhat controversial one (see chap. 18, sec. 18.2), but it is a metaphor that seemed so obvious and apt that it was used by many of the earliest biologists to concern themselves with impacts of non-native species.[43] The terms "invasion" and "invasive" have also been used by biologists in contexts unrelated to non-native species. In medicine, for instance, a treatment can be termed "invasive," and movements of cells or pathogens within an individual organism may be termed an "invasion." An oncologist has even suggested the analogy of a biological invasion as a fruitful framework for understanding and managing the spread of cancer within a body.[44] However, such invasions are not the focus of this book.

More closely related to the spread of non-native species is the gradual or more rapid dispersal of native species as their ranges expand into contiguous areas. This phenomenon is of particular interest today because of climate change (as discussed in chap. 14, sec. 14.4.3). Thus, biogeographers and paleobiologists have sometimes used the term "invasion" to describe the movement of a species into a new and adjacent habitat, or the replacement of one habitat type by another, such as prairie by forest.[45] Also close to the subject of non-native species is the process of ecological succession, unfolding usually over a period of decades, during which native plant species replace other native plant species, as when a bog gradually transforms into a forest. The replacement is sometimes called "invasion."[46] Some aspects of ecological succession so resemble those of invasion by non-native species that Mark Davis and colleagues called for abolishing the entire field of invasion science as a distinct discipline and for non-native species to be subsumed into the study of plant ecology in general, especially succession.[47] Although many similarities exist between ecological succession and invasion by non-natives, important differences abound as well (as detailed in chap. 18, sec. 18.4).[48] Finally, the rapid, continuous spread of a native species in the wake of anthropogenic habitat change is sometimes termed a "native invasion."[49] As with successional changes, similarities exist between these "native invasions" and the spread of some non-natives, but there are important differences.[50] As invasion science has grown, the similarities to successional changes and "native invasions" have increasingly been recognized and brought to bear on research on mechanisms, dynamics, and impacts of biological invasions.

The arrival, establishment, and spread of non-native species is not always termed "invasion," and non-native species have been called many other things, especially "nonindigenous," "exotic," "alien," and "weeds." The terminology has become part of recent controversies about the field of invasion science (see chap. 18, sec. 18.2), but an understanding of the history of recognition and study of non-native species and their possible impacts has not been hindered by the multiplicity of terms.

This book, then, will describe the history of research on how, when, where, and why invasions occur, on what kinds of impacts ensue, and on how invasive species and their impacts can be managed. The amount of research related to these matters and accessibility of relevant literature is biased toward certain developed nations.[51] This imbalance is discussed in chapter 19, section 19.4.

PART I

People Move Species Around and Eventually Recognize Some Impacts

2

The Early Shuffling of the Biosphere

Although biological invasions are a widely recognized phenomenon today, this was not always the case. People introduced species to new areas thousands of years ago, sometimes deliberately, and some transported species not only established populations but also spread and flourished without human assistance. For deliberate introductions by people who left no written record, we have no way of knowing whether their descendants several generations later, seeing them in nature, even recognized that they had been introduced. Nor do we know whether they understood that certain species damaging human activities, such as insect pests of agriculture or shipworms that devastated boats and docks, were caused by non-native species that had arrived with human help. More than likely, in many if not almost all cases, they were seen instead as plagues deities had visited upon society.

By the seventeenth century, certain species were widely understood to have been introduced, especially on islands, although the degree to which the animals and plants of various regions were non-native did not begin to be grasped until the late eighteenth century as trained naturalist-explorers visited many parts of the globe. Reports that some invasions could be damaging also began to appear in the seventeenth century, but the perceived damage, with a few exceptions, was to human endeavors, and it was not until the nineteenth century that the scope of ecological impacts, especially effects on native species and ecosystems, began to be recognized. This growing perception of biological invasion impacts is described in chapter 3, section 3.2.

2.1 Terrestrial Introductions in Prehistory and Antiquity

Many plants and animals were transported prehistorically, deliberately or inadvertently, from their original geographic ranges to more or less distant regions. Some established persistent and even invasive populations. The best-studied examples entail introductions to islands, because in this case the possibility of such movement of terrestrial and freshwater species without human assistance is usually slim. Sometimes direct evidence of introduction is lacking but a cogent case can be constructed based on the biology of the species and geological history, as for the hypothesis that Austronesian sailors prehistorically introduced tortoises to Indian Ocean islands as future food stores, just as early European sailors in historical times did elsewhere with several domesticated animal species.[1] Although most direct evidence for prehistoric introductions is archeological, genetic information, including both contemporary and especially ancient DNA, is increasingly adduced and has modified previous scenarios based on archeology and greatly increased the number of recognized cases of likely early introduction of nonnative species.[2]

One consequential early introduction was the spread of the Pacific rat from Asia through Melanesia and Polynesia. This invasion began early, with transport of the rat by the Lapita people as a food item from islands in Southeast Asia to the Bismarck Archipelago between 3,400 and 3,200 years ago and thence by their descendants, the Polynesians, to the most distant reaches of Polynesia all the way to Easter Island by about 700 years ago.[3] This rat likely played a key role in the extinction of hundreds of endemic island bird species, especially rails, and the deforestation of Easter Island.[4] The popular story of the person cutting down the last tree on Easter Island[5] should probably be replaced by a story of people watching rats destroy all the seeds of their trees. Pacific rats were possibly a major factor in the destruction of much of lowland native forest in Hawaii and the resulting extinction of dozens of species of Hawaiian endemic birds.[6] The Lapita people and Polynesians also spread chickens, native in Asia, throughout the Pacific.[7] By about 2,000 years ago, the Lapita people had introduced dogs to parts of Near Oceania (islands including the main Solomon Island chain) but not beyond, while the dogs introduced to most of Oceania, including Polynesia, also arrived about 2,000 years ago, but from islands in Southeast Asia, not from Near Oceania.[8] The Polynesians also introduced species inadvertently. For instance, the colonists of Hawaii about 1,000 years ago brought seven lizard species, almost certainly as stowaways in cargo, from Asia or islands of the eastern Pacific—four geckos and three skinks.[9]

Humans brought several mammals to various islands in the Antilles during the Late Ceramic Age (AD 500–1492). The guinea pig from South America was probably introduced to Puerto Rico soon after AD 500, and at various times after that from Puerto Rico to other Antillean islands.[10] The agouti and the opossum were introduced early during the Late Ceramic Age from mainland South America to the Antilles, and interisland transport spread then them widely.[11] They disappeared in historical times from several islands; for instance, agoutis no longer exist on Nevis and Carriacou, perhaps because of habitat destruction or competition with other introduced species after Europeans arrived, or both.[12] The Puerto Rican hutia was actually introduced to Puerto Rico from its native Hispaniola.[13] Several other mammal and bird introductions to the Antilles and interisland transfers within the Antilles during the Late Ceramic Age are suspected.[14]

In the North Atlantic, the pine marten was probably deliberately brought to the Orkney Islands in the Neolithic period (5500–4500 BP), as were red foxes between 500 BC and AD 1000, most likely from nearby Scotland, though both disappeared centuries later.[15] The common vole arrived in the Orkneys, probably stowing away in grain or animal fodder, about 5,000 years ago from continental Europe and not from nearby Great Britain.[16] Red deer were introduced to both the Outer Hebrides and the Orkneys during the Neolithic, although not from adjacent Scotland.[17] They were eliminated from the Orkneys at the end of the first millennium AD.[18] Red deer were also introduced to Ireland during the Neolithic, in this case from Great Britain.[19]

Humans dispersed animals around the Mediterranean, to both islands and bordering mainland areas, at least as far back as the earliest years of the Christian era.[20] For instance, the house mouse, inadvertently carried around the world by humans, spread from the area in which its commensal habit evolved (the Fertile Crescent) into the eastern Mediterranean, both mainland and islands, with the spread of early Neolithic farming practices, and then, beginning in antiquity and continuing to the present, through the western Mediterranean.[21] The original large mammal faunas of the five large Mediterranean islands—Cyprus, Crete, Sardinia, Corsica, and Majorca—were completely replaced by anthropogenic introductions, most of them intentional and prehistoric.[22] Some died out after lengthy residences, but most still persist, at least in domesticated form. For instance, on Cyprus, by 8000 BC, goats and pigs were introduced and had become feral, as had fallow deer. Soon they were joined by red foxes, and about 1000 BC by least weasels. The other four large islands received similar introductions mostly in the Neolithic, but somewhat later, and with a few different species—e.g., the badger from Europe and the stone marten on Crete, the edible dormouse on Corsica, red deer on Sardinia, hedgehogs on Crete, Sardinia, and Corsica. Several smaller mammal species that arrived during the Neolithic and Bronze Age were almost surely stowaways.[23]

The house cat is a particularly interesting case.[24] The domesticated form arose from the wildcat of the Middle East and was transported to Cyprus by 7500 BC, probably deliberately as a mouser or pet, or both. Domestic cats reached much of the rest of the world by deliberate transport originally from Egypt, where cat domestication was an enthusiastic activity and the cat became venerated as the official deity of Egypt (the goddess Bastet). Although their export was officially forbidden, by 500 BC cats had been brought to Greece, and by 2,000 years ago, ships carrying grain sailed directly from Alexandria to many sites in the Roman Empire carrying cats on board as ratters. Worldwide, domesticated cats have established feral populations and are a conservation scourge.[25] The Egyptian mongoose was another Roman introduction to Europe, brought to the Iberian Peninsula during the Roman occupation, at least as early as the first century AD, possibly for rodent control.[26]

Although at least as many plants as animals were probably deliberately or accidentally introduced prehistorically, records are often not as clear-cut, particularly of accidental introductions. One problem, especially for herbaceous species, is the difficulty of identifying pollen to species in palynological deposits.[27] Additionally, human activity often rearranges sediment layers; this is also a problem in interpreting animal remains. Nevertheless, unassailable evidence shows that humans have frequently moved plants, often food plants, great distances. During the Archaic Period (8000–1000 BC), early inhabitants of the Antilles brought avocado and sapodilla from Mexico or Central America to the islands.[28] Inhabitants of the Indian subcontinent and eastern Africa exchanged food plants between 2500 BC and AD 100, while humans long spread food plants around the Mediterranean, including cereals to Cyprus at least 10,000 years ago.[29] Polynesians carried the sweet potato from northwestern South America to eastern Pacific islands, including Easter Island, around AD 1200–1300, but these genotypes were replaced much later by other genotypes introduced by Europeans.[30] This plant reached Hawaii in the fourteenth century, brought by Polynesians from western Pacific sites to which Europeans had introduced them.[31] Polynesians also introduced coconut palm, candlenut, breadfruit, cabbage palm, and mountain apple to Hawaii before Europeans arrived.[32]

Early travelers also distributed plants for uses other than as food. Prehistoric Austronesians used paper mulberry, native to temperate and subtropical East Asia, to make barkcloth and probably introduced it to Taiwan between 5500 and 4500 BC. It was later introduced to Japan in AD 610 for papermaking.[33] In both Africa and Australia, humans greatly extended the range of baobab prehistorically.[34] In Australia, Aboriginal populations used boab, a baobab species, not for food but to make string from the root fibers and carvings from the fruits. Prehistoric humans also transported weeds to new regions; again, many islands received such inadvertent introductions.[35]

The use of stored grain, critical in human prehistory in the shift from small populations of hunter-gatherers to sedentary, larger populations, also entailed inadvertent introduction of insect pests.[36] The flightless granary weevil, which probably originated in the Fertile Crescent and is now found only in stored grain, reached Greece by 7,700 years ago, then spread further north and west into Europe; humans almost certainly dispersed it as they carried grain from site to site. Similarly, the saw-toothed grain beetle, of uncertain geographic origin, was found in charred grain at a Greek site dated to around 4,400 years ago, probably spread to Greece by humans from central Europe. The red flour beetle, probably of Indian origin, first appeared on Aegean islands about 3,650 years ago and in Egypt 3,340 years ago.[37] Humans also dispersed their domestic insect pests and parasites at least as early as the Neolithic.[38] The earliest European record of the housefly, possibly native to the Nile Valley, is from 7,000 years ago in the Rhine Valley, and the human flea, probably from South America, reached western Europe by 6,000 years ago, likely carried by people via the Bering land bridge.[39]

None of these prehistoric people left written records about these species, so we have no idea what they thought about the invasions. Even oral traditions indicating knowledge of such events are extremely scarce. According to Maori tradition, the Pacific rat came to New Zealand with the initial Maori colonists from eastern Polynesia with the Great Fleet (1350 AD) as a food item, and this scenario accords with archeological and genetic evidence.[40] Did prehistoric people realize which species were non-native? Obviously the individuals who deliberately introduced these species knew they were non-native, but what about a century or three centuries or a millennium later? And what about people at other locations on the mainland or on large islands; were they aware that certain species they had not seen before had been introduced? And to what extent did prehistoric people recognize that certain non-native species were problems? The Easter Island residents who saw rats destroying the seeds of the palms that made up their forests knew this destruction was a disaster. Did they realize their ancestors had introduced the rats? What about people on Mediterranean islands who saw their forests changing or even disappearing? Did they even realize that feral non-native mammals were the cause? Did they know they were introduced? Did they view it as a problem? Even today, although it is widely known that Kentucky bluegrass is a non-native species introduced from Europe to North America in the seventeenth century that reached Kentucky within a century, Kentucky remains the "Bluegrass State."[41] Probably most of its citizens and those from elsewhere assume it is native. In a preliterate society, how likely is it that knowledge of a common species' introduced status would persist and spread?

2.2 Prehistoric and Ancient Historical Aquatic Introductions

Prehistoric movement of marine organisms, and even some movement in historic times, is difficult to substantiate. James Carlton pointed out that many shelf-dwelling marine animals are found spanning entire coasts of particular oceans, or coasts of more than one ocean, lack substantial fossil records, and could have been moved by early seafaring people, even prehistorically.[42] He thus termed many marine species "cryptogenic"—that is, of unknown native origin—so in any particular location, they could be native or non-native. The naval shipworm is probably native to the Atlantic. It tunnels into wood, including piers and boat hulls, and can easily be spread inadvertently by boats or on floating debris, so its exact geographic origin is difficult to deduce. However, it is believed to have arrived in the Black Sea between 750 and 500 BC with the transport of grain among Greek colonies.[43] Early civilizations certainly knew about the shipworm and its impacts: to stymie the shipworm, Phoenicians coated ship hulls, first with pitch and later with copper sheaths. As early as the fifth century BC, a mixture of arsenic and sulfur in oil was employed, and Greeks used lead sheathing by the third century BC.[44] Another marine mollusc inadvertently transported long distances by ships very early was the soft-shell clam of North America, which reached Europe by the thirteenth century AD, almost certainly carried by Viking ships.[45]

Perhaps the earliest deliberate introduction of a marine species occurred during movement of oysters for culture. Oyster farming had begun by at least the seventh century BC, and by the fourth century BC Aristotle reported Greeks transplanting oysters to previously unoccupied sites on coasts of various islands.[46] We know from Pliny the Elder and Macrobius that the Roman emperor Claudius, in the first century AD, had the prized "scarus" fish (probably the Mediterranean parrotfish) transported from the eastern Mediterranean to the west coast of Italy, a distance of some 1,500 kilometers (932 miles), and protected it from fishing there for five years.[47] Pliny wrote in AD 77:

> At the present day, the first place is given to the scarus, the only fish that is said to ruminate, and to feed on grass and not on other fish. It is mostly found in the Carpathian Sea, and never of its own accord passes Lectum, a promontory of Troas. Optatus Elipertius, the commander of the fleet under the Emperor Claudius, had this fish brought from that locality, and dispersed in various places off the coast between Ostia and the districts of Campania. During five years, the greatest care was taken that those which were caught should be returned to the sea; but since then they have been always found in great abundance off the shores of Italy, where formerly there were none to be taken.[48]

Thus, ships existed by at least this time with tanks that could carry live marine animals a considerable distance; Aristotle in the fourth century BC had described such tanks being used to carry eels.[49]

Even before various forms of aquaculture began several thousand years ago on several continents, it is possible that freshwater species were moved at least short distances to new water bodies, but it is difficult to marshal evidence for specific cases.[50] For instance, it is likely, though unproven, that crayfish were transplanted to new locations.[51] Movement distance would be limited by the ability to keep aquatic organisms alive during transport. With the advent of aquaculture, the combination of moving species to contained settings and escape to nature likely greatly increased the number of introductions to new sites. For example, common carp, the subject of early aquaculture in both Asia and Europe, was almost certainly introduced by humans to central and western Europe as well as Japan.[52] Other aquatic species used by humans for food, such as frogs and turtles, might also have been transported to new sites.

2.3 Early Historical Introductions

In the sixteenth, seventeenth, and eighteenth centuries, travelers to various regions brought species from elsewhere and recorded the presence of species they found and knew or suspected to have been introduced. Deliberate introductions were almost never considered in the context of potential unexpected consequences, and earliest reports of non-native species were mostly simply statements of presence. Occasionally, and with increasing frequency, a traveler remarked on the spread or impact of a non-native species, and several such accounts deplored the spread or impact.

The remote South Atlantic island of St. Helena exemplifies not only the early introduction, deliberate and inadvertent, of non-native plants and animals but also the earliest, halting recognition that such non-native species could be problematic. The first settler on St. Helena after the Portuguese discovered the island in 1502, Fernando López, arrived in 1533 and kept hens, sheep, and goats, all of which reproduced and spread into the island's forests.[53] In 1588, the pirate Thomas Cavendish found many non-native plants as well as populations of pheasants and chukar partridges, both introduced from Asia, as well as thousands of goats.[54] Black rats were inadvertently introduced possibly as early as the fifteenth century and recognized by 1679 as abundant and a pest of food stocks.[55]

By the seventeenth century, the Dutch lamented deforestation caused by the goats and released dogs to kill the goats, and wild dogs were common in 1658.[56] Johan Albrecht von Mandelslo, a German nobleman heralded for an early voyage to India and the East Indies, also visited St. Helena. His accounts of St. Helena and the Azores, both from a 1639 report, make note of established populations of plants, fowl, and mammals in both places that were surely brought by Europeans, but without mention of the fact of introduction except for "poultry" in the Azores and cattle, goats, and unspecified trees on St. Helena. He expressed no concern about impacts.[57] Rather, the gist of the report is the abundance and variety of species useful for people.

Certainly by the late seventeenth century, both goats and rats were seen as obstacles to establishing sustainable agriculture on St. Helena.[58] In 1715, Captain Daniel Beeckman, returning from Borneo, stopped in St. Helena and recorded not only hordes of rats but "vast numbers of Cats, that went away from the Houses, and became wild, living among the Rocks, where they find good Prog, . . . feeding on young Partridges; and will come sometimes and make great destruction among the Poultry, so they are become as great a Plague as the Rats."[59]

Mauritius, in the Mascarenes in the Indian Ocean, similarly saw massive early introductions and the dawning awareness that these could be damaging. Arab ships almost certainly landed at times on Mauritius previously, but, as on St. Helena, Portuguese explorers are often credited with first arrival there, in 1506.[60] Although the Portuguese did not establish a permanent settlement, Portuguese navigator Pedro Mascarenhas in 1512 introduced deer, goats, monkeys, and pigs.[61] In the sixteenth century, black rats, likely inadvertently introduced by the Portuguese, established populations, and by 1606, Dutch sailors reported rats as abundant.[62] In 1598, several Dutch ships blown off course by a cyclone landed on Mauritius, claimed it for Holland, and planted a garden with European plants. The Dutch introduced rusa deer in 1639, and by the 1670s, deer had spread and were apparently valued, as legislation was introduced to control introduced wild dogs, which were harassing them, as well as domestic pigs that had become feral.[63] Whereas Dutch concern in the seventeenth century with impact of non-native species focused largely on damage to introduced cattle and game animals, in 1673, the governor lamented that pigs were eating turtle and tortoise eggs, which colonists collected for food.[64] In the 1690s, François Leguat, a French Huguenot, reported many deer, goats, and pigs in the wild and surmised that predation by pigs was a main cause for the decline of native animals.[65] Leguat also was the first to report house mice on Mauritius, and he lamented the great number of rats on Rodrigues, a smaller island in the Mascarenes (fig. 2.1).[66] In 1761, French astronomer Alexandre Guy Pingré visited Rodrigues and reported that native birds, including the endemic solitaire, were threatened or eliminated by feral introduced cats.[67]

Jacques-Henri Bernardin de Saint-Pierre, a writer and botanist, was briefly superintendent of the Jardin des Plantes in Paris. He spent 1768–71 on Mauritius.[68] As a philosopher and scientist, he held views that were very much in the tradition of a harmonious balance of nature provided by God.[69] His time on Mauritius served to support this conception.[70] But it also led to his observation of how humans damage this harmonious balance through introducing species:

> There are, it is admitted, noxious insects which prey upon our fruits, our corn, nay, our persons. But if snails, may bugs, caterpillars, and locusts,

Fig. 2.1 Depiction by François Leguat of biota of the island of Rodrigues in the Mascarenes, 1691–93. Note the presence of native species but also introduced rats. From Leguat, *Voyage et avantures de Francois Leguat*. Credit: University of Tennessee, Knoxville, Special Collections.

> ravage our plains, it is because we destroy the birds of our groves which live upon them; or, because on transporting the trees of foreign countries into our own, such as the great chestnut of India, the ebony, and others, we have transported with them the eggs of those insects which they nourish, without importing, likewise, the birds of the same climate which destroy them. Every country has those peculiar to itself, for the preservation of its plants.[71]

This is probably the earliest statement of the "enemy release" hypothesis, enunciated two centuries later at the beginning of modern invasion biology as the explanation for the explosive growth of non-native populations (see chap. 15, sec. 15.4) and the foundational hypothesis for biological control as a means to control invasions (described in chap. 17, sec. 17.5).[72] In addition,

Bernardin de Saint-Pierre lamented that Madagascar was "corrupted" by cultivation of introduced rice, specifically because "pestilential vapours are exhaled" from the rotting roots and stalks of harvested rice plants.[73]

The French naturalist Philibert Commerson, who accompanied Louis Antoine de Bougainville on an exploratory circumnavigation, remained in the Mascarenes as the voyage returned to France, studying the biota of Île de la Réunion (then Île de Bourbon) and Mauritius (then Île de France) from 1768 until his death in 1773. Commerson was interested in introducing useful non-native species to both islands and in 1770 proposed a scientific academy for Mauritius whose remit would include research on species from elsewhere that were already present or might be introduced.[74] He died before such an entity was established. As did Bernardin de Saint-Pierre, Commerson recognized the potential role of herbivores and predators in controlling populations they feed on, and he even proposed introducing such species, although only in the case of non-native rats did he explicitly suggest targeting a non-native species by this means. He wrote from Mauritius in 1770 to Joseph-François Cossigny:

> It is a misfortune that we do not have any of those birds that destroy insects. Only this island provides the spectacle of great forests with no woodpeckers. These are the great enemies of termites, ants, and small and large caterpillars. What a service one would render this colony by introducing species of shrikes, "dominicains," tyrant flycatchers, flycatchers, woodpeckers, kingfishers, and other insectivorous birds, which never attack grains; small hawks, butcherbirds, owls to keep the granivorous little birds in check; and also non-venomous snakes to destroy the rats. It would even be worth importing frogs to purge stagnant water of the prodigious quantity of gnat larvae that swarm there.[75]

In North America, the English traveler William Wood, who spent 1630–33 in the Massachusetts Bay Puritan colony, listed several plants used by settlers as food or medicine that were growing without cultivation, including European species.[76] John Josselyn later recorded twenty-two European plant species that had been brought to New England by the English and were spontaneously occurring outside cultivation by 1671 (fig. 2.2).[77] These included plantain, which he noted Native Americans called "*English-Man's foot*, as though produced by their treading," and "sprang up in the Countrey after the *English* came."[78] Several plants Josselyn recorded became invasive, and Richard Mack described the means by which many of them had arrived earlier in the seventeenth century.[79] Josselyn also reported that rats (the black rat) were present and brought from England, as was the honeybee.[80]

The earliest North American laws mandating removal of a non-native plant also concerned damage to agriculture.[81] In the seventeenth century,

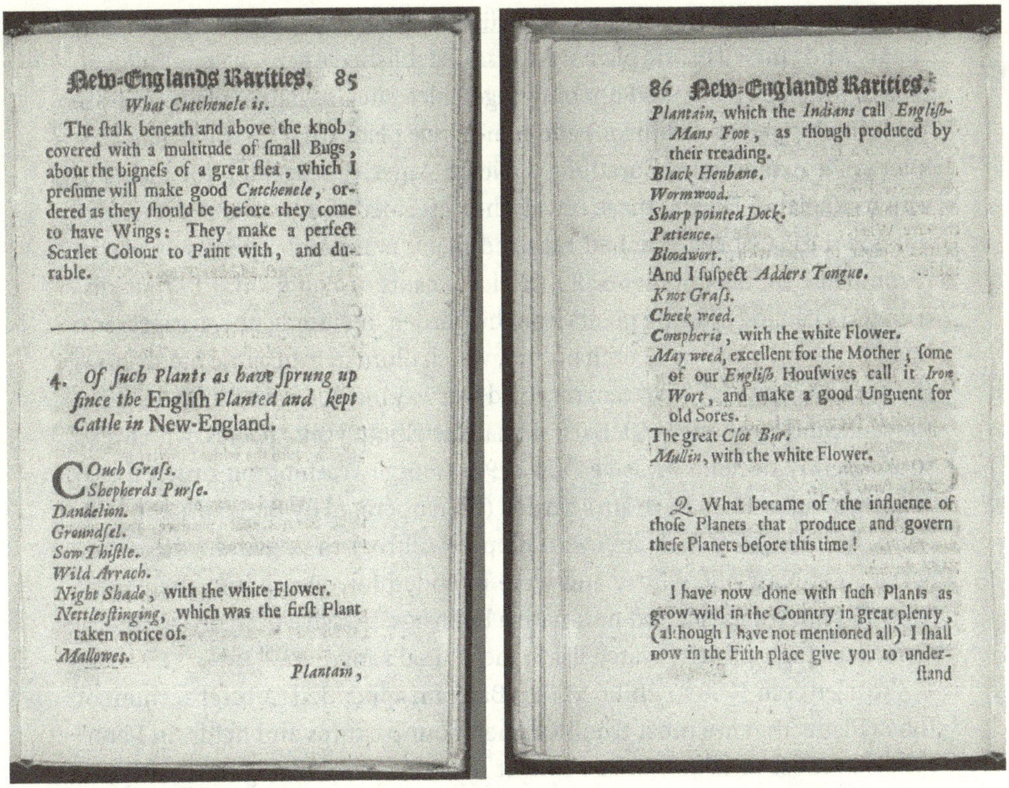

New-Englands Rarities. 85

What Cutchenele is.

The ſtalk beneath and above the knob, covered with a multitude of ſmall Bugs, about the bigneſs of a great flea, which I preſume will make good *Cutchenele*, ordered as they ſhould be before they come to have Wings: They make a perfect Scarlet Colour to Paint with, and durable.

4. *Of ſuch Plants as have ſprung up ſince the* Engliſh *Planted and kept Cattle in* New-England.

Couch Graſs.
Shepherds Purſe.
Dandelion.
Groundſel.
Sow Thiſtle.
Wild Arrach.
Night Shade, with the white Flower.
Nettlesſtinging, which was the firſt Plant taken notice of.
Mallowes.

Plantain,

86 New-Englands Rarities.

Plantain, which the *Indians* call *Engliſh-Mans Foot*, as though produced by their treading.
Black Henbane.
Wormwood.
Sharp pointed Dock.
Patience.
Bloodwort.
And I ſuſpect *Adders Tongue.*
Knot Graſs.
Cheek weed.
Compherie, with the white Flower.
May weed, excellent for the Mother; ſome of our *Engliſh* Houſwives call it *Iron Wort*, and make a good Unguent for old Sores.
The great *Clot Bur.*
Mullin, with the white Flower.

Q. What became of the influence of thoſe Planets that produce and govern theſe Planets before this time!

I have now done with ſuch Plants as grow wild in the Country in great plenty, (although I have not mentioned all) I ſhall now in the Fifth place give you to underſtand

Fig. 2.2 List of European plants that John Josselyn reported in New England in 1672, which he had visited in 1638–39 and 1663–71. Credit: National Agricultural Library, United States Department of Agriculture.

English colonists in North America brought common European barberry as a favored hedge plant; it soon spread autonomously and became widely naturalized. Although the definitive role of this species as the alternate host of wheat stem rust (which also attacks barley, oats, and rye) was not established until 1865, European farmers had long recognized a connection between barberry and wheat stem rust, and in 1660, French farmers in the vicinity of Rouen secured passage of a law to eradicate barberry for this reason. Around the same year, the rust first appeared in North America and quickly threatened the supply of flour; a key role for barberry was either intuited or recalled from Europe. In 1726, the Connecticut Colony passed a law mandating removal of barberry and imposing fines on uncooperative individuals. The Royal Province of Massachusetts Bay in 1754 and Rhode Island Colony in 1766 followed suit with similar laws.[82]

John Bartram, born in 1699 in Pennsylvania, was interested in plants as a child and became a farmer.[83] By 1733, he had begun extensive correspondence and exchange of plant seeds and specimens with Peter Collinson, a London draper who introduced foreign plants into his garden.[84] Bartram shipped boxes of North American plants to Collinson, who distributed them to many clients who fancied exotic plants. Bartram established a garden near Philadelphia in which he cultivated not only vegetables but also native plants (whose seeds he shipped to Collinson) and non-native plants that he sold. He is best known for extensive explorations of North America searching for plants to ship to England. During these travels, he described many species previously unknown to science.[85] He had received many non-native species as partial recompense for his shipments of American plants to Collinson.[86] Bartam distributed a catalog offering plants from his garden, including non-natives, and by 1748 he was growing Scotch broom, which ultimately became highly invasive.[87] In 1756 or 1757, Bartram received Norway maples from Philip Miller in England, and by about 1760, Bartram had distributed this species, which also subsequently became invasive.[88] In 1792, George Washington ordered two Norway maples from Bartram's garden (then operated by Bartram's son William). In 1769, John Bartram even shipped bullfrogs to England, where they survived at least one winter and were intended for release into the wild.[89] They did not establish, and non-native bullfrogs, invasive in many areas, are today on a high-priority watch list in the United Kingdom.[90]

To a letter in 1758 to Philip Miller, Bartram appended "A brief account of those Plants that are most troublesome in our pastures and fields, in Pennsylvania; most of which were brought from Europe," listing ten European species followed by several native North American plants.[91] "Most mischievous" among introduced species was yellow toadflax, followed closely by St. John's wort and stinking chamomile. The recognition that these species had escaped from gardens is striking, although Bartram's specific concerns all relate to difficulties such plants impose on agriculture—mowing meadows or tainting the milk and flesh of cattle.

At Carolus Linnaeus's suggestion, the Swedish Academy in 1747 commissioned Pehr Kalm, a Finnish botanist, to travel to North America in search of useful plants to bring to Sweden. Kalm's voyage, from 1748 to 1751, took him to Pennsylvania, New Jersey, New York, Quebec, and Ontario westward to Niagara Falls. He spent much time with Bartram, frequently acknowledging Bartram's insights in his description of the voyage.[92] Kalm's account is replete with accounts of plants and a few animals that he signaled as introduced from Europe, including several that are now invasive: privet, plantain, English ivy, and a species of *Arundo* (possibly giant reed). Kalm recognized that smallpox was brought by Europeans and devastated Native Americans and noted that Native Americans in their language called the honeybee "English flies because the English first brought them over; but at present they fly plentifully about the woods of North America."[93] He observed, as had Josselyn and Bartram, that the Native American term for plantain translated as "Englishman's foot" because they believed it had never been present previously and followed in Europeans' footsteps. Kalm wrote that several European plants (e.g., privet and *Arundo*) as well as the honeybee were found in natural as well as anthropogenic settings, but he did not evince concern over this fact. Rather, he deplored damage caused by a pest of non-native pea plants—almost certainly the pea weevil, which had to have been introduced, although Kalm did not note this fact.[94]

Brazil received a plethora of introduced species, both animals and plants, beginning with its colonization by the Portuguese in 1532.[95] As on St. Helena and Mauritius, livestock and plants suitable for human use were deliberately introduced, especially from other parts of the Portuguese empire. Inevitably, some of these became invasive. For example, Jesuits introduced jackfruit from Asia near the end of the seventeenth century, which invaded the Atlantic forest in the twentieth century.[96] Notable among introductions were African grasses that were either deliberately imported for forage or, having arrived by unknown means, were then planted for this purpose (see chap. 8, sec. 8.1). For instance, at the end of the eighteenth century, the Portuguese polymath José Correia da Serra was affiliated with Sir Joseph Banks, president of the Royal Society in England.[97] Among many botanical explorations, he studied plants cultivated by the British, Dutch, and French in their colonies, seeking those of possible benefit for the Portuguese. Recognizing that native Brazilian grassland plants were not suited for coexistence with the large, congregating livestock that had been introduced, he arranged for buffel grass, native to Africa and already present in Jamaica, to be sent from that island to Brazil.[98] This species increases fire frequency and intensity to the disadvantage of native species and invades dry forest and grasslands in Brazil.[99]

: : :

St. Helena, Mauritius, and eastern North America exemplify the sorts of observations accompanying accounts by travelers in other areas who reported non-native species, and occasionally their dispersal. To the extent that they noted spread and impact rather than simply presence, the concern was almost wholly with effects on human activities—agriculture, animal husbandry, hunting—or comfort. There was scant concern for anything that we would now call ecology, which is not surprising, since ecology as a distinct, recognized discipline did not arise until the late nineteenth century (see introduction to chap. 4).[100] In this context, Pingré's recognition in the eighteenth century that introductions could cause extinctions of native species, including those not used for anything by humans, stands out. So does Bernardin de Saint-Pierre's citing of the important role of trophic (feeding) interactions in maintaining a sort of "balance of nature" and recognition that non-native species could escape control if species feeding on them are left behind. These insights were not formally recognized as part of the young field of ecology until the early twentieth century.

3

Early Recognition of the Extent of Invasions and Increasing Concern with Their Impacts

Although isolated travelers recorded non-native species in various places beginning at least as early as the sixteenth century, no systematic scientific framework existed that would assemble such observations and allow consideration of the extent and importance of the invasion phenomenon as a whole. Such a framework required a broad knowledge of which species are found in which locations and also a consideration of what it means for a species to be native to a region. The former matter was largely resolved in the nineteenth century with the rapid growth of the science of biogeography, bolstered by trained naturalists who visited many regions that had previously not been substantially explored from a biological perspective. The question of what is meant by saying a species is native to a region is still debated (see chap. 18, sec. 18.2). However, the nineteenth-century development of a science of biogeography, especially a branch known as "historical biogeography," combined with advances in molecular genetics that began in the twentieth century and are still ongoing (described in chap. 11, sec. 11.2), led to strong evidence in many instances that particular species had occupied certain areas for a long time and in other instances that they were relatively recent arrivals, and often even suggested where they had come from.

Many of the naturalist-explorers, though noting the presence of species they presumed to be non-native, did not comment on potential ecological impacts of these species, because either they did not perceive impacts or they considered such matters as secondary to their goal of delimiting biogeographic "provinces" and the presence of such non-native species as obfuscating the patterns they sought. This frequent absence of focus on impact is unsurprising, given that the modern science of ecology did not arise until the late nineteenth century.[1] However, a few of the early and mid-nineteenth-century naturalists commented on and lamented many impacts of non-native species, especially on islands.

3.1 The Science of Biogeography: Which Species Are Where, and Where Are They Native?

Although observations by several travelers in the fifteenth through seventeenth centuries contributed to a growing knowledge of which species were found on which continents and islands, the science of biogeography, which treats the distribution of species and how and when they arrived in their current locations, did not begin in earnest until the eighteenth and especially the nineteenth centuries. At the beginning of the eighteenth century, even the outlines of much of the earth were unknown to Western science; for instance, Captain James Cook was the first European to visit and survey the east coast of Australia in 1770 and to encounter Hawaii in 1778.[2] By the end of the nineteenth century, the geography of the earth was well established, and, partly as a consequence of the voyages establishing that geography, the basic biogeography of the earth was also known—at least which species were where, though the "how" and "when" parts of the field were still being developed.[3]

The comprehensive monographs usually credited as initiating the field of biogeography are Swiss botanist Augustin P. de Candolle's *Essai élémentaire de géographie botanique* (1820) and British biologist Alfred Russel Wallace's *The Geographical Distribution of Animals* (1876), but, as with so many aspects of the history of biology, important antecedents existed.[4] For plants, Alexander von Humboldt and Aimé Bonpland's *Essai sur la géographie des plantes* (1805) contributed importantly with their surveys in Central and South America.[5] The critical antecedent for animals was *Histoire naturelle générale et particulière* (1776) by Georges-Louis LeClerc, Comte de Buffon, who pointed out that mammals of the Old World and New World subtropics and tropics were different species, even though, under reigning views at the time, regions of the two areas with similar physical attributes such as climate would have been expected to have the same species.[6] Von Humboldt's "Sur les lois que l'on observe dans la distribution des formes végétales" (1816) for plants and Pierre André Latreille's "Introduction à la géographie générale des arachnides et des insectes" (1817) for invertebrates extended "Buffon's law," as it was known in the nineteenth century, although de Candolle and Wallace formulated the definitive comprehensive statements.[7]

What the nineteenth-century biogeographers were trying to do was to define distinct regions or provinces containing distinct collections of species. In addition to Buffon, early efforts of this sort were by Johann Christian

Fabricius (*Philosophia Entomologica*; 1778) for insects, Latreille for insects and arachnids, and James Cowles Prichard (*Researches into the Physical History of Mankind*; 1836) for biota in general.[8] The overarching problem that had to be clarified before anyone could comprehensively and intelligently address the matter of non-native species was to recognize biotas in terms of common descent (evolution) and not just assume that regions with similar environments generally had species with certain adaptations to the physical environment (e.g., burrowing behavior of desert species). This recognition was halting and imprecise until Charles Darwin and Wallace developed the full-fledged concept of evolution by descent. It was in this regard that de Candolle, even before Darwin and Wallace, and Wallace, following Philip L. Sclater's "On the General Geographical Distribution of the Members of the Class Aves" (1858), provided the key insights by sketching out large regions of the earth, each with collections of species related to one another by descent but not related to species of other regions.[9]

Given this overarching goal of demarcating regions containing related species, it is unsurprising that non-native species would be at best of little interest and at worst a sort of "noise" obscuring the very patterns biogeographers were seeking to establish. Many naturalist-explorers during the eighteenth and nineteenth centuries construed their mission as contributing to the development of a science of biogeography by describing which species were found in regions of the world that had not previously been systematically examined from a biological perspective. Although several naturalists reported species they surmised had been introduced by humans, these were generally dismissed as insignificant oddities. Pehr Kalm (see chap. 2, sec. 2.3) is in this category. Von Humboldt, whose extensive explorations in South America (1799–1804) yielded sixty thousand specimens and three thousand species new to science, noted in 1805 how "some plants, objects of gardening and agriculture from the earliest times, have accompanied man from one end of the globe to the other," complaining about the resulting difficulty of determining the origins of these species.[10]

Many naturalists of the nineteenth century, and a few of the eighteenth century, explicitly mentioned which species (especially plant species) they believed to have been introduced from elsewhere, separating them in various accounts and tallies of species. I have already noted John Bartram's weed list of 1758 and Kalm's account of 1770, and in the next section I will add William Roxbrough's St. Helena flora of 1816 and Hewett Watson's Azorean flora of 1870. Even earlier, the Swiss anatomist and naturalist Albrecht von Haller in 1742 expressed his "great Abhorrence" of a botanist who "inserts a very great Number of Plants most certainly Exotic, as growing about Paris."[11] Also notable is a list from a paper read in 1832 by Lewis D. Schweinitz tallying all 137 European plant species that he believed had become naturalized in the United States, classified by degree to which they had spread from cultivation or anthropogenic habitats and partially by means of introduction.[12] What is perhaps more striking to a modern reader than the early recognition of non-native species is how few of these naturalists mentioned impact; their interest was biogeographic.

The German botanist Karl Willdenow in 1805 saw anthropogenic dispersal as the single most important force in dispersing plants to diverse regions and included plants in nature as well as those deliberately introduced as crops.[13] He wrote:

> Man, however, more than wind, weather, seas, rivers and animals, contributes to the dissemination of plants. He who commands nature, who changes deserts into beautiful landscapes; lays waste whole Countries, and again brings them to their former state, has in various ways favoured the distribution of a number of plants over our globe. . . . The common thorn apple, the Datura *Stramonium*, which now grows almost throughout all Europe . . . and is thrown out as a noxious weed, came from the East Indies and Abyssinia to us, and was so universally spread over Europe by a set of quacks, who used its seeds as an emetic or cathartic.[14]

In 1832, the geologist Charles Lyell, citing de Candolle and Willdenow, emphasized the role of humans in dispersing seeds and thus complicating the search for the "true" affinities of plants of various regions.[15] Lyell noted Willdenow's *Datura stramonium* example and also cited Linnaeus to the effect that "Canadian erigiron [*sic*], or flea-bane, was brought from America to the botanical garden at Paris, and already the seeds have been carried by the winds, so that it is diffused over France, the British islands, Sicily, Holland, and Germany."[16] British botanist George Bentham, citing the newly published English translation of de Candolle's 1820 work, similarly deplored how non-native plants stymied attempts to establish botanical regions and determine plant origins.[17]

3.2 What About Impacts of Invasions?

Among nineteenth-century naturalist-explorers, three stand out as having recognized and deplored what we would now term ecological impacts of biological invasions: Charles Darwin, Alfred Russel Wallace, and Joseph Hooker. In 1833, Darwin traversed Patagonia and noted cardoon, or artichoke thistle (native to central and western Mediterranean regions):

> Very many (probably several hundred) square miles are covered by one mass of these prickly plants and are impenetrable by man or beast. Over the undulating plains, where these great beds occur, nothing else can now live. Before their introduction, however, the surface must have supported,

> as in other parts, a rank herbage. I doubt whether a case is on record of an invasion on so grand a scale of one plant over the aborigines.[18]

He also recognized milk thistle (native to southern Europe and parts of Asia), which he called "Pampas thistle," as introduced:

> There were immense beds of thistle, as well as of the cardoon: the whole country, indeed, may be called one great bed of these plants. . . . The cardoon is as high as a horse's back, but the Pampas thistle is often higher than the crown of the rider's head.[19]

In the Galápagos Islands in 1835, Darwin reported feral goats and pigs but did not comment on their potential impacts. However, he made perhaps the first attempt to relate evolution to non-native species. In the first edition of the account of his research voyage aboard the *Beagle*, published in 1839, he discussed "a rat which Mr. Waterhouse believes is probably distinct from the English kind; but I cannot help suspecting that it is only the same altered by the peculiar conditions of its new country."[20] In the revised and expanded second edition, published in 1845, he was much more explicit:

> Of terrestrial mammals, there is only one which must be considered as indigenous, namely, a mouse (Mus Galapagoensis), and this is confined, as far as I could ascertain, to Chatham island, the most easterly island of the group. It belongs, as I am informed by Mr. Waterhouse, to a division of the family of mice characteristic of America. At James island, there is a rat sufficiently distinct from the common kind to have been named and described by Mr. Waterhouse; but as it belongs to the old-world division of the family, and as this island has been frequented by ships for at least the last hundred and fifty years, I can hardly doubt that this rat is merely a variety, produced by the new and peculiar climate, food, and soil, to which it has been subjected. Although no one has a right to speculate without distinct facts, yet even with respect to the Chatham island mouse, it should be borne in mind, that it may possibly be an American species imported here; for I have seen, in a most unfrequented part of the Pampas, a native mouse living in the roof of a newly-built hovel, and therefore its transportation in a vessel is not improbable.[21]

On St. Helena in 1836, Darwin adumbrated the issue of non-native species leading to the extinction of natives:

> It is surprising to behold vegetation possessing a character decidedly British. The hills are crowned with irregular plantations of Scotch firs; and the sloping banks are thickly scattered over with thickets of gorse, covered

> with its bright yellow flowers. Weeping-willows are common on the banks of rivulets, and the hedges are made of the blackberry, producing its well-known fruit. When we consider that the number of plants now found on the island is 746, and that of these fifty-two alone are indigenous species, the rest having been imported, and most of them from England, we see the reason of the British character of the vegetation. Many of these English plants appear to flourish better than in their native country; some also from the opposite quarter of Australia succeed remarkably well. The many imported species must have destroyed some of the native kinds; and it is only on the highest and steepest ridges, that the indigenous Flora is now predominant.[22]

Darwin cited the account of former Governor Alexander Beatson stating that enormous herds of free-ranging goats, initially introduced in 1502, had been the main factor in destroying native vegetation on St. Helena, and he noted that by 1724 virtually the entire island had been devastated.[23] Remarkably, Darwin wrote that the loss of native vegetation had doubtless caused all eight native land snail species to become extinct, as well as "a multitude of insects." And he recorded that the non-native English snail *Helix* was likely introduced as eggs on some of the non-native plants and had become common. Darwin did not cite an appendix to Beatson's book by the botanist William Roxburgh, who had lived on St. Helena, entitled "An alphabetical list of plants, seen by Dr. Roxburgh growing on the island of St. Helena, in 1813–14."[24] Roxburgh categorized each species as either indigenous or exotic, and he tallied 56 natives and 295 exotics, in some instances stating the source of the introduction.

Soon thereafter, the *Beagle* landed on Ascension Island, where Darwin climbed Green Mountain (whose largely introduced vegetation became a controversial topic in modern invasion science nearly two centuries later[25]). Among wild non-native vertebrates on Ascension he noted goats, rats (probably the black rat), house mice, "plague numbers" of cats (imported to control the rats and mice), guinea fowl, and chickens. In perhaps his second attempt to relate evolution to introduced species, Darwin observed in the second edition of what has become known as the *Voyage of the Beagle* that "I can hardly doubt that these rats (like the common mouse, which has also run wild) have been imported and . . . have varied from the effect of the new conditions to which they have been exposed: hence the variety on the summit of the island differs from that on the coast."[26]

In his groundbreaking book *On the Origin of Species by Means of Natural Selection*, published in 1859, Darwin adduced non-native species and their impacts at several points to exemplify the "struggle for life" that drives natural selection.[27] For instance, he argued that "if all the animals and plants of Great Britain were set free in New Zealand, a multitude of British forms would in the course of time become thoroughly naturalized there, and would exterminate many of the natives."[28] He extended this argument to suggest that Northern Hemisphere species are generally stronger competitors than those of the Southern Hemisphere: "Hardly a single inhabitant of the southern hemisphere has become wild in any part of Europe . . . the productions of Great Britain stand much higher in scale than those of New Zealand."[29] Darwin pointed to observations by Alphonse de Candolle that established non-native plant species are much more likely to belong to genera new to the region of introduction than to native genera, giving an example: "In the last edition of Dr. Asa Gray's 'Manual of the Flora of the United States' . . . out of the 162 naturalised genera, no less than 100 genera are not there indigenous."[30] He saw this fact as resulting from the likelihood that non-native species unrelated to natives would be less ecologically similar to native species than non-native species related to natives would be. Thus they would be less likely to compete intensely with the natives, a hypothesis subsequently termed "Darwin's naturalization hypothesis" (see chap. 11, sec. 11.3).[31]

During his voyages of biological collecting, discovery, and inspiration to Amazonia and the Malay Archipelago, Wallace made no note of non-native species. However, his book *Island Life* (1880) is replete with many references not only to the presence of non-native animals and plants but frequently also to their impacts on various islands.[32] As had Darwin, Wallace described the devastation on St. Helena wrought by introduced goats, pointing to how they facilitated the establishment of European, American, African, and Australian plants, and he suggested that the "almost total extirpation of a luxuriant and highly peculiar vegetation must inevitably have caused the destruction of a considerable portion of the lower animals which once existed on the island."[33] Wallace also pointed to non-native animals and plants on Bermuda, the Cocos Islands, the Galápagos Islands, the Hawaiian Islands, and New Zealand, even citing European weeds overrunning areas in North America. He detailed the biota introduced by humans to the Azores: rabbits, weasels, rats, mice, goldfish, eels, and at least 101 beetle species. Although travelers had noted the presence of non-native species in the Azores since at least 1639, Wallace was the first to attempt to sketch the extent of introduction of vertebrates, invertebrates, and plants, although his main goal was to determine the affinities and means and dates of arrival of the non-native biota, not to elucidate their impacts.[34]

For plants of the Azores, Wallace stated that, "however, the general character of the whole flora is that of the south-western peninsula of Europe, and as most of the introduced plants have come from the same country, it is almost impossible now to separate them, and Mr. Watson has not attempted to do so."[35] In fact, Hewett C. Watson was at great pains to separate plants introduced by humans from other plants of the Azores, which is unsurprising as he was one of the nineteenth-century biologists who focused most explicitly on distinguishing between species of native and introduced origin, as well as those for which the reason for the presence in a particular place cannot (yet) be determined (as will be discussed shortly).[36] Wallace was referring to Watson's long entry on Azorean plants in a recently published book.[37] Although Watson did not tabulate or enumerate non-native plants, he tried to distinguish among (1) plants introduced by humans from plants that had been present in the archipelago long before humans could have introduced them but were not yet distinguishable morphologically from those of the continents of origin (primarily Europe), and (2) those whose continents of origin could be deduced but had differentiated somewhat from their continental relatives. His species accounts listed twenty-five species as certainly or almost certainly non-native and another twelve as perhaps non-native, while his introductory and concluding sections suggested that these numbers were likely underestimates.

In 1890, Wallace summarized evidence on the theory of evolution by natural selection in light of observations and debate during the three decades since Darwin's publication of *On the Origin of Species*.[38] Wallace's exposition was replete with copious observations on impacts of non-native plants and animals on continents as well as islands: European weeds in North America, New Zealand, and Patagonia; goats on St. Helena; Eurasian livestock in Patagonia, the West Indies, and Mexico; Old World asses in Ecuador; boar and pigs in the West Indies, continental North America, and New Zealand; South African plants in Australia; West Indian lantana in Ceylon (today Sri Lanka); Asian "Norway" rat in Europe; Eurasian insects in Russia, Australia, and New Zealand.[39] Wallace invoked such cases to exemplify principles in the "struggle for existence," notably reprising and extending to animals "Darwin's naturalization hypothesis" (see chap. 11, sec. 11.3.2). Wallace also anticipated what came to be known with the advent of the field of ecology as "top-down" control of populations in nature: the notion that predators and consumers determine the composition of the community or organisms they consume and heavily influence the dynamics of the ecosystem they occupy.[40] And he adumbrated the concept of Charles Elton and later twentieth-century invasion biologists of "biotic resistance" (a concept discussed in chap. 15, sec. 15.4.1), the view that most introduced non-native species fail to establish or spread because they are opposed by native competitors, predators, parasites, and pathogens.

Joseph Hooker's lecture in 1866 discussed in detail the floras of Madeira, the Canary Islands, the Azores, the Cape Verde Islands, St. Helena, Ascension, and Kerguelen, and in passing floras of several other islands.[41] Hooker had explored many islands (including Madeira and Kerguelen) as botanist for the Antarctic expedition of James C. Ross (1839–43). Hooker's main goal in this lecture was to indicate various typical aspects of the floras of remote oceanic islands as reflecting both Darwin's theory of evolution by natural selection and Darwin's insistence on more or less rare instances of long-distance seed dispersal (rather than the existence of lost continents, now submerged) as the means by which plants colonized such islands. Further, he asserted that the fact that aspects of insular floras were rationalized by Darwin's ideas provided strong support for the theory of evolution, which was then controversial. Given these goals, it is unsurprising that Hooker viewed non-native species as obscuring an understanding of the geographic affinities of the different floras and therefore interfering with his defense of the theory of evolution. His interest in non-native species as an ecological force was thus secondary. Nevertheless, he recorded several invaders and commented on their impacts.

Hooker noted that humans and species they introduced, including animals, had likely caused the rarity or extinction of many species. For Madeira, he exulted that the non-native plants were generally easily recognized as such, but he deplored the devastation wrought by deliberately set fires followed by introductions:

> Cultivation and the introduction of European trees followed; and who now can form a conception of the number of peculiar plants that utterly perished in the flames, or that have since been elbowed out by the more vigorous introduced European vegetation? And with regard to Port Santo, in about the year 1418 a mother rabbit and her brood were landed, and increased so rapidly, that they not only consumed the native vegetation, but the cultivated, and actually drove the settlers from the island.[42]

Hooker also remarked on the non-native floras of Australia, New Zealand, and South Africa, and, as had Darwin for New Zealand, he predicted a grim fate for at least some native plants:

> Nor can it be doubted that many of the small local genera of Australia, New Zealand, and South Africa, will ultimately disappear owing to the usurping tendencies of the emigrant plants of the northern hemisphere, energetically supported as they are by the artificial aids that the northern races of men afford them.[43]

Although when this statement is cited, it is generally to support Darwin's view of the inherent superiority of Northern Hemisphere species, in fact Hooker chose his words carefully, including the phrase "artificial aids that the northern races of men afford them."[44] Combined with his comments on deliberately set fires and non-native animals such as rabbits and goats, this phrase suggests that Hooker recognized a key role for anthropogenic habitat change or destruction in the replacement of native plants by non-native ones. He foreshadowed more forceful and direct arguments along these lines in the early twentieth century, especially by New Zealand botanists Leonard Cockayne and Harry H. Allan (more on them in chap. 6, sec. 6.3).

Hooker detailed the impact of non-native plants and animals on the St. Helena flora, and in fact Wallace cited Hooker's account in *Island Life*. Hooker cited the same account of Beatson that Darwin had noted about the impact of goats, and Hooker cited the same sad fact that others had noted, that in 1709 the governor of the island, fearing the utter extinction of native "ebony" trees, petitioned the directors of the East India Company to destroy the goats, receiving the reply that "the goats are not to be destroyed, being more valuable than Ebony." A century later, Governor Beatson recognized the devastation wrought by the goats; in the second book he published on the island, Beatson included the essay "Remarks on the Evil Consequences Which Have Resulted from the Introduction of Goats upon the Island of St. Helena."[45] Beatson's attempt to eradicate goats failed (Hooker erroneously reported it had succeeded) but effected a temporary population reduction. What Hooker added, that Wallace did not, is the important fact that it was Beatson who

> proposed and carried out the introduction of exotic plants on a large scale, and from all parts of the world; these have propagated themselves with such rapidity, and grown with such vigour, that the native plants cannot compete with them. The struggle for existence had no sooner begun, than the issue was pronounced; English Broom, Brambles, Willows, and Poplars, Scotch pines and Gorse bushes, Cape of Good Hope bushes, Australian trees and American weeds, speedily overran the place; and wherever established, they have virtually extinguished the indigenous Flora.[46]

3.3 Classifying Species: Which Are Native and Which Are Not?

Matthew Chew and Andrew Hamilton saw the Cambridge botany professor John Henslow as triggering the classificatory activities of Hewett C. Watson, who worked assiduously to classify non-native species as a separate category in establishing a flora of Great Britain.[47] Actually, it was Watson's publications that spurred Henslow's account.[48] Watson, in turn, was heavily influenced by Nathaniel John Winch, who in 1819 had published a ten-part classification of plants of three British counties, including a ninth category that he listed variously as "Plants which have become naturalized, though originally imported with ballast" and "Plants which have become Naturalized, and are admitted into the British Flora," and a tenth category listed as "Exotics introduced by the same means" and "Exotics found chiefly on the Ballast Hills of Tyne and Wear."[49] The ninth category includes sixteen species listed as "Introduced with Corn" rather than in ballast.[50]

Watson's culminating classification in 1847 "rejected many species commonly admitted into our flora, and has increased the list of naturalised species," Henslow observed.[51] The classification appeared in Watson's massive compendium on British flora, in which he distinguished six categories:

> 1. Native.—Apparently an aboriginal British species; there being little or no reason for supposing it to have been introduced by human agency.
>
> 2. Denizen.—At present maintaining its habitats, as if a native, without the aid of man, yet liable to some suspicion of having been originally introduced.
>
> 3. Colonist.—A weed of cultivated land or about houses, and seldom found except in places where the ground has been adapted for its production by the operations of man; with some tendency, however, to appear also on the shores, landslips, &c.
>
> 4. Alien.—Now more or less established, but either presumed or certainly known to have been originally introduced from other countries.
>
> 5. Incognita.—Reported as British, but requiring confirmation as such. Some of these have been reported through mistake of the species. . . . Others may have been really seen in the character of temporary stragglers from gardens. . . . Others cannot now be found in the localities published for them . . . ; though it is not improbable that some of these may yet be found again. A few may have existed for a time, and become extinct.
>
> 6. Hibernian, *or* Sarninan.—Native, or apparently so, in Ireland, or in the Channel Isles, though not found in Britain proper.[52]

Modern invasion science uses versions of Watson's categories 1, 3, 4, and 5, although with different terms: category 3 does not yet have a stable name, category 4 is increasingly replaced by "non-native," and species in category 5 are now widely termed "cryptogenic."[53]

3.4 A Prescient Person Far from the Mainstream: George Perkins Marsh

One nineteenth-century author who, perhaps more than any figure of his time—even Darwin and Wallace—did sense the global scope and depth of impacts of non-native species was George Perkins Marsh.[54] In 1864, Marsh, a polymath trained as a lawyer who served in Congress and as longtime US ambassador to Italy, published *Man and Nature; or, Physical Geography as Modified by Human Action*, the first thoroughgoing exposition of the many ways humans deliberately or inadvertently have changed the face of the earth. Although not a physical or life scientist, Marsh pointed with examples to staggering past, ongoing, and potential future impacts of human activities, anticipating by a century the initiation in the United States and elsewhere of modern environmental and conservation movements (treated in chap. 7, secs. 7.4 and 7.5) signaled most notably by works of Aldo Leopold and Rachel Carson.[55] Marsh's biographer, David Lowenthal, asserted that

Marsh punctured the then-reigning myth of limitless resources and termed *Man and Nature* the second most influential book of the nineteenth century, after Darwin's *Origin of Species*.[56] He noted ways in which the retitled 1874 edition inspired various subsequent activities, especially forest conservation programs.

Biological invasions were not one of Marsh's main themes. In his extensive "Introductory," he mentioned only the beneficial introduction of new crop plants, and it is indicative that Lowenthal's foreword to the 2003 reprint did not mention invasions among the several important changes that are today of major concern and that Marsh signaled.[57] Marsh dealt with species introductions primarily in a chapter entitled "Transfer, Modification, and Extirpation of Vegetable and of Animal Species."[58] At the outset, he argued, "Whenever man has transported a plant from its native habitat to a new soil, he has introduced a new geographical force to act upon it, and this generally at the expense of some indigenous growth which the foreign vegetable has supplanted."[59] And Marsh recognized that development of a science of biogeography, well underway at the time he wrote, was necessary for one to understand the scope and details of this transformation: "It belongs to vegetable and animal geography, which are almost sciences of themselves, to point out in detail what man has done to change the distribution of plants and of animated life and to revolutionize the aspect of organic nature."[60]

Although the chapter is not mainly about "transfer," Marsh provided many examples of non-native species. In fact, he had previously written two works on such species, one on camels introduced in the United States and the other on artificial fish propagation, the second of these replete with accidental introductions that Marsh reprised in *Man and Nature*: goldfish from Asia in the United States and fish species that had moved between the Hudson River and the Great Lakes through the Erie Canal.[61]

Marsh repeated, without citation, the same statistics that Darwin had cited showing the preponderance of non-native plants on St. Helena, as well as Darwin's description of the invasion of the cardoon in Patagonia.[62] Marsh also noted the invasion of English rivers by Canadian pondweed, which subsequently became, with the advent of modern invasion science, a widely noted challenging case.[63] He incorrectly reported on the introduction of the misnamed Canada thistle into Europe, correcting the account in the 1874 edition to note its Old World origin. He mentioned the then-mysterious parasite attacking grapevines in France without noting that it was not native; in the 1874 edition, he correctly identified it as an American bug, the phylloxera, and noted that resistant American vines were being planted in Europe (as discussed in chap. 4, sec. 4.2). Marsh described the introduction to North America of the Hessian fly, a cereal pest, with mercenary European troops, as well as other European agricultural pests in North America, arguing that their greater impact in their non-native range owed to their "not being checked by the counteracting influences which nature had provided to limit their devastations in the Old World," thus anticipating the "biotic resistance" hypothesis (described in chap. 15, sec. 15.4).[64] In this connection, Marsh cited a preliminary report from the 1860 United States census to the effect that predatory or parasitic insect enemies of some of these pests had recently been introduced from Europe to the United States. Such introductions would have constituted a very early example of the widespread subsequent practice of biological control (detailed in chap. 17, sec. 17.5), but the census report was in error and the proposed introductions did not occur because the enemies were not shipped to the United States.[65]

In a later chapter, discussing the Suez Canal then under construction, Marsh was prescient:

> It will, no doubt, produce very interesting revolutions in the animal and vegetable population of both basins. The Mediterranean . . . is comparatively poor in marine vegetation, and in shell as well as in fin fish. . . . But no one who has not looked down into tropical or subtropical seas can conceive the amazing wealth of the Red Sea in organic life. . . . Most of its vegetables and animals, no doubt, are confined by the laws of their organization to warmer temperatures than that of the Mediterranean, but among them there must be many, whose habitat is of a wider range, many whose powers of accommodation would enable them to acclimate themselves in a colder sea. We may suppose the less numerous aquatic fauna and flora of the Mediterranean to be equally capable of climatic adaptation, and hence, when the canal shall be opened, there will be an interchange of the organic population not already common to both seas. Destructive species, thus newly introduced, may diminish the numbers of their proper prey in either basin, and, on the other hand, the increased supply of appropriate food may greatly multiply the abundance of others, and at the same time add important contributions to the aliment of man in the countries bordering on the Mediterranean.[66]

In fact, the Suez Canal quickly led to invasion by many Red Sea species into the Mediterranean, known as "Lessepsian migrants" after the French diplomat who developed the canal, and far fewer "anti-Lessepsian migrants" from the Mediterranean into the Red Sea.[67] These invasions have indeed caused substantial diverse impacts in the Mediterranean, including declines in some native species, and Lessepsian migrants do support some current Mediterranean fisheries.[68] In his first edition, Marsh mentioned "a mixture of the animal and vegetable life of the two great oceans" if a canal should be built across Panama, without comment on possible impacts, and he did not mention species introductions in discussions of several other potential canals.[69] In the 1874 edition, he added in a short footnote that animal reproduction can be stimulated "by transfer to a foreign clime" and gave the example of European rabbits becoming a "serious evil" on Madeira and in Australia (a case presented in chap. 4, sec. 4.1, and discussed in chap. 11, box 11.1).[70]

His preceding examples are of harmful impacts or were presented without comment on impacts. By contrast, in his first edition, Marsh noted several examples of non-native species beneficial to humans in one way or another, including reindeer in Iceland, many crop plants plus the mountain ash (rowan tree), white poplar, cork oak, and horse chestnut from the Old World in North America, and the silkworm, the cochineal insect, and the honeybee in various new regions.[71] In his second edition, he added ramie from China and Japanese clover as useful additions to the United States flora and lauded the European house sparrow in the United States as "useful by destroying noxious insects and worms not preyed upon by native birds."[72] He favored past or proposed introductions of oysters and fishes, though lamenting that "the intentional naturalization of foreign fish, as I have said, has not thus far yielded important fruits."[73] In 1846, he told the US Congress that "the alpaca, the Thibet goat, and perhaps even the camel" would be useful additions to the American fauna.[74]

Marsh's entire book is not simply a lament about human destruction of nature. Rather, it is a call to recognize that human activities can change natural features and phenomena. The changes can be inimical, but they can also be beneficial, and Marsh pleaded for humans to ensure that the changes benefit human welfare rather than detract from it. Lowenthal argued that, "unlike Thoreau, who also loved nature, but wished to keep it wild, Marsh wanted it tamed."[75] This view pervades all chapters of *Man and Nature*, not just the one on transfers. Thus, for example, Marsh saw many advantages to managing forests stringently rather than leaving them to nature, and he advocated introducing many European trees to North America. He anticipated and heavily influenced Gifford Pinchot, who coined the term "wise use" to capture his sense of the meaning of conservation.[76] Pinchot asserted:

> The first principle of conservation is development, the use of the natural resources now existing on this continent for the benefit of the people who live here now. There may be just as much waste in neglecting the development and use of certain natural resources as there is in their destruction.[77]

As for Marsh's influence on invasion science, it is at most indirect and probably insignificant. Wallace cited him in both of his books treating invasions, but in each not with respect to invasions but to Marsh's suggestion that, but for the presence of goats and camels, many parts of African and Asian deserts would be forested.[78] Major early twentieth-century works dealing with invasions (notably by George Thomson [1922] and Charles

Elton [1927]) did not cite him.[79] Neither did Aldo Leopold in his many writings on non-native species during his earlier years in which he took a utilitarian stance toward them nor later when he viewed them as an environmental problem (see chap. 6, sec. 6.1).[80] Nor did the book by Elton (1958) often credited with initiating invasion science (discussed in chap. 7, sec. 7.1).[81] By the 1980s, when modern research on invasions coalesced into a major science (see chap. 9, sec. 9.1), Marsh was known primarily to historians rather than scientists, unless the scientists were interested in the history of conservation and environmentalism. A 1955 conference inspired by Marsh's book led to a prominent volume, *Man's Role in Changing the Face of the Earth* that included three chapters on non-native species, none of which cited Marsh.[82] Darwin, by contrast, is frequently cited with regard to his insights on biological invasions, including what came to be called his naturalization hypothesis (described in chap. 11, sec. 11.3.2).[83]

: : :

Occasional travelers as early as the sixteenth century noted what they asserted were non-native species, but until the nineteenth century the biogeography of much of the earth was so poorly known that recognizing that a species was not native in a region was challenging and difficult to prove. As naturalist-explorers of the nineteenth century documented which species were found where, their findings made such determinations easier. The task was facilitated by increasing acceptance of Darwin's and Wallace's depictions of evolution by natural selection, which provided a much firmer basis for understanding what it means for a species to be native or non-native in a region. Until the twentieth century, most biogeographers considered non-native species to be impediments to their goal of depicting the native biotas of various regions and their differences. Few concerned themselves with impacts of non-native species, except when these impacts affected human activities such as agriculture. The few who recognized ecological impacts—especially Darwin, Wallace, and Hooker—were especially influenced by observations of species introduced to islands. A mid-nineteenth-century polymath, George Perkins Marsh, remarkably described and predicted many sorts of invasion impacts, including ecological ones. However, Marsh was so far ahead of his time and so out of the mainstream of the thinking of contemporary biologists that led to the modern science of ecology that he had no discernible impact on the subsequent evolution of modern invasion science.

4

Mid-Nineteenth- and Early Twentieth-Century Invasions

Scientists Engage in Management

The middle of the nineteenth century saw the beginning of a rapid development of sciences relevant to understanding and managing invasions. The field of biogeography was well underway and greatly aided scientists' ability to determine which populations were non-native (described in chap. 3, sec. 3.1). Charles Darwin published *On the Origin of Species* in 1859. The rediscovery of Gregor Mendel's laws of inheritance in 1900, along with late nineteenth-century research (especially by Wilhelm Roux, Theodor Boveri, and Walter Sutton) leading to an understanding of the role of chromosomes as bearers of genes, crystallized a science of evolution and genetics that helped to explain aspects of invasions and their management, such as the evolution of resistance to pathogens or pesticides. Although the first modern plant ecology book was published in 1899 and the first animal ecology texts in the early twentieth century, much research throughout the mid- and late nineteenth century relating the biology of organisms to their physical environments constituted an incipient science of ecology that helped early scientists understand why certain non-native species thrived and spread in certain areas.[1] The field of phytopathology developed into a modern science in the nineteenth century, with the first modern textbook published in 1858.[2] This development aided some scientists to understand major pathogen invasions and to recognize appropriate management techniques.

Eight invasions during the mid-nineteenth and early twentieth centuries that attracted enormous attention primarily because of their economic impacts exemplify the engagement of scientists in understanding the scope of invasions and impacts and in attempting to control or even eradicate the invaders.

4.1 European Rabbit in Australia

Humans have intentionally introduced the European rabbit, which originated in the Iberian Peninsula, to many parts of the world beginning at least as early as 1300 BC.[3] Virtually everywhere they were introduced, rabbits came to be seen as a scourge and were targeted for control or eradication.[4] The Australian case is famous and exemplifies how scientists and others attempted to manage this invasion in the nineteenth and early twentieth centuries. Rabbits were brought from England to Sydney with the First Fleet in 1788 and were noted by 1825 as common around houses there, and in 1827 as common on estates in Tasmania.[5] Beginning around 1850, rabbit importation increased, under the aegis of acclimatization societies (as discussed in chap. 5, sec. 5.1) and with government support, but early introductions were of domestic varieties ill-suited to life in the wild.[6] However, in 1859 Thomas Austin, an acclimatization enthusiast and sportsman in Victoria who scorned domestic rabbits and sought challenging prey, imported about a dozen wild English rabbits and released them.[7] Delighted by the new hunting opportunities, he distributed rabbits to many friends. By 1862, thousands of their offspring were damaging trees Austin had planted to benefit non-native game birds. By the early 1870s, the five main rabbit breeding areas in Australia were established, and by the 1880s, farmers and others recognized these and other introduced wild rabbits as terrible pests.[8] The devastation was so great that, by 1879, farmers began to abandon properties (fig. 4.1).[9]

Attempts to control rabbits began by the 1870s and took many forms. Shooting, largely by hired "rabbiters," led to hecatombs of thousands of rabbits, but rabbits compensated for this loss by their rapid reproduction. Neither were wire fences and stone walls effective, even when bordered by deep trenches filled with metal shards and broken glass.[10] Beginning in 1871, Australian states passed various laws aimed at controlling rabbits, often by hiring rabbiters or by employing inspectors and mandating that landowners control rabbits; none of the laws were effective.[11] Poison baits and gassing rabbit burrows killed millions of rabbits but failed to control populations except temporarily and locally. At various times mongooses (presumably the small Indian mongoose), ferrets, and house cats were released as potential biological controls.[12] The mongooses and ferrets failed to establish populations, and, although feral house cats became major threats to native vertebrates, their impact on rabbits was minimal, as was that of introduced foxes, even though rabbits at times constitute at least half the diet of both species.[13] (In New Zealand the government

Fig. 4.1 Rabbits around a water hole during myxomatosis trials on Wardang Island, South Australia, 1938. Credit: National Archives of Australia: A1200, L441286.

funded introduction of ferrets, stoats, and weasels, all of which, but particularly stoats, ultimately devastated native birds but none of which controlled rabbits.[14]) The first "rabbit-proof" fence, constructed in Western Australia in six years beginning in 1901 to keep rabbits from spreading and from reinvading cleared areas, spanned 1,833 kilometers (1,139 miles), and rabbits breached it before it was finished.[15] Two fences farther west, 1,165 kilometers (724 miles) and 257 kilometers (160 miles) long, respectively, also failed, as did a total of 10,143 kilometers (6,303 miles) of fencing in Queensland.[16]

In the late 1880s and 1890s, pastoralist William Rodier proposed controlling rabbits by massively trapping them, killing only the females, and releasing the males. His theory was that, if he could skew the sex ratio sufficiently, the many bucks would so harry the few does that the does would become less fertile, and that in addition the bucks would kill any young the does might produce.[17] Although his plan was endorsed by the prominent British zoologist Philip L. Sclater, and the notion of controlling or even eradicating an invasive species by skewing the sex ratio would recrudesce a century later (described in chap. 17, sec. 17.7), Rodier's plan was impractical on any substantial scale, partly because not nearly enough rabbits could be trapped and released and partly because the males, in well-designed experiments conducted in 1905–6, did not in fact act as predicted.[18]

In 1874, Anthony Willows, New South Wales government veterinary officer, conducted the first experiment on controlling rabbits with an infectious disease by inoculating a rabbit with tuberculosis, but transmission to other individuals was insufficient to control a population and he terminated the experiment.[19] The general idea resurfaced occasionally, with focus on various possible diseases, and in 1918, Brazilian scientist Henrique de Beaurepaire Aragão first suggested using the myxoma virus.[20] The idea was not tested until 1924 in cages, and Australian experimenters, not realizing that insects were needed to vector the virus, ended the experiment. By the time of further testing in 1936–43, the importance of vectors was recognized, but results were inconclusive. Finally laboratory experiments in 1950 led to release of the myxoma virus with spectacular effect in some regions.[21]

However, two related problems limited the success.[22] First, the mosquitoes that vectored myxoma virus were not effective in much of Australia.[23] Second, the virus evolved to be less virulent and the rabbits evolved to resist the virus, so each successive epizootic killed a smaller proportion of the rabbit population. Introduction of more virulent myxoma strains was hindered because mosquitoes more effectively vectored the more benign strains.[24] A more effective vector, the European rabbit flea, had been tested in an Australian laboratory, but it was not understood at the time that these fleas must feed on a pregnant doe to produce eggs, so they were not released.[25] As knowledge of the flea's biology increased, these fleas were tested again in 1967 and released in 1968.[26] They substantially increased the viral impact.[27] However, the continuing evolution of increasingly resistant rabbits and an increasingly benign virus led to the introduction in 1995 (by escape from a field test) of a calicivirus that causes rabbit hemorrhagic disease, which has proven very effective but which will be opposed by the same evolutionary forces that lessened effectiveness of the myxoma virus.[28]

4.2 Phylloxera in France

One of the most damaging nineteenth-century invasions was that of France by the grape root louse phylloxera, an aphid native to the eastern United States. Although American grapevines had been introduced to France at least since the late eighteenth century, the phylloxera apparently could not survive the long voyage until the advent of steamships in the nineteenth century.[29] The phylloxera was observed in England in 1863, but in France the puzzling death of a number of vines observed that year was not recognized as related to this insect. The phylloxera probably reached France in the great number of American vines that were introduced between 1858 and 1862 since they resisted a damaging fungus (oidium) assumed to have arrived on

American vines because it had never been seen in France before American vines were introduced. The phylloxera initially went unrecognized as the cause of the spreading death of grapevines because of its complex life cycle including both aboveground and belowground forms. The belowground form, the primary cause of vine death, was often difficult to find and was not immediately identified as the same species as the aboveground form. By 1868, the connection had been made but was at first doubted by many scientists and viticulturists. In 1869, both English and French scientists finally determined that this insect was a species that had previously been described in its native North America.[30]

Through 1869 the invasion was restricted to the lower Rhone valley, and, owing to an independent introduction, to the Médoc region. However, over the next twenty years, it spread to all French vineyards except in the far north, and by the 1890s, even these were infected, as were other vineyards throughout Europe. The economic disaster spurred a massive attempt to control the invasion. Scientists first turned to sulfur compounds, especially carbon disulfide, because these had been effective against oidium. In certain soil types, annually flooding vineyards with water for an extended period effectively killed phylloxera individuals and eggs but proved too costly for widespread implementation. Scientists who had recognized the significance of the resistance of certain American vines to the phylloxera had quickly suggested either replacing French vines with American ones or grafting French vines onto American rootstocks, but not until the 1880s, by which time sulfur compounds had proven ineffective and immersion too expensive, did grafting spread widely as a management approach.[31] Although grafting largely solved the phylloxera problem in Europe, importation of American rootstocks for this purpose is probably the route by which another pest, downy mildew, arrived in Europe.[32] Also, a phylloxera strain has arisen in California that appears to have surmounted defenses of resistant rootstock.[33]

Remarkably, research by 1877 had shown that, whatever the proximate mechanism of plant resistance to phylloxera, it was hereditary, and this before the significance of Mendel's laws and the basic mechanisms of heredity were understood. Controlled hybridization was then undertaken to produce resistant varieties adapted to local French soils and climate and producing appealing wine, but by the end of the nineteenth century, grafting with and replacement by American vines was largely complete and the industry restored.[34] The genetic research and also development and testing of immersion and testing and rejection of the efficacy of sulfur manifested impressive scientific research by several French scientists. By contrast, it took two or more decades before the scientific community as a whole recognized that the fact that many American vines were resistant, far from exonerating them as culprits in the invasion, suggested that they might well have been responsible for it. However, this was a century before basic genetics and coevolution between hosts and pathogens or parasites were understood. It is ironic that the American vines that probably brought the phylloxera were introduced to stem another invasion, that of oidium, suspected to have come from America. The irony is particularly striking because oidium was believed to have arrived on American vines and American vines were recognized as resistant to this fungus. Finally, it is noteworthy that the phylloxera invasion led several European nations to ban all imports from the United States of nursery stock of any plant species, years before the United States contemplated similarly restrictive measures.[35]

4.3 Cottony Cushion Scale, the Vedalia Beetle, and a Fly in California

In 1868 or 1869, the cottony cushion scale insect was discovered in Menlo Park, California, on Sydney golden wattle trees imported from Australia.[36] By 1886, the scale was established in at least eight widely dispersed counties.[37] The California citrus industry, suffering substantial damage from dense scale infestations, sought advice from Charles V. Riley, entomologist for the United States Department of Agriculture and chief of the Division of Entomology.[38] Riley had been one of the first to notice that American grapevines resisted the phylloxera that was plaguing France, and he had assisted French scientists in research on grafting French vines onto resistant American rootstock. In 1887, Riley addressed a meeting of fruit growers in Riverside, California, and argued that they should "appropriate a couple of thousand dollars for no other purpose than the sending of an expert to Australia to devote some months to the study of these parasites [of the scale] there and to their artificial introduction here."[39] The federal government would not fund such a trip, having specifically inserted a clause in the Department of Agriculture appropriations bill to prohibit foreign trips by employees, in response to Riley's European trips that legislators saw as junkets.

Riley and the California commissioner of agriculture, Norman J. Coleman, effected a political work-around. A US exhibit arranged by the State Department was planned at an international exposition to be held in Melbourne, Australia, in 1888, and the secretary of state was convinced to allocate two thousand dollars for an entomologist to represent the State Department. Riley chose his assistant Albert Koebele, whom he had appointed in 1885 to study insects in California and who had been working on the cottony cushion scale problem there. Koebele's problem upon arrival in Australia in September 1888 was that the scale was uncommon. In Sydney and Melbourne he found none. However, as a representative of the US government, he received much cooperation from local entomologists and located the scale in gardens in Adelaide. The first individual he examined contained nine pupae of a parasitic fly, *Cryptochetum iceryae*. Soon thereafter, while collecting scales in Adelaide to ship to California, Koebele "discovered there, for the first time, feeding upon a large female *Icerya* [cottony cushion scale], the Lady-bird, which will become famed in the United States—*Vedalia cardinalis*."[40] He continued to collect scales with vedalia beetles and *Cryptochetum*, ultimately sending four shipments of these insects to California. En route back to California in January 1889, Koebele stopped in New Zealand, where apparently vedalias had arrived within the last few years and had rapidly spread and destroyed scale populations. In Napier, New Zealand, he collected approximately six thousand individual vedalia beetles and arrived in California in March. These beetles were liberated on caged citrus, as the previous shipments had been, soon after arrival.

The beetles made short work of the caged scales, the cage was removed, and the beetles moved to adjacent trees and were also shipped to citrus growers elsewhere in California.[41] Within a very few years, the beetle had so greatly reduced scale numbers at many citrus plantations that growers viewed it as miraculous. Koebele became a celebrity, and Californians presented him with a gold watch and his wife with diamond earrings in gratitude.[42] Moreover, the beetle became famous, and the success of the project inspired an "infatuation with ladybird beetles" that quickly led not only to introduction of vedalia beetles to many other areas but also to many introductions worldwide of other beetles in the family Coccinellidae.[43] In fact, the majority of non-native ladybird beetles failed to establish populations, and of the minority that did, several became pests by attacking or competing with nontarget insects, massing in people's homes, and even fouling wine with "ladybird taint" by colonizing grape leaves in search of aphids.[44] It remains mysterious exactly why the vedalia was such a successful colonist when so many other ladybird beetles failed, but this project remains one of the great successes of the practice of biological control (as discussed in chap. 17, sec. 17.5).

Meanwhile, the parasitic fly, *Cryptochetum iceryae*, also established populations in California. Although Koebele shipped these flies to California, it is likely that the populations in California arrived by another route, as Koebele's flies, liberated in cages with the vedalia beetles, died in great number.[45] However, in 1887, the California state inspector of fruit pests, W. G. Klee, had corresponded with the entomologist Frazer Crawford in Adelaide seeking assistance in controlling the scale, and Crawford collected *Cryptochetum* and shipped them to Klee, who liberated them near San Francisco in early 1888, before Koebele sailed for Australia. These were probably the progenitors of the current California population. Today, both species together control the cottony cushion scale throughout California. The fly is dominant in coastal areas, including Los Angeles, where the vedalia achieved its initial success, and the vedalia in interior and desert areas.[46] In California, these insects also control cottony cushion scale on host plants other than citrus, such as ornamental *Acacia* and *Pittosporum*.[47]

The arrival of the cottony cushion scale in Florida is a cautionary tale.[48] Having heard of the striking success of the vedalia beetle on California citrus, in 1893 a Florida citrus firm inquired of a California correspondent whether these beetles might be useful against other scales on Florida citrus. To the surprise of the Florida questioners, their California interlocutor sent them a box of vedalia beetles, helpfully enclosing (but not noting on the package) some cottony cushion scales for them to feed on en route. Unaware of the danger, the Florida nurserymen left the opened box near a grapefruit tree. Finding the cottony cushion scale several months later on nearby trees, they cut down and burned the trees, covered the ground with straw, chips, and litter, saturated it with kerosene, and burned it. However, in 1898, the scale was discovered about two miles away on rose bushes, myrtle, and goldenrods, and soon thereafter on nearby citrus. A rapid clearing and burning failed to eradicate the new invasion site, and the scale spread rapidly throughout Florida. The vedalia beetle, though present, did not stem this spread and apparently does not suffice today, as it does in California, to control occasional outbreaks in Florida citrus, which are managed with insecticides.[49] Elsewhere, the vedalia beetle continues to be used successfully as a biological control of cottony cushion scale. In the Galápagos Archipelago, a rapidly spreading invasion first detected in 1982 that infested several threatened native or endemic plants was controlled by vedalia beetles introduced in 2002.[50]

4.4 Spongy Moth in Massachusetts

Étienne Léopold Trouvelot obtained spongy moth egg masses from France to culture in the garden behind his home in Medford, Massachusetts, in 1868 or 1869. It is widely believed that he was attempting to use them in silk production, possibly by hybridizing them with native American moths, but no direct evidence shows this to be the case. Spongy moths escaped from his home and established a population on local vegetation. Though Trouvelot notified local entomologists about the escape and in 1870 an entomologist noted that the species is a pest in Europe and was spreading in Massachusetts, no action was taken to control the nascent population.[51] Initially, the spread was apparently restricted to the vicinity of Medford, and the numbers at first were not great except near Trouvelot's house. By the summer of 1889, the moth had spread far enough and caterpillars had become so abundant in Medford that the extensive defoliation and masses of frass plus caterpillars falling from trees or on the ground generated a public outcry.[52] In 1890, the State of Massachusetts appropriated funds to eradicate the moth from the thirty known infested cities and surrounding areas.[53] The key methods employed teams manually removing or destroying egg masses, destroying caterpillars with burlap tree-banding, cutting and burning vegetation, and

Fig. 4.2 Hired workers removing spongy moth egg clusters from two elm trees in Medford, Massachusetts, 1905. Credit: USDA Forest Service, from *Illustrations of the Ravages of the Gypsy and Brown Tail Moths.*

spraying infested trees with Paris green (copper acetoarsenite) and lead arsenate (fig. 4.2). Several entomologists before and during the ten-year Massachusetts campaign suggested introducing natural enemies of the spongy moth from Europe, especially parasitic insects, but these suggestions were not pursued at that time.

The campaign failed, hamstrung by insufficient knowledge of the moth's biology; insufficient, inconsistent funding from the state so that a large trained workforce could not be maintained; and differing perceptions of the problem by most entomologists, on the one hand, and the public and its representatives, on the other.[54] The public was primarily concerned with immediate relief from the defoliation and the accompanying mess wrought by the moths. Entomologists recognized that, should the moth persist and spread even beyond Massachusetts, agricultural and other damage could be catastrophic. The entomologists thus sought total eradication, elimination of every last spongy moth in North America.

By the time the eradication program began in 1890, whether eradication would have been feasible with the methods at hand, had adequate funding been consistent, is an open question. Even though female spongy moths do not fly, so that, on its own, the species does not disperse rapidly, the original infestation had festered and spread for more than twenty years by the time the program began. Furthermore, humans can easily spread the moth inadvertently in the egg, caterpillar, and pupal stages; through individuals falling into or being attached to wagons (nowadays cars); or by other means. Thus the moth had probably already spread farther than the original target region of a substantial area north and west of Boston (including Medford), even though it was not numerous outside this region. A persistent problem during the eradication program was the discovery of new infestations, some outside the original target region. In addition, the program was not allowed to cut certain potentially or even actually infested trees that probably served as sources for frequent reinvasions of areas that had already been considered cleared of the moth. Probably at least these latter two problems would have defeated the effort even had much more funding been available. However, it gives one pause to know that, between 1920 and 1932, an adequately funded joint state and federal project in New Jersey successfully eradicated a separate invasion of the spongy moth that had spread over an area roughly the size of that treated in Massachusetts using virtually the same methods that had been used in Massachusetts.[55]

In any event, the massive Massachusetts effort locally curtailed spongy moth populations, but this result cut both ways. It allowed project directors to claim progress and to predict ultimate success if only funding were available. However, the public could see the problem as they perceived it to be lessened, and as time went on continued prediction of future success came to be viewed skeptically. In 1899, defoliation was greatly reduced, at least partly owing to the eradication effort, and the state ended the eradication project. By 1905, the spongy moth population had increased rapidly and spread to surrounding states, and in 1906 the federal government appropriated funds to prevent spread of the moth. Federal and state efforts continue to this day.

Since the early twentieth century, the spongy moth range has expanded but slowly, in part owing to the flightlessness of females but also because of effective management.[56] The species now occupies only a third of its potential range in the United States.[57] State and federal governments currently operate a 100-kilometer barrier zone along the entire expanding population front that has greatly reduced spread rates.[58] In uninfested states, nearly one hundred thousand pheromone traps are deployed annually to detect newly established populations founded when egg masses or other life stages are accidentally transported. Once detected, these incipient populations are eradicated, mainly through aerially applied microbial pesticides.[59]

For the most part, spongy moth populations are now controlled, though with occasionally intensive outbreaks. Although knowledge of spongy moth biology is still incomplete, small mammals prey on larvae and pupae and are believed usually to keep moth population levels low. However, the mammals are generalized feeders, and their populations are primarily governed by the size of acorn crops, which vary greatly from year to year. During a year or in a local region with a dearth of acorns, mammal populations plummet and spongy moth populations explode.[60] In some areas, landowners attempt to suppress outbreaks by aerially applying microbial insecticides.[61] Several species of natural enemies were introduced during a flurry of state and federally funded introductions between 1905 and 1915.[62] Others may have been inadvertently introduced, and others may be native. The spongy moth nucleopolyhedrosis virus, accidentally introduced with parasitoids, is host-specific and plays an important role in suppressing outbreaks.[63] Another particularly effective fungal pathogen, *Entomophaga maimaiga*, was introduced in 1909 but not detected with certainty again until 1989.[64] However, it is most effective only during wet periods.

A by-product of early spongy moth management programs concerns the introduction of about forty parasitic insects, of which ten established populations, beginning in 1905.[65] Among these was the tachinid fly *Compsilura concinnata*, introduced in 1906. This parasitoid did not prevent spongy moth spread or outbreaks, and its role in suppressing outbreaks is uncertain, but it apparently is the main reason that the rapid spread through New England of the equally feared introduced browntail moth not only ceased but retracted for many decades to two small coastal populations.[66]

4.5 Small Indian Mongoose in the West Indies and Beyond

The small Indian mongoose was introduced twice to the West Indies.[67] In 1870, several from India were released on Trinidad, and in 1872, four males and five females from Calcutta arrived in Jamaica. The introduction to Trinidad probably did not establish a population, but the ones that sugar planter William B. Espeut imported to Jamaica to control rats thrived. He sold mongooses to other Jamaican planters, and he and they, in turn, sold them to planters in Cuba, Puerto Rico, Grenada, Barbados, and Trinidad.[68] Jamaica was probably the source of mongooses sent to many other islands in the West Indies in the nineteenth century.[69] Mongooses from Jamaica purchased by Hawaiian sugar planters arrived in 1883 at Hilo on the island of Hawaii; they flourished, and their descendants were released and established populations on Maui, Molokai, and Oahu.[70] The small Indian mongoose was also introduced in the late nineteenth and early twentieth centuries to Mauritius, Okinawa, the Fijian islands, and Mljet in the Adriatic. More recently it was introduced to Ngazidja (in the Comoro Islands), several other Adriatic islands, and nearby mainland from Mljet, and to Amami Oshima from Okinawa.[71] These introductions were mostly aimed at controlling rats, especially in sugarcane fields, or venomous snakes.

On most islands to which the mongoose was introduced, rat populations in cane fields fell at least initially, sometimes dramatically, but surviving rats tended to nest more arboreally, rendering them largely inaccessible to the mongoose.[72] Nevertheless, as late as 1890, the mongoose on some West Indian islands was still protected by law because of its perceived impact on rats.[73] However, by ten to fifteen years after introduction to an island, it was usually already considered an agricultural pest, largely because of attacks on poultry, and often a threat to native wildlife as well.[74] Native bird and reptile populations on several West Indian islands were greatly affected, and by 1920, the mongoose was recognized as devastating birds and reptiles on St. Kitts.[75] Elsewhere, in 1899, the Chamber of Agriculture of Mauritius advised against introducing the mongoose, based on reports from other islands, but the government nevertheless proceeded to do so, believing it was avoiding potential problems by introducing only males.[76] However several of the "males" were females, and the mongoose was recognized as a pest there by 1905.[77] Similarly, by 1921, the mongoose was perceived as a plague in the Hawaiian Islands.[78] In the West Indies, several islands, including Jamaica, Barbados, St. Vincent, Antigua, Barbuda, and Trinidad, quickly introduced legislation to destroy the species or prevent its introduction.[79]

Controlling the mongoose proved difficult. Early efforts in the West Indies focused on bounties and organized trapping programs.[80] Bounty programs on Trinidad, St. Kitts, Antigua, Barbados, and St. Vincent often yielded huge numbers of mongooses but at best reduced numbers only locally and temporarily.[81] A similar result obtained on Mauritius, where a bounty was established in 1905, just five years after introduction.[82] The discovery that the mongoose vectored rabies in an outbreak on Puerto Rico in 1950 and similar findings on Grenada led to a substantial effort to find an effective poison and suitable bait, but little recent research has occurred despite the apparent promise of the approach, and livetrapping remains the major method in areas where management is undertaken.[83] A long attempt to eradicate the mongoose from the large Japanese island of Amami Oshima employed first live traps, then lethal traps; bounties, then paid trappers; and dogs trained to detect mongooses. Success was reported in 2024.[84]

4.6 Water Hyacinth in Florida

Water hyacinth from South America was probably introduced to the United States at the Cotton Centennial Exposition of 1884–85 in New Orleans, Louisiana; other possible slightly earlier introductions cannot be confirmed.[85] Visitors distributed it to surrounding Louisiana parishes (the equivalent of other states' counties), and a tourist at the exposition returned with water hyacinth plants to her farm near Palatka, Florida, placing them in a fountain near the St. Johns River.[86] They flourished in the fountain, and excess plants were thrown into the river.[87] By 1893, water hyacinth obstructed boat navigation on the river.[88] By 1896, it had spread throughout the region, partly

Fig. 4.3 Mechanical harvester removing water hyacinth from the Caloosahatchee River at LaBelle, Florida, 1939. Credit: US Army Corps of Engineers.

assisted by ranchers who introduced it from basin to basin, believing it to be a good cattle feed.[89] By the beginning of the twentieth century it had colonized all southeastern coastal states as far north as Virginia as well as California.[90] By 1951, water hyacinth had infested at least fifty-one thousand hectares in Florida.[91]

Navigation problems posed by water hyacinth in Florida quickly led to legislation by the US Congress in 1899 authorizing the US Army Corps of Engineers to remove it from Florida waters, and to similar actions by the Florida legislature.[92] The first army engineer in charge of the project rejected the idea of biological control, cognizant of unintended consequences of earlier introductions, including that of the rabbit to Australia.[93] At first he tried chemicals, including muriatic acid, sulfuric acid, carbonic acid, and kerosene followed by fire, none of which proved effective, while a commercial chemical of unknown composition injured cattle that ate sprayed plants. Thus mechanical approaches dominated the effort for the next forty years. These included boats with roller-mounted knife blades and floating booms and traps along with elevators to remove the masses of hyacinth thus contained (fig. 4.3).[94] The latter approach sufficed to keep river stretches temporarily clear, but the rapid growth and sheer weight of water hyacinth meant that mechanical removal could not produce substantial regional control.[95]

Attention thus returned to the use of chemicals or biological control agents. The Army Corps of Engineers found the herbicide 2,4-D, discovered in 1942, to be effective against water hyacinth and began using it to this end in 1948.[96] In the mid-1950s, the Florida Game and Fresh Water Fish Commission and the Corps of Engineers began working jointly to control water hyacinth using 2,4-D, the new herbicide diquat, and various mechanical harvesters, but they succeeded only in slowing its spread, not stopping it.[97] These two agencies, plus other federal, state, and local agencies that attempted to control water hyacinth for various purposes, managed water hyacinth through the 1960s in uncoordinated fashion, spraying primarily areas when and where water hyacinth was problematic (e.g., by interfering with boating) and leaving some unmanaged waters as ongoing sources for controlled downstream waters. A philosophical change occurred in management in the 1970s: consistent monitoring of all infestations and spraying infestations (primarily with 2,4-D) before they became problematic.[98] This change has produced effective long-term control.[99] Notably, although 2,4-D is not a selective herbicide, the amount used annually is greatly reduced because the ongoing management regime prevents hyacinth from developing immense masses requiring large amounts of 2,4-D.

In the 1970s the Corps of Engineers also began studying possible use of South American insects to attack water hyacinth, releasing the mottled water hyacinth weevil in 1972, its congener, the chevroned water hyacinth weevil, in 1974, and the water hyacinth moth in 1977. The primary impact of these insects has been subtle, consisting of slowing water hyacinth growth and thus rendering the plants more susceptible to other causes of mortality, including herbicides.[100] The subtlety of the impact has led to controversy over whether the combination of 2,4-D and mechanical removal suffices in Florida or whether the biological control insects are necessary for long-term, consistent suppression. Exactly the same debate has played out in Lake Victoria, Africa, where the same two weevils were also introduced.[101] Chemical control and biological control of the same target invader are often viewed as incompatible, largely because the herbicide or pesticide may drive the target population to such a low level that the natural enemy cannot persist, at least if it is host-specific. However, Ted Center and colleagues suggest that, in Florida, the two methods can be integrated, with herbicides generally lowering water hyacinth populations below management thresholds but not so low as to eliminate weevil and moth populations.[102]

4.7 White Pine Blister Rust in North America

White pine blister rust, a fungus native to Asia that attacks pine species in the subgenus *strobus* (white pines or five-needle pines), reached Europe within the last two hundred years, possibly on a tree imported from farther east in Russia, but it was not seen as problematic except on American white pines, which had been introduced to Europe, probably by the sixteenth century. By the middle of the nineteenth century, some of these American white pines were recognized as diseased, and the disease was originally believed to be the same as a known stem rust on European pines.[103] However, white pine blister rust has an alternate host, species in the genus *Ribes* (currants and gooseberries). A new fungus species was first reported on *Ribes* in the mid-nineteenth century, and German phytopathologist and mycologist Heinrich Klebahn in 1899 proved that the fungus on *Ribes* and the fungus on American white pine in Europe were different stages of the same species.[104] Black currant, native to Europe, is both very widely cultivated in Europe and particularly susceptible to white pine blister rust, thus virtually guaranteeing that white pines planted there would become infected.[105]

White pine blister rust arrived in North America around 1900 and was first recorded in 1906 on cultivated currants in Geneva, New York. The decline of white pine stocks in North America in the nineteenth century owing to increased demand for timber and improvement in logging and manufacturing equipment precipitated the introduction of white pine blister rust to North America.[106] This decline became so acute that the industry moved its focus to related Western pines by the late nineteenth century, but white pine was sufficiently valuable as timber that increased planting occurred, and American nurseries could not supply enough seedlings. This demand plus the removal of tariffs and absence of any restrictions led to importation of white pine seedlings from Europe, especially from France and Germany. These shipments included rust-infected seedlings, which are essentially asymptomatic for several years.[107] Although workers destroyed all the rust-infected currants discovered in Geneva in 1906, infected imported pine seedlings continued to be planted in the Northeast. For instance, in 1909, several million white pine seedlings were shipped from Halstenbeck, Germany, where the rust was known to be present, to more than two hundred locations in North America.[108] With hindsight, we can see that someone who had wanted deliberately to introduce a pathogen to devastate white pine in North America could hardly have devised a better plan than to grow American white pine in Europe, allowing it to acquire pathogens present there, and then to ship seedlings from Europe back to America.

Earliest management efforts consisted of destroying infected trees and removing the intermediate host, *Ribes*, from white pine plantations, as Europeans had done.[109] However, beginning in 1915, federal funding became available and control quickly focused on the notion of suppressing the fungus by eradicating *Ribes* within two hundred to three hundred yards of white pine stands, funded by landowners and eventually by many states and localities. Soon various state regulations prevented movement of white pine seedings and gave workers the right to enter private properties, destroy

Ribes, and in some instances compensate owners for destroyed plants.[110] The program morphed into an attempt to eradicate *Ribes* entirely from large regions, ultimately any area one thousand feet or less from white pine. The eradication program entailed two components, cultivated and wild *Ribes*. European black currant, known to be highly susceptible to the rust, was widely cultivated in eastern North America (and less frequently in the West) for its fruit, and destruction of these plants and gooseberries engendered hostility from homeowners.[111] States were slow to authorize black currant eradication (e.g., Connecticut did not pass such a law until 1929), but once the project was authorized, cultivated black currants were quickly eradicated.[112] Importantly, Canada did not conduct a *Ribes* eradication program and so would always constitute a seed source.[113]

Wild *Ribes* was another story entirely, especially in the West, where many susceptible pine stands were in rough, inaccessible terrain containing native *Ribes*. Despite skepticism by some phytopathologists, federal agencies and some states funded an enormous eradication program. Many digging techniques were tried during the program, and crew sizes and marking approaches also evolved. Some *Ribes* species tended to regrow from remaining tissue, leading to use of chemicals, especially sodium chlorate. Streamside *Ribes* in the West were particularly difficult to remove. The Great Depression and early relief programs meant that thousands of men could be employed eradicating *Ribes*, but the onset of World War II removed this labor pool around the time that some program leaders and phytopathologists questioned whether wild *Ribes* could be sufficiently eliminated to justify the eradication program, based especially on accumulated evidence on the distance rust spores can travel.[114] *Ribes* continued to be removed from areas near white pine plantations, but the project to stem white pine blister rust by eradicating its intermediate host over vast areas ended soon after World War II.

Although the *Ribes* eradication program in retrospect seems as if it was destined to fail because of the difficulty of truly eradicating wild species, it is important to recognize that the program did largely succeed in eradicating widely distributed cultivated European black currants and that information on spore dispersal distance accrued only gradually. There simply was not enough known about the biology of the disease for much of the duration of the program.[115] That the concept was not prima facie foolish is suggested by the impact of the program to eradicate common barberry, native to southern Europe but widely naturalized in the United States and Canada. Barberry is the alternate host for wheat stem rust, a pest of wheat undoubtedly introduced to North America inadvertently by early European settlers as they introduced both wheat and barberry. In the wake of a devastating epidemic of wheat in 1916, an enormous barberry eradication program jointly mounted by federal and state agencies ran from 1918 through the late 1970s and was a major factor, along with breeding of resistant wheat, in reducing the stem rust to a minor problem.[116] The barberry eradication program did not eliminate all common barberry bushes, but it removed a huge number in wheat-growing areas, not only reducing the number of aerial rust spores but also greatly aiding the breeding program by reducing the number of races of the rust against which breeders must select for resistance.

When the white pine blister rust eradication program ended, management efforts turned, around 1950, to the possible use of sprayed antibiotics and to breeding resistant Western white pines.[117] The breeding programs, capitalizing on the presence of somewhat resistant genotypes in all Western white pine species, achieved substantial success, to the extent that resistant seed is available for plantation use.[118] However, the antibiotics research, focused on cycloheximide and phytoactin, proved inconclusive and by the late 1960s was largely discontinued.[119]

4.8 Chestnut Blight in Eastern North America

In 1904 Hermann Merkel, chief forester for the New York Zoological Park (now the Bronx Zoo), discovered several American chestnut trees on the grounds dying of unknown cause. From a characteristic lesion on the trunk and branches, he suspected a fungus. A scientist from the US Department of Agriculture (USDA) confirmed it as a fungus but misidentified the species and thus gave bad advice about stopping its ravages.[120] Within a year, virtually all chestnut trees on the grounds were diseased and dying. Skeptical of the treatment advice, Merkel asked a mycologist at the New York Botanical Garden, William A. Murrill, to examine the fungus and trees. Murrill correctly identified the fungus, now known as the chestnut blight, as a previously undescribed species that had by then spread at least to several other states, and he worked out its life cycle and means of attacking chestnut trees.[121]

Murrill was skeptical about stemming the spread of the disease and ultimate death of billions of chestnut trees, but his pessimism was countered by other scientists, including Haven Metcalf, first chief of the recently established USDA Laboratory of Forest Pathology. In 1908, Metcalf established a quarantine zone within a radius of thirty-five miles of Washington, DC, and assigned scouts to destroy every chestnut tree with even the slightest sign of a lesion. His enthusiasm led to early optimism that this scorched-earth approach was working, despite occasional findings of infected trees within the quarantine zone. Largely based on the claimed success and Metcalf's testimony, in 1911 the Pennsylvania legislature allocated $275,000 to implement a similar plan, aiming to eradicate chestnut blight from the state, and especially to keep the western part of Pennsylvania from being infected, if need be by establishing a Maginot Line across the entire state in which every chestnut tree would be removed.[122] Adoption of this program, despite testimony based on the biology of the fungus from Murrill and a New York phytopathologist, Franklin Stewart, that it would fail, was accompanied by a resolution imploring the federal government and other states to implement the same program.[123]

The plan failed, and the Pennsylvania legislature refused in 1913 to continue the program despite the plea of Pennsylvania Chestnut Blight Commission that the program could succeed if maintained. Other states and the federal government had not followed Pennsylvania's example, and within approximately forty years, almost all American chestnut trees were dead or dying. It is difficult to capture adequately the massive change this devastation wrought on the landscape. In many areas, chestnuts had comprised a fourth or even a third of canopy trees and were not only key ecological components but also the economic mainstay (partly through timber) and a cultural icon in some regions.[124] Sucker sprouts continue to grow from the base of ancient stumps, only to be killed by the blight when they reach waist height.

The question of how and from where the blight arrived in New York was immediately posed. Murrill suspected it arose in situ by mutation of a native fungus. Others, noting resistance of chestnuts in Asia, surmised that the fungus was native to China or Japan. Much suspicion settled on the possibility that the fungus had arrived on Japanese chestnuts that had been imported by the thousands in the late nineteenth century.[125] Recent genetic work confirms this hypothesis.[126] A similar chestnut blight invasion of Europe began in Italy on native European chestnut trees in 1938, but the rate of spread quickly slowed and destruction has remained less complete, at least partly because a virus of the fungus decreases its virulence and perhaps also because European chestnuts are more resistant.[127] The European invasion probably has multiple origins from both Asia and North America, but the spread within Europe largely began with the Italian invasion.[128]

Today efforts to resurrect American chestnut focus in three directions. Attempts to introduce the virus that attenuates the European invasion and to induce it to spread have so far failed, but research continues on this approach.[129] Researchers have long sought to hybridize resistant chestnut strains from Asia with American chestnuts to produce a resistant tree, suggested as early as 1914 by Arthur H. Graves, who later became a curator at the Brooklyn Botanic Garden, with expanded work by USDA scientists beginning in 1927.[130] The early research was pioneering, in that Mendel's laws of genetics had only recently been rediscovered and tree generation lengths are long. The early efforts ran into the problems that the genes conferring resistance were unknown and that resistant hybrids that contained those genes also had so many other Asian genes that they did not much resemble American chestnuts. A few genes conferring resistance have now been located, and current research focuses specifically on resistance genes in the

Asian trees so that the resulting tree will be predominantly American chestnut.[131] Finally, a wheat gene that codes for an enzyme breaking down oxalic acid secreted by chestnut blight has been inserted into American chestnut, with some resulting transgenic trees showing strong resistance to the blight. However, the project has become mired in controversy with other transgenic trees manifesting lower resistance and other unfavorable traits.[132]

: : :

Various of these eight cases exemplify frequent difficulties in understanding and combating invasions in the nineteenth and early twentieth centuries, largely stemming from a nonexistent or nascent science of ecology, fuzzy notions of the evolutionary process, and the absence of an effort to draw together the rapidly growing number of recognized damaging invasions into a catalog that would be routinely consulted for suggestions as new invasions were recognized. Nevertheless, scientists in each case had important insights that led to deeper comprehension of the nature of invasion impacts and in most cases to a management approach that diminished the impacts. Sometimes these insights came remarkably quickly (e.g., in the cases of the phylloxera in France and the cottony cushion scale in California) and other times took almost a century (e.g., the rabbit in Australia), but eventually all these invasions except for those of the chestnut blight and mongoose were substantially contained.

A key concept in ecology is the notion of compensatory mortality: mortality in a population caused by some factor (e.g., hunting or predation) that is compensated for by increased reproduction, increased survival, or both in the remaining population, so the mortality does not cause more than a temporary decline in the population.[133] Much effort by modern animal ecologists, beginning with Alexander J. Nicholson in 1933 and continuing through the present, is devoted to determining what factors cause population sizes of particular species in nature to be what they are, and part of that effort consists of determining whether mortality imposed by some factor that kills a fraction of a population is compensatory or not.[134] In other words, does killing a large fraction of a population by certain means actually cause a population to decline? It is now understood that killing many individuals need not have a significant long-term population impact, but this fact was not so clearly recognized in the late nineteenth and early twentieth centuries, and it was not codified in an ecological context that would automatically be considered if one aimed to manage a population. Thus the massive slaughter of rabbits in Australia, the great number of mongooses trapped for bounties in the West Indies, and the huge number of spongy moths killed in the Massachusetts eradication campaign were all optimistically and enthusiastically perceived initially. The target population was temporarily lowered, and only subsequently was it recognized that these deaths were not solving the problem.

Evolutionary thinking was in its infancy in the nineteenth century, and the genetic underpinnings of evolutionary change were not understood. Thus the difficulty of understanding the nature of resistance to pathogens and insect pests hindered management of some invasions. For instance, as French scientists grappled with the phylloxera invasion, the initial failure by many scientists to understand that absence of damage on American vines suggested that North America was the native range of whatever was ravaging the vines slowed adoption of the grafting program that finally salvaged the wine industry. Even as late as the twentieth century, the successful use of myxomatosis to lower rabbit densities was hindered by the unforeseen rapid evolution of resistant rabbits and less virulent viral strains.[135]

Often there was simply insufficient information about the biology and life history of the invading organism to understand why the impact was so great and to begin to fashion an effective response, or at least to rule out ideas that would clearly not work. To some extent, that is often the case today with new invasions, but we now have mature sciences relevant to the various aspects. For white pine blister rust, its life cycle and alternate host (*Ribes*) were finally understood in Europe just seven years before its arrival in North America, and lack of knowledge of spore dispersal distance led an ultimately futile eradication campaign to be maintained for a long time. For the other plant pathogen discussed in this chapter, chestnut blight, the initial identification was incorrect, the correctly identified species had not been previously recognized or studied, and many phytopathologists encouraged a campaign to stem its spread that was doomed because of inadequate knowledge of the biology of the fungus. For the rabbit eradication campaign, failure to understand the need for a vector for myxomatosis, and vector biology, impeded progress. For the phylloxera, the belowground and aboveground stages of the life cycle were not at first recognized as the same species.

The rapid success of the vedalia beetle and the fly *Cryptochetum iceryae* in controlling the cottony cushion scale in California inspired many other biological control projects (as described in chap. 17, sec. 17.5).[136] In many instances, however, attempts to find suitable biological control agents were long delayed because of fear of unintended consequences of introducing a natural enemy. For instance, the Army Corps of Engineers turned immediately to chemical control, with a series of failures, rejecting consideration of biological control because of publicity about impacts of rabbits in Australia and other invasions.[137] Introduction of the mongoose as a biological control agent in many areas was one of several similar nineteenth- and early twentieth-century efforts using vertebrates feeding on many nontarget species that were eventually recognized as serious mistakes, such as the cane toad in Australia, mustelids in New Zealand, and the myna in the Hawaiian Islands.[138] Not until the 1990s did nontarget impacts of introduced biological control agents become a major general concern. Nowadays vertebrates such as those cited in this paragraph are rarely if ever introduced for this purpose, and procedures for predicting nontarget impacts of insects and pathogens proposed for biological control have become far more stringent.[139]

Despite the various delays, failures, and mistakes, it is striking that science was quickly brought to bear on each of these invasions, the biology of the invader was usually fairly quickly understood, and management approaches were fashioned, generally after trial and error, that ameliorated the impact. More importantly for the evolution of a true science of invasions, some of the scientists confronting these invasions knew about previous research on other invasions and took this knowledge into account. Thus, Charles V. Riley, who initiated the biological control project on cottony cushion scale, had earlier assisted French scientists in understanding resistance to phylloxera by American vines and the resulting grafting effort, and he even collaborated with them to introduce an American mite that ate phylloxera.[140] The cottony cushion scale biological control success, especially the role of the vedalia beetle, inspired many subsequent biological control efforts with lady beetles and other insects.[141] For example, the early monograph on the spongy moth, explaining the extensive attempt to introduce natural enemies of the moth in the early twentieth century, began with a description of the striking success of the vedalia beetle in controlling the scale.[142] This nascent development of a recognized, hierarchical literature proceeded in the middle of the twentieth century and was crucial to the development of modern invasion science (see chap. 9, sec. 9.1). What had not yet happened at this point was a view of biological invasions as a global phenomenon with many local manifestations. Rather, each scientist dealing with one of the invasions discussed in this chapter was focused on that invasion as a particular problem to be solved by particular means. The overall increasing flow of invaders globally was not a key point of concern, as it began to be in the middle of the twentieth century.

5

Late 1800s to Early 1900s

The Trickle Becomes a Flood

The invasions reviewed in chapter 4, and others, stimulated awareness of the dangers of introducing species. But they would have remained isolated, though disturbing, lessons had not individuals, private organizations, and government agencies continued introductions. They did so in the face of growing publicity and concern surrounding non-native species that characterized the end of the nineteenth century and first half of the twentieth century, even as many acclimatization societies either disbanded or focused on other activities. Indeed, several introduction programs continued even after Charles Elton's 1958 monograph (discussed in chap. 7) drew increased attention to invasion impacts, right into the modern era of invasion science triggered by the Scientific Committee on Problems of the Environment (SCOPE) program of the 1980s (the topic of chap. 9, sec. 9.1).

Stepping back from those infamous invasions described in chapter 4, we can see a much broader picture of optimistic transplantation. Beginning in the nineteenth century, it seems that no taxon was safe from being hauled halfway across the globe in the name of better hunting, more fishing, better fruits, filling the proverbial empty niche (discussed in chap. 15, sec. 15.5), or simply providing reminders of home for immigrating humans. This chapter reviews highlights of that diaspora, and chapter 6 will describe the stories of some stalwarts who attempted to stem the flow.

5.1 Acclimatization Societies

A nineteenth-century phenomenon that contributed substantially to spreading non-native animals (and a few plants) in various regions was the "acclimatization society." Such societies were formed primarily to introduce species perceived as "useful" to new areas, especially to regions of the globe increasingly settled by Europeans.[1] The striking aspect of these societies from the standpoint of growing concern with invasions and, ultimately, scientific study of them is not so much the blithe insouciance with which they distributed species that often became problematic. Rather, it was the recognition, often delayed but sometimes evident even as the introductions occurred, that such invasions could be troublesome.

Probably the first such formally constituted organization was the Société Zoologique d'Acclimatation established in Paris in 1854 under the chairmanship of Isidore Geoffroy Saint-Hilaire. By 1863, at least thirty-nine regional acclimatization societies in continental France, five in French colonies (Algeria, French Guyana, Île de la Réunion, Martinique, and Guadeloupe), and twenty-four in foreign countries had been established and associated with the Paris society.[2] In 1860, the Acclimatisation Society of the United Kingdom was founded, and several acclimatization societies were formed in Australia from 1861 through 1896.[3] In New Zealand, an acclimatization society patterned after that of the United Kingdom began in Auckland in 1862, followed within a decade by five regional societies.[4] An acclimatization society modeled after the Société Zoologique d'Acclimatation was established in Moscow in 1857, becoming the Imperial Russian Society for the Acclimatisation of Animals and Plants in 1864, with branches in three other cities.[5] In the United States, the American Acclimatization Society was incorporated in New York City in 1871, followed by the Cincinnati Society of Acclimatization in 1873; soon thereafter, similar organizations formed in San Francisco and in Cambridge, Massachusetts.[6] Activities of most such organizations ended by the early twentieth century, partly in response to increasing reports of damage by various non-native species. However, in the Hawaiian Islands, the Hui Manu, established in 1930 primarily by white upper-middle-class bird lovers, competed with the Honolulu Mejiro Club, founded by Japanese immigrants, to introduce bird species.[7] The Hui Manu did not disband until 1968.[8] The Hui Manu and the Honolulu Mejiro Club exemplify societies formed simply to import species for aesthetic or cultural purposes, to remind immigrants of home. Thus the German Song Bird Society was established in Portland, Oregon, in 1889 to introduce European songbirds.[9]

For the most part, these societies consisted of prominent, often wealthy citizens, including civic leaders, and functioned somewhat like community service organizations that also serve a social function, much like Kiwanis and Lions Clubs today in the United States. Stated objectives of the introductions were varied, but all were promoted as providing some sort of societal benefit—carnivores to control pests; fish and game for both food and sport; birds to eat insect pests, as game, or to brighten with colorful or melodious species from their homelands what immigrants perceived as a dull and dreary landscape (especially on islands). For British and French colonials, the latter motivation was often paramount, as was expressed in 1862 by Sir Frederick McCoy, professor of natural sciences at the University of Melbourne in a meeting of the Acclimatisation Society of Victoria, complaining about boys killing songbirds, "those delightful reminders of our early home":

> That those delightful reminders of our English home would even now have spread from that centre over a great part of our colony, and the plains, the bush and the forest would have had their present savage silence, or worse, enlivened by those varied touching, joyous strains of Heaven-taught melody which, our earliest records show, have always done good to man . . . we should, I have no doubt, all have been able ere this to have enjoyed one result of the Society's labours in hearing so many of those songsters wild about us, were it not for the pestilent counteraction of the younger portion of the British public, who, in this portion of Her Majesty's dominions, assert their privileges touching the destruction of such ferae naturae.[10]

McCoy even rejoiced that the European rabbit was "so thoroughly acclimatized that it swarms in hundreds in some localities and can at any time be extended to others."[11]

As noted previously, in the 1850s through the 1870s that constituted the heyday of acclimatization societies, few scientists had recognized harmful impacts of non-native species; McCoy was a prominent scientist, as were members of many other acclimatization societies. Nevertheless, the sheer number of introductions and the fact that potential downsides were usually discussed casually, if at all, is striking. Claude-Auguste Millet, vice president of the Paris society, recommended introducing zander to France while deploring the introduction by the British society of the wels catfish to Great Britain and noting the destruction of native brown trout in Lake Geneva by non-native burbot.[12]

Many introductions planned by societies in Australia and New Zealand were thwarted because the targets did not survive the overseas journey. Most species introduced by acclimatization societies did not persist, at least partly for lack of consideration of the match between their native environment and their intended new homes or the small numbers of individuals introduced, or both (discussed further in chap. 15, sec. 15.2). This is fortunate, as many societies released species that, had they survived, would almost surely have become horrendous pests, as might have happened with the small Indian mongoose in Australia (discussed in chap. 4, sec. 4.5).[13]

However, the societies succeeded in introducing starlings to North America, wels catfish to Great Britain, European rabbits to Australian islands, house sparrows, starlings, and blackbirds to Australia, and opossums, hedgehogs, house sparrows, little owls, and mallards to New Zealand.[14] All these populations are now viewed as pests. Although damage caused by contemporaneous introductions by governments and individuals with similar aims (e.g., ferrets, least weasels, and short-tailed weasels [stoats] in New Zealand to control rabbits[15]) may outweigh that wrought by those of acclimatization societies, the lasting damaging impact of the societies is substantial. Further, some societies collaborated with governments and individuals to introduce species with persistent harmful effects.[16] Several societies targeted plants for introduction, often to aid proposed introductions of animals. Thus George Bennett, physician, naturalist, and first president of the Acclimatisation Society of New South Wales, suggested at an 1863 society meeting that blackberries be introduced to support introduction of the curassow, a large Neotropical turkey-like bird.[17] Blackberries are now a prominent pest plant in Australia and many other areas to which they were introduced. Remarkably, Bennett was well aware of the damage non-native plants could do, and just one year later he warned that "many injurious plants have been imported which afterwards could not be eradicated, which we have experienced in the Scotch thistle, Bathurst burr, also the elegant variegated or milk thistle and the Scotch burr, one of the yellow trefoils."[18]

Scientists at times weighed in on potential dangers of introductions under consideration by acclimatization societies. A review of books on New Zealand birds included this general condemnation:

> We have above hinted at some of the causes of extinction which seem to be at work . . . but there is another, and possibly more powerful cause. . . . This is the silly mania for "acclimatisation" which has been so warmly fostered by many well-meaning though ill-advised persons, both at home and in the colonies, and nowhere more than in New Zealand. The English Acclimatisation Society fortunately came to an end, and before it had time to do any harm here; but its example has been mischievous in our dependencies. In a reckless way animals of extremely doubtless advantage have been transported to the antipodes, and there it seems impossible to deny that they will in a few years be found not only ousting the kinds which are less specialised, and therefore less able to meet them on an equal footing; but, unaccompanied by any of those checks which keep the whole of a natural fauna balanced, the importations will inevitably become the greatest of nuisances. The memory of the patriotic Scot who could not live without his thistles is not exactly blessed by Australians, and among the pilgrim

fathers of New Zealand who will ultimately obtain an apotheosis, the members of their various acclimatisation societies will, we suspect, scarcely be reckoned.[19]

As an example, the suggestion to introduce mustelids (ferrets and weasels) to New Zealand to control non-native rabbits quickly elicited strong objections from both scientists and lay commentators.[20] In 1876, Alfred Newton, professor of zoology at Cambridge, James E. Harting, a noted British ornithologist, and Walter Buller, New Zealand's leading ornithologist, pointed to the hecatomb such mustelids would probably wreak on native animals (especially birds) and cast doubt on the likely efficacy of their control of rabbits. Their plea was echoed the same year by lay letters and editorials in Otago newspapers.[21] These cautions were probably too late, as individuals had already legally released both ferrets and weasels, many more private releases quickly followed, and the government formally approved such releases and even provided legal protection for the mustelids in the Rabbit Nuisance Act of 1881.[22] It should be noted, however, that Newton, Harting, and Buller, though prescient in predicting great harm to native birds, did not cite specific research or data—their arguments and verbiage resembled those of lay citizens who opposed the releases.

By the 1880s, several non-native species in New Zealand were already widely recognized as problematic, and some societies were at pains to deny their role in introducing them. For instance, the Canterbury Acclimatisation Society in 1889 stated, "We most deliberately deny ordering or introducing this questionable bird [the house sparrow]."[23] In Australia, concerns expressed about introductions proposed by the acclimatization societies similarly did not substantially impede their activities in the nineteenth century. Only in the early twentieth century did government agencies act to prevent such introductions, at least partly in response to conservationists' objections.[24]

In general, among factors acclimatization societies considered regarding particular introductions, preventing the decline and possible extinction of native species was not prominent, and in fact was nonexistent until previous introductions were widely perceived as problematic, in particular for devastating native species. Some societies even initially viewed as part of their remit destroying *native* species that were attacking non-native species that societies had previously introduced. For instance, in 1866, the North Canterbury Acclimatisation Society offered a bounty on a native hawk, as did two other New Zealand societies the next year, and in the 1870s at least one New Zealand society paid bounties on a native kingfisher believed to be preying on non-native sport fish. Societies also subsequently introduced species to control species that they had previously introduced that had become problematic, such as the little owl, native to Eurasia, introduced by the Otago Acclimatisation Society in 1909 to control non-native songbirds that were attacking crop plants; within twenty years the owl itself was perceived as a pest and the society was considering control methods. The North Canterbury Acclimatisation Society was instrumental in introducing the black swan from Australia to New Zealand in an attempt to control watercress previously introduced from Europe that was believed to complicate attempts to introduce trout.[25] By the early twentieth century, the swans were viewed as a nuisance. The massive growth of watercress was later said to be partly controlled by competition from non-native willows planted on riverbanks.[26]

As impacts of non-native species became more widely recognized in the late nineteenth century and especially the twentieth century with the advent of a formal science of ecology, acclimatization societies either disbanded or transmogrified into entities supporting hunting, fishing, and bird-watching or similar activities without an acclimatization mission.[27] However, societies in Australia and New Zealand continued to distribute previously introduced salmonids.[28]

5.2 Fish

A federal agency, the US Commission of Fish and Fisheries (often referred to as the US Fish Commission), was established in 1871.[29] From the outset, much US Fish Commission activity consisted of introducing fishes (and invertebrates, such as clams and lobsters) to disjunct areas of the United States where they were not native. The "aquarium car" was a key component. This was the brainchild of Deputy Commissioner Livingston Stone, who in 1872–73 used a modified Central Pacific Railroad car to increase greatly the ability of the commission to introduce fish species of eastern and central North America to Western states, perceived as impoverished in useful food fish.[30] This and similar aquarium cars led to extensive introductions by the US Fish Commission.[31] For instance, between 1871 and 1894, the US Fish Commission introduced to western states twenty-three fish species native to the eastern United States and four Old World fish species that had previously been introduced to the eastern United States.[32] Many of these introductions went first to California, often in cooperation with the California Fish Commission.[33] Of these twenty-seven non-native species, twenty-five survived to establish populations (including all four Old World species—common carp, goldfish, tench, and brown trout), and many were subsequently distributed by the US Fish Commission or by state fish commissions to other states. Except for the goldfish, an ornamental amenity, all US Fish Commission introductions were of species thought to be good food or sport fish. Not only did the US Fish Commission introduce foreign fish to the United States and fish from states where they were native to states where they were not, but it also generously introduced fish to other nations.[34] For instance, in 1882, the US Fish Commission introduced common carp, native to Europe and Asia, to Brazil.[35]

Few of the introductions to western states were considered at the time to pose an environmental or conservation risk. What risks were perceived fell into three categories. The greatest worry was associated with the common carp, on the grounds that its feeding on aquatic plants and muddying waterbody bottoms would destroy major habitat for other fishes and for ducks. Second, for three catfish (brown bullhead, white bullhead, and channel catfish), there was concern that their large populations would deplete native fishes by eating their spawn.[36] Finally, some people feared that the introduced yellow perch simply did not taste good enough.

These worries vastly underestimated the unfolding difficulties. Many of these fishes proved to be problematic both environmentally and to sport and commercial fisheries.[37] For instance, as had been foretold, common carp indeed became perhaps the most environmentally destructive of all fish introduced into North America and the target of many management and eradication campaigns.[38] Their harmful impacts were recognized early, and the muskellunge was introduced by the US and California Fish Commissions partly in the hope that they would control the previously introduced carp.[39] They failed to do so, as had an earlier introduction for the same purpose of sea lions into Lake Merced, a San Francisco reservoir.[40] As for predation concerns, often the newcomers turned out to prey on other prized fish—such was the case for the green sunfish, bluegill, largemouth bass, smallmouth bass, white bass, lake trout, brook trout, northern pike, and muskellunge. Green sunfish and largemouth bass also prey on amphibians of conservation concern. Finally, while hybridization was not a worry in the late 1800s, it has become a mounting problem (discussed in chap. 11, sec. 11.3). Brook trout, by hybridizing with the native bull trout, contribute to the federal threatened status of the bull trout, and several other fishes introduced by either the US Fish Commission or state agencies have raised concern that hybridization with related native species will not only irrevocably change the native species but might do so in ways that are maladaptive.[41]

Given warnings voiced in the nineteenth century, one might have expected diminished twentieth-century interest in introducing fish. Indeed, the US Fish Commission scaled back its introduction activities in the twentieth century, although it did not largely stop them until the Conference on Exotic Fishes and Related Problems, initiated by the American Fisheries Society and the American Society of Ichthyologists and held in 1969 in Washington, DC.[42] However, other federal agencies continued introductions. For example, the US Fish and Wildlife Service imported ohrid trout from Europe around 1965 and helped fish farmers in Arkansas, Missouri, and Mississippi import black carp from Asia, which now may be established in several southeastern rivers.[43] It took the US National Park Service, which began introducing non-native sport fish into several national parks in the late nineteenth and early twentieth centuries, a particularly long time to cease introducing fish.[44] Yellowstone National Park, the first United States national park, was established in 1872. In 1882, cutthroat trout were introduced to fishless lakes in the park, and in 1889, non-native brook trout and rainbow trout were introduced to lakes in the park that had native fish. By the 1920s, Yellowstone had non-native rainbow, brook, lake, and brown trout, and the Bureau of Fisheries (successor to the US Fish Commission) was assisting the National Park Service to introduce non-native fish to several other national parks.[45] Notwithstanding statements in 1922 and 1929 from the National Park Service directorate that non-native species were excluded from parks, non-native trout continued to be introduced to Yellowstone and other parks. Despite objections, including by prominent ichthyologist Carl Hubbs, that these introductions threatened native fish species, the practice continued into midcentury and did not cease definitively until 1968, with publication of "Administrative Policies for Natural Areas of the National Park System," which stated that "nonnative species may not be introduced into natural areas. Where they have become established or threaten invasion of a natural area, an appropriate management plan should be developed to control them, where feasible," and further that "nonnative species of plants and animals will be eliminated where it is possible to do so by approved methods which will preserve wilderness qualities."[46]

As the federal role for introductions diminished in the twentieth century, state agencies took over that responsibility, and the federal government, including the US Fish and Wildlife Service, aided many state efforts to introduce species.[47] Among many other examples of sport and amenity fish introductions between 1900 and 1958, when Charles Elton published the monograph often claimed to have initiated invasion science, the Wisconsin Fish Commission introduced goldfish and rudd, the Nevada Department of Wildlife and Nevada State Fish and Game Commission introduced White River spinedace and speckled dace, the Michigan Board of Fish Commission and the Michigan Department of Conservation introduced rainbow smelt and cherry salmon, the California Department of Fish and Game introduced sargo and orangemouth corvina, the Ohio Division of Conservation introduced coho salmon, the New York Department of Conservation introduced sockeye salmon, the Texas Parks and Wildlife Department introduced red drum, the Hawaii Department of Fish and Game introduced oscar, and State of Missouri personnel introduced rock bass.[48]

Many state agencies continued introducing game fish into waterbodies they had not occupied, even after the advent of modern invasion science during the 1980s shone a bright light on the issue.[49] The Texas Parks and Wildlife Department was particularly active, introducing orangemouth corvina,

Nile perch, spotted seatrout, Atlantic croaker, black drum, peacock bass, and various hybrids of non-native species. Other state agencies introducing non-native fishes for sport or amenities during this period included the Pennsylvania Fish Commission (bowfin and hybrid northern pike ×Amur pike), North Carolina Wildlife Resources Commission (flathead catfish), Georgia Department of Natural Resources (flathead catfish), Arizona Game and Fish Department (flathead catfish), Hawaii State Department of Health (redtail notho), Hawaii Division of Fish and Game (peacock bass and redbreast tilapia), California Department of Fish and Game (longjaw mudsucker), Florida Game and Fresh Water Fish Commission (peacock bass [see box 5.1] and speckled pavon), Delaware Division of Fish and Wildlife (hybrid white perch × striped bass), Louisiana Department of Wildlife and Fisheries (hybrid yellow bass × striped bass), and North Dakota Game and Fish Department (zander). The zander introduction was controversial, as the neighboring states of Minnesota, Montana, and South Dakota plus the Canadian province of Manitoba objected that it would inevitably spread into their waters and might harm native fisheries. However, state agencies develop their own guidelines, and North Dakota proceeded with the introduction in the absence of federal regulation to the contrary.[50] A similar objection by the state of Missouri to the introduction of grass carp by the state of Arkansas was fruitless for the same reason.[51] Grass carp were then legally shipped by Arkansas fish farmers to other states, where they were received illegally, and they were subsequently trucked illegally from several of those states to still other states.[52]

BOX 5.1 Case Study: The Peacock Bass in Florida

A particularly contentious dispute arose in Florida regarding introduction of the peacock bass by the Non-Native Fish Research Laboratory, established by the Florida Game and Fresh Water Fish Commission in 1966. The mission of the laboratory was mainly to develop management procedures for the many non-native species introduced to the state, primarily by aquarium releases and deliberate or inadvertent escapes from breeding ponds. An attempt by the commission to introduce the peacock bass in 1964 had failed.[1] In 1984, Paul L. Shafland, a biologist with the Non-Native Fish Research Laboratory who subsequently became its head, proposed to introduce the fish to south Florida canals, which are mostly box-cut into a coral rock substrate.[2] The proposal argued that four conditions should be met prior to an introduction:

> (1) existence of a justifiable need; (2) presence of a vacant or underutilized niche in a hospitable environment; (3) predictable beneficial effects without an unacceptable adverse impact; (4) availability of an effective means for limiting or controlling the exotic should it become too abundant or have unexpected detrimental effects.[3]

The conditions of justifiable need and presence of a vacant or underutilized niche were stated to be satisfied by the fact that the canals were dominated by non-native fish species, primarily tropical ornamentals released from aquaria. The only substantial predator on these species was the native largemouth bass, and predation by the bass was insufficient to lower the populations of these species enough that native species could flourish. Additionally, the proposal claimed an existing inadequacy of "quality fishing opportunities" in south Florida urban areas. Much of the proposal was devoted to satisfying the third condition, particularly the absence of an unacceptable adverse impact, since the predictable beneficial effects were laid out in satisfying the first two conditions. Three potential downsides were each considered and rejected. First was the possibility that peacock bass could spread from the canals to natural freshwater habitats such as the Everglades. The species cannot tolerate water temperatures lower than 15°C (5°F), so the bass could persist in the canals, but beyond the canals, even mild south Florida winters would kill them. Similarly, they cannot tolerate salt water, so they could not spread to adjacent brackish and marine habitats. A second concern was whether peacock bass could prey on or compete with native largemouth bass, and the proposal cited experience in other locations with non-native peacock bass that indicated such effects were unlikely. The third concern was occasioned by the well-known study showing a massive decline of native forage fish after peacock bass were introduced to Gatun Lake, Panama.[4] The proposal discounted the relevance of this study by citing a report of the Food and Agriculture Organization (FAO) to the effect that the decline was only temporary and additionally by claiming that Gatun Lake, with no previous top-level carnivore, was a special case, that the canals had a top-level carnivore (largemouth bass), and that introduction of the peacock bass to other regions with top-level carnivores did not lead to the decline reported in Gatun Lake.[5] The proposal did not address the fourth condition laid out in its introduction, availability of ways to control the introduced population if unexpected detrimental effects arose.

The 1984 proposal was circulated within and outside the Game and Fresh Water Fish Commission, attracting both favorable and critical comments.[6] Although the proposal suggested that full evaluation of the peacock bass for release would require at least five years, in September 1984 the fish was released into the canals. Fifteen years later, in 1999, Shafland published a four-part review of the outcome, finding evidence for the key predictions of the proposal: the peacock bass had not spread and persisted outside the canals, it did not appear to be competing with largemouth bass, it ate primarily non-native fishes, and it supported an enthusiastic, lucrative sport fishery.[7] With respect to the key prediction that the peacock bass would not disperse into other south Florida habitats, a survey of non-native fishes in several habitats in the Everglades region did not find peacock bass outside urban canals.[8] The survey team believed that low winter temperatures generally were a key factor preventing several introduced fishes from spreading into natural habitats, as Shafland had argued would be the case for peacock bass.[9] Shafland and colleagues reported that mild winters since 2000 had temporarily extended the peacock bass's range north and west but were confident low temperatures would eventually cause a range retraction, citing the bass's previous occasional appearance in the Everglades National Park but inability to establish populations there.[10]

As this debate played out in the 1980s and into the 1990s, Shafland defended the notion that cold temperature would limit the range expansion of the peacock bass in Florida by an appeal to what has now come to be known as the "niche stability hypothesis" (as discussed in chap. 14, sec. 14.4).[11] This hypothesis states that traits such as temperature tolerance limits evolve very slowly. Citing laboratory determination of lower lethal temperatures for fourteen tropical non-native fish established in Florida, he termed as "Lamarckian" suggestions that introduced fish might within decades become adapted to new environmental conditions.[12] Both Walter Courtenay and Joel C. Trexler and colleagues, without specifically suggesting evolution of temperature tolerance limits, pointed to the possibility of delayed impacts of non-native fishes that appeared innocuous in Florida at the time.[13] Meanwhile, Courtenay and Chandler S. Robins observed, "Should adaptation occur to temperature spectra present beyond the sites of introduction in southeastern Florida, it is highly probable that *Cichla* will enter Everglades National Park, compounding problems that this unit is already experiencing with exotic fish invasions."[14] By the 1990s, evidence of global warming was widely cited by ecologists in general and by invasion scientists who predicted that warming would cause ranges of long-introduced species to change, mostly by expansion. The US Fish and Wildlife Service ecological risk screening summary for peacock bass used the species distribution model RAMP to show climatic suitability of United States locations under current conditions but did not address suitability with a warming climate.[15]

The peacock bass introduction program led to increasingly heated disagreements between Shafland of the Non-Native Fish Research Laboratory, which spearheaded the program, and Courtenay, of Florida Atlantic University, located near each other in Boca Raton. In the early 1990s, they occasionally vigorously debated the merits or their positions in a series of teleconferences to discuss development of a state report on the impact of non-native species in Florida; the project directors feared they would come to blows if they had met in person.[16] Courtenay, an ichthyologist, was a prominent herald of the dangers of introducing non-native species, and not only fishes, beginning at least as early as 1969 and continuing to his death in 2014 (chap. 7, sec. 7.2).[17] Starting in 1975, Shafland had a long career with the Non-Native Fish Research Laboratory, much of it spent assessing, publicizing, and attempting to limit impacts of invasive non-native fishes. In 1979, Shafland stated his position:

> In general, biologists have not understood or communicated to the general public the likely environmental effects of these introductions. Virtually all foreign species introductions have had objectionable, if not deleterious, consequences and should, therefore, be made only after an affirmative decision, based on sound ecological principles and past experiences, has been reached.
>
> The introduction of an overwhelming majority of foreign species has proven to have been ill-advised. Some have resulted in environmental catastrophes.[18]

In 1979, Shafland also suggested that the research on the peacock bass introduction to Gatun Lake provided general reasons to be wary of introductions, whereas his proposal five years later to introduce the peacock bass to south Florida canals argued that the relevance to other systems of the outcome of the Gatun Lake introduction was likely minimal, as previously noted.[19]

As time went on, Shafland continued to temper his view about the nearly universal objectionable or deleterious impacts of invaders, at least as concerned the myriad non-native fishes in Florida. Thus, in 1986, Shafland asserted that some of Florida's common non-native fishes may establish and thrive because of previous anthropogenic changes in the environment (the "passenger-driver" effect, discussed in chap. 15, sec. 15.4.5). Although his overall message was that introductions were cause for concern, he stressed, in the wake of the peacock bass introduction that he had led,

> nonetheless, purposeful introductions of exotic fishes could have beneficial effects, especially in highly disturbed man-made habitats (e.g., heated power plant reservoirs or canals) to which native species may be poorly adapted. Finally, introduction of exotics should only be done by professionals who have specialized knowledge of these organisms and the fish communities into which they are to be introduced. Purposeful introduction of exotic fishes will prove beneficial when they are carefully matched to identified needs.
>
> The mere presence of exotic fishes . . . is not necessarily detrimental except in the most puristic of interpretations.[20]

By 1996, Shafland was caustic in criticizing what he termed "unsubstantiated eco-centered dogma" and described his own evolution in terms reminiscent of a road-to-Damascus conversion experience:

> As one who had personally endorsed the "nativism" ideology (i.e., nature without man always knows best, and natives are always superior [Shafland 1976, 1979]), this author found the initial transition to a more critical application of objective science uncomfortable, but ultimately more rewarding and satisfying. The primary purpose of this article is to suggest that much of the current information describing the effects of introduced species is based on scientifically unsubstantiated eco-centered dogma.[21]

He proceeded to argue that catastrophic impacts ascribed to zebra mussels may well be untrue and that their filtering activities are beneficial by virtue of clarifying water and as a food source for commercial important species of fish and waterfowl, that no non-native fish species has caused any native species to become extinct in Florida, that the report of a dramatic impact of the introduced peacock bass in Lake Gatun was premature and exaggerated, and that "nonscientific perspectives masquerading as science in scientific journals" have led to ill-conceived management, research, and legislation.[22] He particularly criticized a document on non-native species in the United States produced by the Office of Technology Assessment of the US Congress.[23] Shafland wrote:

> The OTA's decision to base their recommendations to Congress solely on analyses of presupposed and largely unsubstantiated harmful introductions is prejudicial,

> if not demagogic. Nonscientific dogmatism is now so well established in the scientific literature that the OTA seems unaware it is not science.[24]

It may not be coincidental that the OTA report contained this statement:

> Peacock bass (*Cichla* spp.) were introduced to control other non-indigenous fish and as a game fish in southeast Florida canals (101), but the bass is slowly reducing populations of indigenous bass and bream (73).[25]

Reference 73 is a lay article that addresses the fact that the Game and Freshwater Fish Commission had introduced the peacock bass to reduce populations of non-native fish. In it, Courtenay is quoted:

> The peacock bass cannot tell the difference between exotic fish and native game fish, such as small bass and bream, and is slowly reducing the population of those fish as well. . . . "To use a fish as a bio-control agent is really stupid. They are not very particular about what they eat," Courtenay added.[26]

In fact, Courtenay wrote the contractor's report on nonindigenous fishes on which the OTA assessment is based.[27] It contained a neutral description of the peacock bass introduction into Florida embedded in a document depicting the entire practice of introducing non-native fishes as highly risky and stated that, "with luck," Florida may escape adverse impacts of the peacock bass introduction, although it raised the possibility that natural selection could increase the bass's tolerance of cold weather. In his writings, Courtenay used colorful and often highly pejorative language in large generalities: for example, a statement in a book introduction, "One of the unique natural North American communities, that of tropical Florida, has become a biological cesspool of introduced life," and a section of the same paper entitled "The Florida Travesty."[28] He was particularly creative with paper titles: "Species pollution," "Biological pollution through fish introductions," "Crimes against biodiversity: the lasting legacy of fish introductions."[29] In 1997, Courtenay was quite critical of the peacock bass introduction on the grounds that there was a good probability that it would disperse and survive beyond the initial canals.[30]

1. Courtenay, "Nonindigenous Fishes."
2. Shafland, *Proposal for Introducing Peacock Bass*.
3. Shafland, *Proposal for Introducing Peacock Bass*, 1.
4. Zaret and Paine, "Species Introduction in a Tropical Lake."
5. On the decline as temporary, see Welcomme, *Register of International Transfers of Inland Fish Species*.
6. Courtenay and Robins, "Fish Introductions"; Courtenay, "Nonindigenous Fishes."
7. Shafland, "Introduced Butterfly Peacock (*Cichla ocellaris*) in Florida, I: Fish Community Analysis"; Shafland, "Introduced Butterfly Peacock (*Cichla ocellaris*) in Florida, II: Food and Reproductive Biology"; Shafland, "Introduced Butterfly Peacock (*Cichla ocellaris*) in Florida, III: Length Distribution Analysis"; Shafland

and Stanford, "Introduced Butterfly Peacock (*Cichla ocellaris*) in Florida, IV: Socioeconomic Analysis."

8. Trexler et al., "Empirical Assessment of Fish Introductions."
9. Shafland, *Proposal for Introducing Peacock Bass.*
10. Shafland et al., "Florida's Exotic Freshwater Fishes–2007."
11. Shafland, "Exotic Fish Assessments."
12. Shafland and Pestrak, "Lower Lethal Temperatures for Fourteen Non-Native Fishes in Florida." As reference for the suggestion of rapid evolution, Shafland cited only Philipp and colleagues, who actually made no such claim (Philipp et al., "Point/Counterpoint: Conservation Genetics and Current Stocking Practices"). Rather, they lamented the introduction of individuals from one region of a species' range into other regions, on the ground that local populations are adapted to local environments and hybridization between populations from different regions could threaten that adaptation. Evidence exists that some non-native species have adapted quite rapidly to colder temperatures in their non-native range. For instance, the brown anole of Cuba was introduced to peninsular Florida in the 1940s and has spread northward. More northerly populations now have lower critical thermal minima, although it is not known if this adaptation reflects evolution, phenotypic plasticity, or both (Kolbe, Ehrenberger, et al., "Physiological Variation Among Invasive Populations of the Brown Anole").
13. Courtenay, "Nonindigenous Fishes"; Trexler et al., "Empirical Assessment of Fish Introductions."
14. Courtenay and Robins, "Fish Introductions," 166.
15. United States Fish and Wildlife Service, *Butterfly Peacock Bass (Cichla ocellaris) Ecological Risk Screening Summary*; S. Sanders et al., *Risk Assessment Mapping Program: RAMP*.
16. D. Schmitz, telephone and email communications with the author, 2023. In the end, Shafland did not contribute to the document, produced by Florida's Department of Environmental Protection (Schmitz and Brown, *Assessment of Invasive Non-Indigenous Species in Florida's Public Lands*), while Courtenay wrote two chapters, one of which noted without further comment that the peacock bass had been introduced by the state to provide a new game fish.
17. Weigl, Domey, and Courtenay, "Survival and Range Expansion"; Courtenay, "Florida's Walking Catfish"; Lachner et al., "Exotic Fishes and Other Aquatic Organisms Introduced into North America."
18. Shafland, "Non-Native Fish Introductions with Special Reference to Florida," 18.
19. Shafland, *Proposal for Introducing Peacock Bass.*
20. Shafland, "Review of Florida's Efforts to Regulate, Assess and Manage Exotic Fishes," 24.
21. Shafland, "Exotic Fish Assessments," 124. Shafland cited the following as publications in which he endorsed the "nativism" ideology: Shafland, "Continuing Problem of Non-Native Fishes in Florida"; Shafland, "Non-Native Fish Introductions with Special Reference to Florida."
22. Shafland, "Exotic Fish Assessments," 128.
23. Office of Technology Assessment, US Congress. *Harmful Non-Indigenous Species in the United States.*

24. Shafland, "Exotic Fish Assessments," 128.
25. Office of Technology Assessment, US Congress. *Harmful Non-Indigenous Species in the United States*, 263.
26. McClelland, "Invasion of the Bio-Snatchers," 13.
27. Courtenay, *Pathways and Consequences*.
28. Lachner et al., "Exotic Fishes and Other Aquatic Organisms Introduced into North America," 1, 17.
29. Courtney and Ogilvie, "Species Pollution"; Courtenay, "Biological Pollution Through Fish Introductions"; Courtenay and Moyle, "Crimes Against Biodiversity."
30. Courtenay, "Nonindigenous Fishes."

As noted previously, acclimatization societies introduced many fish species to new locations for sport fishing purposes and sometimes for biological control (or biocontrol, discussed in detail in chap. 17, sec. 17.5). Individuals also often introduced fish species, usually for the same reasons.[53] Many such fish introductions were unrecorded, but the fact and purpose of these introductions can often be deduced easily. In other cases, individuals announced the introductions when they occurred or subsequently claimed them after populations had spread. As various regulations came into play in the late nineteenth century and on into the twentieth century, many introductions by individuals were illegal. Perhaps the best-known perpetrator was Stewart Smith, a British immigrant to New Zealand who outfitted his car with oxygenated aquaria and roamed throughout the nation from the 1960s until his death in 2008 at age ninety-five, liberating more than fifteen thousand individual fish of several species despite at least two arrests.[54] He was motivated by the belief that previous fish introductions, particularly of trout, by acclimatization societies and government agencies were for the benefit of wealthy people, and he wished for others to have similar opportunities with the "coarse fish" (denoted by their rough skin) he had fished for as a boy in England. To this end, he introduced rudd to New Zealand, was perhaps the first to introduce koi, and spread previously introduced perch, tench, and goldfish to new bodies of water. He introduced orfe (ide), which may still exist in New Zealand, and is believed to have introduced gudgeon, which was subsequently eradicated. It is uncertain whether he realized his intention to liberate gambusia and an Australian crayfish (marron); he maintained large numbers of both species in outdoor tanks.

Some illegal fish releases are widely publicized. The capture in 2002 of northern snakehead (native to Asia) in a pond near Washington, DC, and their subsequent spread generated national headlines.[55] The introduction was later traced to a man who had purchased two of the fish in an Asian food market and later released them. In 1994, northern pike were found in

Lake Davis, a California reservoir stocked with rainbow trout, prized for fishing. Because the lake is connected to the Sacramento–San Joaquin Delta and beyond, the pike were viewed as threatening native salmonid fisheries already stressed by other factors, and a state agency attempted to eradicate them with the chemical rotenone in 1997. Whether the attempt succeeded or simply drastically lowered pike populations is uncertain, but pike reappeared in 1999, and local fishers were suspected of both the initial introduction and further efforts after the 1997 operation.[56] A subsequent eradication attempt in 2007 succeeded.

5.3 Birds

Although the United States lacked an analog to the US Fish Commission with a mandate to introduce bird species, the US Biological Survey, established in 1885, performed a similar role.[57] Primarily tasked with studying how diets of native birds and mammals might affect agriculture and conducting research on the geographic distributions of native plants and animals, the Biological Survey was soon charged with enforcing the Lacey Act of 1900; this act and the controversy surrounding it are discussed further in chapter 18, section 18.1. This legislation regulated the importation of birds and interstate transportation of game, and the Biological Survey was tasked with granting permits for importing non-native birds and mammals.[58] Although these roles would today seem more aimed at preventing than facilitating introduction of non-native species, in fact the Biological Survey performed introductions and aided both state agencies and private individuals in introducing many non-native bird species as potential gamebirds.[59] For instance, the US Biological Survey introduced at least four non-native birds to Sapelo Island (Georgia) in 1923 at the behest of automobile and aircraft magnate Howard E. Coffin, who in 1912 had purchased the entire island except for Black communities established by formerly enslaved people soon after abolition.[60] Of these introductions, the plain chachalaca, from Mexico, still persists.[61]

As the US Fish Commission had done with fish, the US Biological Survey helped many state agencies introduce game birds from other nations or from one region of the United States to another, disjunct region, including the ruffed grouse, sharp-tailed grouse, Gambel's quail, elegant quail, bobwhite quail, Chinese quail, Reeves's pheasant, kalij pheasant, golden pheasant, and Pallas's sand grouse.[62] In addition, no bird could be imported into the United States without a permit from the Biological Survey, which was mandated to prohibit introduction of any species believed to have "injurious habits."[63] However, despite the detailed criticism of bird introductions by its biologist Theodore Palmer (more on him in chap. 6, sec. 6.1), the Biological Survey permitted many introductions by individuals, often owners of large hunting estates.[64]

In 1929, Biological Survey biologist Waldo L. McAtee advocated introducing many game birds, arguing that European and Asian species in particular, by virtue of their long association with people, were preadapted to vast areas of the United States that had been altered by humans (e.g., by deforestation and land clearing) and seemed unsuitable for native North American species. He largely discounted potential downsides to such a project in a section entitled "Possible Dangers from Importations," in which he acknowledged harmful impacts from the previously introduced house sparrow and starling:

> A lesson has been drawn from such instances, and now the introduction of practically all kinds of exotic animals and plants that may become pests is forbidden. . . . There is little fear, however, that any of the large and highly edible species classed as game birds will continue for any period as pests. Should they exhibit destructive tendencies their numbers can easily be cut down by extension of the open season and increase in the bag limits. No bird that is widely prized as food is ever likely to become destructively abundant in the United States. . . . Fear has been expressed also that the introduction of foreign game birds might carry with it the introduction of diseases . . . [but] it does not seem an argument to which much weight should be given, in view of the fact that domestic poultry is constantly being introduced . . . and constitutes a source and reservoir of most of the diseases to which our game birds are susceptible. . . . Depletion of the food supply for native birds also has been cited as one possible bad effect of introducing additional species. The food supply for game birds can be increased almost indefinitely, however, if the effort be made.[65]

Remarkably, just four years previously, McAtee had warned:

> Some unfortunate upsets in biotic relations have resulted from introductions even in the case of organisms generally dependent upon man for existence, but when we consider animals and plants not strictly domesticated, successful introductions have almost invariably had regrettable consequences. . . . The process of introduction goes pyramiding with no man able to foretell the eventual results.[66]

The US Fish and Wildlife Service (FWS) in 1940 subsumed the US Biological Survey, and Gardiner Bump succeeded Waldo McAtee in promoting game bird introductions. Bump had served eighteen years with the New York State Conservation Department when he was appointed in 1949 by the FWS, charged by the director of the FWS Branch of Wildlife Service with making

> extensive surveys and studies of major game species in various parts of the world which might be suitable for habitat not currently occupied by native

> game species in the United States. . . . He will be responsible for coordinating and directing current efforts to find new and desirable game birds and mammals that might thrive in habitats not now producing what they should.[67]

While working in New York, Bump had enumerated the twenty-two attempts to introduce or supplement bird populations and twenty attempts to introduce or supplement mammal populations there (over half non-native) and lamented the many failures.[68] He attributed these unsuccessful attempts to insufficient research into biology of target species as well as failure to import enough individuals and to protect them after initial release, and he evinced no concern about harm to native species or ecosystems.[69]

Announcing Bump's appointment to the FWS, the director of his unit had stated:

> Before we transplant any new and untried forms of game birds and animals into the United States we must have assurance, based on biological research, that the newcomers will encounter favorable conditions and will not overwhelm a desirable native fauna. The Service will discourage attempts to introduce any species which might seriously compete with present game populations or which might become a nuisance.[70]

However, in an early presentation as an FWS biologist, Bump did "not seek to resolve the question as to whether or not it is a wise policy to introduce foreign game species," recognized substantial questions, and cited the extended criticisms of the practice by Ralph King (which are discussed in detail in chap. 6, sec. 6.1).[71] Nevertheless, he dismissed the importance or likelihood of two possible harmful impacts, eating crops and introducing disease (the latter in very similar terms to those of McAtee). For a third concern, reducing populations of native species, he simply accepted that such things happen, pointing to the example of the small Indian mongoose. The rest of the presentation consisted of advocating introductions and describing the recently created FWS Foreign Game Investigation Program as an instrument to that end: "the real need for filling vacant habitats."[72] As had McAtee, Bump saw the United States as containing vast areas of anthropogenic habitats unsuitable for native game species. His key message was as in his discussion of New York State Conservation Department introductions—learn the biology of the species, especially habitat requirements, and make sure enough individuals are introduced.[73]

After Bump presented his 1951 paper, the discussion grew testy. An initial question from Herbert Levi, who a year later would write a comprehensive early warning about introducing species (discussed in chap. 6, sec. 6.1), asked about the danger of introducing parasites that might attack native species.[74] Bump responded by saying Levi had asked that question previously and he had answered it: the quarantine procedures were stringent, and more than a million game birds had already been introduced with no evidence of a new pathogen or parasite accompanying them. Antoon de Vos, who in 1956 would coauthor a review of impacts (many harmful) of non-native mammals on native species, expressed skepticism on two points: (1) whether the biology of species planned for introduction would be adequately researched to predict results, and (2) whether an introduction that went awry could really be eradicated, as Bump had indicated.[75] On the second point, Bump responded that the initial introductions were all small and so could be eliminated, but he appeared to assume that any untoward impact would be manifested and recognized quickly (an incorrect assumption, as would later be determined; chap. 18, sec. 18.3, discusses how difficult it is to recognize some sorts of impacts quickly). On the first point, he simply claimed that the few successful game introductions so far had been enormously successful and seemed not to address de Vos's concern about unexpected impacts.

In 1968, Bump reviewed the accomplishments of the Foreign Game Investigation Program, again basing the need for the program on the view that many US anthropogenic habitats (and some relatively pristine ones) were deficient in native game, at which point "something is needed to fill the gap."[76] The program had studied twenty-six species or subspecies of non-native birds that it recommended various states introduce; many states followed the recommendation. Arguing that it was too soon after the introductions to say whether the populations would persist, he worried that for some, the numbers of individuals introduced might have been too small. It was not too soon, however, for Bump to say:

> It is encouraging to note that none of the game bird species introduced have proven detrimental to agricultural crops. Nor is there any record of a disease or parasite, new to North America, being introduced with these species. Nor, again, has their presence, to date, proven inimical to any native game birds that are to be found in coverts within the general vicinity in which some of the introduced species were released.[77]

Acknowledging the controversy surrounding the program, Bump viewed the results as "a long step in the application of science and common sense to the problem of foreign game introductions."[78]

Soon thereafter, in 1970, increasing concern with introductions led to termination of the program. The growing literature on ecological impacts is described in chapter 7. Also, game bird biologists increasingly raised the issue of the program expense and the relatively few successful establishments.[79] However, several states still had non-native birds that the program had provided to them and released them to the wild at least through 1978. Although most of these species failed to establish persistent populations, the legacy of the program includes at least five species established in Hawaii and two in continental North America.[80]

Many private individuals released both game birds and songbirds even as warnings about harmful impacts became increasingly frequent. Most notable was Boston socialite Eastham Guild, who lived with his wife Caroline Guild in Tahiti in French Polynesia from 1923 to 1940.[81] Finding "practically no bird life," and encouraged by the French government, to remedy this perceived impoverishment he acquired and released about seven thousand birds of fifty-nine species from Africa, Asia, Europe, the Americas, Australia, and Pacific islands.[82] Of these, between two and five species survived.[83] One survivor was the silvereye, native to Australia and eastern Pacific islands, which has become the most common land bird on Tahiti and has been instrumental, along with the subsequently introduced red-vented bulbul, in dispersing the most destructive invasive plant on Tahiti, the velvet tree. The tree, introduced to a botanical garden in 1937, began spreading dramatically in the 1970s to occupy 60 to 70 percent of the island.[84] The Guilds also introduced many non-native plants to Tahiti, with Caroline Guild awarded the rank of *Chevalier du Mérite agricole* by the French government for this accomplishment.[85] Despite his low rate of success at establishing non-native birds, Eastham Guild bested Harry Bruins-Lich, a South African gardener who directed public parks in Pretoria for twenty-five years. Posted as a young man to the island of St. Helena between 1928 and 1932 as horticultural officer, Bruins-Lich imported and released sixteen bird species from southern and eastern Africa; fifteen were not present at the time, and none of these established populations.[86]

In the Hawaiian Islands, Alex H. Isenberg and his aunt Mary Dorothea "Dora" Isenberg played a key role in introducing birds to Kauai, where Dora lived.[87] Scion of a prominent sugar plantation family on the island, Alex moved to California, where he eventually established the largest private aviary in the United States in Portola.[88] In 1918, he liberated three non-native bird species on Kauai. In 1931, he reported all had established populations, and he added that he and his aunt had liberated three more bird species.[89] These liberations were despite admonitions by Palmer, McAtee, and Grinnell that introducing non-native birds was highly risky.[90] Isenberg rejected these concerns for Hawaii, for which he was taken to task.[91] Isenberg also complained bitterly that the Hawaiian Board of Forestry forbade his release in 1931 of red-whiskered bulbuls, native to Asia, that he had to take back to California.[92] The Hawaii Invasive Species Council classifies this bird as "injurious to wildlife" and forbids its release into the wild or transport to any island (including Kauai, Maui, and Hawaii) that does not already have it, citing, in addition to harm to native birds, the fact that red-whiskered bulbuls are pests to agriculture and gardens and spread seeds of invasive plants.[93]

The Hawaiian Board of Forestry did not categorically discourage bird introductions, however. In 1937, the president of the Board of Commissioners of Agriculture and Forestry for Hawaii, Frank H. Locey, stated that the Division of Fish and Game "encourages introduction into the Territory of all classes of birds that will add to our natural beauty and happiness, aid economic prosperity and bring no evil with them."[94] He envisioned Hawaii as "a universal aviary, a bird haven where every known species of birds, not injurious to Hawaii, will propagate and thrive. . . . Some day Hawaii will be one grand bird preserve or reservation. We will have here every bird known to the world that will live here to our advantage."[95] This goal was despite his recognition that "the entire subject of bird introduction is very controversial."[96] (See box 5.2.)

BOX 5.2 The Laysan Rail

Frank H. Locey played an ignominious role in the extinction of the Laysan rail that seems antithetical to his stated goal in his 1937 paper.[1] Laysan, a small, remote island in the northwestern chain of the Hawaiian Archipelago, had five endemic bird species. Three were driven to extinction in their native home (i.e., Laysan) by the introduction of rabbits and guinea pigs in 1903 by Max Schlemmer as a failed potential business venture in meat canning and to amuse his children. The rabbits and guinea pigs destroyed the vegetation necessary for the birds' survival, and the last rail on the island succumbed in 1923. The rabbits were eradicated during the same expedition that recorded the last rail, but by then it was too late for the bird. However, in the 1890s and again in 1904, 1905, and 1913, the Laysan rail had been introduced from Laysan to Eastern Island, one of the two main islands of Midway Atoll, and by 1910 Laysan rails had been introduced from Eastern to Sand Island, the other main island, in an attempt to control insects. The rail thrived on both islands and even survived massive construction activities before and during World War II, but black rats colonized Midway around 1943 because of frequent boat landings associated with war activities, devasted the rails, and the last rail was seen in 1945. Massive rat control activities on the Midway islands in 1945 were too late for the rail, as the rabbit eradication on Laysan had been. In 1940, George Munro, an amateur naturalist and bird enthusiast from New Zealand who had emigrated to Hawaii, arranged for the capture of more than twenty rails on Midway, but Locey, in his capacity as president of the Board of Commissioners of Agriculture and Forestry, refused permission for their transfer to Laysan, for no apparent reason. He may have been motivated by the view that the Laysan finch, another Laysan endemic that had been introduced to Midway, came to be viewed as a pest there before it was eliminated by the rats, much as the rail was. Locey's Board had earlier forbidden reintroduction of the finch for that reason, and the US Biological Survey in 1940 had forbidden an attempt to reintroduce the finch and the rail to a bird park on Oahu, citing only the concern with pestiferous behavior by the finch.

1. For his paper, see Locey, "Introduced Game Birds of Hawaii." On Locey's role in the extinction of the Laysan rail, see Hume, "Undescribed Juvenile Plumages of the Laysan Rail."

In Great Britain, many members of the landed gentry played a key role in releasing birds, particularly game birds. In the late nineteenth and early twentieth centuries, the eleventh Duke of Bedford was especially active, importing at least twenty species, releasing a substantial portion of these into the wild, and keeping others in a bird park from which escape was frequent. Many of these introductions failed to establish, including the rufous tinamou and red-crested cardinal from South America, the brush turkey from Australia, the red-crested pochard from mainland Europe and Asia, the American flamingo, the purple gallinule from Africa, and the red-cowled cardinal from Brazil. Whereas the eleventh Duke of Bedford focused particularly on game birds, his son the twelfth Duke of Bedford (whose tenure was 1939–53) was especially interested in parrots and doves, although none of the ten parrot species or five or more species of doves released to the wild appear to have survived. The ornithologist Frank Finn eagerly introduced non-native birds to London, releasing several species of doves and numerous cage birds, taking advantage of foreign research trips and a stint as administrator of the Indian Museum in Calcutta. Included were scarlet grosbeaks and crested larks from India, rosy starlings and red-billed leiothrixes from Asia, black-headed buntings from the Balkans, and orange weavers from western Africa. None of the species Finn liberated established populations.[97]

5.4 Mammals

As with fishes and birds, even as warnings increased about potential harmful impacts and acclimatization societies gradually ended their introductions, private individuals and government agencies continued to introduce mammals. Most introductions were for hunting purposes, although some were of furbearers and a few were for biological control.[98] For instance, Gardiner Bump tallied twenty distinct mammal species introduced into New York State between 1895 and 1939: four from Europe, two from Asia, and the remainder from other areas of North America.[99] All were potential game animals (including nine ungulates): the great majority were introduced by either state agencies or wealthy individuals. Of the foreign species, only wild boar survived, and Bump inveighed against the waste of resources in introducing species unlikely to establish populations. Bump's discussion in 1951, with his general dismissal of risks of harm, is described in the preceding section 5.3 with respect to birds but pertained to mammals as well.

Interestingly, Bump stated, "Anyone widely familiar with . . . the wild boar in Europe would have had a reasonable basis for predicting the undesirable results that have followed their introduction into this country," though the earlier report had no discussion of potential harmful impacts of any non-native mammal, including the boar.[100]

Although private individuals might have been ignorant of the increasing scientific literature on risks of harmful invasion impacts, federal and state agencies must have been aware of them, and in fact Bump's 1951 article cited this literature before largely discounting risks if projects are managed by professional game biologists. However, the ongoing activity inspired a publication by Harry V. Thompson, an English expert on rabbits, preceded by a remarkable note from the journal editor, clearly concerned about continuing game introductions:

> This timely paper was obtained from Harry V. Thompson in order that wildlife administrators and biologists in the United States might be warned of the problems that may result from the introduction of *Oryctolagus* [European rabbit]. Surely our science has reached a point where we are justified in condemning this ignorant and irresponsible gamble.[101]

A 1956 review of the ecological impacts of mammal invasions called attention to a general increase in concerns about risk and consequent decrease in introductions for both North America ("Happily, the short-sighted policy of wholesale introduction on a hit or miss basis is no longer as common as it once was") and South America ("It is gratifying to learn that some South American countries have taken steps to prevent introduction of foreign animals").[102] Sometimes government agencies withdrew proposals to introduce game mammals in the face of public opposition based on information about risks. For instance, the Nevada State Wildlife Agency backed out of a plan to introduce Barbary sheep in the 1960s.[103] Other times, agencies discounted risks.

The US Foreign Game Investigation Program, discussed in section 5.3 with respect to birds, paid relatively little attention to mammals, though they fell within its purview. Bump attributed this lack of interest on the part of the states (which had to initiate an introduction study by request to the program) to his perception that native North American big game fared better than native game birds "under the pressures of advancing civilization" and to concern that grazers and browsers might compete with native game or domestic livestock.[104] However, as Bump also acknowledged, private initiatives, especially by ranchers in the Southwest, had already introduced fourteen such species from Asia, Africa, and Europe, and the introductions into New York State mentioned previously were often by private individuals. For instance, William A. Rockefeller Jr. introduced red deer, sika deer, roe deer, and Siberian roe deer to a vast deer park he had established around 1900.

Western ranchers more than compensated for a paucity of federal activities to introduce big game. Although Bump observed no dearth of native game species in the United States, from at least 1884 through 1940, Texas state resource agencies perceived a shortage of huntable wildlife.[105] More than fifty non-native species were brought to Texas alone in partial response to this ostensible shortage, usually initially to enormous, fenced ranches from which, if they thrived, they were introduced to other ranches or to public land (of which Texas has relatively little), or both. These species included Nilgai antelope in the 1920s, followed by nine species in the 1930s and one each in the 1940s, 1950s, and 1960s.[106] Many of the original propagules came from zoos, although several were shipped directly from their native homes, sometimes in collaboration with US zoos. In some instances, multiple propagules from sources with different genotypes led to hybridization in their Texas range—for example, small dark sika deer from Taiwan hybridized with brightly spotted individuals from Japan, yielding many color variants.[107] Some introduced species hybridized with native domestic livestock, such as mouflon with domestic sheep and boar with domestic pigs.

The collaboration of Texas game ranchers with zoos arose partly from a US regulation for even-hoofed ungulates, such as deer and antelope: imported individuals could go only to quarters such as zoos, where they could not contact native livestock and transmit pathogens. However, their offspring could be sent to ranches or dealers. Thus, for instance, the Fort Worth Zoo and the San Antonio Zoo acquired twenty-two species in collaboration with ranchers, who initially received zoo offspring equal in number to the animals they had arranged to be imported to the zoos. The Texas ranches subsequently were integral to several reintroductions to their native ranges of game species whose populations had dwindled or disappeared there. For example, beginning in 1970, several ranches provided blackbuck antelope to conservation nongovernmental organizations, which arranged their shipment to Pakistan.[108] Many ranches also generated substantial income by offering paid hunting for non-native game species on their properties, a practice that flourishes today.[109]

Several species brought to Texas were also introduced to New Mexico, often earlier than to Texas. In New Mexico, unlike Texas, a state agency—the New Mexico Department of Fish and Game—effected these introductions, and they were released to public land, of which there is proportionally much more than in Texas. States varied greatly in the degree of the enthusiasm of their fish and game agencies about game introductions.[110] Arizona officials, for instance, warned New Mexico authorities that non-native game species crossing into Arizona would be shot.[111] The differing policies of the neighboring states of Arizona and New Mexico demonstrate that a single person can be enormously influential.

In Arizona, the driving force was Charles T. Vorhies, an ecologist who conducted substantial research on both insects and vertebrates. He served as head of the University of Arizona Department of Entomology and Economic Entomology and in 1939 as president of the Ecological Society of America. In 1928, with Walter Taylor, an ecologist and mammal specialist, he drafted a referendum that established the Arizona Game and Fish Commission. Politically active and adroit, he also served as editor of *Arizona Wild Life and Sportsman*, using this position as a soapbox to publicize wildlife issues and advocate for policies. He was particularly concerned with recovering native game animals rather than introducing non-native species as a way to maintain game numbers and game department revenues, and he influenced the Arizona Game and Fish Department to adopt this view as policy, contrasting greatly with its policies regarding fish introductions.[112]

By contrast, New Mexico game management policies were heavily influenced by Frank C. Hibben, a charismatic anthropologist associated with several famous archeological discoveries that subsequently triggered great controversy.[113] Hibben, who became wealthy through marriage, was an enthusiastic hunter who hunted big game in Africa and Asia. In addition to many scholarly and popular archeological writings, Hibben published books about swashbuckling hunting safaris and was featured on the ABC television series *On Safari with Frank Hibben*.[114] He also wielded considerable and somewhat mysterious political influence at state and federal levels.[115] From 1961 to 1971, he chaired the New Mexico Game and Fish Commission, which oversees the New Mexico Department of Game and Fish, and in that capacity led efforts to introduce many species of birds (assisted by the federal Foreign Game Investigation Program), mammals, and plants.[116] The Department of Game and Fish had already introduced Barbary sheep to the wild in 1950, but activity ramped up during Hibben's tenure, with oryx (gemsbok), Siberian ibex, and greater kudu imported for trials in 1962–63.[117] Oryx were released to the wild in 1969 and ibex in 1970, while an unsuccessful release of the greater kudu in 1967 terminated that project.[118] Notably, the oryx release was controversial, with objections based on the risk of ecological damage expressed not only by Arizona authorities but also the National Park Service, but Hibben placed high priority on this introduction (having hunted oryx in Africa's Atlas Mountains).[119] He was strongly supported by William H. Huey, New Mexico's secretary of Natural Resources, who acknowledged that the entire program and particularly the oryx introduction were controversial but minimized the concerns, especially ecological concerns, proclaiming that "the contribution which they might make toward breaking down the balance of nature is certainly among the least important."[120] Subsequently, the spreading oryx population overgrazed both public and private lands, oryx are viewed by many people as a nuisance, and their population, about 3,500, is facilitating shrub encroachment of perennial grassland.[121]

Hibben defended the necessity of the entire program on the grounds that state hunting license fees had plummeted.[122] He also adduced the same justification for the New Mexico introductions as that suggested by non-native game ranchers in Texas: several of these species were threatened in their native ranges, and introduced herds could conserve them for possible reintroduction.[123] However, his main rationale for the program to introduce the oryx, Siberian ibex, and greater kudu (as well as the springbok, whose trialing was planned for later) was a version of the justification that both McAtee and Bump had advanced for introducing non-native birds: conversion to agriculture and other uses had changed North American habitats to the disadvantage of native big game, whose populations were further reduced by hunting.[124] In addition, Hibben viewed the North American climate as unsuitable for many native big game species, particularly by virtue of aridity in the West. Thus, suitably chosen non-native species would be filling a vacant niche.

Interestingly, Frank C. Craighead Jr. and Raymond Dasmann, prominent wildlife biologists and no great enthusiasts about introducing species, accepted the proposition that vacant niches for big game exist in much of North America, though they attributed the vacancy to hunting by prehistoric immigrants rather than land conversion and hunting in historic times (Dasmann's view of non-native species is discussed in detail in chap. 7, sec. 7.2).[125] As did Hibben, they believed certain non-native species might fill these niches without necessarily affecting native species, though they were less sanguine than Hibben that such species could be found and considerably more demanding about research needed to support a proposal to release a non-native species to the wild. Craighead and Dasmann also stressed the importance of adequate methods of control if a non-native species spread to an area where it was unwanted or caused an unanticipated harmful impact. Hibben recognized the need for such an assured control method and pointed to the earlier experience with the Barbary sheep introduction as evidence that hunting could serve this purpose, whereas Craighead and Dasmann were much more skeptical that the Barbary sheep case justified confidence that they could be controlled.[126] Hibben saw no evidence at the time of a harmful impact of the Barbary sheep on native game, while Craighead and Dasmann lamented a dearth of adequate ecological research for any of the big game introductions. Today, although several researchers depict likely harmful impacts of the Barbary sheep on native desert bighorn sheep, adequate research on such impacts is still lacking.[127] Both Hibben and the ranchers importing non-native mammals acknowledged the growing concern about invasion impacts but felt the evidence on non-native big game in the Southwest indicated such introductions were safe.[128]

Mammal introductions in other nations proceeded similarly to those in the United States even as warnings about risk increased. Some introductions were undertaken by individuals and others by government agencies, usually with the goal of establishing game species but sometimes for other purposes. Perhaps the largest effort occurred in the Soviet Union—a harbinger was Boris K. Fortunatov's "General Plan for the Reconstruction of Economically Important Fauna of European Russia and the Ukraine" in 1933, calling for massive introduction of non-native species throughout the European parts of the Soviet Union to fill a hypothesized plethora of "empty ecological niches."[129] The apotheosis of this approach was Soviet leader Joseph Stalin's 1984 "Great Plan for the Transformation of Nature." Acclimatization activities were a part of the concept that vast tracts of forests and other habitats could be molded in a few decades into agricultural and other ecosystems to serve human needs. Soviet biologist Trofim Lysenko's claim that hybridization and inheritance of acquired characteristics could produce new, useful organisms was a pillar of the concept, as were non-native species.[130] Early criticisms of such introductions, particularly at nature protection congresses in 1929 and 1933, were quieted as Stalin consolidated power in the 1930s, highlighted by the imprisonment and death of acclimatization critic Vladimir Stanchinskii.[131]

A wave of introductions occurred throughout the Soviet Union in the 1930s through the 1960s, particularly of furbearers, with the goal of increasing economic yield. Some were failures. For instance, nutria, recommended by Russian zoologist Boris M. Zhitkov in an influential monograph on acclimatization, were repeatedly introduced in many Soviet republics beginning in 1930 but failed in Russia, though they established populations in semicaptive circumstances (e.g., supplemental shelter) in several Asian republics.[132] Ultimately, five species non-native throughout the entire Soviet Union were established: the beaver, raccoon, muskrat, and mink from North America, plus the rabbit from farther south in Europe.[133] Other species native to at least part of the Soviet Union were established in regions where they were non-native.[134] Several of the latter introductions were undertaken to improve the quality of fur of native populations. Thus, for instance, the Teleut squirrel, a southern Siberian subspecies of the widely distributed red squirrel, was introduced repeatedly from 1946 through 1963 in several European republics of the Soviet Union in a failed attempt to improve the fur of local subspecies through hybridization.[135] Other introductions were designed to provide food for furbearers, sometimes themselves non-native. For example, root voles and Norway rats were introduced to the Kuril Islands as food for non-native Arctic foxes and native red foxes.[136]

In the 1950s and 1960s, acclimatization efforts increasingly focused on game animals, especially ungulates, and hunting organizations joined government agencies in undertaking these projects, which proliferated through at least the 1960s.[137] Perhaps because of Stanchinskii's fate, discussion of possible negative consequences, especially ecological consequences, was silenced until de-Stalinization began in the 1950s under Nikita Khrushchev. At a 1954 conference on nature reserves, marine biologist Sergei E. Kleinenberg attacked rampant acclimatization as damaging to nature: "I believe that acclimatization is no game. We must approach this technique with extreme seriousness and acclimatization for the sake of acclimatization must not be permitted."[138] In 1958, game biologist Vasilii N. Skalon observed that "acclimatization was advanced under the slogan of the transformation of nature" and "it could only be assessed favorably; no criticism of it was tolerated."[139] The first prominent concern over harmful ecological consequences was voiced in 1961, when a senior ecologist, Andrei A. Nasimovich, published a remarkable article citing well-known Western critics of introductions: "The number of such examples of ill-conceived acclimatizations could be added to. Many of them are scattered and in the works of foreign authors. From them come the great and justified skepticism about animal acclimatization that is felt in the works of many foreign specialists."[140] He repeated this concern in 1963 at a large Soviet conference on acclimatization, where a few other attendees mentioned the possibility of harmful ecological impacts, although in the context of general support for mammal introductions.[141] Far more presentations simply advocated more carefully planned introductions, sometimes citing the need to fill "empty niches" with species useful to humans.[142] In 1963, Vladimir G. Geptner, another senior ecologist, called for the end "of that harmful idea that acclimatization may be a basic means of increasing the productivity of game resources."[143] (For more on Nasimovich and Geptner, see box 5.3.)

BOX 5.3 Nasimovich and Geptner

Andrei Nasimovich, who specialized in the ecology of mammals, and his colleague Vladimir Geptner, with similar interests, were internationally prominent senior scientists who lived through the Trofim Lysenko–led attempt to implement what came to be termed "Stalin's Great Plan for the Transformation of Nature," a forerunner of the novel ecosystem concept. They were part of a courageous group of scientists who attempted to impede implementation of this plan, which envisioned using the Soviet system of nature reserves as experimental sites for the massive introduction of non-native animals and plants as well as other activities to produce ecosystems that would serve human needs.[1] Nasimovich and Geptner frequently cited foreign publications in support of their views on ecology and conservation at a time when much of Soviet science forswore foreign interactions, and both men were particularly caustic in lamenting the Soviet Union's poor record of nature conservation compared with those of other nations. Geptner is one of several Soviet ecologists who are little known for having anticipated ecological concepts associated with Western scientists, in his case the notion that island biotas, because of their relatively reduced number of species compared with those of mainland, pose weaker "biotic resistance" to biological invasions, an idea widely credited to Charles Elton in his heralded 1958 monograph.[2] Today Geptner and Nasimovich are best known outside Russia for their 1961 comprehensive systematic treatment of Russian mammals, translated into English in 1980.[3]

1. D. Weiner, *Little Corner of Freedom*.
2. Elton, *Ecology of Invasions by Animals and Plants*.
3. Heptner et al., *Mammals of the Soviet Union*, vol. 1, *Artiodactyla and Perissodactyla*.

In Great Britain, aristocrats long introduced deer for hunting, including King James I, who introduced a variety of fallow deer in 1611. In the twentieth century, the self-proclaimed king of Lundy Island, Martin Coles Harman, introduced sika deer in 1927.[144] The eleventh Duke of Bedford, who, as noted previously, introduced many game birds to Great Britain, also introduced at least five deer species, of which only the sika deer, also introduced by others during the same period, established a population.[145]

5.5 Reptiles and Amphibians

The analog for reptiles and amphibians of Livingston Stone for fish or Eastham Guild for birds—a person devoted to introducing a plethora of species—was Giuseppe Despott, curator of the Malta Natural History Museum. The Maltese Islands have only one native amphibian (a frog) and six native reptiles: two snakes, two geckos, a skink, and one other lizard.[146] Around 1910, Despott, attempting to erase this lacuna, introduced eight species of frogs, five salamanders, three lizards, and two turtles, for no apparent purpose other than that he wanted them to be present.[147] None established persistent populations.[148] Globally, unlike for fish, birds, and mammals, there has not been interest to introduce species for sport purposes that would engage government agencies as well as individuals. Nor, with a few exceptions (e.g., some frogs, turtles, and tortoises), have reptiles and amphibians been introduced as potential commercially important food items, as fishes have been. Many non-native populations of reptiles and amphibians arrived in their new homes as hitchhikers on houseplants and other cargo, and others escaped from pet dealers. Malta has populations of two snake species that probably arrived as cargo stowaways.[149] However, many reptiles and amphibians were deliberately introduced in various locations. The common chameleon was intentionally introduced to Malta around 1880 by Jesuit priests, spread throughout the island, and persists to this day.[150] Even in the twentieth century as concern about harmful impacts grew, reptiles and amphibians were frequently deliberately introduced. Some were imported for biological control, such as the cane toad, native from extreme southern Texas through parts of South America and introduced to Hawaii in 1932, Australia in 1935, and many other islands around this time to control insect pests of agriculture.[151]

But many more reptiles and amphibians were introduced to new regions because people simply wanted them around. For instance, Florida has about sixty-three established populations of non-native reptiles and amphibians, of which several were deliberately introduced by people who fancied them for ornamental reasons: the northern curly-tailed lizard from the Bahamas, the crested anole from Puerto Rico, the large-headed anole from the island of Hispaniola, the knight anole from Cuba, the giant anole from Jamaica, and the ringed wall gecko from northern Africa.[152] For these species, unassailable evidence (often reports by the people who introduced them) shows that the introductions were deliberate. For several other species, intentional introduction is probable but uncertain, or one or more deliberate introductions occurred along with some that were likely accidental: Bibron's thick-toed gecko from South Africa, the bark anole from the West Indies, the Tokay gecko from southeastern Asia, the Texas horned lizard, the giant ameiva from tropical South America, and the spectacled caiman from Central America and northern South America.[153] These species were all introduced during the last half of the twentieth century as reports of harmful invasion impacts increased, and at least two (the giant anole and the ringed wall gecko) after the explosion of reports of impacts associated with the advent of modern invasion science.

Hawaii also has many established populations of non-native reptiles and amphibians—about twenty-nine—of which nine were associated with the pet trade in the second half of the twentieth century.[154] Unlike in Florida, little direct evidence exists for most of these that they were deliberately released, but most were illegally imported and their releases were probably intentional.[155] In some cases, an actual record exists. For instance, a student deliberately introduced the gold dust day gecko from the Indian Ocean to Oahu in 1974 near the University of Hawaii, and it was then deliberately spread to other areas on the island.[156] A congener from Madagascar was intentionally introduced in 1996.[157] These introductions were after considerable publicity, particularly in Hawaii, about potential harmful impacts of invasions. A striking aspect of the non-native Hawaiian herpetofauna is the number of species deliberately introduced as biological control agents for pest insects—three frogs and a toad in the 1890s and 1930s, long before concern about nontarget impacts had become an issue (more on biological control and concern with nontarget impacts in chap. 17, sec. 17.5).[158] Four other species of frogs and toads intentionally introduced for the same purpose in the nineteenth and early twentieth centuries failed to establish populations.[159]

Species of reptiles and amphibians intentionally introduced simply as "amenities" were hardly restricted to the United States; examples abound worldwide.[160] For example, the native land and freshwater vertebrate fauna of Great Britain is depauperate relative to that of mainland Europe, partly because of its glaciation history.[161] It includes only one frog, two toads, three salamanders, three lizards, and three snakes.[162] Attempts to remedy this paucity led to attempts to introduce at least sixteen amphibians and thirteen reptiles, of which the great majority did not lead to persistent populations.[163] For almost all of these, there was no stated reason other than to have them present. Many of these attempts were in the wake of Charles Elton's BBC radio broadcasts of 1933 and 1957 and his monograph of 1958 on the risks of unintended consequences from introductions of non-native species (discussed further in chap. 7, sec. 7.1). Although Elton did not use amphibians or reptiles among his examples, his treatment was so wide-ranging that one might have expected heightened concern with any introduction. The naturalist Richard Fitter condemned these releases in a popular article: "The scientist should not, as certain keen herpetologists have been doing in the London area recently, scatter possibly undesirable amphibians about the accessible public sheets of water."[164]

5.6 Plants

Early warnings about impacts of introduced plants were about agricultural weeds. For instance, John Bartram in 1758 listed ten European plants that had become agricultural pests, with yellow toadflax the most noxious. In the nineteenth century, before the advent of a formal discipline of ecology, George Perkins Marsh and the Victorian naturalists, particularly Charles Darwin, Joseph Hooker, and Alfred Russel Wallace, all recognized that non-native plants could replace native ones and perhaps even drive the natives to extinction. (Details on these five men and their work appear in chapters 2 and 3.)

In the late nineteenth century and the first half of the twentieth century, as prominent concerns increased about effects of non-native animals, including many about ecological impacts, similar but fewer and more muted warnings were issued about introducing non-native plants. In New Zealand, Thomas Kirk described introduced plants squeezing out native plants in increasingly remote locations, and George M. Thomson, though stressing the role of such human activities as clearing forests in facilitating invasions, pointed to a number of plant invasions in isolated pristine areas.[165] Charles Marlatt of the United States Department of Agriculture (USDA) warned about the danger that imported nursery stock could carry insect pests, highlighted by his actions to prevent insects and plant pathogens from being introduced to Washington on Japanese cherry trees. (Kirk, Thomson, and Marlatt are all discussed further in chapter 6.)[166] Aldo Leopold's article about the invasion of quackgrass in Illinois and his essay "Cheat Takes Over," both in 1941, were published in journals of very limited circulation, and the latter did not become famous until the belated prominence of his posthumously published *Sand County Almanac* (as detailed in chap. 7).[167] Frank Egler for Hawaii and Harry H. Allan for New Zealand wrote in academic journals about impacts of plant invasions, both emphasizing the facilitating role of anthropogenic activities.[168] Marston Bates, in an early book chapter on invasions (described in chap. 6, sec. 6.1), discussed impacts of plant invasions, citing Allan and Egler in emphasizing the role of anthropogenic habitat change.[169] Henry J. Oosting's widely used plant ecology textbook in 1948 detailed several important plant invasions, such as those of water hyacinth and Japanese honeysuckle in the United States and prickly pear in Australia; a second edition published in 1956 added kudzu, eucalyptus, and cheatgrass in the United States and Canadian pondweed in Europe (more on this textbook in chap. 10, sec. 10.2).[170] Oosting was primarily focused on ecological impacts. Through the 1950s, though many geographers discussed biological invasions (as detailed in chap. 8), focus on ecological impacts was scant, except for Oosting, Andrew H. Clark, and Edgar Aubert de la Rüe.[171]

The aforementioned writings, with the exception of the reprint of Leopold's "Cheat Takes Over" in *Sand County Almanac*, were all in publications targeting scientific audiences. In the 1950s, even before Elton's 1958 book, *The Ecology of Invasions by Animals and Plants*, and the BBC broadcasts that preceded it, two broadly ranging articles appeared in lay-oriented science magazines.[172] Herbert W. Levi described impacts of invasions not only of non-native animals but also of non-native plants, including multiflora rose, prickly pear, Japanese honeysuckle, and Canadian pondweed.[173] Victor R. Cahalane also portrayed impacts of introduced prickly pear, as well as blackberry from North America and European "Canada thistle" in Australia and Canada thistle, Japanese honeysuckle, Scotch broom, Russian thistle, and water chestnut in the United States.[174] (Levi's and Calahane's publications are discussed further in chapter 6.)

Thus, by the first half of the twentieth century, although warnings about harmful consequences of plant invasions were not as plentiful as those about animal invasions, they existed, they grew in number, and they increasingly focused on ecological impacts in addition to agricultural ones. Certainly scientists were aware of them. Yet plant introductions kept mounting. Perhaps gardeners were at best only dimly aware of the issue and, finding it difficult to imagine that an ornamental with pretty flowers or unusual leaves could really be problematic, continued blithely to seek the latest unusual plant, much as Peter Collinson and other English gardeners and horticulturists did who sought a stream of new North American species from John Bartram in the eighteenth century.

More perplexing are government agency scientists who spearheaded plant introductions, similarly to those introducing fishes, mammals, and birds. The United States government perhaps first engaged in this activity in 1825, when

President John Quincy Adams ordered US consuls to send rare plants and seeds to the Department of State for propagation and distribution, and many federal government agencies participated in various ways throughout the nineteenth century in introducing crop plants and trees.[175] In the twentieth century, the most paradoxical agency scientist was Edward H. Graham, who occupied influential positions such as chief biologist and assistant administrator of international programs in the Soil Conservation Service of the USDA, established in 1935.[176] Graham warned of unintended consequences of introductions, extensively describing invasion impacts of non-native animals and plants (as described in chap. 6, sec. 6.1), citing Leopold's "Cheat Takes Over" and other early warnings about invasions.[177] Yet in the 1940s, he testified to Congress in favor of planting non-native species for field borders and contour hedges that subsequently became extremely invasive, and his 1947 book advocated using many non-native plant species to stymie erosion or establish windbreaks.[178] Several later became deplored invaders, such as kudzu, giant reed, Russian olive, and Scotch broom. He particularly touted multiflora rose for erosion control and as a windbreak despite warnings of invasiveness by the 1940s.[179] In sum, the Soil Conservation Service under his guidance was instrumental in introducing and spreading several invasive plants.

It is unsurprising that the USDA should introduce plants, as one of its main charges when it was established in 1862 was "to procure, propagate, and distribute among the people new and valuable seeds and plants."[180] The person who contributed most to fulfilling this role was David G. Fairchild, who in 1898 was appointed director of the newly established Office of Foreign Seed and Plant Introduction in the USDA, which ultimately introduced more than two hundred thousand non-native plant species and varieties.[181] The office established field stations in the United States to test plants sent back by a worldwide team of plant explorers or by more or less formally affiliated associates worldwide. Though emphasis was on edible plants and other "useful" plants, ornamentals were also frequently imported. The Office of Foreign Seed and Plant Introduction contributed not only to introduction of non-native species but also to the work of scientists engaged in hybridization experiments in the early twentieth century, including hybridization between native and non-native species and between pairs of non-native species.[182]

Fairchild, an inveterate traveler, personally engaged in the exploration effort, and beginning in 1908, the Office of Foreign Seed and Plant Introduction issued a *Bulletin of Foreign Plant Introductions* every two to four weeks.[183] Each of the 210 numbers of the *Bulletin* included a section called "New Plant Immigrants" describing seeds or cuttings sent to the office by USDA collectors or foreign correspondents. In 1914, the journal was renamed simply *Plant Immigrants*, and it was produced through 1924. Fairchild listed himself as "Agricultural Explorer in Charge," with a staff of "Agricultural Explorers," "Assistant Agricultural Explorers," and "Plant Introducers." The journal, as well as a more detailed, summarizing annual catalog of new plant introductions issued from 1910 through at least 1922, encouraged citizens to write for seeds from plants propagated from new accessions—and to do so quickly, before the supply was exhausted.[184] A note in a 1911 issue of the *Bulletin* explained:

> Applications for material listed in this bulletin may be made at any time to this Office. As they are received they are filed, and when the material is ready for the use of experimenters it is sent to those on the list of applicants who can show that they are prepared to care for it, as well as to others selected because of their special fitness to experiment with the particular plants imported. One of the main objects of the Office of Foreign Seed and Plant Introduction is to secure material for plant experimenters, and it will undertake as far as possible to fill any specific requests for foreign seeds or plants from plant breeders and others interested.[185]

Although descriptions of new arrivals and suggested acquisitions in the *Bulletin* became quite detailed within a few years, often including locations and habitats where survival and growth would be likely, I have found no example of a discussion of potential harm, especially ecological damage, even for species that subsequently were recognized as widespread damaging invaders. For instance, in 1911, a report on Russian olive based on reports from agricultural explorer Frank Meyer in East Turkestan lauded its remarkable drought resistance, its ability to fix nitrogen, and even the insufficiently valued fruit, with no mention of potential invasiveness.[186] The species was introduced as a hedgerow in the upper Midwest in the nineteenth century by Russian Mennonites and was used in the Great Plains as a windbreak, but its further distribution in the West in the early twentieth century ultimately resulted in a devastating invasion of riparian areas.[187] Fairchild himself was certainly aware of concerns about invasion but viewed them remarkably casually. For example, regarding a species native to eastern Europe and western Asia, showy mullein, that he encountered in Sweden, he commented, "With its seeds I sent a warning that this verbascum might prove a weed but, as it is easily rooted out, we need not be too nervous."[188] (Box 5.4 provides further detail on Fairchild and showy mullein.)

BOX 5.4 David G. Fairchild and Showy Mullein

Contrary to Office of Foreign Seed and Plant Introduction director David G. Fairchild's expectation, *Verbascum speciosum*, known as showy mullein or Hungarian mullein, did naturalize and become a roadside weed in California, though it has not been sufficiently problematic to trigger specific management activities.[1] Fairchild's attitude toward the possibility that non-native plants would become invasive and his appreciation of native plants are enigmatic. In 1902, during a plant exploration trip to Japan, he sent seeds of kudzu back to the office with a note saying it is well known in the United States in private gardens as an arbor plant or simply to produce "tropical effects," estimating that "for this purpose alone, it is a valuable plant."[2] He went on to describe the plant growing wild in Japan and covering hillsides, and he advocated testing it in the United States as a fodder plant in waste places. Later that decade, he tried to establish kudzu at his estate in Bethesda, Maryland, with roots he imported. In the 1930s, Fairchild wrote that these had failed, and, after buying seeds from a nursery, he

> scattered them about rather recklessly. . . . The seedlings all took root with a vengeance, grew over the bushes and climbed the pines, smothering them with masses of vegetation which bent them to the ground and became an awful, tangled nuisance. I spent over two hundred dollars in the years which followed trying to get rid of it.[3]

Immediately after this description, Fairchild observed without comment, "I see by the paper today that the Kudzu vine is now being used by the Soil Erosion Service [later rechristened the Soil Conservation Service] to control gullying and erosion throughout the South."

His earlier experience could—perhaps *should*—have produced a caution. The Soil Conservation Service campaign in the 1930s included hundreds of workers employed by the Civilian Conservation Corps who distributed tens of millions of kudzu seedlings throughout the South. These plantings, combined with efforts by various kudzu clubs and individual evangelists such as Channing Cope, resulted in current occupancy by this plant of three million hectares in the United States.[4] By the 1950s, state highway departments had largely stopped using kudzu to control erosion, and by 1979, the USDA classified it as a weed. In 1997, Congress tagged kudzu as a Federal Noxious Weed. In the United States, kudzu has become a metaphor for a spreading, unmitigated disaster.[5]

It is difficult to reconcile Fairchild's seeming lack of concern over the possibility that an imported plant could become a deplored weed, as well as his ambitious planting of many non-natives at his estates in Maryland and Florida, with his stated enthusiasm for native plants. This latter view was cemented by a visit in 1904 with Ossian C. Simonds, a landscape and garden architect who is credited along with Jens Jensen as a founder of the "prairie style." Simonds was a native plant enthusiast who felt that native plants should be the basis for any landscape, and, according to Fairchild, "this idea sank deeply into my mind and affected everything that I have done in the way of landscape planting."[6]

Fairchild frequently appeared to downplay the possibility that a plant pathogen or parasite imported on a non-native plant might devastate a native plant species, and he periodically lambasted quarantines and inspections instituted to prevent such mishaps because they interfered with his introduction activities. For instance, in 1910, hoping to cultivate and outplant the plethora of non-native plant material arriving at the Office of Foreign Seed and Plant Introduction, he complained that "our task was made even more difficult by the entomologists and pathologists who insisted that every packet of seeds, cuttings or plants must be thoroughly inspected or held in crowded quarantine greenhouses."[7] After the First World War, he hoped his imports would be "given a fair start towards being grown and propagated; and that they did not get smothered by Quarantine regulations and treatments."[8] He saw Charles Marlatt, an erstwhile friend and college classmate at Kansas State University who moved in the same rarefied social circle in Washington, DC, as his leading antagonist.[9] "There were times when I felt that my old friend Charles Marlatt and his associates would gladly have done away with the introduction of plants from abroad altogether," Fairchild recalled. "As long as I was able, I would keep my foot in the doorway and prevent importation from being entirely forbidden." He perhaps even regarded Marlatt as a type of racketeer: "The Plant Quarantine, like many excellent government projects, at times approached dangerously close to what is known as a 'racket.'"[10] Fairchild's antipathy toward quarantines vacillated. Another time, lamenting the changes wrought by the chestnut blight to Appalachian forests, Fairchild "felt like saluting the pathologists who are working to prevent a repetition of this calamity in the future. I regretted any feelings of impatience I may have had towards their quarantines and inspections."[11]

1. R. Preston and Wetherwax, "*Verbascum speciosum*."
2. Fairchild, "9227. *Pueraria thunbergiana*. Kudzu."
3. Fairchild, *World Was My Garden*, 328.
4. Winberry and Jones, "Rise and Decline of the 'Miracle Vine'"; Alderman, "Changing South," Alderman, "Channing Cope and the Making of a Miracle Vine"; Solomon, "Kudzu Is So Much More."
5. Solomon, "Kudzu Is So Much More."
6. Fairchild, *World Was My Garden*, 297.
7. Fairchild, *World Was My Garden*, 374.
8. Fairchild, *World Was My Garden*, 471.
9. Liebhold and Griffin, "Legacy of Charles Marlatt."
10. Fairchild, *World Was My Garden*, 425.
11. Fairchild, *World Was My Garden*, 406.

It was not only in the United States that government agencies led the way in introducing plants with scant concern about possible invasiveness. In Australia, plants suitable for livestock grazing were introduced from many parts of the world into pastures beginning in the 1880s. The earliest efforts were by individuals, but the great majority of introductions were by government organizations such as the Commonwealth Scientific and Industrial Research Organisation (CSIRO) and the Northern Territory Department of Primary Industry and Fisheries.[189] Evaluation focused on whether the plants would survive and thrive under grazing pressure and whether livestock would eat them, with no explicit reference to likelihood of their becoming agricultural weeds or invaders into natural ecosystems.[190] This activity increased in the 1950s, and between 1947 and 1985, 463 non-native grass and legume species or varieties were introduced to Northern Australia, primarily from Africa and Central and South America. Of these, 60 were ultimately listed as weeds, of which 39 were of conservation concern.[191] Finally it was recognized that the very traits being sought in introduced pasture plants were traits conducive to weediness.[192]

: : :

The nineteenth century saw several warnings about harmful impacts of non-native species, often associated with widely recognized particular invasions (described in chap. 4). However, such foreboding had little or no impact in the nineteenth and early twentieth centuries on massive efforts by governments, nongovernmental organizations, and individuals to introduce plants and animals to new regions. Some such activities continued into the mid-twentieth century. Often such efforts were justified on the grounds that the proposed introductions would benefit some human enterprise, such as agriculture, hunting, or fishing for sport or commercial purposes. Other introductions reflected nostalgia by new immigrants for species of their distant homelands. Still other species, especially ornamental plants and songbirds, were introduced as aesthetic amenities to brighten an island or region perceived as biologically depauperate or drab. Most introduction projects during this period failed to establish persistent populations despite commitment of considerable resources. Such failures occasioned complaints that more careful planning was needed—not to avoid impacts, but to increase the likelihood of successful establishment. However, several plants and animals introduced during this period more or less rapidly became lamented invaders. Perhaps the most striking feature of the widespread introduction efforts during this period was the ignorance of or inattention to warnings about potential inimical impacts.

6

Mid-Nineteenth- to Mid-Twentieth-Century Research, Often Forgotten, and Warnings, Largely Ignored

In the nineteenth-century heyday of acclimatization societies, as was observed in chapter 5, occasional authors lambasted the entire project. Others were more narrowly focused, including several scientists who deplored the plan to introduce mustelids to New Zealand on the grounds that they would devastate native animals, especially birds.

By the turn of the century, criticism of introductions in general, or at least of entire groups, rather than of the introduction of a particular species, became more common, and such warnings increased in frequency in many nations through the mid-twentieth century. Scientists of several nations detailed disastrous outcomes of some previous deliberate introductions and called for further research on the phenomenon of spread of non-native species and enhanced regulation of introductions.

6.1 North America

Several scientists in the United States Department of Agriculture (USDA) were early heralds of potential problems caused by non-native species. Theodore S. Palmer, longtime secretary of the American Ornithologists' Union, was a zoologist in the USDA Division of Economic Ornithology and Mammalogy (US Biological Survey) who rose to highly ranked positions in the USDA.[1] He was particularly interested in inimical impacts of non-native birds and mammals.[2] His 1898 paper called attention to the general problem of non-native domesticated animals such as goats and cats becoming feral, citing Alfred Russel Wallace's description of goat impacts on the vegetation of St. Helena (described in chap. 3, sec. 3.2). His main general message, replete with examples such as the rabbit in Australia (chap. 4, sec. 4.1), was that intentionally introduced animals that appear innocuous or beneficial in their native range often become plagues in their new homes, such as the small Indian mongoose in Jamaica, mustelids in New Zealand, house sparrow in the United States, myna in Hawaii, and starling in New Zealand and Australia. Palmer particularly criticized activities of acclimatization societies (as discussed in chap. 5, sec. 5.1). His practical goal with this publication was to inspire federal legislation to limit introductions, and he expressed frustration that a call for such regulations in 1886 by C. Hart Merriam, chief of the US Biological Survey, had not led to action.

The Department of Agriculture yearbook featuring Palmer's paper also included an article by the USDA chief entomologist, Leland O. Howard, "Danger of Importing Insect Pests," which noted the huge cost to agriculture and silviculture imposed by seventy-three non-native insect species already introduced to the United States, including the spongy moth (discussed in chap. 4, sec. 4.4), codling moth, and Hessian fly.[3] Howard predicted that certain other species would become costly invaders if they were introduced; he proved to be prescient for several of these, including the Mediterranean fruit fly, also known as the medfly. He called for a rigorous national program of quarantine procedures and inspection of goods likely to carry non-natives species. Howard tabulated the origins of existing invaders, finding that the majority came from Europe and rather few from the tropics, and he sought possible explanations for these statistics, thus anticipating by almost a century a major contribution to modern invasion science by Reece Sailer.[4] Howard was also an early and persistent advocate of research on biological control of insect pests (a topic discussed in chap. 17, sec. 17.5), especially focusing on the potential use of parasitoids.[5]

Thirteen years after Howard's warning, his colleague Charles Marlatt, later Howard's successor as chief of the Bureau of Entomology, published a USDA bulletin citing primarily data from reports issued by Howard and warning of the risks associated with importing nursery stock infested with the spongy and browntail moths.[6] He recounted his own and Howard's inspection trips to Europe and noted that many plants exported to the United States were likely infested at least by these two moths. He argued more broadly for quarantine and inspection procedures for plant introductions at least as strict as those already imposed by several European nations, which had been triggered by the disastrous phylloxera invasion of Europe (discussed in chap. 4, sec. 4.2).[7] Marlatt had previously detailed the quarantine and inspection procedures implemented by the state of California, advocating similar measures in the eastern United States.[8] That article had focused on several non-native scale insects (including two species from Florida) that had invaded California (including the cottony cushion scale, described in chap. 4, sec. 4.3) and control methods, particularly the promise of introducing insect parasitoids and predators for biological control. In addition to the vedalia beetle, Marlatt noted the effects, or lack thereof, of several insects, especially ladybird beetles, on non-native pest scale insects (he argued that non-native scale insects outpaced all other non-native species except for the phylloxera in France in terms of damage). However, he made no mention of potential nontarget impacts of non-native biological control agents.

Although he was a prominent entomologist with important contributions on periodical cicadas, Marlatt is best known for a 1910 incident involving Japanese cherry trees sent by the citizens of Tokyo to Washington, DC, in a gesture of friendship.[9] The US Army Corps of Engineers intended to plant the trees on what later became the National Mall, but the USDA inspection, led by Marlatt, revealed several pathogens and non-native insects. The trees and their packing material were thus burned, to the dismay of many citizens and amid some skepticism about the real risk. Historian Philip Pauly depicted the incident as a reflection of nativism, although a second shipment in 1912, this time fumigated by the Japanese Imperial Quarantine Service, passed inspection and the trees were promptly planted.[10]

John C. Phillips was a wealthy, nonpracticing physician turned ornithologist who cooperated with the US Biological Survey on an early USDA publication on non-native birds in North America.[11] He viewed non-native birds very much in the tradition of the nineteenth-century biogeographers (described in chap. 3, sec. 3.1), an issue of "faunal geography," a "somewhat neglected field," rather than in terms of general principles or impact.[12] Yet he foreshadowed modern invasion science, as the main matter he addressed was that certain non-native species succeeded in establishing a spreading population while others died out, one of the two main foci of the Scientific Committee on Problems of the Environment (SCOPE) project that later founded the modern field (see chap. 9, sec. 9.1). The bulk of his monograph was an account of which introductions of foreign birds to the United States (and a few introductions of American birds elsewhere) had succeeded and which had failed. Unlike the SCOPE project, Phillips did not usually attempt to explain the different trajectories, and when he did, he appeared to assume that inappropriate climate and other habitat variables were the reason for failures.

Phillips did not focus on impacts, simply pointing at the outset to the differing views at the time of the advisability of deliberate introductions:

> There are at least two schools of thought on the subject of introducing and transplanting birds, and these are widely at variance. One of these, the conservative, represented by such eminent naturalists as Joseph Grinnell of California and many others, believes in preserving at all costs the present or rather the original status of native birds and harmless mammals, and

> points out the great dangers incurred in the importation of new species in other parts of the world, and especially the danger of spreading new diseases. The other school would bring in anything from a button quail to an ostrich without any regard to the general suitability of the species. Most sportsmen and naturalists do not agree fully with either of these views, although the writer sympathizes strongly with the first.[13]

Phillips was referring to a widely read article by Grinnell, a leading zoologist, warning against the possibility that non-native birds, especially game birds, would either outcompete native species or transmit new parasites to them.[14] Phillips also cited Palmer with respect to dangers, but he oddly minimized the likelihood of problems on the grounds that he felt, as did Grinnell, that few non-natives can establish populations.[15] In his accounts of introductions, Phillips cited impacts only for the starling, house sparrow, myna, and crested myna as harmful, and he believed that passage of the Lacey Act of 1900 (covered in chap. 18, sec. 18.1, and highlighted in box 18.2) answered Merriam's and Palmer's pleas for regulation of deliberate introductions.

Yet another USDA scientist who signaled threats posed by non-native species was Edward H. Graham, chief of the US Soil Conservation Service in the USDA Division of Biology and, at the end of his life, chair of the Commission on Ecology of the International Union for Conservation of Nature and Natural Resources. Graham's 1944 book on land use cited Aldo Leopold's essay "Cheat Takes Over," Charles Elton's *Animal Ecology*, Phillips's USDA publication, and a series of papers by Ralph King, and he included an entire chapter entitled "Exotic Things."[16] Pointing to invasions that were well known by then (e.g., chestnut blight in the United States, the small Indian mongoose in the Caribbean, the European rabbit and prickly pear in Australia), Graham broached several subjects later recognized by invasion scientists as key phenomena or issues and given specific names—invasional meltdown, the enemy release hypothesis, the passenger-driver metaphor (more on these in chapter 15). Remarkably, recounting the dramatic success of the South American cactus moth as a biological control agent for non-native prickly pear in Australia, Graham even warned against suggestions to introduce the cactus moth to the American West to control native prickly pear on the grounds that it would inevitably spread and threaten the dozens of native prickly pear species in both the United States and Mexico, where prickly pear is a valued food. In fact, following its arrival in Florida in 1989 after it was introduced for biological control in the West Indies, exactly this threat of spread to native prickly pear in the United States and Mexico is of great concern (detailed in chap. 17, sec. 17.5).[17]

Paradoxically, as chief of the Soil Conservation Service, Graham enthusiastically supported planting non-native species to stem erosion (more on Graham in chap. 5, sec. 5.6), touting them additionally as providing habitat for native birds and mammals. Thus, he advocated using shrubby bushclover and Chinese bushclover, Asian shrubs introduced to the United States in the nineteenth century and now widely recognized as invasive. In testimony to the US House of Representatives in 1946 and 1949, he enthusiastically recommended both bushclovers for use as field borders and in 1949 praised the multiflora rose from Asia as a contour hedge.[18] Graham's 1947 book, *The Land and Wildlife*, went considerably further in addressing erosion at field edges, streamsides, roadsides, and spoil banks, advocating use of giant reed from the Old World, weeping lovegrass from Africa, kudzu and Tatarian honeysuckle from Asia, Scotch broom from Europe, the two bushclovers, and Asian roses, especially multiflora rose.[19] For windbreaks and residential or contour hedges, he suggested Russian olive, Chinese elm, Tatarian honeysuckle, and again multiflora rose; he developed methods of using multiflora rose still further in 1957.[20] For all these plants, he stressed potential use by wildlife; however, all subsequently were recognized as invasive, some of them legendarily so (e.g., kudzu and giant reed). Elton ended his invasion book that warned of the dangers of invasions with a plea to maintain hedgerows in Britain as a seminatural habitat of great use to native animals.[21] (Elton and his work are described in detail in chap. 7, sec. 7.1.) Ironically, he cited Graham's 1944 book to this end and included plates depicting multiflora rose hedgerows in Indiana and Maryland, respectively, lauding their benefits in the captions.[22] To some extent, Graham's failure to anticipate the impacts of many of these plants can be attributed to the lag that many non-native species undergo after establishment but before they spread and become invasive (see chap. 18, sec. 18.3).[23] However, at least in the case of multiflora rose, that is not the entire story. The species was introduced to the United States in the early 1800s, and several scientists were warning of invasive tendencies by the 1940s and 1950s.[24]

Not associated with the USDA but cognizant of work by Phillips and Palmer, Rudolf Martin Anderson was an American-born Canadian zoologist who studied Canadian birds and mammals. He wrote a short, wide-ranging article on invasive non-native animals for the Fifth Pacific Science Congress in 1933.[25] Much as James Ritchie had done for Scotland earlier (as will be detailed in the following section 6.2), Anderson described the impact of introduced vertebrates in North America in the context of a destruction of a putative "largely mythical 'Balance of Nature,'" incidentally noting similar introductions in other parts of the world.[26] Also similarly to Ritchie, Anderson somewhat mistitled his work ("Effect of the Introduction of Exotic Animal Forms"), as he included a section on how non-native plants brought hitchhiking insects, such as the spongy and browntail moths, the European corn borer, and the Hessian fly to North America. For fishes introduced to Canada, in addition to species from the United States, he pointed to the brown trout from Europe and the common carp. Citing Palmer and George

Thomson (profiled in section 6.3), Anderson discussed disastrous impacts of European rabbits in New Zealand and Australia, and he provided more general statistics from New Zealand about the panoply of non-native species there and the role of acclimatization societies in introducing them (as previously discussed in chap. 5, sec. 5.1). In fact, Anderson's treatment of New Zealand introductions was almost as extensive as that for North America. Anderson also described unexpected damage by the small Indian mongoose introduced to Jamaica and Hawaii, the muskrat and gray squirrel from North America to continental Europe and Great Britain, respectively, and the European hare to Ontario and New York. Citing Phillips, he pointed to the harmful impacts of the house sparrow and starling in North America.[27] Anderson also warned of the danger of introducing a pathogen with any non-native species. Oddly, he noted that, at least for vertebrates, Canada was not afflicted to the same extent as many of the other regions he adduced as examples of impact of non-natives, arguing that, at least for mammals, only rats and house mice were consequential and these were not in natural areas. In 1937, focusing just on Canada, he repeated this same point.[28] As he published no other work on non-native species and was not conducting empirical research on them, it is striking that he was sufficiently troubled by the issue, apparently from reading earlier literature, to have written the 1934 article.

Among Americans who were concerned early with potential impacts of biological invasions, Aldo Leopold was the most famous and probably the most influential in sensitizing both scientists and the public to the issue. Leopold, who is often credited, with Rachel Carson, with founding the environmental movement in the twentieth century, at least in the United States (chap. 7, sec. 7.4), deplored introducing non-native species from his earliest writings (fig. 6.1).[29] He published many short papers on this topic in obscure, often lay, journals.[30] He also offered more general treatments in professional journals.[31] However, Leopold's rationale for this dislike changed after meeting in 1931 with a younger Charles Elton, who is often seen as founding modern invasion science and who was long interested in the subject.[32] They became lifelong friends and correspondents on ecological matters until Leopold's early death in 1948.[33] Before the meeting, Leopold's aversion to non-native species was utilitarian—he cited cases in which they were less productive than natives. In addition, he may have had an aesthetic concern—they somehow did not look right embedded in a new setting surrounded by native species. This aesthetic concern was occasionally voiced by other turn-of-the-century scientists, such as Harvard botanist Charles Sprague Sargent (chap. 18, sec. 18.1).[34]

The meeting with Elton catalyzed Leopold's increasing concern with ecological impacts of invasions, rather than utilitarian or aesthetic considerations, and led to major essays on particular invasions and the more general

Fig. 6.1 Aldo Leopold with two carp seined in a dredged borrow pit in Beaver River Refuge, Utah, 1941. Credit: Courtesy of the Aldo Leopold Foundation and University of Wisconsin–Madison University Archives.

phenomenon, culminating in "Cheat Takes Over," published in 1941 but revised for his posthumously published *A Sand County Almanac*.[35] This essay moved from a description of the famous cheatgrass invasion of the American West to discussions of many other invasions by plants, plant pathogens, and animals, and it was one of several other essays in the book focused wholly or partly on non-native invaders, including notably "Good Oak." The evolution of Leopold's views on non-native species is marked by an increasing concern with the long coevolution of co-occurring native species and the absence of such coevolution for a non-native with the inhabitants of its new home. In "The Land Ethic" of *A Sand County Almanac*, perhaps the most explicit statement of Leopold's influential philosophy of conservation, coevolution is front and center as the basis for his conception of ecosystem health and its absence as a source of ecosystemic problems.[36] Leopold's growing focus on coevolution also led to his more sophisticated and explicit "land aesthetic," captured in "Marshland Elegy" of *A Sand County Almanac*. This "land aesthetic" more firmly grounded Leopold's aesthetic objection to non-native species.[37]

In 1942, Leopold's protégé Ralph T. King, a wildlife biologist and long-time faculty member of the New York State College of Forestry at Syracuse University, inveighed against game bird introductions, citing Palmer, Grinnell, Phillips, and Leopold.[38] He listed eight reasons, of which one was utilitarian (such introductions are expensive and most had failed) while the others were ecological (e.g., such species may become pests, they may lead to the decline of native species and deterioration of native habitats, they may inadvertently bring new parasites and pathogens, they can hybridize with native species to the detriment of the natives). Although his focus was narrowly on game birds in the United States, King did not hesitate to adduce invasions of other species to buttress his arguments: the small Indian mongoose in Jamaica, the gray squirrel in Great Britain, the European rabbit in Australia and New Zealand, lantana in Hawaii. And he was one of the first to recognize that pathogens that are innocuous to foreign species owing to eons of coevolution could devastate North American species newly exposed to them. King's arguments were quickly reprinted verbatim in an ornithological journal.[39]

No North American scientist appeared in the first half of the twentieth century to lament the full range of potentially harmful impacts of introduced fish in analogous fashion to King's series of papers on introduced birds. Leopold's first publication on non-native species, in 1918, was on the plethora of trout introductions into rivers of the American Southwest.[40] He opposed these introductions, but not because of ecological concerns. Rather, he foresaw hybridization with native trout and the ensuing production of a population less well adapted to the environment. However, this paper and Leopold's correspondence at the time were enigmatic.[41] In certain circumstances he did advocate introducing species; this appeared to be when they would be well adapted to the habitat of the introduction site and total fish production would thus increase. Yet in both his 1918 paper and a *Game and Fish Handbook* that he assembled in 1915 for the US Forest Service, he said, without explanation, that natives are best: "Ordinarily native and indigenous species are preferable."[42] He repeated the admonition in 1918: "Restock with the best adapted species, the native species always preferred."[43]

In the early twentieth century, many authors favored introducing carp because carp could be fished in waters unfavorable for native fishes, and they questioned its real impact on the natives.[44] Further, as long as fishing was good, concern with non-natives replacing natives was minimal. For instance, Henry Talbott, a law clerk in the US Interstate Commerce Commission with a consuming interest in angling, observed the replacement of native fishes by non-native ones in part of the Potomac River:

> To those who have known the Potomac well it has always been one of the greatest of fishing streams. Strange to say, the fishes for which it was once famous have now nearly disappeared, and it is fished now for an alien stock, for most of its biting population has been imported. . . .
>
> But the river is still a delight to fishermen and on a single day a few weeks ago it was estimated that 5,000 men and women were playing at fishing, lining the rocks on both banks, from Georgetown to Little Falls, and fleets of small boats dotting the stream from Chain Bridge down. . . .
>
> So, in conclusion, the native shad, sturgeon, striped bass, and even herring are flitting, and in their places the river is locally famous for a set of fishes that are imported from other climes and other waters.[45]

However, notes and articles questioning the wisdom of such introductions and suggesting harmful impacts gradually accumulated. For most, the harm was to fishing for native species.[46] Others, however, lamented ecological damage, often to unexpected parts of the ecosystem. Luis H. Rivero described how largemouth bass introduced to Cuba ate native cyprinodont fish that had eaten mosquitoes, thus increasing the mosquito population and triggering an increase in malaria cases.[47] David H. Madsen, supervisor of Fish Resources for the US National Park Service, went even further:

> It would take much time and would be repeating facts already well known to refer to the hundreds, if not thousands, of instances where exotic species have been introduced to the detriment and often the destruction of native species in waters of every State in the Union. Not alone have native species of fish suffered from these mistakes, but certain forms of bird life as well have been driven out of large areas, and valuable aquatic vegetation has been utterly destroyed.[48]

Previously, Madsen had complained about the presence and continued introduction of non-native fishes into the national parks on the grounds that they had adversely changed lake and stream ecosystems.[49]

Leopold's early writings about non-native species were before he met Elton in 1931 and therefore before his growing concern about potential ecological impacts of invasions.[50] His later writings, influenced by Elton and by the accumulating reports of damage by non-native fish and other species, did not focus particularly on fishes, but Leopold cited certain well-known non-native fish, especially common carp, as part of the general invasion problem in that they tend to damage ecosystem stability and what Leopold came to call "land health." For instance, "Cheat Takes Over" and an unpublished manuscript entitled "Conservation: In Whole or in Part?" both cite ecological damage from carp.[51] Leopold was particularly explicit about carp in an essay entitled "Fifth Column of the Fencerow," primarily about quackgrass, native to the Old World, in Illinois but with a condemnation of common carp: "Like the foreign carp which is quietly displacing the native fishes of our lakes and rivers by his superior voracity and reproductive powers, so is this foreign grass [quackgrass] quietly displacing the native plants of our roadsides and fencerows by its superior powers of suckering, seeding, and choking."[52]

If anyone could have provided a comprehensive, influential argument against fish introductions, it would have been George S. Myers, a prominent ichthyologist at Stanford University and head of the Division of Fishes at the United States National Museum. Myers appeared conflicted about introducing fishes. In a 1936 article beginning with a description of the unexpected hecatomb wrought by the small Indian mongoose introduced to Jamaica for rat control (as described in chap. 4, sec. 4.5), Myers nevertheless advised introducing many fish species, especially those in the mosquitofish genus *Gambusia*, for mosquito control, focusing in detail on the need for careful research on which species will thrive when introduced to which new site.[53] However, at the end of this section, he added this caution:

> Finally, it should always be held in mind that the introduction of any foreign animal is apt to cause repercussions on the native fauna in unexpected ways. *Gambusia* has been spread far and wide in anti-mosquito work, frequently in ignorance of valuable native species. *Gambusia affinis* and *G. holbrookii* are, for their size, among the most voracious and destructive of fishes, and although no reports of damage to the young of valuable food fishes in areas into which they have been introduced have appeared, it is possible that introduction into certain places will prove to have been a mistake.[54]

In the same vein, in a 1947 article subtitled "Inadvisability of Recommending North American Fishes Without Careful Appraisal of Foreign Fishes and Ecology," Myers focused on whether introduced North American fishes would thrive in different environmental conditions and whether native fishes on other continents could constitute just as productive food fish at less expense.[55] However, he noted two introductions, one realized and one proposed, for which he perceived what would now be termed conservation or biodiversity concerns. He described common carp introduced from the Old World that had inevitably escaped ponds near the headwaters of the Iguassú River in Brazil, leading to enormous populations downstream and complaints by local fisherman that many species of "better" native fish were becoming scarce; he noted his own efforts to prevent common carp ponds from being located near the Amazon for fear of a similar escape. Myers was also concerned that largemouth bass, a non-native "success" in South Africa, was touted as a potential introduction to the great lakes of central east Africa, although his main worry was that the bass would not adapt well and that these lakes already have enormous numbers of native "basslike" cichlids. This concern preceded by just a few years one of the best-known of all biological invasions, that of the Nile perch into Lake Victoria and other central east African lakes with the consequent extinction at least two hundred native cichlid fish species and devastation of regional societies whose economy was based on them.[56] Myers's paper was inspired by a previous similar effort by Thomas Chamberlain, who also emphasized that introduction of food fish is often inefficient because native fishes would serve the same purpose better, but who also noted in passing an appalling conservation impact: Non-native black bass and sunfish apparently caused the drastic decline of an endemic salamander species, which had long been exploited as food, in an isolated Mexican lake.[57]

In 1965, Myers termed the western mosquitofish, native to southeastern North America, a "fish destroyer," describing decline or disappearance of native fishes in Asia and Africa where this species had been introduced for mosquito control and likening the damage to that of well-known invasions by the house sparrow and the rabbit.[58] Yet the very next year, he edited a book advising on the maintenance of the western mosquitofish as an aquarium fish, celebrating its record in mosquito control with no mention of non-target impacts.[59] Half a century later, much evidence demonstrates great impacts of this species on many native animals.[60] Elsewhere in the same book, Myers advocated introducing South American flame tetra into New Jersey, lauded the introduction of the Mozambique tilapia to the United States, and provided advice for keeping snakeheads and lionfish in aquaria with no mention of potential impact of an escape; both subsequently became heralded invaders. Despite having lamented the distributional confusion engendered by escaped aquarium fishes worldwide in 1940, Myers provided many "hints to fish importers" in *The Aquarium Journal* during the 1950s, recommending new species for aquarium enthusiasts with no apparent concern about potential escape or discarding of live fish.[61]

Frank Egler was an enigmatic American plant ecologist who contributed many papers on key ecological topics of the mid-twentieth century, including several that foresaw developments decades later.[62] In 1942 he published an important paper on the ecology of non-native plants in Hawaii (chap. 5, sec. 5.6).[63] His study was modeled after Harry Allan's paper on the introduced flora of New Zealand (which will be discussed in section 6.3). Casting his argument in terms of debates current at the time about the nature of ecological succession—how plant species colonize and succeed one another on new or newly cleared land (chap. 15, sec. 15.4)—Egler suggested that most of the non-native plants that had escaped cultivation in Hawaii had done so with the aid of human habitat alteration, including deforestation, and introduced mammals. Egler argued that, if human influences were removed, the native Hawaiian flora would probably mostly reestablish itself, though certain invaders would still remain prominent, depending on the specifics of their biology and ecological requirements. Egler subsequently elaborated on these propositions with copious Hawaiian examples.[64] The other locus of Egler's research on introduced species was Aton Forest in Connecticut, for which he published an enormous monograph detailing the invasion history and biology of virtually every non-native plant species as a model for understanding the process of invasion.[65] Egler experimented with use of various herbicides to control unwanted vegetation (much of it non-native), especially on rights-of-way, and was a leading figure pointing to nontarget impacts, some of them persistent, in many papers.[66] Rachel Carson sought him out as an expert on this subject in her research leading to *Silent Spring*, the other founding document, along with Leopold's *Sand County Almanac*, of the modern environmental movement (both discussed in chap. 7, sec. 7.4).[67]

Edgar Anderson, a prominent botanist who specialized in plant genetics, focused on the importance and consequences of hybridization between species. In the 1930s, he published papers on hybridization among North American species of the New World genus *Tradescantia* (spiderworts), including the circumstances under which it occurred and the nature and fate of hybrids. A 1952 book detailed the relationship of this case to plant introductions.[68] In short, the first *Tradescantia* introduced to Europe was probably the Virginia spiderwort, collected in Virginia as early as 1616 by John Tradescant the Younger, who brought it to England. Following his father, Tradescant the Younger became head gardener to King Charles I of England, and the Tradescants had received at least two other spiderwort species by 1640.[69] Later, John Bartram and his son William, plant collectors who operated a garden near Philadelphia in the eighteenth century (we met them back in chapter 2), had brought another spiderwort from farther west to their garden, where it naturalized and, in the artificial, disturbed habitat of the garden, hybridized with Virginia spiderwort.[70] The Bartrams and others sent this second spiderwort to England and may well have sent hybrids. Further, a third spiderwort in the Bartrams' garden, from the American Midwest, had also been sent to Europe, and Anderson demonstrated experimentally that, in the right disturbed environment, all three of these congeners hybridize with one another, concluding that "*Tradescantia virginiana* of European gardens really is . . . a set of mongrels bred out of these three species," either hybridized in Europe, hybridized in gardens in North America before being sent to Europe, or both.[71]

Anderson hypothesized that some European meadow plants may have been introduced prehistorically by the first cattle raisers to migrate from central Asia. Further, he described how both eastern and western North American grasslands are "transported landscapes" dominated in large areas by plants introduced inadvertently from Europe with livestock and crops.[72]

However, despite these descriptions as well as portrayals of both subtle and more obvious landscape-level changes wrought by these mass introductions, Anderson did not depict these changes as ecologically damaging or lamentable. In 1956, he contributed a chapter entitled "Man as a Maker of New Plants and New Plant Communities" to a famous book on human-induced landscape-level change, discussed later in this section.[73] He began with the same general picture of massive movement of plants (and he briefly added animals and microorganisms), initiated prehistorically, as in the 1952 book, but the chapter is primarily on humans as makers of new plant species rather than new plant communities. He detailed several cases like that of *Tradescantia*. For sunflowers, irises, and other genera, cultivated species, or species whose geographic and habitat distribution was shifted by human disturbance of the land, he suggested that the disturbance initiated hybridization among congeneric species and the formation of hybrid swarms or new species. Anderson's interests were not ecological but rather in the vein of evolution and landscape history, for which his approach to invasions resembled that of geographers rather than ecologists. (Geographers and their study of invasions are the topic of chapter 8.)

Henry J. Oosting, 1955 president of the Ecological Society of America, published a plant ecology textbook in 1948 (*The Study of Plant Communities*, described in chap. 10, sec. 10.2).[74] This textbook was not specifically about invasions, and ecology textbooks in those days did not typically include discussions of invasion by non-native species. For instance, the most prominent ecology textbook at the time (*Fundamentals of Ecology*, described in chap. 10, sec. 10.2) included one paragraph on non-native species, simply stating that all potential introductions should be considered with extreme care and referring the reader to the papers of King on birds and a short but broadly ranging paper by Herbert W. Levi, an arachnologist who did not specialize in studies of non-native species (more on him in a moment).[75] However, in a short section, Oosting described great impacts of non-native Japanese honeysuckle, tumble mustard, and water hyacinth, as well as the chestnut blight, in the United States and *Opuntia* (prickly pear) in Australia. Although this was a plant ecology textbook, he also described impacts of the spongy moth and small Indian mongoose. The spongy moth had devastated forest plant communities, but the literature at the time on mongoose impacts was solely on native animals, and this is what Oosting described. In a second edition, Oosting expanded the sections on invaders, adding such famous invaders as kudzu, eucalyptus, the house sparrow, and the starling in the United States, cheatgrass and the Asian shrub halogeton in the US Intermountain West, the muskrat and Canadian pondweed in Europe, and the rabbit from Europe in Australia.[76] He also addressed biological control with four examples: the successful control of *Opuntia* in Australia by the cactus moth (chap. 17, sec. 17.5), the attempt to control the spongy moth in the

United States with introduced natural enemies (chap. 4, sec. 4.4), the initial success in Australia of myxomatosis in controlling rabbits and the waning impact of the virus (chap. 4, sec. 4.1), and the disastrous nontarget impacts of the small Indian mongoose in the West Indies (chap. 4, sec. 4.5).

Oosting's inclusion of all these examples in an otherwise straightforward plant ecology textbook is even more remarkable in light of the fact that he did not study invaders in his research. In his myriad plant ecology papers that focused on the population and community ecology of plants in the United States (especially the Southeast, where he was located for most of his career), he made scant mention of non-natives, even when treating communities heavily invaded by some of the non-native species discussed in his textbook (e.g., kudzu, Japanese honeysuckle). Rather, a typical paper by Oosting and his colleagues would be an exhaustive description of plants, soil, and other physical characteristics, and information on succession, with enormous amounts of quantitative empirical data. The major portion of the textbook consisted of similar material, minus most of the quantitative data. One can only speculate that his acute sensitivity to impacts in several of the systems he studied plus the increasing reports of all sorts of invasions, often with possible impacts, led him to an even broader concern with the role of invasions globally.

A similar case to that of Oosting concerned arachnologist Herbert W. Levi. In 1952, while chairing the Nature Conservancy's Committee on Exotics, he wrote a short paper ostensibly on non-native wildlife (mammals and birds) but that delved into many aspects of invasions and treated invasive pathogens, insects, and plants as well as wildlife.[77] Citing, among others, Elton, Grinnell, Phillips, and Wodzicki, Levi presented examples of various impacts, including competition and hybridization with natives, and he even mentioned phenomena that were not substantially addressed by invasion scientists until much later, such as mutual facilitation by two or more introduced species exacerbating impacts (now termed "invasional meltdown"; more on that in chap. 15, sec. 15.6), and changed behavior by non-native species in their new locations.[78] As with Oosting, Levi's previous and subsequent research did not deal with invasions. Most of his career was spent at Harvard University, where he published on spider systematics, taxonomy, and geography.

Victor R. Cahalane, a mammalogist and ornithologist who long served as a biologist in the US National Park Service, wrote a wide-ranging critique of both animal and plant introductions in 1955 in a lay nature magazine.[79] He particularly attacked the notion that most non-native species have few or no negative impacts (see chap. 18, sec. 18.3), focusing on the underlying anthropocentrism of such claims:

> Most evaluations of the effects of exotics are entirely anthropocentric. We are not biologically truthful when we claim that an exotic has fitted

harmlessly into the environment. In such instances, we ignore or are ignorant of the real consequences which do not directly affect our health or our pocketbooks. It is doubtful whether any introduced species can survive without having a detrimental effect on one or more native species.[80]

He buttressed this argument with a review of well-known examples of harmful impacts of non-native species—vertebrates, invertebrates, plants, and pathogens—citing cases from New Zealand, Australia, many islands, Europe, and especially the United States. Although he pointed to nonecological concerns such as impacts to agriculture and the expense of failed introductions, Calahane's focus was squarely on ecological impacts, and he called for high-level government control, guided by science, in the permitting process for intentional introductions.

In 1956, Antoon de Vos, a Dutch mammalogist (and former student of Leopold's) working in Canada, collaborated with two mammalogists from the United States, Richard H. Manville and Richard G. Van Gelder, to produce the first globally comprehensive report on non-native mammals and their impacts.[81] This was a monumental work, with information from North America assembled by Manville, that from South America by Van Gelder, and most of the rest by de Vos. The bibliography lists 153 sources, most of them about a particular species or group of mammals introduced in one region. Many of these were doctoral theses or publications in wildlife journals. De Vos and his coauthors included introductions from separate locations within large nations and noted many failed invasions. In many ways this report was an abbreviated forerunner of a large book on introduced mammals that appeared during the initial rise of modern invasion science and an even larger and more comprehensive work during the explosive growth of the field.[82] The main shortcoming was a greatly reduced treatment of rats and the house mouse on the grounds that "the history of their spread . . . has been recounted so frequently that it need not be detailed here."[83]

Chapter 3 describes a conference held in 1955 inspired by George Perkins Marsh's book of nearly a century earlier, *Man and Nature; or, Physical Geography as Modified by Human Action.* The resulting volume, *Man's Role in Changing the Face of the Earth,* included three chapters on non-native species.[84] As noted earlier in this section, Edgar Anderson focused on how hybridization creates new species, including hybridization involving non-native species.[85] Andrew Clark, discussed subsequently in section 6.3, reprised the main themes of his New Zealand research and extended his treatment to invasions into New World grasslands in general.[86] The most significant of the three chapters from the standpoint of invasion science was contributed by Marston Bates, an American zoologist primarily known for research on mosquito transmission of disease.[87] His short, broadly ranging chapter treated many types of invasion impacts. Citing Allan on plant invasions of New Zealand and Egler on plant invasions of Hawaii, he stressed how anthropogenic habitat change facilitates invasions.[88] Bates treated the spread of human and animal pathogens, citing, among other works, his own research on mosquito-borne pathogens, especially yellow fever and the role of introduced *Aedes aegypti* (yellow fever mosquito) as vector, and the initial devastation of introduced rabbits in Australia by myxomatosis introduced as a biological control (discussed in chap. 4, sec. 4.1). He discussed biological control more thoroughly in the context of non-native insect pests of agriculture, pointing to the frequent publication of successful biological control introductions, and examples of these from an applied entomology textbook.[89] Bates perceptively added, however, that "I know of no summary of the failures, which would be equally interesting from the biological point of view."[90] Bates thus anticipated by several decades both debate about the overall efficacy of biological control and the recognition that failed invasions can be extremely informative about the invasion process.[91]

Citing Thomson for New Zealand and several Nevada cases as exemplary for the United States, Bates noted the enormous, often successful effort to introduce sport fish without comment on their impact, and he cited the relatively frequent introduction of *Gambusia* species for mosquito control without reports of disasters, though without clear evidence to his knowledge of their efficacy in controlling mosquitoes.[92] Substantial nontarget impacts of *Gambusia* were reported after Bates wrote.[93] For birds, he observed the relatively few surviving species among foreign species introduced to the United States, citing Phillips.[94] For mammals, Bates described the small Indian mongoose introduction to Jamaica for biological control of rats (detailed in chap. 4, sec. 4.5) and the introduction of the North American muskrat to Europe as ecological disasters, while stating that it was too early to render a verdict on the South American nutria in the United States. In the case of the nutria, the species is now recognized as a damaging pest and target of eradication campaigns.[95] For invasions in general, Bates lamented that "no clear principles and no means of predicting the circumstances under which introduction will be successful or the possible consequences of success seem to emerge."[96] On this point he adumbrated the main theme of the SCOPE project thirty years later that initiated modern invasion science (presented in chap. 9, sec. 9.1). Ending his chapter, Bates called for expanded study of many aspects of biological invasions, again anticipating the SCOPE project and, more remarkably, more research on the subsequent changes, including evolutionary changes, in introduced populations. This last matter was not featured in Elton's famous monograph two years later or by the SCOPE project, becoming a prominent feature of invasion science only in the twenty-first century (described in chap. 11, sec. 11.3). Bates also suggested increased study of the role of animal behavior in invasion success and impact, another subject that has received much attention only recently (chap. 19, sec. 19.5.2).[97]

6.2 Europe

In 1920, James Ritchie, a prominent Scottish naturalist, wrote a broadly ranging book, *The Influence of Man on Animal Life in Scotland*.[98] The book was perhaps mistitled because it copiously cited examples from elsewhere.[99] Embedded in a larger treatment focusing on habitat change and destruction are one large chapter and substantial parts of several others discussing impacts of non-native species. This was an astonishingly modern treatment of many aspects of modern invasion science, sixty-five years before the field crystallized (chap. 9, sec. 9.1) and thirty-eight years before Elton's book that is often cited as the founding document of the field (chap. 7, sec. 7.1). After an initial discussion of how humans, especially in Scotland, rely on many non-native species, especially for food, Ritchie detailed how non-natives often massively affect the environment, either directly (e.g., by preying on native species or transmitting diseases to them) or indirectly by habitat change that humans bring about to foster certain non-natives (e.g., forest clearance for agriculture). Although each invasion section eventually focused on Great Britain and especially Scotland, several began with examples from other regions. Included were substantial treatments of such then-current high-profile invaders as the zebra mussel, gray squirrel, and wood wasp *Sirex noctilio* in Great Britain, rabbit in Australia and elsewhere (rabbit detailed in chap. 4, sec. 4.1), Colorado potato beetle in the United States, Norway rat, codling moth, and medfly in many regions, phylloxera in France and California (phylloxera in chap. 4, sec.4.2), and spongy moth in Massachusetts (chap. 4, sec. 4.4). Not only did Ritchie describe both the spread and the impacts of these and many other invaders in Scotland and elsewhere, but he broached several topics not fully treated until decades later, such as reasons for failed invasions and the role of habitat disturbance in facilitating invasions (the role of habitat disturbance in invasions is described in chap. 15, sec. 15.4.4).[100] Elton in fact cited Ritchie's 1920 book, but not for his insights on invasions—rather, for the role of habitat change in species distributions and the spread of capercaillie upon its reintroduction to Scotland after its extermination by hunting and habitat change.[101] Ritchie lauded a 1919 quarantine regulation imposed by the United States on plants and seeds and lamented Britain's tardiness in imposing such a measure.

Ritchie's interest in impacts of non-native species continued. In 1931, he published another book that discussed impacts of the introduced rabbit, Norway rat, gray squirrel, and little owl in Scotland, and in 1932 BBC radio featured Ritchie in a six-part series, "The Changing Face of Nature," drawn from his two books and summarized in the BBC publication *The Listener*.[102] Two episodes focused on non-native species, including impacts of the gray squirrel, rabbit, and little owl in Britain, muskrat in Europe, rabbit in Australia, and house sparrow, starling, and Colorado potato beetle in the United

States. Ritchie closed his BBC series with a plea for Great Britain to prevent further introductions. In his 1933 BBC broadcast on non-native species (chap 7, sec. 7.1), Elton cited Ritchie's broadcasts the previous year.[103]

The twentieth century European of the time before Elton who has received the most attention from modern invasion scientists is Carl H. Lindroth, a Swedish entomologist interested in biogeography and how insect species got from wherever they originated to their current geographic ranges. This interest culminated in a book on animal species (not only insects) held in common between Europe and North America, not counting those deliberately introduced by humans (e.g., domesticated species, biological control introductions, escaped aquarium animals), with speculation for many species on how such a distribution might have arisen.[104] He concluded that unintentional introduction by humans produced nearly half of such introductions. Lindroth also noted that species introduced from Europe to North America are about ten times more numerous than those introduced in the other direction; he was one of the first to comment on a pattern that was subsequently much discussed in the era of modern invasion science.[105] His proposed explanation, at least for soil-dwelling insects and plants, was that ships sailing west from Europe carried soil ballast, whereas ships sailing in the opposite direction did not. Lindroth was not concerned with impact. Five of 344 pages are on "the economic importance of introduced animals," a cursory discussion of agricultural impact and none of ecological impact. Notably, Lindroth cited only Leland Howard among all the authors discussed in this chapter or previously who had treated invasion impacts. His copious citations were rather to biogeographers and systematists.

6.3 New Zealand

New Zealand underwent many introductions of non-native species, both deliberate and inadvertent, even before Europeans arrived and especially after European settlement (see chap. 3, sec. 2). Both acclimatization societies and individuals flooded New Zealand with non-native species in the nineteenth century, occasioning much opposition and discussion, including by scientists, so it is unsurprising that New Zealand would play an outsize role in the invasion literature of the turn of the twentieth century and next few decades. Thomas Kirk, an English immigrant to New Zealand who became a prominent botanist, issued an early, broadly ranging warning in 1895.[106] Kirk addressed the displacement of native New Zealand plants, vertebrates, and invertebrates by human activities, such as logging and grazing, but he focused primarily on invasions by non-native mammals, birds, insects, and plants. Depicting a diverse, massive, and growing threat, he called for the quick adoption of protective measures and was an early advocate of establishing reserves for birds on islands with few or no non-native predatory mammals.

George M. Thomson, a teacher, amateur scientist, and eventually politician, wrote many articles for New Zealand journals on non-native animals and plants, which culminated in 1922 in a massive book, *The Naturalisation of Animals and Plants in New Zealand*, that summarized distributions and often introduction histories of more than twelve hundred non-native species.[107] Only two short chapters dealt with impacts; Thomson noted the absence of much scientific research on impacts at the time he wrote but speculated on them, in some cases along the lines of critics of acclimatization societies (as presented in chap. 5, sec. 5.1). For instance, he attributed the disappearance or severe decline of several native bird species primarily to habitat destruction but with non-native rats, cats, stoats, weasels, and possibly non-native birds playing some role. For insects and fish he surmised a similar story, pointing to even more meager data.

Citing the statement by Charles Darwin that "if all the animals and plants of Great Britain were set free in New Zealand, a multitude of British forms would in the course of time become thoroughly naturalized there, and would exterminate many of the natives,"[108] Thomson adduced examples to support his counterargument:

> Where man does not interfere with the vegetation, the indigenous species [of plants] can hold their own against the imported forms. It is human intervention—either direct or indirect—which completely alters the conditions. The same probably holds good to some extent of the animal life, only the problem is more difficult to follow out.[109]

In his chapter on plants introduced to New Zealand, Thomson rebutted not only Darwin's statement but similar arguments about the dominance of palearctic flora over that of New Zealand by James Hooker and Alfred Russel Wallace, who had cited Hooker.[110] Thomson accepted the thesis of New Zealand botanist Leonard Cockayne, who believed that most non-native plants did not become invasive and those that became invasive did so not because of some inherent "superiority."

As with Ritchie's emphasis on habitat disturbance, Cockayne's and Thomson's early recognition of the frequent role of habitat disturbance in facilitating invasions was a harbinger of subsequent emphasis on this phenomenon by Charles Elton in the first synthetic treatment of the biology of invasions and, in the era of modern invasion science, by Andrew S. MacDougall and Roy Turkington (chap. 15, sec. 15.4.5), who suggested that many if not most plant invasions are "passengers" of physical environmental change, often at the hands of humans.[111] Thomson attributed this view to all New Zealand botanists: "The opinion of all botanists in New Zealand to-day is that when the direct or—to a large extent—the indirect influence of man is eliminated, the native vegetation can always hold its own against the introduced."[112]

Thomson's book received much new attention with the advent of modern invasion science in the 1980s, with hundreds of citations. Primarily, his detailed records of which species were introduced where, when, and in what numbers became grist for analyses of the reasons for introduction success or failure, as in three papers in the 1990s seeking to relate success or failure to propagule size.[113] Subsequently, others have pointed to Thomson's own statements on limitations of the data he could gather to suggest that such analyses are at best inconclusive.[114]

An astounding early twentieth-century publication by a New Zealand sheep rancher, Herbert Guthrie-Smith, is worth mentioning, though its later influence is primarily in the field of environmental history rather than invasion science or ecology.[115] Thomson cited several papers by Guthrie-Smith but not his classic book, possibly because Thomson completed his writing before Guthrie-Smith's book was published, or perhaps because the print run of Guthrie-Smith's first edition was only two hundred copies.[116] Guthrie-Smith established his large ranch on the North Island of New Zealand in the early 1880s, and until his death in 1940 he perceptively recorded environmental changes on his property, particularly the arrival and fate of some 250 non-native species of plants and animals. He often discussed impacts, and he noted phenomena that decades later became foci of invasion scientists, such as the occasional arrival, population explosion, and subsequent dramatic decline of particular invaders.[117] His writing style was idiosyncratic and replete with striking metaphors—for example, on the invasion by the least weasel: "For a short period weasels overran like fire the east coast between Tutira and Poverty Bay, and then like fire died out."[118] He did not provide references, but Guthrie-Smith had clearly read widely; for instance, he cited Hooker's views on competitive advantages of non-native plants. Remarkably, just before his death, in the preface to his imminent third edition, Guthrie-Smith reflected on the many changes he had wrought on Tutira as he introduced livestock and cleared forest and bracken:

> Am I absolutely happy on another score—my substitution of domestic breeds of animals for native lizards and birds; my substitution of one flora for another; my contribution towards more quickly melting New Zealand through erosion into the Pacific—a question of ethics this, of simple right and wrong, one increasingly clamatory in years[?] . . . Have I then for sixty years desecrated God's earth and dubbed it improvement?[119]

Although he cited invasions in other parts of the world, Guthrie-Smith resolutely focused on New Zealand, and especially on the natural history of invasions in one small region of New Zealand. Possibly this is why he is rarely cited by biologists except those from New Zealand, although he is lionized in historical circles. However, he exemplifies an important trend—the early recognition in New Zealand, not only by scientists but by lay citizens, of invasion spread and impacts. Previously I noted the feverish activity in New Zealand of acclimatization societies and concern with their unintended consequences.[120] Although early concerns were primarily on impacts to agriculture, forestry, and hunting, by the late nineteenth century, many critics of both deliberate and inadvertent introductions deplored impacts on native species. For instance, Walter Buller pointed to impacts of Norway rats on native birds and crayfish and later lamented the threat of non-native mustelids to native birds.[121]

In the twentieth century, in addition to treatments of the whole non-native biota of New Zealand by Guthrie-Smith and Thomson, specific components of that biota saw substantial research. Ornithologist Walter R. B. Oliver depicted the myriad harmful impacts of non-native animals on native birds, both through direct predation (e.g., rats, mustelids, and the brushtail opossum) and habitat destruction (red deer and the opossum).[122] Harry H. Allan, Leonard Cockayne's protégé, comprehensively treated the non-native flora.[123] Whereas earlier discussions of impact, such as those by Cockayne, were largely broad-brush generalizations based on observations plus various assumptions and analogies, Allan surmised in detail why particular species had spread, others had spread and then declined, and yet others had not spread beyond the gardens in which they were planted.[124] In doing so he anticipated, as had Ritchie in Scotland and Cockayne and Thomson in New Zealand, the major role many modern invasion scientists grant to physical environmental change.

Another remarkable figure in advancing scientific interest in invasions in New Zealand was Kazimierz A. G. Wodzicki, a Polish expatriate zoologist who arrived in New Zealand in 1941 and published many papers on vertebrates in New Zealand, including impacts of non-native species such as the rabbit. His most important work, published in 1950, was a large monograph on non-native mammals that provided great detail on the history and impacts (both positive and negative) of the introductions, and on their control, with which Wodzicki, as first head of the ecology section of the New Zealand Department of Scientific and Industrial Research, was much occupied.[125] Although focused on mammals, the monograph also discussed history and impacts of many non-native birds and plants.[126] A skilled zoologist, Wodzicki drew exhaustively on the literature from New Zealand and elsewhere (including the books by Ritchie, Guthrie-Smith, and Thomson cited previously, as well as works by Aldo Leopold and Ralph King), and he noted many phenomena now prominent in the invasion science literature, including aspects of control and management. For instance, Wodzicki pointed to disastrous nontarget impacts of introducing and maintaining mammals with generalized diets for biological control of non-native pests, such as mustelids to control rabbits and goats to control weeds. Throughout, Wodzicki emphasized the need for substantial empirical research to determine exact impacts and interactions that he and previous authors had sketched out based on observational evidence. In addition to subsequent short reports based on his 1950 monograph that were published in the 1960s, Wodzicki's foreword to a work by Carolyn King that appeared near the beginning of modern invasion science was published posthumously in 1990.[127] This work was a successor to Wodzicki's 1950 monograph and contained several reports of the sorts of intensive research that Wodzicki had advocated.

Andrew H. Clark, a Canadian historical geographer who spent most of his academic career in the United States, published a monograph on the myriad impacts of non-native animals and plants on the South Island of New Zealand (discussed in chap. 8, sec. 8.2).[128] Clark synthesized the massive literature on the history of these invasions, citing Guthrie-Smith and Thomson, and on various impacts of these introductions, including Allan's and Cockayne's work on non-native plants. He pointed to examples of phenomena that subsequently became important features of invasion science, such as invasional meltdown (described in chap. 15, sec. 15.6) and lag effects (chap. 18, sec. 18.3.1). From an ecological standpoint, perhaps the most striking aspect of Clark's monograph is his description of many complex interactions among various non-native and native species (e.g., non-native gorse facilitating the spread and persistence of non-native rabbits), to an extent foreshadowing for this one island Alfred Crosby's later classic *Ecological Imperialism* (discussed in chap. 8, sec. 8.3), which depicted groups of European species acting together and overrunning native biotas as Europeans proceeded to colonize much of the world.[129] Clark also detailed myriad ways in which non-native species affected people's lives in both positive and negative ways, focusing especially on economic aspects. Although Wodzicki's monograph on non-native mammals in New Zealand appeared the year after Clark's monograph, Clark included Wodzicki's findings in a later publication recapitulating the main themes of his New Zealand work in a broader treatment of invasions into New World grasslands.[130]

6.4 Soviet Union

The person who paid most dearly for warning about potential impacts of non-native species was surely Vladimir V. Stanchinskii, an ecologist who focused on the community and ecosystem levels and served as editor in chief of the first Soviet ecology journal, *Zhurnal Ekologii i Biotsenologii* (Journal of Ecology and Bioecology).[131] As director of scientific research at Askania-Nova, a nature reserve in the Soviet republic of Ukraine, Stanchinskii conducted pathbreaking research on the flow of energy through food webs.[132] He was also a committed conservationist and opposed on scientific grounds an ambitious Soviet program to transform nature to serve human needs (discussed in chap. 5, sec. 5.4), which included introducing many animals and plants from other nations, or from one Soviet region to another. Stanchinskii countered that, not only would many such introduced populations fail to survive, thus wasting the considerable expense of the project, but those that did survive would likely displace and even cause the local or global extinction of native species, and in addition might carry parasites or pathogens that would attack native species. For this objection, probably at the behest of Trofim D. Lysenko and Izaak I. Prezent, who spearheaded the planned transformation, Stanchinskii was sentenced to prison for two years, the research institute he had developed at Askania-Nova was closed, the All-Union Scientific Research Institute for the Acclimatization and Hybridization of Animals was established in its place, and Lysenko and Prezent ordered the compositor to smash the frames of metal type to destroy his book.[133] Barred from teaching after his release, Stanchinskii was able to conduct research at a reserve in Russia but was rearrested in 1941 and died in prison the next year.[134]

: : :

One might ask if the growing number of scientists from the late nineteenth century through the first half of the twentieth century calling attention to the phenomenon of invasions and variety of impacts constituted or at least precipitated a new science of biological invasions. And did these alarms lead to more stringent regulations? In fact, these and similar publications did neither. A modern invasion science arose only in the 1980s, while substantial regulation appeared in the second half of the twentieth century. Why the long delays?

Certainly many early warnings contributed to the growing list of case studies and general literature. Even though most were ostensibly narrowly focused on particular sorts of organisms or on particular regions, they often drew on a broader range of examples. This development foreshadowed the imminent transformation of scientific treatment of invasions from a catalog of disparate one-off cases into a distinct, recognizable discipline—invasion science. Most scientists of this period focused on specific taxa—birds, mammals, birds and mammals, fishes, insects, plants.[135] However, many of these authors cited other organisms to make a general case for invasion impact in support of a specific focus on one taxon. King, for example, writing about potential problems associated with bird introductions, discussed the introduced small Indian mongoose, gray squirrel, rabbit, and lantana, while Myers, concerned with potential impacts of non-native fish, adduced the small Indian mongoose, house sparrow, and rabbit.[136] Leopold, in his essay "Cheat Takes Over" on the scope of impacts of non-native cheatgrass in the American West, cited many unrelated invasions—the house sparrow, chestnut blight, white pine blister rust—to establish the general relevance of invasion impacts.[137]

Others discussed invasions of particular regions—Ritchie focused on Scotland, but he sprinkled each chapter with examples from elsewhere, such as the rabbit in Australia, mustelids in New Zealand, phylloxera in France, and Colorado potato beetle, spongy moth, cottony cushion scale, and house sparrow in the United States.[138] Anderson was primarily concerned with introductions into Canada but mentioned the small Indian mongoose in Jamaica, muskrat in Europe, gray squirrel in England, and (citing Thomson) the plethora of vertebrates imported to New Zealand.[139] Wodzicki, writing on invasions of New Zealand, noted impacts of introduced goats on St. Helena and in Scotland (citing Ritchie).[140] Graham was addressing invasion issues in the United States but cited Australian invasions by water hyacinth and prickly pear.[141]

A mature science is characterized by a recognized, well-defined literature, and contributions to a mature science are expected to cite relevant earlier work. Many of the aforementioned authors manifested considerable familiarity with previous invasion literature. Thomson's work on New Zealand cited not only writings about introductions into New Zealand (including by Cockayne and Guthrie-Smith) but also the main nineteenth-century authors who focused broadly on impact—Wallace, Hooker, and Darwin (chap. 3, sec. 3.2). Treating the South Island of New Zealand, Clark restricted his attention only to works on that location but cited the major previous authors—Guthrie-Smith, Thomson, Cockayne, and Allan. Ritchie, focused on Scotland, cited no specific writings, but the global panoply of cases he advanced showed great familiarity with literature from elsewhere—among others, the nineteenth-century invasions by the rabbit, phylloxera, spongy moth, cottony cushion scale, and vedalia beetle described in chapter 4. King, addressing bird introductions, especially into the United States, cited the three main previous authors about American bird introductions—Palmer, Grinnell, and Phillips—as well as the game management book by Leopold and Leopold's "Chukaremia," a sarcastic critique of the practice of introducing game birds.[142] Cahalane included among suggested reading works by Thomson, Grinnell, Leopold, and King.

Perhaps most indicative of the fact that scientific notice of invasions was rapidly growing by the mid-1900s was the fact that scientists who did not conduct research on invasions nevertheless knew enough about some of them and worried enough about the phenomenon to announce that invasions were a major environmental problem. Herbert Levi was a systematic arachnologist who nevertheless described a range of invasion impacts in an article in a general scientific journal. Even more striking is that Henry Oosting, a plant ecologist, wrote about many invasions by both plants and animals in what was otherwise a focused plant ecology textbook, and his second edition expanded the treatment of invasions.

Authors during this period frequently mentioned very visible impacts of invaders, such as predation on and herbivory of native species, which had been noted by authors in the nineteenth century and occasionally even earlier (chaps. 2, 3). A smaller group pointed to introductions of diseases or their vectors, an impact not broached in the nineteenth century but exemplified by high-profile cases in the early twentieth century—white pine blister rust and chestnut blight (chap. 4, secs. 4.7 and 4.8, respectively).[143]

Several authors featured in this chapter cited other phenomena that had previously been noted occasionally but that subsequently became prominent themes of modern invasion science. Some such phenomena, such as habitat change and destruction wrought by invasive species, were regularly cited from this period onward and quite naturally became foci of modern invasion science.[144] The converse proposition, that habitat change can foster invasions, had rarely if ever been noted previously but became a prominent theme in the first half of the twentieth century.[145] In a particularly early example, Cockayne, discussing the many invasive plant species in New Zealand, observed:

> Many at first sight appear better suited to the soil and climate than are the indigenous species, and over much of the land they give the characteristic stamp to the vegetation; *but this is only the case where draining, cultivation, constant burning of forest, scrub, and tussock, and the grazing of a multitude of domestic animals have made absolutely new edaphic conditions which approximate to those of Europe*, and where it is no wonder that the European invader can replace the original.[146]

Elton, in the first monograph treating most of the scope of modern invasion science, pointed to certain kinds of habitat change that facilitated invasions, and this idea became a commonplace in modern invasion science, culminating in the abiding "passenger-driver" metaphor, stating that many invasions are passengers rather than drivers of environmental change (chap. 15, sec. 15.4.5).[147]

Early twentieth-century authors mentioned several other types of invasion impacts and raised questions about invasions that later became ongoing features of modern invasion science. These matters were not repeatedly addressed and incrementally accrued to the literature on invasions, but rather were rediscovered or reemphasized decades later, often without recognition of the earlier observations. Clark, for instance, gave examples of New Zealand invasions that underwent a substantial lag between arrival and spread, but the notion that such lag times were frequent and the search for generalizations about them did not become prominent until a paper in 1996 that did not cite Clark.[148] Similarly, as had Elton in his 1927 monograph on animal ecology, Leopold, Graham, Clark, and Levi all cited specific cases in which mutual facilitation among different invaders aided the invasion of at least one of them, but this phenomenon did not surface as a prominent subject until well into the era of modern invasion science, under the term "invasional meltdown" (chap. 15, sec. 15.6).[149] King and Anderson were concerned about hybridization between native and introduced populations, a matter that arose prominently as a general issue among invasion biologists only in 1996 (chap. 11, sec. 11.3.2).[150] Guthrie-Smith depicted the striking, apparently inexplicable population crash of a successful invader; the general phenomenon was not raised among modern invasion biologists until 2004.[151] Both Phillips and Bates observed that the fact that some species successfully establish new populations when introduced to new regions whereas many fail to do so is a phenomenon requiring research. Although occasional authors subsequently raised similar questions (e.g., why some plant species become weedy and related ones do not), the general question resurfaced only in the announcement of the SCOPE program in the 1980s.[152] Finally, Bates suggested evolution of invaders as a fertile subject for research. This was the one major thrust of modern invasion science that Elton's famous 1958 monograph almost entirely omitted and that did not become a prominent part of the field until the twenty-first century (chap. 11).

Remarkably, Leopold suggested that many invasion impacts result from perturbation of energy flow and nutrient cycling patterns in invaded communities, anticipating a key component of the concept of ecosystem impact of invasions introduced by Peter Vitousek in the SCOPE program and that subsequently became a leading edge of modern invasion science in the twenty-first century (chap 11, sec. 11.2).[153] Although Elton and others had previously conceived of the notion of food webs and diagrammed several examples, and Elton in 1927 had further juxtaposed the idea that food web relationships normally control animal population sizes with the claim that non-native species can disrupt this control, he had not linked invasion impacts to changed patterns of energy flow.[154] In fact, the concept of "ecosystem" was explicitly defined only in 1935 by Arthur Tansley, and the entire expanding branch of ecology that has come to be known as "ecosystem ecology" was first scoped in Eugene Odum's classic ecology textbook of 1953 (chap. 10, sec. 10.2).[155] The field began to flourish broadly only in the late 1960s and the 1970s.[156] Leopold also broached the notion that some biotas resist invasions and others are more susceptible, an idea elaborated by Elton in his 1958 monograph, where he tentatively suggested that species diversity and food web complexity confer resistance of a biota to invasion.[157] This was one of the three explicit foci of the 1980s SCOPE project that initiated the modern field (chap. 9, sec. 9.1).

Despite the growing number of early heralds and the prescience of some of them with respect to subsequent developments in invasion science, they did not constitute a new field, and when modern invasion science arose in the 1980s, these predecessors were rarely cited. At most, if rediscovered at all, they were viewed as people ahead of their time, rather than early rivulets feeding the mainstream. Part of the reason was the fact that ecology itself was a relatively young field at this time (as described in chap. 1), and virtually all observations on invasions were ecological. In the early twentieth century, most academic ecology consisted of narrowly focused contributions—usually observational—that would serve as case studies for later attempts at synthesis, and a few of these studies concerned non-native species. No journals specialized in invasion research (chap. 10, sec. 10.1). In its first decade, the 1920s, the new journal of the fledgling Ecological Society of America, *Ecology*, published fifteen articles that touched on invasions, but those on specific invasions did not cite literature on other invasions and almost never more general writings such as those cited in this chapter (chap. 10, sec. 10.1). Except for Charles Elton's 1927 textbook, early ecology textbooks gave short shrift to invasions. Not until the 1970s did invasions begin to get substantial attention in ecology textbooks. Many early warnings about invasion related to conservation concerns, yet modern conservation biology did not crystallize until the 1980s, and the first prominent textbook in the field appeared in 1993 (chap. 10, sec. 10.2.2).

Perhaps the advent of a modern invasion science required a continuing buildup of case studies, such as the few published in the first decade of *Ecology*, and the authors featured in this chapter provided others to the growing catalog or commented on the growing number, but the catalog had not yet reached some critical threshold for a new science to emerge. Additionally, many of these early writings traced more or less convincing narratives or scenarios about invasion impacts, rather than framing and attempting to test hypotheses against alternatives, as in a mature science (chap. 15). Few of these efforts entailed controlled, replicated experiments that later became one feature of modern invasion science. As for the fact that many of the early warnings contained explicit pleas to enact policies to slow the arrival and spread of non-native species, little evidence exists to show that they influenced regulations. Although occasional regulations targeting individual non-native species were implemented beginning in the nineteenth century, a plethora of international and national legislation did not arrive until the late twentieth century, and the problem of inadequate regulation persists today.[158]

7

Mid-Twentieth Century

A False Start, and the Lead-Up to Modern Invasion Science

The initiation of modern invasion science is often ascribed to Charles Elton's 1958 monograph, *The Ecology of Invasions by Animals and Plants* (henceforth *EIAP*).[1] Elton is a towering figure in the history of ecology, perhaps best known as one founder of the concepts of the ecological niche, the food web, and energy flow, and for his research on population cycles in nature.[2] His influence on conservation in Great Britain and on conservation biology as a science was also substantial.[3] *EIAP*, written for a lay audience, summarized and expanded a series of three BBC radio broadcasts by Elton in 1957 entitled "Balance and Barrier" and drew also from his 1933 BBC broadcasts "Plagues of Animals" and "Nature Sanctuaries."[4]

As noted in chapter 6, several twentieth-century authors addressed issues associated with non-native species that Elton subsequently explored in *EIAP*. However, none approached the scope and depth of Elton's monograph. Many focused on particular taxa, particular types of ecosystems, or particular regions, whereas Elton treated vertebrates, invertebrates, plants, and pathogens in marine, freshwater, and terrestrial habitats, and he drew on examples from geographically dispersed mainlands, islands, and oceans. Perhaps James Ritchie and Marston Bates came closest to *EIAP* in scope, but Ritchie, though citing invasions elsewhere, always brought his focus back to Scotland, and much of his book is only about Scotland and not very related to the inadvertent sorts of invasions featured in much of Elton's book and modern invasion science.[5] Elton acknowledged Bates's essay in the preface to *EIAP* as having provided useful references and as "the only recent general review of the subject of invasions that I have seen." Though Bates's essay was global in scope, it was a brief overview of the subject with a few well-chosen examples, and many sorts of impacts were not addressed. Although *EIAP* was aimed at lay readers, written in a conversational tone and with popular references (e.g., Buffalo Bill, Charlie Chaplin, Walt Whitman, Peter Brueghel the Elder, Robert Browning, H. G. Wells), Elton's prominence as an academic ecologist and the great number and variety of examples crammed into a short, very accessible book ensured that many scientists would also learn about the book and read it. In fact, this book was certainly a great milestone in the study of biological invasions, especially their impacts, but its true influence on the field, and that of Elton's writings in general, is indirect and probably not great.[6]

7.1 Charles S. Elton and His Invasion Book

Elton's interest in non-native species began early, during his childhood in Liverpool, where he was intrigued by exotic animals in a shop and a small informal zoo.[7] His first major work, written when he was twenty-seven, was one of the first monographs on animal ecology and included important references to impacts of non-native species, including lantana and the myna in Hawaii, the European rabbit in Australia and New Zealand, the spongy moth in North America, the starling and brushtail opossum in New Zealand, and Canadian pondweed in Europe.[8] Elton's take on these invasions was that they constituted disturbances that could act as probes into the determinants of community structure: "Then there are sudden disasters, like fires, floods, droughts, avalanches, the introduction of civilised Europeans and of rabbits, any one of which may destroy much of the existing vegetation."[9]

Elton's concern with non-native species continued through the next decades. As in the animal ecology monograph, he often used articles ostensibly about different or broader issues to focus on invasions, or articles on one invasion to call attention to the general phenomenon. For example, a 1933 newspaper article ostensibly about the need for a census of British animals was headlined "Animal Invaders," while a 1936 book review of a monograph on the Chinese mitten crab in Europe opened with the observation that the species had just arrived in Great Britain, developed ominous scenarios for potential damage it might cause, and closed with a list of Britain's "growing band of invaders," including the gray squirrel, muskrat, French partridge, little owl, willow grouse, rainbow trout, black bass, continental crayfish (signal crayfish), American slipper-limpet, oyster-tingle (drill), Colorado potato beetle, and the cordgrass *Spartina townsendii*, a hybrid between native small cordgrass and North American smooth cordgrass.[10] Two small books he wrote in 1933 contained many references to impacts of invasions.[11] One closed with a chapter largely about non-native species entitled "Plagues of Animals," based on his 1933 radio broadcast with that title.[12]

In 1932, Elton founded the Bureau of Animal Population at Oxford University, which he headed until its closure in 1967 when he retired.[13] During the 1930s and through World War II (1939–45), bureau personnel worked on several species introduced to Great Britain. Soon after the Bureau of Animal Population founding, the British Ministry of Agriculture and Fisheries, as part of an eradication campaign against the non-native muskrat, funded research on the ecology of this species in Great Britain directed by Elton and primarily conducted by Tom Warwick of the bureau.[14] Warwick also studied escapes of non-native nutria, which was later eradicated.[15] Elton also directed early research by Arthur Douglas Middleton on impacts of the American gray squirrel on the native red squirrel and on trees in Great Britain.[16] The gray squirrel later became recognized as a scourge in Great Britain and is now also established in Italy and spreading.[17]

During World War II, Elton directed the entire Bureau of Animal Population program toward aiding the war effort by finding ways to reduce or even eradicate four non-native species that damaged British agriculture: the ship rat, Norway rat, house mouse, and European rabbit. A longtime bureau collaborator and protégé of Elton's, Henry N. "Mick" Southern, made major contributions to control of all four species during this period and for several subsequent years. Those on rats and mice were later summarized in a three-volume work that was widely cited for the remainder of the century.[18]

The only Elton-supervised student whose doctoral research specifically concerned non-native species was Kitty Paviour-Smith (who later married Southern and was a mainstay of the Bureau of Animal Population until its closure). Studying the fauna of fungi colonizing dead and dying trees in Wytham Woods for her doctoral thesis, she found the beetle *Cis bilamellatus*, which she later determined is Australian, to be the main inhabitant of birch bracket fungus, traced its likely recent trajectory into Great Britain, and elucidated its population biology.[19] In the book Elton considered his masterwork, *The Pattern of Animal Communities*, published in 1966, Elton discussed the importance of this beetle in the organization of the Wytham Woods biotic community.[20] In a 1962 talk he had used it to exemplify how a barely noticed invader could have a big impact on an entire ecosystem:

> One last example, to illustrate a principle of our work . . . Kitty Paviour-Smith has been studying a beetle, abundant in the dead stages of the birch-bracket fungus, which has spread from the London area, probably from a specimen of fungus in Kew Herbarium, and is now known to be an Australian invader. . . . So we can see the investigation of single species populations, of predators and prey, the structure and interspersing of whole communities, and of foreign invaders into them, are all connected parts of the animal ecologist's field of study.[21]

This beetle has more recently spread to Ireland, mainland Europe, and Chile.[22]

Paviour-Smith had a long, productive career, even after the closure of the Bureau of Animal Population, conducting research especially on forest ecology and insects. She was, by the 1960s, the key person in the Wytham Biological Survey, a major long-standing bureau project.[23] Although she published on the invasion of sycamore in Great Britain by a pathogenic fungus, she did not focus particularly on invasions, and when Paviour-Smith has been cited with respect to invasions, it is almost exclusively for the research on *Cis bilamellatus*.[24]

Elton heavily influenced two other students who conducted research on biological invasions after their time at Oxford in the bureau.[25] His student William Murdoch studied the population biology of several carabid beetles at Wytham Woods and elsewhere for his thesis, eventually specializing in predator-prey interactions and publishing many papers on this subject.[26] He read *EIAP*, but it was not the inspiration for his later research.[27] He did, however, cite it as one of the four sources for the idea that greater species richness increases ecosystem stability.[28] Murdoch's engagement with invasion science rather originated in his research on biological control, which began in 1982 and included many later papers.[29] His interest in biological control arose from his long interest in predator-prey interactions and their role in population dynamics, and the concern with population dynamics, in turn, was triggered by Elton's 1927 book and discussions at the bureau with Elton and his student Dennis Chitty about small mammal cycles.[30]

Roger Kitching was supervised at the Bureau of Animal Population by Southern but interacted strongly with both Paviour-Smith and Elton throughout his thesis research, which was not on non-native species but on animals living in tree holes.[31] Kitching read *EIAP* while at Oxford, but he was particularly inspired at the time by *The Pattern of Animal Communities*, which Elton had recently published and viewed as the culmination of much of his life's work and that of the bureau.[32] Kitching's interest in and research on non-native species began later, as he studied the introduced sheep blowfly in Australia.[33] This research led him to compile and edit *The Ecology of Pests: Some Australian Case Histories*.[34] Although the book was not specifically about non-native species, eight chapters treated non-native species (none cited *EIAP*). Kitching noticed this fact and subsequently edited *The Ecology of Exotic Animals and Plants—Some Australian Case Histories*.[35]

The *Ecology of Exotic Animals and Plants* project led Kitching to begin thinking about invasion science as a distinct field rather than a collection of particular cases and to seek generalizations among all sorts of invasions.[36] This shift was aided by his being in a new country, Australia, with many high-profile, relatively recent problematic invaders, in contrast to his early years in England, where most non-native species were either ancient and thus almost part of the normal biotic background or else of only local concern. It was at this point that Kitching recalled various insights and cases in *EIAP*, which he had read many years earlier.[37]

The Bureau of Animal Population hosted many long-term visiting scientists, including some of the world's leading twentieth-century ecologists (e.g., Thomas Park, L. Charles Birch, Eugene Odum). One visitor, David Pimentel (who spent part of 1961 there), conducted substantial research on non-native species during his career. He was already an established researcher on non-native species and particularly biological control when he read *EIAP* soon after it appeared in 1958.[38] After receiving his doctorate in 1951, his first post was as chief of the Tropical Research Laboratory in Puerto Rico, where he conducted biological control research but also studied the small Indian mongoose.[39] Pimentel continued to study biological invasions, particularly biological control, and was among early scientists to caution about possible nontarget impacts and other unintended consequences of biological control (chap. 17, sec. 17.5).[40] He also published an early paper on the evolution of both non-natives and natives interacting with them, suggesting that such evolution could attenuate invasion impacts.[41] Elton subsequently expressed tentative skepticism about the efficacy of such a mechanism.[42] Specializing on the intersection of ecology and environmental problems, Pimentel was best known among invasion scientists for attempting to calculate the total economic costs of non-native species in the United States and globally, a project he was inspired to undertake by a report of the United States Congress Office of Technology Assessment estimating costs of a select group of invaders.[43]

It is difficult to assess Elton's influence on Pimentel's contributions to invasion science. Although Elton was an enthusiast of biological control as a mechanism for managing invasions, as witness several examples cited in chapter 4 of *EIAP* and extensive notes for chapter 7 of his planned revision of *EIAP*, he also foresaw in chapter 4 the possibility of damaging nontarget impacts.[44] However, Pimentel was engaged in biological control research before his stint at the bureau and before Elton published *EIAP*. Furthermore, Pimentel's interest in the evolution of non-natives and natives encountering them is one major area of modern invasion science that Elton barely addressed in *EIAP*, touching only on the evolution of insect resistance to pesticides, and Elton's notes for the planned new edition did not expand on this theme.[45] Elton did not discuss the economic cost of invasions, an emphasis of Pimentel's, except in a few particular cases. With the advent of modern invasion science in the 1980s (covered in chapter 9) and concurrent rediscovery of *EIAP* and explosive growth of citations of the book, virtually every invasion scientist was reading (or rereading) *EIAP* with its numerous, striking, and still timely examples of various invasion phenomena.[46] The enthusiasm of many researchers was probably spurred by this book, and this could well be the case for Pimentel. There is no evidence of direct influence on Pimentel's projects.

In the years following publication of *EIAP* and before or at the beginning of the Scientific Committee on Problems of the Environment (SCOPE) project that initiated modern invasion science, three major conferences addressed topics treated in *EIAP* with little or no note of Elton's book.[47] In an international symposium in Asilomar, California, in 1964 about genetics and evolution of colonizing species, including non-natives (chap. 11), only three of twenty-seven authors cited *EIAP*, all with passing references.[48] Although Elton was not an evolutionist and barely addressed evolution in *EIAP*, several conference participants were ecologists, and many of the papers, even by evolutionists and geneticists, addressed ecological topics. A 1969 conference at the Brookhaven National Laboratory on Long Island, New York, assembled primarily ecologists to discuss the proposed diversity-stability relationship.[49] This was the main generalization Elton drew in *EIAP* about why some non-native species establish populations and spread while others disappear, a generalization that was perhaps the key point relating to non-native species in the two chapters of *EIAP* devoted to conservation.[50] Yet none of the nineteen papers cited *EIAP*. In 1984, as the SCOPE project was getting underway, a joint conference at Southampton of the British Ecological Society and the Linnean Society of London addressed the relationship of colonization, ecological succession, and stability; none of the twenty-one contributions cited *EIAP*.[51] Of these conference volumes, those based on the Asilomar and Brookhaven meetings were widely cited for decades, and the Asilomar volume, *The Genetics of Colonizing Species*, even inspired a fiftieth anniversary special issue of the journal *Molecular Ecology*, "Invasion Genetics: The Baker and Stebbins Legacy," and a subsequent book.[52]

Elton's influence on public awareness of non-native species was perhaps greater than his influence on the development of the science, though most of this influence was surely indirect and difficult to measure. Although *EIAP* was written for a popular audience and was based on a series of BBC radio broadcasts that must have been heard by many listeners, I know of no quantitative evidence of a particular uptick in British or international public awareness of invasions and concern with their impacts before modern invasion science crystallized in the 1980s (as shown in table 1.1 in chap. 1). However, Elton strongly influenced the two persons most closely associated with the rise of environmentalism in the mid-twentieth century—Aldo Leopold and Rachel Carson—and particularly sensitized them to problems associated with non-native species. Surely the many readers of Leopold's *A Sand County Almanac* and Carson's *Silent Spring* must have come away with increased awareness of problems posed by biological invasions.

Although Leopold first wrote about non-native species in 1918, in these and other writings of the next decade addressing particular introductions he was concerned especially with utilitarian matters—whether non-native sport fish and game would thrive and provide as good an experience for sportsmen as native species would.[53] Although he mentioned the devastation wrought by introduction of chestnut blight and white pine blister rust to American forests (species discussed in chap. 4, secs. 4.7 and 4.8, respectively), this was a passing comment at the end of an article about the pros and cons of introducing game animals, and ecological impacts were not a focus of Leopold's writings on non-native species at this time.[54] In addition to utilitarian matters, the early writings, as did later ones, also raised aesthetic issues (as described in chap. 6, sec. 6.1).[55] However, Leopold met Elton in 1931 at a meeting on biological cycles at Matamek, Labrador, and began a lifelong friendship and correspondence that entailed a transformation of Leopold's concern with non-native species from mainly utilitarian in focus to one squarely aimed at ecological impacts.[56] Prevention or at least minimization of species introductions became firmly embedded in Leopold's land ethic and his conception of ecosystem health.[57] His concern with non-native species culminated with three essays primarily on invasions in the posthumously published *A Sand County Almanac*, including "Cheat Takes Over."[58] Although *A Sand County Almanac* was not widely read for two decades after its publication, it became a perennial best-seller during the environmental movement heralded by Rachel Carson's *Silent Spring*, and these two books are widely seen as the two most significant environmental books of the twentieth century.[59]

Carson was directed to *EIAP* by Edward O. Wilson, and she thanked him profusely for this tip:

> I am already indebted to you for many things, but perhaps most of all for your references to Elton's "Ecology of Invasions." I had not heard of it until you mentioned it, although I know his earlier work on populations. I found this enormously stimulating. It cuts through all the foggy discussion of insect pests and their control like a keen north wind.[60]

Carson asked Elton to write a foreword to the English edition of *Silent Spring*. Carson, in a gracious response to Elton's declining the invitation, again described its influence on her thinking:

> May I take the opportunity to tell you how much your various books have meant to me over the years? Your recent ECOLOGY OF INVASIONS was especially fine, and I thought it did much to set in proper perspective the problem I have dealt with in SILENT SPRING.[61]

Although not explicitly about non-native species, *Silent Spring* is replete with examples of non-native species and the unintended, often pervasive consequences that arose from attempts to manage them with pesticides and herbicides. In *EIAP* Elton discussed all but one of the non-native species Carson cited—spongy moth, Japanese beetle, prickly pear, codling moth, Dutch elm disease, Hessian fly, St. John's wort, and European corn borer. Carson cited *EIAP* early in her book, also directly quoting Elton's prescient warning on the threat posed by the evolution of resistance: "We are hearing the early rumblings of what may become an avalanche in strength."[62] Carson also cited two examples of successful biological control adduced by Elton, the vedalia beetle (discussed previously, in chap. 4, sec. 4.3) and the South American cactus moth that controlled introduced prickly pear in Australia (discussed further in chap. 17, sec. 17.5). The one case featured by Carson that Elton did not discuss in *EIAP* was the red imported fire ant. Interestingly, Elton was likely to have included this example in the revised edition he long planned but never completed, as his handwritten notes on the proof copy that he used to organize the edition listed three citations of publications on this invasion that appeared the year *EIAP* was published, and detailed notes on E. O. Wilson's article "The Fire Ant" were inserted between the pages of the proof copy.[63]

Although Elton's 1958 book did not lead quickly and directly to a science of invasions, both scientific and popular interest in invasions grew for the next two decades. Three factors in particular spurred this growth. First was the gradual accrual of reports of new invasions whose impacts were visible enough to be noticed in the lay and scientific press and sometimes to inspire research; also, occasional reviews appeared that described particular groups of existing non-native species in specific regions. Second was the rise of the environmental movement, particularly in the United States, and third was the related advent of modern conservation science, both of which recognized non-native species as an important part of a larger, overarching problem, although the recognition was belated for modern conservation science.

7.2 New Invasions, Popular Books, and Reviews of Particular Invasive Biotas

After publication of *EIAP*, several scientists wrote about invasions, especially by particular groups of species in particular regions, yet none of these publications triggered a growth in research on the general phenomenon. For instance, in 1962 Herbert Sukopp described the great number of plant invasions of central Europe beginning with the Columbian interchange (termed "neophytes" by European scientists to distinguish them from species introduced in antiquity), citing *EIAP* as a general introduction to the history of invasions.[64] This work was not followed by a rapid increase in invasion research, but work on Central European plant invasions became a major feature of modern invasion science after its origination in the 1980s.[65]

In 1970, Ernest A. Lachner, C. Richard Robins, and Walter R. Courtenay Jr. published *Exotic Fishes and Other Aquatic Organisms Introduced into North America*, which focused largely on fishes, with special attention to Florida and Hawaii, and mentioned *EIAP* as a general treatment of invasion impacts.[66] Courtenay and Robins later published a paper on non-native aquarium species, especially fishes, established in Florida, and Courtenay and several colleagues discussed non-native fishes in Florida (as described in chap. 5, sec. 5.2 and box 5.1).[67] In 1975 Courtenay and Robins, though still emphasizing Florida and fish, described several aspects of invasion impacts for other birds and mammals and the general issue of non-native species carrying parasites and pathogens.[68] None of these three publications cited Elton. Although they helped alert scientists, especially ichthyologists, to non-native species, particularly in Florida, these papers did not spur the development in the 1970s of a comprehensive, distinct science centered on biological invasions. However, later work on non-native fishes by Courtenay was cited in papers on fishes and Florida during the SCOPE program, and Courtenay was a prominent participant modern invasion science developed.[69] He was active in the field until his death in 2014—for instance, studying the famous invasions of the lionfish along the US. Atlantic coast and the northern snakehead near Washington, DC, earning the sobriquet "Dr. Snakehead" for his expertise and activity related to the latter case.[70]

Florida quickly became and remains a focus of invasion science because of the great number of invasions of plants and animals into the state. In 1971, ecologist Raymond F. Dasmann penned for a lay audience a fiery manifesto aimed at halting the rapid destruction of natural habitats of Florida, with a long chapter on non-native species.[71] Citing *EIAP* for the hypothesis that disturbed habitats are particularly susceptible to invasion by non-native plants (a hypothesis discussed further in chap. 15, sec. 15.4.4), Dasmann provided a litany of examples of invasion impacts. He had previously written about the destruction of California's native grasslands by non-native livestock, which often became feral, and non-native plants, mostly annual grasses and forbs from the Mediterranean area of Europe, and also about non-native game animals on US public lands (discussed previously in chap. 5, sec. 5.4).[72]

Dasmann's books, coming soon after the publication of Rachel Carson's *Silent Spring* in 1962 and, in the case of the one on Florida, after the first Earth Day in 1970, appeared as the environmental movement was rapidly expanding, and they may have sensitized some readers to biological invasions, but they did not influence development of a science of invasions; the chapters on Florida and California in the major relevant SCOPE volume did not cite them.[73] In the 1970s, botanists Daniel F. Austin and Julia F. Morton published several papers on non-native plants in south Florida, and each author summarized observed or suspected impacts.[74] Neither was cited in the relevant SCOPE chapter.[75] However, after the SCOPE project, a comprehensive book about non-native species in Florida appeared, with extensive citations of Austin and Morton, details of many examples cited by Dasmann, and a chapter by Courtenay on fishes.[76]

In 1973, *The Wilson Bulletin* published papers on non-native birds in three regions of the United States—southern California, greater New York City, and southeastern Florida.[77] Of these, only the one on Florida introductions, by Oscar T. Owre, explicitly related to more general aspects of invasions and previous literature not specifically on birds of a particular location, citing *EIAP* as a general reference and as a justification to worry about invasions, John C. Phillips (whom we met in chapter 6) for previous data on non-native birds in the United States, and Lachner and colleagues for other introductions to Florida. However, although a SCOPE program book cited this paper, it received virtually no attention until then but was cited several times in the wake of the SCOPE program of the 1980s.[78] Many other reports of non-native birds appeared in various journals focused on ornithology, some even before *EIAP*. For instance, in 1953, ecologist Gordon R. Williams described how several European birds that had been introduced to Australia and New Zealand had colonized other Pacific islands, for the most part autonomously.[79] Although his main concern was how they reached their new sites, he briefly expressed skepticism that they would compete with native species on at least some islands.

Several books on invasions for lay audiences appeared in the two decades after publication of *EIAP*, and almost all of these either ignored or barely noted Elton. Neither did they ignite substantial further scientific interest in invasions. Lorus J. Milne and Margery Milne, biologists at the University of New Hampshire, wrote many popular books on nature and biology. In 1960, they published *The Balance of Nature*, a series of essays on how various human actions were upsetting the "balance of nature," a paradigm then dominant (and discussed in chap. 2, sec. 2.3).[80] Prominent among these were well-known invasions such as those by the small Indian mongoose, sea lamprey, common carp, red imported fire ant, rabbit, and Dutch elm disease. Although Milne and Milne included detailed references, including scientific ones, in extended notes for "specialists," they did not cite *EIAP*. Nature and outdoor writer George Laycock published *The Alien Animals* in 1966, featuring case studies of twenty animal species introduced to various parts of the world that subsequently caused ecological or economic damage, with passing reference to several others.[81] Less comprehensive in terms of the nature of invasion impacts than *EIAP*, and with far fewer examples, the book was similar in thrust to *EIAP* and cited many scientific publications, but oddly did not cite Elton's book.[82] In 1976 Clive Roots, a biologist who published many books for a popular audience, wrote *Animal Invaders*, which mentioned Elton's estimate of rat densities in British corn ricks but not *EIAP*.[83] As had Laycock, Roots focused on a few dozen examples in various regions (many of the same ones as Laycock), citing many scientific publications. In 1969, farmer and writer Eric C. Rolls published a large, popular book on non-native vertebrates (mostly mammals) and their impacts in Australia, *They All Ran Wild: The Story of Pests on the Land in Australia*; he did not mention Elton.[84] Writer Christopher Lever in 1977 published *The Naturalised Animals of the British Isles*, about non-native vertebrates, for both lay and scientific audiences, consisting primarily of detailed histories of the introductions with attractive line drawings and information on their impacts.[85] This book was the culmination of a decade of interest by Lever in this subject, including popular articles and BBC broadcasts.[86] Although none of the five books discussed in this paragraph were cited in the scientific literature on invasions preceding the SCOPE project, it is possible that they raised public interest in invasions and that this increased interest and recognition of the breadth of invasion impacts contributed to the atmosphere that gave rise to the SCOPE project.

The most scholarly of the lay-oriented books on non-native species in the wake of Elton's monograph and before the SCOPE project focused, like Lever's book, on non-native vertebrates of Great Britain. This was Richard S. R. Fitter's *The Ark in Our Midst: The Story of the Introduced Animals of Britain—Birds, Beasts, Reptiles, Amphibians, Fishes*.[87] Written at Elton's urging and published in 1959, the year after Elton's monograph, the book detailed the history of all the species introduced into Great Britain, including the role of the British Acclimatisation Society (discussed in chap. 5, sec. 5.1). Unlike Lever's book, it also similarly treated both failed invasions and native British species introduced to areas of Great Britain in which they were non-native or had become extinct. Fitter, a versatile naturalist and conservationist who wrote many guides, especially to plants and birds, had previously written a short lay article on non-native vertebrates in Britain and subsequently published a short book partly on the subject.[88] The 1959 book treated impacts to nature and human enterprises. However, despite its exhaustive treatment, it did not stimulate either lay or scientific interest in invasions, although Lever cited it.[89]

All six of these books were almost exclusively on animals, as signaled by the titles of four of them.[90] This focus is unsurprising in books aimed at a popular audience, as impacts of non-native animals, such as grazing and especially predation, are highly and immediately visible and often dramatic, whereas many impacts of non-native plants are more slowly manifested, and some, such as changes in nutrient cycling, are subtle. With the advent of modern invasion science in the 1980s, the scope and importance of non-native plant impacts were rapidly highlighted in both scientific and popular literature. The change in title of Rolls's second edition is indicative. The first edition in 1969, entitled *They All Ran Wild: The Story of Pests on the Land in Australia*, was virtually entirely about vertebrates, with small sections on introduced blackberry and prickly pear, and this was a deliberate choice.[91] The second edition, published in 1984, was otherwise unchanged (including the statement in the preface that "apart from blackberry and prickly pear, weeds have not been dealt with") but was entitled *They All Ran Wild: The Animals and Plants That Plague Australia*.[92] Post-SCOPE, several general books about biological invasions for a popular audience appeared (detailed later in chap. 10, sec. 10.3), and virtually all of them treated plants as well as animals. One, by Australian writer and wildlife biologist Tim Low, explicitly criticized Rolls's 1969 edition for its narrow focus on mammals and neglect of invasive plants.[93]

Throughout the 1960s and 1970s, reports of new invasions appeared routinely in the scientific and popular literatures. For instance, in planning a second edition of *EIAP*, Elton kept notes on articles and collected reprints about various invasions through 1986.[94] Just for the 1960s and 1970s, he recorded invasions of the walking catfish from Asia, pike killifish found from Mexico through Central America, and giant African snail in Florida; a barnacle from Europe in Japan; pink salmon from the US West Coast in Lake Superior; and Japanese wireweed and Texas mud crab in Britain.[95] The walking catfish, giant African snail, and Japanese wireweed were sufficiently high-profile to be reported in the popular press. Many other reports appeared throughout this period, on all sorts of organisms in many places. Examples include the introduction of the monk parakeet from South America in the United States, the spread of the raccoon dog native to Eurasia across much of Europe, the spread of the sirex wood wasp (native to parts of Europe, Asia, and northern Africa) from Tasmania to mainland Australia, the introduction of the giant African snail to the Miami area and its eradication, and the rapid spread of the Asian aquatic plant hydrilla throughout Florida and beyond.[96]

7.3 A Study Ahead of Its Time

One paper stands out in this period as ahead of its time, raising issues that began to be addressed in the SCOPE project and, in some aspects, in the development of modern invasion science in the wake of the SCOPE project. In 1973, Thomas M. Zaret and Robert T. Paine published "Species Introduction in a Tropical Lake" about myriad impacts caused by the introduction of predatory peacock bass from the Amazon Basin into Gatun Lake in what was then the Panama Canal Zone.[97] This fish arrived in the lake in 1967 from the Chagres River, which it had recently reached after escaping into a tributary of the Chagres from an impoundment to which it had been transported in 1965 as a possible sport and food fish. The particularly striking impact that Zaret and Paine depicted was drastic simplification of a Gatun Lake food web within six years, especially via the peacock bass feeding on native fish that feed on zooplankton and nearly eliminating most of them (fig. 7.1).

This simplification of the food web in turn triggered a variety of "second-order effects." Zaret's doctoral research had addressed how, in Gatun Lake, one of these native fish, the Chagres silverside, preys heavily on a cladoceran, *Ceriodaphinia cornuta*, that consists of two morphologically distinct forms, one with "horns" presumed to deter predation.[98] His dissertation did not mention the peacock bass invasion. One second-order effect of the arrival of the peacock bass and consequent near-disappearance of the silverside was that the relative frequency of the horned form plummeted. In areas of the lake where the silverside had largely disappeared, Zaret and Paine also observed absence of two species (Atlantic tarpon and black tern) that feed on the silverside and that remained in parts of the lake where the silverside was still present. They cited similar observations for other birds, such as herons and kingfishers, that feed on small fish. Finally, they presented data on mosquito captures from systematic trapping before and after the peacock bass invasion showing substantially greater abundance after the invasion, and they noted that, in the wake of the invasion, the form of malaria pathogen in victims shifted from *Plasmodium vivax* to *P. falciparum*. A possible explanation is the elimination of mosquito larva–eating fishes such as the silverside, mosquitofish (*Gambusia*), and characins that the peacock bass preys on.

This focus on the whole gamut of invasion impacts—at species, community, and ecosystem levels—was way ahead of its time. Peter Vitousek during the SCOPE project pointed to the existence of ecosystem impacts, but it was not until a decade later that this sort of research became prominent (as described in chap. 12, sec. 12.3).[99] Although Elton's 1958 book had a chapter on "New Food-Chains for Old," he did not sketch concatenated sequences of impacts such as the impact of peacock bass on native silversides and the impact on silversides in turn affecting native cladocerans or mosquitoes (and thus humans via malaria), or the impact of peacock bass on the multiple species that had preyed on the silverside. The closest Elton came to scenarios of this sort was his description in a later chapter of a biological control project in which an introduced parasitic wasp attacked an introduced scale insect to the benefit of introduced citrus trees in California. The concatenated sequence of impacts connecting the peacock bass to malaria instead resembles invasions described much later, such as a case in Flathead Lake, Montana, in which a tiny ghost shrimp was introduced to feed introduced kokanee salmon.[100] Instead, it devastated cladoceran populations, which are also prey for kokanee, and the crash of the kokanee population caused a dramatic decline in populations of bald eagles and grizzly bears that had eaten them. The figures of the greatly simplified Gatun Lake food web depicted by Zaret and Paine could have been a model for the figures in

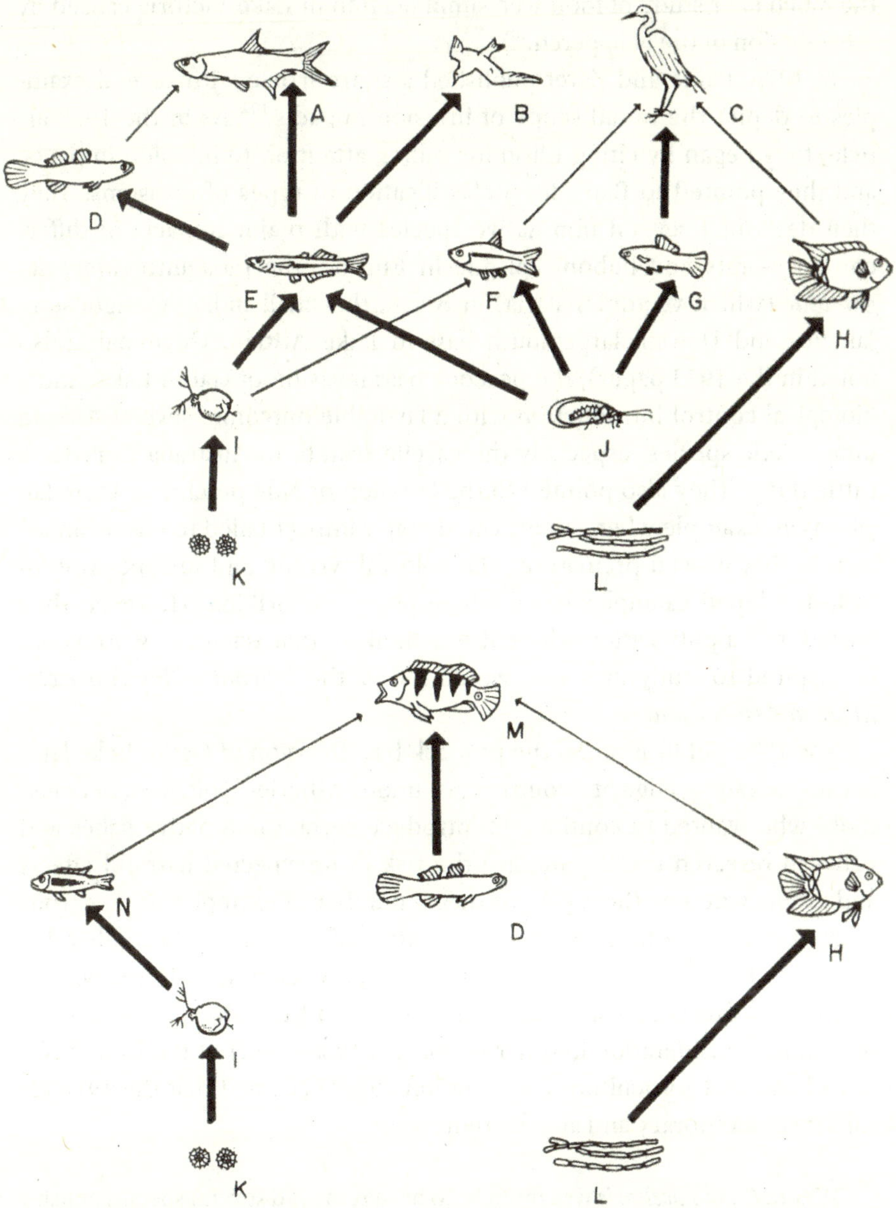

Fig. 7.1 Change wrought in food web by arrival of the peacock bass in Gatun Lake, Panama, 1967. *Upper*: Preinvasion food web. *Lower*: Postinvasion food web. Thick arrows represent important food items; thin arrows represent less important items. A = Atlantic tarpon, B = black tern, C = herons and kingfishers, D = bigmouth sleeper, E = Chagres silverside, F = several species of characins, G = several species of live-bearers, especially shortfin molly and Nicaraguan mosquitofish, H = blackbelt cichlid, I = zooplankton, J = terrestrial insects, K = nanophytoplankton, L = filamentous algae, M = adult peacock bass, N = young peacock bass. Credit: Ned R. McWherter Library, University of Memphis, from T. M. Zaret and R. T. Paine, "Species Introduction in a Tropical Lake," *Science* 118 (1973): 449–55. Reprinted with permission from AAAS.

the much later study of food web simplification in Lake Victoria caused by introduction of the Nile perch.[101]

In 1975, Paine and Zaret published a short article with several examples to depict the broad scope of invasion impacts.[102] As in the 1973 article, they began by citing Elton for calling attention to invasion impacts, and they pointed to Bates for a classification of types of invasions. They then described several non-native species with major impacts of different sorts—rats and bubonic plague in Europe, the mosquito *Anopheles gambiae* (which vectors malaria) in Brazil, the small Indian mongoose in Jamaica and Hawaii, largemouth bass in Lake Atitlán, Guatemala (also noted in the 1973 paper), the peacock bass invasion of Gatun Lake, and a biological control introduction with a favorable outcome: several African dung beetle species, especially the gazelle scarab, to Australia to process cattle dung. They also pointed to the invasion of Nile perch in a Ugandan lake as an example where an expected major impact failed to ensue, an assertion that proved premature. The colorful writing and juxtaposition of well-developed examples in this three-pager are striking. However, they buried it in a publication where it was unlikely that scientists who would be inspired to study invasions would see it: the *Journal of the American Medical Association*.

The 1973 publication on the peacock bass invasion of Gatun Lake later became a leading edge of a controversy among fisheries biologists, between those who wanted to continue to introduce certain non-native fishes and believed research could minimize the risk of unexpected harmful effects and others who saw the rapidly growing number of examples of disastrous impacts as warranting a nearly total cessation of the practice (a conflict discussed in chap. 5, sec. 5.2). By the 1980s, the peacock bass had already been widely introduced as a sport fish (e.g., in Puerto Rico, Brazil, Hawaii) and was under consideration in other regions. A 1988 report of the United Nations Food and Agriculture Organization (FAO) claimed that the 1973 results were temporary and already redressed:

> Where *Cichla ocellaris* was thought to have removed several species of fish from Gatun lake, Panama (Zaret and Paine, 1973), the supposedly eliminated species later recolonized the lake from refugia. . . .
>
> The results of the introduction to Lake Gatun in Panama were documented by Zaret and Paine (1973) who alleged that *Cichla* had eliminated six species of native fish from the lake fauna. It now appears that this evaluation may have been overly exaggerated as a new balance is reported between the predator and the native species, many of which persist in refugia in the mouths of inflowing streams. The situation is further complicated by the introduction of other species, including some *O. niloticus* which provide an alternative prey.[103]

A proposal advocating introducing peacock bass into Florida (discussed in chap. 5, sec. 5.2, and highlighted in box 5.1 case study) went still further in attempting to discount the significance of the 1973 study, claiming, as did FAO report author Welcomme, that the main effect of the introduction had been reversed but also pointing to the absence in Gatun Lake of a top-level piscivore, as had been noted by Zaret, and citing several publications of other introductions of the species to argue that "when introduced in well-adapted fish communities which have top level piscivores, detrimental effects have been absent or insignificant."[104]

However, testing a prediction of Zaret and Paine's that the lake ecosystem would return to some form of equilibrium, and a somewhat different later conjecture by Zaret, Diana M. T. Sharpe and colleagues found in 2017 that the main feature—the dominance of the peacock bass and the dramatic decline of the near-shore native fish community—had persisted for forty-five years.[105] This persistence is despite the subsequent introduction of other non-native fishes, including a fish-eating predator (the jaguar cichlid), an omnivore (the oscar), and three detritivores/algivores (flag cichlid, Nile tilapia, and redbreast tilapia). In addition, the native predatory tiger fish has greatly increased its dietary breadth in the wake of the invasion.[106] It ate primarily live fish before the peacock bass introduction, but much of if its diet today consists of predatory invertebrates as well as scavenged pieces of peacock bass and Nile tilapia discarded by fishers.

Both Zaret and Paine had productive careers, but neither engaged substantially with invasions after this, and neither participated in the SCOPE project. Rather, both sporadically pointed to invasions and devoted their major research efforts in other directions. Zaret, who died at the age of thirty-nine, mainly studied the biology of fishes and aquatic invertebrates, and he occasionally used his work on fish ecology to focus on more general problems such as measurement of the ecological niche and overlap of different species' niches. In 1974, he again discussed the range of impacts of the peacock bass in Gatun Lake, contrasting these to what he perceived as the relative absence of major effects of the introduction of Nile perch into the Great Lakes of East Africa.[107] E. Barton Worthington quickly suggested that it was premature to discount a major impact of the Nile perch introductions, and he was ultimately proved right, as the extinction of many endemic cichlids from Lake Victoria, with important follow-on effects, is widely recognized as one of the greatest of invasion disasters.[108] Zaret also provided data showing that the peacock bass population introduced to the Chagres River, the source of the population in Gatun Lake, had not largely eliminated the native fishes as it had in the lake.[109] In 1980, he used his extensive Gatun Lake data to study the biology of the peacock bass (e.g., growth rate, behavior), independently of its impact.[110] G. L. Swartzmann and Zaret used the Gatun Lake case to test a simulation model in a search for those traits of the peacock bass that most determined its impact.[111] In 1982, Zaret used the Gatun Lake introduction but also results of several other cases in which fish had been introduced to lakes to cast doubt on Aldo Leopold's contention that species diversity confers stability on ecological communities.[112] This latter contribution is more in line with Zaret's occasional invocations of invasions as parts of explorations of other ecological issues. As another example, in his 1980 book *Predation and Freshwater Communities* he discussed several cases in which fish introductions had affected zooplankton communities.[113]

Paine was one of the leading ecologists of the twentieth and early twenty-first centuries, focused often on predation, and is perhaps best known for the concept of the keystone species and elucidating the role of top-down predation in maintaining biodiversity and in trophic cascades.[114] Among his many graduate students, several have focused parts of their research on invasions, including Marjorie Wonham (whose doctoral thesis concerned an invasion), Jennifer Ruesink, and Timothy Wootton; Zaret was Paine's postdoctoral fellow. Paine's subsequent citing of invasions was generally to inform a larger treatment of predation, ecological disturbances, or other impacts on food webs, the unifying thread of his research. In 1995, he included a discussion of invasions as part of an effort to refine the concept of keystone species.[115] In 1998, Paine and two coauthors pointed to cases in which two or more ecological disturbances combined to produce surprising outcomes: a flood plus the Asian overbite clam introduced into San Francisco Bay, and the invasion of the fire tree, a nitrogen-fixing plant native to Atlantic islands, to Hawaii in the wake of a volcanic eruption.[116] A 2002 brief commentary highlighted the role of new resources (in one case, an introduced species) on reorganization of a food web, while a 2011 publication coauthored with twenty-three colleagues used several invasion cases to exemplify impacts of "trophic downgrading," reduction of food chain length owing to loss of top predators.[117]

7.4 The Environmental Movement

The rapid rise of the American environmental movement following Carson's publication of *Silent Spring* in 1962 could not help but attract attention to a phenomenon such as biological invasions perceived as an environmental threat.[118] Although replete with references to effects of non-native species, *Silent Spring* was primarily a call to arms over increasing chemical pollution, not least from pesticides and herbicides employed against biological invaders. The environmental movement as a whole also focused, at least initially, largely on chemical pollutants, though it rapidly grew to incorporate habitat destruction or conversion and biological invasions as prominent concerns.

A comparison of three books reflects the advent of the environmental movement, its early reflection of Carson's emphasis on pollution, and the increasing addition of invasions to the main concerns. George Perkins Marsh, in 1864, was far ahead of his time in calling attention to biological invasions (as described in chap. 3, sec. 3.4) and, in fact, to the whole issue of human modification of the earth. His book *Man and Nature; or, Physical Geography as Modified by Human Action* and its 1874 second edition, *The Earth as Modified by Human Action*, focused mainly on land and water use and concomitant habitat destruction, with one large chapter detailing human introduction of other species and a second chapter mentioning the potential role of canals in such movement.[119] He did not mention chemical pollution. In 1956, a large conference on anthropogenic changes of the earth resulted in a monumental tome dedicated to Marsh, *Man's Role in Changing the Face of the Earth*.[120] Almost all forty-nine of its chapters were on land use change and habitat destruction. Three were on invasions, and two made passing references to them. Only one chapter was on pollution, and the focus of that chapter was how waste disposal triggers growth of pathogen populations, not the key pollution issues Carson was soon to raise. In 1971, nine years after the appearance of Carson's *Silent Spring* and one year after the first Earth Day (the two events commonly said to mark the beginning of the modern environmental movement), Thomas R. Detwyler edited *Man's Impact on Environment*, citing the two books just mentioned as inspirations.[121] The fifty-two chapters of *Man's Impact on Environment* are drawn from previous writings. Of these fifty-two chapters, twenty-one were largely or wholly about pollution (three of these were solely about the insecticide DDT, Carson's main target in *Silent Spring*). Eight were on invasions, and two others made passing reference to them. Only seven were on land use change or habitat destruction, with two others briefly addressing such changes.

No analog to these two efforts, engaging dozens of scientists to assess all types of global impacts of humans, appeared in the 1980s during the explosive growth of invasion science, but the statistics of the previous paragraph suggest that, as the emphasis shifted toward chemical pollution in the wake of *Silent Spring*, invasions were increasingly recognized as part of anthropogenic change. By 2000, when the US National Research Council addressed whether the national research response was adequate to assess known global anthropogenic environmental impacts, four factors and interactions were recognized: (1) climate change, (2) pollution and modification of biogeochemical cycles, (3) land use, and (4) biological invasions.[122]

7.5 A New Conservation Science

Through the early 1970s, the conservation literature focused largely on habitats and the habitat requirements of particular species of interest. Elton, for instance, believed that the reason any particular species was found in particular places and not in others resided largely in its habitat requirements and that understanding why ecological communities consist of particular species and not others can be achieved only by learning the habitat requirements of each species, where "habitat" includes other species in the same location. This was the thesis of the book he considered his masterwork, *The Pattern of Animal Communities*, which consisted largely of synthesizing massive amounts of data Elton and his colleagues and students in the Bureau of Animal Population had gathered on many species in Wytham Woods to characterize ecological communities associated with particular habitats.[123]

With respect to conservation, Elton was not a "preservationist" who believed that a realistic or even desirable goal was to maintain each community in perpetuity; he recognized that human activities would inevitably change too many things.[124] However, for preserving rare species and variety of species (we would now call the latter "biodiversity"), which he viewed as important goals, he believed that the key to success was understanding the myriad factors controlling each species, which he construed as primarily habitat features.[125] Thus, as Elton came to focus heavily on establishing nature reserves as a conservation tactic, his emphasis was on the nature of the habitat with respect to the species in the reserve and the need for management in the face of inevitable ongoing disturbances or changes that would otherwise threaten at least some of those species.[126] Notably, beginning with a small book in 1933, *Exploring the Animal World*, Elton was an enthusiast of small reserves, so long as their habitat was appropriate for particular species of interest and was managed.[127]

In the late 1960s and early 1970s, two areas of research, one in biogeography and ecology and the other in genetics and evolution, arose that generated a very different sort of conservation science by the 1980s.[128] Whereas traditional conservation science focused on particular species and their needs, the new conservation science was concerned not with species identity but with the number of species in a site. In 1967, Robert H. MacArthur and Edward O. Wilson published *The Theory of Island Biogeography*, which suggested that the number of species on an island is an equilibrium between ongoing local extinction of species already present and ongoing immigration from the mainland species pool of species currently absent.[129] Species composition constantly changes, but the statistic of interest is number of species, which stays the same, in "dynamic equilibrium." Importantly, the extinction rate is higher on small islands, which are thus expected to have fewer species than large islands, ceteris paribus, in accord with many observations in nature. MacArthur and Wilson were explicit that "island" in this framework included "habitat islands," such as a forest amid fields, or a lake on land. Inevitably, it was quickly recognized that nature reserves could be construed as islands, and four papers appeared that explicitly aimed to derive rules for the design of nature reserves from the principles derived from the equilibrium theory.[130] Because of the effect of area in the theory, the first rule of refuge design for all authors was that larger reserves are better than small reserves, and other rules also devalued small reserves.

The other pillar of the new conservation science also focused on generalizations about all species, rather than species' specific habitat requirements, and was based on the genetic phenomena of inbreeding depression and genetic drift (as discussed in chap. 11, secs. 11.1 and 11.2). Both inbreeding depression and loss of alleles to drift are greater in small populations. Norman W. Moore and M. D. Hooper first pointed to the potential problems caused by inbreeding depression in small reserves, which would perforce have small populations.[131] During the same period, Otto H. Frankel and Robert J. Berry highlighted the effects of drift on small populations of conservation concern.[132] These genetic impacts were by 1980 combined into a dictum of conservation genetics, that small populations are to be avoided, and the idea of a minimum viable population size, determined by both ecological and genetic factors, became a mainstay of the new conservation science.[133] The focus of much conservation biology shifted from the traditional concern with particular declining populations and how to manage the causes of decline, often in reserves, to a new emphasis on the generic effects of small population sizes on all species, and on numbers of species in reserves, not which species. The paper casting the shift in conservation science focus from a "declining population paradigm" to a "small population paradigm" prominently featured the impacts of invasions under the declining population paradigm, observing as an example that alleviating the threat to the Lord Howe woodhen required eliminating non-native pigs.[134]

The decline of populations of particular species, or threat of a decline, was already ascribed, in a number of cases, to particular invasions by scientists operating in the traditional conservation science mode. The papers cited earlier by Courtenay and his colleagues and Owre in the 1970s are examples. The new conservation science, with its focus on the number of species (biodiversity) and generic problems associated with small population size, did not relate directly to impacts of non-native species. In fact, the growing literature on impacts of invasions failed to support the equilibrium theory of island biogeography, a foundation of the new conservation science.[135] However, the advent of a "new" conservation science led to great excitement not only in the wider conservation community but also among many scientists, and this infectious excitement, arising precisely during the rise of the modern environmental movement, quickly became associated with the environmental movement, with its incorporation of biological invasions as a key focus.[136] This development simply added to gradually growing notice of non-native species impacts in conservation-related literature. For instance, the journal *Biological Conservation* began in 1968 with an editorial statement of the thirty topics intended to be covered; number nine was "Introduction and Acclimatization," and number 10 was "Aliens and Indigenes."[137] Early issues of the journal had occasional papers about non-native species impacts from the traditional conservation science perspective.[138] The inaugural volume of the journal *Environmental Conservation* included a report of non-native African daisy invading national parks in southern Australia and called for increased attention to invasions into native vegetation rather than the predominant focus on agricultural weeds.[139]

: : :

As scientists increasingly reported on invasions in the first half of the twentieth century, many noted earlier reports, and some invaders, especially those with striking impacts, came to be widely known among ecologists: the European rabbit, house sparrow, small Indian mongoose, lantana, starling, water hyacinth, chestnut blight, spongy moth, common carp, Norway and black rats, prickly pear. It seems inevitable that the increasing existence of widely recognized cases would lead to a synthesis of sorts, seeking patterns and commonalities and pointing to puzzles and needed research. Marston Bates pointed in this direction in 1956, but he skimmed the surface, indicating several types of impacts, mentioning a variety of types of invaders, and pointing to some research lacunae.[140] In a sixteen-page chapter, he could hardly cover most aspects of the phenomenon or delve deeply into any of them.

Charles Elton had been thinking about invasions for many years, had often written about particular invasions or about invasions as an ecological phenomenon, had directed his Bureau of Animal Population to study British invasions and their management, and had broadcast a radio feature on invasions in 1933 and an entire radio series about them in 1957. Many of his writings and the bulk of his radio broadcasts were on impacts of non-native species, both impacts on human activities and ecological impacts. It is thus not surprising that, in 1958, he published the first comprehensive treatment, synthesizing and organizing the disparate threads of research and writing on invasions into a unified volume. Though written in an engaging tone and targeting an educated lay audience, *The Ecology of Invasions by Animals and Plants* was a scientific monograph. Many scientists read it, and it has achieved almost biblical status in modern invasion science.

However, *EIAP* did not actually initiate a new invasion science or directly inspire substantial invasion research. Over the next twenty-five years, a continuing and accelerating stream of reports on invasions and their impacts combined with the rapid rise of an environmental movement in the 1960s and the growth of the precursors of modern conservation science in the 1970s to inspire an international series of workshops that established the study of invasions as a distinct, complex science and triggered a phenomenally rapid growth of research with an expanding ambit (chap. 9).

8

Geographers Study Invasions

A Largely Separate Endeavor

Beginning in the nineteenth century, an academic tradition other than biology led to occasional focus on biological invasions and interactions with developments among biologists. The history of this tradition—the study of geography—spans two centuries and intersects with the evolution of invasion science at various times and in various ways. A description of this history and these intersections thus does not fit seamlessly into the chronological sequence I have largely followed for the first half of this book. I have located it here because of the broad overlap with much of invasion science, and I have indicated the relevant other chapters that bear on this overlap.

The discipline today defined as "geography" has several branches, all of which are closely tied to other disciplines. Broadly, those branches that relate to invasion science fall into two categories: cultural/historical geography and biogeography. Practitioners in the first category are concerned primarily with humans as the cause of geographic phenomena, as well as the consequences for humans of such phenomena. Biogeographers, by contrast, focus on patterns in the geographic ranges of other species or classes of species (e.g., plants, birds)—the history, causes, and consequences of these patterns. Furthermore, as a quite recent and heterogeneous academic discipline, geography has been roiled by controversy over its appropriate domain of study, including notably over whether historical and cultural foci and anthropogenic environmental impact are acceptable topics.[1] Cultural/historical geography intersects broadly with history (especially environmental history), sociology, and cultural anthropology. Biogeography has two subbranches, each of which overlaps a branch of biology: historical biogeography is related to aspects of evolutionary biology, and ecological biogeography can also be seen as a branch of ecology.[2]

Scholars in these areas of geography—cultural/historical geography and biogeography—have addressed biological invasions. With few exceptions (primarily in historical biogeography), however, their focus has been on humans, so they have conceived of the landscape as the context for developments in human societies, especially as driven by economics and changing livelihoods. Thus, to the extent that they have treated non-native species, the emphasis of cultural/historical geographers has usually been on species deliberately introduced for agriculture, grazing, and amenities, especially horticultural plants, rather than on inadvertently introduced species, although occasionally they have noted in passing ecological impacts of both sorts of introduction. Historical biogeographers, by contrast, have addressed the spread of species and entire biotas over vast stretches of geological time, built on the nineteenth-century development of biogeography by such pioneers as Augustin de Candolle and Alexander von Humboldt (pioneers discussed in chapter 3), and have usually not been concerned with a human role. Their interest has primarily been in the spread of species by "natural" rather than anthropogenic means, often during periods before human societies and migrations could have played a role. They have thus tended to view non-native species as "noise" obscuring patterns and processes of interest, rather than as subjects of interest in their own right. Their focus has been on geographic patterns, not ecological or environmental phenomena.

8.1 Increasing Interest in Biological Invasions

In 1870, Victor Hehn, a German-Estonian cultural historian, used comparative linguistic data as well as historical records to depict how various domesticated animals and cultivated plants came to inhabit large parts of Europe.[3] Some of these eventually escaped cultivation, but Hehn did not address this fact. He presented evidence of human transport for many plants of Asian origin, correctly identified Spain as the origin of the European rabbit and argued that it was deliberately spread by humans throughout Europe, and pointed to the North American origin of black locust, now invasive in Europe.[4] However, Hehn made no mention of environmental or ecological impact and saw his efforts as filling in a history of the spread of human culture, "for agriculture, horticulture, and domestic economy are only parts of a whole, a mere extract from the history of human culture which accomplishes itself in every part at once."[5]

In 1932, Evgenii V. Wulff, a Russian phytogeographer (plant biogeographer), published a comprehensive introduction to the history of how and when plants achieved their current geographic ranges; in 1943, the book was translated into English.[6] Wulff worked in the tradition of classical plant biogeographers, and his book, especially after translation, generated interest among geographers but was rarely cited by biologists. However, one chapter, entitled "Artificial Factors in the Geographical Distribution of Plants," was devoted to how humans had spread plant species, both deliberately introduced agricultural and horticultural plants and inadvertently distributed weeds.

Almost diametrically in opposition to Hehn and to botanist Edgar Anderson (whom we met in chap. 6, sec. 6.1), Wulff argued that the influence of non-native species on regional floras—that is, which species are found in a region, and their relative abundances—had been greatly exaggerated. In his view, even when many species are brought to a place, the great majority can persist only with continued human cultivation or disturbance of the landscape (chap. 15, sec. 15.4.4), and many simply disappear if forces favoring their presence cease.[7] Thus he minimized the spread of cultivated plants into natural vegetation and argued that evolution that would allow them to invade natural communities is generally very slow (though he gave an example of evolved adaptation to cold in olive trees from Mediterranean nations long established on the southern Crimean coast). Wulff also noted the long cultivation of "tens of thousands of species" in botanical gardens and acclimatization nurseries without substantial invasion as well as the restriction of many non-native weeds spread by railroads to the roadbed itself. He cited Harry H. Allan (whom we met in chap. 6, sec. 6.3) to the effect that only forty-eight of about six hundred non-native plants in New Zealand had become competitive with native species in the absence of disturbance, and he added that, of the approximately nine hundred non-native plant species in Madagascar, "only one, *Adenostemma viscosum*, has gained a foothold in natural plant communities undisturbed by man."[8] Yet throughout the book, Wulff never mentioned ecological impact; rather, his concern was with the extent to which non-native species becloud geographic patterns of floras.

Geographers recognize American historical geographer Carl O. Sauer as the most influential developer of the concept of cultural landscapes, beginning in 1925.[9] Sauer founded the "Berkeley school of geography" focusing on cultural and historical landscape-level patterns and the impacts of humans, the latter focus heavily influenced by his discovery of George Perkins Marsh's *Man and Nature* (discussed in chap. 3, sec. 3.4) in the 1930s.[10] Among many influences of human activities on landscapes, Sauer frequently adduced the introduction of non-native crops and livestock, but he was almost wholly silent on invasions of natural habitats by either deliberately or inadvertently introduced species, unlike Marsh.[11] In 1947, Sauer noted the importance of this phenomenon in an indirect way, pointing to the role of anthropogenic habitat change in facilitating plant invasions:

> The ecological effects included the introduction and establishment of species from other localities, but the most important results are to be found in

> the ways by which man placed certain plants under reproductive disadvantages and increased the opportunity for others. . . . Man's activities were directed especially strongly toward diminishing the shading of the ground. Every leafy tree or shrub that died and was not replaced . . . made room for the colonization of sun-loving herbs. . . . thus the first steps were taken that, as I have indicated elsewhere (Sauer 1944), may have resulted in the great grasslands of the hunting cultures.[12]

However, throughout his copious writings, Sauer rarely elaborated on and never highlighted the role of non-native as opposed to native species in such landscape transformations.

Sauer focused on landscapes at a large scale. He helped organize the 1955 conference that eventuated in the book *Man's Role in Changing the Face of the Earth* (discussed in chap. 7, sec. 4), and in his contribution he emphasized fire, including cultural use of fire, as well as agriculture.[13] He related historic landscape-level alteration almost exclusively to economic forces driving anthropogenic habitat change. His references to invasions were in passing, including a citation of Hehn as describing "Italy as wearing a dress of alien vegetation" but adding that Hehn "carried the theme of plant introductions out of the East too far."[14] Sauer also pointed to massive landscape-level changes (e.g., erosion) in the New World associated with non-native crop plants, but he did not address enormous changes effected by both plant and animal invasions of New World temperate grasslands detailed by another geographer, his student Andrew H. Clark, in the same symposium and earlier (as described in chap. 6, sec. 6.3).[15] Perhaps Sauer's most direct description of an invasion was published in 1958, from a symposium on land use in the humid tropics:

> European livestock, cattle, horses, and dogs ranged over former Indian fields, the New World becoming largely stocked from herds built up first on Haiti and Cuba. These animals contributed to the loss of their own range by disseminating seeds of woody plants through their excrement. The Spaniards complained, for example, of loss of range on Haiti to *guayabales*, thickets of guava (*Psidium*), within a generation of their arrival; these thickets are still extensive.[16]

Notably, *Psidium* is native to Haiti.

Although Sauer's 1958 essay, on tropical America, again emphasized anthropogenic habitat change, especially farming and fire, and minimized the role of non-native species, in the same symposium French geographer Edgar Aubert de la Rüe accorded such species a major role in transforming tropical vegetation.[17] This impact, in his view, was not only by the increasing area planted in non-native crops. He lamented the creation of a "cosmopolitan flora" of ornamentals and weeds, especially with the advent of European colonization, with impacts on islands but also on continents. He noted the effects of goats on several islands, including St. Helena, but singled out several plants as damaging invaders of mainland as well is insular vegetation: cogon grass, lantana, white leadtree, acacia, and water hyacinth.

Among several members of Sauer's "Berkeley school of geography," his student James J. Parsons continued similar lines of research. His academic home throughout his long career was Berkeley, and he chaired the Department of Geography. Parsons detailed the introduction, spread, and ultimate dominance of several perennial African pasture grasses in tropical and subtropical regions of Latin America, transforming large regions from forest to grassland.[18] However, he did not address the ecological consequences of this massive transformation, as his interest was in its impact on human enterprises, particularly agriculture:

> The African grasses stand up better to grazing and have higher nutritive values than native American species. In this respect, the invasions can be considered advantageous, although botanists may mourn the disappearance of native members of the flora that it may cause.... The largely unnoticed but massive invasions of the vigorous African species [are] at the base of the new hope for the development of a viable commercial livestock industry in the low latitudes of the New World Tropics.[19]

Parsons cited another member of the Berkeley Geography Department, Brazilian native Hilgard O. Sternberg, for having similarly noted the vast changes in parts of South America wrought by African grasses, part of what Sternberg referred to as the "Africanization" of large regions of the continent.[20] Sternberg's main focus was Amazonia, but his research encompassed the entire continent, and although his dominant concern was human activities, his perspective was mainly on the changes these activities wrought on the landscape, and, aside from whatever benefit these changes may have conferred on their perpetrators, the consequent ecological damage. In addition to the non-native grasses, Sternberg pointed to vast plantings of non-native trees, especially southern blue gum from Australia and Monterey pine from North America, in reforestation schemes, and he did not think that only botanists would mourn the loss of native species.[21] Sternberg also stressed that many introductions were not intercontinental but rather from one region of South America to another, and that some of these introductions might have preceded European colonization. He pointed to the role of non-native livestock in environmental degradation:

> Although changes in the biota operated for and by bovines represent the most widespread consequences of livestock introduction, sheep and goats

> have also played a significant, if more localized ecological role. Deterioration of the landscape in South America's arid Caribbean shores . . . can be blamed on the presence of goats, as has [*sic*] an impoverishment of the flora and fauna of the Galápagos Islands off the coast of Ecuador. In some other areas of South America, sheep rather than goats have been the main cause for dilapidation of the environment, notable examples being found in Patagonia and Tierra del Fuego, as well as in the Andean rangelands.[22]

Carl Sauer's son, Jonathan D. Sauer, majored in history as an undergraduate at the University of California, Berkeley. He then entered the geography department at the University of Wisconsin, where the chair, Andrew Clark (more on him in the next section, 8.2) and another faculty member, Frederick Simoons, had completed their doctoral research under Carl Sauer. After service in World War II, the younger Sauer conducted doctoral research in botany at Washington University, St. Louis, under Edgar Anderson, his father's friend at Berkeley, subsequently landing a joint faculty appointment in the geography and botany departments at the University of Wisconsin. In 1971, he became a professor in the geography department at the University of California, Los Angeles, where he remained for the rest of his academic career, focusing on biogeography.[23]

As had Carl Sauer, Jonathan Sauer focused on landscapes, but at a more granular level, and he was particularly interested in the role of dispersal, both recent and in the distant past, in establishing the plants found at a particular site. And, much more than Carl Sauer had, Jonathan Sauer emphasized the frequent role of humans in dispersing plants, both deliberately and inadvertently. His master work on the topic—*Plant Migration* (1988)—detailed numerous plant communities worldwide.[24] In many locations, non-native species were prominent, and often they dominated. In several locations, Sauer identified natural or anthropogenic disturbances that facilitated the dominance (cf. chap. 15, sec. 15.4.4). His background in plant evolution from his doctoral years was evident in descriptions how evolution and dispersal interacted, and Sauer noted examples in which hybridization between native and non-native species, or between two non-native species, produced new invasive species. Whereas Carl Sauer was not particularly concerned with invasion of natural habitats by non-native species, such invasions were a main feature of Jonathan Sauer's research.

One might have expected Jonathan Sauer's biological expertise and attention to the detailed biology—especially evolution—of many plant species to have caused his copious writings to have heavily influenced biologists studying invasions. However, his work was infrequently cited in mainstream invasion science literature. This relative neglect was partly because his biological interests were primarily evolutionary and not ecological, whereas ecological impacts of invasions were the main focus of biologists through the twentieth century. Thus, although Jonathan Sauer reported many invasions of natural landscapes by non-native plants, he rarely addressed their impacts beyond noting their presence and abundance. Similarly to Carl Sauer, Jonathan Sauer almost never wrote about the role of non-native animals except for livestock as they trampled, ate, and (for Jonathan) dispersed non-native plants. Perhaps a more fundamental reason for Jonathan Sauer's lack of interaction with the developing modern science of invasions in the late twentieth century was his distrust of grand theories that subsumed a great diversity of observations into a general framework, whereas most ecologists of the time were driven to cast ecology in exactly such a framework (as described in the introductory part of chap. 15).[25] He was particularly noted for fierce criticism of the dynamic equilibrium theory of island biogeography.[26] This criticism greatly intrigued and motivated ecologists precisely during the period in which research on non-native species coalesced into a modern science.[27] This same skepticism about general models and focus on particular cases characterized Clark's declining influence among geographers (as discussed in the next section, 8.2) and the relative obscurity of Charles Elton's last book, which he considered the culmination of his research (as discussed in chap. 7, sec. 7.1).[28]

Elmer D. Merrill, an American botanist specializing in taxonomy, published a biogeographic history of plants of the Pacific islands in 1945 more in the vein of geographers' treatments than that of early twentieth-century authors concerned with invasion impacts, such as Aldo Leopold and Henry J. Oosting (chap. 6).[29] His chapter on weeds acknowledged the prevalence of non-native species, and, as a biologist, he mentioned several themes prominent in ecological literature, especially how, by modifying habitat, humans create favorable circumstances for weeds. He traced the likely invasion routes of some weeds, much as Hehn had, and he even anticipated an important hypothesis later advanced by Elton in his 1958 monograph, *The Ecology of Invasions by Animals and Plants* (discussed in chap. 7, sec. 7.1). This hypothesis has come to be known as the "enemy release" hypothesis: non-native species thrive, in part, because they arrive in their new homes without the natural enemies (plant pathogens and phytophagous insects, for Merrill) that suppress them in their native range (more on this hypothesis in chap. 15, sec. 15.4.2). And Merrill pointed to several plant and a few animal invasions in other regions to show similarities to invasions of Pacific islands, particularly those of dominant invaders that he characterized as having "Napoleonic ambitions" (citing an unnamed botanist in the Dutch East Indies).[30]

However, what Merrill did not do in the chapter on weeds and a chapter on cultivated plants (both agricultural and ornamental species, many non-native) was to address ecological impacts on native species or ecosystems. Though he suggested landscape-level changes in the weed chapter, the focus was on descriptions of particular species and the human motivations and actions that led to their introduction and facilitated their survival and spread.

In 1968, geographer Donald W. Meinig published a comprehensive "historical regional geography" of the Great Columbia Plain in the Pacific Northwest in the nineteenth and early twentieth centuries, emphasizing what he termed "ecology"—human relationship with the physical environment.[31] He described massive land use change—primarily destruction of the native perennial grasses by overgrazing, in many areas to be replaced by desert shrubs. Meinig briefly reported on the invasion by several non-native weeds, especially wild oats, Russian thistle, China lettuce, and Jim Hill mustard—the mustard named after a railroad magnate, signifying the perception that it arrived in the region with the railroad. However, his treatment of these weeds focused on their effect on tillage methods and fallowing rather than on ecological impact, and he did not discuss the invasion by non-native bromes, seen as problem weeds by the 1890s.[32] In particular, he omitted the invasion of the region by 1889 by cheatgrass, recognized as a damaging weed in certain areas by 1903.[33] Cheatgrass spread rapidly beginning around 1905 to become the region's worst weed, dominating and transforming thousands of hectares with major ecological impacts on many native species and on entire ecosystems and serving Aldo Leopold as a symbol for the massive effects of impacts in general in his 1941 essay "Cheat Takes Over" (discussed in chap. 6, sec. 6.1).[34]

8.2 Two Geographers Interact More Strongly with Biologists

Several other geographers treated invasions at least partly from the same perspective as contemporaneous biologists focused on impact and were occasionally cited in the developing invasion science literature, though there is scant indication that they influenced the development of the field. Andrew Clark, a Canadian student of Carl Sauer's at the University of California, Berkeley, published a doctoral dissertation focused on non-native plants and animals of the South Island of New Zealand.[35] This dissertation was the basis for his book, *The Invasion of New Zealand by People, Plants, and Animals: The South Island* (discussed in chap. 6, sec. 6.3).[36] Both the book and the dissertation were primarily concerned with economics and human movements and activities (especially agricultural and pastoral activities), as would be expected for a cultural/historical geographer. Thus, the discussion in both volumes was primarily about species deliberately introduced for agriculture—the introduced animal discussed most was sheep, and the featured introduced plant was wheat.

However, Clark also exhaustively dredged the substantial previous biological literature on invasions in New Zealand. He frequently noted ecological impacts as well as phenomena that later came to be prominently studied by modern invasion scientists, especially lag effects and complex

Fig. 8.1 Illustration from doctoral dissertation of geographer Andrew H. Clark (1944) exemplifying dominance of parts of the South Island, New Zealand, landscape by European species. Credit: Library of the University of California, Berkeley.

interactions among non-native and native species. His focus on impacts of non-native species included such environmental impacts as erosion caused by rabbits and effects of non-native red deer and gorse on native vegetation. The dissertation had many more photographs than the book, and often these were of environmental impacts. Some were ironic, such as his "Fig. 372," which captured the notion of entire transplanted landscapes: "English sheep grazing English grasses under the spreading limbs of English oaks at Te Waimati, South Canterbury" (reproduced here as fig. 8.1). However, Clark was primarily concerned with both positive and negative impacts of invasions on people's lives—for example, the impact of rabbits on raising sheep or of gorse on pastures. In the 1955 symposium on large-scale human impacts on the earth, Clark developed the key points of his dissertation and 1949 book on New Zealand and briefly explored their relevance and relationship to other temperate grasslands.[37]

In 1959, Clark published *Three Centuries and the Island: A Historical Geography of Settlement and Agriculture in Prince Edward Island, Canada*, "the second major regional effort in what is planned as a series," of which the

New Zealand book was the first.[38] His aim was to understand landscape-level changes of "new worlds" colonized by Europeans. He planned to investigate in similar fashion Australia and South Africa, but he did not live to do so. Rather, his third major region of concentration was Nova Scotia, on which he published one volume in 1968 but died in 1976 with a planned second volume uncompleted.[39] The Prince Edward Island book made no direct mention of non-native species as non-native, much less of their impacts, though Clark discussed non-native species in various sections: dogs, cats, sheep, timothy (for hay), and so on. Perhaps the closest Clark came in this volume to impacts of non-native species was his condemnation of unnamed weeds left by the French in the late eighteenth century among "curses of the settlers of the next two centuries" and of feral swine killing domestic fowl and sheep in the early nineteenth century.[40] However, these impacts were agricultural, not ecological, and the book is fundamentally a geographic history of the economy and human demography, and their relation to livelihoods, especially agriculture.

Clark's third book, *Acadia: The Geography of Early Nova Scotia to 1760*, resembled the volume on Prince Edward Island in its focus on the history of human demography and livelihoods, in this instance particularly the early French settlers, their interactions with Native Americans, and their ultimate expulsion by the British.[41] Clark described Nova Scotia as "one among innumerable instances in the history of mankind in which an area has been invaded by an alien people . . . and their retinue of exotic plants and animals."[42] However, the only explicit mention of an invasion is a complaint about the "grey rat" for which Clark cited Marc Lescarbot, who visited Nova Scotia in 1606–7: "We had ample provision of them, and made unceasing warfare on them. The savages had no knowledge of these animals before our coming: but in our time they have been beset by them."[43] As he had for Prince Edward Island, Clark noted the early presence of crop plants and livestock from Europe, but his descriptions of habitats, especially forests and salt marshes, included no non-native species.

Michael Roche attributed the major shift away from invasions in Clark's focus between the New Zealand book and the two on Canada primarily to an evolution of his thinking on the role and proper subject of academic geography, which was a controversial matter in a relatively new discipline during Clark's lifetime, as noted previously.[44] It seems possible, however, that this shift in focus from Clark's treatment of New Zealand to that of Prince Edward Island and Nova Scotia was instead largely because invasions into natural areas occurred much more rapidly and dramatically in New Zealand than in Atlantic Canada. Clark himself pointed to this difference:

> At one extreme we can see, as in the case of New Zealand in the two centuries since the first visit of James Cook, a vast alteration in the face of the

> earth. In contrast, it is arguable that, except for a few hundred square miles which the Canadians farmed (and most of that, reclaimed tidal marshland), the landscape, flora, fauna, and pre-European culture and economy of Acadia had altered very little by 1760.[45]

Perhaps had Clark lived to write the second planned volume on Nova Scotia, he would have elaborated on this difference. Prince Edward Island had very little natural habitat remaining by the time Clark wrote about it. In fact, Graeme Wynn believed Clark may have chosen Prince Edward Island for his second book because it was much simpler ecologically than Nova Scotia and the influence of culture rather than ecology was therefore clearer.[46] Whatever his reason for choosing Prince Edward Island, Clark depicted how it quickly became predominantly agricultural, and I can find no early accounts lamenting invasions into the rapidly dwindling natural areas by either escaped agricultural plants or inadvertently introduced weeds. By the middle of the nineteenth century, the island was almost entirely given over to agriculture. Possibly this is why Prince Edward Island produced no analog to Leonard Cockayne, Herbert Guthrie-Smith, or George Thomson (all of whom appeared in chap. 6, sec. 6.3), whose early twentieth-century writings aided Clark's investigation of shifting New Zealand landscapes. It was not until the twentieth century that game animals were introduced to Prince Edward Island and non-native marine plants were remarked upon.[47]

Nova Scotia is an entirely different matter, and much forest remains even now. The earliest Acadian settlers introduced not only crop plants and livestock but also several tree species, medicinal plants, and ornamentals, as well as weeds.[48] However, even though certain of these have been tagged as major invaders elsewhere, it was not until the rise of modern invasion science that invasions became a substantial concern in Nova Scotia (and Prince Edward Island), and the concern has largely been focused on marine rather than terrestrial invaders.[49]

Clark was a dominant figure in American cultural/historical geography in the 1950s and 1960s, and his work remains well known among geographers.[50] However, by the 1970s, his leadership role had waned because his approach was frankly idiographic in a period when the maturation of the field led many geographers toward a nomothetic goal (as discussed in the introductory section of chap. 15).[51] It was not that Clark did not believe generalization was possible or a useful goal:

> In the historical geography of the overseas expansion of Europe there are scores or hundreds of such studies to be made and very few yet completed. Thus the next stage, or purpose, of historical investigation, to attempt some broad generalizations, may be premature—to say the least—for historical geographers.[52]

In his chapter on New World temperate grasslands in *Man's Role in Changing the Face of the Earth*, Clark emphasized the great variation in trajectories after European colonization, the intricate, complex sociological and ecological idiosyncrasies of the various systems, and the dearth of detailed research, concluding that "few generalizations can be drawn."[53]

In short, Clark believed that the comparative method was the appropriate way to produce generalizations, that geographic studies of European colonization were too few at the time to allow useful generalizations, and that perhaps the generalizations were not as interesting as the underlying variation. But he did not eschew the goal of generalization. Meinig implied that Clark's interests and goals shifted and narrowed as he proceeded through his three books.[54] As I suggested previously, however, at least with respect to non-native species, the different foci of the books may simply reflect the different ecology and history of the three islands. In any event, Clark was not often cited as modern invasion science arose in the 1980s and is rarely recognized today in the invasion science literature. As for his diminished status in geography, it is striking that it reflects a similar aspect of Elton's influence. Because Elton wrote so early in the development of ecology, he is widely remembered for his elaboration of concepts that remain important today, such as food chains and webs and the ecological niche, as well as his early recognition of the huge global impact of invasions.[55] However, as discussed in chapter 7, section 7.1, what Elton considered his main project for much of the end of his active career—directing the development and methods of community ecology—foundered, and the book he considered his most important *The Pattern of Animal Communities*, was not well received and is barely remembered by ecologists today, largely because, just as did Clark, Elton believed that generalizations must be preceded by production of a catalog of exhaustively studied individual cases.[56] Wytham Woods was the case he presented in his book. As community ecology matured in the 1960s and 1970s, generalizations became the main goal, and they remain the main goal today, with the inductive approach suggested by Elton (and Clark) often criticized.[57]

Archibald Grenfell Price was an Australian historical geographer whose early academic career concerned Australian issues. In 1939, he published a more expansive book on the impact of white settlers in the tropics generally, and in 1949, he focused explicitly on their interactions with natives.[58] In the 1949 volume, Price pointed to diseases introduced by settlers as a key devastating impact, and in 1964, he even published a small book on this impact, *The Importance of Disease in History*.[59] None of these writings addressed other non-native species and how they affected humans and the environment. However, in 1963, Price published *The Western Invasions of the Pacific and Its Continents*, replete with substantial attention to impacts of all non-native organisms. One entire chapter—"The Moving Frontiers of Animals and Plants"—begins with "the effects of the invasions on flora and fauna."[60] Acknowledging George Perkins Marsh's *Man and Nature* (discussed in chap. 3, sec. 3.4) as the pioneering work on these effects and lamenting that humankind had not heeded Marsh's warnings, Price proceeded to detail several examples. He was particularly explicit on impacts in New Zealand, citing Thomson, Allan, Clark, and Wodzicki. A geographer, Price cited relevant works by many geographers, including Wulff, Clark, and Carl Sauer. Price had also familiarized himself with the writings of some biologists, including Merrill, Anderson, and Marston Bates. A footnote to this chapter stated, "Since this book has been with the publishers my attention has been drawn to Charles S. Elton, *The Ecology of Invasions by Animals and Plants*, London, 1958, 1960. This is a valuable work, containing references to important biological changes in the Pacific and its continents."[61]

Price treated invasions as an important component among several anthropogenic impacts following European and other colonists' arrivals, irrevocably changing Pacific societies and their ecosystems: "Amongst these changes none were more striking than the changed landscapes which the Westerners effected by their purposeful or accidental movements of animals and plants."[62] Human disease was the most prominent driver of massive cultural change in his view, but he detailed the roles of agriculture, weaponry, and even steel as opposed to stone implements.

Similarly to Clark, at least for invasions, Price viewed the substantial research in New Zealand as a model case and believed that other cases would be required to permit broad generalizations: "If one had several lives it would be a fascinating task to examine on similar lines the entry of exotic life into regions such as the United States, Canada, and Alaska (where there is a great wealth of wild life material), Latin America, and the Pacific Islands, but such a review would require a team of expert scientists."[63] Price's influence on modern invasion science was probably minimal and at best indirect. Although Clark provided a prominent favorable review in *Geographical Reviews*, Price's *The Western Invasions of the Pacific and Its Continents* is infrequently cited.[64] Of forty-nine citations in a Google Scholar search (on June 10, 2019), only four were related to Price's descriptions of movements of plants and animals, and three of these were by geographers and historians. However, one of these was by Alfred Crosby in a book that I will discuss next and that was widely read by invasion scientists—*Ecological Imperialism*, in which Crosby cited Price for information on both weeds and animal invaders introduced to Australia.

8.3 A Twentieth-Century Luminary Recognized by Invasion Scientists

The geographer best known to invasion scientists is Alfred W. Crosby Jr., whose title at his death in 2018 captures the heterogeneity of the academic field of geography: professor emeritus of History, Geography, and American Studies at the University of Texas, where he spent much of his academic career. Often referred to as a founder of the field of environmental history, Crosby published two widely read books, partly based on a series of previous papers, that traced the geographic dispersal of humans, animals, plants, and pathogens from Europe to islands and other continents, and the sociological and ecological consequences of this spread.[65] More than any geographer except Clark in his treatment of New Zealand, Crosby detailed the spread and impacts of other organisms along with those of humans, and also ecological impacts as well as socioeconomic ones. However, again as had Clark, he treated the ecological impacts primarily as they affected humans and their enterprises: "the fundamental elements of the human story."[66] Although these are scholarly books reflecting an astounding breadth of knowledge, Crosby, especially in the second book, wrote in a relaxed, popular mode, so the educated lay public, part of his target audience, could easily grasp the key points.

The earlier book, *The Columbian Exchange: Biological and Cultural Consequences of 1492*, was restricted to the European colonization of the Americas but presented all the themes of the later volume, *Ecological Imperialism: The Biological Expansion of Europe, 900–1900*, which extended the analysis to other regions. In *Columbian Exchange*, Crosby began, as had other geographers, with a focus on agricultural plants and animals carried to the Americas by Christopher Columbus and those who followed him, later elaborating on their impacts when they became wild. But he quickly expanded his ambit to include weeds, either deliberately introduced before their weediness was recognized or inadvertently carried as hitchhikers, as well as accidentally introduced invasive animals such as insect pests and the black rat. However, perhaps the chief novel emphasis of this book was the focus on diseases, not only diseases that colonizing humans spread to Indigenous peoples but also livestock diseases infecting native mammals, as well as plant pathogens such as chestnut blight and Dutch elm disease. In this emphasis Crosby acknowledged the influence of historian William McNeill's book *The Rise of the West*, which advanced the idea that human history is driven particularly by interactions between previously separate societies.[67]

In *Columbian Exchange*, Crosby also foreshadowed the idea of "invasional meltdown" (a concept discussed in chap. 15, sec. 15.6), the main theme of his second book, juxtaposing Darwin's observation of massive invasion of the Uruguayan pampas by a Mediterranean thistle, the cardoon, with descriptions of great herds of wild cattle grazing and trampling on native plants and thus aiding the cardoon invasion. The last chapter of *Columbian Exchange* described the continuation of the exchange in the twentieth century, again emphasizing the impact of non-native plants, animals, and pathogens on both humans and nature. This book preceded by a decade the rise of modern invasion biology, but Crosby cited Elton's *Invasions* as well as Carl Lindroth's *Faunal Connections* (discussed in chap. 6, sec. 6.2) in developing examples.

Ecological Imperialism is Crosby's work that invasion scientists know best. It was an expansion and elaboration, replete with copious references, of his short 1978 paper, "Ecological Imperialism: The Overseas Migration of Western Europeans as a Biological Phenomenon."[68] It extended to other regions, especially Australia, New Zealand, and the Canary Islands (termed "Neo-Europes"), the hypothesis of species from Europe (which he termed a "portmanteau biota") working jointly to produce massive socioeconomic and ecological impacts, and the hypothesis was much more explicitly framed as a version of what later came to be known as "invasional meltdown":

> One of the most important factors in the success of the portmanteau biota is so simple that it is easy to overlook. Its members did not function alone, but as a team. . . . Sometimes the mutual support is obvious, as with Europeans importing honeybees to pollinate their crops. Sometimes the connection is obscure: In the Great Plains, the whites and their hirelings killed off almost all the buffalo—encouraging the spread of venereal pathogens, some of which were certainly immigrants. . . .
>
> The success of the portmanteau biota and of its dominant member, the European human, was a team effort by organisms that had evolved in conflict and cooperation over a long time. The period of that co-evolution most significant for the success overseas of this biota with sails and wheels occurred during and after the Old World Neolithic, a multispecies revolution whose aftershocks still rock the biosphere.[69]

Ecological Imperialism supplied many examples of invasions and their impacts but contained only one reference to the flurry of publications by biologists that spurred the Scientific Committee on Problems of the Environment (SCOPE) project that initiated modern invasion biology. The initial prominent SCOPE book appeared the same year as *Ecological Imperialism.*[70] Discussing invasions of California grasslands, Crosby cited Clark's contribution to *Man's Role in Changing the Face of the Earth.*[71] Oddly, an extensive chapter on invasions into New Zealand did not cite Clark's book on the South Island.[72] Both *Columbian Exchange* and *Ecological Imperialism* are credited by Charles C. Mann as the inspiration for his popular book *1493*, which is more a series of detailed vignettes than a comprehensive history, as Crosby aimed for.[73]

Although the portmanteau biota comprises animals (including humans), plants, and pathogens, and Crosby gave many examples from all three groups, the greatest focus, as in *Columbian Exchange*, was on the pathogens—diseases, especially of humans. By the time Crosby published *Ecological Imperialism*, the historian William McNeill had published *Plagues and Peoples*, a greatly expanded treatment of the role of diseases in human history that McNeill had briefly introduced in his earlier book that influenced *Columbian Exchange*.[74] In the acknowledgments for *Ecological Imperialism*, Crosby particularly singled out McNeill's encouragement, and both *Plagues and Peoples* and *Ecological Imperialism* are known among invasion scientists for their emphasis on the great influence of introduced pathogens.[75] In fact, both McNeill and Crosby cited an earlier work by US Army physician Percy M. Ashburn as a key source on this point, as had Price.[76] Crosby described the impact on his career of discovering Ashburn's book:

> My venture into epidemiological history quickly carried me to a more general subject: ecological history, the history of all organisms pertinent to human history and their (our) environment. Yes, germs are important, and so, it turned out, were insects, fungi, weeds, crops, and domesticated and wild animals. Humanity turned out to be the purposeful but often drunken ringmaster of a three-ring circus of organisms. Humanity affected the organisms, and they affected humanity.[77]

In *Ecological Imperialism*, Crosby repeatedly asked why the European portmanteau biota constituted a steamroller crushing native biotas, but other biotas—for example, that of North America—when transported elsewhere did not similarly dominate native ecosystems. A version of this question has been frequently asked for specific parts of the biota—for example, for insects.[78] Charles Darwin claimed as a fact that British species were stronger competitors than New Zealand species, proposing this as a special case of a hypothesized competitive superiority of European species over those of the Southern Hemisphere (as described in chap. 3, sec. 3.2). However, aside from the question of whether either of these claims about competitive superiority is true, Darwin did not suggest an explanation. What seems novel about Crosby's depiction is his contention that European introduced species aided one another. He attributed the greater teamwork and therefore greater impact upon introduction to:

1. the much longer presence of dense human populations on the Eurasian landmass than in the devastated Neo-Europes (thus evolved resistance to human pathogens),
2. the far greater number of human pathogens dispersed to the Neo-Europes from the Old World than vice versa,
3. the elimination of most large herbivores and carnivores in the Neo-Europes by the relatively recent arrival of humans (the "Pleistocene overkill" hypothesis[79]) allowing European livestock to spread easily,
4. the early domestication of many mammal species in Eurasia but not the Neo-Europes,

5. the close association of humans and these domesticated mammals in Eurasia (facilitating transfer of pathogens), and
6. the coevolved adaptation of Eurasian weeds to grazing and trampling Eurasian livestock that devasted native plants of the Neo-Europes.

This last factor, coevolution of weeds and livestock, was not explicit in either the book or the earlier paper but was implicit in the observation that trampling and grazing by the non-native livestock devastated the native plants and thus facilitated the spread of the non-native weeds.

When Crosby wrote *Ecological Imperialism*, at the beginning of modern invasion science, existing data did not contradict any of his hypothesized explanations for the devastation by European biota of Neo-European ecosystems and absence of a reverse phenomenon, and the various proposed explanations had varying amounts of supporting evidence. Similarly, certain well-studied particular cases exemplified his idea of mutually adapted and cooperating members of a portmanteau biota, but not many data bore on the overall significance of this phenomenon in facilitating invasions and exacerbating their impacts. However, the important point here is not the veracity of Crosby's idea but the extent to which it played a role in developing invasion science.

Ecological Imperialism remains a widely read book, with a second edition in 2004 and 4,625 citations on Google Scholar (accessed June 5, 2019). Most of these are in works in the social sciences and humanities—154 of the first 200 in the Google Scholar search. However, 46 of the first 200 were in biological writings, of which 30 were about invasions. But Crosby's book may have substantially influenced (and continues to influence) modern invasion science indirectly. The SCOPE synthesis volume, *Biological Invasions: A Global Perspective*, was one of two key SCOPE books that helped initiate modern invasion science (as described in chap. 9, sec. 9.1).[80] The first chapter, by Francesco di Castri, entitled "History of Biological Invasions with Special Emphasis on the Old World," addressed particularly the success and impact of European species in other regions, such as the Americas and Australia, and sought an explanation, exactly as Crosby had.[81] His tentative explanation was that the long association of European species, including weeds and parasites, with large human populations, as well as practices of European agriculture, led to their enormous impacts on introduction to other regions. The main facet of Crosby's hypothesis that di Castri did not include was the role of human diseases.

At the outset of his exploration of the question, di Castri cited two sources. He especially expressed a debt to Fernand Braudel's monumental, three-volume 1979 work, *Civilisation Matérielle, Économie et Capitalisme*, and then added, "From another historical point of view, it is also worth reading Crosby's book (1986) on European 'Ecological Imperialism.'"[82]

Braudel's book, though a staggering display of erudition, is almost wholly about the economic drivers of human history. The first volume contains a short section on human disease, citing McNeill's 1976 book *Plagues and Peoples*, but with respect only to the arrival of diseases in Europe. Nowhere else did Braudel mention diseases or weeds, nor did he explore the ecological impact of non-native animals, except for the role of the black rat in spreading bubonic plague in Europe. In any event, di Castri's chapter has been cited 474 times, according to a Google Scholar search (on June 5, 2019), virtually always by biologists. Published in one of the most influential books precisely when modern invasion science was rapidly growing, di Castri's chapter was surely widely read by invasion scientists. Some may have followed up his citation of Crosby, and many may indirectly have absorbed Crosby's idea about the role of long interactions with humans producing highly invasive species. In addition to di Castri's opening essay, the next two papers in the SCOPE synthesis volume both cited *Ecological Imperialism*, perhaps another indication of its influence on modern invasion science.

Even more likely to have indirectly spread Crosby's influence among invasion biologists was a paper Betsy Von Holle and I published in the first issue of the journal *Biological Invasions* in 1999, entitled "Positive Interactions of Nonindigenous Species: Invasional Meltdown?"[83] The paper began with a short introduction about the prominence at that time of focus on "biotic resistance"—competition, predation, and parasitism—in determining whether an invasion succeeds or not (chap. 15, sec. 15.4.1). We then turned to our main theme:

> However, a diametrically opposite conception of invasion success is also possible, one in which nonindigenous species, instead of interfering with one another, facilitate each other's establishment and/or continued existence. Thus, for example, Crosby [in *Ecological Imperialism*] depicts the colonization of the Americas, Australia, New Zealand, and the Canary Islands by European animals, plants, and pathogens as a mutualistic process. The European species, including European humans, are seen as biological allies that together constituted a synergistic juggernaut crushing native peoples and their ecosystems because the European species had coevolved with humans and with one another. For instance, European weeds, having coevolved with pigs, sheep, and cattle, were adapted to their activities, while North American plants were devastated by them.[84]

This paper has been widely cited (1,835 times according to a Google Scholar search on June 5, 2019), and the concept of invasional meltdown became a prominent theme in invasion science. A follow-up paper also citing *Ecological Imperialism* has been cited 419 times (according to a Google Scholar search on June 5, 2019).[85] Many researchers might have learned of Crosby's idea through these papers, and some may have been inspired to read his book.

Crosby later published *Germs, Seeds, and Animals: Studies in Ecological History*, a collection of previous writings (some newly edited) on the themes of *Ecological Imperialism*.[86] Almost all emphasize the role of introduced human pathogens in environmental history. This book appears not to have influenced invasion scientists; virtually all 160 citations found through a Google Scholar search (on June5, 2019) were in publications in the social sciences (particularly anthropology) and humanities. However, it may have inspired the title of a more widely cited work (discussed in the next three paragraphs). *Germs, Seeds, and Animals* was also the only place I have found in which Crosby cited a product of the SCOPE project.[87]

A later book by a geographer that attracted an even greater readership than *Ecological Imperialism* is *Guns, Germs, and Steel* by Jared Diamond.[88] Although introduced pathogens—the "germs"—played a prominent role in this expansive treatment, subtitled *The Fates of Human Societies*, non-native plants and animals, especially accidentally introduced ones, were very subordinate, and the sorts of ecological impacts featured by Crosby and Clark were entirely absent. This absence at first blush is surprising, because Diamond had previously termed invasions one of the "evil quartet" causing recent species extinctions, along with overkill, habitat destruction and fragmentation, and secondary extinctions.[89]

Diamond, who earlier had published many ecology papers, turned to historical and cultural geography in the 1980s, and in 2002 he became a professor of geography at the University of California, Los Angeles, where he is now emeritus. In *Guns, Germs, and Steel*, he perceived the ultimate cause of the European spread and dominance discussed by Crosby to be geography—(1) the dominant east-west geographic axis of Eurasia versus the north-south axis of the Americas, and (2) the highly indented coastline and high mountains of Europe versus the relatively smooth coastline and less formidable peaks of China. The more proximate causes, in his view, were dense human populations, written languages, and abundant plant and animal food, and the immediate causes were guns, germs, and, to a lesser extent, steel implements. With the exception of written languages, the proximate and immediate causes had previously been advanced by earlier authors—Price, McNeill, Crosby. Thus the seeds and animals of Crosby's title are embedded in the food abundance in Diamond's scenario. However, whereas for Crosby, the seeds and animals were of both agricultural organisms and (often) inadvertently introduced non-native species, Diamond wholly omitted the inadvertently introduced non-natives, though he cited both *Columbian Exchange* and *Ecological Imperialism* in suggested further readings.

In addition to the invasions in the "evil quartet" that Diamond exemplified with a few examples (e.g., the elimination of endemic island bird species by rats, cats, and the brown tree snake), Diamond attributed many recent extinctions of birds and mammals, mostly on islands, to predation and competition with non-native species.[90] He also occasionally attributed observed recent extinctions of the Melanesian birds he studied to invasions.[91] Further, the ecological work for which Diamond is best known is a series of "assembly rules" for ecological communities that is based entirely on geographic patterns and that posited the existence of stable communities, constructed according to the rules, whose stability is manifested by the fact that they resist invasion.[92] Given the concern with invasions in his writings on ecology and conservation, one might have expected Diamond to have acknowledged some role for invasions in the fates of human societies. Except for introduced germs—and here he did cite McNeill's two books as well as Crosby's—no such role was indicated. Thus, although *Guns, Germs, and Steel* was and remains widely read and has elicited much discussion and controversy among geographers, historians, anthropologists, and sociologists, it has not influenced modern invasion science.[93]

8.4 Geographers Continue to Mine Biological Invasions

As biological invasions have become the focus of an active modern science and also a topical issue in popular media (as discussed in the introductory section of chap. 1), geographers have occasionally studied particular invasions from various perspectives. Derek Alderman, for instance, has depicted the history of the introduction of Asian kudzu into the United States, its spread through the South, and its eventual status as a cultural icon with widespread metaphoric use, citing it as one example of what he hoped would become a "cultural geography of exotic species."[94] He advanced the suggestion as part of a special issue of *Geographical Review* on the culture and politics of invasive species.[95] Another example is the adoption of the highly ecologically damaging beaver, introduced from North America, as a mascot by citizens of Navarino Island (Tierra del Fuego, Chile), who see it as a symbol of their own identity and display a cartoon version of a smiling beaver on stickers and on a map showing island highlights.[96] Similarly, the recent invasion of Colombia by descendants of hippopotami purchased from zoos by drug lord Pablo Escobar and released to the wild after his death in 1993 has led to a regional tourist industry featuring hippopotamus statues and signs, despite substantial evidence of growing ecological impact of the invasion.[97]

Geographers have also used particular invasions to explore other questions that have spatial or geographic underpinnings. Emily Bukowski, for instance, used the geography of the spread of and ultimately failed attempt to control Dutch elm disease in Syracuse, New York, to explore the effects of mixed property ownership on the efficacy of pest management strategies.[98] Anders Halverson holds a biology doctorate based on research on a wood frog but has since worked in the area of environmental history, including a popular book on widespread introduction of rainbow trout outside their range, where they have become highly invasive with many impacts on native species.[99] His focus was primarily on the people who implemented this invasion (to improve sportfishing) and the geography of their introduction activities. All of these recent authors writing about particular invasions from the standpoint of cultural or historical geography in the wake of the explosive growth of modern invasion science have dredged the biological literature as part of their research, though their focus is primarily on the people associated with the invasive species rather than the species itself. Some of their writings have become well known among invasion scientists, at least those working on the same or related species. Most citations of Halverson's book, for example, are by biologists, many of them studying invasions (according to a Google Scholar search June 6–11, 2019). However, although there is a persistent and growing call for invasion scientists to pay more attention to the full scope of human interactions with invasive species (as discussed in chap. 16, sec. 16.3, and chap. 18, sec. 18.6), there is no evidence that these recent geography-based writings have influenced the development of invasion science.[100]

Geographers have also been among the social scientists who dominate a recurring criticism of invasion science and management, especially challenging the entire concept of "nativeness" and exploring the analogy of antipathy toward non-native species with xenophobia (as discussed in chap. 18, sec. 18.1).[101] Particularly noteworthy on this topic have been Lesley Head of Australia and Charles R. Warren of Scotland, who related their general critique to invasions of their respective nations.[102] As with geographers in general, Head and Warren focused particularly on human attitudes toward non-native species and invasion management. Perhaps the most expansive conception along these lines was that of American geographer Paul Robbins, who argued that the status of a species as non-native or invasive is always determined at least partly by the cultural and political context and that particular invasions are not by species alone but by networks of people, institutions, and species. In his view, it is these networks that are the proper unit of geographic research.[103] Though Robbins is frequently cited by other social scientists discussing invasions, including both Head and Warren, his approach has not transformed the methods and goals of invasion science.

: : :

Overall, then, increasing engagement of geographers with biological invasions, although roughly synchronous with the crystallization and maturation of modern invasion science, constituted more a parallel effort with a different focus—impact on humans rather than impact on ecology and the environment. Many of the geographers who treated the issue read and cited key papers in the biology literature. And geographers pointed to themes that were addressed by biologists even before the SCOPE project and increasingly in the years since. Wulff and Jonathan Sauer suggested a key role for physical disturbance in plant invasions, while Wulff, Carl and Jonathan Sauer, Merrill, and Crosby noted how changed habitat facilitated many plant invasions. Merrill, Jonathan Sauer, and Crosby anticipated the invasional meltdown hypothesis, and Merrill suggested the importance of enemy release, just as Elton did in the next decade. Price and Crosby stressed the role of non-native human pathogens as a major one in drastic landscape changes associated with invasions. Diamond viewed human pathogens as a key proximate cause of the spread, subjugation, or disappearance of human societies, although he did not relate societal change to other biological invasions. Clark even adduced examples of a lag effect in impacts of some biological invasions well before such lags became a prominent feature of invasion science in the wake of SCOPE.[104] However, except for Crosby, there is little evidence that geographers substantially influenced the development of invasion science.

9

The SCOPE Project Jump-Starts Modern Invasion Science

In 1980, ecologists convened in the coastal town of Hermanus, South Africa, for the Third International Conference on Mediterranean-Type Ecosystems. Invasions were not the specific focus of the conference, although several speakers discussed them. The conferees could easily see non-native trees invading the fynbos vegetation on the mountain slopes above the town. This crystallizing observation spurred intense discussion about invasions during the conference.[1] Harold A. Mooney, a Stanford University plant physiological ecologist, and Fred J. Kruger, an ecologist at the South African Forestry Research Institute, evolved a plan to seek support from the Scientific Committee on Problems of the Environment (SCOPE) for a program of workshops and conferences on plant invasions into Mediterranean-type ecosystems.[2] Mooney had already edited, with Francesco di Castri, *Mediterranean Type Ecosystems: Origin and Structure*, in which several chapters dealt with invasions, and Kruger had published a report on research related to environmental problems in the fynbos biome that asserted "the most critical problem in fynbos conservation remains that of invasive woody plants."[3] SCOPE had been established in 1969 as an arm of the International Council of Scientific Unions (ICSU, an umbrella organization founded in 1931) to assemble scientific expertise to address societal impacts on the environment and to generate responses to environmental problems. Most SCOPE projects at the time were about aspects of chemical pollution and biogeochemical cycles (a trend that continues today), although some had been on other topics (e.g., fire, Saharan dust).

On May 31, 1982, an eleven-member international working group with representatives from Australia, Canada, Czechoslovakia, France, India, South

Africa, and the United States met to draft a proposal to the SCOPE general assembly, including a recommendation that Mooney chair the scientific advisory committee that would guide the project. The working group included Ralph Slatyer, an Australian ecologist who was president of SCOPE at the time, and he requested that the project be expanded to a global scope, aligning with the SCOPE mandate to address environmental problems of global significance.[4] At the SCOPE general assembly meeting in Ottawa in June 1982, the project thus expanded was approved, and it was announced in the March 1985 edition of the *SCOPE Newsletter* as the SCOPE Programme on the Ecology of Biological Invasions, with Mooney as chair.[5]

9.1 The SCOPE Programme on the Ecology of Biological Invasions

The announcement described national programs under the auspices of SCOPE developed by committees in Australia, South Africa, the United Kingdom, and the United States and anticipated additional contributions from scientists in Germany, Norway, the Netherlands, and New Zealand. Three specific questions were to guide the project:

1. What factors determine whether a species will be an invader or not?
2. What site properties determine whether an ecological system will be relatively susceptible, or resistant, to invasion?
3. How should management systems be developed to use the knowledge gained from answering questions (1) and (2)?

These questions closely tracked questions framed by the working group, with one exception. The working group had included a fourth question: "What are the implications of biological invasions to ecosystem-level processes such as water balance and nutrient cycling?"[6] As it happened, although questions of this nature were posed in two of the volumes emerging from the SCOPE program, this focus did not emerge as a leading edge of invasion science until a decade later (as described in chap. 12, sec. 12.3).[7]

In addition to the aforementioned national efforts, the program established five international working groups to commence work in 1984–85 to address specific problems believed to demand quick international attention and that would, in turn, provide input into a projected final synthesis activity. These working groups were to target (1) impacts and control of invaders in nature reserves and protected areas, (2) mathematical models of the invasion and spread of organisms, (3) invasibility of stressed ecosystems, such as forests affected by acid rain, (4) invasions into Mediterranean-type ecosystems (the original conception of conferees at the 1980 South Africa conference), and (5) invasions into tropical ecosystems. A synthesis workshop was planned for November 1986 in Stellenbosch, South Africa, near where the 1980 conference that had led to the SCOPE invasions program had been held. The synthesis workshop was to focus on three areas: (1) biogeography (determine which areas were invaded by which species when and by what means), (2) prediction and quantitative assessment (analyze and quantify processes involved in invasions and explore models to explain or predict invasions), and (3) assessment of invasion consequences and costs and benefits of control programs.

The SCOPE Programme on the Ecology of Biological Invasions generated enormous enthusiasm and directly engaged numerous scientists in addressing invasions, including many who had not worked on non-native species previously but had expertise in allied fields, such as genetics and mathematical modeling. Workshops and symposia under the auspices of the program were held in Great Britain, the Netherlands, France, the United States, Australia (see box 9.1), South Africa, and India. Six books and special issues of two journals were published as products of the program, as were many papers in other journals.[8] Just the six books and two special issues contained 173 papers by 251 authors from seventeen nations. Even these numbers do not capture the reach of the program. For instance, the Dutch workshop led to a special issue of the *Proceedings of the Koninklijke Nederlandse Akademie van Wetenschappen* containing 10 papers by 21 authors.[9] However, about 100 scientists participated in a symposium associated with the workshop, and 25 participated in the workshop.[10]

BOX 9.1 Studying Invasions in Australia

The nearly simultaneous appearance of the Australian SCOPE volume and two books on invasion in Australia edited by Roger Kitching (discussed in chap. 7, sec. 7.1) is noteworthy.[1] The SCOPE papers tended to pose general questions or focused on groups of species, such as mammals or aquatic animals, though with details of particular invasions; in addition, eleven posters summarized in the Australian SCOPE book largely focused on single species.[2] Kitching compiled his 1986 book in 1984 and was then unaware of the SCOPE program.[3] Furthermore, of the twenty chapter authors in the SCOPE book and the sixteen chapter authors in Kitching's *The Ecology of Exotic Animals and Plants: Some Australian Case Histories*, only one (Angela H. Arthington, an ecologist specializing in rivers and fish) is shared between books. Thus, these were truly independent efforts.[4] An Ecological Society of Australia symposium held in Adelaide entitled "Exotic Species—Their Establishment and Success" preceded by a decade both Kitching's books and the SCOPE program.[5] Several of the fourteen published papers sought generalizations about invasions, and others detailed specific invasions. Only two cited Charles Elton's monograph. Clearly, both the public and scientists in Australia were keenly aware of invasions, especially by species seen as pests, by the 1970s. The popular book by Eric Rolls (described in chap. 7, sec. 7.2) had recently been published, and, as noted by Kitching, Australia had many highly visible problematic invaders.[6]

1. Kitching and Jones, *Ecology of Pests*; Kitching, *Ecology of Exotic Animals and Plants*.
2. Simberloff, "SCOPE Project."
3. Roger Kitching, email message to the author, 2008.
4. Simberloff, "SCOPE Project."
5. D. Anderson, *Exotic Species in Australia*.
6. Rolls, *They All Ran Wild: The Story of Pests on the Land in Australia*.

According to a Google Scholar search, the two most widely cited books, the North America/Hawaii volume and the synthesis volume, began to be heavily cited in 1995, with a sharp spike in 1999 leading to approximately 100 citations annually of the synthesis volume and 80 annually of the North America/Hawaii volume between 2001 and 2006.[11] The SCOPE program struck a highly responsive and still resounding chord, as the trickle of reports of invasions had increased through the 1960s and 1970s (as discussed in chap. 7, sec. 7.2), and the SCOPE program finally achieved the recognition of invasion science as a distinct, important science primarily by two means. First, it cemented in the minds of many researchers working on understanding or managing particular invasions the notion that they were all working on aspects of a single phenomenon of global extent—biological invasions—and that they could benefit by learning about one another's projects. Second, it engaged in a broad exploration of the subject many people who had not previously been especially interested in invasions per se, not only ecologists but evolutionists, geneticists, mathematicians, economists, sociologists, and others.

For the most part, the ambitious SCOPE invasions program goals were met. National and regional programs were developed, as planned, by committees in Australia, South Africa, the United Kingdom, the United States, and Europe, with workshops, symposia, and resulting books focusing on Australia, South Africa, United Kingdom, United States, and Europe.[12] Three of the four anticipated other national contributions materialized, albeit with substantial delays in two instances. The Netherlands contribution, including a workshop, symposium, and published proceedings, was described earlier in this section.[13] A SCOPE-sponsored Norwegian conference on invasions triggered the development of a sequel to the SCOPE invasions program, the Global Invasive Species Program (GISP), described subsequently in section 9.2.[14] A workshop in Bonn, Germany in 1999 was part of that sequel.[15] As for New Zealand, there is an odd lacuna in its participation in the SCOPE project, indicated to me by Richard H. Groves.[16] As discussed in chapter 6, section 6.3, many New Zealand citizens and scientists recognized invasion impacts even in the nineteenth century. Charles Elton cited George Thomson's massive tome on New Zealand invasions in his early monograph on animal ecology, and in his classic 1958 monograph, the chapter on impacts of invasions on islands featured New Zealand.[17] And New Zealand, a relatively small nation, continues to produce a disproportionate amount of scientific research on invasions, especially management and eradication of invasives.[18] However, the sole direct New Zealand contribution to the SCOPE invasions program was a paper by Robert E. Brockie and coauthors from a SCOPE program working group on invasions in nature reserves.[19]

The synthesis workshop was held as planned in 1986, but it was moved from Stellenbosch to Honolulu after objections from some participants to holding a conference in South Africa during apartheid. The objection led to substantial debate, but a short note from Desmond Tutu, who had been asked his opinion on the matter, was decisive. Five of the twenty-two authors who attended the conference were nevertheless South African. The synthesis meeting did address the three topics originally indicated—(1) biogeography, (2) prediction and quantification, and (3) invasion consequences and costs and benefits of control.[20] However, the papers dealt with many other topics as well—almost the full range of issues raised in the various SCOPE program activities through 1986.

The five international working groups described in the SCOPE invasions program announcement produced synthesis documents, except for the one on invasibility of stressed ecosystems. A special issue of *Biological Conservation* treated impacts of invaders into nature reserves (though not, as anticipated, control of these invaders in depth).[21] The British national contribution entailed a special issue of the *Philosophical Transactions of the Royal Society of London (B. Biological Sciences)* on mathematical models of aspects of invasions.[22] Di Castri and colleagues encompassed Mediterranean-type ecosystems in their book on invasions in Europe and the Mediterranean basin, focusing largely on the European Mediterranean, which was followed by a book featuring invasions in all Mediterranean-type ecosystems, including Australia, South Africa, Chile, and the United States.[23] During the synthesis session in Hawaii, several attendees noticed a major gap—a dearth of information on invasions in the tropics.[24] This observation led to a SCOPE-sponsored workshop in 1989 in Nainital, India. As what was probably the final formal published product of the SCOPE program, in 1991 an edited volume appeared based on that workshop.[25] Other publications on these topics arose from activities conducted as part of the SCOPE program, including several papers on invasions in nature reserves and mathematical models of the invasion process.[26]

9.2 A Gap in the SCOPE Program, and GISP, a Sequel

In his summary of the British contribution to the SCOPE invasions program, Martin Holdgate asserted that, despite the third objective of the program, "very little consideration is given to management questions."[27] His perception was correct. Of the 142 papers in the six books plus the British and Dutch special journal issues described previously, in section 9.1, only 31 treat management. The South African book alone substantially addressed management of invaders as opposed to aspects of their biology and impacts, as heralded by its title: *The Ecology and Management of Biological Invasions in Southern Africa*.[28] Eleven of the 25 contributions in this volume dealt with management.

This relative paucity of papers on management was widely recognized. As articulated by Mooney, the program "did not offer much to land managers except to inform them that they were not alone in the world dealing with these problems."[29] To fill this gap, a new SCOPE program was developed in the wake of the SCOPE-supported Norway/UN Conference on Alien Species, held in Trondheim, Norway, in 1996.[30] In 1997, a collaboration of SCOPE, the World Conservation Union (IUCN, for International Union for Conservation of Nature), and the Commonwealth Agricultural Bureau International (CABI) began a new program that evolved as the Global Invasive Species Program (GISP), designed to engage many constituents trying to address the problem from various perspectives, including natural and social scientists, legal scholars, and resource managers.[31]

GISP was an ambitious effort with a far greater ambit than the initial SCOPE invasions program, including many elements seen as contributing to "the comprehensive approach that is needed for dealing with invasive species."[32] These included academic research on the ecology of invasive species that was a key focus of the initial program, as well as how global change (especially climate change) will affect invasion impacts, plus analysis of economic consequences of invasions and especially development of tools and approaches to manage invasions, the third, mostly neglected goal of the initial program. With the growing recognition of the breadth and costs of invasion impacts, initial funding for GISP was ample, with support from the Global Environment Facility, the United Nations Environment Programme, the United Nations Educational, Scientific, and Cultural Organization, National Aeronautics and Space Administration of the United States, ICSU, La Fondation Total, the John T. and Catherine D. MacArthur Foundation, and the Norwegian government.[33]

A California meeting led to a 2000 book sponsored by SCOPE and GISP on the interaction between invasions and other global changes, such as climate change and altered nutrient cycles.[34] Charles Perrings and colleagues produced a treatment of economic aspects of biological invasions intended as a "first major step towards formulating and codifying" advice on economic aspects of invasions as called for by GISP.[35] A meeting in Cape Town in 2000 synthesized the progress of the various GISP projects until that date and led to a SCOPE synthesis volume including a ten-point global strategy to confront the various problems posed by non-native species, slightly developed from an earlier GISP document endorsed by SCOPE.[36] GISP in collaboration with SCOPE produced other notable products, including a *Toolkit of Best Prevention and Management Practices* aimed especially at underdeveloped nations (including islands struggling with invasion impacts).[37] However, as a partnership between agencies with differing missions, GISP suffered recurring funding problems and eventually closed in 2011.

9.3 A Controversial Analogy: Genetically Modified Organisms

The SCOPE program engaged in a development that initially appeared likely to increase greatly public and scientific interest in risk assessment for planned introductions of non-native species and in research on impacts of invasions generally. In the late 1970s and the 1980s, developments in "genetic engineering" technologies led to intense debate about the safety of releasing genetically engineered organisms (GEOs), also known as genetically modified organisms, especially those featuring transgenes, into the environment.[38] Ecologists and evolutionists quickly recognized an analogy between such releases and biological invasions, and a SCOPE program paper by Philip J. Regal developed the notion that this analogy could inform the promulgation of risk assessment procedures for releasing GEOs into the environment.[39] A study by ecologist Frances E. Sharples commissioned by the United States Environmental Protection Agency (EPA) and published in 1983 in a journal targeting geneticists working on recombinant DNA technology reached a similar conclusion, as did a group of ecologists who participated in a workshop convened by the EPA with academic geneticists, scientists in the rapidly growing biotechnology industry, and federal agency representatives.[40]

In fact, risk assessment for non-native species was at best a nascent field in the 1980s (as described in chap. 16, sec. 16.4), and ecologists conceded that impacts of introductions were extremely difficult to predict, a situation that persists today.[41] Thus deliberate introductions into the United States were not stringently regulated, risk assessment tools were primitive, and many invasion scientists hoped that the analogy to release of GEOs would inspire a ramped-up effort to improve risk assessment methods for invasions. However, the aptness of this analogy was promptly contested by scientists engaged in genetic engineering, defended by ecologists, and again rejected.[42] Opponents of the analogy feared that association in the minds of the public of GEOs with well-known invaders with highly visible consequences such as kudzu, the starling, and the spongy moth in the United States would lead to regulations that would cripple development of the technology and associated industry. They suggested generic reasons, detailed by Regal, why GEOs would be unlikely to generate such unexpected environmental impacts.[43] Probably the most comprehensive defense of the usefulness of the analogy was by Simon A. Levin and Mark A. Harwell, who enumerated many types of impacts of non-native species on native populations, communities, and ecosystems and described how, improbable as this might seem, GEOs could unexpectedly produce similar impacts.[44]

In practical terms, the impact of the controversy subsided when ecologists generally conceded the validity of the contention of defenders of the technology that the technology producing GEOs does not inherently generate more environmental risks than does traditional breeding of plants and animals for desirable traits, so that the focus of risk assessment should be on the traits of the engineered organism rather than how they were produced.[45] This was also the main recommendation of a committee appointed by the US National Academy of Sciences to study risks associated with field-testing and release of GEOs.[46] In 1976, ICSU had initiated a Committee for Genetic Experimentation (COGENE) with a working group on risk assessment for recombinant organisms and a meeting in 1987 in Italy, cohosted by SCOPE and COGENE.[47] Mooney and James Drake contributed a chapter to the resulting book summarizing the findings of the SCOPE invasions program and asserting, with qualifications, the relevance of the non-native species analogy to risk assessment for GEOs.[48] Ecologist and SCOPE invasions program participant Simon A. Levin also recognized qualified relevance of biological invasions research to risk assessment for GEOs.[49] Nevertheless, the take-home message of the SCOPE-COGENE working group was similar to that of the US National Research Council: "While much attention has been focused on the methods used to modify organisms, it is the products of these technologies and the uses to which they will be put that should be the objects of attention, rather than the particular techniques employed to achieve those ends."[50] With the imprimatur of Harold Mooney, the main driver of the SCOPE invasions program, the view became dominant that GEOs were no riskier than products of traditional breeding and warranted no additional assessments and safeguards.

Thus, whatever the degree of aptness of the analogy of GEO releases to biological invasions, given the undeveloped state of risk assessments for invasions at the time, it was not onerous for the GEO community to be subject to scrutiny no greater than that accorded to most non-native species. Many scientists and some policymakers have called for increased research on biological invasions and on risk assessments for invasions in particular, both during the controversy on GEOs and subsequently. Regal explicitly motivated this call by stressing the need to understand invasion processes and impacts better to assess risk of GEOs better.[51] However, I have found no evidence that the claimed equivalence of risks from biological invasions and GEOs released into the environment has spurred increased research on invasions or on invasion risk assessments.

The same controversy recrudesced in the second decade of the twenty-first century with the advent of gene editing, particularly through RNA-guided gene drives such as CRISPR-Cas9. Ironically, in this incarnation of the controversy, one of the main goals of geneticists undertaking the modification is to control invasive non-native species.[52] As in the previous incarnation, much of the opposition to release of genetically modified organisms comes from conservationists, highlighted by a statement issued by thirty prominent environmental advocates at the World Conservation Congress in 2016: "A Call for Conservation with a Conscience: No Place for Gene Drives in Conservation."[53] More recently, early enthusiasts of the technology have issued cautions about releasing gene-edited organisms into the environment.[54] The United States National Academies released a report on this iteration of the controversy with the main recommendation—cautious, controlled field-testing—remarkably similar to the recommendation of the analogous report on the earlier iteration by the National Research Council and that of a United Nations treaty.[55] Neither the reports nor the treaty quite detailed how field tests could be sufficiently controlled. Recent research on applying gene editing and other genetic techniques to invasion science and management are discussed in chapter 17, section 17.7.

9.4 What Does "Global" Mean, Really?

Most participants in the SCOPE program were anglophones, and in fact, a great preponderance of those voicing concern about non-native species from the mid-nineteenth century through the SCOPE program were from anglophone nations—Great Britain, the United States, Australia, New Zealand, Canada, and South Africa.[56] Perhaps this preponderance arose primarily from the sheer size of the British overseas empire, although the remarkable contributions of James Ritchie (described in chap. 6) and Elton (presented in chap. 7, sec. 7.1) were from residents of the homeland rather than imperial outposts. Also, the SCOPE program occurred during a period when the quickly increasing use of English was a key factor in what Mooney termed the "globalization of ecological thought."[57]

The dominance of English in the SCOPE workshops and publications persists in the mainstream published invasion literature and is increasingly recognized as a hindrance in advancing invasion science as a science and in improving invasion management.[58] As the field has grown explosively in the wake of the SCOPE program, an increasing literature in other languages remains relatively inaccessible and is often untapped in reviews and meta-analyses of invasion issues (such meta-analyses are presented in chap. 13, sec. 13.1). This relative neglect is partly because search engines such as Web of Science and Google Scholar do not readily find many articles in journals published in languages other than English, especially if they lack an English abstract, and partly owing to the huge gray literature in invasion science, much of which consists of government and similar reports not in English. Elena Angulo and colleagues provided a striking example of the problem in their attempt to estimate the economic cost of invasions (discussed in chap. 19, sec. 19.4).[59] Their English-language search turned up 838 published documents yielding 2,396 estimates of costs of particular invasions, while an analogous search in ten other languages produced a total of 356 published documents that included 5,212 cost estimates.

The issue of English dominance intersects with a similarly growing concern with the preponderance of a few highly developed nations in providing authors of mainstream journal publications.[60] The SCOPE program, ostensibly global, was dominated by scientists from North America, Europe, Australia, and South Africa. Although an increasing number of scientists and managers now focus on invasive species in nations all over the world, the mainstream literature still has major underrepresentation by most developing nations and most of the Global South.[61] These issues—dominance of the English language and by developed nations—are not unique to invasion science.[62] But for a discipline focused on the movement of species to and from all regions of the globe, these disparities may be particularly debilitating.[63]

Peder Anker included Elton as a key member of "The Oxford School of Imperial Ecology," but he focused on how Elton's ecological contributions, particularly on population cycles and his conception of the food web, aided Great Britain's imperial ambitions and enterprise, and he did not address the issue of biological invasions.[64] The geographers Andrew H. Clark and especially Alfred W. Crosby (who were discussed in chap. 8, secs. 8.2 and 8.3) depicted in detail how certain British activities in their overseas empire entailed many impactful invasions, although they portrayed these invasions as by-products of imperialism rather than a conscious imperialistic strategy. Jonah Peretti, CEO of BuzzFeed and cofounder of the Huffington Post (also known as HuffPost), saw the entire SCOPE program as a nativist, xenophobic venture, tainted at the outset by its origin at a conference in South Africa and heavily influenced by what he described as the overrepresentation of South African scientists at the SCOPE conferences and in resulting publications.[65] Peretti did not mention impacts of non-native species in South Africa as possible reasons for South African concern with invasions. He was one of several authors from the social sciences and humanities who perceived invasion science and management as inspired by nativism rather than ecological or environmental concerns (more on this idea in chap. 18, sec. 18.1).

: : :

In less than a decade, the SCOPE program crystallized an increasing concentration of concerns about specific invasions, precipitating the formation of a recognizable, distinct science out of what had largely been considered specific cases related to aspects of ecology. The SCOPE format—an international series of workshops—served this function well for two main reasons. First, it assembled scientists who had individually been working on particular invasions and thereby fostered the realization that they could benefit from exploring one another's research. Second, it engaged experts in a variety of disciplines aside from ecology, many of whom had previously been at best tenuously interested in invasions, and demonstrated the relevance of their fields to a large issue and the availability of interesting, relatively understudied problems that would constitute fruitful research loci.

Despite the plethora of products arising directly from the SCOPE program, one of its main goals was largely unaccomplished: how to integrate developing knowledge on the causes and consequences of individual invasions in effective management programs. Also, one aspect of the underlying philosophy and strategy of SCOPE—that environmental problems are international in extent and require international scientific collaboration to manage them—has yet to be adequately realized in invasion science (as discussed further in chap. 19, sec. 19.4). The explosion of research activity triggered by the SCOPE program is global in nature, but the global integration of invasion scientists and managers into an interactive, equitable enterprise remains a work in progress.

10

Invasions Invade the Scientific and Popular Literature

As the field of invasion science matured in the wake of the Scientific Committee on Problems of the Environment (SCOPE) project, its findings increasingly appeared in scientific journals and textbooks of allied fields, especially ecology, and were featured in books for a mass readership. In conservation biology, which crystallized into a modern science in the 1980s, around the time of the SCOPE project, the research that contributed to this crystallization also surfaced in journals and textbooks of related fields, in this case especially ecology, evolution, genetics, and wildlife management. When journals and textbooks specifically devoted to modern conservation biology appeared, reports of invasion impacts and management were prominent features from the outset. Inevitably, the rapid growth of invasion science (chronicled in chapters 11–17) led to journals and textbooks focusing specifically on this field. The existence of these journals and textbooks did not lead journals and textbooks in related fields to cease covering invasions. Articles on invasions continued to flourish in journals that had already published such reports—such as *Ecology* and *Conservation Biology*—and in new journals not specifically treating invasions—such as *Molecular Ecology* and *PLoS One*. Leading ecology and conservation biology textbooks deepened coverage of invasions apace with the wealth of new research findings and recognition of the global scope of invasions and their interaction with other global environmental changes. Increasing media coverage of invasion-related phenomena also led to many books for lay readers.

10.1 Newsletters and Journals

One manifestation of the burst of activity spurred by the 1980s SCOPE project (the topic of chap. 9, sec. 9.1) was an increasing flow, beginning in the early 1990s, of scientific journal articles focusing on aspects of invasions—impacts, management, the invasion process, and simply records of new species arriving in various locations. As noted in chapter 7, section 7.2, between Charles Elton's 1958 monograph and the SCOPE project, several articles on invasions appeared in journals devoted to ecology or particular taxa (e.g., birds or fish), and even in leading general science journals.[1] The evolution of a new sort of conservation science during this period (discussed in chap. 7, sec. 7.5) was also accompanied by invasion articles in conservation journals.

Even before Elton, in the early twentieth century, scientific journals occasionally published articles on non-native species, but these rarely addressed the sorts of questions posed by modern invasion scientists. For instance, fifteen articles mentioning non-native species appeared in the new journal *Ecology* in the 1920s, but few related to invasions as a common phenomenon. One mentioned in passing that many plant species in Hawaii are introduced while focusing on geographic affinities of native species.[2] Four concerned the biology of a recently introduced agricultural pest, the European corn borer, and two similarly addressed the newly invading Mexican bean beetle.[3] None of these related the focal invasion to other invasions or to more general aspects of invasions. Two other articles concerned managing plants in the United States that are alternate hosts of non-native plant pathogens ravaging species of commercial concern—barberry from Europe (alternate host of wheat stem rust) and three native species of *Ribes* (currants or gooseberries) serving as alternate hosts of white pine blister rust (see chap. 4, sec. 4.7, for detail on white pine blister rust).[4] Three articles described particular invasions but minimized their impacts on native species or ecosystems. Both Harry H. Allan in passing and Henry A. Gleason in a review of Herbert Guthrie-Smith's *Tutira* (chap. 6, sec. 6.3) observed that native species had replaced several non-native plants in New Zealand, and Gleason even used this observation to counter Joseph Hooker's earlier descriptions of invasive takeover (chap. 3, sec. 3.2).[5] Robert C. Miller predicted that the naval shipworm, then recently introduced to San Francisco Bay from the Atlantic coast, would not displace any of the four resident wood-boring species.[6] The most striking example of the lack of concern with invasion impacts at the time was an experimental study of 139 plant species—some non-native—deliberately introduced to new habitats to see which would thrive and to determine reasons for success or failure. This publication, by an advocate of using experiments to study evolution, Daniel T. MacDougal, mentioned no potential impacts.[7] No plantings were subsequently removed, and a list was provided because "some of the species may yet escape, and unless their existence in the region were authenticated might become a source of confusion to the plant geographer."[8]

Only two of the 1920s *Ecology* papers addressed problems that subsequently became leading-edge invasion science topics. Everett F. Phillips observed that the honeybee is native to Asia. Using data on plants that bees were believed to be using in producing marketable honey in the United States, and admitting shortcomings of these data, he noted that three-fourths of marketable honey must come from plant species (native to both the United States and elsewhere) that the honeybee could not possibly have coevolved with, and from this observation drew the remarkable conclusion that "these incomplete and in some regards unsatisfactory data yield little comfort to those who desire to see in nature the beautiful mutual adaptations on which the theory of organic evolution was at one time supposed to lean heavily."[9] A century later, the relationship of native pollinators to non-native plants and vice versa is a topic of substantial evolutionary and ecological interest (discussed further in chap. 13, sec. 13.2.3).[10] Alvin R. Cahn described the utter transformation of a small Wisconsin lake by non-native common carp, with wholesale elimination of the previously dominant native plant and fish species, anticipating by many decades the rapid increase in research on community- and ecosystem-level impacts of non-native species, as well as the concepts of keystone or transformer species (concepts discussed further in chap. 12, sec. 12.3).[11]

The SCOPE project precipitated a rapid rise of publications in leading ecology journals, exemplified by *Ecology* (the main journal of the Ecological Society of America), the *Journal of Ecology* and *Journal of Animal Ecology* (venerable journals of the British Ecological Society), and *Oecologia* (a commercial journal established in 1968 in Germany and strongly international after it became an English-language journal in 1977). A similar rapid increase occurred in the leading conservation biology journals, *Conservation Biology* (the main journal of the Society for Conservation Biology) and *Biological Conservation* (a commercial journal established in 1968). Table 10.1 shows the increase in number of articles in these six journals, in ten-year increments, from 1981 to 2020. In total, coverage of invasions in scientific journals rose nearly monotonically from 3 articles in 3 journals in 1980 to 2,535 articles in 495 journals in 2020, with abrupt increases in 1990–91 and 1997–98.[12]

Authorship also changed greatly as the field grew. As in scientific journals generally, the proportion of articles written by a single author declined and that of multiauthored papers rose.[13] In 1980, two of the three articles published were single-authored and the third had two authors. By 2020, only 3.9 percent of articles were single-authored, and 48.3 percent had five or more authors.[14] One article in 2020 had a whopping 642 authors, another had 126 authors, and many articles had more than ten authors.[15] Several

Table 10.1 Articles on Biological Invasions in Leading Ecology and Conservation Biology Journals, 1981–2020

	ECOL	*JE*	*JAE*	*OEC*	*CONS*	*BICO*
1981–1990	3	0	0	3	3	7
1991–2000	35	8	4	29	49	43
2001–2010	144	79	22	123	126	164
2011–2020	181	139	43	180	84	181

Note: *ECOL*: *Ecology*; *JE*: *Journal of Ecology*; *JAE*: *Journal of Animal Ecology*; *OEC*: *Oecologia*; *CONS*: *Conservation Biology*; *BICO*: *Biological Conservation*.

factors have combined to increase multiauthorship of invasion articles.[16] The increasing complexity of phenomena studied as the field of invasion science matured contributed, as larger skill sets were required to address increasingly complex questions. Increasing access to online conference platforms surely facilitated interactions within teams of authors. The increasing complexity of research questions, accrual of "big data," and generally increasing interest in multidisciplinary research led to establishment of synthesis centers beginning in the 1990s.[17] Several such synthesis centers created working groups on invasion topics, such as the Invasions Working Group supported by the National Center for Ecological Analysis and Synthesis (established in 1995 by the United States National Science Foundation and the State of California), which produced an influential early approach to quantifying invasion impact (discussed in chap. 16, sec. 16.1) written by 11 authors.[18] Changed authorship patterns also reflect the increasing role of women in invasion science, with the proportion of articles single-authored by women or, for multiauthored articles, with women listed as first author rising from about 15 percent in the early 1990s to 38.9 percent in 2020.[19]

10.1.1 Newsletters and Informal Journals

As modern invasion science rapidly expanded following the SCOPE project, newsletters and similar publications arose targeting particular audiences, such as scientists and managers interested in a particular non-native species, or particular types of non-native species, or invasions into a particular region or type of ecosystem. Some of these evolved into journals, though not quite academic journals in that they were oriented more toward news, survey information, meeting notices, and the like rather than scientific research, and standard peer review was not routine and could be informal or omitted entirely. Several such newsletters and informal journals for a time became major venues for transmitting information, with articles cited in the standard scientific literature, but all were ephemeral, replaced in the 1990s

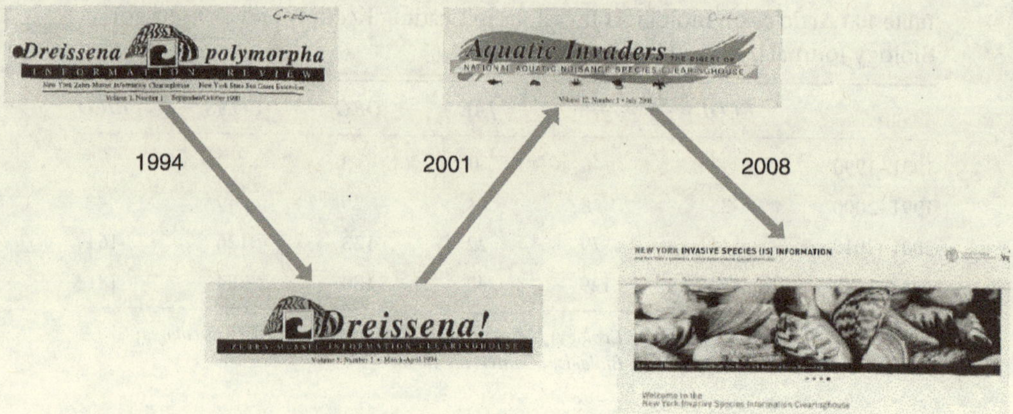

Fig. 10.1 Sequence of newsletters featuring impacts and management of zebra mussel invasion of North America, and the newsletters' ultimate replacement by a website. Credit: James Carlton.

or the early 2000s as such information migrated to websites operated by government agencies and nongovernmental organizations (NGOs), and as rigorous, standard scientific journals focusing specifically on invasions were initiated (described subsequently in subsection 10.1.2).

An example of such a newsletter was the *Dreissena polymorpha Information Review*, which began in 1990. *Dreissena polymorpha*, the zebra mussel, was introduced from the Ponto-Caspian region of Eurasia to Lake St. Clair between Ontario and Michigan in the late 1980s, probably either in ballast water or as settled adults on an anchor. It spread rapidly in North America, causing ecological and economic damage; previously it had dispersed through Europe, largely through the extensive canal system.[20] This newsletter was published bimonthly by the New York Zebra Mussel Clearinghouse, an NGO that soon became a project of the New York Sea Grant, funded by the United States National Oceanographic and Atmospheric Administration. In 1994 it morphed into *Dreissena! (The Digest of National Aquatic Nuisance Species Clearinghouse)*, and after four issues of volume 11 in 2001, *Dreissena!* became *Aquatic Invaders*, a quarterly beginning with the four issues of volume 12 in 2001 (fig. 10.1). The initial newsletter had consisted primarily of a few summaries of research projects (usually lacking references), short abstracts of relevant scientific publications, and conference announcements, virtually all related to zebra mussels. By volume 13 in 2002, *Aquatic Invaders* was publishing substantial research articles by international authors, fully referenced and apparently peer-reviewed, and most were about aquatic invaders other than the zebra mussel. Short abstracts of articles in the scientific literature were still included, as were announcements. However, *Aquatic Invaders* ceased publication after a double issue in

2008 (volume 19), with some information that would have been published in subsequent issues migrating to the website of the New York Invasive Species Information Clearinghouse.[21]

Wildland Weeds was a magazine published beginning in 1997 by the nongovernmental Florida Exotic Pest Plant Council (FL-EPPC). Its goal was to provide "a forum for individuals to convey the impacts non-native plant species have on native biodiversity and native plants species" as well as to provide "a mechanism for educating resource managers and the public about identification, biology, distribution and management of invasive exotic plants."[22] All articles were about plant invasions in Florida, and the magazine also published reviews of books about invasions. In 1999, the Southeast Exotic Pest Plant Council (SE-EPPC) was established as a coalition of the exotic pest plant councils of several states, including Florida, and *Wildland Weeds* became a publication of both FL-EPPC and SE-EPPC. Accordingly, although Florida invasions were heavily represented, articles began to appear about invasions in other southeastern states as well as elsewhere in the United States and even internationally. Each volume had four issues through 2010, while the 2011 and 2012 volumes had three issues, none appeared in 2013, a single issue was published in 2014, and the last issue during the winter of 2015–16. Until the last issue, *Wildland Weeds* had been a glossy, colorful hard copy, and the articles, though factual and often containing important information not readily available except perhaps in gray literature reports, were written in a conversational mode accessible to the public. However, the final issue contained an editorial announcing that *Wildland Weeds* would henceforth be a digital-only publication, though it retained the colorful, nontechnical style.[23] The editor at the time, newly appointed, tried to get help to continue publishing the magazine but failed, and the 2015–16 issue was the last.[24]

Some newsletters focused on a single non-native plant. *Chromolaena odorata* (Siam weed) is a perennial herb native to subtropical and tropical regions in the Americas that was introduced beginning in the nineteenth century to Asia, Africa, and Australia, often as an ornamental. In many parts of its non-native range, it damages both agriculture and natural habitats, and it appears on the list of 100 of the world's worst invaders published by the International Union for Conservation of Nature (IUCN).[25] At an international workshop in Thailand in 1988 on biological control of Siam weed, the workshop attendees recommended producing a *Chromolaena odorata Newsletter*, and the first issue appeared later that year. Through issue number 16, the newsletter was published at irregular intervals by the Agricultural Experiment Station of the University of Guam. Beginning with number 17 in 2008, publication shifted to the Agricultural Research Council of South Africa, and the last issue, number 19, appeared in 2014. The editorial board still issues calls for articles, but potential authors appear to be publishing in academic journals instead.[26]

In 1998, the New Zealand Ministry of Agriculture and Forestry (MAF) started a new glossy, amply illustrated publication for a lay audience, *Biosecurity*. Published every six weeks with input from other government departments and ministries, it aimed to be "of special interest to all those with a stake in the protection of New Zealand's economic, environmental and social assets from the dangers posed by pests and diseases."[27] Although this magazine published articles on other topics, such as animal welfare, it was heavily focused on excluding and managing non-native species, particularly as regards New Zealand. The last issue, number 103 in 2011, contained no announcement or explanation of the termination, although the previous issue had contained a feature—"Have your say about *Biosecurity* magazine"—asking whether readers were satisfied with the content and format, inviting suggestions for improvement, and providing a link to an online survey on these matters.[28] In 2012, MAF was incorporated in a new Ministry of Primary Industry (MPI). In 2015, MPI initiated an online newsletter, *The Border Space*, which has since appeared every two months. The initial issue contained no announcement of target audience. However, issues of the newsletter, though shorter than those of *Biosecurity* and written in a breezier style, cover much the same material, though in less detail, and are similarly copiously illustrated.[29]

In 1995, the newly formed Invasive Species Specialist Group (ISSG) of the IUCN launched a newsletter, *Aliens*, published in New Zealand, base of the ISSG chair. The founding editor, Sarah Lowe, described the aims of the publication: "We would like this to be very much a NEWsletter, with short items focussing on new research and control strategies, upcoming meetings, and other 'current awareness' information related to invasive species research and management."[30] She also announced that production of further copies would depend on support from both individuals and possible sponsors. The first issue had eighteen articles, some highly detailed, about impacts and management of animal, plant, and microbe invasions worldwide. Interest was high, and the newsletter appeared twice yearly through 2008, with many articles by international invasion scientists, announcements of meetings and publications, and a new editor in 1998, Maj De Poorter, still in New Zealand. In 2008, a new chair of ISSG was appointed, Piero Genovesi. In 2009, publication of *Aliens* moved to Italy with a new, glossy format and color figures, and began publishing one issue each year, each issue about twice as long as earlier ones. Genovesi and Riccardo Scalera assumed the editorship. However, support for the publication was insufficient to sustain it as a paper publication beyond issue 33 in 2013, and an online successor has been discussed but not published.[31]

The aforementioned are a small sample of the many newsletters and informal journals that have arisen and, for the most part, eventually disappeared. A web search reveals dozens, some extant, most extinct. A few, such as *Aquatic Invaders, Wildland Weeds,* and *Aliens,* published articles not greatly dissimilar from those in academic journals. However, in addition to increasing use of the web to disseminate information more rapidly and less expensively than in printed publications, the advent of an expanding new field of modern invasion science meant that it was only a matter of time before academic journals specifically devoted to invasions appeared, and as they did, they became the venue of choice for many articles that might, in somewhat different form, have previously appeared in newsletters or informal journals.

10.1.2 Invasion Science Journals

10.1.2.1 *Biological Invasions*

First off the block was *Biological Invasions,* with an initial volume in 1999. In 1995, René Mijs, then acquisitions editor for the Dutch publisher Kluwer Academic Publishers, attended a meeting of the International Society of Limnology in São Paulo, where he and several attendees discussed the concept of a journal focused on biological invasions.[32] During a North American acquisitions trip in 1996, Mijs mentioned the idea to various scientists, and a leading marine invasion scientist, James Carlton, learned of this initiative as he was about to depart the United States for a meeting in Poland on invasions convened by the International Council for the Exploration of the Sea. Feeling that the possibility of such a journal would interest the attendees, Carlton attempted to contact Mijs to express his interest and willingness to serve in an editorial capacity. He was unable to reach Mijs, but a Kluwer office in the United States forwarded his email to Mijs, who did not respond at the time.[33]

In 1997, Carlton met in London with Bob Carling, an editor at Chapman & Hall, which had published the paperback reprint of Charles Elton's monograph. He mentioned that a publisher was considering starting an invasions journal, and Carling informed him soon thereafter that Chapman & Hall had concluded that invasions were "probably too narrow a subject to sustain a journal" but that Kluwer was conducting market research on the possibility.[34] Soon thereafter, on behalf of Kluwer, Mijs contacted Carlton to inquire if he would consider being senior editor of an invasion journal, as the concept was advancing in the publishing house. In November 1997, Mijs told Carlton that Kluwer wanted more evidence on the types of papers that would be published and the importance and relevance of the journal. Carlton responded with examples of journal articles on invasions during the previous two years and a statement about the interests of potential subscribers to the journal and likely readers. In December 1997, Kluwer approved the project, and the first issue appeared in March 1999, with Carlton as editor in chief.

Biological Invasions began with four issues per year, and volume 1 included 32 articles. Starting in 2004, it grew rapidly; the 2020 volume contained twelve issues with 249 articles. Through 2020, the journal had received 9,838 submissions and published 3,478 articles from seventy-six nations.[35] Similarly, the editorial board initially included twenty members at institutions in ten countries, but a rapid increase in size beginning in 2010 led to more than one hundred editors from institutions in twenty-five countries in 2020. From the inception of the journal, the editorial board has been dominated by scientists working in the United States (35–56 percent), and by 2020, 9 percent of board members were Canadian.[36] All editors in chief were United States citizens through 2024, when two Argentinian citizens assumed that role. Of published articles, 45 percent were senior-authored by scientists working in North America and 28 percent by scientists at European institutions.[37] Relative to their population sizes, New Zealand and Australia have been particularly highly represented on the editorial board and among senior authors, surely at least partly because of the long histories of concern with and research on non-native species in those nations (as described in chapters 5 and 6) but also likely because they are anglophone nations and the journal language is English (though with abstracts also in Spanish and an optional third language beginning in 2021).

Articles analyzing demography of authors and editorial board members identified several geographic disparities, particularly the relative dearth of editors and authors from Asia, South America, and especially Africa (aside from South Africa) and pointed to the anomaly of a journal reporting on a global phenomenon with large areas of the globe poorly represented in authors and editors.[38] The fact that the journal language is English was identified as contributing substantially to this pattern. In ecology in general, the difficulty of nonanglophone scientists publishing in English-language journals and problems of using results of literature searches conducted only in English to draw general conclusions are increasingly recognized (as discussed further in chap. 19, sec. 19.4).[39]

10.1.2.2 *NeoBiota*

Unlike *Biological Invasions*, which is a commercial journal, *NeoBiota* is a society journal, published initially by NEOBIOTA, a German organization of scientists founded in Berlin in 1999, and then by the same entity, with the added tag "The European Group on Biological Invasions."[40] "Neobiota" is a neologism, defined subsequently as "organisms, independent of their taxonomic rank, which occur in a region outside their natural range due to human agency or which evolved from such taxa."[41] The founders of NEOBIOTA envisioned the organization as helping to resolve what they perceived as a highly fragmented effort in Germany, with "virtually no exchange of information between plant and animal ecologists" and "few attempts to integrate the studies on terrestrial, freshwater and marine ecosystems."[42]

NeoBiota lacks volume numbers, publishing numbered issues instead. The first issue appeared in 2002, with twenty-five papers in German and seventeen in English, and contributions were drawn from the first conference convened by NEOBIOTA, in 2002 in Berlin, plus additional invited papers. The conference drew contributions from six European countries but with the focus mainly on Germany. Other papers from this conference were published in English, along with several invited papers, as a special issue of *Biological Invasions* in 2003.[43] Subsequent issues of *NeoBiota* were in English, and the first eight issues, through 2009, drew from the proceedings of the subsequent meetings of NEOBIOTA, which were held in English every two years (following Berlin, in Halle, Germany; Bern, Switzerland; Vienna, Austria; and Prague, Czech Republic) with additional invited papers. During the sixth NEOBIOTA conference in Copenhagen in 2010, conferees agreed to initiate an international journal, named *NeoBiota*, as a continuation of the earlier NEOBIOTA publications, so the first issue of this version of the journal was numbered 9 and appeared in 2011.[44] The stated rationale for a new journal was the increasing importance of invasions and the development of new publishing technology allowing open access. Publication was assigned to Pensoft, a publisher in Sofia, Bulgaria, pioneering in open-access technology.[45] The number of articles per year through 2018 varied between fifteen (in three issues) and twenty-four (in five issues), jumping rapidly to sixty-one (in twelve issues) in 2019 and eighty-four (in ten issues) in 2020. The initial issue in 2002 had two editors, both German. When the journal began publishing through Pensoft, fifteen countries were represented by the thirty-five editors, of whom twenty-four (69 percent) were European (seven of these were German), and six (17 percent) were North American. By 2020, the editorial team consisted of a board of three Germans plus sixty-four subject editors representing twenty-one countries. Of these sixty-four editors, thirty-four (53 percent) were European (eleven of these were German) and fourteen (21 percent) were North American.[46]

NEOBIOTA, originally a German working group, quickly grew to encompass all of Europe, with a council established in 2008 consisting of twenty-three countries.[47] That number had expanded to twenty-nine countries by 2020. The journal similarly rapidly published papers from Europe and, especially with the changed format beginning in 2011, from around the globe.

10.1.2.3 *Diversity and Distributions*

Slightly before the launch of *Biological Invasions*, a journal focused on the geography of biodiversity arose and subsequently became a prominent outlet for invasion science research. The publisher Blackwell in 1993 initiated the journal *Biodiversity Letters*. The last volume appeared in 1995, having published twenty-three or twenty-four articles annually, none of which dealt with biological invasions. In 1998, a successor journal appeared, *Diversity and Distributions*, publishing its first issue (volume 4, number 1) in 1998. The editor in chief was David Richardson of South Africa, a leading invasion scientist. In an inaugural editorial, Richardson portrayed *Diversity and Distributions* as part of a trio of Blackwell journals (the others were *Journal of Biogeography* and *Global Ecology and Biogeography Letters*) that he described as "essential reading for ecologists and biogeographers."[48]

After a brief rationalization of the need for the new journal, Richardson described the focus as "all facets of geographical range and biodiversity," listing seven categories of suitable topics. He ended the editorial with two paragraphs describing how the new journal would feature biological invasions to help "sharpen the 'game rules' for invasion ecology—to clarify and develop techniques of observation, measurement, modelling, and extrapolation."[49] He detailed how invasions could contribute to a better understanding of species' geographic ranges, confronted the occasional claim that invasions are so idiosyncratic that prediction and generalization are nearly impossible, and foresaw the use of advanced modelling techniques focused on the role of interactions between life-history traits and environmental features in determining invasion dynamics.

Diversity and Distributions started off slowly, with twenty-two or twenty-three articles in each of the first four years. However, after one article on invasions in the first year, the journal published an increasing number of invasion papers, reaching seven of the total twenty-two in the fourth year. Beginning with volume 10 (2002), the total number of articles increased rapidly, so that by 2010, every volume included more than ninety articles, and those treating invasions always composed at least 25 percent. Three special issues, in 2004, 2008, and 2011, boosted the fraction of invasion articles for those years to about 50 percent, and most volumes from 2004 through 2015 included more than thirty papers on invasions. Also, between 1998 and 2006 (when the journal ceased publishing book reviews), *Diversity and Distributions* published twenty-three reviews of books about biological invasions. Two events then coincided with a lessening of the invasion focus. In 2016, Janet Franklin, a prominent American geographer, replaced Richardson as editor in chief. Her own research foci included species distribution models (discussed in chap. 14, sec. 14.4.4) but did not particularly feature invasions, and her main focus for the journal was conservation geography.[50] Volume 22 in 2016 had a similar percentage of papers on invasion papers to those of the preceding decade, but thereafter the percentage fell to below 20 percent through at least 2020. Then in 2018, Wiley, which had merged in 2007 with Blackwell and become the publisher of *Diversity and Distributions*, announced that the journal would become an open-access publication beginning in 2019. Many members of the editorial board objected, primarily because the fee to publish an article would be beyond many authors' means. Franklin resigned as editor in chief, as did the majority of the editorial board, including invasion scientists. Of the new editors in chief, Kevin C. Burns, Luca Santini, and Aibin Zhan, only Zhan had a substantial research program on invasions, and the new editors, in an initial editorial attempting to reestablish rapport between academics and Wiley, described the journal as a biogeography journal.[51] In 2020, they were joined as editors in chief by Céline Bellard, whose research features aspects of invasion science.

10.1.2.4 *Aquatic Invasions*

At a 2005 invasion science workshop in Florence, Italy, two biologists specializing in marine and freshwater invasions, Vadim E. Panov, a Russian, and Stephan Gollasch, a German, suggested the need for a new journal on aquatic invasions. The next year saw the first issue of *Aquatic Invasions* with Panov and Gollasch as editors in chief. The journal focused on invasions of the inland and coastal waters of Europe, and the lead editorial termed it a "new European journal of applied research on biological invasions in aquatic ecosystems." Among topics particularly sought were first records of invaders, technical reports, large data sets, and other accounts "not publishable in regular scientific journals."[52] It was established as a journal of a society (International Association of Theoretical and Applied Limnology) and published on the behalf of the society by the Regional Euro-Asian Biological Invasions Center, a largely virtual institute that also publishes *BioInvasions Records* and *Management of Biological Invasions*, discussed subsequently. It is also affiliated with the European Research Network on Aquatic Invasive Species, developed between 1999 and 2002, and the journal received start-up support from the European Commission.

Aquatic Invasions evolved in both its subject coverage and its geographic focus. The first volume, in 2006, contained fifty-two articles (thirty-six research articles, fifteen short communications, and one technical report), of which forty-eight were first records of non-native species in particular sites or similar topics (e.g., range extensions of invasive species in their invaded range), and all but three articles focused on invasions in Europe or contiguous Asia. In 2020, volume 15 contained thirty-nine articles (all classified as research articles), of which only eighteen focused on invasions into Europe and contiguous Asia; others treated invasions in North America, South America, and distant regions of Asia. The board of forty editors in 2020 included scientists from twenty-one nations on six continents; only seventeen were European. Only one article reported a new record, as would be expected given the initiation of the new companion journal *BioInvasions Records* in 2012 (see subsection 10.1.2.8). Volume 15 included a large range of topics dealing with invasions, similar to topics featured by *Biological*

Invasions and *NeoBiota*, including impacts, invasion routes, and traits of non-native species. Management was largely absent, in accord with the 2010 establishment of the companion journal *Management of Biological Invasions* (see subsection 10.1.2.7).

10.1.2.5 *Russian Journal of Biological Invasions*

The *Russian Journal of Biological Invasions,* published by the Pleiades Publishing Company, was initiated in 2008, with ten articles in Russian plus additional English abstracts. This format persisted through a second issue in 2008 and two issues in 2009. Beginning in 2010, the journal published a volume with four issues each year in English, distributed by Springer Nature, with Russian versions available on a separate website operated by Pleiades. The 2010 volume was designated volume 1. From the outset, the focus has primarily been on invasions in Russia and nearby regions, and by 2020 the editorial board consisted of a Russian editor in chief and deputy editor in chief plus fifteen other Russian scientists, three from the United States, and one each from five other nations.

10.1.2.6 *Invasive Plant Science and Management*

The term "weed" is defined variously, but all definitions capture the notion of a plant growing in a place where it is unwanted. Historically, "weed" has connoted a plant interfering with agriculture.[53] However, with the advent of modern invasion science, non-native plants affecting natural environments have been designated "weeds," as epitomized by the informal regional journal *Wildland Weeds* (discussed in the opening subsection 10.1.1). In general, research and publications on weeds have not distinguished between native and non-native plants—the key is that the plant is found where it is unwanted. With the more recent publications on unwanted plants in natural environments, the term "weed" has the more restrictive connotation of a non-native species, but the general weed literature makes no such distinction. As early as 1879, the American botanist Asa Gray remarked that most weeds are non-native: "In the United States, and perhaps in most parts of the world, a large majority of the weeds are introduced plants, brought into the country directly or indirectly by man."[54] By the time a field of weed science was coming into existence around the beginning of the twentieth century, it was widely recognized that many weeds, or even most, were non-native.[55] For instance, perhaps the first weed science book, published in England in 1920, observed that

> the weeds that are ubiquitous in this country are chiefly cosmopolitan, and they either occur under arable cultivation in other countries or they colonise rapidly if they are introduced. In parts of Australia the bush has been cleared comparatively recently and the land brought under cultivation.

> Where this has occurred the native flora has disappeared entirely, being unable to put up with the interference and the new conditions, but a weed flora has appeared composed of species that are common as weeds of cultivation in America and the Old World. The introduced species have adopted their new home to such an extent that they threaten to spread in most injurious fashion. . . . A similar thing has happened in Canada. The list of Canadian Farm Weeds consists chiefly of species that have been introduced from Europe.[56]

In the United States, organizations concerned with weed management formed in several states in the 1930s, soon thereafter coalescing into regional organizations, culminating in the founding of the Association of Regional Weed Conferences (ARWC) in 1949. In 1951, AWRC decided to publish a journal, *Weeds*, and the first volume appeared that year with six editors, all American men. In 1957, the ARWC renamed itself the Weed Society of America (WSA), which continued to publish *Weeds* through 1967.[57] In line with the general understanding of what constituted weeds, *Weeds* published papers on both native and non-native plants, with emphasis on chemical control in agricultural settings. In 1967, the WSA renamed itself the Weed Science Society of America (WSSA) and changed the title of its journal from *Weeds* to *Weed Science*, with the last volume of *Weeds* (number 15) appearing in 1967 and the first volume of *Weed Science* (16) in 1968. The focus continued to encompass chemical control of agricultural weeds, both native and non-native, but included other sorts of research—for example, on the biology of particular species—that was not directly applicable to management. In 1987, the WSSA initiated a new journal, *Weed Technology*, focused specifically on management, although *Weed Science* continued to publish management-related articles.

After several years of discussion, in 2008, the WSSA initiated a new quarterly journal, *Invasive Plant Science and Management*, reporting exclusively on non-native plants. The journal adopted the definition of "invasive" in President Bill Clinton's Executive Order 13112 of 1999—any "alien" (non-native) species not native to its environment or ecosystem introduced deliberately or inadvertently by humans and that is likely to harm the economy, the environment, or human health. The ambit of the journal was invaders into nonagricultural communities. Thus, articles would be published on species that overlap between two types of ecosystems, but not invaders strictly into cropland or urban environments. Although published by a United States society, the objective was to "provide the best possible information to facilitate better communication and coordination in the development of invasive plant management programs throughout the country and the world."[58] However, in 2020, thirteen members of the fifteen-member editorial board were based in the United States, and, of the thirty-two articles published that year, lead authors of twenty-eight were at US institutions.

10.1.2.7 *Management of Biological Invasions*

In 2010, the Regional Euro-Asian Biological Invasions Center initiated a new journal, *Management of Biological Invasions*, to complement *Aquatic Invasions*. After the establishment in 2016 of the International Association for Open Knowledge of Invasive Alien Species, *Management of Biological Invasions* became an official journal of that organization. The main goal of the journal was to provide managers with access to both rigorously reviewed management information and "pure" science relevant to invasion management.[59] The journal started slowly, with the two initial volumes each consisting of a single issue, with five and nine articles, respectively. The number of articles grew rapidly beginning in 2013, and all subsequent volumes consisted of four issues; by 2011, volume 11 published forty-eight articles. Although *Management of Biological Invasions* consistently published more papers with a clear management focus than, say, *Biological Invasions* and *NeoBiota* (both of which publish relatively few management-oriented papers), from the outset it also published papers on more purely academic research tenuously related to management. However, it is difficult to think of any research on biological invasions for which one could not imagine a potential indirect link to management. *Management of Biological Invasions* is an international journal—its website subtitles the journal "International Journal of Applied Research on Biological Invasions." Although founded in Europe, the editorial board in 2020 had editors in chief from Australia and the United States, and the editorial board of thirty-nine members included twenty-two North American scientists and only seven from Europe.

10.1.2.8 *BioInvasions Records*

In 2012, *BioInvasions Records* was established as the third journal of the Regional Euro-Asian Biological Invasions Center, as a continuation of a former electronic supplement to *Aquatic Invasions*. The focus was new records of both terrestrial and aquatic non-native species, but related applied manuscripts on invasions were welcomed.[60] The emphasis on first records and range extensions, not necessarily accompanied by information on biology, impact, or means of arrival in the non-native region, filled a gap in the international literature, as invasion journals such as *Biological Invasions* and *NeoBiota* published few such accounts. The *Russian Journal of Biological Invasions* published such articles, but it was a new journal, not widely read, and in any event focused on one region, whereas *BioInvasions Records* was explicitly international in scope. The upshot was that, at the time of the founding of *BioInvasions Records*, only *Aquatic Invasions* published substantial numbers of accounts of first records and range extensions of non-native species, and only for aquatic invaders. Even for these, as well as for terrestrial invaders, most such reportage ended up in regional journals, taxon-based journals, or the gray literature, often not known by or accessible to authors seeking to synthesize aspects of the invasion literature.

BioInvasions Records satisfied a perceived need, and it grew rapidly, publishing forty-five papers in its first year and ninety-one in 2020 (volume 9), the great majority on first records and range extensions, representing all regions of the world. The editorial board grew from an initial eighteen members (including eleven Europeans) to seventy-nine members, with Europeans constituting 47 percent and the United States the largest national component (thirteen members).

10.1.3 Biological Control Journals

Journals focused on biological control to manage pests perforce publish articles on impact and management of non-native species. Technically, "classical biological control" is the introduction of natural enemies from the native range of a non-native pest to control populations of that pest (discussed further in chap. 17, sec. 17.5), and in the invasion and ecology literatures the term "biological control" usually refers to classical biological control, although the biological control literature also reports on research involving both native pests and native natural enemies.[61] Thus biological control journals, although they report primarily on research on non-native species, also publish articles unrelated to biological invasions. Another feature of this literature is that it frequently addresses the use of microorganisms and nematodes (both native and non-native) to control non-native species, and the invasion literature on microorganisms and nematodes is notably scarce compared to that on plants and animals other than nematodes. Finally, the great majority of publications in biological control journals focus on the use of the technology in agriculture and other human endeavors rather than on the "natural" ecosystems that are the concern of much of the invasion science literature. The term "conservation biological control" in the technical literature is a misnomer, as it is at best tangentially related to conservation biology concerns. It refers to the collection of practices used in agricultural settings to facilitate the survival and reproduction of natural enemies of agricultural pests—both native and non-native—to benefit agriculture.[62]

10.1.3.1 *Journal of Economic Entomology*

The *Journal of Economic Entomology*, an early journal that published occasional biological control articles, began in 1908 as the official journal of the American Association of Economic Entomologists.[63] This society had formed in 1889 and was focused on managing insects as they affect economic interests. It was thus dominated by research on agricultural pests, with occasional contributions on insects affecting forestry or public health. From its inception the *Journal of Economic Entomology* featured chemical control.

Several papers reported the arrival of new insects of economic (usually agricultural) concern, and many articles reported on impacts and chemical management of prominent non-native species such as the housefly, codling moth, Argentine ant, spongy moth, browntail moth, and Colorado potato beetle.[64] A few reports treated biological control projects, notably parasitoid wasps and flies of the spongy moth and browntail moth.[65] However, decade after decade, even as research on biological control broadened and proliferated, the journal focused on chemical control. In 1953, the American Association of Economic Entomologists was absorbed by the Entomological Society of America, an organization that was initiated in 1906 because the focus of the older society was perceived as too narrow for the ambit of entomology as it was developing as a science. The journal from 1953 became an official journal (one of several) of the Entomological Society of America, but its character—especially its heavy emphasis on chemical control—remained. Beginning with volume 85, number 5, in 1992, the *Journal of Economic Entomology* divided its articles into sections, one of which was "Biological and Microbial Control." This section includes not only articles on classical biological control—natural enemies introduced from elsewhere—but also augmentation of native enemies as well as other technologies such as the use of pheromones and release of sterile insects (discussed further in chap. 17, sec. 17.6). Still, the section is usually a small part—often a single article—of a journal dominated by research on chemical control.

10.1.3.2 *Entomophaga/BioControl*

The earliest journal specifically focusing on biological control, *Entomophaga*, was the journal of the International Organization for Biological Control (IOBC) and, despite its name, reported from the outset (1956) on biological control of plants as well as arthropods. In 1998, the journal title changed to *BioControl*, and Springer became the publisher for the IOBC, replacing Lavoisier. Throughout its history, the great majority of papers in *BioControl/Entomophaga* have described research focused on particular potential control agents for particular pests. However, as controversy arose over nontarget impacts of non-native species introduced for biological control (discussed in chap. 17, sec. 17.5), an editorial by the editor in chief and the lead article presaged sporadic articles on that issue.[66] Thereafter, an occasional article treated nontarget impacts, highlighted in 2008 by a special issue with nineteen articles on impacts and management of the multicolored lady beetle, used in biological control, that had become highly invasive in North America and Europe.[67] This beetle competes with and preys on other insects that eat aphids, attacks soft fruit, fouls wine, and masses in homes and other buildings. Articles appeared that focused on more general ecological topics relevant to invasion, but tailored to biological control—for example, on indirect effects of a non-native predator on native predators.[68]

A 2012 special issue (volume 57, number 2) treated biological control in natural rather than agricultural settings, with several articles discussing nontarget impacts.[69] In 2021, a special issue contained fourteen articles on the risks of introducing generalized predatory insects, such as the multicolored lady beetle.[70] Notably, many articles were couched in the framework of invasion science, such as analysis of interaction networks (covered in chap. 13, sec. 13.2), consideration of what species traits and phylogeny can reveal about invasion impacts (chap. 15, sec. 15.3), and assessment of the multicolored lady beetle in the context of the enemy release hypothesis (chap. 15, sec. 15.4.2).[71] As biological control specialists increasingly lamented the burden of considering risk of nontarget impacts for their projects, *BioControl* published a trenchant article depicting declining use of the technology, particularly for conservation purposes.[72]

10.1.3.3 *Biological Control*

In addition to *BioControl*, the commercial journal *Biological Control*, started by Academic Press and soon acquired by Elsevier, is the other major academic journal in the field. It was initiated in 1991—that is, as modern invasion science was getting underway.[73] As is the case for *BioControl*, most papers in *Biological Control* narrowly focus on particular potential control agents for particular pests. Also as in *BioControl*, microbes and nematodes—relatively rare in the invasion science literature—are frequently discussed as potential biological control agents. And most papers in *Biological Control* address pests of agriculture and other human enterprises. Nevertheless, some papers treat topics squarely in the mainstream of invasion science or closely related to issues in the invasion science literature. Early in the journal's history, Lester E. Ehler commented on similarities and differences between invasion science and the research on non-native pests and their natural enemies in biological control, citing, for instance the tens rule and the lag effect (both discussed in chap. 18, sec. 18.3.1) from invasion science literature and assessing their utility for biological control.[74] On the whole, he was less sanguine than George E. Heimpel and Nicholas J. Mills, who wrote two decades later, about the ability of invasion science to aid biological control research.[75] In 2010, *Biological Control* published a special supplemental article on the use of biological control to protect natural ecosystems.[76]

As might be expected in light of the emergence of the journal just as concern over nontarget impacts of biological control grew rapidly among invasion scientists, conservation biologists, and the public (as discussed further in chap. 17, sec. 17.5), *Biological Control* has published several papers on this issue, including an early commentary on risks associated with the possible use of transgenic biological control agents (discussed earlier in chap. 9, sec. 9.3).[77] A special issue on risks of biological control of weeds appeared in 2005, including an early attempt to assess the risk of biological control in particular cases and an important discussion of possible nontarget indirect effects.[78] Other papers on risk assessment followed, including a discussion by Barbara I. P. Barratt and colleagues in 2010 expressing optimism that improved risk assessment could moderate the tension between conservationists and biological control practitioners.[79] This view contrasted with the pessimistic assessment of Barratt and associates in 2018 attributing a decline in the use of biological control to the persistence of this tension, as Vincent C. Moran and John H. Hoffmann had similarly lamented.[80] Other authors found documented nontarget impacts to be scarce, and many papers in *Biological Control* detailed protocols for testing for the likelihood of nontarget impacts before a natural enemy is released or reported results of such tests.[81]

10.2. College Textbooks

As scientific notice of biological invasions grew in the twentieth century, accelerated in the wake of Charles Elton's 1958 monograph, and crystallized into the modern field of invasion science in the 1980s, college courses in the fields generating the most research on invasions evolved to reflect the burgeoning research effort. Most scientific publications concerning invasions during this period addressed ecology, as indicated by the title of Elton's book: *The Ecology of Invasions by Animals and Plants.* Ecology as the subject of college courses began in the late nineteenth century and grew to major status throughout the twentieth century, and thus textbooks appeared before Elton's monograph, they multiplied during the approximately thirty-year period between his monograph and the SCOPE project, and new textbooks as well as revised editions of earlier ones abounded after the SCOPE project. Modern conservation biology arose much later, around the time of the SCOPE workshops, and widely adopted textbooks for conservation biology courses did not appear until the 1990s, but even during this shorter time span, some conservation biology textbooks evolved through subsequent editions and new ones appeared. Soon thereafter, the first textbooks for courses in invasion science appeared.

Initially, if ecology textbooks mentioned non-native species at all, it was to cite specific invasions to exemplify particular phenomena or processes. Certain prominent invasions were cited by many authors—for example, the chestnut blight to exemplify the impact of a plant pathogen on a host plant population and the spongy moth as an example of herbivory impact (both discussed in chapter 4). The roster of such examples tended to increase in successive editions of the same textbook as new research findings appeared or existing examples became better known. Invasions as a general phenomenon were not featured until the late twentieth century, and even then they rarely merited their own section or chapter, with an overview of several examples. As concern with global changes grew throughout the late twentieth and early twenty-first centuries, some ecology textbooks identified invasions as a global change and assessed its magnitude with respect to other global changes, and sometimes its interactions with other global changes, such as climate change. Conservation biology textbooks, by contrast, arriving in college courses much later, from the outset treated substantial numbers of non-native species, but in this case they were discussed as threats to the existence of particular species, communities, or ecosystems rather than as examples of ecological phenomena or probes into mechanisms of ecological processes. Finally, the late twentieth and early twenty-first centuries saw the first appearance of textbooks targeting courses specifically in invasion science and attempting to cover the entire field.

10.2.1 Ecology

10.2.1.1 *Animal Ecology*

For ecology, Elton's 1927 textbook, *Animal Ecology*, was the first textbook on animal ecology and included references to the establishment of non-native species (including plants) that Elton viewed as disturbances, the results of which could shed light on the forces structuring communities (as discussed in chap. 7, sec. 7.1).[82] For instance, for the introduction from Mexico to Hawaii of the ornamental plant lantana, Elton showed how the non-native myna ate its berries and thus increased its own population while dispersing the plant. The myna in turn fed on non-native army worm caterpillars, suppressing this pest of sugarcane plantations and grasslands. However, several insects were introduced to control lantana, in an early biological control effort, and succeeded to the extent that populations of the myna declined, leading to renewed outbreaks of the army worm. Elton cited other invasions, such as those of the spongy moth in North America and the rabbit introduced from Europe into Australia (cases examined in chap. 4), to illustrate how interactions between species affect their populations and the communities in which they are embedded.

10.2.1.2 *Bio-Ecology*

The dominant ecology textbook in the first half of the twentieth century was *Bio-Ecology*, published in 1939 by two towering American figures in the early years of ecology, plant ecologist Frederic E. Clements and animal ecologist Victor E. Shelford (cf. chap. 12, sec. 12.2 and chap. 14, sec. 14.4.1).[83] Although drawing examples heavily from the United States, *Bio-Ecology* was conceived as a general ecology textbook. The first substantial effort to synthesize plant and animal ecology plus terrestrial and aquatic ecology, it achieved an international readership. Although the index contained no entries on introductions, invasions, or non-native species and the book featured no discussion of invasions as a topic, *Bio-Ecology* cited several well-known invasions to exemplify particular phenomena (e.g., rabbits from Europe in Australia for herbivory impacts, starlings and house sparrows from Europe in North America for adaptation to new habitats, chestnut blight from Asia in North America for impacts of pathogens on plants, cheatgrass from Asia in eastern Washington state for the concept of a disclimax community). Clements and Shelford cited Elton's 1927 book, but not with regard to invasions. The authors were almost surely well aware of the potential scope of invasions, but this awareness was manifested only in a short description of one example, the impact of non-native grazers combined with non-native plants that had coevolved with them—the sort of "ecological imperialism" later publicized by geographer Alfred Crosby (whom we met in chap. 8, sec. 8.3). Discussing one such case, the transformation of Palouse prairie in eastern Washington, they noted "an increasing body of knowledge as to the food habits of domestic cattle, sheep, and goats, all of which are unfortunately of European or Asiatic origin."[84]

Perhaps most telling in terms of their downplaying invasion impacts was Clements and Shelford's citation of a remarkable paper by Alvin R. Cahn in 1929. Cahn, who also published and guided early research anticipating species distribution models (discussed in chap. 14, sec. 14.1), described the wholesale transformation of the animal and plant communities of a small lake and a connected pond in southern Wisconsin after introduction of Old World common carp.[85] Small individuals were thrown overboard by fishermen who had been using them as bait for prized native fish such as largemouth bass, northern pike, and walleye. Although Cahn noted the non-native status of the carp, Clements and Shelford initially omitted this fact entirely, simply citing the case as an example of a species acting as a "dominant"—that is, controlling community character and composition. Later in the book they reprised the case and mentioned the introduction to suggest, following Cahn's proposal, that native fishes may have the same impact but that the "experiment" had never been performed.

10.2.1.3 *The Study of Plant Communities*

Henry J. Oosting, in both editions (1948, 1956) of his plant ecology textbook, *The Study of Plant Communities*, had a section on invasions that included not only impacts of such non-native plants as water hyacinth and Japanese honeysuckle in North America and prickly pear in Australia as well as a plant pathogen (chestnut blight), but also non-native animals such as the spongy moth in eastern North America and the small Indian mongoose in the West Indies (several of these discussed in chap. 4).[86] For most examples, he stated the impact in general terms; for instance, water hyacinth "choked canals" and "destroyed wildlife." Invasions were not used to exemplify processes of general ecological importance. The second edition added further examples, again including animals, as well as biological control introductions. As noted in chapter 6, section 6.1, Oosting's emphasis on invasions is mysterious, as invasions were not prominent in the literature at the time, and his own research did not address invasions. Neither did Oosting cite notable early publications on invasion. Except for Clements and Shelford, the only other direct citation was to an obscure conference proceedings paper by Rudolf M. Anderson.[87] Oosting's invasion descriptions were more an appeal to his readers to note this aspect of the scope of human destruction of natural systems, akin to Rachel Carson's *Silent Spring* and Aldo Leopold's *A Sand County Almanac*, both of which targeted lay readerships (as discussed in chap. 7, sec. 7.1), while Oosting's book was a college textbook. The invasion examples were barely integrated into the rest of the textbook, which consisted mostly of descriptions of plant communities and their physical environments.

10.2.1.4 *Fundamentals of Ecology*

The work that superseded *Bio-Ecology* as the dominant ecology textbook was *Fundamentals of Ecology*, by Eugene Odum, first published in 1953.[88] Odum was the leading figure in the development of ecological research at the ecosystem level (as discussed further in chap. 12, sec. 12.3).[89] His *Fundamentals of Ecology* ranked first in a survey of members of the American Institute of Biological Sciences as the book that had most influenced their career training.[90] The book went through five editions (1953, 1959, 1971, 1983, and 2005) and was translated into at least twelve languages.[91] The timing of *Fundamentals of Ecology* is relevant, as is Odum's main research focus of ecosystem-level ecology. The first edition was published before Elton's 1958 popular book that anticipated much of modern invasion science (described in chap. 7, sec. 7.1), and the second edition was surely written before Elton's book appeared and was published almost contemporaneously with it. The third edition appeared before the SCOPE project, the fourth edition was published as the SCOPE project was getting underway, and Odum and Gary Barrett wrote the fifth edition as the modern field was rapidly expanding in the wake of SCOPE. As for Odum's focus on ecosystems, invasion impacts at the ecosystem level were rarely addressed before the SCOPE project, but several contributions to the project depicted ecosystem impacts, and Peter Vitousek, both within SCOPE books and in other publications during this period, pointed to the breadth and importance of such impacts (more on these observations in chap. 12, sec. 12.3).

The index of Odum's first edition had no entry for invasions or introduced species but had one for "exotic game, introduction of, 328." This entry referred to a short paragraph that transcends game species, beginning thus:

> We may well close this brief review [of wildlife management] with a comment on the introduction of exotics, that is, species not native to the region.

> Since species introduced into a new environment often either fail entirely or succeed so well that they become pests (rabbits in Australia, for example), all introductions are to be considered with extreme care. In general, if there is a suitable species native to the region, it is better to concentrate on management of this species rather than to attempt to introduce a substitute.[92]

Odum then listed pheasant from Europe as a "good" non-native game species because pheasants were adapted in their native range to grain fields resembling those replacing native grasslands in North America, and he noted Mexican quail as a non-native game species introduced to the southeastern United States as an example of utter failure.

The other aspect of invasions Odum treated in his first edition was the devastation wrought by some introduced predators, parasites, and pathogens, exemplified by a description of the impact of the chestnut blight in North America (cf. chap. 4, sec. 4.8). He ended this section with a warning:

> In fact, if one makes a list of the diseases, parasites, and insect pests which cause the greatest loss to agriculture . . . the list will include a large number of species which have recently been introduced into a new area, as in the case of the chestnut blight, or acquired a new host or prey. The lesson for man, of course, is to beware of new negative interactions and to avoid sponsoring new ones any faster than is absolutely necessary.[93]

Odum devoted several pages to the use of climographs to depict factors that limit different populations, referring to Arthur Twomey's use of temperature-humidity climographs to explain the reasons for survival of non-native "Hungarian partridges" (grey partridge) in certain locations in North America but not in others (cf. chap. 14, sec. 14.4.1). This discussion of climographs continued through all five editions, with the figure of climographs related to the partridge introductions reproduced in all but the last one. The last edition, though published when species distribution models (SDMs) were becoming common to predict ranges of non-native species (cf. chap. 14, sec. 14.4.4) and climographs were passé, simply mentioned that climographs can be useful in determining suitable locations for proposed introductions, with no mention of SDMs.

Odum's second edition, six years after the first, repeated the invasion-related points from the first and added the invasion of California chaparral by eucalyptus from Australia as an example of non-native species thriving if their new habitat resembles that in their native range. The third edition, twelve years later, maintained all these examples but included several notable additions. Elton's invasion book, which had appeared almost simultaneously with Odum's second edition, was cited in the third edition, and Odum quite possibly broadened his view of invasion impacts by having read Elton.

Odum in the third edition described two invasions Elton had detailed.[94] One was the hybridization in England of North American smooth cordgrass with native small cordgrass to produce a hybrid that was invading English mudflats and transforming them into poorly drained marshes.[95] The other was the devastation wrought by the introduced parasitic sea lamprey in the North American Great Lakes, to which Odum added information on invasions of the Great Lakes by the alewife and coho salmon. Odum also added a brief discussion of an important paper that had recently been published and that remains infrequently cited in ecology textbooks, by John L. Brooks and Stanley I. Dodson.[96] They found that, in a Connecticut lake to which a near relative of the alewife had been introduced, all large species of zooplankton were replaced by much smaller species. The fact that the species composition and characteristics of an entire trophic level could be controlled by one predator surely struck a responsive chord with Odum, as much of his interest in ecosystems was focused on trophic interactions and patterns of the resultant energy flow through an ecosystem.

The fourth edition, in 1983, added a small section on biological control (cf. chap. 17, sec. 17.5), with the admonition that attempts to use large, generalized predators as control agents are unsuccessful, citing the small Indian mongoose on Caribbean islands, which failed to control rats and attacked native ground-nesting birds (cf. chap. 4, sec. 4.5). Odum's fifth edition, published posthumously after a lapse of twenty-two years with the assistance of a colleague who had been his graduate student, might have been expected to expand coverage of invasions still further, as it was written after the SCOPE project and during a period of rapidly expanding invasion research.[97] However, there was some retrenchment, perhaps signaled by the removal of Elton's invasion book from the bibliography, as well as examples that might have been inspired by Elton, such as the invasions of the sea lamprey and the hybrid cordgrass, plus the small section on biological control. In sum, Odum certainly recognized the massive damage invasions could inflict—his evolving treatment of chestnut blight through all editions is perhaps the best example. However, although he briefly warned of possible consequences of invasions, he never synthesized this information into a substantial treatment of invasions. Only the third edition had an index entry for "invasions," and none had an entry for "introduced species." The "exotic game" entry disappeared after the third edition. Rather, Odum cited particular invasions to exemplify particular phenomena or, in the earlier editions, to note their prominence in particular biomes. For instance, the lone index entry in the third edition was to a single invasion, that of hybrid cordgrass, as an example of sympatric speciation.

10.2.1.5 *The Experimental Analysis of Distribution and Abundance*

The dominance of *Fundamentals of Ecology* waned in the 1970s, when prominent ecologists published a spate of new textbooks.[98] One that achieved substantial distribution was *The Experimental Analysis of Distribution and Abundance* by Charles J. Krebs, which went through six editions (1972, 1978, 1985, 1994, 2001, 2009) starting well before the SCOPE project and extending into the explosion of invasion research in the twenty-first century.[99] That the first edition was dedicated to Charles Elton and Dennis Chitty (Krebs's doctoral advisor, whose own doctorate was advised by Elton) telegraphs substantial treatment of invasions. Krebs adduced them frequently, primarily to exemplify particular ecological phenomena or problems—for example, Himalayan tahr in New Zealand for life tables, reindeer on islands for exponential population growth, sea lamprey for predation, two introduced ant species on Bermuda for competition, spongy moth and browntail moth for various population controls. Krebs also discussed reasons for successful and failed invasions, presenting quantitative data for several species, and he treated evolution of non-native species, including coevolution in Australia of the European rabbit and myxomatosis (chap. 4, sec. 4.1) as an example. Many of the chapter-ending questions for students addressed invasions, including one referring to general invasion patterns proposed by Elton.

Krebs extensively treated applied problems, and he featured one of these, biological control, in all six editions both as a general topic and with specific cases used to exemplify other phenomena—e.g., three parasitic wasps of the genus *Aphytis* introduced to control California red scale (actually native to China) in California to show the mechanism of interspecific competition and myxoma virus introduced to Australia to control European rabbits as an example of evolution of non-native species. Krebs noted the importance of using biological control agents with narrow host or prey ranges as part of a description of the successful introduction to California of the Klamath-weed beetle, native to Europe and North Africa, to control Klamath weed from Eurasia, an example in all editions but the sixth. He tended at first to focus on well-known successes such as the non-native cactus moth controlling prickly pear in Australia (cf. chap. 17, sec. 17.5) and the vedalia beetle controlling cottony cushion scale in California (cf. chap. 4, sec. 4.3), and he included a large table of successes beginning with the second edition. Krebs paid little attention initially to nontarget impacts. This narrow focus was exemplified by his citing in all six editions the introduction of the myna from India to Mauritius in 1762 to control red locust, which he termed a success and for which he made no mention of potential nontarget impacts. The highly omnivorous myna is one of seven species on the IUCN list of one hundred of the worst global invaders that were introduced for biological control (chap. 17, sec. 17.5).[100] On Mauritius, it competes for nest sites with the highly endangered endemic echo parakeet and preys on the endemic macchabé skink.[101]

After nontarget impacts of biological control became a major issue in ecology and conservation biology in the 1990s (as discussed further in chap. 17, sec. 17.5), Krebs featured them prominently.[102] In the fifth edition (2001) he cited a general portrayal of the problem and described nontarget impacts of the small Indian mongoose (cf. chap. 4, sec. 4.5) and rosy wolf snail.[103] The sixth edition (2009) cited Francis Howarth's review of the issue and added the then-recent claim that the multicolored lady beetle, introduced to the United States from Asia to control aphids, threatened the monarch butterfly.[104]

In successive editions, Krebs increasingly focused on invasions, with new examples of impacts as these appeared in the literature. Also, without increasing the size of his book, he expanded treatment of several he had included earlier. Krebs's orientation toward ecology was resolutely at the population and community levels, rather than the ecosystem level as it was for Odum, and most of Krebs's treatment of invasions reflected this orientation. However, he did not neglect ecosystem-level impacts, introducing in his second edition the study by Brooks and Dodson of the impact on routes of energy flow through the highly modified community of a lake to which a predatory fish had been introduced. In the fifth edition he included an essay on recent research on impacts of eutrophication on nutrient cycling and energy flow in a Finnish lake and the role of non-native pikeperch in redressing them.[105]

10.2.1.6 *Ecology* and *Ecology: The Economy of Nature*

Soon after Krebs, Robert Ricklefs in 1973 published an ecology textbook entitled simply *Ecology*, a 966-page behemoth.[106] As Krebs had, Ricklefs published a series of editions extending through the SCOPE project and into the period during which the field of invasion science expanded rapidly, ending with a fourth edition in 2000 coauthored with Gary L. Miller.[107] In addition, beginning in 1976, Ricklefs published an abridged version of *Ecology*, at half the length (455 pages), entitled *Ecology: The Economy of Nature*.[108] This book, targeting less advanced general ecology courses, went through nine editions, acquiring as coauthor Rick Relyea, an ecologist specializing in aquatic amphibians, for the seventh and eighth editions (2014 and 2018, respectively).[109] Relyea became the sole author in 2021 for the ninth edition.[110]

The first edition of *Ecology* had no index entry for invasions or introduced species but used non-native species to exemplify several phenomena, frequently omitting mention of their non-native status. For instance, among evolutionary processes—a particular emphasis of Ricklefs's books—the myxoma virus and European rabbit in Australia were used to exemplify rapid evolution, with reference to the fact that both were introduced. By contrast, Ricklefs described the rapid evolution of geographically differentiated populations of the house sparrow in North America (chap. 11, sec. 11.1) with no mention of its introduction. Similarly, he outlined how sympatric speciation could occur with the example of apple maggots in

North America without noting the crucial role played by the non-native status in North America of one of the host plants, apple (chap. 11, sec. 11.3.2). The second edition, published in 1979, resembled the first, with a few added uses of non-native species as examples, such as competition among parasitoids introduced as biological control agents for agricultural pests in both California and Hawaii, with an index entry for "biological control."[111] A photograph of water hyacinth, native to South America, choking a waterway in Louisiana that was used to exemplify freshwater habitats did not note its non-native status. Like the earlier editions, the third edition, in 1990, lacked an index entry for invasions or introduced species.[112] Examples of ecological phenomena using non-native species (e.g., how herbivores can control plant populations) were largely maintained. A striking addition was a series of maps depicting the spread of the starling after its arrival in North America, followed by a short paragraph, citing two papers and a book published after the second edition, describing how some introductions fail, including earlier starling introductions. By the aforementioned fourth and last edition, "introduced species" merited its own short index entry, although the list of new emphases for this edition did not include invasions.[113] This edition added several invasions, such as a section on a range of impacts of the introduction of the Nile perch on Lake Victoria and its cichlid species (cf. chap. 7, sec. 7.3). However, several cases from earlier editions were omitted, such as the introduction of prickly pear cactus to Australia and its control by the introduced cactus moth (cf. chap. 17, sec. 17.5).

The abridged version of Ricklefs's text, *The Economy of Nature*, underwent a greater transformation, as might have been expected, as it persisted for longer, running through nine editions from 1976 to at least 2021, including a much longer span of the dramatic increase in invasion research than did the larger book. The first edition, published just three years after the first edition of *Ecology*, contained most of the invasion examples of the longer predecessor plus others, such as the introduction of parasitic wasps for biological control in California and Hawaii, which did not appear in *Ecology* until the second edition.[114] In addition, the evolution of genetically based geographic variation in the house sparrow introduced to North America was treated as a sequel to its introduction. Most remarkably, Ricklefs discussed how the effects of introducing the peacock bass to Gatun Lake rippled through the entire community and reorganized the food web (as discussed in chap. 7, sec. 7.3), an early demonstration of the potentially wide-ranging effects of a non-native species not cited in any edition of the larger book.[115] The photograph of water hyacinths in Louisiana with no mention that it was non-native, depicted in the second edition of *Ecology*, made its first appearance here.[116]

Subsequent editions of *The Economy of Nature* saw a gradual increase in use of invasions as examples of ecological and evolutionary phenomena. By the sixth edition in 2008, both "introduced species" and "invasions" had index entries, and a new feature, "Ecologists in the Field," portrayed several invasions and their impacts: the Nile perch in Lake Victoria, the global decline of amphibians in the wake of the chytrid fungus, and the rapid evolution of calling behavior in an introduced cricket species in Hawaii upon subsequent introduction of a parasitic fly (chap. 11, sec. 11.3.2).[117] The introduction to this edition pointed to increased emphasis on global changes and noted that some sections associated with these changes were written by Rick Relyea, including several on non-native species. There was extensive treatment of SDMs to predict the spread of non-native species (more on SDMs in chap. 14, sec. 14.4) as well as explicit discussion of invasions as an economic factor and a cause of extinctions. Relyea was listed as a coauthor of the entire seventh edition, which included a new feature, "Biology Today: Connecting the Concepts."[118] Two examples were the invasion of North America by the emerald ash borer and the invasion of Macquarie Island by rabbits and cats, including a controversy over the impacts of eradicating the rabbits.[119] Many examples of other processes involved non-native species, although not all were indicated as such. Relyea became senior author of the eighth edition. The ninth edition, with Relyea as the sole author, included still more biological invasions, both as examples of ecological and evolutionary phenomena and as a major ongoing global change.[120] A new feature, signaling sections of the book as pertaining to "Global Change," designated many invasions: water hyacinth in several parts of the world, chestnut blight in North America and attempts to produce blight-resistant genotypes, species distribution modeling of invasion spread, the introduction of the brown tree snake to Guam, the introduction to California of Klamath weed and its control by an introduced beetle, the Macquarie Island saga, the invasion of the United States by mosquito-borne West Nile virus, the impact of non-native garlic mustard in North America on mutualistic mycorrhizal fungi of native plants, and non-native species as a cause of declining biodiversity.

10.2.1.7 *Evolutionary Ecology*

Not all new ecology textbooks first appearing in the 1970s and progressing through several editions featured invasions even in the later, post-SCOPE editions. Ecologist Eric Pianka first published *Evolutionary Ecology* in 1974, and the last print edition, the sixth, appeared in 2000. He published a seventh online and ebook edition in 2011. As had Ricklefs, Pianka focused his ecology text on evolutionary underpinnings of ecological phenomena, as signaled by its title. In line with the evolutionary focus, the first edition noted that the house sparrow was introduced to North America and thus has a separate gene pool from native European populations (as discussed further in chap. 11, sec. 11.1). The only other mention of invasions was the suggestion that the extinction of several Hawaiian birds and Australian marsupial mammals was resulted from competition with non-native species.

Immediately following this hypothesis was this statement: "Elton (1958) discusses many other examples of ecological invasions among both plants and animals."[121]

Subsequent editions did not refer to invasions much more frequently. The second edition, published in 1978, added the exponential growth and subsequent dramatic crash of reindeer introduced to St. Paul Island, in the Bering Strait, citing Krebs's first edition.[122] The myxoma virus/European rabbit case in Australia that exemplified coevolution of parasites and their hosts for Krebs and Ricklefs in their respective first editions first appeared in Pianka's third edition, published in 1983.[123] The fifth edition, published in 1994, cited apple maggots in North America to exemplify sympatric speciation and followed Ricklefs's first edition in omitting the non-native status of apple as a critical feature (as discussed further in chap. 11, sec. 11.3.2).[124] This omission persisted in the sixth edition (published in 2000) but was finally redressed in the seventh (published in 2011) with explicit discussion of the importance of the fact that apple was non-native in North America.[125] This was the only change relevant to invasions from the sixth edition. All seven editions reproduced Twomey's climograph showing the degree of overlap of climate between the native range and two areas in North America considered for introduction of the grey partridge, citing Odum's second edition, and none mentioned SDMs to address climatic relationships between native and non-native geographic ranges.[126]

10.2.1.8 *Ecology: From Individuals to Ecosystems*

A later popular ecology text that appeared first while the SCOPE project was underway and went through five editions between 1986 and 2021 was *Ecology: From Individuals to Ecosystems* by ecologists Michael Begon, John L. Harper, and Colin R. Townsend.[127] Unlike the authors of the aforementioned textbooks, who were all Americans, these authors were British. Also unlike most previous authors, Townsend, who subsequently relocated to New Zealand, conducted substantial research on invasions, and he published many papers on various impacts of non-native brown trout in New Zealand.[128] However, although they cited Elton's invasion book, Begon and colleagues' treatment of invasions was not extensive until the fourth edition. In the first edition they mentioned biological control, discussing the cactus moth/prickly pear project in Australia and the reduction of the European rabbit population in Great Britain after the myxoma virus was introduced, although without noting that the rabbit itself was introduced to Great Britain by Normans in the twelfth century. There was passing mention of two other examples. In several places in which research on non-native species was cited to exemplify interspecific interactions, no mention was made of the non-native status of one or both species.[129] The second edition, published in 1990, had increased treatment of biological control, with detail not only for the cactus moth/prickly pear case but also that of the cottony cushion scale and vedalia beetle (discussed in chap. 4, sec. 4.3) and water hyacinth and the mottled water hyacinth weevil (discussed in chap. 4, sec. 4.6). The use of non-native species to exemplify general ecological phenomena without noting their non-native status continued, with addition of the potato blight in Ireland as a plant pathogen.[130] Little changed in the third edition, published in 1996. By this time, concern with nontarget impacts of biological control was widely discussed, but Begon and colleagues did not mention the issue.[131]

However, invasions received substantially more attention in the fourth edition, published in 2006, with several new examples, including the red imported fire ant in North America, a section on costs of invasions featuring zebra mussel and yellow star thistle in North America, extensive treatment of predicting spread of invaders (including in response to climate change) and invasional meltdown (chap. 15, sec. 15.6 discusses invasional meltdown). Indirect effects of invaders (chap. 17, sec. 17.5) were featured, including the brown tree snake in Guam maintaining high populations by eating lizards and thus eliminating birds by preying on them even as bird populations declined. Also, nontarget impacts of biological control made their first appearance with the case of the musk thistle head weevil, from Eurasia and North Africa, introduced to North America to attack non-native musk thistle, threatening imperiled native thistles (chap. 17, sec. 17.5).[132]

The fifth edition, published fifteen years later in the wake of the rapid expansion of invasion science, had still greater coverage of invasions to exemplify ecological phenomena as well as in a small section near the end on invasions as one of the major ongoing global changes. This edition, in line with the ever-increasing concern with global warming, extensively treated how global warming will affect ecological processes and patterns. For the intersection of global climate change with invasions, there was treatment of SDMs (specifically MaxEnt, discussed further in chap. 14, sec. 14.4) to predict how non-native ranges will shift (and often expand). Invasions as a major factor in species extinction received substantial coverage, as did invasions as examples of indirect effects of one species on another. Invasions were frequently incorporated in detailed, illustrated boxes, termed "Applications," of which an early one depicting homogenization was "Application 1.5"—"Human History and Species Invasions." The first explanation of use of SDMs for invasions was "Application 2.1."[133] A notable difference from earlier editions was that, when invasions were used to exemplify an ecological phenomenon, they were almost always identified as invasions, a feature that contributed to the greater emphasis on invasion as a major ecological force.

10.2.1.9 *Ecology*

Prominent ecology texts initially written later, during the increasing interest in invasions in the twenty-first century, generally featured invasions at the outset and increased coverage of the topic in subsequent editions as the field developed. A good example is *Ecology* by Michael L. Cain, William D. Bowman, and Sally D. Hacker, which has gone through six editions (between 2008 and 2023) so far.[134] The first edition had a short index entry for "invasive species" but treated invasions in many places not listed in the index. Specific non-native species were often cited to exemplify ecological phenomena: for instance, spotted knapweed for interspecific competition, golden apple snail for herbivory, foxes for predation, rabbits and myxoma virus and also chestnut blight for parasitism, trout introduction to New Zealand for a trophic cascade of impacts on entire communities. The invasion of the Mediterranean Sea by the "killer alga" (discussed subsequently) was detailed, including not only impacts but the eradication of the species after its establishment in California (discussed further in chap. 17, sec. 17.2). A chapter on conservation biology[135] mentioned that invasions can have ecosystem impacts (discussed further in chap. 12, sec. 12.3) and also how an ecosystem-level process can facilitate invasions—in this case, prescribed fire to maintain longleaf pine ecosystems in the southeastern United States favoring invasion by fire-adapted Asian cogon grass.[136] Biological control was exemplified by the prickly pear-cactus moth project and the rabbit-myxoma case, both in Australia. Notably, Cain and coauthors introduced the fact that both invaders and native species can evolve in response to their interaction, using the example of a native bug in Florida evolving to use an introduced host plant more effectively (discussed further in chap. 11, sec. 11.3.2).[137]

The second edition, published in 2011 had more extensive index entries plus a larger index entry for non-native species and made several additions to the aspects and cases of invasion treated previously. It included a discussion of the evolution of a native North American apple maggot to use a non-native host (apple) more effectively (chap. 11, sec. 11.3.2) and also an expanded treatment of ecosystem impacts. To the example of Asian cheatgrass modifying nutrient cycling in western grasslands (discussed further in chap. 12, sec.12.3), it added the notion of nutrient enrichment with the example of impacts of nitrogen fixation by introduced kudzu in the southeastern United States (discussed in chap. 6, sec. 6.1).[138] The third edition, published in 2014, increased treatment of invasions further, notably with a box describing how predation by non-native rainbow trout and brook trout in western lakes led to a decline in the population of the native mountain yellow-legged frog.[139] Another noteworthy addition was a discussion of the possibility that competition with the native eastern tree-hole mosquito in the United States could impede the spread of the tiger mosquito, recently introduced from Asia, then in the news because of the possibility that the tiger mosquito could vector diseases such as dengue fever.[140] The third edition also addressed one of the rising controversies about invasions (discussed further in chap. 18, sec. 18.3.2), citing Martin A. Schlaepfer and colleagues about potential conservation benefits of invading non-natives.[141] The fourth edition, published in 2017, qualified the statement about benefits by prefacing it with "In rare circumstances . . ."[142]

The fourth edition opened with a statement that the myriad impacts of climate change will be signaled by a "Climate Change Connection" label in the text, and for non-native species there was now a detailed treatment of the connections, with an example of how climate change will affect impacts and management of the European green crab in North America.[143] (In the third edition this example appeared only in an online supplement.) The fifth edition, published in 2021, resembled the fourth with respect to invasions. Overall, this textbook had substantial, increasing coverage of biological invasions.[144]

10.2.2 Conservation Biology

10.2.2.1 *Essentials of Conservation Biology*

Modern conservation biology crystallized in the 1980s, just as modern invasion science developed (as described earlier in chap. 7, sec. 7.5).[145] Thus, unlike for ecology, there were no modern conservation biology courses at universities until near the end of the twentieth century, nor were there modern conservation biology textbooks that antedate invasion science or that have undergone a series of revisions traversing the rise of invasion science. The first prominent conservation biology textbook was botanist and conservation biologist Richard Primack's *Essentials of Conservation Biology*. The first edition was published in 1993 during the initial rise in invasion research following the SCOPE project, and the sixth and last edition appeared in 2014, when the explosive rise was well underway.[146] From the outset, Primack featured invasions prominently as one of the main challenges for conservation biologists. The first edition had an extensive index entry for "Exotic species" and many examples of the range of threats invasions pose to native species. One emphasis was how habitat destruction and change tend to benefit non-native species.[147]

Subsequent editions largely maintained the invasion material from previous ones, modifying or replacing examples in light of more relevant recent research. For instance, the second edition (1998) added material from the 1993 report of the United States Congress Office of Technology Assessment and from a 1997 book on invasions in Florida, as well as a two-page box on the invasion of the Nile perch and water hyacinth in Lake Victoria.[148] A new feature was discussion topics, several of which treated invasions. In the third edition, published in 2002, Primack greatly modified his treatment of invasion impacts on human health. He also first noted nontarget impacts of biological control with the example of the parasitic fly introduced to control the spongy moth but that attacked saturniid moths of conservation concern, an impact widely publicized by a paper that had recently appeared (discussed further in chap. 17, sec. 17.5).[149] Possibly driven by a related motivation, Primack removed a section from the second edition in which he had used *Eucalyptus* trees in North America to exemplify non-native species that can be seen as beneficial. The fourth edition, published in 2006, added the example of the musk thistle head weevil as a biological control introduction with nontarget impacts inimical to conservation (more on biological control in chap. 17, sec. 17.5).[150]

Primack's fifth edition, published in 2010, retained most of the previous examples of invasion impacts and added new ones, such as the killer alga in the Mediterranean (and its eradication before it became widespread in California). Primack enhanced his treatment of zebra mussel impacts in the Great Lakes with added information on the quagga mussel invasion and a map showing the continuing spread of both species throughout the contiguous United States.[151] The sixth and final edition, published in 2014, resembled the fifth edition. As in each successive edition, Primack noted a few new invasions that rose to public prominence since the previous edition. He added discussion of the invasion of the Caribbean and parts of the western Atlantic Ocean by the lionfish of the Indo-Pacific region. The chapter on overexploitation, invasive species, and disease opened with a dramatic photograph of three people holding a large Burmese python captured in Florida.[152]

10.2.2.2 *Principles of Conservation Biology*

In 1994, the year after Primack's first edition of *Essentials of Conservation Biology* appeared, Gary K. Meffe, C. Ronald Carroll, "and contributors" published a competitor, *Principles of Conservation Biology*.[153] As did Primack, Meffe and Carroll provided many examples of invasive species impacts on conservation, including a chapter by Meffe, Carroll, and Stuart Pimm entitled "Community-Level Species Interactions, Disturbance Regimes, and Invading Species" with a table of ostensible characteristics of invasive non-native species and invadable ecosystems. Notably, this chapter included an essay by Ariel Lugo downplaying invasion threats, entitled "Keeping an Open Mind on Exotic Species." The essay was immediately preceded by a statement that many invasives have disastrous impacts for conservation and that Lugo's perspective is "a minority opinion."[154] A later chapter contained a short section on "Management Principles in Practice: Controlling Invasive Species" outlining approaches for melaleuca, Brazilian pepper, ring-necked pheasant, common carp, and brown trout. However, most such methods (cf. chap. 17) were not addressed, including biological control. The second edition, in 1997, which included several contributors as chapter authors, was little changed from the first with respect to invasions.[155] Perhaps most noteworthy was a new essay by Lloyd Loope on Hawaii as a laboratory for addressing conservation problems posed by invasions. Lugo's essay remained, again immediately preceded by a statement on disastrous impacts of invasions, but Lugo's perspective was no longer stated to be a minority one.

The senior author of the third edition, published in 2006, was Martha J. Groom, and it included, as part of a section on "primary threats to biodiversity," a chapter on invasions by invasion scientist Marjorie Wonham.[156] The chapter also contained the Lugo essay from earlier editions, a box on impacts by invasion scientist Ingrid Parker, and several detailed case studies of particular invasions. These features complemented a battery of more briefly noted invasions exemplifying many aspects of invasion impacts and management. Invasions cropped up more cursorily in several other parts of the book, particularly in a chapter on protected areas and another on ecological restoration. Lugo's essay was included in Wonham's chapter, described as listing "some of the benefits of introduced species" but embedded in a section describing many damaging impacts on conservation.[157]

10.2.2.3 *Conservation Science: Balancing the Needs of People and Nature*

A still later book, by Peter Kareiva and Michelle Marvier, signaled a new perspective on conservation, focused on the impact of conservation policies and actions on human activities: *Conservation Science: Balancing the Needs of People and Nature*, the first edition of which was published in 2011.[158] Kareiva and Marvier, both ecologists, had conducted, often together and occasionally with Groom and Parker, research on invasions or closely related phenomena.[159] Although their orientation toward invasions through the early part of the first decade of the twenty-first century had more or less aligned with those of most invasion scientists, on the larger issue of conservation in general they had become iconoclasts. They published a manifesto radically revising traditional conservation science, and the textbook outlined the new science and steps to implement it.[160]

Although the manifesto did not focus on invasions, several of its features (e.g., the surprising resilience of ecosystems subjected to disturbance, the many benefits for humans of modified ecosystems, the universal and unavoidable footprint of human activities on nature, the dominant role of human wants and welfare) can easily be recognized as difficult to reconcile with traditional views of invasions. Kareiva was chief scientist of the Nature Conservancy at this time, and his perspectives caused great consternation among Nature Conservancy scientists and, with respect to invasions, Nature Conservancy land stewards, many of whom had been vigorously fighting non-native species seen as threatening the refuges they were managing.[161] Kareiva had by then articulated a vision of invasion management that cast their efforts into doubt, citing such critics of invasion management as Mark Davis and Emma Marris (more on them in chap. 18, sec. 18.3):

> There will be increasingly large expanses of ecosystems that are infested with invasive species, some of which might seem like "pretty shrubs and flowers," and that support a lot of animal species—including threatened and endangered species. Because of the expanding human footprint, these infested and novel ecosystems will be valued, not shunned as somehow unworthy. This is something Davis hints at, although it is perhaps better developed in Emma Marris's new book, *Rambunctious Garden*. So on one hand, the toll taken by invasive species will be more evident in 10 years, but the world will have faced so much change and human impact that some highly invaded ecosystems will be valued as "natural systems" in their own right.[162]

One might then have expected the new conservation biology textbook to approach invasions very differently from the books by Primack and by Groom and colleagues. Kareiva and Marvier did, in fact, emphasize different points. They saw impacts on humans of non-native species and their management as equal in importance to their impact on other species, the traditional focus of conservation biologists. And Kareiva and Marvier discussed economic costs and cost-benefit analysis more frequently and in greater detail than the previous books. Particular invasions and their impacts were mentioned throughout the book, as in those of the other authors, but most treatment of invasions was in a chapter entitled "Getting Practical About Introduced Species," the title telegraphing the message that much policy and management on invasions was outdated, ineffective, and impractical—which accorded well with their manifesto about conservation science. Kareiva and Marvier did not neglect negative conservation impacts of some invasions, citing prominent examples—for example, zebra mussel, cheatgrass, rats, various trout species. They also discussed how invasions can change ecosystem processes, citing the fire tree example in Hawaii, and they pointed out that the lag effect means that currently innocuous established non-native species may still be risky and that the tens rule (discussed in chap. 18, sec. 18.3) probably underestimates the fraction of damaging invasions because indirect or subtle effects may go unnoticed. These features did not differ greatly from other conservation biology textbooks.

However, the notion that non-native species can be beneficial, for humans and perhaps for traditional conservation targets, was stressed, often buttressed by an appeal to economics. In the section "The Economic Consequences of Introduced and Invasive Species," the first subhead was "Intentional Introductions Have Benefits and Costs." This subsection pointed out that several economic sectors (e.g., horticulture, the pet trade) routinely import non-native species for profit and advocated a cost-benefit analysis to determine whether advantages (from a human perspective) outweigh disadvantages. Kareiva and Marvier also adopted a suggestion of Marris's that minimizes the likely impact to humans of the Nile perch introduction on the Lake Victoria ecosystem, and they included a box entitled "Has Tamarisk Been Wrongly Vilified?" supporting the controversial view that the salt cedar invasion of the US Southwest has important benefits, such as providing habitat for a bird species of conservation concern, and that the negative impacts have been overblown.[163]

Kareiva and Marvier's second edition of *Conservation Science*, published in 2017, maintained substantial treatment of negative impacts of non-native species and updated some examples but included new examples tending to downplay negative conservation impacts and to accentuate benefits.[164] For example, the section comparing benefits and costs "to humanity" of increased trade included the statement that plant species diversity has increased in Hawaii and California owing to non-native species. Similarly, citing publications that had appeared after the first edition, they stated, "Moreover, even when species are lost due to human activities, nonnative species commonly invade, so that the net loss in local diversity is minimal."[165] The next year, Kareiva and Marvier, joined by Brian Silliman, edited *Effective Conservation Science: Data not Dogma.*[166] The only two chapters dealing substantially with invasions were by Schlaepfer ("Introduced Species Are Not Always the Enemy of Conservation") and Richard J. Hobbs ("Novel Ecosystems: Can't We Just Pretend They're Not There?"), a prominent advocate of the benefits for humans of novel ecosystems incorporating non-native species.[167] A passing reference to introduced green crabs in New England in a chapter on coastal restoration claimed they aid marsh restoration with no mention of much research suggesting damage to both human and conservation interests.[168]

10.2.3 Invasion Science

Many scholarly books pitched at scientists in general and invasion scientists in particular appeared in the wake of the SCOPE project. Several focused on invasions by particular groups of species, and others discussed particular aspects of invasions.[169] Many of these were edited volumes featuring chapters by authors specializing in particular species or invasions. Several were used as the basis for invasion science courses, usually at the graduate level and often discussion-type courses, but none were organized as introductory, comprehensive textbooks covering the field as a whole.

10.2.3.1 *Biological Invasions*

As the explosive growth of the field was getting underway, the first book to fill this void was written by British ecologist and invasion scientist Mark Williamson, a leading participant in the SCOPE project. Williamson's *Biological Invasions*, published in 1996, was pitched at advanced undergraduates and graduate students but was introductory in the sense that it assumed at least some readers have only cursory knowledge of the rapidly expanding field.[170] Williamson organized his book around eleven "conceptual framework points," each a general pronouncement about which species invade and how they reach their new locations, what types of environments are invaded, or what the impacts are. For instance, the third point was "All communities are invasible, perhaps some more than others," while the sixth was "Most invaders have minor consequences (tens rule)." Discussion of the principles addressed almost all major topics in invasion science, including evolution of invaders and native species confronting them, a topic Elton had ignored. Williamson exemplified the principles with many classic cases adduced in the earlier ecology textbooks, such as rabbits and myxomatosis in Australia, muskrats in Europe, gray squirrels in Great Britain, and stoats in New Zealand. He also drew from cases not often discussed in the contemporary invasion literature, such as viral diseases. Although the organization in terms of general, short dicta would help students grasp the nature of the field as a science rather than a long catalog of one-offs, Williamson undercut this take-home message by pointing out that all of the "rules" have exceptions: "Individual invasions can do almost anything, and can behave in many strange ways."[171]

10.2.3.2 *Invasion Ecology*

As growth of the field accelerated in the new century, invasion scientists Julie L. Lockwood, Martha F. Hoopes, and Michael P. Marchetti published *Invasion Ecology* in 2007.[172] For advanced undergraduates or graduate students, this textbook treated the major themes of the field. Despite the title, Lockwood and colleagues included a chapter on evolution of invaders. Their biggest focus was invasion impacts, and a final chapter addressed management. The authors included even more examples than Williamson had, as might have been expected in light of an additional decade of intensive invasion research, and the new knowledge allowed them to make their points without several of the previously obligatory old examples, such as the Australian rabbit/myxomatosis and prickly pear/cactus moth cases. Although Lockwood and colleagues were frank in pointing to weaknesses in regnant theories and paradigms in the rapidly growing field, these were generally in the realm of technical details and inadequacy of empirical data. The preface promised to convey the complexity of current debates, but the book did not address a fundamental controversy roiling the field at this time—the charge that the entire enterprise was tainted by xenophobia (a controversy addressed further in chap. 18, sec. 18.1).

As might be expected in a rapidly expanding science, a second edition soon followed, in 2013, and it was 50 percent longer than the first.[173] The bibliography was also 50 percent longer, and the additional number of references was largely accounted for by publications appearing after the first edition.[174] In addition to updating examples in light of new research, the authors added new chapters on eradication and on the impact of global changes—especially climate change—on invasions. The global changes chapter began with an entire primer on the nature of most major ongoing global changes. Although again promising to convey the complexity of current debates, the authors still did not directly address the charge of nativism and xenophobia, which had become even more prominent in the intervening six years, although they enlarged the section on terminological issues and included the related debate on whether language in the field is needlessly militaristic (a debate discussed in chap. 18, sec. 18.2).

10.2.3.3 *Invasion Biology*

In 2009, two years after the first edition of Lockwood and colleagues' *Invasion Ecology*, invasion scientist Mark Davis published *Invasion Biology*.[175] Davis considered Williamson's 1996 *Biological Invasions* textbook to be outdated in light of the massive amount of new research and *Invasion Ecology* to be introductory and targeting new students, whereas he saw his primary readership as graduate students and professional biologists. Although forswearing an attempt to be comprehensive, Davis cited more than a thousand references and covered to greater or lesser extent all the topics one might expect in an introductory invasion science course. Although Lockwood and colleagues often provided more explanatory background than Davis did when discussing similar topics and had many more figures, for some subjects (e.g., the spread of newly arrived introduced populations) their treatment was the more advanced. Davis used a wealth of examples, including many not often cited, to address how non-native species get introduced, why some species tend to be more frequent invaders, whether certain types of habitats are more or less likely than others to be invaded, the wide array of types of impacts on native species and ecosystems, and how invasions are managed.

Remarkably, his last four chapters detailed a litany of complaints against the field, culminating in a plea to abolish a science focused on invasions and instead to fold it into a broader science of biotic changes associated with the movement of species, whether by humans or under their own steam. His underlying complaint was the dichotomy that invasion scientists draw between native and non-native species. Instead, he perceived a continuum of nativeness, arguing that the dichotomy misleads scientists and managers into wasting resources fighting innocuous non-native species. Davis also assailed military metaphors common in invasion science (discussed in chap. 18, sec. 18.2) on the grounds that they exacerbate the dichotomous perspective. Though he did not explicitly imply that invasion scientists and managers are xenophobes, as he did in a publication two years later, he recognized "distinct nativism tendencies" in the field.[176]

10.3 Popular Books

Several books targeting a popular readership followed Charles Elton's *The Ecology of Invasions* (1958) in the two decades leading to the SCOPE project (as discussed in chap. 7, sec. 7.2). The explosion of interest in invasions following SCOPE was reflected in a great increase in newspaper articles on invasions (as discussed in chap. 1) and also a continuing flow of books for lay readers. Several have been broad-brush depictions of the invasion phenomenon as a whole, focusing on a particular set of cases to examine the larger gamut of impacts and issues associated with invasions.[177] Their subtitles telegraph the scope and urgency of the problem: *How Invasive Species Are Transforming the Planet—and Ourselves*; *The Growing Threat of Species Invasions*; *Bio-Invasions and Their Impacts on Nature, the Economy and Public Health*; *Un danger pour la biodiversité*. Their covers often feature newsworthy invaders: bullfrog, cane toad, giant African snail, giant hogweed, Norway rat, sea walnut, brown tree snake, boar.

As certain invasions became widely known, they featured in books for the broad public, including on the cane toad in Australia, the Nile perch in Lake Victoria, the killer alga in the Mediterranean, the brown tree snake on Guam, the northern snakehead in Maryland, the chestnut blight in eastern North America, and the common wasp in New Zealand.[178] Other authors focused on the entire range of invasions in a particular region: Australia, Patagonia, the Laurentian Great Lakes.[179] Carolyn King described invasions of predatory mammals in New Zealand, while Kim Todd detailed a range of non-native species into North America and the motives of the people who introduced them.[180] Still other popular books depicted successful eradication campaigns—William Stolzenburg briefly treated many examples but focused on the eradication of non-native rats and foxes from several Aleutian islands, and Eric N. Muñoz described the eradication of the non-native killer alga from two sites in California.[181] By contrast, several authors described disastrous failed eradication campaigns for a general readership: Joshua B. Buhs recounted the long effort to eliminate the red imported fire ant from North America, Robert J. Spear similarly described the twelve-year attempt to eradicate the spongy moth in Massachusetts, and Jordan D. Marché recapped the more recent invasion of the American Midwest by the emerald ash borer and the state and provincial debacles that attempted to eliminate it.[182] Children's books on invasions have also appeared, including Morris Gleitzman's *Toad Rage* and Patricia Seibert's *Toad Overload* on the cane toad in Australia and a discussion of many invasions in Italy by Piero Genovesi and Sandro Natalini.[183]

Inevitably, as critics of invasion science fed controversies over invasion causes, consequences, and management (such controversies are the subject of chap. 18), man-bites-dog accounts of invasions surfaced for lay readerships. In these volumes, rather than wreaking havoc on nature and human endeavors, non-native species are for the most part portrayed as harmless, often beneficial, and efforts to exclude or eradicate them as misguided. As do writings in the scientific literature, these popular books, though focusing on one of several criticisms on invasion science and management, adduce other criticisms to greater or lesser extents. Most at least hint that the entire enterprise is tainted by xenophobia (as discussed further in chap. 18, sec. 18.1). Among such efforts are those by Garry Hamilton, Emma Marris, Martha L. Molnar, Fred Pearce, Jacques Tassin, and Ken Thompson.[184] Other lay-oriented books, such as those by Tao Orion and Sonia Shah, include similar material in particular chapters while building toward a related theme.[185] Several suggest that non-native species will somehow "save" nature, as announced by subtitles such as *Saving Nature in a Post-Wild World* and *Why Invasive Species Will Be Nature's Salvation.*

: : :

In addition to the profusion of journals explicitly focused on biological invasions or those that, though not explicitly so focused, published primarily on non-native species, many journals in other fields (e.g., ecology, evolution, global change science, conservation science, biogeography) published an increasing number of articles on invasions in the wake of the advent of modern invasion science. Both invasions-focused journals and invasion articles in other journals particularly increased in the new millennium. Because the rapid growth of invasion science coincided with the period during which the internet matured and became a dominant form of communication, many newsletters on invasions briefly flourished in more or less polished printed form but then disappeared. Their raison d'être—transmission of information on invasions—evaporated with the advent of invasion journals and websites on the science and management of non-native species.

Throughout the growth of invasion science, a split has persisted between literature aimed at scientific understanding per se of invasions—where, when, and how they occur, what impacts they cause—and that aimed at managing them, through both policy and actions on the ground. An entire field devoted largely to managing invasions—economic entomology—arose in the mid-nineteenth century, and the *Journal of Economic Entomology*, discussed previously in section 10.1, remains a leading journal in that field.[186] The focus is still how to destroy or manage insects causing economic damage, a focus distinct from those of journals more broadly treating entomology or subdisciplines such as insect physiology or social insects. Many newsletters that proliferated in the late twentieth century mainly dealt with management, sometimes management of particular non-native species, such as the

Chromolaena odorata Newsletter, other times management of large groups of non-native species, such as *Wildland Weeds*. The two biological control journals, *BioControl/Entomophaga* and *Biological Control*, perforce focused on management from their beginnings. And many of the websites that replaced newsletters feature management of non-native species.

As academic journals arose specifically devoted to invasion science, their focus and most of their articles were resolutely on "pure science"—the biology of non-native populations, temporal and geographic trajectories of invasions, invasion impacts, and the like. However, the founding editors of these journals, hoping to bridge a recognized gap between academic scientists and managers, explicitly solicited manuscripts on management.[187] Nevertheless, the academic journals published rather few management-oriented papers, leading to the initiation of journals specifically devoted to management: the Weed Society of America journals (*Weeds*, *Weed Science*, *Weed Technology*, *Invasive Plant Science and Management*) and *Management of Biological Invasions*. However, the latter journal from its initiation and continuing until today publishes some papers on the science of invasions quite similar to those of the *Biological Invasions* and *NeoBiota* and not closely tied to management.

Another trend has been the shift in publication of first records of non-native species in particular places. Initially these appeared in regional or taxon-focused journals not specifically focused on non-native species. As newsletters proliferated, many of these reported first records as well as range extensions of previous invaders. As newsletters were superseded by websites, many of the websites reported first records and some recorded all locations in the non-native range where a species has been found. At its founding in 2006, only *Aquatic Invasions* among academic journals on invasion science routinely published first records and range extensions, but with the advent in 2012 of its sister journal *BioInvasions Records*, designed as an outlet for such reports, the number of such reports in *Aquatic Invasions* rapidly declined. *BioInvasions Records* continues primarily to serve this purpose.

The trajectory of concern with invasions and research on them, and especially the coalescence of research into the modern field of invasion science beginning with the 1980s SCOPE project, is reflected in college ecology textbooks. Textbooks through the first sixty years of the twentieth century, if they mentioned invasions at all, cited few of them and did so simply to exemplify ecological processes or principles. An exception, a midcentury plant ecology textbook by Henry J. Oosting, cited many invasions, but for the most part he did not use these as examples to explore ecological phenomena. Rather, he pointed to them as cautions about one way—not widely noted at the time—in which human activities were greatly changing many natural ecosystems. Most modern ecology textbooks arriving on the scene in the early 1970s devoted more space to invasions as examples of ecological phenomena. This increase probably reflected partly the increasing number of consequential invasions reported in both the scientific and popular literatures (as discussed in chapters 5 and 7) and perhaps also recognition of Charles Elton's 1958 catalog of types of invasion impacts and synthesis of impacts as a global change, not simply a series of unrelated cases. As ecology textbooks underwent revision, invasions played an increasing role, and after the SCOPE publications, invasions often merited their own sections in ecology textbooks. In the twenty-first century, as global changes (especially climate change) became leading topics in both popular media and scientific programs and literature, ecology textbooks often tagged invasions as an important global change.

College textbooks in conservation biology appeared much later than ecology textbooks because modern conservation biology as a distinct science evolved in the late twentieth century. Leading textbooks incorporated invasions at the outset and increased coverage through successive editions. However, the emphasis differed between ecology and conservation biology textbooks. Whereas ecology texts used invasions to exemplify ecological phenomena at population, community, and ecosystem levels, conservation texts pointed to them as threats to biodiversity and were thus largely focused at the population level—for example, non-native species that change critical habitat for an imperiled native species, that prey on or hybridize with imperiled species, and the like. This focus was maintained even when a particular non-native species may have had community-level and ecosystem-level impacts, as was sometimes the case for invaders that changed habitat. Finally, as invasion science grew in scope and prominence and invasions received increasing media attention, textbooks appeared that focused on the science itself and its various facets, not just particular invasions to exemplify ecological phenomena or conservation problems.

Although many invasions were cited through the years, famous ones were repeated in various ecology, conservation biology, and invasion science textbooks and in successive editions of evolving textbooks, such as rabbit impacts and rabbit interactions with myxoma virus in Australia, the control of prickly pear cactus by the cactus moth in Australia, and chestnut blight in North America. As other impactful, highly publicized invasions occurred later, they were also included and often highlighted in multiple textbooks. Examples include the invasions of the zebra mussel in North America and the killer alga in the Mediterranean and California.

The proliferation of journals treating biological invasions, the increasing prominence of invasions in ecology and conservation biology textbooks, and the snowballing number of newspaper articles about invasions all were matched by a flurry of books targeting a lay audience. Some focused on prominent invasions, others on invasions in particular regions, and still others on successful or unsuccessful eradication projects. Controversies over aspects of modern invasion science arose as the field exploded in the wake of the SCOPE program, and several books aimed to capture a general audience by adopting and elaborating on contrarian views of the field propounded by various critics.

PART II

A Rapidly Growing Science Expands and Evolves

11

Invasion Science Embraces Evolution and Genetics

Although evolution and the genetics that underpin evolution are now a major part of modern invasion science, a delay occurred between the Scientific Committee on Problems of the Environment (SCOPE) program of the 1980s that initiated the modern field (as discussed in chap. 9, sec. 9.1) and the substantial integration of evolution and genetics into invasion science. We must be clear on the biological meaning of "evolution." "Descent with modification," as Charles Darwin defined it, is accurate, but in his day the mechanisms of descent were not understood.[1] The rediscovery of Gregor Mendel's laws of heredity in 1900 established that genes are the bearers of heredity. This has led to textbook definitions of "evolution" such as "changes in the genetic constitution of a population with the passage of each generation," which, though also correct, may not immediately capture what biologists have in mind when they say a change is evolutionary.[2] The important point is that some change over several generations in morphological, physiological, or behavioral traits must have a basis that is at least partly hereditary in order to be evolutionary, and not simply be a phenotypic change that the environment could have induced. Importantly, recent research on invasions suggests that some adaptive changes do not entail changes in frequencies of certain genes (as detailed subsequently in section 11.3.2).

In the late nineteenth and early twentieth centuries, observations on non-native species were adduced as evidence both for and against Darwin's conception of natural selection as the force driving evolutionary change. For instance, in 1887, Wilhelm Detmer argued that Lamarckian inheritance of acquired characteristics explained why deciduous sour cherry from

Europe became evergreen when planted in what is now Sri Lanka.[3] Similarly, Edmond Bordage in 1910 attributed to inheritance of acquired characteristics the facts that deciduous peach transplanted as saplings or planted as seeds from Europe to Île de la Réunion in the Indian Ocean became semideciduous or evergreen, and that the semideciduous or evergreen habit persisted in trees produced by seed from transplanted trees.[4] Evolutionary changes in non-native species were cited in the evolution and genetics literatures in the early twentieth century; such accounts became more common by midcentury, and they were quite frequent around the time of the SCOPE program. However, this research was peripheral to the developing science of invasions, which was basically ecological. Rather, research on non-native species was used to exemplify and elucidate aspects of the evolutionary process, especially the potential speed of evolution. As an early example, Darwin suspected a rat population in the Galápagos Islands had evolved from non-native Norway or black rats, and he similarly suggested that rats on Ascension Island were introduced and had evolved (as described in greater detail in chap. 3, sec. 3.2).

11.1 Early Research on Evolution of Non-Native Species

In 1919, Joseph Grinnell pointed to a recently arrived population of Old World house sparrows in Death Valley, California, as "a particularly convincing 'experiment' already under way, of just the sort called for by certain critics of the work of the systematist and distributionist, which in time would test the question of the evanescence versus the relative permanence of characters of the category commonly viewed as subspecific."[5] John B. Calhoun in 1947 responded to Grinnell's suggestion, measuring morphological traits in house sparrows collected at different times in different regions of the United States, where the sparrows had been introduced from England in the 1850s. He found that American sparrows had quickly become larger than their European forebears, that this change was hereditary—that is, genetically based—and that birds in different parts of the United States differed from one another.[6] Richard F. Johnston and Robert K. Selander in the 1960s and 1970s added much detail to an analysis of American house sparrow morphological variation (chiefly size variation, based on skeletal traits) and concluded that the variation was at least largely driven by evolution and that the evolution was adaptive, rather than a consequence of a random phenomenon.[7] They wrote:

> Clearly, our thinking must not exclude the possibility of animals attaining to extremely rapid rates of evolution at the racial level.[8]
>
> . . . size variation in these birds is adaptively organized.[9]

The alternative explanation that Johnston and Selander were rejecting, that the change was random, referred to the fact that small populations, such as those that often initiate a species introduction, may differ greatly genetically from the native source population or evolve to differ genetically, or both. The initial difference is often referred to as a *bottleneck* or *founder effect*, and the evolution leading to differences is known as *genetic drift*. These phenomena can all be shown mathematically to derive from small population size.[10]

In the early 1950s, anatomists E. H. Ashton and Solomon Zuckerman studied the African green vervet monkey on the Caribbean island of St. Kitts, where it was introduced as a pet with the slave trade in the seventeenth century and quickly escaped and established a population. Using cranial and dental traits, they found strong evidence of substantial evolution in less than three centuries, and they argued that the changes reflect natural selection, not drift or a founder effect.[11] They emphasized the great speed of this evolution: "The speed at which the changes in the St. Kitts monkey have occurred appears to be from 5,000 to 10,000 times the estimated rate of evolution in the teeth of horses over a period of from 5 to 16 million years."[12]

Herbert G. Andrewartha and L. Charles Birch, Australian ecologists, described apparent evolution in a prominent insect pest, the European corn borer, after its introduction to North America.[13] In its European native range, it has one generation per year in northern areas and two generations per year in the southern region, and this pattern ("univoltine" vs. "bivoltine") persists if individuals are reared in different climates, implying a genetic basis that likely arose by natural selection operating over long periods. In North America, the corn borer was first reported as an introduced pest in 1917 near Boston, where it has two generations per year.[14] Soon thereafter, it was found in the state of New York, where it has one generation per year. Insects experimentally moved from New York to Massachusetts remained univoltine for four years, suggesting a genetic basis. As the species gradually spread through the Midwest beginning in the 1920s, at first populations were strictly univoltine, but the proportion of the population that was bivoltine gradually increased, reaching 75 percent in some regions. From these observations plus experiments, Andrewartha and Birch concluded that the corn rootworm was introduced twice to North America from two sources, that the bivoltine form was never introduced to the Midwest or else failed to survive there, and that the bivoltine form there now had evolved there from the univoltine form. Subsequent research suggests that the voltinism differences may be partly maintained by the fact that the species in North America has two forms with different sex-attractant pheromones.[15] Andrewartha and Birch suggested a similar basis—evolution in the non-native range—for North American voltinism patterns in another European species, the spruce sawfly, introduced in the late nineteenth century.[16]

Elwood C. Zimmerman, an entomologist in Hawaii, exploited a different sort of evidence to support the notion of rapid evolution.[17] Polynesians colonized the Hawaiian Islands in the thirteenth century, bringing plant and animal species (as discussed in chap. 2, sec. 2.1).[18] Among these were bananas, of which no native species or relatives grow within thousands of kilometers of Hawaii. Since bananas arrived, several native Hawaiian moth species of the genus *Hedylepta* have evolved to feed on this plant and no other. Zimmerman traced the origin of these banana-feeding moths to a native Hawaiian congeneric species that feeds on an endemic Hawaiian palm that had been common but was largely eliminated in lowland areas of Hawaii by human activities such as fire and cultivation. Zimmerman hypothesized that, as the palm population dwindled, moth species that had fed on them increasingly fed on non-native banana and coconut palm and that natural selection had favored individuals that survived and reproduced best on the new host plants. Over time, this process led to species that feed only on banana.

British geneticist R. J. "Sam" Berry conducted copious research on the genetics and evolution of island species, especially mice, and several of his studies featured rapid evolution of non-native species. House mice were introduced in the 1890s from the nearby mainland to the 100-hectare (247-acre) island of Skokholm, about three kilometers (just under two miles) off the Welsh coast. Berry studied skeletal traits, weights, and allozymes (different versions of an enzyme caused by mutational changes in the underlying DNA coding sequence) of Skokholm and mainland mice and found substantial differences between the island and mainland populations.[19] The Skokholm mice were much larger than those of the mainland. The skeletal traits were not measurements, such as skull lengths, but rather thirty-five "nonmetrical" characteristics such as presence or absence of a particular tooth or a foramen in a bone; these are assumed to have a genetic basis (though the connection of a trait to the underlying gene or genes is not as clear or certain as for allozymes), whereas size measurements such as weights and that of some bones are known in many instances to be affected by the environment. Thus, at least some differences between mainland and Skokholm mice are evolutionary. Although Berry thought that aspects of the evolution were adaptive, he attributed many characteristics to a founder effect and believed that natural selection had acted quickly to increase genetic variation.[20]

Berry also studied house mice of the Faroe Islands, midway between Iceland and Norway.[21] These mice were long known for their immense size. For instance, William Eagle Clarke reported that, "indeed, they are veritable giants, being considerably larger than the type, and, I believe, also than of any of its numerous geographical races hitherto described."[22] They are also famous for having evolved very rapidly.[23] They were brought to the archipelago, doubtless inadvertently, probably by the Vikings no more than a thousand years ago, and, for certain islands that had not been colonized initially, possibly as recently as two hundred years ago from previously colonized islands.[24] Remarkably, the separate islands all have distinct "races" based on measurable and nonmetrical skeletal traits and allozyme frequencies.[25] Berry believed that, even though some of the distinctive features may have evolved by natural selection, the initial populations were very distinct from mainland populations because of a random founder effect.[26]

The British First Fleet brought rabbits to Australia in 1788, and the population was supplemented with about a dozen wild English rabbits in 1859 (as detailed in chap. 4, sec. 4.1; see also box 11.1). They spread by 1920 to most of Australia south of the Tropic of Capricorn.[27] By the 1970s, regional variation in color pattern, morphology, allozyme and blood group frequencies, and resistance to the disease myxomatosis were recognized and attributed at least partly to genetic change—that is, evolution.[28] The remaining question for Australian rabbits is not whether the changes are evolutionary but the extent to which they are adaptive (driven by natural selection) or simply reflect a founder effect or genetic drift, but for most of these changes an adaptive explanation is plausible.[29]

BOX 11.1 Rabbits in Australia

Joel M. Alves and colleagues, using several molecular genetic techniques, recently showed that, despite several other introductions, including by the British First Fleet, rabbits in Australia today are overwhelmingly descended from those few wild individuals introduced in 1859 by Thomas Austin, an English settler hoping to establish a wild population for hunting.[1] The other introductions were of domestic rabbits, but Austin's were trapped in the wild in southwest England. The small number of individuals Austin introduced produced a severe genetic bottleneck—probably between three and five females founded almost the entire Australian rabbit population. Furthermore, the initial introduction from mainland Europe to Great Britain in the thirteenth century had already reduced the genetic diversity of the British rabbit population relative to those in the native range. Alves and coworkers suggest that the genetic features of the introduced wild rabbits adapted them for invasion in some way that was not available in the genomes of the domesticated rabbits, but what that adaptation was remains undetermined.

1. Alves et al., "Single Introduction of Wild Rabbits."

Striking among the aforementioned examples of interest in evolution of non-native species is that all were focused on questions about evolution and its genetic basis—especially the speed of evolution—rather than on ecological impacts of invasions, the subject of Charles Elton's 1958 monograph (discussed in chap. 7, sec. 7.1) and the SCOPE program (the topic of chap. 9).

This was even true when the subject of the research was well known for its ecological impacts, such as rabbits in Australia or the European corn borer in North America. Elton in 1930 wrote a small book focused on the interaction of ecology and evolution, but he was an ecologist and rarely addressed how evolution could affect invasions or vice versa.[30] In his 1958 monograph, *The Ecology of Invasions by Animals and Plants* (*EIAP*), he mentioned evolution only with respect to the rapid evolution of resistance to insecticides in non-native pest insects. David Pimentel suggested a general genetic mechanism by which evolution would lessen impacts of invasive predators, parasites, or herbivores on their prey or host populations, to which Elton responded with skepticism (as detailed in chap. 7, sec. 7.1).[31] Marston Bates, in his contribution to *Man's Role in Changing the Face of the Earth*, advocated expanded research on evolution of non-native populations, but this paper had little if any influence on the development of modern invasion science (as mentioned in the introduction to chap. 7).

Evolutionists and geneticists interested in evolution of non-native species were not closely associated with ecologists' developing interest in invasions outlined in chapter 6. An international workshop held at Asilomar, California, in 1964 assembled evolutionists, geneticists, and ecologists "to present facts and exchange ideas about the kinds of evolutionary change which take place when organisms are introduced into new territories."[32] Of twenty-seven papers, only three authors cited Elton's *EIAP*, two ecologists and one evolutionist.[33] Edward O. Wilson included Elton among authors who found few generalizations about non-native species. Birch pointed to the fact that *EIAP* dealt with impact but not evolution. And evolutionist Ernst Mayr simply noted the presence of several bird invasions in *EIAP*. Few authors, and almost no evolutionists, cited earlier ecological literature on invasions of the sort noted in chapters 4 and 6, although the ample discussions following each chapter, recorded from the workshop, delved into ecological as well as evolutionary matters.

Even though several authors in the books from the SCOPE program cited the book arising from the Asilomar workshop, it took over a decade for research along evolutionary and genetic lines to become well integrated into invasion science, but to understand this delay and subsequent integration requires a foray into the rapid, ongoing development of molecular genetics that began in the second half of the twentieth century.[34]

11.2 Advances in Molecular Genetic Technologies

In the examples presented in section 11.1, observed changes in non-native species were phenotypic: body size and shape, voltinism, host plants; even allozymes are phenotypic, as discussed in the following paragraph. Researchers who were claiming that such changes were evolutionary had to cite other evidence that these traits were, in fact, at least partly determined genetically and did not simply reflect phenotypic plasticity—an environmental influence on the phenotype. Often breeding experiments provided such evidence. For some traits, such as animal body size as reflected by body weight, the environment during development was shown to cause differences. This was why some studies aiming to show evolution of body size used skeletal traits, and especially nonmetrical ones, rather than weight.

Allozymes (slightly differing forms of the same enzyme, coded for by different alleles) were discovered in 1957.[35] In 1966, gel electrophoresis was used to separate allozymes: an electric field applied to a gel allowed different molecules to migrate over the gel at different rates, depending on their size or electrical charge.[36] The results reported in these 1966 papers revealed extensive amounts of previously unrecognized genetic variation, revolutionizing research on many aspects of evolution and population genetics, including evolution of non-native species. Allozymes are phenotypic traits—they are not the genetic material itself—but they are immediately and directly formed from the genetic material, and they provided a greatly expanded collection of traits for evolutionary studies. An explosion of electrophoretic research followed, including research on non-native species.[37] For instance, in 1972, Michael Clegg and Robert Allard studied allozyme frequencies at five enzyme loci in slender wild oat, a Mediterranean plant probably introduced to California around 1800. Comparing fourteen California populations to similar data from Mediterranean sites, they concluded that this plant was brought to California several times from a number of European sites.[38] Their work anticipated a later major use of molecular genetic information in invasion science: identifying source regions for particular invaders and the likelihood of multiple introductions.[39] Clegg and Allard also cast doubt on the early assumption that most allozyme variants of the same enzyme were selectively neutral with respect to one another.

In 1973, Selander and Donald Kaufman examined allozymes coded for by twenty-five loci in the decollate snail, introduced from the Mediterranean to North America near Charleston, South Carolina, in the early nineteenth century and which reached the West Coast by 1966. They found no genetic variation within or among American populations, compared to considerable variation among native populations in Tunisia and France, demonstrating that a widespread and apparently successful non-native species succeeded in the apparent absence of genetic variation.[40] They interpreted their finding to mean that adaptation to various environmental conditions occurred primarily or wholly by phenotypic plasticity and not evolution. Peter Brussard in 1975 examined allozymes in eastern North American populations of the grove snail, native to Europe and likely introduced in the eighteenth or nineteenth century, as well as three populations from Wales in the native range. Unlike with the decollate snail, American populations of the grove snail had as much genetic variation as the Welsh ones, which Brussard attributed to both multiple introductions and aspects of their life history (monoecy, outcrossing, multiple matings, viable sperm storage) that allow a single snail to distribute several individual genomes while reproducing.[41] He also used the geographic distribution of particular allozymes to suggest European source regions for particular American populations.

In 1973, William Klitz examined some twenty allozymes in North American house sparrows from thirteen scattered populations and found they were genetically identical both between and within populations.[42] This genetic depauperization contrasted with much genetic variation in native house sparrows in Europe.[43] The relative genetic impoverishment of the North American populations also contrasted with the substantial morphological variation described previously. Similarly, a subsequent allozyme study showed that Australian and New Zealand non-native house sparrow populations have many fewer alleles per locus than native European populations, and New Zealand birds tend to be less heterozygous.[44]

These early forays were harbingers of many studies using allozymes to compare genetic variation in native and non-native populations.[45] Just among birds, for instance, similar allozyme research was conducted on native populations of starlings in Great Britain and non-native populations in New Zealand; on native populations of the myna in India and non-native populations in Australia, New Zealand, Fiji, Hawaii, and South Africa; and on native chaffinch populations in Europe and non-native populations in New Zealand.[46] Sébastien Lavergne and Jane Molofsky used allozymes from the European native range and North American introduced range of reed canary grass to show that the species had been introduced multiple times from different European locations, that the various European genotypes had recombined in North America to produce genotypes not found in the native range, and that some of these genotypes were particularly invasive.[47]

If allozymes are considered the first generation of molecular genetic techniques applied to biological invasions, new molecular genetic techniques to examine DNA directly were rapidly developed, primarily driven by uses in medicine and agriculture, and were quickly applied to non-native species and largely superseded the use of allozymes. First off the block among new techniques was use of restriction fragment length polymorphism (RFLP), in which enzymes are used to break a DNA sample into fragments, which are then separated by size by gel electrophoresis.[48] The development by Kerry Mullis and colleagues in 1985 of the polymerase chain reaction (PCR), which permits rapid multiplication of DNA fragments, gave great impetus to the new genetic techniques.[49] The combination of PCR and RFLP was used to study non-native species. For example, Albert Weathersbee III and collaborators combined PCR and RFLP to distinguish egg masses of a destructive weevil pest of horticulture and agriculture (*Diaprepes abbreviatus*), introduced to Florida from the Caribbean, from those of the relatively innocuous native blue-green citrus root weevil.[50]

Another method of breaking DNA into fragments and comparing the fragments by electrophoresis to reveal polymorphisms was proposed in 1990—random amplified polymorphic DNA (RAPD).[51] This method was soon applied to a non-native species, the Argentine stem weevil, introduced from South America to Australia and New Zealand.[52] The analysis suggested that the source of the Australian and New Zealand introductions was probably the Buenos Aires region or nearby Uruguay and that possibly either Australia or New Zealand was invaded first and then served as the source for the other nation. However, RAPDs were often found to be poorly reproducible and normally detected only presence of a locus, not whether the individual is heterozygous or homozygous.[53]

RFLP and RAPD may be considered second-generation molecular genetic technologies. Soon thereafter, however, yet another method of fragmenting DNA and comparing the fragments by electrophoresis to find polymorphisms was proposed in 1995: amplified fragment length polymorphism (AFLP).[54] This method had technical advantages over RFLP and RAPD, especially repeatability, though, as with RAPD, AFLP primarily revealed only presence of a locus and not heterozygosity or homozygosity.[55] Again this approach was quickly used in research on non-native species. Thus Laurent Amsellem and colleagues studied an invasive Asian weed, giant bramble, introduced to Indian Ocean islands, to show reduced genetic diversity in the islands relative to the native range and to propose an initial introduction to Madagascar (possibly several introductions), which became the source of introductions to the other islands and to Australia.[56]

Even before the advent of RAPD, yet another molecular genetic technique using PCR and electrophoresis arose that may, with AFLP, be considered a third-generation approach. Microsatellites, also known as simple sequence repeats (SSRs) or short tandem repeats (STRs), are short pieces of DNA repeated several times in tandem. First discovered in the mid-1980s, they quickly became popular for many purposes, including invasion research, and generally replaced RAPDs after the 1990s.[57] A striking paper by Neil Tsutsui and colleagues reported using microsatellites of the highly invasive Argentine ant in its native South American range and in California to propose that one possible reason for its competitive dominance in California is reduced genetic variability, probably associated with a bottleneck at introduction, and reduced aggression between nests, possibly owing to the relative genetic uniformity.[58] The reduced aggression allows the formation of a massive "supercolony" stretching over 1,000 kilometers (621 miles). Just two years later, two enormous European supercolonies of the Argentine ant were identified, and in this case, microsatellites showed a bottleneck, but not a severe one, suggesting that reduced genetic diversity is not the only or possibly even main cause of invasive supercolony formation.[59] Microsatellites were also employed to demonstrate that a notorious marine invader, the warty comb jelly, invaded the Black and Caspian Seas from within or near the Gulf of Mexico, whereas the source of the later invasion of the North and Baltic Seas was New England.[60]

Another third-generation technique using PCR involves mitochondrial DNA (mtDNA), discovered in the mid-1960s as part of the research showing that mitochondria are endosymbionts (organisms that live within the body or cells of another organism, usually in a mutualistic relationship).[61] By the 1990s, mtDNA became the most widely used type of genetic marker in ecological and evolutionary studies and was frequently used in invasion research, not least because its high mutation rate allowed inferences about population histories over relatively short periods.[62] A mitochondrial gene, cytochrome *c* oxidase I (COI), became the key gene whose sequence is the basis for the "Barcode of Life" that is increasingly used to identify and distinguish species, including non-native species.[63] Among much research using mtDNA to infer sources and routes of invasions, Jason Kolbe and colleagues found that brown anoles from different parts of Cuba have almost exclusively mtDNA of single haplotypes (groups of genes that tend to be inherited together) and that these haplotypes differ from site to site.[64] However, populations in most areas of Florida, where the brown anole has been introduced, have several haplotypes. Thus it appears that multiple introductions from Cuba to Florida have occurred, so populations in Florida, far from being genetically depauperate, have more genetic variation than populations in the native range, contradicting conventional wisdom. This study and others showing multiple introductions were major factors in at least partly resolving the "genetic paradox of invasions," the observation that, despite what was assumed to be reduced genetic variation of any non-native species in its introduced range (e.g., the aforementioned papers implicating a bottleneck effect), some non-native species not only persist but thrive and become invasive, whereas genetic poverty should enhance inbreeding depression and maladaptive genetic drift.[65]

The Human Genome Project was the main force behind development of yet another type of genetic marker, single-nucleotide polymorphisms (SNPs) which consist of sites in the genome in which two (usually) or more nucleotides are present in non-negligible frequencies (say, > 1%) in a population.[66] The Human Genome Project revealed hundreds of millions of SNPs, and once various "next-generation" DNA sequencing methods became accessible and relatively inexpensive shortly after the turn of the millennium, SNPs became an increasingly prominent tool for inferring aspects of population history in many areas of evolution and ecology, partly because they are so much more numerous than microsatellites.[67] Eventually they were used to investigate invasions. For instance, Benjamin Fitzpatrick and colleagues used SNPs to show rapid fixation of alleles from a non-native tiger salamander introduced into the range of the native California tiger salamander and hybridizing with the native one.[68] Oriol Vidal and associates employed SNPs to build on previous research, which used mtDNA, to show that European populations of a North American mosquitofish probably arose from several introductions, had undergone bottlenecks, and were differentiated into five distinct population groups.[69] And a team led by Katrien Vandepitte used SNPs to show that quick genetic adaptation of the Pyrenean rocket, a plant introduced from northern Spain and southern France to Belgium in the early nineteenth century, was likely instrumental in its abrupt spread into the Netherlands beginning in the 1980s.[70]

In fact, the rapid development of accessible next-generation sequencing methods and their application to markers such as SNPs, mtDNA, and microsatellites may be considered the fourth generation of molecular genetics technologies.[71] Some have argued that previous methods using gel electrophoresis will be increasingly rendered obsolete by improvements in sequencing techniques, although a tendency for many researchers to use the methods they have familiarity with combined with the remarkably rapid evolution of ever more powerful molecular genetic technologies ensures that, at least in invasion science if not in medical sciences, many older methods will continue to be used for some time, alone or in combination with newer approaches.[72] Carvalho and Avise summarized advantages and disadvantages of using the various molecular genetic techniques.[73]

In sum, in about forty years, rapidly evolving molecular genetic technologies driven primarily by medical and agricultural interests provided a new, powerful way to study many aspects of evolution and ecology. Invasion scientists quickly explored each new approach, often collaborating with molecular geneticists who had not previously engaged with invasion science but were eager to use their expertise to explore new and interesting questions.

11.3 Evolution and Invasion Science in the Modern Era

Research on evolution of non-native species dramatically increased with the advance of molecular genetic techniques. Interaction between students of evolution and students of invasions in the modern era became two-way. Just as non-native species were often studied before the 1980s to elucidate evolution (especially the speed at which evolution could occur), they continued to be studied by many evolutionists not primarily for insights into invasions per se but for insights into the evolutionary process. However, additionally, both alone and also often in collaboration with evolutionists and geneticists, researchers interested in ecological aspects of invasions increasingly sought patterns and evidence in genetic and evolutionary data. Many researchers ranging from those with backgrounds ranging from genetics to ecology even saw themselves as practitioners of "invasion genetics," a new subdiscipline of invasion science.[74]

11.3.1 Insights into Evolutionary Processes and Patterns

Research on non-native species to elucidate evolution expanded well beyond examples of evolutionary rates, using morphological and other phenotypic data, mathematical models, and, increasingly, empirically derived data from molecular genetics. The breadth was captured in a 2005 book, *Species Invasions: Insights into Ecology, Evolution, and Biogeography*.[75] Part 2, "Insights into Evolution," contained six chapters addressing a variety of evolutionary questions. Raymond Huey and collaborators examined rate and predictability of evolution using examples from *Drosophila* fruit flies and from salmon.[76] A key point was the implication that observed rapid evolution in these cases can be adaptive rather than simply reflecting a bottleneck effect or genetic drift, or both. For *Drosophila*, Huey and colleagues pointed to evolution of chromosomal inversion frequencies and wing size in non-native populations, citing earlier research showing not only rapid evolution of wing size of non-native populations of *Drosophila subobscura* but that different parts of the wing evolved rapidly in non-native populations in North and South America to produce similar overall wing size clines.[77] The similarity of wing size clines in different *Drosophila melanogaster* populations had been noted as early as 1975, but the fact that some populations were non-native was not used to address the implications with respect to the adaptive nature of the rapid evolution.[78] The salmon examples cited by Huey and colleagues relied on morphological traits and on egg number and also depicted rapid evolution in non-native populations that could largely be interpreted as adaptive.

Robert Ricklefs's chapter in *Species Invasions* used phylogenies of Antillean land birds based on molecular genetics plus observed species ranges to support the validity of the concept of a "taxon cycle," the notion that island species pass through predictable phases of geographic expansion and contraction.[79] Stephen Novak and Richard Mack cited their own data and that of other authors generated through various molecular genetics methods to explore the degree to which non-native plant species undergo a bottleneck.[80] They found many examples of bottlenecks, but the plant mating system and the introduction history (especially whether multiple introductions had occurred) greatly influenced the degree of reduction of genetic diversity. John Wares and associates reviewed much molecular genetic research to ask how introduction affects genetic diversity, evolution of phenotypic traits, genetic fitness, and population differentiation.[81] They were particularly interested in how research on invasions could elucidate the relative importance of natural selection and drift, and the circumstances in which population differentiation leads to speciation. Robert Holt and colleagues used mathematical models to suggest that the trajectories of biological invasions could help test evolutionary theory on the concept of "niche stability," the notion that it is difficult for a population to evolve to use an ecological niche substantially different from its current one (a concept discussed further in chap. 14, sec. 14.4).[82] William R. Rice and Dov Sax described how invasions could elucidate the process of sexual selection (in which members of one sex choose mates of the opposite sex and compete with members of their own sex for access to mates), the evolution of reproductive isolation, and the long-standing question of the selective advantages of sexual as opposed to asexual reproduction.[83] They cited existing studies (and proposed new ones) using morphological and life-history traits, molecular genetics, and biogeographic range data on non-native populations to argue that invasions as uncontrolled "natural experiments" could expand the inferences derived from small-scale, controlled, manipulative laboratory and field experiments by virtue of their natural setting and larger spatial and longer temporal scales.

This group of contributions to *Species Invasions* exemplifies the scope of evolutionary questions to which research on invasive species was applied in the modern era of invasion science, but they do not exhaust the application of research on invasions to elucidate evolution. For instance, Emily V. Moran and Jake Alexander explored the extent to which evolution of non-native species can serve as guides to evolutionary changes that native species might undergo in response to global climate change.[84] Studies of invasions of several distinct plants and animals have played a prominent role in a flurry of research on the role of phenotypic plasticity, and particularly its importance relative to rapid natural selection of adaptive genotypes in a new or drastically changed environment. Among non-native species for which phenotypic plasticity appeared to play a role in invasion establishment, persistence, and spread are salt cedar and tufted knotweed from Asia, spotted knapweed from Eurasia, and the brown anole from Cuba in North America; three non-native aquatic plants in New Zealand lakes and rivers; common selfheal, native to the Northern Hemisphere, in Chile; the New Zealand mud snail in the western United States; and the Mediterranean fruit fly in South Africa.[85] Insights from such studies have deepened understanding of the nature of phenotypic plasticity and its potential importance when a population is confronted with a changing environment, as happens, for instance, with global warming.[86] An important point with respect to invasions is the observation that, to the extent that limited genetic variation owing to the bottleneck effect constrains adaptive evolution when a small propagule invades a new site (but see the following subsection 11.3.2), adaptive phenotypic plasticity can allow a population to persist until mutations or gene flow diversify the gene pool.

As another example of new insights into the evolutionary process, research on the western corn rootworm, a devastating beetle pest of maize (corn) in the United States, shows that it has evolved in several ways that have adapted it to a new host plant and thereby complicated management. The species is native to Central America and spread northward with the cultivation of maize, probably reaching northern US states after AD 700 and New England by 1400.[87] During most of the twentieth century, the species was controlled by rotating maize and soy, an Asian plant, because rootworm eggs, laid on or in the soil, were not laid outside cornfields and larvae did not survive on soy roots.[88] Almost all eggs remain dormant for only one year. However, in the mid-1990s, major damage to maize from western corn rootworm occurred in crop-rotated fields. This damage was due to the evolution of a genotype that is attracted to soy, feeds on soy, and then oviposits in soy fields.[89] The eggs then hatch the next year, when the fields are planted with maize, and the larvae attack the maize roots. This "rotation-resistant" genotype thrives on soy, which is normally a much poorer host than maize is for western corn rootworm, at least partly by virtue of a shift in its bacterial microbiota, an interesting example of adaptation between a host and its microbial symbionts.[90]

The aforementioned examples are but a sample of the ways in which the recent great enhancement of research on evolution of invaders has provided insights into the processes of evolution in general, and new findings on this active research front appear every year. Spencer Barrett and associates, reprinting a special issue of the journal *Molecular Ecology*, and the contribution to that special issue by Dan G. Bock and colleagues, provided a sampling of other aspects of evolution now studied through invasion research.[91]

11.3.2 Insights into Invasion Processes, Impacts, and Management

As noted previously, although evolutionists participated in the SCOPE program, research discussing evolution was scarce in the invasion literature for over a decade. The first monograph on the subject did not appear until 2004.[92] The journal *Biological Invasions* was initiated in 1999, publishing 33 articles that year (as discussed in chap. 10, sec. 10.1). Of these, only 3 featured evolutionary or genetic issues. By contrast, a decade later, the 2009 volume with 190 articles (plus one special issue concerned with a nonevolutionary subject) had 29 articles featuring evolution or genetics. In 2018, the journal published 246 articles, of which 35 dealt with evolution or genetics. These data reflect not only the explosive growth of modern invasion science but also the increase in research on evolution and genetics of invading populations within the discipline. Only 1 of the 3 evolution/genetics articles the first year reported using molecular genetic techniques, while the great majority in 2009 and 2018 relied on molecular genetic data. The increase in such research is understated by these statistics, because much research on evolution and genetics of non-native populations was published elsewhere. For instance, the journal *Molecular Ecology* was first published in 1992 with 35 articles, none of which were on invasions. The 2018 volume contained 309 articles (not counting two special issues inappropriate for invasion-related research), of which 17 dealt with invasions.

Important insights into the processes and impacts of invasions derived from genetic and evolutionary considerations abound. First, the research showing multiple introductions of the brown anole from Cuba into Florida, endowing the Florida populations with great genetic diversity, was just an early example of an increasing number of cases in which molecular genetic techniques have shown multiple introductions, thus resolving, at least for those cases, the genetic paradox of invasions.[93] The same researchers subsequently showed that many invasions by various anole species derive from multiple sources and thus have enhanced genetic variation.[94] Further, in several instances, this wealth of genetic variation from different native sources contributed to the invasive nature of the non-native population, as was demonstrated in the reed canary grass example cited in section 11.2 and in the case of the bladder campion, a plant native to Europe, introduced to North America.[95] The management implication of potential evolution of a non-native species owing to multiple source populations is clear: even if a non-native species is not spreading after initial establishment and appears relatively innocuous, this does not necessarily imply that preventing subsequent deliberate or inadvertent introduction of propagules is pointless, because such introductions could bring or permit evolution of new genotypes that could allow range expansion and increased impact. The green crab, native to the European and North African Atlantic coast, invaded the Atlantic coasts of New York and southern New England in the nineteenth century, spread to the Gulf of Maine in 1905, and was restricted to this region through the 1990s, when its range abruptly expanded northward through the Canadian Maritime provinces. This spread was attributed to newly introduced genotypes from northern parts of its native range.[96]

The term "hybridization" can refer to mating between individuals from different populations at any taxonomic level—populations or subspecies within species, or species within genera. The previous paragraph included cases of intraspecific hybridization—that is, mating between conspecific individuals deriving from different source populations (e.g., in the reed canary grass and brown anole examples). Another major finding in the wake of advances in molecular genetics is the far greater frequency than had previously been suspected of hybridization between individuals of non-native species and individuals of related native species and also a number of cases of hybridization between two different non-native species.[97] Such hybridization is a further source of greatly increased genetic variation and has led to evolution of more impactful genotypes.[98] In 2009, *Biological Invasions* even published a special issue entitled "Ecological and Evolutionary Consequences after Invaders Hybridize," with contributions detailing the evolutionary history of invasive hybrids between native species and non-native species, such as common cordgrass, which originated in the 1890s in England by hybridization between smooth cordgrass introduced from North America and native small cordgrass and became invasive initially in England and then when introduced elsewhere.[99] Hybridization between pairs of introduced species is exemplified in the Old World genus *Tamarix* (salt cedar) in North America.[100] Hybridization has even resulted in reproductively isolated invasive new species such as common cordgrass as well as the marbled crayfish, which arose in the German aquarium trade from hybridization between individuals from distantly related populations of the slough crayfish from North America.[101]

An ever-increasing number of cases of interspecific hybridization have been documented because molecular methods are more sensitive to hybridization than traditional morphological approaches and thus reveal hybridization where none had been suspected or show the extent of hybridization to have been more massive than surmised. The mallard from the Northern Hemisphere was introduced to New Zealand in the 1860s. By 1922, some individual ducks appeared morphologically to be hybrids with the native grey duck.[102] In 1985, an index of hybridity of each individual bird based on plumage, leg and bill color, and size traits suggested the fraction of its makeup from each parental species.[103] Judith Rhymer and colleagues used mtDNA in 1994 not only to assess the hereditary background of each individual but also to demonstrate the previously unsuspected fact that hybridization between these two species was leading to two-way gene flow, not only grey duck DNA into the mallard population, as had been assumed, but vice versa.[104] This gene flow imperils the existence of the New Zealand grey duck as a genetically distinct population. A similar case involves the endangered white-headed duck from the Palearctic region, whose existence is threatened in its last western European redoubt in Spain by the ruddy duck, introduced from the New World in 1948 as an amenity to Great Britain, whence it escaped in the early 1950s and spread southward, eventually reaching the range of the Spanish population of the white-headed duck.[105] An aggressive campaign by Spanish authorities to cull ruddy ducks and hybrids relies on a battery of morphological criteria to distinguish the hybrids. However, a molecular analysis employing noncoding nuclear markers (introns), microsatellites, and mtDNA showed that, although the culling program has been generally effective, some individuals are misidentified based on morphological grounds. The culling program and an aggressive effort to eradicate the non-native ruddy duck population in Great Britain to stem the source of hybridizing individuals have thus far staved off the loss of the Spanish white-headed duck population as a distinct species.[106]

Interest has recently arisen about a role of the hereditary system in invasions that does not entail the arrival of new genes by mutation, gene flow, or hybridization. One such factor consists of what are known as epigenetic changes in gene expression—that is, alterations in gene expression that usually entail processes at the level of transcription from DNA to messenger RNA and are not caused by changes in the DNA sequence.[107] One such epigenetic change is that of methylation, a chemical process that regulates gene expression by switching genes to the "off" position—that is, silences them, so they are no longer doing whatever they had been doing. Christina L. Richards and colleagues found, using AFLP, that Japanese knotweed that had invaded a variety of different habitats in the northeastern United States differed little among habitats in DNA sequences but much more among which genes had been inactivated by methylation. Whether these differences are adaptive or not remains to be determined.[108] However, it is known that novel or stressful environments induce epigenetic changes.[109] This finding leads to the possibility that epigenetic changes such as those found in Japanese knotweed might help an invasive population persist in its new location.

Another feature of the genetic system is the existence of transposable elements (TEs), DNA sequences that change positions in the genome, often creating long sequences of a segment repeatedly copied. TEs make up a large fraction—often approximately one-half—of the genome of many species. TEs are greatly affected by stressful environments, including in ways that might adapt a non-native species to a new environment.[110] For instance, the arboreal sneaking ant, native to Southeast Asia, has been widely introduced globally. In non-native populations of this ant from Japan and Brazil, a large fraction of its remarkably small genome consists of accumulations of TEs, termed "TE islands."[111] These regions of the genome evolve faster than other regions in terms of SNPs. Furthermore, gene expression is lower in TE islands, and methylation of genes is greater in TE islands. These features could facilitate adaptation to new environments.

Molecular genetic tools also permitted the recognition of myriad examples of "cryptic invasions"—that is, previously unrecognized invasions of new genotypes from non-native sources into native populations, non-native species that are very similar morphologically to native ones, and multiple non-native species where only one had been assumed present.[112] Perhaps the best-known example of the first sort of cryptic invasion was demonstrated in 2002 by Kristin Saltonstall, who sequenced chloroplast DNA and showed that the dramatic spread in North America over the previous 150 years of common reed, a wetland plant, had resulted from the invasion into the previously unremarkable native population of the species by an invasive European strain that has even replaced the native population in several areas.[113] The invasive strain likely arrived in ship ballast, and its dispersal was probably facilitated by the construction of railroads and highways in the late nineteenth and early twentieth centuries. This previously unrecognized invasion has had such enormous impacts that it was the subject of an entire special issue of *Biological Invasions* in 2016.[114] Similarly, the genetic constitution of native common carp in Japan has been greatly changed by introductions of non-native domesticated strains from elsewhere in Eurasia.[115] On the west coast of North America, the native bay mussel was replaced in the southern part of its range by the mid-twentieth century by the cryptic invasion of the morphologically similar Mediterranean mussel, as demonstrated by analysis of mtDNA from specimens collected from the late nineteenth through the late twentieth century.[116]

Among several examples in which what had been assumed to be a single native species turned out to be a native species plus one or more morphologically nearly identical non-native "cryptic species" is the sweetpotato whitefly in South Africa.[117] This insect devastates many crop plants by transmitting plant pathogens, and a sudden transition in the late twentieth century in several regions worldwide from relatively minor pest to agricultural scourge is now recognized as having resulted from cryptic invasions of similar non-native species.[118] Among assumed single invasions that turned out, upon molecular genetic analysis, to consist of multiple non-native species, European marine green crabs of the genus *Carcinus* constitute a good example. For both Japan and South Africa, prominent invasions that had been attributed to single species—the green crab—were shown, in fact, to consist of two: the green crab and the Mediterranean green crab.[119] Introduction of the predatory rosy wolf snail to the Hawaiian Islands to control the giant African snail resulted in inadvertent release of a morphologically similar second species, subsequently identified by molecular means.[120]

The previous examples address the matter of evolution of invaders in their new range. A subject of increasing research has been the evolution of native species confronted with a new invader. The corn rootworm case described earlier in this chapter is just one example in which research on a non-native species has shed light on how insects evolve to eat particular host plants. The classic example of evolution of use of a new host is research, begun before the advent of modern invasion science, on how a species native to North America, the apple maggot, adapted to the arrival of a new potential host. Its ancestral host is hawthorn, but around the mid-1850s to mid-1860s, it was found feeding on domesticated apple in the Hudson Valley, and this apple "race" quickly spread. Apple originated by hybridization and cultivation in central Asia and was then brought during the Neolithic to Europe, from which it was introduced to North America by settlers in the sixteenth and seventeenth centuries. In 1969, Guy Bush found differences in ovipositor length, body size, number of bristles, and time when the larvae leave the fruit to pupate in the soil and suggested that the hawthorn and apple races are in the process of evolving to be separate species.[121] In the era of molecular genetics, other differences between flies on apple and hawthorn were found, first in allozymes and a chromosomal inversion, later in mtDNA and nuclear gene sequences, and finally in SNPs.[122] This work continues to inform research on how speciation can be sympatric—that is, two species arising from one without physical barriers obstructing mating.[123]

Another corpus of research on evolution of a native species in response to arrival of a non-native concerns Edith's checkerspot butterfly, a North American species used in much research on evolution and ecology.[124] In 1971, Michael Singer and colleagues began research on the evolution of its host plant preference.[125] Among host plants on which some populations oviposit is ribwort plantain.[126] This plant was introduced from Europe to the western United States in the 1820s. The ability to thrive on the invader apparently existed in at least some individuals of most or all of the butterfly populations before the invader arrived, but the propensity to oviposit on it (and thus for caterpillars to be able to use it as a host plant) was present or arose only in certain populations, and, when present, frequency of oviposition on ribwort plantain rose after the plantain was introduced.[127] Growth rates and survivorship of individuals on different host plants often differ.[128] Analysis of allozymes, nuclear genes by AFLPs, mtDNA, and microsatellites showed that populations can be distinguished, but individuals adapted to different hosts do not form distinct clusters in genetic trees produced by these methods.[129]

Scott Carroll and colleagues studied the evolution of native soapberry bugs in Florida and Australia in response to newly introduced host plants.[130] In south Florida, the red-shouldered soapberry bug uses a native host, the balloon vine. In central and north Florida, the bug also uses the goldenrain tree, introduced from Asia in the 1950s; in many areas, both trees are common. Populations on the different host plants belong to different "host races" that were determined experimentally to be genetically distinct, and they differ greatly in beak length, with beaks shorter for the population on the goldenrain tree, which has much smaller fruits.[131] Individuals of each host race also develop faster on their host species. The new host race also quickly evolved different host preference, physiology, and life-history traits.[132] Analogously, in Australia the native red-eye soapberry bug feeds on the native woolly rambutan but also on showy balloonvine from tropical South America, which became invasive in the 1960s. This balloon vine has much larger fruits than the native host, and the bugs evolved 5 percent to 10 percent longer beaks in just thirty to forty years.[133] Many other native insects use newly available non-native host plants, although in some instances the evidence for evolutionary adaptation is not strong.[134]

Plants have also evolved resistance to non-native herbivores, and prey species have evolved antipredatory behavior and morphology in response to non-native predators.[135] The famous invasion of the cane toad from the American tropics into Australia in 1935 triggered massive impacts on native biota and much research into possible evolutionary responses of various native species to the invasion.[136] Among impacts are mortality of many native predators of the toad, which is highly toxic. The redbellied black snake and green tree snake have become longer, and their gape size has diminished; both changes are probably evolutionary rather than resulting from phenotypic plasticity.[137] The increased body size should decrease effect of the toxin, while the smaller gape prevents the snakes from swallowing the toads. The redbellied black snake also increasingly avoids the toad as prey and is increasingly resistant to the toxin.[138] Another snake, the death adder, is undergoing selection for avoidance behavior toward the toad and for decreased gape size.[139]

A remarkable example of evolution in response to a non-native species involves the Pacific field cricket, native to Australia and western Pacific islands and first recorded in Hawaii in 1877.[140] In its native range, almost all males of this species produce a loud, characteristic call that attracts females as part of a mating sequence; a small percentage are silent.[141] In some cricket populations in the Hawaiian Islands, however, the great majority of males are silent. A North American fly, *Ormia ochracea*, parasitizes several cricket species in its native range. The female flies locate hosts by detecting the male mating call, then deposit larvae that burrow into the cricket, eating the host and ultimately emerging and pupating. This fly, absent from the native range of the Pacific field cricket, was introduced to Hawaii, probably in some introduced American cricket in the nineteenth or twentieth century; no native Hawaiian crickets are sufficiently large to support the growing larval parasite.[142] In Hawaii, and especially on Kauai, where the density of the parasitic fly is greatest, over 90 percent of male Pacific field crickets are silent.[143] The silence is attributable to a mutation favored by natural selection because the silence prevents the lethal parasitism, even though silence makes it harder for females and males to find one another. The mutation, "flat-wing," occurred in a single gene and spread through the population between the late 1990s and 2003, eliminating a morphological structure on the forewings of males that was used to produce the call.[144] Mutant males mate by intercepting females attracted to one of the few remaining singing males.

Much recent research on rapid evolution in short time spans rather than geological eons falls under the rubrics of "eco-evolutionary feedbacks" and "eco-evolutionary dynamics."[145] In these categories are cases in which the ecological context of a population of a species selects for changes in one or more traits of that species, and these changes, in turn, affect the ecological context. The examples given so far in this chapter feature rapid evolution of a non-native species in response to the new ecological context in which it finds itself or of a native species in response to a non-native species to which it is newly exposed. However, the follow-on consequences for the rest of the ecosystem have usually not been emphasized or even explored. As research on evolution became a key component of modern invasion science, several cases of rapid evolution were explored that demonstrate eco-evolutionary dynamics with impacts on much of the ecosystem (discussed further in chap. 12, sec. 12.3), not just the species evolving rapidly and one or a few other species with which it interacts.

One striking example involving evolution of a native species concerns the introduction of the signal crayfish in the 1990s into Enos Lake in British Columbia, Canada, one of several lakes containing a pair of fish species both classified as three-spined sticklebacks. However, species in these pairs, which have evolved over the past few thousand years and can still hybridize and produce viable offspring, differ morphologically in ways that adapt them to different feeding ecologies. One member of each pair is termed "benthic" and feeds on the lake bottom, while the other is called "limnetic" and feeds on plankton in the water.[146] The non-native crayfish in Enos Lake has caused the population of the limnetic form to plummet (perhaps by disrupting the mating behavior of limnetic males), which in turn has led to massive hybridization between the two forms and the population dominance of an intermediate form much more like the benthic form than the limnetic form.[147] This "reverse speciation" has led to a huge shift in the entire zooplankton community, with a cladoceran species normally rare in lakes with both forms now comprising over half the zooplankton individuals in Enos Lake. Total zooplankton density is increased, phytoplankton density is decreased, and decomposition rate is decreased. In an example featuring evolution of non-native species, cutthroat and golden trout introduced in the United States during the first half of the twentieth century to fishless lakes in Wyoming greatly altered the planktonic community.[148] Zooplankton populations crashed and phytoplankton biomass surged. Densities of larger zooplankters progressively declined, and those of smaller species increased. Zooplankton species composition underwent massive changes, with cladoceran and rotifer densities increasing nearly twenty- and thirty-fold, respectively.[149] The plankton community changes in turn led to rapid evolution in total gill raker number in the trout, a trait that is heritable and not phenotypically plastic in these species. Gill raker number increased in both species, constituting an adaptation to the changed plankton community. The increased feeding efficiency on small zooplankton afforded by this evolutionary change is likely, in turn, to change the planktonic community back in the direction of the preintroduction state.

Finally, among Darwin's many writings relating to invasions, his "naturalization hypothesis" is perhaps best known.[150] In *On the Origin of Species*, he pointed to Alphonse de Candolle's statement that far more non-native species belong to new genera than to genera already present in the native flora.[151] Darwin gave an example: "In the last edition of Dr. Asa Gray's 'Manual of the Flora of the United States' . . . out of the 162 naturalised genera, no less than 100 genera are not there indigenous."[152] Curtis Daehler thus coined the term "Darwin's naturalization hypothesis" for the notion that non-native species less closely related to native species are more likely to survive upon introduction.[153] Following two studies that simply examined the genera of non-native and native plants in particular floras, Daehler performed a similar examination for plants introduced to Hawaii and found they conflicted with Darwin's idea.[154]

However, Daehler's 2001 publication, coming just as rapidly developing molecular genetic techniques were revolutionizing systematics, struck a responsive chord with many biologists using molecular data to produce new phylogenies or revise existing ones for innumerable taxa. This inspired a flurry of publications examining whether Darwin's naturalization hypothesis accorded with phylogenies of particular taxa, virtually all constructed recently with molecular genetic data. A Google Scholar search on September 29, 2019, returned thirty-eight publications with "Darwin's Naturalization Hypothesis" or "Darwin's Naturalization Conundrum" in the title, plus another twenty-seven publications citing it in the abstract. Among these, the most widely cited after Daehler is an elegant study of introduced grasses in California using a phylogeny of grasses in California to show that non-native grasses classified as pests are more distantly related to native grasses than are non-native grasses not classified as pests.[155] The authors thus reframed the hypothesis as "exotic taxa less related to native species are more invasive" and noted that the California grass data are consistent with this formulation. The many new publications referencing Darwin's naturalization hypothesis construe the hypothesis in various ways, some using data only on survival of non-native species, others on aspects of spread or impact. Some interpret their data as according with the hypothesis and others see them as in conflict.

Another hypothesis about invasive species and evolution that inspired many publications is the evolution of increased competitive ability (EICA) hypothesis (detailed in chap. 15, sec. 15.4.3). Observing reports of non-native plant species that were larger in their non-native ranges than in their native ranges, Bernd Blossey and Rolf Nötzold suggested in 1995 that the non-native populations, released from pressure from at least some of their natural enemies (herbivores and pathogens), evolved to shunt more resources away from defenses (e.g., chemicals that repel herbivores or make the plant less digestible) and toward increased growth and reproduction.[156] Certainly reports of extraordinarily large non-native plants are often striking. Henry N. Ridley, for instançe, cited a letter from W. T. Locke Travers in New Zealand to Joseph Hooker in England: "You would be surprised at the rapid spread of European and foreign plants in this country. All along the main lines of road through the plain the knotgrass (*Polygonum aviculare*) grows most luxuriantly, the roots sometimes 2 feet in length, and the plants spreading over an area of from 4 to 5 feet in diameter." Ridley noted that "this plant in England is usually but 1 foot across at the most."[157] Blossey and Nötzold buttressed their suggestion with an experiment with purple loosestrife, planting seeds from a site in their native Eurasian range and a site in North America, where the species is introduced, in a common garden. They found that plants grown from North American seeds were taller than those from Eurasian seeds, but that a Eurasian weevil that feeds on purple loosestrife roots performed better on plants from North American seeds.

Because purple loosestrife had been in North America for only about two hundred years, EICA elicited great interest as it suggested a general trend for fast evolution in non-native plants. The inspiration for the idea—the large size of introduced plants—turned out not to be a general phenomenon.[158] Although many plant species are, in fact, larger in their non-native ranges than in their native ranges, at least as many are larger in their native ranges than in their non-native ranges. Those that were larger in the non-native range might have been larger for the reasons suggested by EICA or for other reasons (e.g., phenotypic plasticity). EICA thus spurred several dozen experiments on various plant species to test the idea, and even meta-analyses of the sort described in chapter 13.[159] Results have been mixed, even for the meta-analyses, with one review finding substantial evidence of evolution of non-native plants in twenty-seven studies, though not in the manner predicted by EICA, while another found fairly broad support among thirty studies.[160] It is difficult to get a clear picture of the prevalence of EICA because studies have examined various combinations of growth, reproduction, and defense traits in various ways, rendering it difficult to interpret the causes of demonstrated evolution.[161]

: : :

Well before the advent of modern invasion science, some evolutionists recognized that non-native species elucidated aspects of evolution, especially in the early and mid-twentieth century. The rapid development of an array of molecular genetic technologies beginning in the mid-twentieth century was quickly followed by evolutionists exploiting these techniques to study non-native species in research on evolutionary phenomena. Invasion scientists lagged in using molecular genetic methods to understand particular invasions or invasion phenomena in general. Part of the delay reflected the fact that most leading invasion scientists during and in the wake of the SCOPE program were mainly interested in ecological questions and impacts—as almost all earlier students of invasions had been. However, as molecular genetic tools became prominent in many fields, including ecology, invasion scientists recognized their utility in addressing ecological questions. In addition, as the field matured, evolutionary aspects of invasions were increasingly recognized as interesting in their own right. Also, by the early twenty-first century, graduate students in all areas of biology, including ecology, became familiar with molecular genetic methods in a way that most in the preceding generation had not. These forces combined to produce a rapid growth of evolutionary and genetic study of invasions, and such research became a leading edge of modern invasion science.

12

Impacts, from Populations and Communities to Ecosystems

Through the nineteenth century, observations of non-native species' impacts mostly focused on effects on human endeavors. In the seventeenth century, concern over non-native mammals on St. Helena and Mauritius centered on impacts to food stocks, game animals, and especially agriculture. John Bartram in the eighteenth century deplored the unfortunate effects of introduced plants on agriculture. (Such early historical introductions are discussed in chap. 2, sec. 2.3.) Impacts of nineteenth- and early twentieth-century invasions that aroused great interest (described in chap. 4) were on viticulture (the phylloxera, sec. 4.2), agriculture (the small Indian mongoose, sec. 4.5), forestry (white pine blister rust and chestnut blight, secs. 4.7 and 4.8), the citrus industry (cottony cushion scale, sec. 4.3), and navigation in inland waterbodies (water hyacinth, sec. 4.6).

Far less frequently remarked were impacts that we would now term "ecological," surely at least partly because ecology as a recognized discipline did not exist before the late nineteenth century. Noteworthy in the seventeenth century were François Leguat's suggestion that predation by non-native feral pigs caused native animal populations on Mauritius to decline and Alexandre Guy Pingré's observation that, on Rodrigues, feral introduced house cats were driving native birds, including the endemic solitaire, to extinction (both described in chap. 2, sec. 2.3). In the nineteenth century, while most naturalist-explorers focused on the geographic distributions of non-native species, Charles Darwin, Alfred Russel Wallace, and Joseph Hooker pointed to ecological impacts of various non-native populations, often to exemplify the concept of a "struggle for existence" driving natural selection. They all noted impacts of introduced goats on St. Helena in this regard (in chap. 3, sec. 3.2). As unintended consequences of the activities of acclimatization societies became clear in the nineteenth century, most concern continued to focus on impacts on human enterprises, but a few individuals pointed to impacts on native species, such as Claude-Auguste Millet for sportfish introductions in Europe and Alfred Newton for introductions of mustelids to New Zealand to control rabbits (both described in chap. 5).[1] Among the prominent invasions of the late nineteenth and early twentieth centuries highlighted in chapter 4, those of the spongy moth, chestnut blight, and small Indian mongoose elicited concern about environmental impacts as well as impacts on human activities.

12.1 Ecological Impacts at the Population Level

Many of the impacts noted through the mid-twentieth century affected particular species, such as the solitaire on Rodrigues or the native birds of New Zealand. Theodore Palmer in 1899 lamented the destruction by introduced cats and dogs of two endemic rail species in the Chatham Islands east of New Zealand, as well as the devastation by the small Indian mongoose of several endemic mammals, birds, and reptiles in the West Indies.[2] Later reports described how non-native bass and sunfish caused the decline of an endemic salamander in a Mexican lake and how several species of non-native fishes in three separate water bodies in Nevada had eliminated one local race of a native fish species and threatened populations of other native species by virtue of predation, competition, or both.[3] Charles Elton noted that non-native rats, perhaps aided by feral cats, eliminated the endemic "coot" of Tristan da Cunha.[4]

Predation by a vertebrate—especially a terrestrial mammal like a cat or a stoat or a ferret—features in many of these early cases of a non-native predator preying on native species. Predation is often dramatic and readily observed. One can easily imagine that predation might devastate the population of a native prey species, especially if population decline of that native species had been reported. However, if the individuals killed by a predator would have died anyway by other means before reproduction, or if their death leads to greater reproduction by surviving members of the population, the predation, no matter how dramatic and obvious, would not have caused a population decline. Such mortality is called "compensatory mortality" because the population compensates for the loss (as discussed at the end of chap. 4). This fact first became clear in the mid-twentieth century, especially through theoretical work by Alexander J. Nicholson and extensive consideration of data on predation of vertebrates by Paul Errington.[5] Today many scientific papers on impacts of non-native predators on some particular prey species acknowledge the importance of demonstrating a population impact, and they adduce evidence of varying sorts, such as declines in the prey species correlated with the arrival and population growth of the predator, as for the impact of the brown tree snake on forest birds of Guam or of the Burmese python on mammals in Florida.[6] Media reports targeting a lay audience are often not so cautious and imply a likely population impact.[7]

Competition for resources, such as food for animals or light for plants, is not as dramatic as predation, and whether it is occurring is often difficult to establish. If a type of food used by a native species is not in short supply, then the observation that a non-native species is eating that food need not imply any population effect whatsoever. One would have to establish that the consumption of the food item by the native population is lowered to the extent that its survival or reproduction is affected. Does the native species simply switch to some other food? "Competition" is a term often used very loosely. Darwin hypothesized the general competitive superiority of Northern Hemisphere species over those of the Southern Hemisphere (as discussed in chap. 3, sec. 3.2), though he did not suggest the competitive mechanism(s). Similarly, Hooker noted the competitive superiority of English, African, Australian, and American plants over those of St. Helena but not the nature of the competition (also in chap. 3, sec. 3.2).

Thus, in the twentieth century, competition was not invoked as frequently as predation as a means by which a non-native species was threatening particular native species. Joseph Grinnell feared that introduction of the "Hungarian partridge" (grey partridge) would lead to extinction of the native California quail by means of competition for food or cover.[8] Citing Grinnell and several other authors, Herbert Levi depicted several cases in which non-native species were believed to have diminished or eliminated populations of native species by virtue of competition: American crayfish replacing European crayfish, rainbow trout replacing native trout in the American West, and, citing Elton, the gray squirrel from North America replacing the native red squirrel in England.[9] Actually, Elton had found the replacement quite mysterious:

> At the present time it is well known that the American grey squirrel is replacing the native red squirrel in various parts of England, but it is entirely unknown why this is occurring, and no good explanation seems to exist.[10]

By the time Elton wrote his invasion monograph thirty years later, he still found the replacement puzzling:

> Our ignorance of the nature of competition is also illustrated by the history of the red and grey squirrels in England. The American grey squirrel, *Sciuris carolinensis*, has replaced our native red one, *Sciuris vulgaris*, in the Midlands and part of the South of England. This sometimes happened

> quite quickly. But often it has taken up to fifteen years for the process, whatever it is, to be complete. For we still have no notion what happens, except that there is a change.[11]

In fact, the devastation wrought by the gray squirrel on the red squirrel in England is partly caused by a disease, squirrelpox virus, that was introduced with the gray squirrel, which is resistant, and causes massive mortality in the red squirrel. Several early researchers on the invasion of Great Britain by the gray squirrel, including Arthur D. Middleton and Monica Shorten in Elton's Bureau of Animal Population (discussed in chap. 7, sec. 7.1), had speculated that a disease vectored by the invader contributed to declines of red squirrel populations, and F. B. Edwards reported local elimination of red squirrels by a disease in 1962.[12] However, the role of squirrelpox virus vectored by gray squirrels in the red squirrel decline was doubted until late in the twentieth century:

> Poxvirus disease was sporadically and locally a catastrophic mortality factor in the red squirrel. Coupled with the relative isolation of local populations, it could have threatened population persistence in larger areas. Grey squirrels seem unlikely to have been a source, and are not implicated as a vector, of red squirrel disease.[13]

It is now widely recognized that squirrelpox, habitat transformation, and competition for nuts have caused the continuing great decline and range contraction of the red squirrel in Great Britain.[14] A recent examination of an enormous corpus of literature on invasion impacts finds far more extinctions caused by predation than by competition.[15]

Impacts on particular species or small groups of them (e.g., all species in a genus or family) remain a prominent focus of invasion science. Many highly publicized non-native insects and pathogens that attack plants fall in this category. Thus, for instance, the hemlock woolly adelgid from Japan attacks both eastern North American hemlock species, eastern hemlock and Carolina hemlock, the emerald ash borer from Asia attacks all North American ash species it has encountered to date, and thousand cankers disease and the walnut twig beetle that transmits it arrived in Knoxville, Tennessee, around 2010 and have since spread to seven eastern states, threatening the black walnut, native to eastern North America.[16] Introduced animal pathogens may act similarly. For example, a new strain of rabbit hemorrhagic disease (RHDV2) that arose in France in 2010 reached North America in 2018 and now threatens not only the non-native European rabbit, often kept as a pet, but also several native North American lagomorph species.[17] In principle, classical biological control aims to introduce a non-native species that will attack only a single non-native target species, although occasionally such introductions attack nontarget native species, including those of conservation concern (as discussed further in chap. 17, sec. 17.5).[18]

12.2 Community-Level Impacts

Several early authors went beyond deploring the impact of a non-native species on one native species or a small group of them. For instance, in the nineteenth century, Darwin, Wallace, and Hooker all lamented how goats had devastated the native plants of St. Helena. Darwin and Wallace used the term "vegetation" to signify the affected entity, while Hooker most frequently referred to "flora," though he occasionally used "vegetation" and "forest" in a similar manner. As used by scientists, these terms today have technically different meanings and had similar connotations in the nineteenth century, but these were perhaps not so rigorously adhered to then. "Flora" denotes the list of species (often specifically native species) of a region or island and is more widely used in discussions of biogeography and phylogeny than of ecology. Thus, one often sees discussions of the phylogenetic relationships of an island's flora, and that was frequently Hooker's subject for plants. For instance, lamenting the fact that many endemic plants of St. Helena had been eliminated by goats before they could be collected or studied, he wrote, "From such fragmentary data it is difficult to form any exact conclusions as to the affinities of this Flora, but I think it may be safely regarded as an African one, and characteristic of Southern extra-tropical Africa."[19] Although Hooker sometimes used the word "vegetation" interchangeably with "flora," "vegetation" more customarily refers to the concrete entity—the plants on the ground (or in the water)—than the list of species names and thus is usually the term ecologists would use for an undifferentiated mass of plants of various species. "Forest" is simply a type of vegetation, dominated by trees.

The fact that ecology was not a formally defined science when these Victorian scientists wrote may have contributed to a laxness in Hooker's usage, but the important point was that all three scientists were referring not to an impact on one or several species but to the entire plant component of the biota. The term that ecologists use today for the entire biota of a location is "community," and the term in this sense did not exist in the mid-nineteenth century. The concept of a biotic community developed through the nineteenth century, with various antecedents and much argument, including whether plants and animals constitute separate communities and whether the assemblage of all species is characterized simply by the fact that they are all together in one place at one time or whether they acquire the characteristics of a single living organism, a "superorganism."[20] In the early days of the science of ecology, the question of the importance of studying communities was a heated one. Thus Victor E. Shelford, a leading figure of early ecology in the United States, focused his research on communities.[21] He proclaimed:

> Ecology is the science of communities. A study of the relations of a single species to the environment conceived without reference to communities and, in the end, unrelated to the natural phenomena of its habitat and community associates, is not properly included in the field of ecology.[22]

Today, debates about the utility of research at the community level persist, but they do not threaten the field of ecology or even research at the community level.[23] Another continuing issue is that, even though almost all ecologists would today agree that, in principle, an ecological or biotic community contains all species of all taxa (though disagreement persists on the matter of organismic properties of a community so defined), as a sort of shorthand, many biologists even in scientific publications continue to refer to some subset of the biota as a community, such as all birds or all fishes. Complaints about the looseness of usage are old and legion. Thus, for instance, in 1953, J. B. Cragg complained:

> Perhaps the reviewer is not alone in asking his students to look at least twice at any statement in which the word "community" occurs. Its meaning is limitless, for it ranges from a piece of shorthand denoting an assemblage of organisms to something endowed with the attributes of organization.[24]

As concern with invasion impacts built in the twentieth century, an increasing number of observers focused on impacts on entire communities or large components of communities. Aldo Leopold's famous essay "Cheat Takes Over" lamented the replacement of whole communities of native Western plant species by cheatgrass from Asia and follow-on impacts on animals.[25] Elton in his 1958 book *The Ecology of Invasions by Animals and Plants* described the impact of the sea lamprey on many Great Lakes fishes, the Japanese beetle on over 250 plant species in the United States, the Argentine ant on many native ant species in the United States, the defoliation of many North American tree species by the European spongy moth, and other community-level impacts. Raymond F. Dasmann, depicting the destruction of natural habitats in Florida (detailed in chap. 7, sec. 7.2), indicted water hyacinth from South America and paperbark from Australia for replacing entire native communities. Several participants in the Scientific Committee on Problems of the Environment (SCOPE) project focused on invasion impacts on communities. Thus G. J. Breytenbach described impacts of many kinds of invaders, but especially plants, on communities of plants and of mammals in the southwestern Cape region of South Africa, and Harold A. Mooney and colleagues showed how perennial grassland communities in California were largely replaced by non-native annual plants.[26] Richard Mack depicted the replacement of entire temperate communities featuring native densely clumped grasses by non-native plants aided by anthropogenic habitat changes associated with agriculture and grazing livestock.[27]

12.3 Ecosystem Impacts

These impacts on ecological communities were largely described in terms of which native species were still present in the communities, which had declined in number or disappeared, what non-native species had been added, and how many species were present. Changes such as these in community features are often characterized as changes in "community structure," by which is meant which species are present and the relative abundances of each. However, the 1980s SCOPE project included a few contributions with a different focus—dynamic aspects of a community, such as who eats whom—and the interactions between the community and its physical environment, such as the flow of nutrients like carbon and nitrogen through both the species comprising the community and the physical surroundings, such as soil and water.

This sort of impact is known as an ecosystem impact, but the term "ecosystem" was coined only in 1935 by Arthur Tansley:

> But the more fundamental conception is . . . the whole system, . . . including not only the organism-complex, but also the whole complex of physical factors forming what we call the environment . . .—the habitat factors in the widest sense. Though the organisms may claim our primary interest, when we are trying to think fundamentally we cannot separate them from their special environment, with which they form one physical system. It is the systems so formed which, from the point of view of the ecologist, are the basic units of nature. . . . Our natural human prejudices force us to consider the organisms . . . as the most important parts of these systems, but certainly the inorganic "factors" are also parts—there could be no systems without them, and there is constant interchange of the most various kinds within each system, not only between the organisms but between the organic and the inorganic. These ecosystems, as we may call them, are of the most various kinds and sizes.[28]

The groundwork for the concept of the ecosystem uniting the biological community with the physical environment had been laid in the late nineteenth and early twentieth centuries, and, as with the community concept, the ecosystem concept has been riven by conflicts over the degree to which ecosystems have the properties of living organisms that need not concern us.[29] The key features of ecosystems that have guided research at that level, including research on invasions, are the inclusion of both the biological and physical components and the concern primarily with dynamic processes rather than simply community composition and "structure" in the aforementioned sense.

Early research on the ecosystem impact of an invasion was conducted by Thomas M. Zaret and Robert T. Paine in their study of the impact of the non-native peacock bass into Gatun Lake, Panama. They cast their chief result as the modification of the food web—that is, the routes by which energy flowed through the system. As discussed in chapter 7, section 7.3, this result was ahead of its time, and little similar invasion research was published until the 1990s. Compared with research on invasion impacts at the population and community levels, there is still a relative dearth of studies on ecosystems.[30] However, beginning with the SCOPE project, there has been an uptick in such research, especially in the twenty-first century.

Publications of the SCOPE project included several papers on ecosystem impacts, particularly by Peter Vitousek.[31] Vitousek pointed to recent literature examples showing that non-native species had affected the flux of energy, nutrients, or water through an ecosystem, as well as such ecosystem-level processes as productivity and decomposition.[32] In particular, he described the introduction of an Atlantic nitrogen-fixing shrub, fire tree, to regions of the young volcanic island of Hawaii lacking native nitrogen-fixers. He hypothesized that nearly monospecific stands of this shrub could therefore change productivity, nutrient cycling, and plant successional pathways and could also facilitate invasion by other non-native plants currently limited by the nitrogen-poor soil.[33] Mooney and colleagues and Vitousek's team added changes in the disturbance regime, such as the rate and intensity of fires, as causes of ecosystem impacts.[34] For instance, in Hawaii Volcanoes National Park, highly flammable non-native grasses in the genera *Andropogon* and *Pennisetum* greatly increased fire size and frequency.[35] Non-native grasses cycles in many other ecosystems worldwide have wrought similar changes in fire regimes (as discussed in chap. 8, sec. 8.1, and chap. 14, sec. 14.3).[36]

In 1990, Vitousek aimed to define more formally the ways in which a non-native species could have an ecosystem impact:

> I suggest invaders can change ecosystems where they 1) differ substantially from natives in resource acquisition or utilization; 2) alter the trophic structure of the invaded area; or 3) alter disturbance frequency and/or intensity.[37]

By "trophic structure," Vitousek meant the topology of the food web—who eats whom. In fact, most of the early literature on ecosystem impacts focused on impacts on processes, or even explicitly defined "ecosystem impact" to mean an impact on a process or process pathway (such as trophic structure for energy flow and nutrient cycling). Similarly to Vitousek, the leading invasion science monograph of the 1990s, *Biological Invasions* (described in chap. 10, sec. 10.2), stated, "Invaders that affect ecosystem processes are those that are said to have ecosystem effects."[38] Perhaps this focus reflects the fact that modern ecosystem-level ecology, often dated to a paper by Raymond Lindeman in 1942, has been primarily concerned with processes, especially energy flow and nutrient and water cycling.[39]

However, another SCOPE contribution had described a different way in which an invader can affect an entire ecosystem: alteration of geomorphological processes, such as stabilization and reshaping of sand dunes by the non-native European beachgrass on the coasts of California and Oregon, or the transformation of mudflats into salt marshes by common cordgrass (a hybrid of native small cordgrass and non-native smooth cordgrass) in England.[40] Vitousek and colleagues added this fourth mechanism to the previous three in 2010, and a subsequent review provided many examples of geomorphological impacts by non-native species.[41] In 1991, I pointed to a "keystone species" as another sort of species likely to have an ecosystem impact if introduced beyond its native range.[42] Mary Power and associates sought to distinguish a keystone species from an ecological "dominant" (a species, such as a tree species in a forest, whose great biomass and abundance give it enormous influence over an entire ecosystem and its community, often by constituting the physical structure of the system) by defining a "keystone species" to be one whose impact on a community or ecosystem is disproportionately large relative to its abundance.[43] An example of a non-native species serving this role is the periwinkle, introduced from Europe to New England within the last few centuries.[44] This small snail has completely changed the character of high tide pools and other coastal areas in New England by consuming massive amounts of algae and marsh grasses and bulldozing sediment.[45]

Other classes of species have been designated that play particularly large roles in ecosystem physical structure and function, some of which have been subsumed into an expanded concept of keystone species. "Foundation species" are "species that have roles in the maintenance of the community disproportionate to [their] abundance or biomass."[46] A "transformer species" is defined as one whose activity or promotion of a phenomenon or process (such as fire) changes an entire ecosystem physically.[47] Dominant species, simply by virtue of constituting a great fraction of the physical structure, affect the entire ecosystem. The most widely used such concept is that of the "ecosystem engineer," a species whose activity determines the physical structure of an ecosystem in a way that modifies the availability of biotic or abiotic resources to other species.[48] Many non-native species, both plants and animals, generate ecosystem impacts by virtue of engineering.[49]

Perhaps Vitousek and colleagues captured the notion of ecosystem impact most comprehensively with a metaphor: "Where an exotic species changes ecosystem characteristics, it changes the rules of the game for all native and exotic species."[50] Several important points with respect to invasions and all these classifications of species with particularly great impacts on the entire community or ecosystem became clear as, after a lag of about a decade, an increasing amount of research on ecosystem impacts followed Vitousek's early papers.[51] First, many non-native species produce ecosystem impacts in ways similar to that of fire tree in Hawaii—by changing nutrient pools and flows, and not only of nitrogen. For instance, two non-native plants in phosphorus-poor parts of Western Australia have increased soil phosphorus to the detriment of native dominant plants.[52] Second, any species that invades an ecosystem and greatly diminishes the population of a particularly important species can affect the entire ecosystem, as exemplified by the non-native periwinkle snail removing dominant plants on the New England coast.[53] Third, a non-native species that provides a new and radically different physical structure changes the whole ecosystem, as have non-native cordgrasses replacing coastal mudflats in England and the paperbark, an Australian tree, replacing meadows with forests in south Florida.[54] Fourth, the fact that a species plays an outsize role in its native ecosystem or when introduced in one ecosystem does not automatically mean it will do so when introduced elsewhere; ecosystem impacts are highly context-dependent. For instance, the sour fig forms dense, dominant stands in parts of its native range in South Africa, is highly invasive and causes massive ecosystem impacts in some regions where it has been introduced, such as California, yet has minimally affected a dune region of Portugal.[55]

As interest in ecosystem impacts of invasions grew, several invasions that had previously been found to have generated population or community impacts, or both, were shown by further research or reconsideration of the original observations to have affected entire ecosystems. For example, in the heyday of the explosion of interest of invasions precipitated by the SCOPE project, the brown tree snake introduced to Guam around 1949 was widely heralded as having caused the precipitous decline of all forest bird species after an incriminating paper by Julie Savidge.[56] Two decades later, research showed that the loss of birds rippled through the entire forest ecosystem because many native plants had depended on birds for pollination or seed dispersal and were suffering reduced recruitment.[57] Subsequent similar research on the loss of insectivorous birds on Guam showed a manyfold increase in populations of spiders, likely because of reduced predation by birds on the spiders, possibly aided by reduced competition with birds for insect prey.[58] Parsing the exact influence of the snake on the various changes in Guam populations and ecosystems is complicated by other major anthropogenic factors, including habitat destruction and pollution, but there is no doubt that the snake played a key role in devasting the birds.[59]

12.4 Ecosystem Services and Ecosystem Impacts

Aside from purely scientific interest in the nature of invasion impacts, concern with ecosystem impacts increasingly arose with respect to conservation of biodiversity and environmental damage. In 1981, as modern invasion science was getting underway, Paul and Anne Ehrlich coined the term "ecosystem services" to capture the various benefits that intact natural ecosystems provide to humans.[60] Their definition was formalized by Gretchen Daily:

> Ecosystem services are the conditions and processes through which natural ecosystems, and the species that make them up, sustain and fulfill human life. They maintain biodiversity and the production of ecosystem goods, such as seafood, forage timber, biomass fuels, natural fiber, and many pharmaceuticals, industrial products, and their precursors.[61]

Mooney and Ehrlich trace the concept to works of George Perkins Marsh (described in chap. 3, sec. 3.4) and Aldo Leopold (discussed in chap. 6, sec. 6.1), who wrote before the advent of modern ecosystem ecology, and to a 1970 report on environmental problems, but the concept became widely discussed only in the 1980s, leading to a SCOPE project beginning in 1991 and a widely read comprehensive edited volume on the array of such services in 1997.[62] It was a short step to attempt to relate ecosystem impacts of biological invasions to their effects on ecosystem services.

An enormous international project, the Millennium Ecosystem Assessment (MEA), convened hundreds of scientists from 2001 to 2005 to assess the state of ecosystems and, in particular, how various phenomena, including invasions, are affecting services ecosystems provide to humans. The widely cited resulting synthesis reports expanded the definition of ecosystem services to include ways in which human-constructed or modified ecosystems, and not only natural ones, benefit humans.[63] The other main way in which the MEA influenced subsequent research and discussion of ecosystem services was to classify services into four categories:

1. *Provisioning services*, which provide material goods such as food and water to humans.
2. *Regulating services*, which moderate and control natural phenomena to the benefit of humans, such as flood control or climate control.
3. *Cultural services*, nonmaterial benefits that contribute to the development and cultural advancement of people, such as recreation and interactions with artistic creation.
4. *Supporting services*, fundamental processes that usually do not directly benefit humans but rather support the conditions that underpin the other types of services. Examples include nutrient cycling and soil formation.

Several parts of the MEA addressed the impact of invasions on ecosystems and their services. For example, a figure depicted the five main direct drivers of change in biodiversity and ecosystems—habitat change, climate change, invasive species, overexploitation, and pollution—and whether the impact of each driver on thirteen types of ecosystems was believed to be increasing, decreasing, or unchanging.[64] For invasions, the figure showed invasion impact increasing in seven ecosystem types and unchanging in six types.[65] However, although the MEA gave examples for some impacts, the depictions of impacts on ecosystems and of these ecosystem impacts on services were largely in general terms, and quantification and methods of determining whether an impact was increasing or decreasing, or low, moderate, or high, were not provided.

Soon after publication of the MEA, Heather Charles and Jeffrey Dukes explored, within the MEA framework, ways in which certain documented ecosystem impacts of invasions affected particular ecosystem services, pointing to the general problem posed by assigning a "value" (usually in economic terms) to a service to quantify the magnitude of an ecosystem impact in terms of its effect on that service.[66] Although they provided economic values of certain services and magnitude of impacts of particular invasions that affect a service, they suggested that generalizations were limited if not impossible at that time. In fact, they could not cite evidence that species causing the greatest ecosystem impacts when introduced have the greatest effects on ecosystem services, even though the literature on ecosystem impacts had grown enormously since the SCOPE program. They included a table of their impressions of which ecosystem services were most at risk of damage from non-native species in several types of ecosystems. Also working within the MEA framework, Liba Pejchar and Harold Mooney estimated the economic costs of non-native species to several provisioning, regulating, and cultural services.[67] Soon thereafter, Montserrat Vilà and collaborators exploited the large DAISIE database of invasions in Europe, finding 1,094 documented ecological impacts and 1,347 documented economic impacts on ecosystem services, across all four service categories.[68] At least for this data set, ecological and economic impacts are highly correlated, with terrestrial invertebrates creating greater economic impacts than ecological impacts and terrestrial plants having greater ecological impacts.

In 2017, an edited volume, *Impact of Invasions on Ecosystem Services*, included nineteen chapters detailing many impacts of non-native species on ecosystem services in each of the four MEA categories.[69] A review of this book observed that, although several chapters acknowledge positive impacts of purposeful introduction of species on ecosystem services (e.g., food crops as a provisioning service), the emphasis is almost wholly on how invasions are detrimental to ecosystem services.[70] However, the synthesis chapter pointed to the sorts of problems generated by different perceptions of ecosystem services and impacts on them by different stakeholders:

> Most non-native species introduced by humans have value to one or more sectors of society. As a result, conflicts arise in the estimates of the net value of non-native species to ecosystem services. . . . Addressing the impacts on ecosystem services will require understanding of environmental conflicts and policy tools that can better capture impacts of commercially important non-native species.[71]

In fact, the burgeoning literature on ecosystem services relates only partly to the concept of ecosystem impacts, although it appears that many authors tacitly assume that an impact of a non-native species on an ecosystem service operates through an impact at the ecosystem level. Charles and Dukes observed that impacts of non-native species on all four categories of ecosystem services can broadly be seen as economic, environmental, or social, and that only those they term "environmental impacts" are ecosystem impacts, affecting ecosystem structure and function. Thus, for instance, invasions that affect the aesthetics of nature are often seen as affecting cultural ecosystem services in negative or positive ways, but they may or may not affect ecosystem structure or function.[72] Another confounding factor is that many deliberately introduced species that provide an ecosystem service (e.g., food crops), though non-native, are not invasive in the sense that they do not invade and perpetuate themselves in natural habitats. Even those that do became invasive may not have a substantial ecosystem impact. A search for evidence for or against impacts on provisioning, regulating, and cultural services (several services classified by the MEA as supporting were included with regulating services) among publications on myriad non-native tree species found, amid a great deal of context dependence, many examples of impacts on all three service types.[73] Certain of these impacts derived from an impact on ecosystem function, such as changed fire regime or soil formation, while other impacts on services, such as some examples of aesthetic and cultural heritage services, arose via impacts on particular native species.

To an extent, the interest in accounting for positive impacts of invasions on ecosystem services (discussed in chap. 16, secs. 16.1 and 16.3) overlaps with the controversial literature on conservation benefits of non-native species (discussed in chap. 18, sec. 18.3). For instance, most of the claimed conservation benefits of non-native species listed by Martin Schlaepfer and colleagues accrue to particular species of conservation concern, but they cited three unintentional introductions and one planned introduction as favoring ecosystem services and one planned introduction (a failed biological control project) as damaging a desired ecosystem service.[74] Also, the concern with whether a non-native species is acting through the ecosystem as a whole in providing an ecosystem service, rather than simply affecting a native species that is providing a service, is closely tied to a long-standing debate about the importance to ecosystem function of each individual species within a community. This debate began with an oft-quoted section of Charles Elton's 1958 invasion monograph arguing that diversity of species (nowadays "species richness" [number of species], as is often meant by the term "biodiversity") causes a community to be more "stable," and that one manifestation of this stability is resistance to invasions. This notion of greater species diversity conferring community stability probably traces at least partly to Elton's interactions with Aldo Leopold, whose writings in 1942 and 1944 proposed the same hypothesis.[75] This hypothesis, generally attributed to Elton, stirred a vigorous debate as the conservation and environmental movements gathered steam in the late twentieth century (as discussed in chap. 7, secs. 7.4 and 7.5), because, if true, it would provide a rationale for conserving every species.[76] This controversy transmogrified in the twenty-first century into a heated and unresolved argument about whether biodiversity per se contributes to ecosystem function and inspired research on the roles of particular species in ecosystems.[77]

12.5 Ecosystem Impacts, Microbial Ecology, and Soil Communities

Finally, much of the research on ecosystem impacts of invasions, including Vitousek's seminal SCOPE project paper, focuses on aboveground-belowground interactions, often mediated by microorganisms.[78] The ecology of belowground systems was long a stepchild of academic ecology, as was the study of interactions between the belowground and aboveground components of terrestrial ecosystems. Soil science and the ecology of soil communities were largely studied separately from the ecology of aboveground communities until the turn of the twenty-first century.[79] The SCOPE project barely addressed belowground matters. Aside from Vitousek's contributions, the main mention was a description of the impact of non-native plantation trees on aspects of soil chemistry in Africa, similar to previous research.[80] This low status of belowground ecology was epitomized by a special issue of *Science* in 2004 entitled "Soils—The Final Frontier."[81] Other authors in this period who pointed to the importance of the belowground community and its interaction with the aboveground community include Wim Van der Putten and colleagues and Richard D. Bardgett and David A. Wardle.[82] One of the articles in the special issue reviewed how invasions affect both belowground ecology and belowground-aboveground interactions.[83] Previously research had explored effects of invasions on aboveground-belowground interactions in particular ecosystems.[84] Bardgett and Wardle described the history of the rise of an integrated aboveground-belowground ecology.[85]

Much of the impact of invasions on the belowground component of ecosystems concerns microbes, either non-native microbes or impacts of other non-native species, such as plants or herbivores, on native microbes. Research on the ecology and biogeography of microbes in general, and of soil microbes in particular, made a quantum leap in the late twentieth century owing to the rapid development of molecular genetic tools (described in chap. 11, sec. 11.2), which greatly increased the ability both to detect and to identify microorganisms. Even the number of species of microorganisms had been vastly underestimated.[86] Early reviews of new findings on how ongoing global changes affect microbial ecology focused on changes in climate and land use, but invasions quickly became part of the focus.[87] To a large extent, the role of microbial invasions could barely be studied before this explosion of microbial research because the ranges and native regions of most microbes could not be determined. Thus, for instance, the SCOPE project treatment of microbes dealt virtually exclusively with pathogens of humans, other animals, and plants, often because the non-native status of a pathogen could be deduced from its impact on hosts or because of its association with a vector, such as a mosquito species, known to be non-native. It was widely assumed that many microorganisms were completely cosmopolitan—the "everything is everywhere" precept formalized by Lourens G. M. Baas-Becking—so that the concept of a non-native microbe would rarely apply.[88] That assumption turned out to be very incorrect, and microbial distributions appear to be similar to those of plants and animals and determined by the same forces.[89] As microbial identities and distributions were increasingly determined, much research ensued on the introduction and impact of non-native microbes.[90]

: : :

The earliest alarm over invasion impacts was almost wholly focused on how they affect human enterprises such as agriculture. When concern first arose for what we now consider ecological impacts, it was largely over impacts, usually caused by introduced predators, on particular species or small groups of them. Well before the advent of a science of ecology, a few scientists, such as Charles Darwin, Alfred Russel Wallace, and Joseph Hooker in the mid-nineteenth century, described impacts at the community level of ecological organization. These impacts were essentially changes in the population sizes or proportional representation of particular native species in the assemblage of species at a site or in a region. The concept of "ecosystem," formally defined in 1935, trained ecologists' attention on dynamic processes such as the flow of energy or nutrients through an ecosystem or modifications wrought by disturbances on both dynamic processes and the physical structure of an ecosystem. Except for a 1973 paper on the impact of non-native peacock bass on the topology of a food web, before the advent of modern invasion science, students of invasions largely ignored impacts at the ecosystem level until the SCOPE program of the 1980s.

Peter Vitousek, participating in the SCOPE program, described the ecosystem impact of a non-native nitrogen-fixing plant in a part of Hawaii lacking native nitrogen-fixers and called for increased research on ecosystem impacts. This call was answered after a delay of about a decade, and such research became a major component of invasion science. Study of ecosystem impacts of invasions often entails research on microbial interactions with other species in the ecosystem, and this research has been a major part of the dramatic increase in research in microbial ecology triggered by the use of newly developed molecular genetic tools to identify microbial species. Much of the surge in microbial ecology research, including that featuring non-native microbes, non-native plants, or both, focuses on belowground interactions and thus has contributed to a major expansion of soil science.

The concept of ecosystem services—benefits provided to humankind by ecosystems—was proposed and developed almost simultaneously with the advent of modern invasion science and the explosion of invasion research. Ecosystem impacts and ecosystem services do not map directly onto one another—an ecosystem impact need not directly affect an identified ecosystem service, and invasion impacts at the population or community level can affect ecosystem services. Nevertheless, as research on both ecosystem impacts and ecosystem services has increased, many potential ecosystem impacts on various ecosystem services have been identified, although confirmation of such effects is often still lacking.

13

Invasion Science Catches Two Ecology Waves

As invasion science matured, developments in other disciplines found their way into the new field. Sometimes this accretion was delayed, as with genetics and evolution (the topic of chap. 11), and sometimes the infiltration into invasion science happened in more timely fashion, as with research on ecosystem impacts. Generally, the process began when scientists working on an invasion-related issue recognized the relevance of developments in another discipline, as Peter Vitousek had for research on ecosystem impacts (as detailed in chap. 12, sec. 12.3). Vitousek had studied nutrient concentrations in a set of streams for his doctoral dissertation and focused on nutrient cycling in his early career.[1] A native of Hawaii, he quickly undertook ecosystem-level research there, discovered a non-native plant greatly affecting nitrogen cycling, and recognized that many invasions probably trigger similar impacts and that this would be a fertile area of research.[2]

Two relatively new approaches to portraying and analyzing scientific data arose in the mid-twentieth century and eventually found their way into ecology. Then, around the turn of the century, ecologists recognized the possibility that each approach could elucidate aspects of biological invasions. These approaches—meta-analysis and network analysis—became increasingly popular in ecology and, after a relatively short lag, similarly spread in invasion science.

13.1 Meta-Analysis Enters Invasion Science

In 1976, education researcher Gene V. Glass coined the term "meta-analysis" to mean "the statistical analysis of a large collection of analysis results from individual studies for the purpose if integrating the findings."[3] The goal of a meta-analysis is to combine all analyses of the effect of a particular "treatment" on a specific outcome. In one of Glass's examples, the treatment was glutamic acid and the outcome was effect on intelligence of intellectually disabled children. In another example, the treatment was television instruction and the outcome was learning. Although Glass intended the term "meta-analysis" to include any statistical procedure with this aim, he favored a particular approach.[4] Glass suggested that for each individual study, one should calculate the "effect size" of the treatment as the mean difference in the outcome variable between treated and untreated subjects, divided by the standard deviation of the outcome variable for the untreated subjects. Later Glass reluctantly advocated using the pooled standard deviation for both groups, and this convention has become standard.[5] Many other measures of effect size have been proposed, but the basic approach of estimating the mean difference between treatment and control subjects remains.[6] I will term meta-analyses of this type "formal meta-analyses" and use "broad-sense meta-analyses" to indicate other statistical approaches.[7]

Formal meta-analyses began to appear in the late 1970s in the social sciences and medicine, and over the next two decades they became common in those disciplines.[8] The approach became so popular that entire textbooks on meta-analysis pitched at social scientists and medical researchers appeared within a decade.[9] In both fields, numbers of publications employing meta-analyses grew rather slowly through the 1980s but increasingly rapidly through the 1990s, reaching nearly 6,000 in medicine and 600 in the social sciences in 2010.[10] Entry into ecology and evolution was delayed by about a decade. In 2001, Jessica Gurevitch and colleagues identified the two first meta-analyses in those fields as being published in 1991, with a total through the 1990s of 100 papers either discussing or performing meta-analyses (including several criticizing particular meta-analyses or uses).[11] However, the next decade saw a bandwagon effect, with about 550 such articles published in 2010 and with the first meta-analysis textbook aimed at ecologists and evolutionists, an edited compendium of papers, appearing in 2013.[12]

13.1.1 Matters of Interpretation

Issues with interpreting meta-analytic results also arose quickly in both medicine and social sciences. The first issue is based primarily on the validity of combining results from different, independent studies undertaken by different methods in different settings based on samples that may have differed in some important way that could have influenced the treatment outcome.[13] In medicine, many meta-analyses seem relatively unproblematic on this point. For instance, often meta-analysis is used to test to what extent a pharmaceutical is effective in treating a particular disease, in cases in which several very similar small studies have been conducted with somewhat varying results. As an example, suppose the question is whether, compared with a placebo, a particular drug reduces the risk of heart attack in males, aged forty through sixty, with no prior history of heart attack, an initial cholesterol level of 250–300, a dose of the drug between 10 and 15 mg/day; with patients tracked for at least a year; with a dropout rate of 5 percent or less; with appropriate, identical randomization and blinding criteria; and with patients under the care of a primary care physician for the entire study.[14] In this case, if all individual studies met all these criteria, one would probably not question a claim that combining all the results in a properly organized meta-analysis is a legitimate approach to answering the question. On the other hand, if the question is about whether television instruction affects learning and, if so, by how much, it would probably be far more difficult to amass a set of independent studies that were similar, even in the nature of the outcome that was tested. One could, as Glass proposed, find a number of studies on the same general question and attempt to compute an effect size for each and an overall mean effect size, but the interpretation and implications of these effect sizes and especially the mean effect size would likely be open to question.

A second widespread concern with interpreting meta-analyses results (in fact, in any type of literature review), even if the independent studies included seem very similar, is whether all relevant independent studies have been included. Gray literature reports and student theses, for instance, are often not as readily located as journal articles. Another aspect of this issue is widely known as the "file drawer problem," because results that are not statistically significant or striking might well end up in a file drawer rather than a publication.[15] This "publication bias" is a well-known problem in medicine, the social sciences, and the natural sciences, including invasion science. Various statistical procedures have been developed to estimate the extent of the file drawer problem and to correct, at least partially, for it in meta-analyses.[16]

13.1.2 Meta-Analysis in Invasion Science

In invasion science, perhaps the first published formal meta-analysis, with calculation of effect sizes as outlined at the beginning of section 13.1, was by Marjorie Wonham and colleagues in 2000, who sought patterns in documented impacts of invasive plants, insects, marine invertebrates, and fishes from ninety field studies.[17] Next was an effort by Jonathan M. Levine and associates in 2004 to assess the importance of biotic resistance (discussed further in chap. 15, sec. 15.4.1) by the native species community to invasion by non-native plants as revealed by data from fifty-two published studies.[18] A slow but steady increase of meta-analyses of aspects of invasions

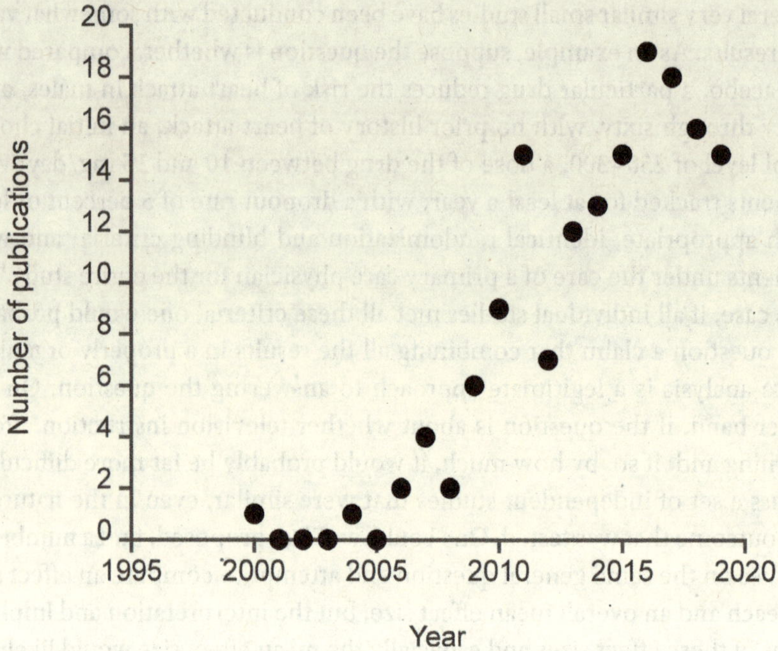

Fig. 13.1 Number of meta-analyses reported in invasion science literature, from Web of Science search on February 21, 2021, with search term combination ((“non-native” OR “nonnative” OR “exotic speci*” OR “alien” OR “invas* speci*” OR “biology* invas*”) AND (“network*” OR “meta-anal*” OR “metaanal*”)).

ensued, reaching fifteen in 2011 and fluctuating around that number for the rest of the decade (fig. 13.1).[19] By 2021, meta-analysis had been so widely employed in invasion science that a review assessed seventy-two meta-analyses, mostly formal ones, of either trait differences between native and non-native species or impacts of non-native species on native species and the environment.[20] The great majority of meta-analyses in invasion science have been formal meta-analyses entailing calculations of effect size, though a few have used and defended other statistical approaches, such as vote counting.[21] Most authors employing formal meta-analyses have examined invasion impacts from various perspectives, such as whether degree of impact varies with the type of non-native species or the ecosystem in which it occurs, or the relative sizes of impacts on prey by non-native predators and native predators. Others have sought differences between traits of invasive non-native species and either native species or noninvasive non-native species, while yet others have asked if degree of phenotypic plasticity in non-native species exceeds that in native species. Several have addressed evolution of species after introduction. Another theme has been effectiveness of management approaches, particularly biological control.

Doubtless the timing and trajectory of meta-analyses in invasion science reflects the rapid growth of the field beginning in the late 1980s (as discussed in chap. 9, sec. 9.1). A meta-analysis cannot be performed unless a collection of studies exists on which to perform it, and a large enough collection probably would not have been available to address the aforementioned questions credibly before the first decade of the twenty-first century. Furthermore, to undertake a formal meta-analysis requires studies that report certain types of quantitative data, or from which quantitative data can be gleaned from graphs. Several early efforts along these lines were abandoned because such data were unavailable from enough studies. For instance, Montserrat Vilà and colleagues, seeking information from published experiments on competition between crops and non-native weeds, forswore formal meta-analysis because few studies reported measures of variation such as standard deviation or standard error, so instead they resorted to simple descriptive statistics plus an analysis of variance for a subset of the studies.[22]

To assess the contribution of meta-analyses to invasion science, one must consider what questions they aimed to answer. In data analysis in general and meta-analysis in particular, there are two types of goals. One is to test a null hypothesis and either reject it or not. In the previously mentioned case of the drug that may or may not reduce heart attack risk, the null hypothesis would be that it does not reduce that risk, and one either rejects it or not, based on some probability level—for example, there is a 5 percent or lower probability that the result indicated by the data arose by chance. The other type of goal is to estimate the size of an effect of interest (which may be zero). A problem with several meta-analyses in invasion science is that they are planned as hypothesis tests and reject (or not) the tested hypothesis based on mean effect sizes, even though individual studies within the meta-analysis vary, and some differ in sign (positive or negative) from the mean effect size. For example, Amy M. Davidson and colleagues tested the hypothesis "that invasive species show greater phenotypic plasticity across various growth, morphological, physiological and fitness traits compared with co-occurring non-invasive species" in a formal meta-analysis of seventy-five pairs of species from forty-six separate studies.[23] They concluded, based on a mean effect size, that invasive species are, in fact, more phenotypically plastic than noninvasive species. However, the separate studies examined different species in different settings, and in several of them the invasive species did not show greater phenotypic plasticity. The conclusion from this study exemplifies the concern expressed earlier about combining results from different, independent studies undertaken by different methods in different settings based on samples that may have differed in significant ways among the studies.

Some meta-analyses correctly reported the effect sizes of individual studies and stressed that they differ, even in direction. For instance, Mirijam Gaertner and colleagues asked whether invasion by non-native plant species reduces the number of native plant species in Mediterranean-type ecosystems and analyzed forty-seven studies from four continents.[24] In their formal meta-analysis, they found an "overall significant decline of species richness after invasion" but depicted and discussed great variation among the studies, five of which had greater species richness after invasion. The remaining forty-two had lower native species richness, although for twenty-nine of these the effect size did not differ significantly from zero. The authors proceeded to analyze these differences. Many other meta-analyses of invasions also discussed variation of results in studies framed as hypothesis tests. A problem, however, is that often the abstract of a paper categorically states an overall result without qualification or recognition of the variation among individual studies. For instance, in their abstract, Gaertner and colleagues described the hypothesis being tested, then stated, "Our results confirm a significant decline in native species richness attributable to alien invasions," with no mention of cases depicting the opposite effect or of the fact that, in the majority of cases, the effect size did not differ significantly from zero.[25] Many readers probably read or remember only the abstract, and the result of such blanket statements in abstracts is probably to give the impression that invasions adhere to certain general rules, when in fact they are remarkably context-dependent.

Other meta-analyses of invasions did not aim to test a specific hypothesis but rather to examine the array of certain relationships between native and non-native species, or between invasive and noninvasive non-native species. For example, a broad-sense meta-analysis was used to ask how often non-native plants used more water than co-occurring native plants of the same growth form.[26] For each pair of plant species, for various measures of water use the authors calculated the percentage difference in water use as ([non-native - native]/native) ×100%, and they depicted the entire array of results: how many pairs differed in each direction, and by what amount. The abstract stressed the variation in results for all measures. Vilà and colleagues conducted formal meta-analyses of twenty-four types of impacts of non-native plants, based on 135 different plant species in 1,041 studies, and asserted in the abstract, based on whether the mean effect size for each type of impact differed significantly from zero, a "significant effect" for several impacts. This statement implied, for example, that non-native plant species affect animal fitness. However, the authors immediately undercut this interpretation by pointing out that "alien species impacts are heterogeneous and not unidirectional even within particular impact types."[27] In fact, the wide variety of non-native animals and plants and surrounding ecosystems in the 1,041 studies meant that no single prediction could be made accurately. Some non-native plants in some systems decrease fitness of some animal species, other non-native plants in other systems increase fitness of some animal species, and in some cases there is no unambiguous effect. The survey of meta-analyses described in the first paragraph of this subsection 13.1.2 found that 44 percent of the seventy-two meta-analyses had mixed results and that the majority found mean effect sizes to be insignificant.

A further complication is that most individual studies included in meta-analyses in invasion science treated species as undifferentiated entities. In fact, from a genetic standpoint, species are structured in a nested way: different genotypes within a population and different populations within a species. Because this nested genetic variation was unaccounted for in most studies, effect sizes tended to be inflated over effect size if the nested variation had been accounted for because recorded variation between species may actually reflect unexamined variation between genotypes or populations, or both. One study suggested that they are likely inflated by at least 20 percent, and quite possibly by more.[28] This inflation beclouds the interpretation of degree of variation among the individual studies, and it substantially increases mean effect size.

The net effect of the rise of meta-analyses on invasion science and perceptions of the nature and importance of invasions has probably been mixed. To the extent that researchers recognize variation among both the nature of the systems and the size and direction of the outcomes among the individual studies, a meta-analysis can give an overall perspective on whatever effect is studied—for example, whether phenotypic plasticity in non-native species tends to exceed that in co-occurring native species. The perspective will be still more accurate if the reader bears in mind the file drawer problem, the likelihood that any study found an adequate fraction of relevant studies, the fact that a mean is an arithmetic average only, and the finding that many effect sizes are inflated because of untallied nested genetic variation.

However, the fact that some meta-analysts have emphasized the significance or lack thereof of a mean effect size and downplayed the degree of variation among individual studies suggests that, among both scientific and lay readers, some will see the take-home message as some sort of universal effect or "rule." This interpretation must be very tempting in a science that is so highly context-dependent and has fought from its inception, as has ecology in general, against the claim that it is a collection of case studies with no unifying principles.[29]

That many abstracts in the published literature in particular noted the mean effect size and relegated discussion of the variation in effect sizes to the body of the paper increases the likelihood that readers have concluded that some general rule exists about an effect, a conclusion that may be propagated by subsequent authors. For instance, the study by Davidson and colleagues concluded that invasive non-native species are more phenotypically plastic than co-occurring noninvasive non-native species, and the abstract said flatly that "our analysis shows that invasive species demonstrate significantly higher phenotypic plasticity than non-invasive species," with no mention of variation. Andrew Hendry, in a paper about phenotypic plasticity in general, stated, "However, a more recent quantitative meta-analysis that examined 75 invasive/noninvasive plant species pairs came down decisively in favor of greater plasticity in invaders (Davidson et al. 2011)," which is true in one possible interpretation of their paper but misleading in another, given that the individual studies analyzed by Davidson and colleagues showed much variation.[30] Similarly, Ayub M. Odour and colleagues asked whether non-native vertebrate herbivores aid non-native plants. Their meta-analysis indeed showed that mean effect size supported the hypothesis, although the supplementary information on the individual studies showed this was far from a universal effect. However, their abstract stated that "we found that exotic vertebrate herbivores have positive effects on exotic plant performance and population sizes, and no significant effects on native plants," leading Peter Bird and associates, studying European rabbit impact on native seedling survival in Australian coastal shrubland, to assert that "exotic herbivores damage native plant communities, in part, because they preferentially graze on native versus exotic plant species, thereby increasing exotic plant abundance, survival, flower and seed production, canopy cover, height and species richness in plant communities (see multispecies reviews by Parker et al. 2006; Oduor et al. 2010)."[31]

Another convention that may mislead interpretation of meta-analysis results is the frequent presentation of mean effect size for a collection of studies with a 95 percent confidence interval, calculated as described by Julia Koricheva and colleagues and interpreted to signify a positive or negative effect of the treatment if the confidence interval does not overlap 0.[32] This depiction, even in studies that reveal and emphasize variation among effect sizes of individual studies, suggests the existence of some true effect of the treatment, with its value within the confidence interval with 95 percent probability, and that deviation of effect sizes of individual studies from this true value represents a type of "error," as in standard statistical analyses. However, if the individual studies are not essentially replicates of one another, as is usually the case, it is more likely that the deviation of the effect size of an individual study from the mean effect size simply represents a different effect, even if the mean effect size is positive and the effect size of the individual study is negative, or vice versa.

13.2 Networks Are Increasingly Depicted and Analyzed

By the twenty-first century, much research on community-wide impacts of non-native species became incorporated into a new science, called "network science," that developed rapidly and became a major aspect of many sciences, including ecology.[33] This incorporation transformed theoretical ecological and invasion research from an endeavor that primarily entailed exploring systems of equations to one that seeks patterns in the topology of pictorial representations of the "generalized community matrix" (more on that in subsection 13.2.2). Understanding this transformation requires some discussion and definitions.[34]

13.2.1 A Plethora of Terms and Statistics

A set of species can be represented as a collection of "nodes" (or vertices) in space called a "network," and if two species interact, they are connected by lines called "links" or "edges." Links can be "undirected" or symmetrical, or they can be "directed," and the direction can be represented by an arrow. In addition, networks can be quantified. A link can be weighted; for instance, if a link represents predation, its width could be proportional to the biomass of the prey consumed by the predator, relative to biomass consumed by this and other predators eating this and other prey species. Also, nodes can be weighted, with weights represented by the size of the node; weights could be associated with population size, or total population biomass, or other measures. Importantly, a network can also be represented as a square matrix, with each node represented by both a row and a column, with zeros at the intersection of two nodes that are unlinked and ones at the intersection of two linked nodes. The ones can be replaced by other numbers if the links are weighted, and the entries can have positive or negative signs. Thus, use of networks is twofold: (1) visualization of the system as a set of nodes and lines, and (2) an analytic tool to study patterns and dynamic processes of a community.

A menagerie of statistics characterizes networks. These can be divided into statistics that characterize a network as a whole and statistics that characterize the network role of particular species.[35] Table 13.1 lists those most commonly calculated in network research in invasion science.

13.2.2 Species Interactions Depicted in Networks

By the early nineteenth century, many authors had verbally described the trophic relationships among species in a community—who eats whom. The first person to diagram a food chain is uncertain.[36] However, the first clear generalized statement of the concept was by English naturalist Richard Bradley in 1718.[37] Several authors united food chains to represent biological communities of organisms as food webs in the late nineteenth and early twentieth centuries. Perhaps the first was Lorenzo Camerano in 1880, although Charles Elton (discussed in chap. 7, sec. 7.1) was probably primarily responsible for spreading the food web concept into the developing science of ecology, having depicted the food web of Bear Island, in the Barents Sea (fig. 13.2), embedded in a diagram of the nitrogen cycle, and discussed food webs in his early animal ecology textbook in 1927.[38] Importantly, trophic

Table 13.1 Statistics Frequently Calculated for Network Research in Invasion Science, and Terms Often Used to Characterize Parts or Aspects of Networks

Statistic or Term	Definition
Network-wide statistics	
connectance	number of actual links in a network divided by the number of links possible if each node were connected to all other nodes
compartmentalization	degree to which a network is divided into subsets of nodes (compartments) not connected with nodes in other compartments
modularity	degree to which a network is divided into subsets of nodes (modules) connected to more nodes within the module than to nodes in other modules
nestedness	degree to which nodes with few links are linked to nodes that are subsets of the nodes to which other nodes are linked, rather than distributed randomly throughout the network
evenness	degree to which nodes throughout the network tend to be similar in the number of nodes to which they are linked (node degree)
degree distribution	how often nodes with different numbers of links occur in the network
Individual node statistics	
centrality	any of several statistics intended to capture the "importance" of a particular node relative to that of other nodes in the network[1]
node degree	number of links connecting a node to other nodes
dependence of node A on node B	in quantitative networks, the number of interaction events between a node (A) and one of its linked nodes (B) relative to the number of interaction events between that node (A) and all its linked nodes[2]
importance or node strength	the sum of dependences of all linked nodes on a particular node
Terms	
motif	particular type (topological shape) of subnetwork embedded within a network that occurs more frequently than expected
hub or connector	node with exceptionally large number of links, or a node that connects two compartments

1. A central node is one that most paths from one node in the network must travel through to reach another node, while fewer paths must travel through a less central node to reach another node.

2. Number of interaction events is the most common measure of dependence, but other measures (e.g., energy flow) could be employed.

interactions in a food web constitute a network, and, as can any network, they can also be characterized as a matrix. If the resource species are represented by rows and the consumer species as columns, then the intersection of a row and column in the upper triangular matrix can be a black square (or simply a 1) if that consumer eats that resource and a white square (or simply

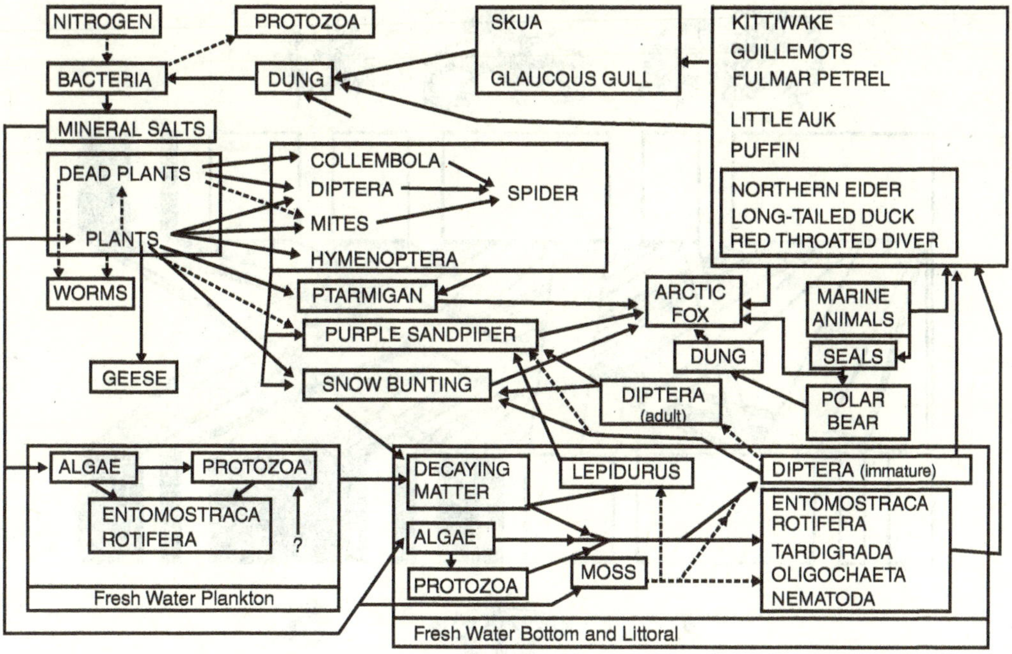

Fig. 13.2 Food web and nitrogen cycle on Bear Island, southernmost island of the Norwegian Svalbard Archipelago, Barents Sea. Credit: From Summerhayes and Elton, "Bear Island," 232. Courtesy of the Oak Street Library, University of Illinois at Urbana-Champaign.

a 0) if it does not.[39] Or the intersection can be a number if the interaction is weighted by how much of the resource the consumer eats.

Another form of species interaction within communities that can be represented by a matrix is competition. Competition between pairs of species has often been depicted by the Lotka-Volterra competition equations, and this approach was generalized by Richard Levins to a "community matrix" in which each species is represented by both a row and a column.[40] In the matrix, the value at the intersection of a row and a column representing competing species is the competition coefficient between the two species (usually the pairwise Lotka-Volterra competition coefficient, but other underlying models are possible). A group of species whose sole relationships with one another are competitive does not lend itself readily to a pictorial representation, as do trophic interactions, so the developing literature on competitive communities through most of the twentieth century did not feature pictures in the same way that food webs were common in research on trophic interactions. Though focusing primarily on competition, Levins also envisioned predator-prey pairs as part of the community represented by the matrix. Competitive and predator-prey interactions are often lumped together, along with parasitism and pathogenic disease, as "antagonistic" interactions.

Fig. 13.3 Major pollinators and plants in pollination network in 2000 on Flores Island, Azores. *Upper row*: plants; *lower row*: pollinators. *Black boxes*: non-native species; *white boxes*: native species. Size of boxes is proportional to number of pollination visits. Light gray connector indicates interaction between two native species; dark gray connector indicates interaction between a native and a non-native species, or between two non-native species. Credit: Data from Olesen et al., "Invasion of Pollination Networks on Oceanic Islands"; Aizen et al., "Invasive Mutualists Erode Native Pollination Webs."

Another type of interaction within a group of species is mutualistic, such as interactions between plants and pollinators or plants and their seed dispersers. These interactions have typically been portrayed as a particular kind of network, known as a "bipartite network" (fig. 13.3), with the associated matrix called a "bipartite graph" or a "bigraph." In a bipartite network, the nodes can be divided into two sets—say, A and B—in which each link connects a node in A to a node in B.[41] This is perhaps the simplest type of multispecies network and has been depicted and analyzed in many disciplines, including biology.[42] In invasion science, it has primarily represented plant species and their pollinators, first in 2002 in tabular form by Jens M. Oleson and colleagues and soon thereafter in network portrayals by many others.[43]

Although Robert May generalized the community matrix to incorporate not just competition and predation but all pairwise interactions, including mutualism, most research through the twentieth century focused on one sort of interaction at a time.[44] A few researchers studied multiple types of interactions, such as Michael S. Ringel and colleagues, who embedded a mutualistic two-species interaction (pollination) in a larger system including antagonistic interactions (predation, nectar theft), and Sophia R.-J. Jang, who modeled a three-species system consisting of a plant, an herbivore, and a pollinator.[45] Most research along these lines sought to understand the conditions under which a modeled system was "stable." Stability in ecology has been defined in multiple ways, but research in ecology and invasion science has largely focused on the circumstances under which a perturbation, such as introducing a new species or removing an existing one, leads to one or more species of the new community (including the newly introduced species in the case of introduction) disappearing from the system (population of zero) according to the equations governing the dynamics of interactions among the species.[46]

Importantly in invasion research invoking network science, "stability" in the sense of species loss following disturbance has rarely been studied directly in nature. Rather, when networks are depicted in matrix form with equations (e.g., Lotka-Volterra equations) characterizing the links, stability has usually been sought mathematically: did the mathematical equations lead to one or more species having a population size of zero, and, if so, how long did it take for a species to go extinct? With networks depicted as a collection of nodes and links in space, stability has been studied by simulation. Species were sequentially removed from the network according to a preselected algorithm (e.g., least connected nodes removed earliest) and whenever removal of a node caused a linked node to have no further links, the unlinked node was removed.[47] Again, stability was represented by how many species went extinct after how many steps in the simulation.[48] In either case—the equations and a matrix or simulation and a network—whether extinctions would ensue in nature as predicted by the equations or the simulations was not determined in this sort of research. Nico Blüthgen pointed to several assumptions in the construction of networks from field-collected data that made him skeptical about whether, in fact, predictions about changes in network structure or extinctions based on a simulated change in such a network would be accurate if the change were actually to happen in the real system on which the network was based.[49] An interesting experiment seemed to support this skepticism.[50] The pollination network for a meadow was constructed based on substantial field observations. A node representing one of three hub plant species in the meadow was removed from the network, and subsequent network changes were determined by a method commonly used to analyze changes in networks.[51] Most network statistics remained relatively unchanged. The species represented by that node was then experimentally removed from the meadow. A surprising result was that a previously minor plant species became a hub species.

Early consideration of stability of ecological systems led to the view that "complexity" (large numbers of species with many connections between them) conferred stability. Charles Elton is often cited as the key exponent of this view.[52] He listed greater species diversity as one explanation for why certain types of communities seem less readily invaded than others (a form of stability).[53] May greatly changed the standard view, showing with models based on his generalized community matrix that, in fact, randomly increased diversity and complexity lowered the likelihood that the resultant community was stable.[54] This result precipitated a plethora of research aimed at exploring the basis of stability in mathematical models of ecological communities.[55] Because introduction of a new species is construed as a destabilizing disturbance, it was inevitable that, amid this research, efforts would arise seeking to understand the impact of invasions on existing communities by studying the dynamics of the equations governing their interactions. For instance, Ted Case explored the outcome of invasions into a community of competitors interacting according to the Lotka-Volterra competition equations, and Steven J. Mithen and John H. Lawton similarly examined the outcome of invasion by a new predator or new prey into a food web governed by the Lotka-Volterra predation equations.[56] In both cases, the immediate question was whether the new, expanded community is stable in the aforementioned sense. However, such efforts were quickly augmented and even overtaken by explorations undertaken by new methods associated with network science.

13.2.3 The Integration of Network Science into Invasion Science

In the late twentieth and early twenty-first centuries, network science quickly entered many disciplines, particularly those focused on systems considered "complex." Ecology with its myriad species and their interactions at the community and ecosystem levels was no exception.[57] Perhaps the first application of network science to ecology was in 1961, and the number of articles rapidly increased to 3,063 in 2012, with the percentage of ecology papers in this category increasing from less than 1.5 percent in 1991 to over 5 percent in 2012.[58] Invasion science lagged by about a decade.

A Web of Science search through 2019 showed the first such publication in invasion biology in 2002, followed by a trickle of papers until four papers in 2008, fifteen in 2009, and nine or more annually beginning in 2013 (fig. 13.4).[59] Of 146 publications identified in the search, 67 were of bipartite pollination networks, consisting of a row of pollinators (usually insects), a row of plants, and a line between a particular pollinator species and a particular plant species if that animal pollinates that plant (depicted previously in fig. 13.3). Tiffany Knight and coauthors described a massive shift in pollination research in general (and including the influence of invasions) since the beginning of the twenty-first century from focus primarily on pairwise interactions between particular pollinators and particular plants to focus on entire communities conceived as networks.[60]

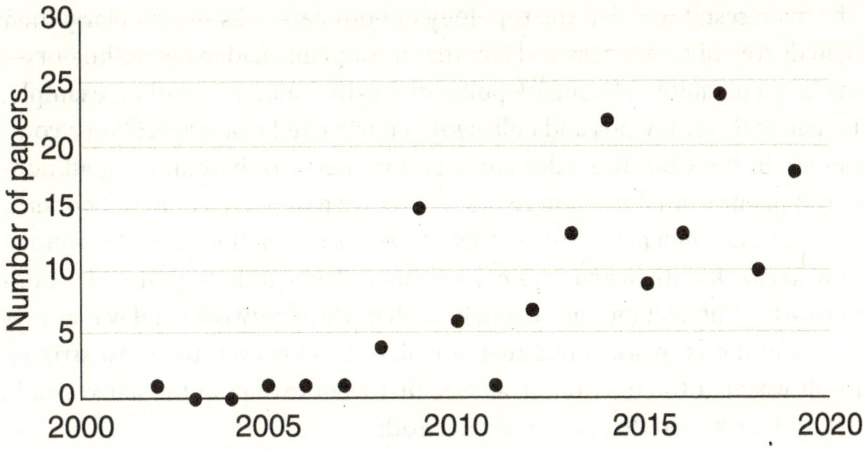

Fig. 13.4 Number of network analyses reported in invasion science literature, from Web of Science search on February 21, 2021, with search term combination (("non-native" OR "nonnative" OR "exotic speci*" OR "alien" OR "invas* speci*" OR "biology* invas*") AND ("network*" OR "meta-anal*" OR "metaanal*")).

The great majority of the 67 pollination network publications detected by the search examined networks based on observed systems in nature, tallying various network-wide statistics such as connectance or compartmentalization. Often these efforts compared an uninvaded pollination network with one similar in species composition but containing one or more introduced plants or pollinators. For instance, Christopher Kaiser-Bunbury and colleagues examined two otherwise similar heathland sites on Mauritius, one heavily invaded by strawberry guava and other invasive plants, whereas these had been removed from the other.[61] For each pollination network, they calculated, among other things, connectance and evenness as well as mean node degree for animal species and for plant species, although these network statistics did not reveal nearly as much about the impact of the invaders as did statistics based on particular species (e.g., which animal species visits which plant species how often). Diane L. Larson and collaborators studied the bipartite pollination networks of prairie plots in South Dakota that were invaded or not invaded by Canada thistle.[62] They calculated both network-wide and species-level statistics, finding lower modularity of networks in invaded plots and that certain species are hubs or connectors.

Several studies first constructed pollination networks from field-collected data and then manipulated the networks hypothetically. For instance, Jane Memmott and colleagues constructed bipartite pollination networks from data on flowers and their visitors from two sites in North America, then sequentially removed pollinator species from the networks according to three different algorithms and determined which plants would be left without pollinators—these would be assumed to go extinct in nature.[63]

The main result was that the topology of both networks—particularly their high degree of nestedness and fact that most plants had many pollinators—made them quite tolerant of pollinator extinction. In another example, Fernanda S. Valdovinos and colleagues constructed a network from a community in the Chilean Andes and a second network by removing all non-native plants from the first network.[64] They then subjected both networks to evolution through a dynamic model of species interactions and determined how network statistics such as connectance changed through time for both networks. Connectance increased in both types of networks and was always higher in the network containing only natives. However, the most striking result was that the fraction of species that went extinct in the natives-only network exceeded that in the real network.

Yet other research constructed a pollination network from field observations, then, based on the network, experimented on the underlying community. An experiment by Martha E. Lopezaraiza-Mikel and associates consisted of constructing a pollination network for a plant community in Great Britain heavily invaded by Himalayan balsam, then physically removing Himalayan balsam during its flowering season from certain plots and leaving it in others.[65] The comparison between the network of the control (invaded) plots with that of the removal plots allowed them to assess the invader impact on pollinator visits to native plants and pollination of them. However, the researchers did not calculate network statistics from the networks they constructed. Laura Russo and colleagues simulated adding non-native plants to the network based on the removal plots of Lopezaraiza-Mikel and colleagues, having the added plants interact with the native plants according to various models, and compared six network statistics of these simulated networks to that of the real balsam-invaded network.[66] A key result was that a non-native plant that is a "generalist" (interacts frequently with many pollinators) will have a larger impact on the topology of the network than will a non-native plant that interacts with few pollinators. To test the generality of this finding, Russo and associates introduced the very generalized non-native plumeless thistle into plant communities in Pennsylvania constructed of five native plants, with control uninvaded plots, and examined various statistics of the resulting networks.[67]

Other studies produced hypothetical pollination networks, not based on real systems, and then subjected them to hypothetical invasions. For instance, Valdovinos and colleagues created many hypothetical pollination networks with various numbers of species and degrees of connectance and nestedness, with dynamic interaction equations governing their trajectories through time. They then "invaded" them with hypothetical additional pollinators and asked which features of the invaders and invaded networks rendered the networks most likely to sustain various sorts of impacts.[68] The most striking results reflected properties of individual species, both members of the original networks and invaders. Knight and her team suggested that most of the twenty-first-century literature on pollination networks in general has been descriptive and characterizes interactions as static, and they argued that progress on employing networks to understand functions and predict responses to various perturbations (including invasions) will require sufficient data to model the network dynamically yet realistically, and also to integrate pollination networks with other networks such as food web networks.[69]

Of the 146 publications returned by the search for networks and invasions, 39 were on food webs. As with pollination networks, most of these illustrated networks described from nature and simply calculated both network-wide and species-level statistics. Several of these efforts attempted, in one way or another, to compare invaded networks with uninvaded ones. For example, the Newfoundland terrestrial mammal community of thirteen native species was progressively invaded by ten non-natives beginning with human settlement in the late fifteenth century. One study used present-day information on trophic relationships of the various species to construct the hypothesized food web networks that arose as the invasions proceeded, finding an increase in connectance powered by an increasing number of species at intermediate positions between top predators and prey species.[70] Another project experimentally added the non-native seaweed Japanese wireweed to a community of native algae, while maintaining a control without the non-native, and then characterized the food web networks of the species colonizing the invaded and uninvaded communities.[71] As in the Newfoundland study, the invaded network had increased connectance.

Several researchers first constructed food web networks from field data, then added hypothetical invaders, and compared network statistics between invaded and uninvaded networks. Thus Núria Galiana and colleagues constructed food web networks from data on Pyrenees terrestrial vertebrate communities and then introduced virtual species with various characteristics. The result was that invasions generally lowered the number of species and led to fewer links per species that remained.[72] They found by various measures of complexity that more-complex networks resisted invasion more strongly (contradicting May's theoretical result) but that invasions succeeded even in networks with relatively high connectance. Other studies used wholly hypothetical networks. An early effort entailed constructing 150 hypothetical networks with varying levels of connectance and then invading them with species with various traits.[73] In general, invasions of networks with low connectance were more likely to succeed than invasions of high-connectance networks, although the exact collection of traits of the hypothetical invaders was important in predicting success or failure. By contrast, in a study that constructed hypothetical food web networks with three levels of connectance, lower connectance characterized networks more resistant to invasion, in line with May's finding.[74]

Network research has been undertaken on other sorts of invasions in addition to invasions of pollination and food web networks. For instance, one team studied the effect of a non-native plant on a network based on a natural system. The network consisted wholly of plants in which two species were connected only if pollen was transferred by any means between them.[75] Six studies focused on the impact of one or more non-native plants on bipartite networks based on field observations of plants and animals that disperse their seeds. Among other research on networks constructed from field data, scientists examined a network consisting of both native and non-native ants and plants with extrafloral nectaries visited by the ants; networks of marine algae, some with a non-native kelp, wakame, and epiphytic organisms that settled on them; and the food web network of a lake invaded by two non-native fish and five non-native parasites associated with them.[76] A network based on observations of forty-four plant species and fifteen bird species that disperse their seeds in Hawaii was of particular interest on two grounds.[77] First, the great majority of both birds and plants at the seven studied sites were non-native (thirteen birds and twenty-nine plants), yet the various network-wide statistics seemed not very different from analogous statistics from forty-two published networks of predominantly native plants whose seeds are dispersed by predominantly native birds. The researchers suggested that this fact implies that perhaps highly coevolved interactions, at least involving avian seed dispersal, are not as important as traits such as beak, seed, and fruit sizes of species, whether or not the species are native, in determining network topology.[78] The second striking result was that networks of the seven individual sites were topologically similar though the interactions were often quite different between sites, even when the sites largely shared the same species.

Some network science studies in ecology qualify as "metaecology" research. For instance, Stuart Borrett and colleagues constructed networks of topic diversity and scientific collaboration in ecology in scientific publications through 2012.[79] For topics, the network depiction was based on "coword frequency" (similar terms used in the titles, abstracts, and keyword records of publications), while for collaboration among researchers the network was constructed based on joint authorship. Striking was that fact that "invasion" was not among the sixty-three most frequently cited ecological concepts or methods in their search, although several of these concepts are frequently associated with invasions and have been the topic of invasion network research (e.g., "food webs," "predator-prey interactions," "[diversity or biodiversity] and stability," "competition and species exclusion"). Neither did the term "invasion" appear in the depiction of the topic network, nor did the depicted networks of authors indicate particularly prominent students of invasion networks. Perhaps this low visibility reflects the relative recency of the invasion of invasion science by network science; as noted previously, the substantial contribution of network publications to invasion science began in 2013.

Other network publications fall in the "meta–invasion science" category. In one, a network was constructed based on publications on invasions relating to the plant family Poaceae to examine the extent to which authors integrated various ecological and evolutionary concepts, using titles, abstracts, and keywords of relevant papers.[80] The major findings were that ecological concepts were more frequently studied and better connected than were evolutionary concepts and that certain connections, generally involving evolutionary concepts, were not found at all, suggesting important research gaps. A more recent study focused on collaboration networks in the South African Working for Water program as a case study on managing biological invasions.[81] The main finding was the key role of a few "core" authors in shaping research directions, mediating collaborations, and maintaining network cohesion. Topic analysis suggested these authors stressed ecological research, leading to an underrepresentation of social science research.

: : :

In ways that simple vote counting cannot, formal meta-analysis with correctly calculated individual effect sizes can provide information that can aid understanding of the scope and magnitude of various impacts wrought by invasions as well as of patterns of features of non-native species relative to those of native species. However, invasions and their impacts are highly idiosyncratic and context-dependent, and the calculation of mean effect size in a meta-analysis connotes that invasions are close replicates of one another, as are replicates in highly controlled medical experiments. In fact, this is rarely the case. Publication of a mean effect size, even with error bars, can thus lead readers to overlook differences among particular invasions and impede understanding of their characteristics and impacts and can suggest incorrect policy and procedures for management of invasion impacts. These problems can be ameliorated by careful attention to interpretation and wording in reports of meta-analyses.

Although it is recent, the integration of network science into invasion science has already generated reviews for food web networks and pollination networks and a book on the topic as a whole.[82] If one were to ask what we have learned from the application of network science to invasions, it seems that the main take-home messages so far largely reside in the specific traits of particular species, either the non-native invading ones or particular natives, and the pairwise interactions between them, rather than the network-wide topology as represented by network-wide statistics. Thus, for instance, degree of connectance varies among studies of real networks, even within the same category, such as food web networks, and those researchers who have explored particular network-wide findings, such as differing connectance between invaded and uninvaded networks, have usually appealed for explanation to the specifics of particular species in the networks or their interactions. The conclusion of Knight and colleagues regarding pollination networks—that the findings so far are largely descriptive and may set the stage for more explanatory research—pertains to the network invasion literature in general.[83] The increasing availability of software packages that calculate network statistics from the underlying matrix data has made it easy for authors to add network statistics to research on particular species and their interactions or on particular systems, or simply to see whether the resulting networks suggest generalizations or interesting follow-up research questions.[84]

It is widely recognized that these are early days for use of network analysis to study invasions.[85] Several authors have called for expanding network analysis to include not just one sort of interaction.[86] Knight and colleagues observed that pollination networks are embedded within larger communities and advocated combining pollination networks with food web networks and plant-microbe networks.[87] Sonia Kéfi and colleagues termed networks combining different sorts of interactions "multiplex networks," depicted a striking rocky Chilean intertidal example, and suggested that examining such multiplex networks will generate important new insights.[88] Though focused on food webs, Patrice David and colleagues pointed to the importance of other interactions interacting with trophic interactions when invasions occur.[89] A theoretical treatment combining mutualistic and antagonistic interactions indicated that the network statistics nestedness and modularity differ greatly in networks combining both sorts of interactions from those representing just one or the other.[90]

To advance the contribution of network science to ecology in general and to invasion science, several authors have suggested avenues beyond consideration of multiplex networks. Perhaps most frequently stressed has been amassing quantitative data on interactions so that network links are not simply lines of equal width, or of different hypothetical widths, but are proportional to their relative importance, and also so that temporal and spatial variation can be incorporated. Such data—how much of species A does species B eat, what fraction of the pollination of plant species C is accomplished by insect species D, and the like, and how do these things vary over time and location?—are difficult to gather, but the proliferation of molecular techniques such as stable isotope analysis and various molecular genetic methods may increase their accessibility.[91] A particular challenge is posed by "ecosystem engineers," species that create new habitats that can greatly affect existing species as well as facilitate persistence of new non-native species. Among non-native species whose engineering has generated many impacts are beavers from North America in Tierra del Fuego and zebra mussels from the Black, Caspian, and Azov Seas introduced to North America.[92] To document and quantify the many network changes wrought by the establishment of an ecological engineer will require extensive research. Rapid evolution of invaders or native species in response to invaders (discussed in chap. 11) may also substantially modify even an initially well-documented network. In 2002, Ron Milo and colleagues introduced the concept of "motifs"—topological patterns of links that appear in an observed network more frequently than in similar randomized networks—in systems biology, although versions of the concept had long been applied in ecology.[93] This approach—seeking recurring motifs in observed matrices—has since become a major component of network analysis in many sciences but has rarely surfaced in invasion science and may feature more prominently in future research.[94]

Perhaps the biggest advance in network treatment in ecology in general and in invasion science in particular would consist of turning networks into dynamic, mechanistically driven representations of communities rather that static, phenomenological snapshots.[95] In addition to the quantitative data described in this chapter, this transformation will require flexible and accurate functions representing pairwise interactions. If this transformation is achieved, network analysis could increasingly become an important predictive tool rather than a search for patterns, which would facilitate prediction of impacts of invasions and of management responses.[96]

As for what new insights might be generated by continued network-based explorations of invasions, the main one is probably a better understanding of the full range of impacts and the variety of types of interaction—direct and indirect—associated with an invasion. With an approach focused narrowly on an invader and one or a few species it interacts with directly (for instance, by predation), the entire scope of an invasion and its potential effects was subordinate if not invisible. A network with all interactions included at least makes the possible impacts evident and may focus empirical research on effects, particularly indirect effects, that would otherwise have been ignored. A substantially quantified network might also aid managers in pinpointing critical locations in the network to reduce the population and impact of a targeted invader.[97] The problem of unintended effects arising with management or eradication of an invader could be reduced by examining the entire network, particularly a mechanistically driven one.[98] In general, the topology of a network alone should at least suggest possibilities, but firm predictions will require quantifying the interactions, as noted earlier in this chapter.

Finally, as network science expands, a variety of different sorts of networks have been proposed that might be usefully applied to invasions. Multiplex networks are an example. So far, invasion scientists have largely been restricted to bipartite networks of pollinators and plants (or similar bipartite networks, such as plants and seed dispersers) and networks such as food web networks in which each node is a species or population and the nodes are connected by one type of interaction, such as a trophic interaction. The study of plant species connected only if pollen is exchanged between them employs a "unipartite" network, in the sense that the pollinating species in an underlying bipartite network are not represented.[99] Still other sorts of networks could be used, such as individual-based networks and multilevel networks.[100] Similarly, new approaches to analyzing effects of invasion on networks have recently been proposed as this book is being written. Edgar F. Cagua and colleagues applied control theory to several pollination networks based on observations in paired invaded and uninvaded communities, and Colin Campbell and colleagues used a model of community assembly for mutualists to study invasion of existing hypothetical pollination networks by entire communities of plants and pollinators.[101]

14

How Will Climate Change Affect Biological Invasions and Their Management?

Biological invasions are one of the four major global environmental changes of great concern to ecologists and conservationists—the other three are (1) land use changes entailing habitat destruction, (2) pollution, and (3) climate change.[1] Invasions and climate change were relative latecomers; land use change was first off the block. It was highlighted in George Perkins Marsh's prophetic nineteenth-century works (described in chap. 3, sec. 3.4) and dominated the volume *Man's Role in Changing the Face of the Earth* in 1956, perhaps unsurprisingly, as the book was dedicated to Marsh (chap. 7, sec. 7.4).[2] Brought to public attention by Rachel Carson's *Silent Spring*, pollution quickly replaced land use change as the focus of environmental and conservation attention in the 1960s.[3] Marsh had not addressed pollution, and *Man's Role in Changing the Face of the Earth* barely mentioned it. However, the first Earth Day in 1970 featured prominent speeches and other activities about pollution, and the next year the book *Man's Impact on Environment*, inspired by Marsh and the 1956 volume, signaled the rapid rise of pollution as an issue.[4] Of its fifty-two chapters, twenty-one were about pollution and only seven about land use change.

Biological invasions as a scientific and public issue percolated in the 1960s and 1970s and emerged as a modern science, major conservation concern, and public issue with the Scientific Committee on Problems of the Environment (SCOPE) project of the 1980s (discussed in chaps. 9 and 10). Climate science had some early antecedents (similarly to invasion science) and then matured rapidly in the 1980s. By the twenty-first century, global warming had become the leading environmental and political issue and a major and increasing conservation concern.[5] The intersection of global warming with invasions also became a prominent part of invasion science. It seems obvious that, say, giraffes would not survive in Canadian boreal forests or hippopotami in deserts, and early students of non-native species recognized the general principle that introduced species need suitable environments to survive. For instance, during the nineteenth-century heyday of acclimatization societies (discussed in chap. 5, sec. 5.1), John Gray, keeper of zoology at the British Museum, doubted that yaks from Tibet or zebu cattle from India would survive in Great Britain, averring that successful acclimatization required a physical environment resembling "as much as possible their natural proclivities."[6] Climate change—and global warming in particular—followed a similar timeline to that of biological invasions as a conservation issue and a public concern. Thus, interactions among climate change, species distributions, and invasions deserve particular attention.

14.1 Climate Science Recognizes Anthropogenic Global Warming

Climatology coalesced from a number of more or less independent efforts into a cohesive science in the second half of the nineteenth century.[7] A key development was efforts by Norwegian physicist and meteorologist Vilhelm Bjerknes, beginning in 1903, to forecast weather and to generate predictive climate models based on the laws of physics. Although his models could not be adequately developed and implemented for several decades because computing speed was too slow and necessary observational data were too scarce, Bjerknes's equations underpinned developments beginning in the 1950s that led to modern general circulation models now used to model and predict global and regional climates.

With respect to global warming, a major component of climate change, the key development was an understanding of greenhouse gases, especially carbon dioxide (CO_2), and the human role in their production. The principle was perhaps first enunciated in 1824 by French mathematician and physicist Joseph Fourier, who proposed that the earth was kept warm by its atmosphere acting as a greenhouse, with some of the sun's energy entering the earth being trapped by the atmosphere.[8] The recognition that CO_2 plays a key role in trapping that energy has usually been credited to Irish physicist John Tyndall.[9] However, Eunice Foote, an American scientist and early campaigner for women's rights, in fact was first to demonstrate the role of CO_2 in 1856.[10] In 1896, the Swedish physical chemist Svante A. Arrhenius argued that increased burning of coal was increasing atmospheric CO_2, that variation in atmospheric CO_2 concentrations was the main determinant of past climatic changes, including temperature changes, that increased temperature in turn would increase the amount of atmospheric water vapor, and that the increase in atmospheric water vapor would exacerbate the effects of the increased CO_2.[11]

In the mid-twentieth century, Charles D. Keeling, an American geochemist, played a major role in the subsequent explosion of concern about climate change. Keeling installed infrared gas analyzers at a meteorological station on Mauna Loa, an active volcano on the island of Hawaii. In 1960 he described cyclical seasonal variation in atmospheric CO_2 concentration in Hawaii, California, and Antarctica as well as a small but perceptible longer-term increase in Antarctica, which he hinted might be due to fossil fuel exhaust.[12] In 1965 a similar slight increase at Mauna Loa over four years was diagrammed, though the seasonal aspects were not readily gleaned from the figure.[13] With colleagues, Keeling continued to study geographic and temporal variation in CO_2, and in 1976 they depicted both the strikingly regular, saw-toothed seasonal cycles on Mauna Loa and an undeniable and inexorable regular increase in the annual mean for fourteen years.[14] This figure, which came to be known as the "Keeling curve," caught the attention not only of scientists but also of the public and of journalists and helped galvanize the developing environmental movement and focus it on climate change, which had previously been a relatively minor concern.[15] Versions have been featured in textbooks in many fields, including ecology and conservation biology; in newspapers; and in the opening chapter of *Earth in the Balance*, a book by then-Senator Al Gore, who soon became US vice president and whose book, along with the related film *An Inconvenient Truth*, went far to publicize many impacts of global warming.[16]

Advances in modeling, rapidly increasing computing power, and increasing data on greenhouse gases such as CO_2 led to a burst of activity. In 1977, the United States National Research Council assessed the models available, noted the variation in the predictions they produced, and determined that

> the span of agreement embraces a fourfold to eightfold increase in atmospheric carbon dioxide in the latter part of the twenty-second century. Our best understanding of the relation between an increase in carbon dioxide in the atmosphere and change in global temperature suggests a corresponding increase in average world temperature of more than 6°C, with polar temperature increases of as much as three times this figure.[17]

In 1979, the United States National Research Council concluded that concentrations of atmospheric CO_2 would probably double by 2030, that this would lead to an average temperature increase of 3°C, and that resolving any of the uncertainties in the modeling effort would be unlikely to change that prediction substantially.[18] American climate scientist James E. Hansen contributed much to the modeling effort and has become, somewhat controversially, a highly visible climate activist. In 1981, Hansen and colleagues reported that their climate model showed atmospheric CO_2 leading to warming more quickly than had been suggested.[19] In the 1980s, temperatures began to rise sufficiently that it became increasingly difficult to deny the change, although the causes—and especially the human contribution—remained controversial in certain circles. The issue began to dominate public discussions of the environment.

In 1986, SCOPE published a book on predicted ecosystem impacts of climate change focused primarily on agricultural ecosystems.[20] In 1988, the United Nations Environment Programme and the World Meteorological Organization formed a United Nations body, the Intergovernmental Panel on Climate Change (IPCC), charged with periodically assessing relevant research on anthropogenic climate change, including literature on impacts. The IPCC issued its first report in 1988 and quickly became the most authoritative and cited source on climate change. Successive reports largely confirmed earlier predictions of the United States National Research Council and Hansen and his colleagues. Its sixth report in 2021–22 issued a dire warning that, without immediate action to reduce greenhouse gas emissions, an increase in the range of 2°C will occur in this century. The section on impacts included a prediction of massive impacts on biodiversity.[21]

14.2 Ecological and Biodiversity Impacts of Global Warming

Because global warming causes massive coastal flooding, increased droughts, heat waves, and other extreme climatic events, as well as major agriculture failures and migration of "climate refugees," it is unsurprising that most of the concern with its effects as recognition of the issue grew was focused on impacts on human endeavors or well-being and not on how warming will affect ecological phenomena or species in nature except as these are clearly linked to human uses. Biodiversity and ecology remain secondary concerns, but recognition and discussion of impacts of this sort have grown rapidly.

As the relationship of enhanced atmospheric CO_2 to global climate change was confirmed and refined, American ecologist George M. Woodwell led efforts to understand the full scope of sources of the CO_2. In particular he pointed to a major contribution of land use change—felling forests.[22] This was already becoming a major concern of conservationists and conservation scientists from another angle—extinction of forest species.[23] Although this impact of deforestation and other forms of habitat destruction and land use change continue to be a huge conservation issue, the direct effects of increased atmospheric CO_2 on species and the concomitant warming on species persistence have gradually become an equal or even greater concern.

Disentangling these effects on species is not a trivial matter.[24] CO_2 and warming can each affect species independently of their joint effect. An international network of free-air CO_2 enrichment (FACE) sites in several ecosystems, including twenty-four sites focusing on noncrop plant species, was established beginning in the 1990s partly to address that problem.[25] Pure CO_2 or air enriched with CO_2 was released into open-top plots so that the concentration was maintained at a constant, elevated level. Generally, results showed that elevated CO_2 interacted with other factors—not only temperature but also hydrology and changed soil nitrogen dynamics—so that response to increased CO_2 varied between species. Certain species tended to grow better while others did not. The FACE project did not specifically address the response of non-native species, but some results suggested that enhanced CO_2 could greatly affect their spread and impact. For instance, Travis Belote and colleagues studied response of the understory plant community at a FACE site in East Tennessee consisting of a planted native sweetgum monoculture.[26] The understory was dominated by two non-native species, Japanese stiltgrass and Japanese honeysuckle, both highly invasive in the region. Japanese honeysuckle productivity was consistently 2.5 to 3.3 times greater in elevated CO_2 plots than in control plots, and its cover was three times greater in the wet year. By contrast, Japanese stiltgrass productivity was 1.7 times greater and its cover was 1.4 times greater in the control plots than in the elevated CO_2 plots in the wet year, but no differences between control and elevated CO_2 plots appeared in the dry year.

Direct and indirect effects on plants of CO_2 elevated to the degree anticipated by climate models have often been found. Demonstrated direct population effects on animals are so far few, if any. However, indirect effects may be common, even aside from the contribution of elevated CO_2 to warming. For instance, for insects that chew on vegetation, a general trend is that when they are fed on foliage grown at elevated CO_2 they grow more slowly, take longer to mature, and have lower survival, largely because of the impact of elevated CO_2 on plant nitrogen. Aphids suck up phloem rather than chew leaves, and among five aphid species, only two responded to elevated CO_2 even though leaf nitrogen concentration increased in the host plants.[27]

There is no dearth of evidence that increased temperature independently of CO_2 concentration can affect performance of many species. Most ecology textbooks include a section on Shelford's law of tolerance, which states that, for any physical factor—and temperature is the paradigmatic factor—every species has a maximum and a minimum compatible with survival and reproduction and an optimum value somewhere between these two extremes (fig. 14.1).[28] The law has been qualified by corollaries—especially that different physical factors can interact, so that, for instance, the maximum, minimum, and optimum temperature might depend on the humidity level. In the context of the law of tolerance, researchers have conducted laboratory studies on many kinds of species in which temperature is varied, CO_2 is constant among treatments, and other factors are controlled.[29]

For non-native species, a meta-analysis of performance by both animals and plants found native and non-native species responding similarly

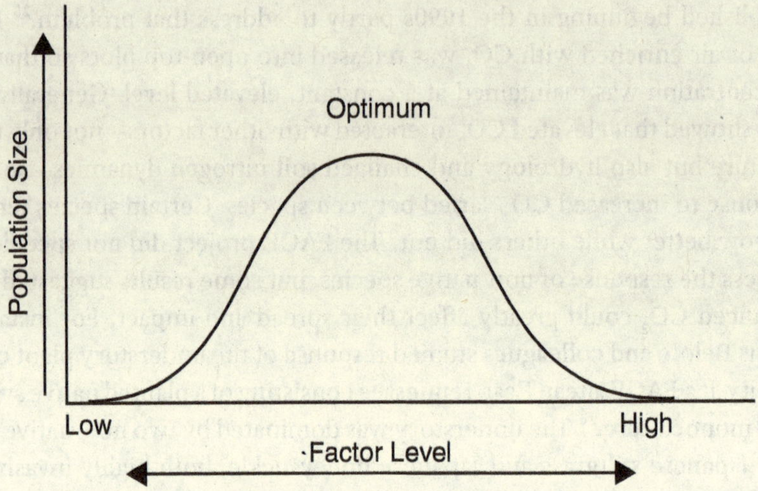

Fig. 14.1 Shelford's law of tolerance, schematically depicting range of suitability of any physical environmental factor for a given species, with minimum and maximum values compatible with survival and reproduction as well as optimum value. Cf. Shelford, "Physiological Animal Geography."

to varied temperature, CO_2, and precipitation in terrestrial (mostly plant) systems but a different result in aquatic (mostly animal) systems.[30] In the aquatic systems, increased CO_2 and temperature often reduced aspects of fitness in native species, whereas non-native species frequently responded positively to these changes. A scan of studies of freshwater invertebrates and fishes found a tendency for non-native species to tolerate increased temperature better than coexisting native species.[31] A meta-analysis of the response of marine plants and animals to increased temperature found that non-native species tended to respond positively and native species negatively.[32]

14.3 Changes in Native and Non-Native Species' Ranges and Effect on Invasiveness

With just a few years of data, it is difficult to determine whether a change in a species' geographic range is simply a short-term reflection of small fluctuations that must constantly occur at most range boundaries.[33] However, abundant evidence shows that many native species are shifting their ranges unidirectionally in response to climate change.[34] A good example is the wasp spider of Europe, which, beginning in 1930, spread from the Mediterranean region all the way to Norway, Sweden, and Finland, initially because warming permitted persistence in new regions and subsequently because the spread brought together formerly isolated genotypes, which interbred and produced new genotypes adapted to colder environments.[35] For non-native species, data are increasingly reported that strongly suggest warming-induced range expansions. For instance, for Great Britain, Philip Hulme found evidence that thirty-one non-native species (mostly insects and plant pathogens) had expanded their ranges in response to a warming climate.[36] One of them, the southern green stink bug or green shield bug, a major agricultural pest, is native to Ethiopia and has been widely introduced accidentally throughout much of the world. It has recently extended its non-native range in several regions, including Great Britain, Japan, Hungary, and Germany.[37] Strong evidence suggests that these expansions reflect adaptations to a warmer climate.[38]

A widely noted large range expansion is that of blue tongue virus, which causes a disease in ruminants, especially sheep and some deer species.[39] Before 1998, the virus occurred only occasionally in Europe, restricted to parts of the Mediterranean basin. It was transmitted there and in Africa, where it has long been present, by a midge species, *Culicoides imicola*, believed to be limited to tropical and subtropical climates. Beginning in 1998, the midge spread throughout the Mediterranean basin, carrying the virus. However, in 2006, an outbreak of blue tongue disease occurred in Belgium, the Netherlands, Germany, and France, far beyond the known range of the midge. The warming European climate allowed the virus to survive the winter where it could not have previously, allowed the midge to expand farther north, and, in the northernmost regions where blue tongue virus now persists beyond even the expanding range of the original midge vector, it is being transmitted by a congeneric native European midge species.

Several other studies have implicated climate warming as contributing to increased ranges of non-native species. For example, on the European Atlantic coast, increasing seawater temperature over fifty years caused a major northward expansion of the range of the non-native Pacific oyster, from southern France to Norway.[40] Other research has demonstrated increased impact within current ranges. For instance, in coastal Connecticut waters, warming caused earlier recruitment of three non-native invertebrates: the European sea squirt, the chain tunicate, and another tunicate, *Diplosoma listerianum*. Warming also increased these species' growth rates and recruitment relative to those of native species.[41] On the North Island of New Zealand, a native shrub, monoao, is disadvantaged by the impact of global warming on the flowering period of non-native heather.[42] Warming greatly increases the flowering overlap of monoao with heather, and competition with heather for insect pollinators decreased pollination of monoao in areas with high densities of heather flowers. On Maui, the devastating fires of 2023 were fueled and exacerbated by highly flammable non-native grasses, such as fountain grass, buffel grass, Guinea grass, and molasses grass. These had been brought to the Hawaiian Islands as forage for cattle beginning in the late eighteenth century and became highly invasive in many settings.[43] On Maui, they dominated sugarcane and pineapple plantations abandoned in the late twentieth and early twenty-first centuries, and warming and increased drought conditions facilitated the 2023 fires. Scientists had warned for several years that failure to manage the abandoned grasslands combined with changing climate would lead to disastrous conflagrations, but management by landowners in the vicinity of the 2023 Maui fires—primarily the State of Hawaii, an educational trust, and a commercial land development company—was inadequate or perfunctory.[44]

A key feature of global warming is an increased frequency of episodic extreme events, such as droughts, catastrophic fires, and hurricanes. Many impacts on biodiversity of such extreme events have already been documented.[45] One such impact that has not been emphasized is the likelihood of increased long-distance dispersal of species, introducing them to non-native regions. Such events are already well known with respect to particular invasions.[46] For instance, soybean rust, a fungal pathogen of both soy and also other legumes (including kudzu), was first reported in 1902 in Asia and soon thereafter in Australia. By the mid-1990s, it had spread to Africa, and then, probably dispersed by wind-borne spores, it appeared in South America in 2001, quickly spreading widely there. It first appeared in North America in 2004 in the wake of Hurricane Ivan near the landfall of that storm.[47] Severe flooding has allowed several aquatic species to reach new locations by escaping from ponds or bypassing dams.[48] The Japanese earthquake of 2011 produced a tsunami that tore many items from the coasts into the Pacific Ocean. These rafts carried numerous organisms—by 2017, 289 coastal marine animal species had landed on Hawaiian and North American shores.[49] Extreme ocean warming events have permitted dispersal of many marine organisms to distant new locations, such as Xantus's swimming crab, a warm-water species whose range extended 400 kilometers (about 250 miles) northward along the California coast during an extreme high-temperature event in 2016.[50]

Beginning around the turn of the millennium, predictions have greatly exceeded direct observations and experiments implicating climate change in expanded range of both native and non-native species. These have usually been based on statistical methods (outlined in the following section 14.4) that predict that ranges of certain native and established non-native species will shift (usually expand poleward) and that non-native species that had been unable to gain a foothold in a particular region because of a stressful climate will now be able to establish populations. For instance, a team led by Bethany Bradley conducted such analyses for non-native cheatgrass, kudzu, cogon grass, privet, Chinese privet, leafy spurge, yellow star thistle, spotted knapweed, and salt cedar in North America.[51] A tour de force of such analyses predicted the trajectories of 896 non-native terrestrial plant species in the continental United States, pointing to range contractions as well as range expansions.[52] Such forecasts are now common. These include a prediction that climate change will help the bullfrog from North America invade many South American protected areas, and another that the yellow-legged hornet native to Asia will expand its range in Europe owing to climate change and that it will increasingly invade much of the United States.[53] Céline Bellard and colleagues applied climate models and statistical approaches relating climate to species' suitability to the widely cited list of one hundred of the world's worst invaders.[54] She and her colleagues predicted various future changes, with ranges of invasive amphibians and birds often shrinking and invertebrate ranges often expanding.[55] To explore the history of thinking about climate change and invasions, I must digress to trace the history of how scientists studied the problem, after which I will resume assessing the empirical evidence as well as the predictions.

14.4 Predictions of Range Expansion

By the early twentieth century, many scientists had concluded that expansion of non-native species in their new ranges was limited by weather, especially temperature and precipitation. In 1908, E. Dwight Sanderson summarized literature of this sort on ten insect species introduced into the United States plus one (the boll weevil) that had entered and spread in the United States after previous restriction of its native range to Central America and Mexico.[56] He detailed his own research on the browntail moth, native to Europe and nearby Asia and Africa, which was accidentally introduced near Boston, first discovered in 1897, and which rapidly spread to occupy much of New England and adjacent Canada.[57] Sanderson compared minimum temperatures in the European range of the moth with those at the then-current northern limits of its new American range and also recorded mortality of their caterpillars at many American sites. He concluded that the minimum temperature was indeed the main limit to its further northward expansion but suggested that the species is kept in check in northwestern Europe by high humidity, which favors a pathogenic fungus. In general, from these observations and the other cases, he concluded that the key weather statistic was not the absolute minimum temperature but the average annual minimum temperature.[58]

14.4.1 Climographs

An advance in depicting climates set the stage for a more comprehensive treatment of climatological limits to non-native species establishment and spread. In 1910, climatologist John Ball depicted what he termed "climatological diagrams." With temperature on one axis and relative humidity on the other, for any particular site (in his example, Egyptian and Sudanese locations) he plotted twelve points, one for each month, and connected them sequentially.[59] Apparently independently, T. Griffith Taylor produced the same innovation but used the term "climograph," which persisted in subsequent literature.[60] The only difference was that Ball had used dry bulb temperature whereas Griffith used wet bulb temperature, which depends partly on humidity. Climographs came to be used to describe conditions (e.g., temperature, humidity, rainfall) at sites occupied by a given species and thence to predict the future range of that species.[61] In 1925, William C. Cook, having previously applied climographs to understand aspects of native species' ecology, was perhaps the first to use climographs to predict where a recently arrived invader would eventually spread in its new range.[62] The alfalfa weevil, native to Europe, south-central Asia, and North Africa, was first detected in North America near Salt Lake City, Utah, in 1904. When Cook wrote, it had spread beyond Utah to parts of Oregon, Idaho, Wyoming, Nevada, and California, and Cook asked what the North American limits would be. To address this question, he constructed climographs from the native range (locations in France, Turkestan, Denmark, Russia, and present-day Georgia and Armenia), comparing their amount of overlap with climographs at points where it had established in Utah and in Idaho and Colorado near the edge of its then-current range. He concluded that the alfalfa weevil had already approached the limits of its climatological niche. In fact, the alfalfa weevil later spread throughout the East, but that was from a separate invasion in Maryland, and, in any event, Cook did not consider climate change.[63]

Cook later suggested that his climograph method should be used to predict not only future range of any recently arrived non-native insect but also its abundance in different parts of the future range.[64] This paper, published in 1931, served as an introduction to research conducted as a thesis for a bachelor's degree under Cook's direction by Claude M. Gjullin on the probable range in the United States of the Mediterranean fruit fly ("medfly"), which had recently arrived in Florida.[65] Gjullin constructed climographs (using rainfall rather than humidity) for the medfly in several regions to which it had been introduced and had survived: the Hawaiian Islands, France, Spain, Australia, and South Africa.[66] He then drew climographs of several locations in the United States and superimposed them on a large area in a two-dimensional space (temperature and rainfall) composed of all points occurring in one or more of the climographs from the previously invaded regions. From this superimposition, he deduced locations in the United States where the medfly could survive, suggesting that severe infestations should be expected in southern Florida and southern Texas. In fact, medflies have subsequently been found in Florida, Texas, and California but eradicated, although whether the eradication campaign in California succeeded is controversial.[67] Foreshadowing an issue that would arise many years later regarding predicting ranges, Henry J. Quayle quickly questioned

Gjullin's result on the grounds that presence of suitable fruit hosts would allow the medfly to flourish in areas Gjullin had concluded were at best marginally suitable.[68] Andrew D. Hopkins, using a different method of assessing climatological factors, predicted that Florida summers were too hot and rainy for the medfly.[69]

In 1936, Arthur Twomey used climographs of temperature and relative humidity (or rainfall) to try to understand the fates of several bird species introduced to the United States.[70] The grey partridge was of particular interest, as a game bird native to western Eurasia that had been introduced to the United States from the late eighteenth century through the early twentieth century and had flourished in some states but failed in others. Twomey compared a climograph of its native range in central Europe to those of two North American sites where partridge introduction had failed and three where the species had thrived. Large parts of the climographs of the two failure sites were outside that of the European site, while climographs from the three regions with successful populations largely coincided with the European climograph.[71]

Twomey similarly used climographs to explain why Texas bobwhite quail introduced to Illinois had failed to establish a population and why the skylark from Europe did not persist in New York and San Jose, California, as well as why it survived on Vancouver Island.[72] Alvin R. Cahn, who had guided Twomey's study as a graduate project, used the same method to predict the outcome if three non-native gamebirds (grey partridge, ring-necked pheasant, and chukar partridge) were introduced to the Tennessee Valley and vicinity (fig. 14.2).[73] Superimposing climographs of locations in or near the Tennessee Valley on climographs of the native ranges, he suggested that the grey partridge could not persist except perhaps in the vicinity of Boone, North Carolina. He also said the pheasant would probably fail throughout the region and the chukar partridge could not possibly persist, based on the fact that not a single month in this area approached conditions in its native Indian range. In fact, pheasant introductions in this region have failed. No record exists of attempts to introduce the two partridge species.[74]

Climographs (sometimes called "climograms") are subjective, visual assessments. They continued to be employed in ecology for a while, but their use as predictors of invasion success and ultimate range has waned, and they have been superseded by statistical methods outlined in the following subsections 14.4.2, 14.4.3, and 14.4.4. Rather, they have occasionally been used as a compact way to display the overall climate in a place and to compare climates between places, and they sometimes motivate or rationalize field or laboratory research on the effect of particular physical factors, especially temperature, on specific life-history stages of a species.[75] As noted previously, various physical factors have been used as climograph axes, and occasionally climographs have depicted three dimensions. A few authors have

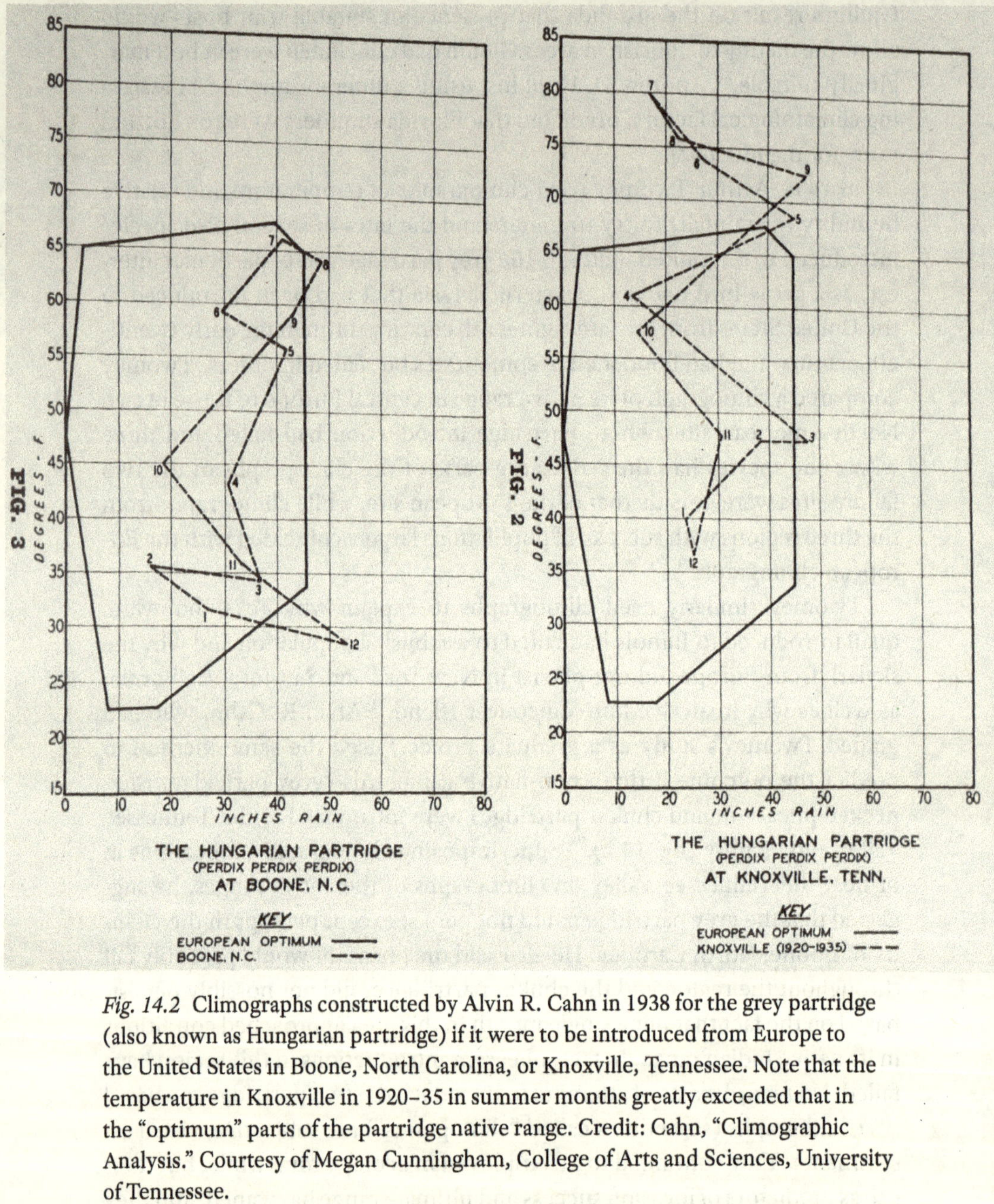

Fig. 14.2 Climographs constructed by Alvin R. Cahn in 1938 for the grey partridge (also known as Hungarian partridge) if it were to be introduced from Europe to the United States in Boone, North Carolina, or Knoxville, Tennessee. Note that the temperature in Knoxville in 1920–35 in summer months greatly exceeded that in the "optimum" parts of the partridge native range. Credit: Cahn, "Climographic Analysis." Courtesy of Megan Cunningham, College of Arts and Sciences, University of Tennessee.

analogized the climograph to an ecological niche. What no one seems to have done is to have extended the concept to some sort of *n*-dimensional solid similar to the conception of a "fundamental" ecological niche determined by physical factors as developed by Joseph Grinnell and G. Evelyn Hutchinson.[76] Thus, the choice of variables included in the climograph, and on which predicted range was based, was limited and determined by somewhat arbitrary means—for example, Shelford's "general knowledge."[77] An advance beyond this approach began in the 1950s with "geographic information systems."

14.4.2 Geographic Information Systems

Although a species' range can be mapped, climographs that were used to predict the ultimate range of non-native species were generally not depicted on a map. The range prediction was generated by constructing a series of climographs at different locations and comparing each to the climograph from a location in the native range. Gjullin did produce a map, and he did so indirectly from climographs of individual locations.[78] Further, climographs used only two types of climate data, usually mean temperature and mean rainfall or relative humidity, or rarely three, whereas many distinct aspects of climate, alone or in conjunction with one another, can affect survival and reproduction, and therefore the range of a species.[79] Thus, for instance, Twomey suggested that the weather in May through September, not annual mean temperature, is the key factor determining persistence of grey partridge in North America.[80] Similarly, minimum temperature lasting for a particular period may be crucial for an insect, rather than mean winter temperature.[81]

An innovation that began in 1950 greatly changed the approach to predicting species ranges, not only of spreading or proposed introduced species but also of native species in the face of climate change. Although landscape architects and planners had been using map overlays since at least the early twentieth century, pioneering town planner Jaqueline Tyrwhitt was the first person to describe overlaying transparent maps upon one another, each map with different information: relief, rock type, hydrology, soil drainage, and so on.[82] The method was publicized by landscape architect Ian McHarg, who used it to indicate land suitable or unsuitable for various uses.[83] Such overlaid maps came to be known as *geographic information systems*, or GISs. During the 1960s, United States government agencies created databases of topography and land cover, and computer programmers wrote software allowing a user to trace landscape features (including climatic data) on maps and to transform the maps into a digital format.[84] Such databases quickly became increasingly detailed and comprehensive. Today, for instance, global data on climate at a fine resolution are available on WorldClim and on hydrology and topography on HYDRO1k.[85] Inevitably, the advent of GISs and availability of massive amounts of climatic and geographic data led to new hypotheses regarding likelihood of establishment and potential geographic range for non-native species. An important innovation for GIS use in modeling species ranges was the ability to interpolate estimated climates from points where climatic data were actually measured to points where they were not.[86] This innovation was pioneered by Grace Wahba and James Wendelberger in 1980 and improved through the 1980s by Michael F. Hutchinson, whose programs were used to produce the WorldClim database.[87]

GISs can be used directly to predict the ultimate, or at least the possible, geographic range of a non-native species. For instance, Steven A. Dewey and colleagues used a GIS consisting of sixty spectral land cover classes (reflectance as a function of wavelength, as detected by satellite) on 30-by-30-meter (just under a quarter acre) land units covering a large part of a national forest in Utah to predict the potential distribution of a Eurasian weed, dyers woad.[88] They used a chi-square test to determine that certain land cover classes had more units occupied by dyers woad than would be expected if existing occurrences were simply randomly strewn over the area. Finding ten cover types highly associated with the plant, they considered all uninfested land with those cover types to be potentially invasible. Most subsequent uses of GIS data to predict non-native species ranges use many more types of data layers and much more complex statistical models of two general types.

14.4.3 Mechanistic Approaches

Most of the metrics used in climographs and GIS layers do not act directly to limit organisms' ranges. For instance, mean minimum temperature, mean precipitation, and other averages are unlikely to cause an animal or plant to die or fail to reproduce, although a specific cold temperature might, or a specific drought. That is, the presence of a species at any specific site may be limited not by climatic mean values but rather by particular extreme events, such as a long drought or an anomalous freeze.[89] Neither does elevation per se kill an individual, though a cold night associated with high elevation might. The statistical relationship of some average climatic variable or other physical factor to a species' absence could generate a hypothesis about a specific, direct cause of that absence. The hypothesis could be tested directly by laboratory or field tests, or both, of how some climatic event, such as exposure to a specified temperature for specified length of time, affects life-history features (survival, growth, reproduction) that directly determine whether or not the population persists. Such an approach is often termed "mechanistic," in the sense that it is based on the biological mechanisms underlying survival and reproduction.

Mechanistic approaches were used to predict range and abundance of non-native species well before the advent of GISs and continue to be generated by a variety of means. Research in 1954 on three non-native fruit flies (including the medfly) was conducted in growth chambers set to simulate climate in eighteen locations in the continental United States.[90] Scientists in the eighteen locations provided weather data, and the flies' growth rates, life cycle lengths, and reproduction were measured for three generations. Because simulated winter temperature was determined to be the key determinant of whether the fly populations grew, a map of the United States was then constructed with regions with suitable and marginal conditions indicated for persistence of these species were they to be introduced. Many researchers have attempted to determine which weather variables are critical to particular species or groups of species and to combine multiple weather variables into simple indices that indicate persistence or abundance of species in general, not only non-native species.[91] This research, which flourished particularly in the middle of the twentieth century, was often termed "bioclimatics" and included examples of application to non-native species and empirical research on some species supporting bioclimatic predictions.[92]

Several biologists have explored the relationship of the physiology of particular native or non-native species to physical environmental features, often with at least one goal being to rationalize or predict the geographic range of that species. For instance, Warren P. Porter and David M. Gates modeled the energy budgets of terrestrial vertebrates, pointing to the various climatic variables—especially air temperature—confronting an animal and determining where a population can persist.[93] For seven species they then constructed "climate spaces" that were essentially *n*-dimensional climographs, though they did not explicitly relate these climate spaces to geographic ranges. Porter and colleagues developed a computer program to calculate the energy budget of any animal species given many directly measured input variables, which was used to predict geographic range limits of various species, such as the non-native cane toad in Australia.[94] However, the program was a patented product until recently, perhaps limiting its more widespread use in such research.[95]

The number of environmental factors that could potentially contribute to determining range is enormous, and the complexity of the way in which these factors could affect the range is forbidding; C. Richard Tracy provided an example for reptiles.[96] This is why much effort has gone into combining variables into a small number of indices and also into focusing on particular variables suspected from laboratory or field experiments to contribute to limiting geographic range. For instance, snow persistence time appears to be critical to persistence of whitebark pine in western North America, and snow persistence time is influenced in turn not only by temperature and precipitation but by wind, slope, aspect, local topography, and freeze-thaw cycles.[97] A few of these efforts to understand range limitations by focusing directly on species traits (particularly physiological traits) and physical environmental factors involve non-native species, such as famine weed from the American tropics and musk thistle from Eurasia, both in Australia; North American ragweed in Europe; and the cane toad, a subtropical species that had spread, after introduction in the Miami area, to much colder regions.[98] These studies did not use GIS datasets.

The first use of a GIS database to assess the potential survival and growth of non-native species was probably by Henry Nix and collaborators, who estimated the potential for growing certain non-native legume species in Australia with a mechanistic model using a climatic data set amassed by Nix.[99] This estimate was produced by comparing climatic data in parts of Australia with a model (GROWEST) of the effect of temperature, moisture, and light indices, determined on a weekly basis, on plant growth, the plant growth based on laboratory and field observations. In 1985, Robert W. Sutherst and Gunter F. Maywald provided a widely cited computer package, CLIMEX, for applying this approach to any species for which the relevant life-history data can be estimated, based on the GROWEST model.[100] Using Nix's climatic database, Sutherst and Maywald demonstrated use of CLIMEX with several arthropod pests of livestock, asking such questions as why population sizes of the horn fly, native to Europe, differ greatly among locations in the North American introduced range. Similarly, Susan P. Worner used CLIMEX to assess the likelihood of establishment and spread of the medfly and the Colorado potato beetle in New Zealand.[101]

CLIMEX has undergone a series of updates and has been widely employed to predict the ultimate invaded range of recently arrived non-native species, especially insects.[102] It has also been tested in the short term against subsequently gathered field data, displaying general accuracy at large scales, but it has frequently been found deficient at very local scales, likely because underlying GIS data often cannot adequately capture variation in local conditions.[103] However, a test of the prediction of the ultimate range can be attempted only once the range has ceased expanding, and criteria for such cessation have not been established.

Not all GIS-based mechanistic models are versions of CLIMEX. For example, Michael Kearney and colleagues studied the Central American cane toad in Australia, constructing a mechanistic model using GIS climatic data and direct data on physical factors and on several aspects of the toad's biology, not information from mapped locations of the toad.[104] They sought two sets of conditions—those presently suitable for the toad and those expected to be suitable in 2050 following predicted climate change. They compared their results to the then-current distribution of the toad in Australia plus four earlier predictions: one produced by a different mechanistic model, one generated by CLIMEX, one produced by an expanded version of CLIMEX using data from the invaded as well as native ranges plus data on road coverage and urban area, and one constructed by simply mapping four kinds of weather features viewed as inimical to the toad. The main result of interest was that their mechanistic model showed substantial parts of southern

Australia to be unsuitable for the toad whereas the CLIMEX-related models showed some of these regions to be suitable. The authors attributed this difference to the fact that some of the areas in question have combinations of environmental conditions not found in the native or introduced range of the toad (known as "non-analog climates," discussed further in the next subsection, 14.4.4), so the predictions based on CLIMEX relied on extrapolation rather than empirical findings.

Physiologists have asserted that physiological models are underused in conservation-related ecology, including invasion research (as discussed further in chap. 19, sec. 19.5.1).[105] The relative dearth of mechanistic models of species ranges compared with the wave of statistical, correlative methods described in the following subsection 14.4.4 is part of this trend. The reason is surely the myriad environmental and other factors that can affect a species survival, growth, and reproduction and the complexity of their interaction, so that comprehensive data from field or laboratory observations are scarce, and even nonexistent for many species.[106] Even Kearney and colleagues, who used many sorts of data on many aspects of cane toad biology, were far from comprehensive in this regard, focusing on those aspects of the life cycle that, in their view, were most likely to be critical in determining the species' range limits. Just as climatologists sought to reduce the complexity of climate data by combining various types of data into a small number of indices that in some sense captured the key aspects of climate, physiologists have sought features that integrate many aspects of species' physiology, such as thermal performance curves and knockdown times, and that could therefore reduce the complexity of predictive models.[107]

14.4.4 Species Distribution Models

Species distribution modeling is a different approach to predicting ranges of non-native species. Species distribution models (SDMs) do not attempt to relate climate or other physical factors directly to a species' traits or performance. Rather, they assume that the native range is somehow determined by its physical attributes, but the precise mechanism is irrelevant in predicting the range in the invaded region. They make full use of GIS databases of climate and other features (e.g., topography) to predict invaded range statistically, based on the premise that the physical factors in the native range determine that range and that the same physical factors will determine the invaded range. Such models are often termed "phenomenological" or "correlative" or "inductive" for contrast with mechanistic models. In the most basic version, first, locations in which the species is known to be present in the native range (as determined, for example, by museum or research records or through a publicly available database, the Global Biodiversity Information Facility [GBIF], that amasses such information) are recorded, and climatic and other variables associated with each of those points are collected from the layers of a GIS database (e.g., WorldClim).[108] Next, various statistical methods are used to find points in the region of the introduction whose physical environment, as listed in the GIS, contains climatic and other features similar to those associated with presence in the native range. These points or some solid figure surrounding them are interpreted as the predicted ultimate range in the newly invaded region. The terms "species distribution model" and "environmental niche model" (ENM) are often used interchangeably, and the Hutchinsonian conception of the ecological niche (discussed later in this subsection) underpins the use of SDMs to characterize or predict the geographic ranges of non-native species.[109] However, the main interest of invasion scientists is to predict and understand the distribution of an invading species, not to characterize its niche per se, so I will refer to this type of model as a species distribution model.[110]

The first SDM of this sort was BIOCLIM, developed by Nix and first published by Hew D. V. Prendergast and Paul Hattersley and by Trevor H. Booth, though only Booth was concerned with non-native species: an Australian *Eucalyptus* species in Africa.[111] The program has been updated since its introduction.[112] BIOCLIM uses the latitude, longitude, and elevation of points where the species is present in its native range and then interpolates values of maximum temperature, minimum temperature, precipitation, and similar variables. From these it calculates many derived variables, such as mean annual temperature, mean maximum temperature of the hottest month, mean rainfall of the driest quarter, and so on, and then identifies the range of each of these derived variables in the native range of the species to construct the "climate envelope" of conditions in which the species can persist. It then scans the region in which the species was introduced (or will be introduced) to find all locations that fall within that envelope, and these locations are the predicted range.

Soon after its publication, BIOCLIM was used to predict the potential range of non-native species, such as three non-native plants whose importation is forbidden in New Zealand, Madagascar ragwort in Australia, and heath-leaved banksia from Australia in South Africa.[113] It has been used even more often to predict changes in ranges of both native and non-native species with climate change. However, it has largely been replaced by various other SDMs. Several statistical frameworks were used in the wake of BIOCLIM to transform the physical factors in GIS datasets into predictions, at first a genetic algorithm for rule-set prediction (GARP) and more recently, the Mahalanobis distance method, generalized linear models (GLMs), generalized additive models (GAMs), random forests, and classification and regression trees.[114] One advantage of these methods is that, unlike BIOCLIM, which uses only presence data, these other methods also can take account of locations where the species is known or believed to be absent, which aids in determining nonlinear relationships between a species' presence or absence and the environment.

GARP is termed a "genetic algorithm" because it is a machine-learning technique that entails using a set of classification rules (in the case of species distributions, the classification would generally be either present or absent) that "evolve" by a process similar to mutation and natural selection. That is, the rules make a prediction, that prediction is tested against real data, and rules are changed (evolve) to try to make a more accurate prediction. Such genetic algorithms were first developed in the 1970s and were apparently first applied to biological invasions by A. Townsend Peterson in a series of papers in the early 2000s, using GARP.[115] Peterson and David A. Vieglais showed that the then-current distribution of the house finch, native to the western North America and introduced into Long Island, New York, in 1940, was quite well approximated by the GARP prediction.[116] They also predicted the spread to most of the eastern and midwestern United States of the Asian longhorn beetle (ALHB), accidentally introduced to North America in wooden packing material, first detected in 1996 in Brooklyn, New York, and subsequently in disjunct locations in several other US states and the Canadian province of Ontario. It is the target of aggressive eradication campaigns, but its ultimate range is still unclear. Finally, they predicted that, if an ALHB congener, the Japanese whitespotted citrus longhorn beetle, were to establish a population in the United States, it would persist in many areas; several incursions have been detected since 1999, but not establishment. Similarly, Peterson and C. Richard Robins predicted that, without aggressive management, the endangered northern spotted owl in the Pacific Northwest would likely be largely or wholly replaced by the barred owl from eastern North America, which arrived in the West in the late twentieth century, facilitated by landscape plantings and fire suppression across the Great Plains.[117] Recent observations suggest that this is happening.[118]

The SDM that gradually became dominant in invasion science (and also in the literature on range changes owing to climate change) since it became available in 2004 is MaxEnt (named for a machine-learning technique known as "maximum entropy").[119] Although subject to criticism, MaxEnt was assessed as generally more accurate than other SDMs and has been applied to many biological invasions.[120] For instance, Gentile F. Ficetola and collaborators asked which regions of the world not yet invaded by the bullfrog would support populations of this species.[121] Darren F. Ward sought the expected ranges of six non-native ant species long present in New Zealand.[122] Notably, Ward compared performance by BIOCLIM and MaxEnt on this task, finding MaxEnt superior.

For non-native species that have not been present for long, one would expect that their range in their new region might still be expanding. Thus some locations in the new region that they do not (yet) occupy might be perfectly suitable in terms of the climatic and other physical factors in the GIS accessed by an SDM; they simply have not dispersed there yet. Indeed, one goal of predicting range is to determine whether the species will persist in such a location if it disperses there. It was quickly realized that, for this reason, location data in the new region should be used with caution if at all in determining which combination of physical factors allow a species to persist.[123] However, in certain circumstances such data might be useful—for example, when evidence suggests that the introduced population has different physical requirements than the native population.[124] Several authors addressing the possibility of range expansion owing to climate change have attempted to integrate SDMs with models of dispersal based on relevant data for the species of interest.[125] Such a combination of an SDM and a dispersal model of the spread of non-native ragweed in Austria performed better than did the dispersal model or the SDM separately.[126] Mark C. Urban and colleagues suggested a statistical method to update existing SDM predictions based on evidence that a non-native species (the cane toad, in their example) has either evolved new tolerances of environmental extremes or had access in its introduced range to environments not available in its native range.[127]

Another issue with applying SDMs to predict ranges of non-native species and of native species under climate change is that a species' range may be affected by the presence or absence of other species. For instance, non-native fig species in south Florida did not spread until years after their introduction, when obligate pollinating wasps from their native range also arrived.[128] Several cases are known in which the presence of a non-native species has greatly reduced the range of a native species. For example, the gray squirrel native to eastern North America, first introduced to Great Britain in 1876, has reduced and fragmented the range of the native red squirrel, primarily by competition and by spreading a pathogen, squirrel pox, to which the invader is resistant but the native is highly susceptible.[129] But it is difficult to find examples in which one non-native species has reduced the range of another non-native species. One case with strong evidence for exactly such a range reduction is that of the browntail moth, native to the Old World, whose substantial range in northeastern North America was reduced to two small, isolated regions by a subsequently introduced parasitic fly.[130]

The effect of other species on a species' geographic range is related to Hutchinson's conception of the ecological niche, which underpins the use of SDMs.[131] Hutchinson conceived of *n* physical factors, including climatic features, determining an *n*-dimensional solid representing the "fundamental niche," the set of all physical conditions that would permit a species to persist. In nature, he perceived the fundamental niche as usually being reduced to a smaller "realized niche" as various natural enemies (e.g., predators, competitors, pathogens) excluded the species from parts of its fundamental niche (a mutualistic species could actually cause the realized niche to exceed the fundamental niche, at least in certain dimensions).[132] Much of the literature on predicting ranges of non-native species is cast in terms of predicting the species' realized niche in the new region.[133] This goal is not identical to predicting its range (more on this point subsequently).

Ecologists have long recognized that, at least on a small scale, species ranges could depend on other species through direct interactions such as competition, predation, mutualism, or disease or indirectly through availability of a needed food item.[134] Nevertheless, researchers recently studying the possible influence of climate change on species ranges have generally assumed that physical factors such as climate usually determine large-scale ranges, while, within these large-scale ranges, interactions with other species may affect local presence or absence.[135] Several scientists studying possible effects of climate change on species' ranges have suggested ways of incorporating the presence or absence of other species in SDMs.[136] If strong evidence exists that one other species (say, a predator or food source) is likely to affect the range of the species of interest, one straightforward way to include this factor in an SDM is simply to include the locations of that other species as a data layer in the GIS used by the SDM. Then whatever equations characterize the interaction between the two species are applied to the resultant range predicted by the physical factors. As climate changes, different species' ranges may shift in similar or different directions and at different rates. Additionally, a species of interest, such as a non-native species, may come into contact with different sets of species, native or non-native, or both. Whether these new potential interactions will modify a previous prediction generated by an SDM can sometimes be addressed by experiments.[137] Other times researchers may have to make educated guesses. Miguel B. Araújo and Miska Luoto used an SDM approach (in this case, GAM) to predict the range of a native butterfly under climate change both with and without the presence of its host plant as a variable, finding that the host plant distribution is likely to be limiting the butterfly's range.[138]

However, the potential interactions of the focal species with other species may not be well understood, and several important interacting species may be important, not just one; in fact, that is more likely to be the situation with a non-native species in its new range. The complications that arise when one aims to predict the spread and impact of a non-native species under climate change are exemplified by alligator weed, an aquatic South American plant that invaded several other regions and is regarded as a major pest in North America, China, and Australia. Alligator weed is well controlled in major parts of its non-native range by the non-native alligator weed flea beetle.[139] However, alligator weed tolerates cold better than the flea beetle does, so the flea beetle does not persist in colder parts of the plant's introduced range (as discussed further in chap. 17, sec. 17.5).[140]

Because alligator weed itself is limited by low temperature, efforts using CLIMEX and MaxEnt sought to predict where it would spread in its non-native range under current climate conditions.[141] As concern with climate change increased, others used MaxEnt to ask where it would go under predicted climate change.[142] And because the flea beetle is also limited by low temperature, CLIMEX was also used to ask if the beetle range could expand in the current climate.[143] The general question is whether the beetle range can increase apace with the alligator weed range increase predicted to occur as climate warms.

Physiological experiments also suggest important factors to consider. For instance, in China, experimental warming in the field caused alligator weed to compete better against the main co-occurring native plant species, and it has adapted to colder weather in the northern part of its Chinese range through phenotypic plasticity (discussed in chap. 11, sec. 11.3.2)—in this instance, modified gene expression.[144] For the beetle, lab experiments showed that increased CO_2 concentrations have direct effects and indirect effects through eating alligator weed grown in increased CO_2 concentrations.[145] These changes would all seem to increase the beetle population. In sum, it is still not possible to say whether the beetle will gain or lose ground against the plant under climate change, and the details of species interactions appear to be important in determining the outcome.

One approach to dealing with the complexity of species interactions is to find some surrogate variable believed to be correlated with interaction intensity and include a GIS layer with that surrogate variable.[146] Although the same SDM models have been used to predict ranges of non-native species as have been used to predict the effect of climate change on range, inclusion of other species in these models has largely been restricted to questions of climate change, not invasion. However, Sutherst and colleagues combined CLIMEX, which in its later iterations is actually a hybrid of a mechanistic model and an SDM, with an expert system to incorporate presence of potential host plants in assessing the likelihood that a new arrival in Europe, the Colorado potato beetle, could establish and spread.[147] Another approach consists of using community-level models that attempt to fit a species co-occurrence matrix to environmental variables to predict both community composition and distributions of individual species.[148]

As interest in the effects of global climate change increased in the 1990s, several scientists predicted its possible effects on non-native species using SDMs. In 1995, David J. Beerling and associates used an SDM incorporating regression methods to predict the future range of Japanese knotweed in Great Britain, and Trond Rafoss and May-Guri Sæthre used CLIMEX to predict the effect of climate change on the range of the Colorado potato beetle from the New World and the codling moth from southeastern Europe in Norway.[149] As noted in section 14.3, Bethany Bradley and colleagues used

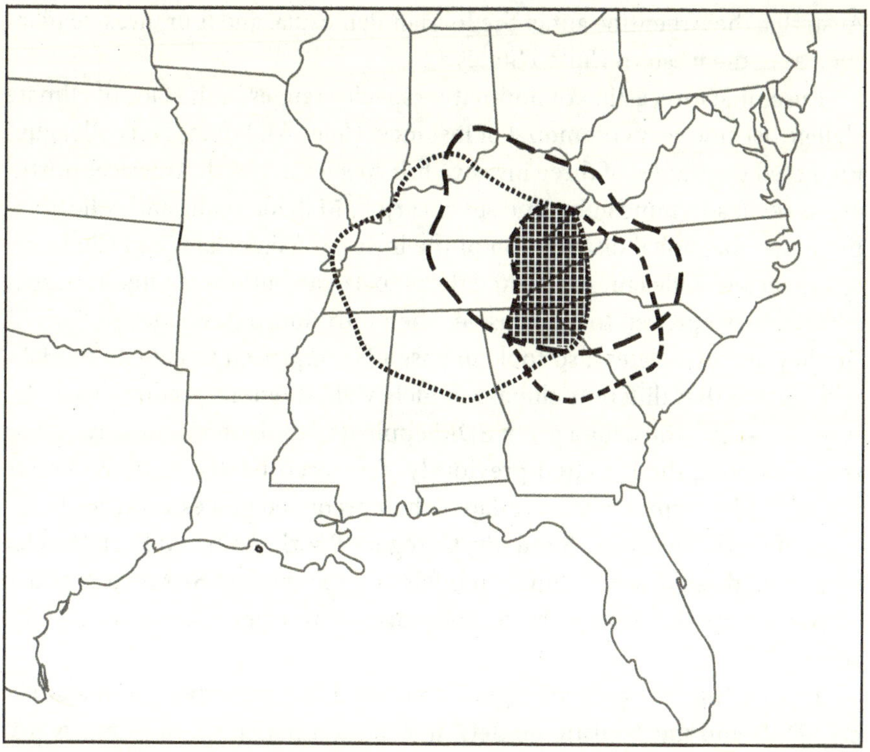

Fig. 14.3 Hypothetical ensemble prediction of future range of a spreading non-native species—the area of overlap of predictions generated by three different species distribution models.

SDMs with several different climate projection models to predict the range under climate change of several non-native weeds in the United States—yellow star thistle, salt cedar, cheatgrass, spotted knapweed, leafy spurge, kudzu, privet, Chinese privet, and cogon grass.[150] Soon thereafter, use of SDMs combined with climate projection models to predict ranges of non-native species became very common.

The fact that different SDMs and different climate change models predict different ranges of species subjected to climate change led to the idea of an "ensemble" or "consensus" method—that is, use several different SDMs and climate change models and construe the overlap of their predicted ranges as the best estimate of the predicted range (fig. 14.3).[151] Wilfried Thuiller and colleagues published a software package that automatically applies an array of SDMs to any given species along with a variety of different climate models and seeks the overlap of their predicted ranges under climate change.[152] And Mathieu Marmion and coauthors discussed how exactly to arrive at the ensemble map—count each SDM prediction equally or give different weights to the predictions of different SDMs.[153] Ranges of several non-native species were predicted by ensembles of SDMs: olive from Europe in southern

Australia, the Argentine ant in the Iberian Peninsula, and four invasive plant species in the western United States.[154]

Ensemble forecasting of non-native species ranges in the face of climate change became very common. For instance, Helen M. Poulos and colleagues predicted the ranges of three invasive fish species in North America: northern snakehead, round goby, and silver carp.[155] Mafalda Gama and colleagues predicted the global distribution of the invasive Asian clam, and Cleo Bertelsmeier and colleagues predicted the global distributions of fifteen ant species that have proven to be invasive when introduced in various regions.[156] Bradley and associates also took an ensemble approach to climate models, recognizing that different climate models yield different predictions of future climate.[157] Thus, for a given SDM applied to each of four invasive weed species among the list cited previously, they used twelve distinct climate models and interpreted the overlap of their predicted ranges as the best estimate of predicted range. Similarly, Grzegorz Buczkowski and Bertelsmeier employed three distinct climate models and ten distinct SDMs to produce consensus predictions for the future ranges of thirteen non-native termite species.[158]

Lauren Buckley and colleagues compared the performance of statistical SDMs and mechanistic models, not on non-native species but on two native species—a butterfly and a lizard—asking how their ranges would change with a temperature increase of 3°C.[159] They used MaxEnt and GLM as the SDMs and chose three different mechanistic models. Two of these models explicitly modeled population dynamics, while the third calculated environmental thresholds based on the species' physiology. The SDMs and mechanistic models performed similarly in identifying current distributions, but the mechanistic models predicted more drastic range changes in response to climate change. A similar comparison of predictions generated by an ensemble of SDMs and by a mechanistic model for climate-induced range changes for fifteen North American boreal and temperate tree species found, by contrast, that the SDM ensemble predicted greater likelihoods of extinction and greater range changes.[160]

Species distribution models do not account for evolution, and their utility in predicting the geographic range of a non-native species assumes that evolution does not cause a species' ecological niche to change rapidly even when the species is exposed to a new set of circumstances after introduction to a new region—a concept termed "niche stability." Much of the research on niche stability had been conducted by paleobiologists studying long spans of time in the distant geological past. However, with the advent of concern with both non-native species and rapid climate change, the concept became important to ecologists. Defending the utility of SDMs for predicting non-native species' ranges, Peterson argued for the generality and pervasiveness of niche stability, citing the paleobiological literature and what he perceived as statistical errors in applying SDMs to recent invasions.[161] However, a growing wave of recent research on rapid evolution in ecological rather than geological time—termed "eco-evolutionary dynamics"—challenges this view.[162] An increasing number of cases have revealed that evolutionary change can adapt a population to environmental change even in the short term. These cases include several involving non-native species (as described in chap. 11).

The frequency with which evolutionary change, or phenotypic plasticity until evolution occurs, would cause a population's non-native range to differ from that predicted by an SDM remains an open question. Many striking examples of rapid evolution (or at least rapid phenotypic change) concern interactions between a non-native species and another species, either native or non-native, with the evolution of either the focal non-native species or the species it interacts with, or both, serving to increase or decrease the invasion impact. For instance, Richard Lankau and colleagues showed that non-native garlic mustard from Asia evolved reduced allelopathic impacts on competing native plants in North America over just fifty years, and at least one native species (Canadian clearweed) evolved increased resistance to garlic mustard.[163] Thus, the invasion impact was reduced. By contrast, the many cases now known of hybridization between a non-native species and a native congener (discussed in chap. 11, sec. 11.3.2) include several in which the resultant hybrid is more invasive than the non-native parent, such as the event that produced common cordgrass.[164] It seems possible that such interactions could affect the range of an invading species, and it is unclear how incorporating the presence of one or more native species in an SDM could allow prediction of the evolution spurred by interactions.

Increasing concern with global warming impacts has led to much research on the degree to which species can rapidly evolve to tolerate increased temperature. Such research is relevant to the reliability of SDMs in forecasting non-native species' ranges, as statistics associated with temperature are the most basic feature of SDMs. David K. Skelly and collaborators pointed to research showing rapid evolution of tolerance of increased or decreased temperature in a variety of species.[165] Subsequent research provided several other examples, including the three-spined stickleback, the water flea *Daphnia magna*, and the acorn ant.[166] These and most other studies concerned species in their native ranges, but there is no reason to think that such research would not aid prediction of the range of a species introduced to a region warmer than its native home. The chromosomal inversion frequencies underlying the convergence of wing size clines in non-native populations of the fruit fly *Drosophila subobscura* toward the cline in its native range (as described in chap. 11, sec. 11.3.1) may be an example.[167] This evolution in both the native and non-native ranges occurred in response to global warming, presumably resulting in adaptation to the increased temperature. For the most part, these are species with short generation times and large populations, and rapid evolutionary response would likely arise more readily in such species.[168] Nevertheless, both phenotypic plasticity (discussed in chap. 11, sec. 11.3.1) and a variety of evolutionary changes in timing of life cycle events could aid survival of longer-lived species, particularly vertebrates, faced with warming.[169] Similar phenomena could foster range change in an invaded range that would not have been predicted by an SDM based on conditions in the native range. For instance, the cane toad established a population in extreme south Florida in 1955. It has spread northward gradually, and both phenotypic plasticity and evolution by natural selection have produced cold-tolerant genotypes allowing the spread.[170]

The range of a species introduced to a new region may fail to be predicted by an SDM based on its range in its native region because of a modified niche that may have nothing to do with evolution. Consider a new region that has areas with a set of weather features (temperature, humidity, etc.) that is not found anywhere in its native region—often referred to as "non-analog climates" because they are not analogous to any set of conditions in the original region.[171] Predictions about whether a non-native species that reaches this region will be able to occupy such sites without evolving based on the locations it occupies in its native range will be limited.[172] Precisely this problem caused the SDM "random forest" to fail to predict the range of Douglas fir, native to North America, in Europe, where it has been present since the mid-nineteenth century.[173]

In sum, the generality of niche stability over short- and medium time spans is unresolved, with studies on various groups of non-native species showing frequent or infrequent niche shifts.[174] This is an active area of research. With respect to using SDMs to predict the ultimate range of a non-native species in its new region, the issue boils down to whether SDMs adequately capture the fundamental Hutchinsonian niche of a species in its native range, and, if they do, whether this representation adequately predicts its range in the new region. For the first step, most observers accept that the fundamental Hutchinsonian niche in the native range is adequately represented by modern SDMs using data-rich GISs. Whether this representation allows adequate prediction of the realized niche in the new region, which in turn would constitute a huge step forward in predicting the new range, is open to question and must depend on context, primarily because of (1) possible effects of other species, (2) possible evolution, and (3) possible non-analog climates.[175] Although researchers are generally aware of these issues, they have usually been ignored or downplayed in most studies applying SDMs to non-native species.

Although managers would want to predict the speed and direction with which an invading species spreads in its new location as well as the ultimate range, such dispersal has rarely been integrated with SDMs—research on ragweed noted previously is an exception.[176] A rich literature exists on the mathematics of the spread of a newly invading species, most of it tracing to a reaction-diffusion model of random dispersal proposed by John G. Skellam in 1951 that he fit to observed data on the spread of the muskrat after its introduction to Europe.[177] Several authors fit this model to invasions of particular species, and others elaborated and expanded this model.[178] However, Michael E. Moody and Richard N. Mack suggested that this model, which envisions an invasion spreading continuously from an initial starting point, greatly underestimates the spread of many non-native plant species, because for many species, a few individuals will spread far beyond an advancing continuous invasion front and establish new populations ("nascent foci") that then spread independently.[179] Perhaps because parameterizing both the reaction-diffusion equations and Moody and Mack's "jump-dispersal" model is challenging, these dynamic spread models for invasions have developed largely independently of SDMs.

14.4.5 Combining Physiology and Species Distribution Models

As observed earlier in this section, the array of physical and biological factors that can affect a species' biology is enormous, so data of the sort needed for an adequate mechanistic model often do not exist. Thus, for instance, CLIMEX, originally conceived as a mechanistic model that uses GIS-based climate data, evolved to use physical factor data from the existing range supplemented where possible by field and laboratory observations, in effect becoming a hybrid of a mechanistic model and an SDM.[180] Kearney and Porter have suggested ways in which physiological and other data that inform mechanistic models could be incorporated into GIS databases on climate and terrain, thereby producing mechanistic SDMs.[181]

Lauren Buckley and colleagues proposed hybridizing mechanistic models and SDMs by including output from a mechanistic model as part of the input to an SDM, along with the physical factors usually employed by an SDM.[182] They tested this approach using MaxEnt combined with experimentally determined estimates of the lower developmental threshold (temperature at which development stops) for six native British butterflies and found modest improvements in predictions of response to recent climate change. Xiao Feng and Monica Papeş suggested modifying the input maps of species presences and absences (presumably in the native range) with physiological data to remove presences that simply reflect survival, or survival and growth, but not reproduction, and to eliminate absences that simply reflect failure to sample at a site rather than physiological inability to survive and reproduce there.[183] They explored the feasibility of this approach by scanning available empirical literature on the effects of temperature and salinity on survival, growth, and reproduction of nine non-native freshwater bivalves in the United States, finding substantial gaps in the literature that would hinder implementing this method. The presence of other species, physiological limitations, and evolution are but three of several factors that will determine changes in ranges of species, including non-native species, as climate changes.[184]

The tension between using mechanistic models or statistical models to predict ranges of non-native species (or of any species subjected to global warming) mirrors a controversy spawned in the early 1990s by Robert H. Peters. Peters claimed that ecology was being led badly astray by the dominance of mechanistic models in predicting values of various factors (such as population size) based on presumed causal relationships between the factor of interest and certain other things. He argued that the primary goal of ecology should be statistical prediction and extrapolation of observed relationships—through multiple regression—rather than mechanistic explanations of relationships, stating that "the cause of such predictive success is irrelevant."[185] This argument elicited angry denunciations and claims that such a course would reduce the science of ecology to "regressing one variable against another," "presents a very empty view of our discipline," and would render ecology "engineering without science."[186] Tellingly, the authors of all three brickbats agreed with Peters that the mechanistic models dominating ecology at the time had not led to substantial progress in the field. Their objection was rather to his view that such models would never succeed, and they saw the correct goal of science as not just prediction but understanding the relationships that yield correct predictions. This sentiment and the aforementioned challenges to predicting ranges by SDMs seem to underpin the desire of physiological ecologists and other scientists to go beyond predicting a species range from a purely statistical, correlative model.[187]

14.5 The Overall Picture

The general tenor of the mass of empirical evidence of range shifts and statistical and mechanistic predictions has been that, on average, global warming will exacerbate invasions. The exacerbation may entail a newly invading species persisting where otherwise it would have died out. Or warming might extend the range of an established non-native species poleward, as has been predicted for many weeds. Or it might cause a previously innocuous species to have a greater impact, as with the chain tunicate or heather. Extreme events that will occur more often frequently liberate resources, such as space or food, that favor non-native over native species (cf. chap. 15, sec. 15.4.6).[188] Warming has opened new routes across the Arctic that have already greatly increased shipping between previously isolated regions, with the prospect that marine species will exploit this new pathway and foster new invasions—for instance, by means of ballast water.[189]

The precise feature that facilitates survival and/or range expansion of a non-native species in a warming climate may vary. For instance, Eurasian yellow star thistle, highly invasive in California, extends its growing season into summer, avoiding competition for water from native winter annuals.[190] In eastern North America, non-native forest shrubs and vines drop their leaves later than native species do, giving them additional time to photosynthesize.[191] Flowering time has often been shown to differ between native and non-native invasive plants, with the non-natives flowering longer, earlier, or later than the natives in situations where the difference favors the non-natives.[192] A formal meta-analysis (as discussed in chap. 13, sec. 13.1) of native versus non-native species found that non-native plants almost always senesce later than native plants (usually leaf senescence, as in the aforementioned forest shrubs and vines). For native versus non-native populations of particular plant species, the non-native populations were found more often to reproduce earlier than the native populations.[193] In Massachusetts, plants are now leafing out earlier in the spring than they did in the 1850s, and the earlier leafing out is particularly pronounced for non-native shrubs.[194] These and other timing differences in response to climate change appear to play a major role in many plant invasions and suggest that global warming may often favor new invaders or facilitate existing ones by responses in timing.[195] Far fewer studies of this sort exist for animals.

Not all invasions will be exacerbated by climate change, and as ranges of particular non-natives shift, some areas that had previously been invaded are predicted to be rid of particular non-native species (predictions subject to the limitations of SDMs described earlier in this chapter). For example, bat white-nose syndrome fungus, introduced from Europe to New York State in 2006 and now spread all the way to the west coast of North America, has killed millions of hibernating bats. This fungus has adapted to cold temperatures, and its range is predicted to contract and its impact to lessen in the wake of global warming.[196] In the study of 896 non-native plant species in the United States described in section 14.3, two areas were predicted to have fewer non-native plant species as warming proceeds.[197] A region in the southwestern United States known as the southern semiarid highlands and also the Mediterranean climate region of California both were predicted to have 7 to 8 percent fewer non-native plant species in 2050 than they have now. For three major invaders of the American West—cheatgrass, spotted knapweed, and leafy spurge—Bethany Bradley and colleagues predicted both range contraction and reduced invasion risk in many areas.[198] For instance, cheatgrass currently dominates at least 400,000 square kilometers (154,441 square miles) in the states of Nevada and Utah alone, precipitating fires that devastate native vegetation. However, climate models combined with SDMs predicted that large areas of southern Nevada and southern Utah now occupied by cheatgrass will become less suitable because of increased summer precipitation or higher temperature.[199] Such cases suggest that the predicted retrenchment of some non-native species will provide opportunities to restore some places to a preinvasion state.[200] However, the treadmill effect (defined in chap. 15, sec. 15.4.6) is always a threat following retreat of a non-native species. For example, if efforts to control cheatgrass succeed, it might be replaced by red brome, a congeneric species native to Europe, western Asia, and North Africa that is present in the region, often dominated by cheatgrass.[201] Red brome tolerates high temperatures better than cheatgrass does.[202]

Just as climate change will shift the geographic ranges and performance of many non-native species, so will it have similar impacts on ranges of many native species. For some native species, their ranges are predicted to expand to regions where they will be co-occurring with other native species with which they have never had contact or have not had contact for thousands of years. These are exactly the conditions in which non-native species become invasive and impactful, and, in fact, such native range expansions have precipitated a further discussion about whether to restrict the term "invasion" to non-native species (described in chap. 18, sec. 18.2). Franz Essl and coauthors, pointing to the phenomenon of native species expanding their ranges rapidly because of anthropogenic influences, especially climate change, proposed the term "neonative" for such species whose range expansion dates from later than 1950, conventionally viewed as the beginning of the Anthropocene.[203] The phenomenon of native species expanding their ranges owing to climate change has led to a new controversy about non-native species, even encompassing the term "neonative."

14.6 A Controversy over Preserving Biodiversity in the Face of Climate Change

During climate changes in the geological past, abundant paleontological evidence shows that many species shifted their native ranges long distances, thereby staying within the climate envelopes to which they were adapted.[204] That might be impossible today, for two reasons. First, increasingly sophisticated climate models predict a rapid speed of climate change far exceeding the speed of past climate change. Second, because of anthropogenic changes in the landscape, species that might have dispersed apace with the changing climate in the past may no longer be able to do so.[205] Huge forested areas have been leveled, major highways dissect the landscape, enormous conurbations have no suitable habitat for many species. One can imagine that a landscape that previously would have provided a relatively easy path for dispersal to shift a range gradually would now pose a formidable barrier.

This possibility elicited the suggestion of "assisted migration" (also known as "assisted colonization" or "planned relocation")—human movement of individuals of a species (especially, imperiled species) out of a region predicted by climate models and SDMs to become unsuitable for them to a region predicted to contain suitable habitat decades in the future. The idea was first proposed in 1985, and by 2014 a search revealed 868 publications discussing the concept.[206] In addition to conservation concerns, the forest industry contributed to this literature because of the likelihood that climate-induced forest change would invalidate long-term harvest plans.[207]

Suggestions of assisted migration triggered alarm at the prospect that such a project—moving a species to an area it had never occupied, or at least had not occupied for many millennia or longer—could lead to an impactful invasion.[208] It was also criticized as at best a "techno-fix" that would at most solve a small fraction of the conservation problems wrought by climate change.[209] Many advocates of assisted migration acknowledged the concern over the possibility of harmful invasions but either discounted it as overblown or argued that careful planning could avoid such events.[210] Jillian M. Mueller and Jessica J. Hellmann, generally enthusiastic about assisted migration, argued that the likelihood of a harmful impact is low but conceded that, if one occurs, the impact could be great.[211]

The flood of publications on assisted migration is notable for the fact that almost all are opinion or perspective pieces, often frameworks or decision trees to apply to proposed projects.[212] Several authors even treated ethical considerations.[213] Few reports of empirical research exist that might indicate the magnitude of the risks. No doubt this disproportion reflects at least partly the fact that global climate change seems to be such a pervasive, complicated, and intractable phenomenon, and individuals, including scientists and conservationists, can easily despair about how to counteract it or mitigate its impacts. Thus they grasp at straws that at least suggest some possibly useful action. However, the simple repetition of the idea can tend to establish it as a proven, relatively risk-free technology in the absence of real evidence. Anna Bucharova captured the situation well:

> Despite the lack of clear evidence, journals keep publishing reviews and perspective papers where authors state that assisted migration is a promising and possibly necessary approach . . . leading to propagation among practitioners and incorporation into guidelines. . . . Since anything repeated often enough becomes "the truth," the putative benefits of assisted migration are often perceived as facts.[214]

To an extent, assisted migration is already happening despite the misgivings. For example, the Torreya Guardians, an informal group of private citizens, began translocating the endangered Florida torreya tree in 2008, initially from its native range in north Florida and south Georgia to North Carolina and subsequently to at least eight other states with no published risk assessment or permitting procedures.[215] Another tree species involved in assisted migration is the Douglas fir (see box 14.1). In 2014 in Glacier National Park, Montana, personnel from federal agencies, including the National Park Service, introduced bull trout, threatened by competition from non-native lake trout and hybridization with non-native brook trout, into a stream-and-lake network that had never contained bull trout.[216] This assisted migration was performed after researchers constructed a risk assessment, despite apparent federal policy forbidding such actions.

BOX 14.1 Assisted Migration of Trees

Perhaps the most frequent proposals for assisted migration concern trees, either to safeguard imperiled species or to aid the forestry industry.[1] Proposals and projects involving Douglas fir exemplify the issues. An early activity directly related to assisted migration of this species occurred in 1975, when Keith Illingworth, a Canadian forest geneticist, planted seedlings from sixty-four regions, as far south as central Mexico, in the Okanagan Valley of British Columbia. Forty years later, massive differences in growth characterized the different provenances.[2] This result supported findings that species distribution models (SDMs) based on data that do not account for the variety of genotypes can yield highly inaccurate results.[3] In addition, the European range of non-native Douglas fir differs greatly from that predicted by an SDM based on the native range in North America, contradicting the hypothesis of niche stability described in chapter 14.[4] Despite the uncertainty, assisted migration of Douglas fir is already underway in British Columbia, with regulations changed to allow seed from any particular region to be moved up to 300 kilometers (about 186 miles) farther north.[5] Seeds from fifteen species, including Douglas fir, were planted between 2009 and 2012 at forty-eight sites from northern California to southern Yukon.

1. On safeguarding imperiled species, see, e.g., C. Palmer and Larson, "Should We Move the Whitebark Pine?" On aiding the forest industry, see, e.g., Pedlar et al., "Implementation of Assisted Migration in Canadian Forests"; Klenk and Larson, "Assisted Migration of Western Larch."
2. Mullard, "Run, Forest, Run."
3. Bayliss et al., "Aggregate Population-Level Models Informed by Genetics."
4. Boiffin et al., "Species Distribution Models May Misdirect Assisted Migration."
5. On the assisted migration, see Government of British Columbia, "Assisted Migration Adaptation Trial." On the change in regulations, see G. O'Neill et al., *Assisted Migration to Address Climate Change in British Columbia*. See also Government of British Columbia, "Chief Forester's Standards for Seed Use."

In the United States, the Endangered Species Act of 1973 was amended in 1982 to permit federal agencies to declare a population of a federally listed endangered species as "experimental." This designation, in turn, would allow the population to be released into apparently suitable habitats outside its current range but within its probable historic range. In 2022, the Fish and Wildlife Service, one of the federal agencies charged with implementing the Endangered Species Act, proposed to allow such an experimental population to be translocated to locations *outside* the historical range if the ability of the habitat within the current and historical ranges "to support one or more life history stages has been reduced due to threats, such as climate change or invasive species."[217] An accompanying press release elaborated on the rationale:

> Scientists have already observed wildlife responding to the effects of climate change, with some species and ecosystems losing habitat due to increased temperatures, altered rain and snow patterns, sea level rise, and greater frequency and intensity of drought and wildfires. These species include the Mt. Rainier ptarmigan in Washington state, Montana stoneflies and the emperor penguin, found in the Antarctic. Climate change has also exacerbated existing threats to plants and wildlife, such as greater threats from disease and invasive species. In Hawaii, increased temperatures are driving the spread of avian malaria among some of the world's most endangered birds, as mosquitoes move upslope.[218]

In 2021, Jedediah F. Brodie and colleagues called for a greatly increased effort using assisted migration at the international level, with the Convention on Biological Diversity playing a leading role by creating a technical committee to develop a protocol that all nations could implement.[219] In fact, the convention in its 2015 Conference of Parties had acknowledged the possibility of assisted migration in the face of climate change, but with "a precautionary approach in order to avoid unintended ecological consequences including, for example, the spread of invasive alien species."[220] Brodie and colleagues advocated using assisted migration to move species beyond their historical range and proposed expanding the new term "neonative" to incorporate non-native species.[221] Their proposal was criticized on the grounds that it minimized the likelihood of inimical consequences of such introductions to previously unoccupied sites.[222]

: : :

Climate change will have enormous impacts on where species live, how well they thrive, how they interact with other species, and how they affect ecosystem-level processes and phenomena. This is as true for non-native species as it is for native species. There is a long, sporadic history of predicting, based on climate, where introduced species will spread to once they are established in a non-native range, and such prediction increased enormously in the late twentieth century as geographic information systems led to a rapid increase in mechanistic and especially species distribution models for invasions. Despite several limitations for SDMs, SDM-based predictions for non-native species have proliferated, while mechanistic models have been developed less frequently. This entire prediction literature should be taken with a huge grain of salt, however. With very few exceptions, tests of these predictions of range changes are lacking. The fact that ranges of so many species, both native and non-native, are shifting, producing and eliminating interactions among them, particularly challenges prediction. Typically, the prediction, with or without predicted climate change, and whether or not an ensemble method is used, is for several decades in the future, and it is far too soon to assess predictive success. This caveat is often omitted from publications.

Direct observation of short-term range shifts or abundance changes that can be plausibly linked to climate change have slowly accrued, more for plants than for animals. Although most researchers believe climate change will, on average, increase the number of invasions and exacerbate invasion impacts, some non-native species will be less likely to establish or persist in certain areas. The precise, underlying reason why a species has been favored by climate change is often unknown, even when the fact that it has been favored seems incontrovertible. However, research on the underlying mechanisms is gradually increasing. It suggests that many impacts of climate change on invasive plants appear to be due to shifts in the timing of life cycle events. Research of this sort on animals is scarce.

15

Hypotheses Explaining Biological Invasions Proliferate

Every scientific discipline begins with observations. As relevant observations increase, practitioners try to formulate general rules or laws that explain the growing number. So it was with biological invasions. Reports of impactful invasions piled up in the twentieth century, each described as a singular event. As they coalesced into modern invasion science in the 1980s (as described in chap. 9), a key element of the transition was studying invasions as part of a global phenomenon, rather than each invasion as a one-off, singular event. Inevitably, scientists began to seek commonalities among invasions and their consequences. They also sought to understand apparent idiosyncrasies and paradoxes. For instance, for some species, some populations became invasive behemoths, spreading widely and with obvious impacts, while other populations of the same species or apparently very similar species failed to establish populations when introduced. Other puzzling observations included established, thriving small non-native populations that failed to spread, and established populations that spread widely but failed to generate apparent major impacts. In 1894, the philosopher Wilhelm Windelband distinguished between what he termed "nomothetic" knowledge, consisting of general laws governing the structure and workings of nature, and "idiographic" knowledge, which consists of portraying particular events in a coherent narrative that emphasizes features unique to each event.[1] Physical scientists have long worked in the nomothetic tradition, identifying universal laws such as the laws of thermodynamics. Biologists have sought with greater or lesser success to emulate them while studying phenomena with complex, varying contexts that seem to demand an idiographic perspective.[2] Ecology is a leading edge of this tension.[3]

Invasion science has yet to produce a theory that purports to explain all aspects of all invasions, but it has spawned many hypotheses that aim to explain groups of invasions that share certain features. Some had antecedents well before the advent of a distinct invasion science, while others arose directly from rapidly growing scientific literature on invasions. An early catalog, published in 2009, described twenty-nine such hypotheses.[4] Later efforts progressively upped the number to thirty-nine.[5] Various hypotheses are so closely related to one another that they might be viewed as variants of the same hypothesis.

15.1 Invasions Pass Through a Sequence of Stages

A growing recognition that invasions must pass through several stages greatly clarified understanding of why some species are often invasive and others are not, why some introductions lead to population establishment and others do not, and why some established non-native populations spread and become invasive while others remain restricted or innocuous. Invasion scientists and managers now generally recognize four stages in the trajectory of a non-native species as it becomes invasive—(1) transport from the native range to a distant site, (2) initial survival at the new site, (3) population establishment at the new site, and (4) spread from the initial location. Some variation exists in how this partition is characterized.[6] For instance, stage 1 can be divided into (a) the non-native species becoming entrained in a transport mechanism (such as a cargo container on a ship) and (b) actually surviving the journey. However, the important point was the growing recognition that one must understand each stage of the process to be able to explain why or how an invasion proceeded as it did or to predict whether an invasion will occur.

During the Scientific Committee on Problems of the Environment (SCOPE) program, with its emphasis on traits of species that predisposed them to be invasive and ecosystem features that rendered them susceptible to invasions (discussed in chap. 9, sec. 9.1), few contributors explicitly associated these issues with specific invasion stages. Several authors featured diagrams of the various stages and focused their efforts on these stages, while the analyses by Richard H. Groves were stage-specific.[7] The 1990s saw increasing attention to invasion stages. Geerat J. Vermeij, for example, proposing a research agenda for invasion biology, called for studying the factors acting at each stage, rather than looking for universal explanations for causes or features of invasions.[8] Mark Williamson organized his early monograph on invasions according to a set of stages (as detailed in chap. 10,

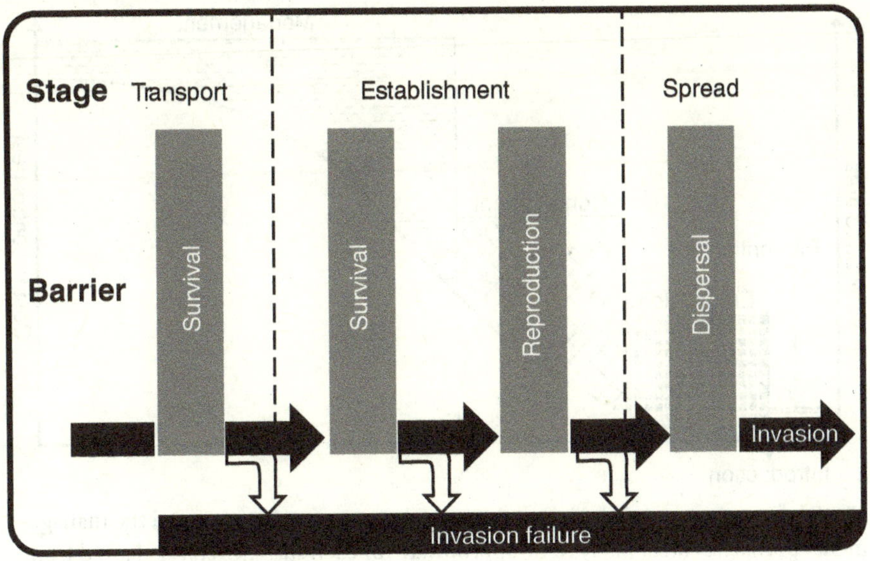

Fig. 15.1 Any non-native species must overcome barriers to pass through a series of stages before becoming invasive. At any point along this trajectory, an impassable barrier can cause the invasion to fail.

sec. 10.2.3.1).[9] By the first decade of the twenty-first century, many invasion scientists observed that explanations for success or failure of invasions should focus on the stage at which the explanatory mechanism was believed to act.[10] By the 2010s, this modus operandi had largely been adopted (fig. 15.1; see also box 15.1). The most widely cited reference for how to partition the invasion process into stages is by Tim M. Blackburn and colleagues.[11]

BOX 15.1 The Invasion Curve

In policy circles and in outreach to the public, the recognition that invasion management methods should be tied to the targeted invasion stage became well established in the 2010s, with many documents and media presentations featuring versions of what came to be known as the "invasion curve," perhaps first proposed in 2010 by the State of Victoria's Department of Primary Industries (fig. B1).[1]

1. State of Victoria, Department of Primary Industries, *Invasive Plants and Animals Policy Framework*, 14. See also Harvey and Mazzotti, *Invasion Curve*.

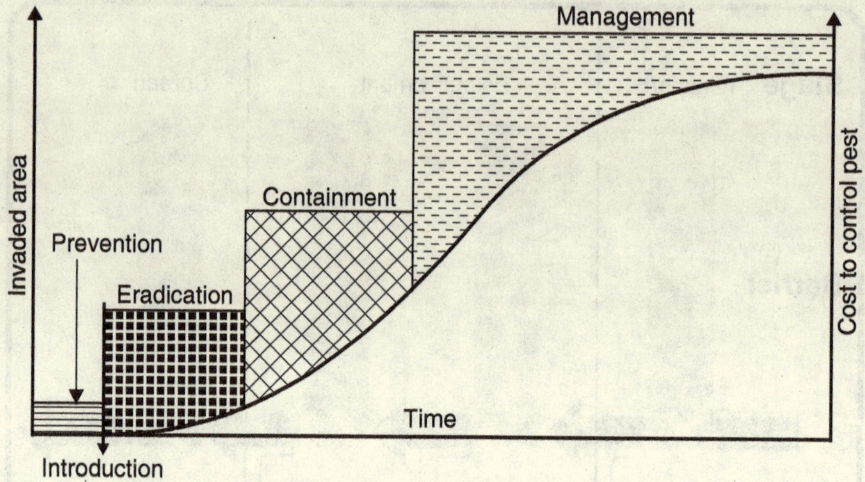

Fig. B1 "Invasion curve" depicting the typical stages of an invasion and the management approaches most likely to be appropriate for each stage. Credit: Adapted from US Army Corps of Engineers, Jacksonville District, "Invasion Curve."

15.2 What Hypotheses Are Obvious?

Two features of any introduction of a non-native species that seem obviously relevant to whether it will lead to population establishment and subsequent spread are how similar the climate of the new location is to that of the species' native range and how many individuals are introduced. Their potential roles in the outcome of particular invasions had been remarked on occasionally before the advent of modern invasion science but did not receive much attention. This remained the case during the SCOPE program, perhaps because the relevance seems so obvious. Both factors have since been widely noted and today are explicitly recognized by invasion scientists.

15.2.1 Climate

The most evident hypothesis for the fate and impact of introductions would seem to be whether the climate is suitable—one would not expect toucans to establish a population in Norway. Acclimatization societies in the nineteenth century (described in chap. 5, sec. 5.1) often blithely ignored apparent climate mismatches, which surely partly accounted for many of their failed introductions. They occasionally pointed retrospectively to climate as a cause—as, for instance, Lord Powerscourt attributing the failure of eland, native to Africa, to establish in Ireland to the combination of cold and dampness there.[12]

In the SCOPE program, a few papers mentioned climate as an important factor in determining whether a non-native species would establish a population or what impact it would have.[13] Most participants ignored it, probably because it seemed so obvious. Even the few who mentioned it did not detail how climate might limit invasions but rather in one or a few sentences acknowledged a role. For instance, Herbert Baker, whose SCOPE contribution was seminal in the history of attributing invasion success and impact to species traits (discussed further in subsequent sec. 15.3.1), concluded his paper by stating that "the available facts about successful invaders suggest: 1. Climate of source and reception area must be similar . . . ," although he had nowhere discussed a role for climate.[14] Similarly, Michael J. Crawley in 1989, in a section on "Failure to Establish," wrote that among "all of the factors identified in the data base as likely causes of failure to establish (climate, predators, parasitoids, disease, competition; see Crawley, 1986) . . . bad weather is both the most frequent cause of failure, and the chanciest of factors."[15] However in neither his 1989 publication nor the one he cited for 1986—his effort in the British contribution to SCOPE—had Crawley actually discussed climate. Almost certainly, the relative dearth of attention to climate during the SCOPE program was because it occurred about a decade before the heyday of species distribution models and growing recognition that climate change would affect species' ranges (as discussed in chap. 14, sec. 14.4). Had the program developed a few years later, surely many participants would have acknowledged a role for climate, and several would have featured it in their contributions.

Aside from SCOPE, by the late twentieth century, enterprises that purposefully introduced species such as forestry and biological control routinely and systematically assessed potential introductions in terms of climatic suitability.[16] Foresters, for instance, quickly employed BIOCLIM, the first species distribution model (SDM; detailed in chap. 14, sec. 14.4.4), to aid tree selection for planted forests. And the widespread effort in the twenty-first century to predict whether a non-native species will survive and, if so, where it will spread based on SDMs assumes that climatic factors are determinative. As discussed in chapter 14, two factors that challenge this effort are (1) the possibility that species rapidly evolve to tolerate new climatic features, and (2) species may be introduced to areas climatically unlike anything in their native range, so there is little basis for prediction. An early monograph in the modern era of invasion science pointed to several examples of the second sort in which a non-native species thrived, such as rabbits in interior Australia, which is hotter and more arid than any part of their native European and North African range.[17]

In sum, climatic matching seems to be a rough but useful rule of thumb as an explanatory hypothesis for invasion outcome. It is now widely employed in planned introductions. When a planned introduction nevertheless fails, or at least does not succeed well enough to perform whatever function was anticipated for it, one possibility is a subtle detail of climate that had not been accounted for, but usually the explanation lies in an entirely different aspect of the receiving environment.

15.2.2 Introduction Effort ("Propagule Pressure")

The idea that introducing very few individuals should be less likely to lead to population establishment than introducing very many seems intuitively clear, but it was not considered as a broadly applicable hypothesis to explain success or failure until the late 1990s. The term "propagule pressure" to characterize the number of individuals was coined by Mark Williamson in 1996, but previously the general idea had been suggested occasionally for particular invasions.[18] For instance, Bryan P. Beirne in 1975 observed that, for insects introduced to Canada for biological control purposes, the probability of population establishment increased substantially with total number of individuals introduced; he did not indicate whether these were all introduced at one time or in several batches.[19] Participants in the SCOPE program occasionally speculated on the role of propagule pressure in initial establishment of particular non-native species, but far fewer than considered the roles of, for example, species traits or habitat disturbance.[20]

A trickle of similar observations appeared in the biological literature through the late 1990s, after which a rapidly increasing number of publications focused on the role of propagule pressure in establishing non-native populations.[21] For inadvertent introductions as well as some deliberate ones, the actual numbers of propagules introduced is unknown, so many authors estimated propagule pressure by assuming a correlation with a variable that could be measured, such as numbers of nursery catalogs and other indicators of horticultural activity, aquarium trade data, and boat and shipping traffic.[22] For two sorts of invasions, numbers of introduced individuals were often recorded more or less accurately: introductions of insects for biological control (discussed further in chap. 17, sec. 17.5) and introductions by acclimatization societies. Several authors examined records for birds introduced to New Zealand by such societies, as had Alan E. Newsome and Ian R. Noble in Australia, all finding that, on average, likelihood of establishing a population increased with numbers of individuals introduced.[23] For biological control, in addition to retrospectively examining establishment where numbers of individuals introduced had not been systematically varied, as Beirne had done, several authors experimented with different numbers of introduced individuals or different numbers of introduction events with fixed numbers of individuals, or both. For example, in New Zealand, Jane Memmott and colleagues released different numbers of individuals of a thrips species that attacks invasive gorse during different numbers of releases, and Fritzi S. Grevstad in the United States released different numbers of two beetle species that attack purple loosestrife.[24] Such experiments, combined with observational and correlational information of the sort previously described, generally confirmed that greater propagule pressure often plays an important role in establishing a non-native population. Robert I. Colautti and coauthors suggested that insufficient propagule pressure should be considered a null hypothesis for failed non-native establishment before explanations invoking species traits or habitat features are invoked.[25]

By the end of the first decade of the twenty-first century, invasion scientists were far less likely to neglect a possible role for propagule pressure in seeking causes for establishment failure or in planning releases. Much of the research underpinning this attention traces back to another literature, that of population modeling. The most influential early work was a 1967 book by Robert H. MacArthur and Edward O. Wilson, who explored the probability of population (non-native or not) establishment depending on the numbers of individuals introduced to an unoccupied site and on demographic parameters of the species.[26] This approach led to formulation of a phenomenon known as "demographic stochasticity," the random fluctuation of population size by virtue of the population's having a finite number of individuals. It incorporates random fluctuation in demographic traits such as birth rate, death rate, and sex ratio. In a population consisting of a very few individuals, the probability that all will die in the next interval, or that all offspring will be of one sex, is far greater than in a large population. This phenomenon, in turn, is a component of what came to be known as the "minimum viable population size," a frequent feature of planning for management of threatened populations.[27] The relationship of this concept to the role of propagule pressure became apparent once invasion scientists focused on propagule pressure. Some researchers seeking specific causes of establishment failure, especially in perplexing cases such as that of the house sparrow in North America in which a spectacularly successful establishment and spread had been preceded by several failures, have invoked demographic stochasticity associated with small propagule size.[28]

15.3 Do Species Traits Predict Invasion Scope and Impact?

As noted in chapter 9, section 9.1, the SCOPE project was charged with addressing two distinct questions: (1) What factors determine whether a non-native species will be invasive? and (2) What site properties determine whether an ecological system will be relatively susceptible to, or resistant to, invasion? As hypotheses proliferated with the growth of the field, two hypotheses and several variants attributed invasion success or impact largely to features of the non-native species.

15.3.1 The "Ideal Weed"

In an influential book in the run-up to the SCOPE project, *The Genetics of Colonizing Species* (discussed in chap. 7, sec. 7.1), several authors proposed that particular plant traits preadapted them to become weeds.[29] Not all weeds are non-native (as discussed in chap. 10, sec. 10.1.2.6), but very many are. Thus, factors that predispose a plant to become a weed—for example, thriving in a wide variety of habitats—are relevant to the study of invasions. The set of traits proposed by Herbert G. Baker, a British botanist who spent most of his career at the University of California, became known in the invasion literature as the "Baker List," and he slightly modified it in 1974.[30] Earlier research by Baker on the importance of self-compatibility in allowing long-distance dispersers to establish populations was a key inspiration of this exercise, and there were still earlier antecedents in the nineteenth century.[31] For instance, in 1879, British botanist George Henslow argued that self-fertilization was a key factor in the spread and dominance of non-native plants, eliciting a rebuttal from American botanist Asa Gray.[32] Despite Baker's acknowledgment that few if any weeds would possess all the traits on the list, the approach was criticized by invasion scientists on exactly these grounds.[33] Nevertheless, the list was frequently cited through the 1990s as predictive in petitions to release genetically modified organisms into the environment.[34]

In the SCOPE project, Baker himself downplayed the ideal weed approach to explaining outcomes of plant introductions to North America, focusing instead on features of the receiving ecosystem.[35] The only species traits he pointed to in 1986 as conducive to successful establishment were certain reproductive features, especially apomixis (the ability to self-fertilize, without sexual reproduction). By the beginning of the twenty-first century, it was widely recognized that the list was at best a rough guide for predicting plant invasion trajectories, although amassing such trait lists through systematic examination of growing numbers of publications continues to be a holy grail in the search for predictive tools.[36] The dominant approach in regulating plant introductions became that pioneered by the Australian Weed Risk Assessment procedure, which was also often adapted to other settings.[37] Such risk assessment tools typically include various species traits but also many other factors, such as features of the potential receiving location and historical observations.

Similar attempts to produce lists of species traits facilitating invasion have appeared periodically, though virtually every one acknowledges the likelihood of exceptions to whatever set the authors advocate. In the SCOPE program, Paul Ehrlich considered pairs of species that were apparently similar, but one member was a successful invader and the other was not.[38] In 1986 for animals and in 1989 for vertebrates only, he produced lists very similar in form to those of Baker, one column of traits "good" for invasion and the other matched traits not good for invasion. By contrast, Noble examined the Baker List approach for terrestrial vascular plants and was not optimistic, viewing it as, at best, a checklist of potential warning signs.[39] Scanning features of invasive plant populations worldwide, Jacques Roy came to the same conclusion.[40]

Other efforts in this vein were more narrowly focused. A notable example entailed examining twenty-four pine species, twelve of which had proven invasive upon introduction and twelve of which had not.[41] Using statistical discriminant analysis, the researchers created a function that clearly separated the invasive from the noninvasive species based on only three traits: seed mass, minimum juvenile period, and mean interval between large seed crops.[42] The same team extended the analysis from pines to all conifer species (seventy-nine species, of which thirty-six are invasive) and found that the same three traits continued to discriminate well between invasive and noninvasive species.[43] Subsequent research along these lines with pines found that high relative growth rate (mass of new growth/mass of plant/day) was also a key feature separating invasive from noninvasive species.[44] The trait contributing most to relative growth rate was specific leaf area (leaf area/leaf mass).

Without seeking lists of traits applicable to large groups of species, others have studied the role of particular traits in the success of one or a few invaders to predict the invasion trajectory of another species based on those traits. One example concerns traits common to zebra mussels and golden mussels. Ladd Johnson and James Carlton attributed the dramatic spread and impact in North America of zebra mussels from Eurasia primarily to two traits: byssal threads for attachment to hard substrates and highly dispersive veliger larvae.[45] Anthony Ricciardi subsequently observed that the golden mussel from China, recently introduced to South America, had similar traits.[46] He predicted that it would spread widely and rapidly there and was proven correct.[47] Most risk assessments, such as the aforementioned Australian Weed Risk Assessment, include questions related to species traits that have seemed to be associated with past invasions, whether or not research has explored potential mechanisms for the association.

Several other hypotheses are variants of the "ideal weed hypothesis," depicting species traits as the key or sole determinant of the outcome of introducing non-native species. Among these are hypotheses termed "novel weapons," "habitat filtering," "adaptation," "human commensalism," "plasticity," and "ecological imbalance."[48] Recently the idea of using sets of features of a species to explain introduction outcome has recrudesced as part of a surging interest among ecologists in "functional traits."[49] Functional traits are measurable features of individuals that may affect fitness, and they can be physiological, morphological, behavioral, or temporal (e.g., flowering time). Since it is, in fact, these traits that determine where a species can persist, some invasion scientists have conceived of species as collections of individuals with particular sets of traits and sought general explanations of persistence of a non-native species as well as impact in terms of traits.[50] This approach might be seen as a version of Baker's ideal weed hypothesis updated in light of half a century of intensive research on invasions.

15.3.2 Functional Response

A particular functional trait that has spawned a prominent general hypothesis is known as the "functional response." In 1949, ecologist Maurice E. Solomon explored what controls animal populations in nature.[51] With regard to the possible role of predators and parasites, he distinguished between two ways that a predator or parasite might respond to a population increase of its prey or host. The "functional response" is how an individual predator or parasite deals with this increase. Does it attack more individuals of that species in total? If so, does its rate of attack increase linearly with the increase in the prey or host population? Or does its rate of attack accelerate with the rate of increase of the prey or host population, perhaps because the predator or parasite specializes in attacking that species among others it might attack? Is the response increased by decreased time required to handle each prey or host individual? The "numerical response," by contrast, is the possible increase in the predator or parasite population in response to the increased prey or host population, generated by increased survival or reproduction, or both, owing to the functional response.

Solomon's paper was not often cited for about a decade, until others, especially ecologist Crawford S. Holling, pointed to the importance of this distinction and used it in planning and interpreting results of their own research.[52] For instance, Holling studied how three native small predatory mammals—masked shrew, northern short-tailed shrew, and deer mouse—responded to the introduction and population fluctuations of the non-native European pine sawfly.[53] Noting that Solomon's distinction between the two kinds of responses had been "largely ignored in the literature," Holling nevertheless adopted it and found that the functional responses of the three predators differed greatly. All three increased their attack on sawfly cocoons as sawfly population densities increased, though the northern short-tail shrew did so much more sharply. By contrast, the numerical response of the masked shrew greatly exceeded that of the other two species. Publications by Holling and others led to the functional and numerical responses becoming part of the working lexicon of ecologists, featured in the population dynamics or predation sections of textbooks—though often attributed to later authors rather than to Solomon. It took longer for invasion scientists to embrace this approach.

Researchers working on biological control of agricultural pests (a topic covered in chap. 17, sec. 17.5) studied the role of functional response well before other invasion scientists did. For instance, in 1984 Susan E. Riechert and Tim Lockley summarized empirical and theoretical data on functional and numerical responses of spiders to prey densities in an exploration of circumstances under which these predators would be useful in biological control.[54] Phil J. Lester and colleagues examined the functional responses of a predatory mite used in biological control of an agricultural pest—the red spider mite—to see what method of applying an acaricide least affected the biological control agent.[55] However, Lester and Rudolf Harmsen also performed experiments on systems of two predatory mite species and two prey mite species that attack apples. They concluded that results of simple laboratory experiments determining functional and numerical responses of a single predator species attacking a single prey species should be applied very cautiously to predictions of the effectiveness of biological control in the field.[56] Similarly, in several cases in which non-native parasitoid wasps were used as biological control agents and functional response data were gathered, functional response alone did not effectively predict successful control of targeted pests.[57] Nevertheless, functional response with or without numerical response has been studied to aid prediction of biological control efficacy by species such as the multicolored lady beetle and several predatory mites.[58]

The first use of functional response to attempt to explain impact of a non-native species outside the biological control literature was reported in 2008 by Loic Bollache and associates, who studied the killer shrimp, an amphipod native to the Caspian region that had invaded parts of western Europe.[59] This team presented individuals of the killer shrimp and three native amphipod species that it is replacing with differing densities of juvenile isopods, a prey species commonly found in nature by all four amphipods. The killer shrimp had a substantial greater functional response than those of the natives to increasing prey density—the rate of consumption increased more rapidly with prey density and the maximum rate of consumption was higher. This result inspired a growing amount of research on the possibility that functional response could predict the impact of a potential invader and that it might explain differences among impacts of different invaders (fig. 15.2). Much of this research was conducted by Jaimie T. A. Dick and his colleagues and included studies to deepen understanding of some prominent invaders, such as the lionfish and the Pacific oyster.[60] Dick and colleagues suggested that comparing functional responses between native and non-native species or between two non-native species would be an effective way of predicting impacts and testing hypotheses about invasions.[61] They proposed that a simple statistic based on functional response that could be determined by laboratory experiments could be used in predictions.[62] The statistic, "functional response ratio," is simply the attack rate divided by the

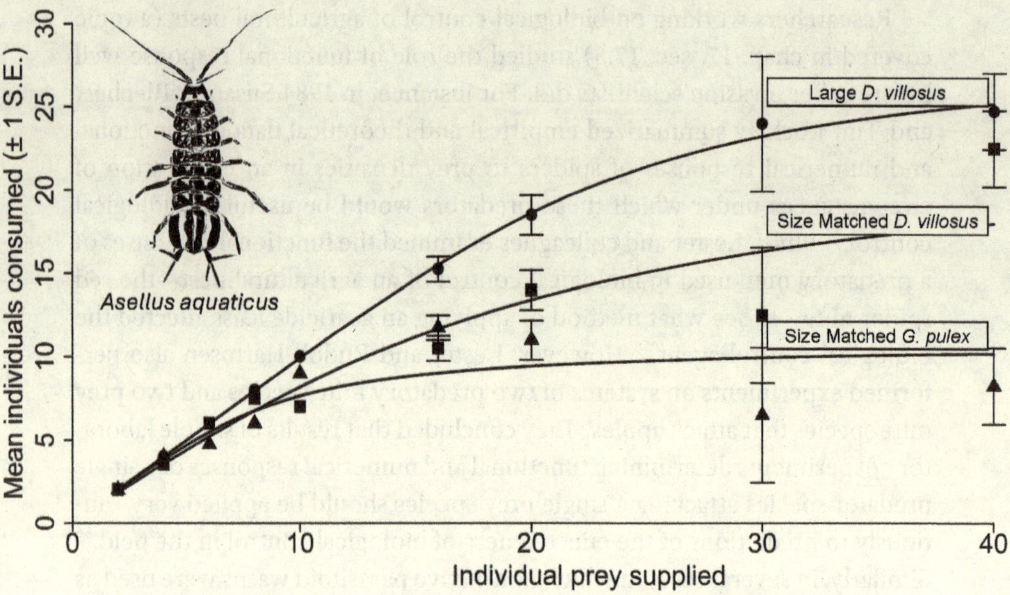

Fig. 15.2 Functional response of predatory native amphipod (*Gammarus pulex*) and non-native killer shrimp (*Dikerogammarus villosus*) to varied densities of prey isopod (*Asellus aquaticus*). In nature, killer shrimp tend to be larger than the native amphipod, but the experiment run with individuals of the two species selected to be the same size (lower two curves) demonstrated that the functional response of the non-native still exceeded that of the native. Credit: Jennifer A. Dodd; cf. Dodd et al., "Predicting the Ecological Impacts of a New Freshwater Invader."

handling time. In nature, however, values of both of these features could differ greatly from those tallied in the laboratory, depending on both the physical environment and which other species are present.

Although, as formulated by Solomon and popularized by Holling and others, functional response would seem applicable only to animals that consume other species (e.g., predators, parasites, herbivores), in fact one could use it to characterize aspects of competitive interactions—say, by comparing functional responses of two or more predatory animals, as Bollache's team had done. It might also be used to compare competitive interactions between plant species by assessing the rates at which plant species use resources such as soil nutrients.[63] David Tilman used the "functional resource acquisition and/or utilization response" of two species of planktonic freshwater algae in laboratory experiments to attempt, with some success, to understand outcomes of competition between them under varied resource conditions.[64] Anna G. Aguilera suggested that feedback mechanisms operating between plants and the associated soil communities could be modeled in terms of functional responses of two or more plant species to determine the outcome of competition or an invasion.[65]

Dick and colleagues, lamenting the wealth of competing hypotheses in invasion science, suggested that in the great majority of them—in fact, twenty-eight of the twenty-nine listed by Jane Catford and collaborators—resource use is a central mechanism.[66] Dick and associates also noted that, in cases where the necessary data have been gathered, non-native species with high impacts consistently have higher maximum feeding or resource use rates than similar native species, and this relationship appeared to obtain for plants, animals, vertebrates, invertebrates, terrestrial species, and aquatic ones.[67] Thus, they argued that functional responses can "unify" invasion science in the sense that these responses will reveal commonalities among a wide range of invasions and permit improved impact predictions. James Vonesh and colleagues, agreeing that functional responses are important quantitative descriptors of consumer effects on resources, nevertheless have argued that Dick and colleagues greatly overestimate the likelihood that comparisons of functional responses can unify invasion science.[68] Their chief concern is that the simplified, relatively short-term laboratory experiments used to characterize functional responses ignore the complex network of interactions and the physical environment that any non-native species confronts in nature and thus will often yield incorrect predictions. They also noted that many high-impact invaders, such as the brown tree snake on Guam and the Burmese python in Florida, invade systems in which there are no more or less similar species to use in comparisons of functional response. A team including Dick appeared to back off somewhat from the notion that comparing functional responses alone would produce excellent predictors of invasion impact, suggesting instead using them in combination with proxies for numerical response.[69]

15.4 What Features Make Ecosystems More or Less Prone to Impactful Invasions?

The stated SCOPE mission included the goal of understanding what site properties make certain systems more susceptible to invasions than others. This question led to several hypotheses, some of which antedate SCOPE, or at least were assumed to be true even if not stated as hypotheses. Although each has distinctive features and they are treated in recent literature as separate hypotheses, many are related to one another but emphasize different aspects of the same underlying phenomena.

15.4.1 Biotic Resistance

In his 1958 monograph *The Ecology of Invasions by Animals and Plants*, Charles Elton proposed that many—perhaps most—invasions fail because of what he termed "ecological resistance," opposing forces exerted by native species. Elton conceived of this force as competitive, with native species occupying so much of what is now termed "ecological niche space" that not enough is left to support a new species. Elton attributed the observed dearth of successful invasions or invasion impacts in Wytham Woods—object of long-term research by his team—to ecological resistance. He particularly emphasized its operation in discussing why islands appeared far more susceptible than continents to invasions. In his view, the greater number of species on mainland posed an ecological resistance to invasion that was lacking on species-poor islands.[70] (Elton and his work on ecological resistance are discussed in greater detail in chap. 7, sec. 7.1.)

This hypothetical resistance by native species to invasion came to be termed "biotic resistance." Elton's adduced examples of apparently invasible species-poor sites were buttressed by theoretical treatments, especially by ecologist Robert H. MacArthur. MacArthur described model communities of three or four trophic levels as being more "stable" in the face of perturbation the more linkages connected the species, and he portrayed model communities of a single trophic level structured by competition between the species as less easily invaded the more species are originally present.[71] Several authors in the SCOPE program suggested versions of biotic resistance for particular failed invasions, citing Elton.[72] However, Jonathan M. Levine and Carla M. D'Antonio, reviewing the literature on biotic resistance, cited observational, empirical, and theoretical research that contradicted the claimed pattern of increased diversity conferring resistance to invasion.[73] Several observational studies even showed the opposite pattern—more species-rich communities containing more non-native species.[74] These and similar observations led Jason Fridley and colleagues to proclaim an "invasion paradox."[75] They suggested at least a partial resolution of the paradox might reside in scale-dependence: in small sites, more native species would mean greater resistance to invaders, while over large areas, whatever environmental features render an area suitable for large numbers of native species also render it suitable for many non-native species. But a recent meta-analysis of papers reporting a relationship of numbers of non-native and native plant species to the physical size of studied areas found no paradox, and in fact no scale at which the number of non-native species was negatively related to the number of native species.[76]

Elton focused only on competitive exclusion and total number of native species (biodiversity) as the causal factors in resisting invasion. Most of the literature addressing the hypothesis has taken this tack.[77] Almost all the theoretical literature on the topic assumes just one trophic level, interacting through competition.[78] Similarly, several small-scale experiments testing the biotic resistance hypothesis, in which potential invaders were added to systems with variable numbers of established resident species, used a single trophic level, such as plants or certain aquatic invertebrates, and interpreted the results, whether supporting the hypothesis or not, as reflecting competitive interactions.[79] However, some studies found herbivory or predation as the force causing a native community to resist invasion: native species ate the invader.[80]

Much of this research addressed biotic resistance to invasion in a more general framework of the importance of biodiversity. However, several studies found that, although a native community indeed resisted invasion, the resistance was caused not by the total number of native species, as Elton had imagined, but rather by the action of single native dominant species. For instance, Anna Henriksson and collaborators examined 184 Swedish lakes into which Arctic char had been introduced.[81] Char had failed to establish in 126 of these, and the number of native fish species was unrelated to whether or not char established populations. However, for the great majority of failures, northern pike—a native species that preys on fish—was present, and char almost never established populations where pike were present.

15.4.2 Enemy Release

Elton also helped lay the groundwork for a widely cited hypothesis to explain successful establishment of non-native populations, particularly those that generate large impacts. This hypothesis, known as the "enemy release hypothesis," is in a sense an extension of the biotic resistance hypothesis.[82] It has two components. First is that biotic resistance by native species at the new site has failed—an invader has established a population. Second is that the invader has been introduced without a coterie of natural enemies that suppress its numbers in its native range.[83] Perhaps the first clear statement of this idea, with respect to plants, was by Swiss botanist Albert Thellung.[84] In 1958, Elton speculated that the prominence in Britain of non-native common rhododendron, native to the Iberian Peninsula, might be due to the absence of its native natural enemies. Similarly, he suggested that the South American cactus moth was able to control non-native prickly pear in Australia (discussed further in chap. 17, sec. 17.5) because the moth arrived from its native South America without its natural enemies. Decades earlier, he had attributed the dramatic spread of the spongy moth in North America (detailed in chap. 4, sec. 4.4) to the absence of parasites that he believed limits its numbers in Europe.[85]

The entire field of classical biological control (the topic of chap. 17, sec. 17.5), an approach Elton endorsed enthusiastically, is based on the premise that non-native "pests" are pests precisely because they have been released from the enemies of their native ranges, so introducing those enemies will reestablish control.[86] Classical biological control projects date back to the mid-nineteenth century, but early practitioners tended to focus on particular non-native pests and not on general phenomena applicable to many invasions. Versions of the enemy release hypothesis as a principle in biological control of insects arose under other names in the 1930s and for biological control of weeds surfaced in the 1940s and 1950s.[87] Successful biological control projects have been cited as strongly supporting this hypothesis.[88] Some modern biological control textbooks envision much of biological control as part of invasion science and discuss the enemy release hypothesis in that context.[89] Versions of the enemy release hypothesis were standard explanations for some successful invasions in many of the first invasion science monographs and textbooks in the wake of the SCOPE project (discussed in chap. 10, sec. 10.2.3).[90]

Although the enemy release hypothesis is often cited, the suggestion that it explains the fate of particular invasions has been subject to criticism on various grounds. Several authors have pointed to the variety of forms of evidence adduced in support of the enemy release hypothesis and observed that, in most putative examples of enemy release based on a single type of evidence, other explanations are possible for a particular invader's thriving.[91] A review of 176 studies of the enemy release hypothesis and the types of evidence reported by each found as many instances where the evidence does not strongly support the hypothesis as instances where it does.[92]

Perhaps the most basic challenge to the hypothesis is its underlying assumption that it is natural enemies that keep a non-native species from being very common in its native range. What forces control populations in their native or non-native ranges is part of a much larger, controversial subject beyond the scope of this book, including the relative importance of predators or herbivores ("top-down" forces), resources ("bottom-up" forces), and physical factors.[93] Suffice it to say that, with respect to invasions, the evidence on how frequently natural enemies control non-native species in their native ranges is mixed.[94] In his early summary of the principles and accomplishments of biological control of plants, Carl B. Huffaker recognized that this assumption may often be unwarranted:

> While aggressive and abundant alien pests are always good prospects for biological control, it does not follow and is ecologically unsound to postulate that such aggressiveness and abundance in the new land is necessarily due to an absence of natural enemies. There are far too many other reasons why a pest may be relatively innocuous in its native land yet aggressive and troublesome in environments new to it.[95]

An alternative to enemy release as an explanation for an observed spread of a non-native species that has established without the apparent natural enemies of its native range is that the physical environment of its new location differs from that of its native range in a way that favors the species' survival and reproduction.[96] This hypothesis will be discussed under the term "disturbance hypothesis," because the great majority of appeals to this alternative explanation cite anthropogenic disturbances such as nutrient enrichment or modified fire regime as producing the relevant environmental difference.

15.4.3 Evolution of Increased Competitive Ability (EICA)

Robert Colautti and colleagues pointed to the fact that enemy release can fail to be manifested over the short term, even if a species leaves the natural enemies of its native range behind, yet can precipitate an evolutionary change in a non-native species that ultimately increases its impact.[97] Their scenario is related to the fact noted previously that, at least in the short term, some species are not controlled by natural enemies but rather by other forces such as limited resources or stressful physical factors. It is well established that many plant species either resist herbivores by virtue of defensive chemicals or tolerate a degree of herbivory without having reduced survival or reproduction.[98] Thus, in their native range, they may not be controlled by herbivores. However, if production of defensive chemicals or the mechanism of tolerance exacts a metabolic cost, release from herbivores would seem to generate natural selection to reduce that now unnecessary metabolic cost and to use the freed energy for other purposes, such as survival or reproduction. In fact, that was the hypothesis of Bernd Blossey and Rolf Nötzold, the evolution of increased competitive ability (EICA).[99]

The EICA hypothesis, discussed in chapter 11, section 11.3.2, suggests that enemy release triggers an evolutionary change in established non-native plants—namely, the shunting of metabolic energy away from traits providing defense against natural enemies (which are decreased or absent in the invaded range) and toward growth or reproduction, or both. Such a shift would be expected eventually to lead to larger populations of these non-natives and thus perhaps to greater impacts. However, the patterns that Blossey and Nötzold perceived that inspired the hypothesis—greater size or greater seed production of plants in their non-native ranges than in their native ranges, or both—turned out not to be true in general.[100] Certainly some plant species are, in fact, larger in their non-native ranges, and some plants do produce more seeds in their non-native range, but it has proven difficult to link these features in specific cases to the mechanism envisioned in EICA. Several reviews and meta-analyses have yielded mixed evidence on the frequency of the proposed phenomenon in nature.[101]

15.4.4 Disturbance

As nineteenth-century botanists increasingly recorded non-native plants in various floras, some noted their association with disturbed habitats. For instance, in 1879, Asa Gray, having observed that most plants considered weeds in the United States had been introduced from elsewhere (as mentioned in chap. 3, sec. 3.2), observed, "We may for present purposes consider weeds to be plants which tend to take prevalent possession of soil used for man's purposes."[102]

Elton focused primarily on the importance of the biological component of an ecosystem in resisting invasion. He rarely addressed the potential role of physical disturbance in facilitating invasion, though he surely recognized its importance for non-native plants. This relative neglect was probably also because of his predominant focus on animals, whereas the role of disturbance in invasions has mainly been claimed for non-native plants.[103] Notably, where Elton wrote about a role for disturbance, he was usually referring to plants: "But by far the greater part of our alien plants live in habitats drastically simplified by man, including of course our crop plants: in arable land, waste dumps, railway tracks, walls, and so forth"; ". . . invasions or outbreaks most often happen on cultivated or planted land—that is, in habitats and communities very much simplified by man."[104]

Aside from its potential role in invasions, disturbance is the initiating force in the process of ecological succession, by which communities replace one another in more or less orderly fashion. The forces driving succession have been a prominent theme in the science of ecology from its inception.[105] Although there were several precursors, the formal depiction of succession as an ecological process began in 1899 with Henry C. Cowles and was dominated in the first half of the twentieth century by a controversy over the degree to which communities are organized holistic entities and succession is truly a deterministic community-level process. This view was most strongly advanced by American ecologist Frederic Clements and contested by Henry A. Gleason.[106] The literature on invasions was largely separate from that on succession, with the notable exception of American ecologist Frank Egler (discussed in chap. 6, sec. 6.1) and Australian ecologists Ian R. Noble and Ralph O. Slatyer, and the field of invasion science has been criticized for exactly this reason (as discussed in chap. 18, sec. 18.4).[107]

Yet another focus of research on disturbance has been its relationship to biodiversity. This relationship has long been studied independently of invasions, with recognition of the fact that extremely stressful conditions can keep all but a few species from occupying a site.[108] In the context of succession, Eugene Odum (discussed in chap. 10, sec. 10.2), in his 1963 textbook, proposed that "the greatest diversity occurs in the moderate or middle range of a physical gradient" of many physical factors, using salinity of coastal bays as an example.[109] This idea was formalized by British ecologist J. Philip Grime in 1973, but without using the term "disturbance."[110] Grime's idea subsequently was often credited to American ecologist Joseph H. Connell, who studied a similar phenomenon in coral reefs and tropical rain forests and named it the "intermediate disturbance hypothesis" (IDH).[111] In a nutshell, the IDH, as formulated by both Grime and Connell, states that, in communities with numerous species, unless an occasional disturbance intervenes, one or a few species will eventually outcompete the others and thereby lower diversity, but if disturbances are too frequent, few species will be able to tolerate the disturbance regime.[112] Observations of such a pattern, though not formulated as a general hypothesis, appeared as early as the eighteenth century and were common by the 1940s.[113]

But what exactly is a disturbance? Grime, focusing on plants, thought of disturbances such as grazing, mowing, trampling, or burning. Connell added herbivory and, for animals, predation. In both conceptions, there is some blurring of the boundaries between episodic disturbances and persistent stressful regimes, and of course a regime of sufficiently frequent disturbances would be stressful. As with so many terms in ecology and invasion science, the word "disturbance" has long been used either without definition or with definitions that may conflict. For instance, one definition is "any relatively discrete event in time that disrupts ecosystem, community, or population structure and changes resources, substrate availability, or the physical environment."[114] By contrast, Grime construed disturbance as a process removing or damaging biomass: "disturbance, which may be said to consist of the mechanisms which limit the plant biomass by causing its partial or total destruction."[115] Some of the disagreements that surfaced in later reviews and discussions of disturbance as a general explanation for invasions can be ascribed to absent or varying definitions of the term. Similarly, definitions of "intermediate" are usually unstated in discussions of the IDH, again accounting for differences of opinion about its importance in ecology in general and in understanding invasions in particular.

Disturbance, then, has been viewed in the modern invasion science literature as a double-edged sword: as a phenomenon that can facilitate invasions or exacerbate their impacts but also in other contexts as an event or process necessary to maintain natural successional processes and biodiversity. In the decade preceding the SCOPE project, several authors noted an association between disturbance and invasion by non-native plants. For instance, in 1977, Ronald L. Amor and Colin M. Piggin observed that few non-native plants have established in Australia without prior disturbance by such phenomena as cultivation, grazing, fire, and engineering works.[116]

The books resulting from the SCOPE project included several chapters focusing on disturbance. Many described particular species or groups of species in particular invaded locations. Per E. Christensen and Neil D. Burrows discussed how fire can facilitate plant invasions in Australia.[117] Donald E. Swincer asserted that "usually invaders occupy disturbed habitats more easily and most habitats are disturbed to some extent in Australia, if only by fire."[118] Herbert Baker pointed to the susceptibility to plant invasions of many types of human-disturbed habitats in North America.[119] Peter Moyle noted that human-modified freshwater systems are particularly frequently invaded by non-native fish in North America.[120] Richard Mack pointed to a role for disturbance in plant invasions in the intermountain western United States, as did Harold Mooney and colleagues for North America and John J. Ewel for South Florida.[121] Most participants in the Dutch contribution to SCOPE cited disturbed or anthropogenic habitats, or both, as particularly heavily invaded by various taxa in the Netherlands.[122] Several participants in the European contribution and the global SCOPE synthesis volume similarly claimed a major role for disturbance in invasions of particular taxa and regions.[123]

Other SCOPE participants attempted to look more broadly at various types of disturbance as generally important phenomena in invasions. Marilyn D. Fox and Barry J. Fox examined several types of disturbance of natural communities and their relationship to invasions by various plant and animal species in Australia, concluding, "There is no invasion of natural communities without disturbance."[124] Gordon Orians asserted that research on a number of taxa since the publication of Elton's 1958 monograph implicated disturbance as facilitating invasions by several non-native plant and bird species and suggested that the next advance will be to understand what specifically about disturbance plays this role.[125] In 1990, Jan Kornaś, building on his earlier work, showed that, in central Europe, heavy human disturbance characterizes the first sites invaded by many non-native plant species.[126] These sites are then the origin for seeds that colonize less heavily disturbed sites, and still later, at least for some species, invasion occurs in relatively undisturbed sites. Fred Kruger and colleagues distinguished between "endogenous" or natural disturbance regimes—entailing natural processes and phenomena such as rainfall, windstorms, and earth movements—and disturbance events and regimes generated directly or indirectly by humans.[127] They concluded that endogenous disturbance had little to do with invasions of the various Mediterranean ecosystems but that human-caused disturbances were of great importance. Richard Hobbs described several experiments on the role of disturbance in invasions, pointed to potential generalizations that might be drawn from them, and called for a greatly expanded experimental approach to elucidate how disturbance facilitates (or does not facilitate) invasions.[128] During this period, even aside from SCOPE participants, biologists cited many examples to highlight the key role of disturbance in facilitating invasions—usually plant invasions—in particular habitats, such as boreal forest and islands.[129]

As invasion research proliferated rapidly after the SCOPE project, reviews of the expanding literature on the role of disturbance inevitably appeared. Hobbs and Laura Huenneke pointed to the challenge posed to conservation by the fact that, for plants, disturbance, on the one hand, often facilitates invasions but, on the other, at least at intermediate levels, increases native species diversity.[130] Reviewing literature of several types of disturbance, they concluded that it is often modification of an existing disturbance regime rather than the disturbance regime per se that facilitates invasion (anticipating the "fluctuating resource hypothesis," discussed in subsequent subsection 15.4.6) and that interaction between disturbances (e.g., between grazing and fire) is often decisive.[131] In 1997, Joseph D. Lozon and Hugh J. MacIsaac conducted two literature reviews, using BIOSIS to locate hundreds of papers about disturbance, from which they extracted information on the relationship of disturbance to successful (established) invasions.[132] They then sought patterns—for instance, (1) whether disturbance is essential to establishment, and (2) habitats in which disturbance facilitated establishment. For each pattern, they depicted differences between animals and plants and discussed published suggestions bearing on the pattern. In 1999, Carla D'Antonio and colleagues provided examples of invasions fostered by many specific types of disturbance, elaborating on earlier calls for research on the specific mechanisms by which disturbances facilitate invasions.[133] They emphasized that an invasion can itself constitute a disturbance that facilitates further invasions. In particular, an invasion can disrupt natural disturbance regimes—for instance, by enhancing or suppressing fire or interfering with geomorphological disturbances such as scouring of flooded riverbanks.

By 2007, an early invasion science textbook could assert that the notion that disturbances facilitate invasion is "one of the most commonly accepted truisms in the field of invasion biology."[134] The 2000s have seen a spate of review papers and broad consideration of the role of disturbance in invasions. Angela T. Moles and colleagues, examining the rapidly accumulating literature on plant invasions, assessed research findings on several topics (e.g., community and habitat susceptibility or resistance to invasions, the role of invader traits, evolution of non-native species) and focused on the hypothesis that disturbance facilitates invasion.[135] They observed that then-recent empirical and theoretical research had found mixed support for the hypothesis. This result, in turn, suggested to them that changes from the natural disturbance regime—a form of disturbance itself—might be the key way in which disturbance facilitates invasion, but a test of the hypothesis from two hundred sites in eight nations found that, although the degree of change in a disturbance regime does correlate with aspects of invasion more strongly than disturbance frequency or intensity alone, none of these variables explained much of the variation in number of non-native species or percent cover by non-native species. Focusing on the IDH, Jane Catford and coauthors suggested management approaches to deal with the problem that disturbances can be critical to succession of native plant communities yet also facilitate invasions.[136] Miia Jauni and colleagues performed what was perhaps the first formal statistical meta-analysis of the hypothesis that non-native plants benefit from disturbance, in terms of number and abundance of non-native species, using a WoS literature search to identify seventy-three relevant articles.[137] They also conducted a similar meta-analysis for native plants. Results from individual cases varied, but in terms of mean effect size (a concept discussed in chap. 13, sec. 13.1) they found that non-native plant species on average benefited from disturbance and native plant species did not. They also detected differences in mean effect size depending on type of disturbance and type of habitat and that these features were the major determinants of disturbance impact. In addition to reviews, experimental research has continued on the effect of various kinds of disturbance on invasion of particular communities by particular invaders.[138]

15.4.5 Passenger-Driver

In 2005 Andrew S. MacDougall and Roy Turkington proposed a hypothesis for the forces leading to a successful invasion that might be considered a corollary of the disturbance hypothesis: the "passenger-driver hypothesis."[139] Experimenting on plant invasions in a series of sites in British Columbia, Canada, they observed that many occurred in previously degraded or disturbed sites, so that the invasions were not driving an environmental change but rather were passengers of the degradation or disturbance. This metaphor quickly resonated with many invasion scientists.[140] Jonathan T. Bauer observed that a non-native species need not be either a passenger or a driver but rather might be, in his extension of the metaphor, a "back-seat driver": a species that benefits from a disturbance in establishing a population or spreading but that in addition may exacerbate the impact of the disturbance by directly harming natives or doing so indirectly by altering ecosystem properties.[141] He presented as an example the invasion of eastern North America by garlic mustard, facilitated by overbrowsing by native deer and activities of non-native earthworms. However, garlic mustard releases chemicals that suppress mycorrhizal fungi on which many native plants rely, thus further changing the invaded habitat.

The passenger-driver hypothesis became widely recognized as crucially important to the project of ecological restoration, the attempt to restore a degraded ecosystem to some previous, desired state or long-term trajectory.[142] A large fraction of the effort to restore degraded ecosystems has entailed removing non-native species.[143] But if the underlying cause of the degradation is some anthropogenic disturbance and the non-native species are only symptoms, then simply removing the invaders without addressing the underlying cause may lead simply to reinvasion by the removed species or their replacement by other non-native species (the "treadmill effect").[144] The hypothesis was also frequently cited by critics of invasion science and management in the controversy over whether the threat to biodiversity and the environment posed by invasions is overblown (as discussed further in chap. 18, sec. 18.3).[145]

15.4.6 Fluctuating Resource

In 2000, Mark Davis and colleagues proposed a "general theory of invasibility" for plant invasions. They posited that any disturbance that causes an increase in resources—for example, light, nutrients, or water—is likely to spur an invasion if propagules of non-native species are nearby.[146] The "fluctuating resource hypothesis" (FRH) was founded on J. Philip Grime's early influential early assertion that competition among plant species is reduced in the wake of disturbance because the native species are temporarily prevented from monopolizing resources.[147] Davis and colleagues, in turn, had extended this idea by suggesting that competition intensity is inversely related to the amount of unused resources.[148] Importantly, Davis and colleagues later saw the FRH as explaining why attempted tests of the biotic resistance hypothesis yielded inconsistent results—there would not necessarily be a relationship between the number of native species and available resources in the wake of a disturbance. Nor would high productivity per se facilitate invasion because native species would monopolize it unless disturbance prevented them from doing so. Davis's team also emphasized that, to the extent that fluctuating resources explain how certain plant species replace other plant species, the FRH also applies to native plants replacing native plants (e.g., when native tree species encroach on prairies in the wake of decreased fire frequency) as well as to non-native plants doing so.[149] This explanation was part of their broad-based argument that invasion science is founded on an incorrect premise that non-native species somehow differ from native ones and should be studied separately from the rest of plant ecology (as discussed further in chap. 18, sec. 18.4).[150]

The FRH quickly found favor with plant ecologists and, increasingly, with invasion scientists working with other taxa.[151] In many studies, the hypothesis seemed to describe the actual mechanism precipitating an invasion.[152] An example is the research on invasive versus noninvasive pines discussed previously, in which two traits that distinguished invasive from noninvasive species—high relative growth rate and high specific leaf area—appeared to be particularly advantageous when a disturbance, such as a tree fall, produced a rapid increase in light. In some circumstances, the mechanism initiating an invasion did not appear related to a flush of resources.[153] For instance, an experiment in New Zealand found that the FRH failed to explain invasion of grasslands by a Eurasian forb.[154] By contrast, in grasslands of Montana in the United States and of Argentina, colonization by both native and non-native plants was indeed facilitated by increases in resources generated by disturbance. However, many studies showed an advantage of non-native over native plants in such invasions, contradicting the claim that FRH implies no categorical difference between colonization of sites by native and non-native plants.[155]

The FRH came to be embedded in a larger framework considering the role of resource pulses in many ecological phenomena, such as population explosions of many species, not only non-natives.[156] Perhaps its main contribution to invasion science, in addition to elucidating a number of specific invasions, is that it helped explain the tremendous amount of puzzling variation in the outcome of apparently similar situations in which a small non-native population established a temporary foothold in a new area. As Davis and his coauthors observed, the great variation in causes, magnitudes, and details of disturbance-induced flushes of resources, combined with variation in which species will have propagules in the vicinity when a resource flush occurs, guarantees varied outcomes.[157]

15.5 How Do Ecosystem Features and Species Traits Interact to Affect Invasions?

It has long been recognized that the fate of all non-native species depends to an extent on both species traits and features of the receiving ecosystem. It is likely that most if not all scientists focusing on one of these factors believed that the other was important, though many did not mention the fact. During the SCOPE program, with its mission focused on either species traits or ecosystems, few authors explicitly argued that the interaction of the species and the receiving site was important.[158] However, several general hypotheses to explain the outcome of invasions, often related to one another or to hypotheses discussed in section 15.4, have explicitly invoked an interaction between species and site, including "reckless invader," "limiting similarity," "new associations," "habitat filtering," "Darwin's naturalization hypothesis," and "opportunity windows."[159] Some of these antedate modern invasion science.

One such venerable hypothesis is that of the "empty niche" or "vacant niche." The concept of the "ecological niche," though usually attributed by ecologists to Grinnell, Elton, and Hutchinson in the twentieth century, was understood by nineteenth-century authors even before the field of ecology coalesced into a science, and the term "niche" was used by several of these authors to capture versions of the concept formalized by the twentieth-century ecologists.[160] The idea of a vacant niche that might be filled by a non-native species with appropriate traits is closely related to the biotic resistance hypothesis described in section 15.4.1. At least for the version of the biotic resistance hypothesis that focused on competition as conferring the resistance, one might imagine that the more species are present, the smaller the amount of empty "niche space" available for an invader to occupy.

Although a long-standing controversy in ecology questions the very existence of such a thing as a vacant niche, many authors have attributed specific successful invasions to the presence of a vacant niche.[161] Hutchinson pointed out that many invasions seem attributable to vacant niches, but he raised the possibility that what appears to be invasion of a vacant niche might often result from a disturbance rendering a location more invasible: "The rapid spread of introduced species often gives evidence of empty niches, but such rapid spread in many instances has taken place in disturbed areas."[162] In his 1958 invasion monograph, Elton attributed the invasion by the black rat into forests on Cyprus to its exploitation of an empty niche.[163]

A few participants during the SCOPE project ascribed particular invasions to the existence of an empty niche. For instance, Adrie J. Verkaik explained the rapid spread of the muskrat, native to North America, throughout the Netherlands by aspects of its own biology in the tradition of "ideal weed" explanations plus the finding that the few potential competitors make slightly different use of the environment and are not well adapted to an aquatic habitat, leaving a niche unoccupied.[164] Several invasions in the modern invasion science era have been ascribed to the presence of a vacant niche in the invaded ecosystem, although in many cases the evidence for this hypothesis could equally well support the hypothesis that some sort of disturbance has changed the habitat to favor a non-native species if one were to arrive or already be present.[165]

15.6 A Hypothesis That Stands Apart: Invasional Meltdown

Several hypotheses that have been advanced to explain the outcomes of certain types of introductions are neither so obvious as not to warrant specific mention (e.g., climate), nor do they fall cleanly into categories of species traits, environmental features, or the interaction between the two. One of these hypotheses has attracted growing attention. Whereas the biotic resistance hypothesis implies that each invader makes it more difficult for the next potential invader to establish itself, the "invasional meltdown hypothesis" posits that two or more non-native species can facilitate one another, thereby increasing the likelihood that one or both will establish populations and spread or enhancing their impact.[166] Geographer Alfred Crosby anticipated the concept of invasional meltdown when he depicted how groups of coadapted European species—livestock, weeds, insect pests, and pathogens—acted in concert to dominate and largely replace biotas of other regions (as discussed in chap. 8, sec. 8.3).[167] Francis G. Howarth also adumbrated the idea, suggesting synergisms among non-native species in Hawaii and also that the presence of a non-native species can create new niches that might be filled by either natives or new non-natives.[168] This kind of facilitation often entails a non-native species being reunited with another species introduced from its native range with which it has a coevolved mutualistic interaction. For example, the laurel fig, native to Asia and Australia, was present in Florida from the early twentieth century onward but did not begin spreading until the 1970s when its obligate pollinating wasp, *Eupristina verticillata* was introduced.[169] However, two species introduced to the same location from different regions, thus lacking any coevolution, can produce a meltdown. For instance, the Argentine ant from South America, first recorded in California in 1908, protects the red scale, native to Asia but introduced to California from Australia in the nineteenth century, from natural enemies, leading to an increase in scale density on citrus, a facilitation noted by Elton in his 1958 monograph (discussed in chap. 7, sec. 7.1).[170] Such facilitation could occur in many ways, including, for instance, a non-native species dispersing seeds of another non-native species or changing habitat in a way that favors another non-native species.[171]

The invasional meltdown hypothesis attracted much attention but quickly became controversial, largely on two grounds: (1) the degree to which the apparent facilitation has a substantial enough population impact to affect survival, spread, and impact, and (2) just how frequently meltdown occurs relative to other factors affecting invasions, such as biotic resistance.[172] Despite the continuing paucity of instances in which population impacts are proven, examples that suggest that meltdown is occurring have continued to accumulate since the hypothesis was proposed, and it has featured in all reviews of general hypotheses to explain invasions. In fact, an early review cited meltdown as the most strongly supported among six prominent invasion hypotheses.[173] Later, Raul R. Braga and colleagues found broad support overall in 876 publications addressing the hypothesis, but they suggested that further research should focus on three distinct subhypotheses: (1) where one non-native species facilitates another, but the facilitation is not reciprocated; (2) where there is mutual facilitation between two non-native species; and (3) where more than two non-native species interact directly or indirectly.[174]

The range of possible examples was expanded by the observation that an interaction between two species that does not appear to be facilitatory might in fact be so.[175] Rather, the interaction might be strongly asymmetric in terms of population impact. One can imagine, for example, that a nonnative predator's impact on a non-native prey species might have minimal population impact for the prey but be crucial for the predator population. Few empirical studies on particular systems have explored this possibility. Finally, Peter T. Green and associates further expanded the range of possible interactions that might fall under the general hypothesis of meltdown.[176] Their research showed how a mutually facilitating interaction between two non-native species has suppressed a native species, and that suppression then facilitated the spread of a third non-native species (fig. 15.3). On

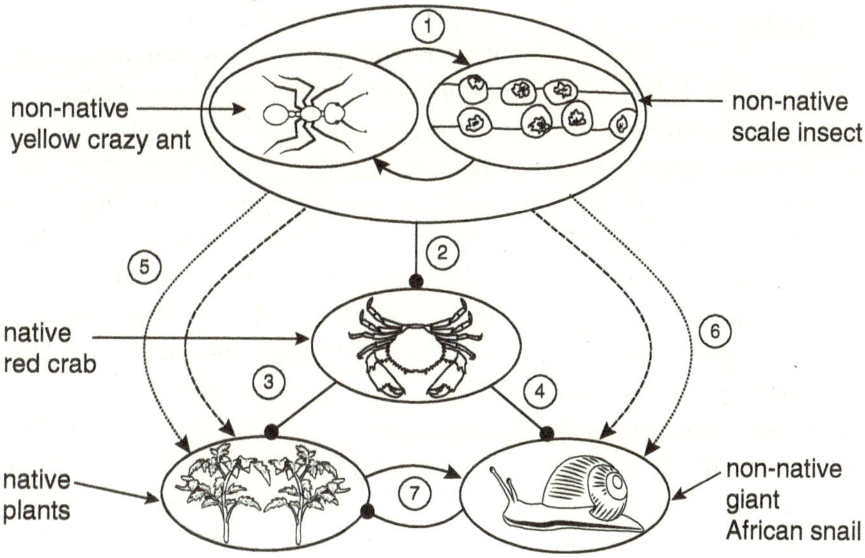

Fig. 15.3 Invasional meltdown on Christmas Island (in the Indian Ocean) in which interactions among non-native yellow crazy ants, non-native scale insects, and native red crabs aid the establishment of a non-native snail, the giant African snail, to the probable detriment of native plants. Solid lines indicate direct effects. Dotted and dashed lines indicate indirect effects. Circles indicate negative impacts, and arrows indicate positive impacts on the adjacent species. Credit: Courtesy of Peter Green, redrawn by Joseph Antoine, Newgen, from P. Green et al., "Invasional Meltdown: Invader–Invader Mutualism."

Christmas Island, in the Indian Ocean, non-native yellow crazy ants, native to Asia or West Africa, arrived in the early twentieth century but were rare until 1989. Their population then exploded, spurred by the introduction of scale insects—*Tachardina aurantiaca* and *Coccus* sp.—that secrete honeydew. As in many such instances, the ant feeds on the honeydew and protects the scale insects.[177] The greatly increased ant populations devastated the population of native red crabs, which had preyed on the giant African snail, thereby suppressing its population.[178] The population crash of the crab permitted the snail population to grow and spread.

Mutualistic interactions of invaders with one another or with natives have attracted far less attention than negative interactions such as predation and competition. One more narrowly focused hypothesis related to invasional meltdown has come to be known as the "enemy-of-my-enemy hypothesis," in which the buildup of enemies of an invader (which can include other non-native species, such as those from the invader's native range) leads to greater harm to native populations than to that of the invader, thus indirectly aiding the invader.[179]

: : :

The proliferation of explanatory hypotheses and modifications and variants of older ones is a consequence of the maturation and growth of modern invasion science. Sciences originate with observations in nature that sooner or later are recognized as relevant to the same phenomenon or question, and they coalesce into what is known as a descriptive stage, in which increasing numbers of observations are amassed and noted, whether or not the science has even been named. The accumulation of observed invasions and their impacts in the seventeenth through nineteenth centuries (chronicled in chaps. 3 and 4) and especially their proliferation in the era of naturalist-explorers and biogeographers beginning in the late eighteenth and early nineteenth centuries (detailed in chap. 3) constitute this descriptive stage for invasion science. As the catalog of observations continues to accumulate, some individuals go further and seek explanations for the observations, explications of the phenomenon to which observations appear relevant. However, the explanations initially consist of more or less convincing verbal narratives or scenarios rather than testable hypotheses and generally do not include proposals for further research aside from amassing more and more confirmative case studies.[180] Observations that do not accord with the hypothesis are usually simply ignored. The nineteenth-century claims by George Henslow about the key role of self-fertilization and by Asa Gray on the importance of disturbance in plant invasions are in this category, as is Charles Elton's 1958 proposal of biotic resistance.[181]

At some point, at least in the physical sciences, as observations and narratives pile up, practitioners recognize the need for criteria to decide among competing hypotheses or at least to try to understand their relative importance. Often, but not always, these criteria or tests of hypotheses entail experiment. The method of strong inference and the importance and limitations of testable hypotheses and attempting to refute rather than confirm them are associated with this stage.[182] Controversies over the applicability of this approach to biological sciences such as ecology and evolution are beyond the scope of this book.[183] Joel Cohen labeled aspects of biologists' attempts to formulate useful general laws "physics-envy" and "the curse of biology."[184] Suffice it to say that the myriad complexities and idiosyncrasies of invasions alone would render it difficult for invasion scientists to seek anything analogous to the general laws of physics, and they also make replication difficult if one is considering historical invasions as experiments. Thus, the modern field has developed largely in an idiographic tradition, and the many proposed hypotheses, at least in the modern era, have been accompanied by explicit recognition that they apply to a limited domain (e.g., plant invasions, invasions of species-poor islands). Furthermore, a component of the rapid growth of modern invasion science is the frequent use of controlled and replicated experiments, and these are often employed specifically to test a proposed hypothesis with a specified, limited domain.

The proliferation of hypotheses reflects not a science mired in a narrative stage but rather the nature of the subject and the astounding amount of research in the short period since the inception of the modern field. This corpus of research has permitted the focusing and fine-tuning of quite general hypotheses and the proposal of new hypotheses, thereby increasing our understanding of the patterns and mechanistic bases of invasions and their impacts.[185]

16

Measuring, Ranking, and Predicting Invasion Impacts

As invasions were increasingly described in the nineteenth and early twentieth centuries (in chaps. 4 and 6), most publications focused on their perceived impacts, often their impact on some human activity. Impacts were recounted in descriptive terms. Thus, for example, accounts from that time described in great detail the devastation of farming areas by rabbits in Australia and of the vineyards in France by the phylloxera (in chap. 4, secs. 4.1 and 4.2, respectively). Environmental impacts were occasionally detailed as well, as in Darwin's description of the domination of the Argentinian pampas by two non-native plant species (in chap. 3, sec. 3.2). The growing awareness of the scope of invasion impacts led to a rapidly expanding array of proposals and methods to eradicate non-native species or to manage them (as discussed in chap. 17). However, the vast catalog of types of impacts challenged attempts to compare impacts and thus to prioritize species for management in the face of limited resources and uncertainty about the likelihood that various management procedures would be effective. Some kind of quantification was needed.

Absence of an accepted method of quantification with which to rank impacts of diverse non-native species exacerbated the problem that different stakeholders view the degree of impact differently, or even whether an impact is positive or negative (as discussed further in chap. 18, sec. 18.3). Even invasions with highly visible impacts could be judged very differently by different observers.[1] For instance, non-native salmonids such as trout, char, and salmon have inimical consequences for native fishes worldwide, yet anglers enthusiastically travel great distances and spend money to pursue them, so many observers rate their beneficial effects as greater than their negative ones. Another challenge of quantifying impacts is evident when one considers that, to rank invaders in terms of impact, a method must compare species as different as rainbow trout, chestnut blight, kudzu, Norway rat, Burmese python, and spotted wing drosophila.

Developing a universal measure of impacts is further complicated by their variety: ecological/conservation, economic, public health. As modern invasion science grew rapidly, monetary cost became almost the default currency. For instance, the United States Congress Office of Technology Assessment (OTA), in an influential 1993 report aiming specifically to inform US government activities regarding invasions, acknowledged public health and ecological impacts but emphasized economic impacts, quantifying estimated economic cost of selected invasions.[2] This document also provided economic cost-benefit analyses for management of eleven invaders and one invasion pathway (import of logs from Siberia). The Global Invasive Species Program (GISP), the sequel to the Scientific Committee on Problems of the Environment (SCOPE) program (discussed in chap. 9, sec. 9.2), was intended to provide tools to address issues raised by invasions, especially how to manage them—which, as noted in chapter 9, section 9.1, was downplayed in the original SCOPE project. One part of this effort was an analysis of the economic consequences of invasions, led by economist Charles Perrings and ecologist Mark Williamson. An early book produced under the auspices of GISP, *The Economics of Biological Invasions*, initiated an entire subdiscipline of invasion economics.[3] Several chapters in this volume and subsequent efforts acknowledged that they were considering only economic aspects but nevertheless cast decisions about management of invasions solely in terms of models of economic costs and benefits.[4] As the scope and details of invasion impacts were increasingly elucidated, the international nature of the issue led inevitably to explorations and recommendations on national and international trade policies, with strong emphasis on economic aspects.[5]

Montserrat Vilà and colleagues explored a database of over ten thousand non-native species in Europe; for approximately 13 percent of these, information on economic impact was available, while such information on ecological impact was available for only 11 percent.[6] The authors attributed this difference to the greater ease of recognizing economic impacts and the likelihood that they will therefore be reported more frequently and promptly. An additional factor tending to focus management decisions on economic impacts was a series of widely publicized cost estimates, beginning with that in the 1993 OTA report, which pegged the cost of seventy-nine especially damaging non-native species in the United States at $97 billion over the previous eighty-five years.[7] In 2000, David Pimentel and collaborators, expanding the scope of analysis for the United States to approximately fifty thousand species, estimated the annual cost to be $137 billion.[8] In 2002, Pimentel's team revised this estimate for the United States to $34 billion annually, and for six nations plus continental Europe in total, Pimentel estimated a cost of $314 billion annually.[9] In 2011, Pimentel revised the estimate for the United States to $219 billion annually, updated much of the information for the previously discussed six nations (although without estimating a comprehensive annual cost), and added data from New Zealand.[10] A recent effort has increased focus on economic costs of invasions, beginning with the assembly of a public global database of these costs—InvaCost.[11] Using this database, Christophe Diagne and colleagues estimated a global cost of at least $1.288 trillion between 1970 and 2017.[12] A workshop in 2019 in Paris based on this database resulted in 2021 in a special issue of *NeoBiota* with nineteen articles, each focused on costs in a particular nation or region and, in 2022, a special section of *Biological Invasions* with eleven articles on various aspects of these costs, such as stakeholder perceptions, modeling approaches, and pathways of particularly costly invasions.[13]

16.1 Quantifying Environmental Impacts

Because much of the earlier impetus for the rise of modern invasion science came from ecologists (as described in chap. 6) and the intersection of the rise of invasion science with that of modern conservation biology (as detailed in chap. 7, sec. 7.5), a purely monetary quantification of impact to guide management seemed inadequate to many people in the field. The absence of a market in a species often makes assigning a cost to an invasion a highly subjective exercise. What is the economic value of an uncommon wildflower of no commercial value or a salamander species whose existence is threatened by a non-native species? This conundrum helped drive the fervent, controversial quest to show that biodiversity per se—simply the number of species, independently of their identity—was crucial to aspects of ecosystem function and thus to the services that ecosystems could provide to humans.[14] But even if at least some types of ecosystem services could be assigned a monetary value, that value fails to capture the total value that biodiversity has for humankind.[15] One can question the entire approach of attempting to quantify invasion impact at the ecosystem level by estimating its economic impact on ecosystem services.[16] (Chap. 12, sec. 12.4, provides a more detailed discussion of ecosystem services and ecosystem impacts.) One response from economists is that the value of a non-market entity, such as a species with no commercial value, can be estimated by asking people how much they would be willing to pay to preserve it.[17] However, many ecologists, following philosophers such as Paul W. Taylor and Nicholas Agar, believe that other species have intrinsic value independently of human assessment.[18] In any event, whether answers to a survey question adequately reflect what people are truly willing to pay for something has frequently been questioned.[19]

Perhaps the "ecological" impact of an invasion that is easiest to quantify is extinction of one or more native species. In fact, extinction was the sole impact of an invasion tallied in many mathematical models of invasion. For instance, Ted Case explored invasion into communities modeled by the Lotka-Volterra competition equations, Steven J. Mithen and John H. Lawton similarly used Lotka-Volterra predation equations to model invasions, and R. Daniel Morton and Richard Law employed generalized Lotka-Volterra equations of species interactions to study the outcome of invasions (as mentioned in chap. 13, sec. 13.2.2).[20] In each study, the "impact" of the invasion was characterized by the extinction (or lack thereof) of one or more native species. Invasion scientists have increasingly represented communities visually as networks—collection of nodes (species) and links between them (interspecific interactions) in space. A popular approach to exploring the likely impact of an invasion has been to simulate the result of an invasion by manipulating the network according to a specific algorithm, see which nodes cease to have any links, and consider the species represented by these nodes as extinct (discussed in chap. 13, sec. 13.2.1). Philosopher Mark Sagoff, a persistent critic of invasion science and management (whom we'll encounter again in chap. 18, sec. 18.5), described extinction of native species as perhaps the only one of the various invasion impacts that can be viewed as an ecological "harm" without qualification.[21] He then doubted the evidence suggesting that invasions have caused many extinctions, thereby reprising his skepticism about environmental impacts of invasions. However, society as a whole clearly accepts other environmental invasion impacts short of species extinction as harmful, as witness innumerable efforts and some laws designed to protect habitats and ecosystems, such as wetlands, from various threats, including non-native species.

In 1999, plant ecologist Ingrid Parker and colleagues suggested an approach to quantifying invasion impact to help inform management priorities.[22] This approach served as a springboard for many subsequent efforts. They conceived of the total ecological impact of a non-native species (I) to be determined by the geographical range in the invaded area (R), the average abundance (A) of the invader within that range in terms of numbers of individuals or biomass per unit area, and the effect (E) per individual or per biomass unit of the species. The effect on the invaded community and its species was seen as occurring at the levels of the individual (e.g., in their behavior), population (e.g., in population size or growth rate), species (e.g., evolution of various traits), community (e.g., number of species or trophic structure), or ecosystem (e.g., nutrient cycling, primary productivity). They combined these three factors multiplicatively, so that $I = R \times A \times E$, an equation presented as a starting point to discussions of impact in subsequent invasion science textbooks.[23] This equation is very general and, as conceded by the authors, would be challenging to apply in any particular instance; they enumerated five reasons:

1. The equation assumes that impact, I, increases linearly with an increase in any of the three independent variables. This may not be the case—for instance, a threshold abundance or population density (A) may exist below which the impact is nil and above which it increases, either linearly or nonlinearly.
2. The multiplicative form that is assumed would be violated if any pairs of the three factors were correlated, as would be the case, for example, if abundance affected the per capita impact.
3. The great variety of impacts even of a single invasion and the fact that they may be manifested at more than one of the aforementioned levels complicate the ability to combine them in one factor, E.
4. Many sorts of impacts vary temporally or spatially, or both, which could greatly complicate calculating overall impact. For instance, spatial variation could invalidate the use of single variables such as A to capture areal range and E to represent per capita impact.
5. The entire approach treats invasions as single events involving just one invading species. In fact, different invading species could interact with one another to affect their impacts. The phenomenon of invasional meltdown (discussed in chap. 15, sec. 15.6) is but one example of such interactions.

Anthony Ricciardi modified this equation to emphasize the species composition of the invaded region: $I = A \times F \times C$.[24] Here A and I were as in the original equation, F was defined as the invader's ecological function, and C was the species composition. If the impact were to be calculated over the entire invaded area, C would vary spatially, depending on which species were present in which areas, thus accounting, to an extent, for the fourth and fifth listed issues. However, as with the original equation, this modified equation is very general, and it would be difficult to gather the data needed to quantify F, which was intended as analogous to E, the per capita impact in the original equation, as such quantification would include determining all the ways the non-native species affects all the individuals, populations, communities, and ecosystems in the invaded range. Possibly for some local part of the invaded range, such information could be gathered, but it seems impossible for the entire range of any widespread invader. In 2010, Jan Thiele and colleagues took a different tack, modifying the equation of Parker and coauthors in two ways.[25] First, observing that any invader will have different impacts in different habitats, they divided the invaded range into habitats—$R_1, R_2, R_3, \ldots R_i \ldots$ etc. Second, they attempted to account for the possibility that per capita impact in each habitat, E_i, is not independent of abundance, A_i, but rather a function of the abundance. The total impact over the entire range would then be a far more complicated expression than the simple product of Parker and colleagues. As Thiele and colleagues observed, actually measuring both abundance and per capita impact would remain challenging.

Guillaume Latombe and coauthors in 2022 proposed a different expanded version of the method of Parker and colleagues: GIRAE (Generalized Impact = Range size × Abundance × Per-unit Effect).[26] They suggested this version would apply not only to ecological impact but also to other impacts, such as economic costs, and also that it could be applied not only at the species level but also for multiple species. The particular proposed improvement revolves around the fact that the equation $I = R \times A \times E$ assumes that impact I increases linearly with range R and abundance A. Since some measures of impact (e.g., loss of native species) appear in at least certain cases to be nonlinear functions of abundance or range, or both, Latombe and colleagues suggested a more general equation, $I = R^{\alpha} \times A^{\beta} \times E_0$, where E_0 is defined as the per-unit effect of a species for one unit of range size and abundance. Different values of the exponents α and β allow the equation to accommodate nonlinearities between impact, on the one hand, and range size and abundance, on the other. For instance, values of α and β greater than 1 indicate an acceleration in the increase in impact with range size and abundance. However, Latombe and colleagues questioned whether adequate data exist to parameterize the model for many species. Applying the model to multiple species in one region would therefore require assuming that the relationship among I, R, and A is the same for all species.

16.2 The IUCN Red Lists for Imperiled Species: A Template for Quantifying Invasion Impacts?

Since the mid-2010s, an increasingly dominant approach to quantifying invasion impacts for use in prioritizing management actions has arisen, based on the International Union for Conservation of Nature (IUCN) Red List classification for quantifying the degree of threat facing imperiled species.[27] The Red List has been enormously influential.[28] It is the document the Global Environment Facility—the funding mechanism of the Convention on Biological Diversity—uses to develop its resource allocation strategies, and it is widely used in formulating the details of multilateral environmental agreements such as the Convention on International Trade in Endangered Species and the Convention on Migratory Species. It is instrumental in activities of the Intergovernmental Science-Policy Platform on Biodiversity and Ecosystem Services, including those on non-native species (discussed in chap. 18, sec. 18.3).

Craig Hilton-Taylor outlined the history of the IUCN Red List.[29] The opening chapter—a list of fourteen mammals and thirteen birds in danger of extinction—resulted from a 1950 IUCN meeting. In 1959, a card index was established with details of the threats and species status for mammals (by then thirty-four species) considered threatened. Between 1960 and 1962, data for the card index were transcribed into data sheets in red-covered loose-leaf binders. The data sheets were continually updated, and in 1964, they formed the basis of a proto–Red Book published as a supplement to the *IUCN Bulletin*. In 1966, this content was formally published as a two-volume set (one on mammals, one on birds) entitled *Red Data Books*. Volume 3 on reptiles and amphibians and volume 4 on freshwater fishes appeared in 1968 and volume 5 on flowering plants in 1970. All these *Red Data Books* were updated in loose-leaf versions through 1981.

Throughout this period, the *Red Data Books* used several qualitative terms and categories to capture degree of endangerment. In 1969, these were standardized to four terms: "endangered," "rare," "depleted," and "indeterminate." They were revised in 1972 to "endangered," "vulnerable," "rare," "out of danger," and "indeterminate," and other categories were later added, such as "extinct," "insufficiently known," "commercially threatened," and "not threatened." These categories were used by IUCN for animals until 1994 and for plants until 1998. Modifications were made as the lists were used over the years. In 1984, an IUCN symposium sought to analyze and synthesize the aspects of the extinction process to have a better, more clearly defined set of categories. In 1989, the IUCN commissioned Georgina Mace and Russell Lande to outline a new approach to this goal, and Mace and Lande proposed three categories (plus "extinct") based on population viability analysis, a technique then undergoing rapid development:[30]

Critical: ≥ 50% probability of extinction within 5 years or 2 generations, whichever is longer
Endangered: = 20% probability of extinction within 20 years or 10 generations, whichever is longer
Vulnerable: = 10% probability of extinction within 100 years

Mace and Lande's list also included the criteria that would have to be met for a species to be placed in each category.

After widespread consultation, a modified version of this system was adopted by the IUCN in 1994, with categories critically endangered (CR), endangered (EN), vulnerable (VU), lower risk (LR), and data-deficient (DD).[31] The criteria were altered and made much more specific than those suggested by Mace and Lande, but the general approach was adopted.[32] This system was used in a new book on threatened birds, the 1996 *IUCN Red List of Threatened Animals*, and *The World List of Threatened Trees*.[33] Continual review of the system led to a revised version in 2001.[34] Guidelines for use are continually updated, and there are occasional minor revisions.[35] The key features remain use of as many quantitative data as possible; explicit, transparent publication of all evidence leading to an assignment; and the assignment of degree of threat to five categories rather than a single numerical score. Despite occasional disputes about particular categorizations, the Red List has become the gold standard in assessing risk of species extinction.[36]

16.3 Red List Analogs for Invasive Species: EICAT, EICAT+, and SEICAT

Development of a method for quantifying and prioritizing non-native species impacts analogous to the IUCN Red Lists for imperiled species traces primarily to a 2010 paper by Wolfgang Nentwig, Elfi Kühnel, and Sven Bacher, who attempted to rank the non-native mammals in Europe in terms of both environmental and economic potential impact.[37] Theirs was a Delphi method—that is, it relied on expert opinions; in this case, those of the three authors.[38] They dredged the literature on all non-native mammals in Europe, then each author assigned a score, to each species, for each of five types of environmental impacts (e.g., predation, competition) and five types of economic impacts (e.g., agricultural loss). Independently, each expert assigned scores from 0 (no detectable impact) to 5 (maximum impact) for each type of impact. Each expert then summed the scores for all five types for environmental impact and, separately for all five types of economic impact. To account for invaded area (*A* in the equation of Parker's team), they multiplied the median of the three summed scores (one for each expert) by the percentage of European 50 × 50 km grid squares (about 965 square miles each) occupied by the species. Thus, for each species, they ended up with an overall numerical score for environmental impact and another for economic impact, and these could be ranked. Sabrina Kumschick and colleagues applied this method to compare impacts of different non-native bird and mammal species in Europe.[39]

Kumschick and colleagues developed a more complicated Delphi method incorporating a version of the scoring approach advocated by Nentwig and colleagues.[40] A decision-maker—perhaps a government agency—would appoint a panel of experts to determine and score impact types, as in the previous method. The decision-maker would also appoint a second panel that would also undertake a scoring, but of a different kind. This panel would comprise "stakeholders," individuals perceived as having a stake in the potential management of the invader. These could be scientists but also individuals affected by the invasion impacts. Members of this panel would score the relative importance of the various types of impacts scored by the first panel. The impact scores of the first panel would then be weighted by importance scores of the second panel to produce the final impact scores that the decision-maker would use. With final impact scores for all non-native species whose management was under consideration, the decision-maker could decide how to allocate resources to manage these species. The impact scores were again in two categories, environmental and socioeconomic, the latter in place of the economic category of Nentwig and colleagues. Importantly, Kumschick's team conceived of impacts being considered as possibly positive as well as negative. Thus, for instance, in a socioeconomic example well known to many invasion scientists, in Australia the non-native weed Paterson's curse is viewed by beekeepers as a boon because it increases honey production but by ranchers as a plague because it is toxic to livestock.[41] The relative importance of environmental benefits and damage caused by the invasion of the paperbark tree, native to Australia, in Florida is similarly contested. Some scientists prize the paperbark for providing habitat for the native snail kite in the Florida Everglades, while vast resources have been spent attempting to control it because the dense stands with sparse understory vegetation disfavor many native species adapted to the wetland and prairie habitat paperbark replaces.[42]

An international working group sponsored by the German Research Foundation transformed these Delphi-process quantification procedures into a version aligned with the IUCN Red Lists. For quantification and ranking of environmental impact, it proposed the Environmental Impact Classification for Alien Taxa (EICAT).[43] EICAT retains the foundation of a panel of experts scoring environmental impacts but expands the five types of such impacts to thirteen impact types divided among three levels—individual, population, or community. For each of these thirteen types, if adequate data exist, an assessor would give the impact one of five scores: massive, major, moderate, minor, or minimal. A sixth category, where insufficient information is available to estimate impact, is "data-deficient," as in the Red Lists. An estimate of uncertainty (three levels—high, medium, and low confidence) is appended to each assessment of impact, analogously to the Red List approach. Whereas Nentwig and colleagues and Kumschick and colleagues summed the scores for the different impact types to generate an overall impact score, EICAT does not sum impact scores; it just presents the scores of the impacts for each impact type. Charlotte L. Hawkins and coauthors elaborated the framework for assigning impact scores in EICAT.[44] They proposed adding parallel scores for the maximum impact ever recorded by an invader in the area of interest as well as the current impact, and named the scores: massive = MV, major = MR, moderate = MO, minor = MN, and minimal = MC. In 2020, Kumschick and colleagues clarified that the overall impact score for a species is taken to be the maximum impact score for all types of impacts.[45] A decision-maker could construct a different algorithm to combine these into an overall score—for example, by assigning 5 points to an MV score, 4 to an MR score, and so forth, and then summing these scores.

EICAT was next subjected to a consultation process by IUCN, resulting in several revisions to the framework and guidelines.[46] These included (1) clarifying the distinction between impacts scored as massive and those scored as major, (2) clarifying how uncertainty is evaluated in impact assessment, (3) expanding the definition of the impact "interaction with other invasive species" to capture indirect effects (discussed in chap. 17, sec. 17.5), (4) refining and clarifying the criteria for scoring the impacts "transmission of disease" and "hybridization," and (5) removing the assessment of a species' current impact, leaving maximum recorded impact as the impact to be scored. On September 15, 2020, IUCN formally adopted EICAT as its global standard for classifying the severity and type of impacts caused by non-native species.[47]

In 2022, Giovanni Vimercati and colleagues followed up suggestions to incorporate positive as well as negative impacts of non-native species by proposing EICAT+ for environmental impacts.[48] As does EICAT for negative environmental impacts, EICAT+ lists several types of positive environmental impacts.[49] In close parallel with EICAT, EICAT+ has five possible scores for each impact: massive positive (MV+), major positive (MR+), moderate positive (MO+), minor positive (MN+), and minimal positive (ML+). Also as in EICAT, a possible score is data-deficient (DD) if information is insufficient to assess an impact, and each estimate is to be accompanied by a rating (high, medium, low) for how confident the assessor is in the score. Unlike EICAT as adopted by IUCN, the authors of EICAT+ did not propose an algorithm for producing an overall positive impact score, although a decision-maker could easily fashion one (as in the previous EICAT description) or simply adopt the IUCN approach for EICAT. Nor did they suggest a way to combine EICAT and EICAT+ scores in a sort of overall impact score for a species.

Rather, the goal appeared to be to render transparent what is known about the full range of environmental impacts of non-native species, as Jean Vitule and coauthors had advocated.[50] Interestingly, and perhaps with an eye to the controversy over impacts (discussed in chap. 18, sec. 18.3), Vimercati and colleagues stated explicitly that EICAT+ should not be used to offset or to understate negative impacts. Although EICAT+ as presented would not appear to understate negative impacts as assessed by EICAT, it is difficult to see how it would not be used to offset negative impacts if EICAT and EICAT+ are to be used together to help decision-makers consider the full range of impacts in deciding which non-native species to attempt to manage and how.

For socioeconomic impacts, in 2018 Sven Bacher and colleagues proposed Socio-Economic Impact Classification of Alien Taxa (SEICAT) as an analog to EICAT.[51] In most ways, SEICAT is completely analogous. There are the same five possible scores for each impact—MV, MR, MO, MN,

MC—and a data-deficient score, DD. Uncertainty is similarly estimated as one of three levels. Similarly to EICAT, SEICAT does not account for positive impacts. Whereas EICAT enumerates thirteen specific impact types, SEICAT lists four, termed "constituents of human well-being" (following the Millennium Ecosystem Assessment)—(1) safety, (2) material and immaterial assets, (3) health, and (4) social, spiritual, and cultural relations.[52] For each, several examples are provided, but the authors state that "there are no precise guidelines [*sic*] how to aggregate activities meaningfully" and that "different activities are differently valued in different contexts."[53] Unlike for EICAT, SEICAT has no specific method for combining scores for different impact types into an overall score. Bacher and colleagues compared EICAT and SEICAT scores for non-native amphibians and appear, for the purposes of that comparison, to have followed the EICAT convention that the overall impact score for a species is taken to be the maximum impact score for all types of impacts. Interestingly, the EICAT and SEICAT scores for non-native amphibians were generally poorly correlated. In both EICAT and SEICAT, most species were scored data-deficient for many types of impacts. A similar exercise for non-native fish in the Mediterranean also showed low correlation between EICAT and SEICAT scores.[54]

16.4 Risk Analysis

A major impetus for efforts to quantify and rank impacts of invasive species arose from a desire to regulate the arrival of new species that will become impactful—that is, to be able to analyze risks of various impacts to inform a permitting process. As impacts were increasingly enumerated, and especially as estimates of economic impact proliferated, attention turned to risk analyses in the service of permitting.[55] For example, the aforementioned 1993 report of the OTA specifically called for formal risk assessment procedures both for planned introductions of particular species and for pathways that might facilitate unplanned hitchhikers. This report also acknowledged similar concerns surrounding the potential release of genetically modified organisms (discussed in chap. 9, sec. 9.3) and consequent research into risk analysis that might inform such analyses for non-native species. Even independently of the advent and growth of modern invasion science, attempts were underway to assess risks associated with movement of living organisms. For instance, a 1994 multilateral agreement on sanitary and phytosanitary (SPS) measures led the Food and Agriculture Organization (FAO), through its associated International Plant Protection Convention of 1997, to require that attempts to ban import of a product on grounds that it threatens human, animal, or plant life be accompanied by an objective assessment of risks.[56] In 1997, the Harvard Center for Risk Analysis, citing the requirement of the North American Free Trade Agreement and the General Agreement on Tariffs and Trade for risk analyses to support quarantine decisions for imported living organisms, convened a workshop to develop general principles for such analyses.[57] The SPS agreement inspired the Society for Risk Analysis and the Ecological Society of America to hold a similar workshop in 2001.[58]

Risk assessments were already being conducted by various US agencies, especially the US Department of Agriculture (USDA), for pathways that might introduce species.[59] A similar assessment in 1996 targeted the black carp as a potential invader in its own right, not as a pathway for pathogens.[60] A shortcoming of the USDA assessments, in the view of the OTA, was the heavy focus on economic impacts relative to environmental ones. The US Environmental Protection Agency (EPA), independently of the USDA, published a proposed framework for ecological risk assessment in 1992, based almost exclusively on risk assessment of chemical "stressors."[61] Responding to concern about applicability of such a framework to biological stressors such as non-native species and genetically modified organisms, the EPA in 1993 solicited issue papers for a potential revision or expansion of the framework, including a paper on biological stressors. That paper emphasized that random aspects of the biology of living organisms (especially evolution), the complexity of species interactions, and absence of important information on the ecology of almost all species that do not affect agriculture or human health make risk assessment much more challenging than that for chemicals, casting into doubt the utility at that time of highly quantitative risk estimates.[62] The 1998 successor to the 1992 framework, however, was still strongly analogous to that for chemicals, acknowledging the concerns raised by the issue papers and readers of a draft version of the successor document but relegating them to "future detailed treatment."[63] It presented one of the USDA pathway examples, logs from Chile as potential vectors of non-native insects and plant pathogens, as an example for defining management goals but without comment on the actual method of assessment.[64] In 1999, President Bill Clinton's Executive Order 13112 ("Invasive Species") mandated a unified, risk-based federal approach to invasions, and the initial federal management plan developed to respond to that executive order called for the formulation of risk assessment instruments to that end.[65]

The OTA report lamented the limited usefulness of early risk assessments because they "do not quantitatively predict with high degrees of confidence either the likelihood that a new species will become established or its impacts."[66] The risk assessments at the time, such as the aforementioned early USDA efforts, were explicitly products of a Delphi process. Experts answered a series of questions about the likelihood of establishment of a nonindigenous species (or several of them, in cases of pathway risk analysis) and likely degree and nature of impact. Similarly, a risk assessment tool targeting potential non-native plant invaders into fynbos vegetation in South

Africa relied on expert responses to a series of questions based on previous analyses of traits conducing to tree invasion (described in chap. 15, sec. 15.3.1).[67] Several other early risk assessment methods for specific types of invasions in particular ecosystems or regions likewise entailed a Delphi process.[68]

Risk assessment tools for specific proposed species introductions and for pathways that might inadvertently bring non-native species still rely on a Delphi process, so the quantification is not of risk per se but of expert opinions of risk. The most widely used such instrument is the Australian Weed Risk Assessment (AWRA), developed in 1995, which has been adapted for use in several other nations.[69] The AWRA consists of forty-nine questions, most of which are answered by "yes," "no," or "unknown"; a few permit a range of numerical scores. The USDA developed a weed risk assessment tool (PPQ WRA) for specific species rather than pathways; it includes twenty-three questions designed to determine risk of establishment and eighteen questions focused on risk of major impact if establishment were to occur.[70] In the United Kingdom, the Non-Native Species Secretariat Risk Assessment Scheme, targeting both animals and plants, currently poses more questions (seventy-one) than the AWRA and PPQ WRA and also asks more questions that do not receive "yes," "no" or "unknown" answers but request a probability (e.g., "How likely is it that establishment will occur?").[71] Both the PPQ WRA and the British system aim to make some of the uncertainties explicit by asking assessors for their degree of confidence in answers to each question.

To the extent that experts can differ in their answers to questions, these Delphic risk assessments are subjective.[72] Probably the greatest degree of uncertainty, even for experts, arises from the point raised previously—the lack of knowledge of the ecology of the great majority of species.[73] For pathways rather than proposed introductions of a single species, the situation is compounded because the pathway can bring many species, and it is a judgment call which species are most likely to be entrained in the pathway. The nature of the Delphi process as implemented in the examples presented in this chapter yields a quantification, but the quantification does not address the concern of the OTA about lack of quantified predictions because the quantification is of the summed judgment of experts and the variance around that judgment. Experts estimate various probabilities (e.g., that a species disperses to a new location, establishes, spreads, generates various kinds of impacts) as low, medium, or high, after which an ad hoc algorithm converts these verbal guesses to numbers and combines them across assessors to yield a number.[74] Another entity that actually regulates introductions then takes that number into account in determining permitting or other activities relevant to the species or pathway under consideration. Although the rapid growth of modern invasion science has helped fill some of the striking knowledge lacunae hindering quantitative risk assessment, it is questionable how much confidence a regulatory entity should have today in a quantitative risk assessment of a planned introduction or a pathway, even aside from whatever other factors (e.g., sociopolitical) are taken into account.[75] An important point is that, even if subsequent events show that a risk assessment had correctly predicted the great majority of species that in fact became problematic, it is likely also to have incorrectly predicted that several species that did not become problems were going to become pests. The costs of such an error relative to the cost of failing to have predicted a species that subsequently becomes problematic is not something assessors will consider, but it is surely a factor that a regulatory entity will bear in mind.[76]

: : :

Quantifying impacts and ranking non-native species for management efforts will never be exact sciences. The myriad types of impacts at several levels will stymie any attempt to quantify the totality of the effect a non-native population has on the resident ecological community and on human well-being and endeavors. Similarly, the dynamism of biological systems will be a constant impediment—even if the total impact of some non-native population were grasped at one point, various of its components would inevitably shift as populations fluctuate, new species arrive, existing ones dwindle or disappear, climate changes, and so forth. Rankings for management are also complicated by the fact that different stakeholders view impacts of one and the same species differently. Although invasion scientists will never achieve the holy grail of a single number characterizing impact, and despite ongoing efforts to modify the equation of Ingrid Parker and colleagues to that end, the field appears to be coalescing around an approach similar to that of the IUCN Red Lists for conservation—that is, assigning a category of degree of impact rather than a number, with the category assignment based on very explicit consideration of all available information and acknowledgment of uncertainty. The rapid expansion of the science means that, at the least, more and more relevant data will be available on which to base estimates of impacts by species already classified and also to reduce the large class of species classified as data-deficient and therefore not assessed.

An important point with respect to rankings and decisions as regards management is that, even if a species achieves the highest impact ranking and no one disputes that the impact is negative and great, a management approach that could possibly reduce that impact may not currently exist. Furthermore, the likelihood that any management technique will succeed in any particular case is an uncertain matter. Even venerable approaches such as biological and chemical control often fail. Also, assessments of impacts do not include assessments of potential nontarget impacts associated with any management method. In sum, decision-makers will benefit from the evolving methods of categorizing degrees of impact, but such information will remain one of several inputs they must consider.

Similarly, risk assessments of impacts of a particular planned introduction of a species or of a pathway that might bring invaders are another input that decision-makers will consider, but they will be only one input, and the predictions embedded in risk assessments are simply best guesses of individuals seen as experts on the particular species or pathway. Furthermore, the estimates of these experts are combined in arbitrary ways. The most recent assessment methods, the results of Delphi processes similar to earlier risk assessment tools, attempt to lay bare the variation among assessments by different experts and also the degree of confidence they place in particular estimates. These features should aid decision-makers in determining how much weight to put on a risk assessment.

17

Management of Biological Invasions

As biological invasions affected various human activities, people fought them using a staggering variety of techniques. The prominent invasions in the late nineteenth and early twentieth centuries detailed in chapter 4 capture this array. For the European rabbit invasion of Australia alone, early attempts included massive shooting, wire fences and stone walls (some with moats filled with metal shards and broken glass), poison baits, gassing of burrows, introduced predators (ferret, cat, and small Indian mongoose), and inoculation with tuberculosis, as well as trapping individuals and killing females while releasing males. None of these methods provided control, but subsequent introduction of the myxoma virus along with its vector, the European rabbit flea, achieved at least temporary success. Similarly, early futile efforts to control water hyacinth in Florida employed numerous chemicals, fire, and many mechanical approaches, including use of boats mounted with roller-mounted knife blades, and floating booms with elevators to remove the plants. Later introduction of herbivorous weevils also failed to control the invasion, which was ultimately well managed with the herbicide 2,4-D. In the United States, large government teams were employed in fruitless attempts to stem the invasions of white pine blister rust by removing its intermediate host plant and of the spongy moth by destroying its egg masses and pupae and burning its caterpillars (described in chap. 4, secs. 4.7 and 4.4, respectively).

These early management efforts included the three main classes of technologies still used against non-native species: mechanical and physical control, chemical control, and biological control. All three of these methods have been used successfully in some instances, often in combination with one another or with other techniques. All three have also failed utterly in other cases. Importantly, for each of these approaches, the technology has evolved, often in response to lessons learned from previous uses, including failed cases. However, the failures, as well as various ancillary effects of both successful and failed approaches, have led several critics of invasion science to suggest that management is too often futile, is sometimes damaging, and should usually not even be attempted until harmful impacts of the invasion are clearly manifested.

Mark Davis and colleagues argued that managers confronting non-native species should cease trying to "achieve the often impossible goal of eradicating them or drastically reducing their abundance."[1] Indeed, eradication and management failures abound, and these have induced pessimism about the entire enterprise of confronting non-native species. For instance, failure to eradicate invasive non-native blackberry in the Galápagos Archipelago led the head of restoration efforts there to acknowledge the damage but call for ending the effort and declaring it a native despite its harmful effects.[2] US eradication campaigns targeting three non-native species in the late nineteenth and early to mid-twentieth century—the spongy moth, white pine blister rust, and red imported fire ant—are recognized by scientists and the public as spectacular failures and have contributed to the sense that trying to eradicate an invader is a losing battle.[3] In each case, not only were massive human efforts and economic resources wasted, but nontarget species and various human endeavors were harmed. The white pine blister rust campaign entailed ripping out countless patches of currants and gooseberries (intermediate hosts of the rust) over wide swaths of the United States. The fourteen-year campaign to eradicate the red imported fire ant from the Southeast cost over $200 million and featured widespread use of broad-spectrum chemicals that affected many species and probably exacerbated the ant's invasion by killing many of its natural enemies. Conservation biologist Edward O. Wilson termed this fiasco "the Vietnam of entomology."[4]

Although the Scientific Committee on Problems of the Environment (SCOPE) project included management as one of its three goals, this goal was greatly de-emphasized and barely addressed in all but one of the resulting major publications (as discussed in chap. 9, sec. 9.2). This relative de-emphasis was part of a separation of the literature about invasion mechanisms and impacts, on the one hand, and on managing invasions, on the other. The former literature is dominated by academic scientists publishing in academic journals, while the latter includes many resource managers and government agency scientists and often consists of gray literature—annual reports, white papers, government agency guides, and the like. Today, several academic journals focus largely on management (especially by biological control or chemicals). Others, such as *Biological Invasions*, specifically invite papers on management as well as invasion mechanisms and impacts (as described in chap. 10, sec. 10.1.2.1), but the substantial separation of management from academic "science" in various cultures and venues remains a problem.[5]

17.1 Early Detection, Rapid Response (EDRR)

It has long been a mantra of invasion managers that "an ounce of prevention is worth a pound of cure," leading to widespread efforts to implement early detection, rapid response (EDRR) programs. Despite the relative deemphasis of management during the SCOPE program, several authors discussed the importance of EDRR. A joint South Africa–United States symposium on best management practices for excluding and controlling non-native species included a workshop on EDRR that described campaigns to manage five local invasions.[6] One of these (the eradication of the giant African snail in south Florida in the 1960s and 1970s) succeeded, and one is still underway today (witchweed, described in the next paragraph); in both of these early detection and rapid response were seen as key.[7] Three failed, two for want of rapid response and one because the campaign was terminated for lack of funding. The Global Invasive Species Program (GISP), sequel to the SCOPE program (discussed in chap. 9, sec. 9.2), published a global strategy for research on management of invasive non-native species that featured the importance of early detection, and the GISP handbook on best prevention and management practices included a chapter on early warning.[8]

Perhaps the first such program in the United States was the Witchweed Eradication Program, initiated in the 1960s, to eradicate an introduced root parasite of several crops in the Carolinas, a coordination between federal and state agencies that has progressively reduced the area of infestation but not (yet) eradicated it.[9] In 2001, the US General Accounting Office, responding to a congressional request, identified lack of a national system as the overarching obstacle to a rapid response and found inadequate detection and inadequate resources as significant problems.[10] In 2004, a federal interagency committee began developing a plan to facilitate EDRR of invasive plants through multiple partnerships with state and local government agencies as well as nongovernmental organizations (NGOs).[11] A special issue of the journal *Biological Invasions* reviewed progress toward effective EDRR in the United States.[12] Despite decades of calls by various federal agencies for greater attention to EDRR and enormous expenditures toward management of non-native species, progress has been limited.[13] In Europe, the Bern Convention on the Conservation of European Wildlife and Natural Habitats of the Council of Europe, effective in 1982, produced a strategy on biological invasions in 2004.[14] Included in the goals was an EDRR system; some details of the early detection component were published in 2010.[15]

It has also long been recognized that the early detection part of EDRR is a weak link in such programs, particularly for unplanned introductions, because the human labor required to monitor all land and water effectively in the search for what must initially be small populations would be staggeringly expensive. That citizen scientists could aid in this effort was quickly apparent, and some programs were initiated to train volunteers to undertake specific reconnaissance chores—for example, the Invasive Plant Atlas of New England, created in 2001, and the Victorian Weed Spotters, begun in 2008.[16]

However, as useful as such efforts are, organized and trained volunteers will never come close to generating enough early detections, especially in nations that are not wealthy. Interest thus exploded in developing apps to use the camera function of smartphones that any person could use, to integrate geolocated photographs with appropriate websites, and to encourage the public to use these apps. For instance, various state agencies and NGOs in Florida collaborated with EDDMapS, a system launched in 2005 by the University of Georgia to document the distribution of invasive species in the United States and Canada, to launch the IveGot1 iPhone application in 2010, expanded to an IveGot1 Android app in 2012.[17] Other mobile apps that are increasingly used by the public include iNaturalist (initiated in 2008) and eBird (begun in 2002).In addition to EDDMapS, the Global Biodiversity Information Facility is a database that receives such information.[18]

17.2 Eradication

Eradication is the complete removal of all individuals from a discrete population, not contiguous with other conspecific populations. Despite the prominent failed eradication campaigns mentioned in section 17.1, the early twentieth century saw several successful campaigns, particularly on islands.[19] Examples include a tsetse fly from Principe in 1911–14 and again in 1956; the citrus blackfly from Key West in 1934–37; the New World screwworm, a species of blowfly, from Curaçao in 1954–55; feral cats from Aride in 1932, and goats from Ascension right after World War II.[20] Well before Charles Elton published his 1958 monograph *The Ecology of Invasions by Animals and Plants*, in 1944 he reviewed a book on the successful eradication in northeast Brazil of the African mosquito *Anopheles gambiae*, an efficient malaria vector.[21] He was optimistic about eradication, citing two successful earlier campaigns in the United States as well as the eradication of the muskrat in Great Britain, to which his Bureau of Animal Population had contributed. Elton even went so far as to predict success in the then-current eradication campaign against Norway rats and black rats in Great Britain and other campaigns against these rodents that he foresaw elsewhere in the world.[22]

By 1958, the rat eradications in Great Britain had failed, and Elton was considerably less sanguine, describing the *Anopheles gambiae* eradication as "a very rare event. Usually, if an invasion has got really going it can only be dealt with by keeping the numbers within bounds, that is by *control*."[23] Nevertheless, various eradication campaigns continued to be undertaken, some succeeding and others failing. A great many of these targeted mammals—especially rodents and carnivores—are on islands, and enthusiasm increased in the new millennium. The literature also showed an increasing degree of synthesis and collaboration. Previously, most publications about eradication of non-native species were one-offs, often in gray literature or journals of restricted circulation, with the exception of a 1989 book, *Eradication of Exotic Pests*.[24] The new spirit of collaboration was evident when the International Union for Conservation of Nature (IUCN) convened a series of international conferences on eradication of invasive species on islands, in Auckland in 2001 and 2010 and in Dundee, Scotland, in 2017.[25] These conferences attracted hundreds of participants from many nations, who contributed many accounts of eradication campaigns (primarily on islands but, especially in the 2017 conference, occasionally on continents) and discussed factors associated with success or failure.

The second IUCN conference included the announcement of the development of a publicly accessible database—the Database of Island Invasive Species Eradications (DIISE)—intended to centralize information on island eradication campaigns for vertebrates other than fishes and to allow syntheses of various sorts (methods, regions, taxa, etc.).[26] The database, managed as a partnership between the University of California at Santa Cruz, the University of Auckland, the IUCN Invasive Species Specialist Group, Landcare Research (a New Zealand government resource management agency), and Island Conservation (an NGO whose mission is eradicating invasive vertebrates on islands), is continually updated.[27] By 2019, DIISE had tabulated over two thousand eradication attempts in more than sixty nations and territories. The number of campaigns increased starting in the 1980s, and the success rate for mammal eradications since the mid-1990s is around 90 percent.[28] Projects have targeted increasingly challenging situations, including multiple invaders, larger islands, and islands with larger human populations. Indeed, the title of the 2017 IUCN eradication conference—*Island Invasives: Scaling Up to Meet the Challenge*—captured the thrust of developments in eradication technology and targets, featuring many presentations on simultaneous eradications of multiple species, often on large islands.[29] Probably the most ambitious eradication project is Predator Free 2050, a New Zealand campaign announced in 2016 that is engaging the government, NGOs, and the public in an effort to eradicate all invasive non-native mammals by 2050, not without controversy (as noted in chap. 18, sec. 18.6).[30]

For insect pests, plant pathogens, and a smattering of other species (nematodes, molluscs), the Global Eradication and Response Database, initiated around 2012, is continually updated and by 2025 had recorded information on 1,048 eradication programs, beginning in 1869, against 329 taxa in 108 countries.[31] As with vertebrates, the eradication campaigns proliferated in the late twentieth century and on into the twenty-first. More than 80 percent of the campaigns against insects have succeeded, owing to improved technologies and increasing experience.[32]

No existing database attempts to amass information on all campaigns to eradicate non-native plant populations, a lamented lacuna.[33] Plant eradication has been attempted less frequently than animal eradication.[34] Invasion scientists and managers have occasionally expressed great skepticism about the prospect of successfully eradicating plant populations established beyond very small areas, partly because soil seedbanks can persist for a long time—up to centuries in certain cases.[35] Nevertheless, many invasive plant populations have been eradicated, mostly but not only in very small areas.[36] Several ongoing campaigns targeting multiple populations employ technological advances that may allow more plant populations to be targeted.[37]

Eradication of aquatic invasions, especially in the marine environment, has not seen the recent progress reported for terrestrial taxa, at least partly because aquatic invasions are always or intermittently underwater, thus harder to see and often entailing more complicated management procedures.[38] A review tallied seventy-seven non-native freshwater fish eradication campaigns for thirty-nine species with the majority reporting success but with inadequate supporting evidence.[39] Non-native crayfish populations have also been targeted, with a few successful eradications and many more failures. Many very small populations of invasive aquatic plants have been eradicated through use of herbicides, hand-pulling, suction pumps, bottom-shading, biological control, and draining or drawdown of small water bodies, but reinvasion has been a common problem.[40] Although most attempts at eradicating marine invaders have failed, two noteworthy successes inspire hope that other campaigns may succeed. In Australia, an established population of the Caribbean black-striped mussel was eradicated in 1999 by massive treatment with liquid bleach and copper sulfate, and in two California locations the non-native "killer alga" was eradicated with bleach and chlorine released under tarpaulins covering patches of the alga.[41]

For many invasive populations, either an eradication campaign fails or none is undertaken, but subsequent activities have attempted to control the population at low densities, a process termed "maintenance management." Often the same methods used in a failed eradication attempt are employed in maintenance management. The three main technologies are (1) physical and mechanical control, (2) chemical control, and (3) biological control, also known as biocontrol. Both successes and failures abound for all three approaches. Other methods have been applied to a more limited array of invaders, and several new methods based on molecular genetics are currently in various stages of development or have recently been deployed in particular cases.

17.3 Physical and Mechanical Control

Physical and mechanical control—physical removal of invasive plants or animals with or without use of machinery—has included many methods, from hand-pulling or digging out plants with shovels or root-wrenches and shooting animals with guns to using large earth-moving equipment and hunting animals from aircraft.[42] Although projects and methods have been described in both gray literature and academic journals such as *Invasive Plant Science and Management* and *Management of Biological Invasions* (journals described in chap. 10, secs. 10.1.2.6 and 10.1.2.7, respectively), these projects are highly idiosyncratic, and no journal serves as a repository of information on the outcome of past and current projects, unlike the situation for chemical and biological methods. Efforts using such techniques go back many years. In the late nineteenth and early twentieth centuries, rabbits were shot in Australia, spongy moth egg masses were manually removed in Massachusetts, mongooses were trapped in the West Indies, crusher boats and elevators were employed against water hyacinth in Florida, intermediate host plants of white pine blister rust were ripped out in the United States, and chestnut trees were felled to control chestnut blight in Pennsylvania (all as discussed in chap. 4). All these efforts failed, but by the late twentieth century, physical methods used in conjunction with other techniques had contributed to long-term control of both animal and plant invaders. For instance, the Canadian province of Alberta in 1950 established a program that has successfully maintained Norway rat populations at low densities through the Alberta Rat Patrol, which employs guns and anticoagulant baits.[43] In coastal California, federal refuge personnel, public works laborers, and inmate work crews have kept European beachgrass at low levels since the early 1990s in large dune areas primarily by cutting and uprooting plants.[44]

17.4 Chemical Control

The use of salt to control weeds in agricultural fields and gardens dates back at least to the sixteenth century in England, and some of these weeds were probably non-native.[45] In the United States in the nineteenth and early twentieth centuries, salt was used against non-native orange hawkweed and field bindweed, and copper sulfate, iron sulfate, sulfuric acid, and nitric acid were tested during this period against various weeds in Europe and North America.[46] Sulfuric acid was employed through at least the 1930s, and copper sulfate continues to be widely used as an algaecide. In the early twentieth century, arsenicals came into wide use, such as sodium arsenite for water hyacinth in Louisiana. Petroleum distillates began to be used at this time in the United States for weeding. A major shift in herbicide use began with the discovery of 2,4-D in 1942–44, which initiated what has been called the "chemical era" of herbicide development, featuring organic chemicals.[47] In 1962, the same year that Rachel Carson published *Silent Spring* (discussed in chap. 7, sec. 7.1), more than 200 million pounds of herbicides were used to control weeds, many of which were non-native.

Plant pathogens, especially fungi, were also targeted by chemicals. By the mid-nineteenth century, sulfur dust was being used against various powdery mildews, and in the late nineteenth century, French vintners fought grape downy mildew, a fungus introduced from North America, with "Bordeaux mixture," consisting of copper sulfate and lime.[48] By the 1940s, development of several chemical fungicides was in full swing, and the 1960s saw introduction of many synthetic fungicides, including some, termed "broad spectrum," that attacked many fungi.[49]

As for plants and fungi, chemical control of non-native animals, especially insects, became common in the nineteenth and early twentieth centuries.[50] When the Colorado potato beetle (native to Mexico; see box 17.1) reached the North American Great Plains, farmers applied "Paris green" (a copper acetoarsenite) to kill it. By the late nineteenth century Paris green was applied to many crops, becoming the first widely used insecticide, employed in the eradication of the mosquito *Anopheles gambiae* in Brazil (discussed previously in sec. 17.2). Other arsenical insecticides, such as calcium arsenate (used in the United States against the boll weevil) and lead arsenate (used to control the spongy moth, discussed in chap. 4, sec. 4.4), were largely replaced by DDT in the 1950s because, at least in typical exposures, DDT appeared to cause little or no damage to plants or vertebrates.

BOX 17.1 Colorado Potato Beetle

The Colorado potato beetle is not native to Colorado, and it did not originally attack potato. When Spanish conquistadores invaded Mexico in the seventeenth century, this beetle was living on the central Mexican plateau and feeding on burweed, a potato congener. The Spaniards introduced cattle to Mexico in the late seventeenth century and began driving cattle north to markets in Texas. Cattle carried burweed seeds north, where the plant established and its seeds attached to bison. The plant moved northward in the early eighteenth century, reaching the Great Plains in the early nineteenth century. Meanwhile, the potato did not reach North America directly from its native region, the Andes. The first introduction to Europe is unknown, but the potato reached France by the late sixteenth century. In 1719, potato seed stock was sent from Ireland to New Hampshire, and by 1820, just as the Colorado potato beetle reached the Great Plains (on burweed), so did the potato. By 1859, the beetle was attacking potato, and, after a time lag, the beetle population exploded. From North America, the beetle reached Great Britain by the 1870s, was in France by 1921, and then spread through Europe and into Asia, reaching China in the 1990s. The beetle has now evolved different biotypes adapted to many species related to potato, including tomato and horse nettle. During World War II (1939–45), the Germans charged the Americans with dropping Colorado potato beetles on their potato fields, and a similar charge was leveled by the Soviet Union during the Cold War.[1]

1. Kahn, "Staffs of Life II: Man Is What He Eats."

DDT was synthesized in 1874, but it was not until 1939 that it was developed as an insecticide by Paul Hermann Müller at Geigy in Switzerland. Geigy patented DDT in 1940, and it was widely used in the Second World War to control body lice and disease vectors, especially mosquitoes. Use increased in the 1950s. Müller won the Nobel Prize for Medicine in 1948 for discovering and developing its insecticidal properties. Starting in the 1940s, chemists developed a plethora of organochlorine and organophosphate insecticides, as well as new synthetic herbicides and fungicides.[51] However, Rachel Carson's *Silent Spring*, published in 1962, triggered not only the environmental movement but also—as one of its leading edges—a continuing controversy over chemical management of non-native species and problematic species in general. Carson detailed acute and chronic pesticide poisoning of wildlife and humans, as well as ecosystem impacts, and chemical pollution was the main initial focus of the environmental movement.[52]

A legacy of *Silent Spring* in some circles is a lingering "chemophobia" about chemical control of invasive species.[53] Nowadays, herbicides and pesticides are usually designed to target particular species and to minimize nontarget impacts if they are used properly (not always the case), but the full range of impacts of long-term use of many chemicals remains unexamined. In some cases, herbicides and pesticides have controlled invasions well at least in the medium term, and they have averted several imminent disasters.[54] In the United States, the herbicide fluridone has often controlled small infestations of hydrilla, an aquatic invader from Asia, Africa, and Australia.[55] On Christmas Island in the Indian Ocean, the insecticide fipronil drastically reduced the population of the non-native yellow crazy ant and thus at least temporarily stalled a precipitous decline in the population of the endemic red crab, a keystone species whose actions affect the entire island ecosystem.[56] The anticoagulant brodifacoum has long been used to eradicate or control non-native rodents and other mammals, and the piscicide rotenone has been widely used to eradicate or manage non-native fish populations.[57]

However, chemical control has been bedeviled by four major concerns. First is the main issue raised by Carson—nontarget impacts. Use of fipronil to control yellow crazy ants has been criticized because it affects many arthropods, and use of the anticoagulant brodifacoum has spurred repeated controversies partly because of nontarget impacts, such as in a project to eradicate house mice from the Farallon Islands of California.[58] Brodifacoum impacts are exacerbated because predators of targeted species can be affected through the process of biological magnification (progressive increase of an ingested toxin in higher levels of a food chain) emphasized by Carson. Nontarget impacts of rotenone in aquatic systems have often been publicized as well.[59]

Evolution of resistance is the second major problem.[60] Resistance was recognized early, particularly in two cases. In 1916, both entomologists and crop fumigators in California recognized that red scale from Asia was increasingly resisting cyanide fumigation of citrus, and in 1928, in Virginia and Colorado, larvae of the codling moth (native to Europe or Asia) were able to enter apples sprayed with lead arsenate. Although the 1940 presidential address of the American Association of Economic Entomologists described these and a few other early cases, the dramatic success of DDT in controlling many insect populations during and soon after the Second World War led to great enthusiasm about chemical control, the aforementioned proliferation of new classes of organic pesticides, and a willful neglect of increasing reports of resistance.[61] Eventually, so many cases of resistance surfaced that the problem could not be ignored. For instance, because hydrilla in Florida is a female clone reproducing only vegetatively, it was assumed that resistance to fluridone would be slow to evolve. However, in 2004, hydrilla was found to resist fluridone by virtue of three separate somatic mutations that arose in three separate lakes.[62] Rabbits in southwestern Australia rapidly evolved resistance to sodium monofluoroacetate (1080) used to control them.[63] Rodents in several locations have become resistant to anticoagulants such as brodifacoum, and mustelids (martens and weasels) who ingest these anticoagulants when they prey on rodents that have consumed them have also evolved resistance to them.[64] Insects, because they have short life cycles and often enormous populations, are especially likely to evolve resistance quickly. For instance, the diamondback moth, probably native to Europe, is an invasive crop pest in many regions and was targeted early by pesticides, including DDT. It was the first crop pest to evolve resistance to DDT, and it has since evolved resistance to every chemical deployed against it.[65]

A third problem is that chemicals are often expensive, for the most part produced for profit, and cost increases as increasing amounts are used in response to evolved resistance. The increased cost for agricultural crops can be passed on to the consumer. But when there is no direct market—as is often the case for management of an invasive species for environmental purposes—costs are borne by the government or an NGO, and these have many competing demands.[66]

Finally, a fourth problem is that, when eradication or maintenance management of non-native invasive mammals and occasionally birds is attempted, animal welfare and animal rights advocates often object on the grounds that pesticides are inhumane (as discussed in chap. 18, sec. 18.6). One issue is that sodium monofluoroacetate, which inhibits enzymes involved in producing energy under aerobic conditions, often leads to a slow and painful death.[67] Similar reports abound of suffering of brushtail opossums and rodents caused by brodifacoum.[68] Such concerns have engendered efforts to produce pesticides that would cause death in a more humane way, such as PAPP (para-aminopropiophenone), which acts similarly to carbon monoxide poisoning.[69]

17.5 Biological Control (Biocontrol)

Classical biological control—introduction of a natural enemy from the native range of an introduced target species—has successfully maintained many populations of pestiferous non-native insects and plants at low levels for extended periods, as with the introduction of the predatory vedalia beetle to control cottony cushion scale on California citrus in 1889 (detailed in chap. 4, sec. 4.3). Probably the first international introduction of a predatory arthropod targeting an invasive insect was that of the mite *Rhizoglyphus phylloxerae* from the United States to France in 1873 for control of the phylloxera (described in chap. 4, sec. 4.2), followed quickly by introduction of the eleven-spotted lady beetle from England to New Zealand in 1874 as a biocontrol for aphids and of the parasitic wasp *Cotesia glomerata* from England to the United States in 1883 as a biocontrol for the insect known as imported cabbage worm.[70]

For plants, perhaps the first deliberate attempt to control a non-native weed by introducing an insect occurred in 1863, when the cochineal insect (native to Brazil) was introduced to control drooping prickly pear (native to South America) in southern India. This project succeeded.[71] Another early success was the introduction of the cactus moth from South America to Australia, where in the 1920s its caterpillars devastated common prickly pear cactus, introduced from the Americas, thereby freeing up enormous areas for livestock grazing.[72] The Australian success inspired introduction of the cactus moth to South Africa in 1933 for the same purpose—controlling pest prickly pear species. It succeeded to some extent there, though it did not spread as widely or rapidly as in Australia.[73] Many other projects have employed classical biological control to attack non-native cacti, and a few projects were not "classical" because they targeted native cacti viewed as pests.[74] Non-native species threatening public health were another early biological control target, particularly disease-vectoring mosquitoes, for which many insectivorous fishes were introduced in the early twentieth century.[75] By the 1920s, western mosquitofish and eastern mosquitofish had been widely distributed from various North American sources and were generally considered effective at controlling mosquitoes.[76]

The late nineteenth century biocontrol introductions presaged a deluge of such efforts for several decades. By 1936, in the continental United States alone, seventy-three non-native insect parasites and predators of insects from every continent but South America and Antarctica were established; the analogous figure for the Hawaiian Islands was ninety-four.[77] By then, the vedalia beetle, first introduced to California in 1889, had been introduced to more than fifty countries, and a parasitic wasp from the northeastern United States, *Aphelinus mali*, that attacks the woolly apple aphid had been introduced to about forty nations, beginning in 1920.[78]

Most early biological control projects aimed to reduce populations of non-native pests of agriculture or pastoralism, such as the cottony cushion scale or prickly pear cacti, and some, such those involving mosquito fish, served public health. Threats to the environment or native ecosystems and species were rarely targeted at the outset. An exception was the spongy moth in the northeastern United States, deplored for its devastation of nonproduction forests as well as for its general nuisance characteristics (as discussed in chap. 4, sec. 4.4). It was targeted by several unsuccessful biological control introductions between 1905 and 1915. By the late twentieth century, biological control was routinely attempted in the service of environmental goals with several successes, especially against non-native aquatic plants.[79] Perhaps the first successful campaign of this sort began in 1964 in the southeastern United States against alligator weed (native to South America), which was well controlled in southern areas by the introduced alligator weed flea beetle.[80] However, the beetle does not consistently control alligator weed in more northerly locations (as noted in chap. 14, sec. 14.4.4) and in some terrestrial habitats, a striking example of the context dependence of impacts of non-native species (more on context dependence in chap. 19, sec. 19.6).[81]

As with mechanical and chemical methods, biological control has seen many failures as well as successes. Although criteria for success vary among authors, a recent estimate is that 10.1 percent of insects introduced to manage insect pests have conferred control.[82] For classical biological control of plants, several estimates suggest that 50 to 70 percent of projects have achieved partial control, and 25 percent complete control.[83] It is important to recognize that a single project may have entailed introducing several species, with successes having been preceded by species that failed to establish or species that established populations but failed to confer adequate control. In some cases, none of several introductions conferred substantial control.

Many more non-native species introduced for biological control purposes thus established populations than actually controlled the targeted pests. For insects introduced to control insects, 32.6 percent (2,007) of those introduced established populations, while 10.1 percent (620) conferred satisfactory control.[84]

Biological control has long been touted in popular media as well as scientific literature as a "green" alternative to chemical control. In 1962, Rachel Carson devoted part of the closing chapter of *Silent Spring* to classical biological control, citing the successful use of the vedalia beetle to control cottony cushion scale and lamenting the relative lack of government and societal support for biological rather than chemical control. Charles Elton, in *The Ecology of Invasions* (discussed in chap. 7, sec. 7.1), was also enthusiastic. He too cited the vedalia beetle/cottony cushion scale case, and he also mentioned the prickly pear/cactus moth example and the introduction of predatory snails to Hawaii to control the giant African snail. He particularly contrasted the control of the red scale from Asia on California citrus by the introduced parasitic wasp *Aphytis chrysomphali* with the repeated use of pesticides that killed both the wasp and various predatory insects that were enemies of the scale.[85] In unpublished notes probably intended for use in a planned second edition of *The Ecology of Invasions*, Elton also discussed many successful or ongoing biological control projects for invasive plants, based on comments by the Australian entomologist and population biologist Alexander J. Nicholson.[86]

However, Elton foreshadowed a later controversy about biological control, citing the possibility that species introduced to control a previously introduced invasive species might devastate a nontarget native species, even driving it to extinction.[87] The controversy was primarily triggered by two reviews in the 1990s of a number of instances in which classical biological control in the modern era was suspected of threatening populations of native nontarget species.[88] In the nineteenth century, two well-known conservation debacles entailed the deliberate introduction of carnivores: the small Indian mongoose introduced in 1872 to control rats in Jamaica and subsequently in many other locations (detailed in chap. 4, sec. 4.5) and the stoat introduced from England to New Zealand in 1883 to control European rabbits (chronicled in chap. 4, sec. 4.1), the stoat introduction accompanied by objections that the predator would attack native birds (as described in chap. 5, sec. 5.1).[89] Both species attacked native species of great conservation concern and contributed to the extinction of several of them. The mongoose and the stoat were not the only early introductions for biological control that went drastically awry by threatening native species. Perhaps the worst case in terms of sheer numbers of native species negatively affected is that of the predatory rosy wolf snail from the southeastern United States, widely introduced to Pacific islands beginning in 1955, when it was taken to Hawaii from Florida in a failed attempt to control the non-native giant African snail, an agricultural pest. Since then, the rosy wolf snail has globally extinguished over fifty endemic island snail species, including in French Polynesia and Hawaii.[90] Remarkably, despite widespread knowledge of this disaster, another generalized predator, the New Guinea flatworm, was introduced to several islands, such as Bugsuk in the Philippines in 1981 and the Maldives in 1985 for the same purpose—control of the giant African snail.[91]

The IUCN list of one hundred of the world's worst invaders includes seven species introduced for biological control. These include the western mosquitofish, small Indian mongoose, stoat, rosy wolf snail, and New Guinea flatworm, all discussed previously.[92] The other two are the cane toad, native to South and Central America and introduced for insect control to Hawaii in 1932, Australia in 1935, and many other locations; and the Indian myna, introduced to control insects in the early eighteenth century to Madagascar, in 1755 to Île de la Réunion, in 1762 to Mauritius, in 1862 to Australia, in 1865 to Hawaii, and to many other sites for reasons that are unclear but may well have been insect control.[93] Except for that of the flatworm, these are all old introductions, and other old introductions for biological control purposes have surely also had damaging nontarget impacts. For instance, the attempt to control the spongy moth in New England included a spate of biological control introductions between 1905 and 1915. Among these was the parasitic European tachinid fly *Compsilura concinnata*. Although this fly did not control the spongy moth, it established a widespread population that now plays a major role in the decline of three nontarget native silk moth species, one of which is state-listed as threatened.[94] On the plus side, the same fly is believed to have stemmed the burgeoning invasion of the feared browntail moth in New England.[95] The barn owl was introduced in the 1950s to both Hawaii and the Seychelles to control introduced rats damaging agriculture, after failed introductions for similar purposes to New Zealand in 1899 and to Lord Howe Island from 1922 to 1950.[96] In both the Hawaiian Islands and the Seychelles, the owl threatens imperiled seabird populations.

Modern biological control practitioners take nontarget impacts seriously, for the most part acknowledge that several early introductions such as those of the small Indian mongoose and giant African snail were disastrous, and advocate procedures that would minimize the likelihood of unforeseen nontarget impacts.[97] Key among these procedures are forswearing the use of species that do not narrowly focus their attacks on the particular target species and host-testing procedures that determine the range of host or prey usage.

Two introductions of biological control agents with substantial nontarget impacts have often been cited, and, although the conservation threats each poses are widely acknowledged, some effort has been expended to exonerate biocontrol practitioners in each case. In 1957, the Commonwealth Institute of Biological Control released the cactus moth on the island of

Nevis to control prickly pear cactus.[98] It island-hopped to many other islands in the West Indies, and in 1989 it was found in the Florida Keys attacking a native cactus of conservation concern. An attempt was made to exculpate the Commonwealth Institute on the grounds that the horticulture industry likely brought the moth to the United States on imported plants.[99] Helmuth G. Zimmermann and associates suggested that transportation of plants rather than autonomous flight by the moth might have been responsible for much of the spread in the West Indies as well as that to Florida, but they conceded that the cactus moth in the continental United States now threatens a number of cactus species of conservation and cultural concern and that a release such as the one on Nevis should not be sanctioned today because of the risk to native species.[100]

The other widely discussed case is that of the musk thistle head weevil, introduced in 1968 from France to North America by Agriculture Canada to control musk thistle (native to Eurasia) and then dispersed throughout the United States by federal and state agencies.[101] It was discovered attacking at least twenty-two native thistles in the genus *Cirsium*, including some of conservation concern, and the threat of extinction of at least one of these species is acknowledged in the biological control literature.[102] Nevertheless, the responsibility of the introducers has been minimized.[103] One such argument is based on the fact that Agriculture Canada and the Commonwealth Institute of Biological Control had conducted host-testing to determine whether nontarget native plants would be attacked. The tests were insufficient to detect risk to native plants, as they included only one native North American species of *Cirsium* out of at least eighty-eight native species. Even though nontarget species were attacked, researchers concluded the weevil would not damage native species because in choice tests the weevil chose to oviposit and feed on the target musk thistle, larval mortality was relatively high on the few *Cirsium* species tested, and adults that did mature on those *Cirsium* were smaller than those maturing on the targeted musk thistle.[104] An additional argument was that, even had the tests implied that native species might be damaged, at the time of testing, damage to native plants of no economic value was not seen as problematic.

For biological control of plants, host-testing protocols have become more rigorous, and neither the cactus moth introduction to Nevis nor that of *Rhinocyllus conicus* to North America would likely be undertaken by major biological control agencies.[105] It is widely recognized that narrow host ranges of non-native natural enemies are necessary to minimize risk, both by lowering the probability of attack of nontargets and by limiting the likelihood that the non-native natural enemy might maintain large populations on other species even as the population of a threatened nontarget species declines. However, biological control practitioners have argued that very few nontarget impacts have been documented and complained that, driven by undue influence of the cactus moth and *Rhinocyllus* cases, unnecessarily strict procedures have been implemented that limit use of this entire approach, at great cost to both conservation and human enterprises.[106] Critics have responded that estimates of very low numbers of nontarget impacts are misleading, largely because of absence of research on nontarget insects and plants of no market value, and that strong regulation of the approach and much more monitoring of results are warranted.[107]

Another issue that has received increasing attention is indirect effects of biological control agents on nontarget species and even whole ecosystems.[108] One example concerns two Old World gall flies, the knapweed gall fly and the UV knapweed seedhead fly, introduced in North America to control spotted knapweed, also native to the Old World. Though they proved ineffective on the targeted weeds, gall flies flourished and supported greatly increased populations of a native deer mouse. Gall flies comprised 85 percent of the mouse's winter diet in knapweed-infested areas. Possible impacts of this mouse population increase include consumption of native plant seeds, competition with other small mammals, predation on and competition with insects, and even increased transmission of hantavirus to humans.[109] Several kinds of indirect effects of biological control introductions might be predicted by analysis of diagrams of pairwise interactions among community members including the proposed biological control agent.[110] However, even prominent biological control practitioners acknowledge that, given the variety of possible interactions within any community and the potentially complex interaction chains linking species, as in the gall fly/deer mouse case, many indirect effects probably cannot be predicted.[111]

17.6 Other Technologies

Although versions of physical/mechanical, chemical, and biological control have all been applied against many non-native species of many types, other technologies have often proven useful against a limited range of target species. The sterile male technique (also known as SIT—sterile insect technique) was initially developed by Edward F. Knipling and Raymond C. Bushland in the early 1950s and successfully eradicated a particular fly, the New World screwworm, in 1954–55 on the island of Curaçao, where it was native, then Florida and the southeastern United States, to which it had been introduced.[112] It was subsequently widely employed against dipterans, especially non-native fruit flies, on islands and mainland worldwide.[113] SIT consists of mass releases of males sterilized by radiation or chemicals. The sterile males compete with wild males to mate with wild females, reducing the number of offspring and causing population decline or even extinction. A huge number of males is usually released because the sterilization procedure often reduces competitiveness of the sterilized males, and various ancillary procedures are often employed to increase competitiveness of treated males.[114] In addition to Diptera, a few beetles and moths have been targeted, and preliminary research on using the technique on non-native crayfish has been conducted.[115] A project using sterile males to target invasive sea lampreys in the Laurentian Great Lakes was terminated after ambiguous results of a field test.[116]

Another venerable technology is mating disruption by pheromones. This method, developed in the 1970s, has been deployed against non-native populations of several lepidopterans and, more recently, at least one non-native beetle.[117] The use of pheromones has not spread beyond insects, although preliminary research has been conducted for the sea lamprey and two non-native crayfish species.[118] The underlying idea is to synthesize versions of a sex-attractant pheromone that is then used to disrupt mating—for instance, by setting out attractant traps to capture males seeking females, or simply to saturate an area with the pheromone so that males can no longer locate females. Among non-native species in North America, the method has helped control the pink bollworm (an Asian moth pest of cotton), the spongy moth, the codling moth, and the Oriental fruit moth, an Asian pest of many stone and pome fruits.[119] Although the pheromones are species-specific and not believed to threaten humans or other nontarget species as some pesticides do, a controversy erupted in California over use of the method in an attempt to eradicate the light brown apple moth from Australia, a pest of many fruit crops.[120] Concern about human health impacts of the inert products in the formulation of the aerial spray on which the campaign was based caused the program to be terminated.

Still other technologies are narrowly focused on one or a few non-native species. For instance, zebra mussels, present as non-natives in both Europe and North America, avoid being poisoned by toxic chemicals by closing their valves when they detect trace amounts. "BioBullets" are tiny beads of a fatty substance coating and masking a core of toxic potassium chloride.[121] As the mussels filter water, they ingest the beads, the coating dissolves, and the potassium chloride kills them. In nature, the method would also kill nontarget molluscs, but it is used in water treatments plants and industrial facilities. Several other techniques are employed against particular invasive populations, including acoustic detection by smartphones; dogs; mechanical odor-detecting "e-noses"; drones; and remotely operated underwater vehicles.[122]

Most improvements in the dominant technologies—mechanical/physical, chemical, and biological control—as well as more narrowly focused efforts can be seen as incremental: for instance, a slightly different type of chemical bait that is more attractive or toxic to the target organism or less attractive or toxic to nontarget species, or an improvement in testing prospective biological control agents for possible nontarget impacts. Some, like the BioBullets, are completely new. Other times, a modification of a venerable method seems more like a quantum leap. An example is use of "Judas goats" to eradicate invasive goats on islands. The term derives from "Judas steer," applied at least as early as the nineteenth century to animals used to lead cattle as a group to slaughter. A Judas animal is one that can be tracked and that finds or attracts other individuals, thus permitting hunters to find them. The method is potentially useful against any invasive animal species in which normal behavior includes individuals aggregating. Perhaps the first application to an eradication campaign against non-native goats was in the 1980s. Although a bell can be used to enable finding a Judas goat, a major innovation derived from the use of radio telemetry to track ungulates.[123] This technique was used in the early 1980s to eradicate a non-native goat population in Hawaii Volcanoes National Park that had defied previous efforts.[124] A subsequent innovation consisted of the production of a "Mata Hari goat" by sterilizing female Judas goats and inducing them into long-term estrus, thus luring otherwise wary male goats for long periods.[125] This innovation permitted eradication of large, refractory, non-native goat populations from large islands in the Galápagos Archipelago.[126] Versions of the Judas animal technique have been used to eradicate or greatly reduce non-native populations of other social animals such as pigs, dromedary camels, and wild turkeys.[127] The Mata Hari technique has also been used with pigs.[128]

17.7 New Methods Based on Genetics

Several new technologies based on genetics to control or eradicate invasive populations promise to add to the managers' armamentarium but have also spawned controversies. The antecedents of a phenomenon that came to be called "gene silencing" consisted of several experiments performed in the late 1980s and early 1990s. By the late 1990s, it had been determined that double-stranded RNA (dsRNA) rendered messenger RNA molecules unstable or inaccessible in plants, nematodes, insects, and protozoa, thereby "silencing" whatever gene the messenger RNA corresponds to by preventing production of protein. The advent of gene silencing spurred great optimism that it could be used to treat human disease and to increase agricultural productivity.[129] Research on control of non-native agricultural pests quickly became part of a wave of gene-silencing publications.[130] At first, the goal was to spray a formulation containing the gene-silencing chemical (usually dsRNA) on leaves or other surfaces and hope the target pest would ingest it and enough would get from the pest's digestive system to the targeted cells, where it would silence a critical gene, killing the pest.[131]

However, Monsanto received approval from the Canadian Food Inspection Agency in 2016 and the US Environmental Protection Agency in 2017 to distribute corn produced by a project that combined gene silencing with a genetic manipulation.[132] The project, which largely flew under the public radar, introduced a transgene to corn that silences genes in the western corn rootworm when the insect attacks the plant. The rootworm, a beetle native to North America, was introduced in Europe. Whereas simply spraying a gene-silencing formulation on a surface for a targeted species to eat would not constitute a genetic change in any species, the transgene in corn is heritable, and any technology that causes a heritable change in a species engenders controversy and often substantial regulatory hurdles. DuPont applied to patent a similar method using gene silencing to attack stinkbugs, including the brown marmorated stinkbug, an Asian invader in the United States.[133] A remarkable recent gene-silencing project entailed genetically modifying the genome of a symbiotic bacterium (*Snodgrassella alvi*) in the gut of the honeybee to produce dsRNA sequences for genes of the varroa mite, a damaging Asian parasite of honeybees in much of the world, including North America, where the honeybee is non-native.[134] The dsRNA silences genes in the mite, killing it. Gene-silencing projects under development target other non-native species, including an introduced genotype of common reed in the United States and crustaceans used in aquaculture.[135]

Several other projects entail either a version of the aforementioned sterile male technique or a version of an approach based on sex ratios that was foreshadowed in a classic paper by evolutionist William D. Hamilton in 1967.[136] He aimed to explain why sex ratios in nature were usually close to one male per one female and why a few populations had highly skewed sex ratios. Hamilton noted that any heritable process that caused the sex ratio to be skewed toward one sex (he envisioned meiotic drive as such a process) could drive a population to extinction and that such a process might be useful in controlling pest insects. At that time, no technology was imagined that could generate this effect. But several decades later, research suggested two approaches to producing the skewed ratio envisioned by Hamilton. One approach directly follows Hamilton's idea of a heritable change, while the other is more in the spirit of the sterile male technique—that is, not employing a heritable trait but instead repeatedly flooding the target population with individuals whose matings will be fruitless.

The second approach followed a 2006 proposal to use a technology that had been developed in the 1970s and was already widely used in aquaculture to produce fish that are all phenotypically of one sex, no matter which sex chromosomes they contain, by means of treatment with a sex hormone (androgen or estrogen).[137] The proposal combined such treatment with a mating scheme that would produce phenotypic females containing two copies of the normally male-determining Y chromosome. In this instance, a Y chromosome in a female was termed a "Trojan Y." Continued release of such YY females would eventually cause the population to have no wild-type females for males to mate with, and a YY female can produce only male offspring. At this point, the population would be extinguished, as no females would remain. A team in Idaho developed this approach, devising a breeding scheme for brook trout that, combined with an estrogen treatment, produced YY individuals, both phenotypic males (termed "supermales") and phenotypic females.[138] Supermales were released into four streams in Idaho, where brook trout are not native, combined in two streams with electrofishing to reduce the initial wild population.[139] The experiment is ongoing. Other fish, including some with non-native populations, have been treated with sex hormones in aquaculture to produce all male progeny, and a combination of ultraviolet light and estrogen treatment has produced supermales and XY females of common carp, often considered the world's most invasive fish.[140]

Another method in the spirit of the sterile male technique, developed by the commercial enterprise Oxitec, used a transgene in the yellow fever mosquito, which transmits yellow fever, dengue, chikungunya, and Zika viral pathogens to humans. Transgenes to control invasive species have been discussed since the beginning of the twenty-first century, but tests of the transgenic Oxitec mosquito in nature triggered criticism.[141] The transgene used by Oxitec renders females flightless, which would be lethal in nature, but if individuals are reared on a diet with tetracycline, the gene is inactivated.[142] This is a heritable genetic change, although in principle it would not be transmitted in nature because females bearing it would die without reproducing. Oxitec planned to rear masses of transgenic mosquitoes in the presence of tetracycline, discard the females, and flood wild populations with transgenic males so that wild-type females would be likely to mate with the modified males rather than wild ones. This scheme is analogous to releasing sterile males: even though females mating with the modified males have progeny, their female offspring die. Reports of a field test of this method on Grand Cayman Island spurred controversy over lack of regulation and potential unintended consequences.[143] However, as Zika virus spread in the Americas, resistance to releasing Oxitec mosquitoes yielded to the urgency of countering the virus.[144] Oxitec claimed great success in drastically lowering the yellow fever mosquito population in Piracicaba, Brazil—a Zika hot spot. However a further controversy arose with the discovery that genes from the transgenic mosquito had been recovered in wild yellow fever mosquito populations in Brazil, apparently showing that at least some offspring carrying the transgene had survived.[145] This finding has led to greater concern about the possibility of unintended consequences of the method, although the publication reporting this discovery was subsequently severely criticized.[146] Nevertheless, Oxitec is developing similar versions for several other dipteran and lepidopteran agricultural pests.[147]

Approaches directly in the spirit of Hamilton's 1967 proposal, entailing a genetic modification intended to be transmitted, have been under development as part of the burst of activity surrounding the advent of CRISPR-Cas9 gene-editing technology. Research on this "gene drive" technology rapidly evolved in the 1990s, marked by the award of the Nobel Prize in Chemistry in 2020 to two lead researchers, Emmanuelle Charpentier and Jennifer Doudna.[148] Although most excitement centers on medical and agricultural applications, the possibility was soon recognized that gene editing could control or eradicate populations of invasive species under certain circumstances.[149] Two projects well underway target malaria vector *Anopheles* mosquitoes (Target Malaria) and invasive rodents (GBIRd).[150] Rapid improvements in gene-editing technology for mosquitoes and mice are key to progress in these projects.[151] The US Bureau of Reclamation is sponsoring a project to control zebra mussels by introducing a transmissible cancer found in soft-shell clams, potentially by using CRISPR-Cas9 to silence a tumor-suppressing gene.[152]

The swift movement toward using gene drives to manage or eradicate non-native species has increased concern about unintended consequences, highlighted in 2016 by a manifesto issued by prominent conservation scientists at the World Conservation Congress: "A call for conservation with a conscience: no place for gene drives in conservation."[153] Both the United States National Research Council and a United Nations treaty acknowledged potential unintended consequences, such as the transfer of a driven gene into an untargeted species, but sanctioned cautious continuation of testing, including field-testing.[154] Several geneticists have suggested that natural selection will act rapidly to impede the spread of a driven gene, lessening the likelihood that it would actually drive a population to extinction.[155] However, some mathematical models predict that, at least in small populations, the race between population reduction to extinction and evolved resistance to the gene drive would frequently end in extinction.[156] Concern about the possibility of nontarget impacts and the irreversibility of an extinction caused by a gene drive has stimulated second thoughts by several previous enthusiasts of using gene drives to attack non-native species and has led to calls to develop methods to halt or call back a gene drive observed to be generating an unintended consequence.[157] Heightening worry about unintended consequences is the fear that the increasing accessibility of gene drive technology could lead to freelance efforts by well-meaning individuals lacking the substantial scientific and ethical oversight of projects such as Target Malaria and GBIRd or, worse yet, by ecoterrorists.[158]

A method using DNA has recently revolutionized detection of non-native individuals.[159] Environmental DNA (eDNA) consists of short DNA fragments that an organism leaves in water, air, or sediments, or on surfaces, found in skin, mucus, urine, feces, or other tissues, products, or remains of living organisms. It has even been found on plastic and textile marine litter.[160] Environmental DNA has been used since the 1980s to detect bacteria in marine sediments and since the 1990s to track phytoplankton blooms.[161]

In 2008, eDNA was first used to detect a non-native population, the bullfrog (native to the United States) in ponds in France, through sampling of pond water.[162] Soon thereafter, the method was used to detect eDNA from silver carp and bighead carp from Asia in several locations in the Chicago area waterway, a river and canal complex that has served as a pathway for invasions into the Laurentian Great Lakes.[163] DNA of both species was detected upstream of what had been believed to be the invasion front based on traditional fishing methods. In the 2010s, eDNA was widely used not only to detect invasive aquatic animals but also to detect low-density populations of species of conservation concern, to assay biodiversity or community structure, and to study trophic interactions.[164]

A series of technical advances has led to using eDNA for an increasing range of conservation and biodiversity studies.[165] A literature search showed an increase from just one publication on the topic in 2008 to over two hundred in 2019.[166] Detection and monitoring of non-native species quickly became a major use. Among non-native species detected by this means are several aquatic plants, the aquatic weed hydrilla, Burmese python, New Zealand mud snail, Brazilian waterweed, rusty crayfish, red swamp crayfish, spiny-cheek crayfish, and signal crayfish, and feral pigs.[167] Although initially almost all targets were aquatic animals, both plants and terrestrial animals have been detected by eDNA, including the brown marmorated stinkbug from leaf surfaces.[168] Honeybee pollen pellets contain plant eDNA, suggesting another avenue for detecting terrestrial plants.[169] Several researchers have now begun sampling eDNA in air, aiming to develop a method to survey large volumes rapidly, as is routinely done with water.[170]

The major advantage of eDNA over previous methods of collecting specimens and then identifying them morphologically or molecularly is that it is faster and less labor-intensive.[171] Many limitations exist, but methods are developing so rapidly that some will likely be surmounted.[172] The rates of false positives and false negatives must be estimated and reduced, and contamination is an issue. Environmental DNA degrades, but the rates at which it does so in various environments is not well known for most species, making assessment of an apparent absence difficult. Various technical challenges with amplifying DNA must be addressed. And dead individuals also shed DNA into the environment, so a positive detection need not mean a living individual was present in the sampled area.

: : :

By the late nineteenth century, as biological invasions damaging to agriculture were increasingly recognized, efforts to eradicate or at least control them proliferated. Although attempts to eradicate non-native species have long been stymied by failure to detect them early enough, non-native mammals have been eradicated from many islands. Several technological advances have led to increasingly ambitious campaigns against mammalian island invaders, especially beginning in the late twentieth century. Some attempts to eradicate non-native fishes and plants from freshwater locations have succeeded, although the targeted areas have usually been small. Attempts to eradicate marine invaders have rarely succeeded, especially in areas distant from shorelines. Many non-native insect populations have been eradicated, mostly from relatively small regions. Eradication techniques have usually entailed physical, mechanical, or chemical means, or combinations of these, although other methods, such as the sterile insect technique, have been employed for particular taxa. Several newly developed methods relying on molecular genetic techniques are being rapidly developed that may permit eradication in circumstances that would previously have rendered such efforts infeasible. In addition, the advent of programs linking citizen scientists armed with cell phone cameras to entities that can verify and organize responses to invasions is enabling earlier detection of some invasions, as is rapid development of technology to capture and identify eDNA.

When eradication fails or is not attempted, efforts to control non-native species have generally employed the same methods as those used in eradication campaigns, plus classical biological control—the introduction of natural enemies of the targeted invader from its native range. All these methods have been marked by improvements during the twentieth and twenty-first centuries, and all have achieved successful long-term control over various invaders. However, all these methods have also failed to control other invaders, and, for certain invaders of multiple sites, the same method has achieved control of some populations but not of others, exemplifying the context specificity of invasions. Impacts on nontarget native species have been documented for both chemical and biological control, leading to controversy over their use in particular projects and also generating efforts by their practitioners to establish procedures that minimize such impacts.

Although efforts to eradicate non-native species are most likely to succeed soon after the invasion occurs, while the population is small and localized, eradication has been attempted and has sometimes succeeded for long-standing, widespread invasions, particularly on islands. However, when a non-native species has long been established, especially when it is a prominent member of the ecological community, it is likely to have established patterns of interaction with one or more native species or other non-native species. Removing such a species, or even greatly reducing its population over the long term, may engender ripple effects by virtue of loss of one or more interactions. Potential unintended consequences of such losses may sometimes be avoided by careful consideration of the ecology of the community in which the targeted invader is embedded. For instance, several successful eradications of rats on islands have led to explosions of house mouse populations that the rats had formerly suppressed, leading managers undertaking rat eradication to consider simultaneously attempting to eradicate mice.[173] However, the complexity of ecological communities and myriad interactions among its members will inevitably mean that certain consequences will not be predicted.

18

Controversies Abound

Invasion science and the practice of managing non-native species have been criticized on several grounds, with critics who focus on one issue often mentioning others in passing.[1] For instance, an article arguing that most non-native species are not problematic and that management efforts are often futile and costly has a title suggesting a different criticism, that categorizing some species as invasive is xenophobic—"Don't Judge Species on Their Origins."[2] Several of these criticisms have led to the argument that the relatively young modern field of invasion science as a separate entity should be abandoned, with these studies incorporated into ecology or other research programs, a proposal vigorously rebutted.[3] Despite linkages among them, the main controversies fall into seven distinct categories: (1) the charge of xenophobia or nativism; (2) terminology and categories; (3) invaders perceived as beneficial; (4) "native invaders" and ecological succession; (5) invasive species denialism; (6) animal rights, animal welfare, and compassionate conservation; and (7) the utility of management. Many of the controversies feature a split between biologists and nonbiologists, the nonbiologists primarily in the social sciences.[4] Issues surrounding the utility of management are discussed in chapter 17. The remaining six controversies are treated in the sections that follow.

18.1 The Charge of Xenophobia or Nativism

The most heated controversy has raged over the contention that invasion science is tainted by nativism and xenophobia.[5] This charge was occasionally leveled by scholars in the social sciences and humanities in the 1990s, and it became a more persistent drumbeat, in popular media as well as in biological literature, in the twenty-first century. An example is a series of papers by the German garden architects Joachim Wolschke-Bulmahn and Gert Gröning, who pointed to the vigorous Nazi program to "cleanse the German landscape of unharmonious foreign substance [plant species]" as an ecological analog of Hitler's campaign to rid the German *Volk* of foreign elements.[6] Gröning and Wolschke-Bulmahn perceived similar motivations among native plant enthusiasts of their time who deplored non-natives.[7]

Gröning and Wolschke-Bulmahn's writings were widely cited with varying degrees of approbation by many critics of aspects of invasion science, including Jonah Peretti, who subsequently cofounded Huffington Post and BuzzFeed; geographers Lesley Head, Pat Muir, Charles Warren, and Juliet J. Fall; anthropologist David S. Trigger; philosophers Isis Brook and Eric Katz; and biologist Jacques Tassin.[8] The American journalist Michael Pollan, citing Wolschke-Bulmahn and Gröning, pointed to the Nazi "Rules of the Design of the Landscape," including exclusive use of native plants, and the campaign of the Reich's Central Office of Vegetation Mapping to eradicate a well-established small herb, small-flowered touch-me-not, because it is native to Asia. He then hinted that the rise of a "natural gardening" movement in the United States featuring native plants might be similarly motivated:

> Am I implying that natural gardening in America is a crypto-Fascist movement? I hope not. I mention the historical precedent partly to suggest that the "new American garden" is neither as new nor as American as its proponents would have us think. . . . It's hard to believe that there is nothing more than scientific concern about invasive species behind the current fashion for natural gardening and native plants in America—not when our national policies are rife with anxieties about immigration and isolationist sentiment. The garden isn't the only corner of American culture where nativism is in flower just now.[9]

Evolutionary biologist Stephen Jay Gould, citing Wolschke-Bulmahn's publications, expressed concern that well-meaning native plant enthusiasts, indulging a "kindly romanticism," could slide into nativist terminology and perhaps viewpoints.[10] Donna Haraway, a poststructuralist historian of science and technology, raised the same specter of fascism in modern ecological restoration because of what she viewed as similarities with the Nazi native plant project.[11] This juxtaposition of invasion science and management with Nazi enthusiasm for native species reached its apogee in a 2003 book by David I. Theodoropoulos that flatly declared the origin of invasion science to be in the Third Reich, indicted invasion biologists for failing to recognize this origin, and outlined what he perceived to be parallels between activities of invasion managers and a series of Nazi horrors such as gas chamber "selections."[12]

Several critics of invasion science have similarly drawn a connection between South African apartheid and invasion science and management (e.g., anthropologists Jean Comaroff and John L. Comaroff and biologist Jacques Tassin).[13] The link was most strongly claimed by Peretti, who saw the Scientific Committee on Problems of the Environment (SCOPE) program and goals that triggered modern invasion science (covered in chap. 9, sec. 9.1) as shaped by South African scientists influenced by apartheid:

> A more recent and subtle example of this can be found in South Africa in the 1980s. This is where the initial proposal for the Scientific Committee on Problems of the Environment's (SCOPE) invasive species project was proposed. 'South African scientists were instrumental in preparing the project proposal that was put to SCOPE for approval in 1982' (Macdonald et al. 1986).... South African scientists have had a great deal of influence over the SCOPE project.... Why are scientists from South Africa especially concerned with biological invasion? Like Nazism, apartheid thinking is concerned with separating the pure from the impure. Even anti-racist scientists living in an apartheid culture may be influenced by this sort of purist, xenophobic, and racist way of thinking. It is not surprising that SCOPE's hard-line biological nativism has roots in South Africa.[14]

The claimed Nazi link has drawn more attention. Authors in many disciplines have alluded to a connection between invasion science and Nazism, but the controversy has been particularly acrimonious in Wolschke-Bulmahn and Gröning's field of landscape and garden architecture and the related science of restoration ecology.[15] Wolschke-Bulmahn and Gröning saw the origin of the Nazi antipathy toward non-native species in the work of the early German landscape architect Willy Lange (1864–1941) and Irish garden architect William Robinson (1838–1935). Both Lange and Robinson favored native plants but advocated planting non-native species in certain circumstances. Although Lange's writings contained nationalistic statements, direct evidence that either man was xenophobic or racist is scant.[16] Wolschke-Bulmahn looked to Lange's and Robinson's milieus to determine their mindsets: "Both Lange and Robinson were members of a conservative bourgeoisie in which attitudes like nationalism, biologism, and anti-Semitism were widespread."[17] Wolschke-Bulmahn went even further in the same article, tracing Lange's views at least partly to the influence of Alexander von Humboldt, who believed in innate mental characteristics of people of different geographic origins, supposedly owing to the climatic conditions of their homelands, as he wrote in the early nineteenth century.[18] This view, in turn, meshed with Lange's early twentieth-century belief that native plant gardens are suitable for the temperaments of people native to a region. The notion that aesthetic tastes are conditioned by the landscape that individuals experience in their homelands was common in gardening and landscape circles at the time Lange wrote, as seen in the writings of his contemporary Neltje Blanchan, a prominent and prolific gardening author (see box 18.1), and is still occasionally voiced.[19]

BOX 18.1 Neltje Blanchan

Neltje Blanchan is often referred to as "Nellie Doubleday"(she married Frank N. Doubleday in 1886), although she published as "Neltje Blanchan." Born in Chicago, she frequently asserted that Americans, at least during her lifetime, are fundamentally "Teutonic" and thus not suited to produce or appreciate the formal European gardens of the age epitomized for her by Italian gardens:

> It is painfully true that we Americans, like the English, are too Teutonic to be an artistic people.[1]
>
> Our Teutonic blood prejudices us, as a people, toward a more general love for nature than for art.[2]
>
> What are the special claims for naturalistic treatment of our home grounds? It accords with our racial temperament; therefore it is destined to become the dominant style of gardening here, for the same reason that the English language prevails on the continent over every other tongue. People of Latin blood have carried art to the very highest perfection, but our strong Teutonic strain predisposes us toward nature and naturalistic methods.[3]

1. Blanchan, *American Flower Garden*, 38.
2. Blanchan, *American Flower Garden*, 41.
3. Blanchan, *American Flower Garden*, 74–75.

A particular object of Wolschke-Bulmahn's scorn was Danish landscape architect Jens Jensen, who moved to the United States in 1884, becoming a leading exponent of using native plants and a founder, along with Ossian C. Simonds, of the "prairie style" of landscape architecture (as mentioned in chap. 5, box 5.4).[20] Jensen's language made him an easy target; he wrote:

> The gardens that I created myself shall, like any landscape design it does not matter where, be in harmony with their landscape environment and

> the racial characteristics of its inhabitants. They shall express the spirit of America and therefore have to be free of foreign character as far as possible . . . the Latin and the Oriental crept and creeps more and more over our land, coming from the South, which is settled by Latin people, and also from other centers of mixed masses of immigrants. The Germanic character of our race, of our cities and settlements was overgrown by foreign [character]. Latin has spoiled a lot and still spoils things every day; and only sometimes might it be good and enrich our world of forms.[21]

However, landscape architects Dave Egan and William H. Tishler, examining this quote and the entire corpus of Jensen's writings, argued that his animus was not specifically against Latin or any other particular presence, but rather toward the potential obliteration of distinct native North American (especially midwestern) plant assemblages in a process subsequently termed "biotic homogenization."[22] This concern anticipated the more recent worry about the cosmopolitization or "McDonaldization" of nature, which seems immune to the charge of nativism.[23] In any event, the worry may fade as human tastes change.[24] The desire to retain intact native assemblages seems quite analogous to attempts to retain distinct human cultures, as codified by the Council of Europe and UNESCO, and need not be linked to nativism.[25]

The argument against using non-native species in landscaping can be framed in wholly aesthetic terms, as exemplified by a dispute between landscape architect Fredrick Law Olmsted and botanist Charles Sprague Sargent, director of the Arnold Arboretum of Harvard University, over landscaping a restoration of the Muddy River and the surrounding forest.[26] Olmsted advocated using non-native species, while Sargent insisted on natives, resulting in the Brookline side of the river being planted exclusively with natives and the Boston side with many non-natives.[27] The stated basis of the dispute was entirely aesthetic. Sargent conceded that "it is not easy to explain why certain plants look distinctly in place in certain situations and why other plants look as distinctly out of place" and asserted that non-native species "inevitably produce inharmonious results."[28] Olmsted allowed that "planting far-fetched trees with little discrimination has led to deplorable results" but was unwilling "to taboo all trees coming from over the sea."[29]

Aesthetic preferences can be founded on an appeal to the idea of the existence of racial or national temperaments, as for Blanchan, but they need not lead to as strict a prohibition of non-native species as Jensen propounded. Blanchan frequently expressed a preference for native American plants and urged gardeners to use them more frequently, but she also advocated planting non-native species in particular circumstances.[30] Jensen's motivation for preserving native assemblages may not have been founded in a belief in racial temperaments.[31] Philosopher Isis Brook recognized that limiting garden constituents to native plants can be perceived as xenophobic but defended the practice, at least in one's native region, on the grounds of aesthetic appeal, and she justified use of non-native garden plants by expatriates on the basis that they help create a sense of home.[32] She thus espoused the same idea noted previously that aesthetic preferences are conditioned by people's experiences of landscapes in their homelands, but this time in order to justify introducing non-native species. By contrast, art historian Anne Helmreich described the naturalist-explorer Alfred Russel Wallace (whose work was described in chap. 3, sec. 3.2) as a nationalist and an imperialist on grounds that, in a letter from the Malay Archipelago to his sister in England in 1861, he wrote, "A hill of gorse, or of heath, a bank of foxgloves & a hedge of wild roses & purple vetches surpass in beauty anything I have ever seen in the tropics."[33] Historian Peter Raby observed, however, that Wallace variously proclaimed European superiority or lack thereof.[34] Rather than an aesthetic judgment based on racism, Wallace's appreciation of English plants, at least at this time when he had been traveling for seven years, frequently ill and often in straitened circumstances, might simply have been an expression of homesickness.[35]

Several other authors have charged xenophobia or nativism, or both, without invoking a link to Nazis or apartheid, most prominently historian Philip Pauly, who interpreted early twentieth-century activity against non-native species in the United States as a component of a pervasive nativism during the Progressive Era (1896–1916) that was also reflected in increasingly restrictive immigration laws.[36] He wrote:

> Attitudes towards foreign pests merged with ethnic prejudices: the gypsy [spongy] moth and the oriental chestnut blight both took on and contributed to characteristics ascribed to their presumed human compatriots. . . . It should be clear that attitudes about foreign and native organisms were intimately linked, through both everyday experience and analogies of policy, to views on "alien" and "native" humans.[37]

Pauly depicted as nativists three United States government scientists discussed in chapter 6 who presciently recognized and publicized threats that non-native species posed both to native species and to human enterprises: Theodore Palmer, Charles Marlatt, and Leland Howard. When apprised of the absence of direct evidence of nativist tendencies for any of these men, Pauly assumed that they shared the prevailing cultural view.[38] He argued:

> Scholars of the Progressive Era are well aware of the pervasiveness of prejudicial attitudes at all levels of American society during this period. Genteel scientists, however, seldom displayed their prejudices in print.[39]

By contrast, Pauly perceived the very genteel scientist David G. Fairchild (a contemporary of Palmer, Marlatt, and Howard) as the enlightened antagonist of nativist government scientists. Member of a prominent family established in Connecticut in the seventeenth century, son of a college president, and son-in-law of Alexander Graham Bell, Fairchild was a swashbuckling naturalist-explorer and founding director of the US Department of Agriculture's Office of Foreign Seed and Plant Introduction, which introduced over 200,000 non-native plant species and varieties to the United States. He blithely dismissed concerns about invasion (as discussed in chap. 5, sec. 5.6, and detailed in the accompanying box 5.4).

The linchpin of Pauly's argument that nativism was a key factor motivating early twentieth-century American activity regarding invasions was the approximate synchrony between certain California and federal regulations and laws regarding non-native species and increasingly stringent federal immigration requirements. These immigration requirements are widely accepted as having been influenced by a contemporaneous rise in nativism and xenophobia in response to massive immigration from southern and eastern Europe, culminating in the institution of national quotas in 1921 and the Immigration Act of 1924.[40] However, unacknowledged by Pauly was the fact that each piece of legislation on non-native species was explicitly motivated by observed impacts of certain non-native species.[41] For instance, the 1885 California act "to prevent the spreading of fruit and fruit-tree pests" by authorizing inspection of incoming fruit from other states or nations was motivated by devastation of the citrus industry by the cottony cushion scale (detailed in chap. 4, sec. 4.3) and similar agricultural damage, such as that wrought by the San Jose scale from Asia.[42] The main relevant federal legislation, the Lacey Act of 1900 (see box 18.2), was primarily drafted to stem the decline of native game bird populations by prohibiting certain forms of interstate transport of wildlife; prevention of import of specific non-native animal species was added almost as an afterthought. Secretary of Agriculture James Wilson, citing Palmer's admonitions about the impacts of the non-native small Indian mongoose and the house sparrow, convinced Representative John Lacey to include a section on importing non-native species.[43]

BOX 18.2 The Lacey Act

The Lacey Act of 1900 has been widely depicted as ineffective at preventing invasions.[1] However, this reputation neglects the fact that many species listed under the act were already present when listed, and some were widely distributed (e.g., the starling and the house sparrow). The Lacey Act has undergone many amendments, and listing responsibility was transferred from the Department of Agriculture to the Department of the Interior in 1940.[2] Despite amendments in 1948 that actually weakened the authority to prevent a species from being imported, the act has in fact been quite effective for at least seventy years in dealing with species not

Fig. B2 Poster advising opposition to a 2022 proposal to amend the Lacey Act. Credit: Debbie Goodrich for the American Federation of Aviculture.

already present when listed.[3] This accomplishment was aided by an amendment in 1960. Of 288 species listed for reasons of potential invasiveness before they were actually present, none have established a population in the United States. All 19 species that were already present when listed remain established; 12 have remained within the state or territory they occupied when listed, and 7 have spread to new states. Figure B2 depicts a poster arguing in 2022 against a proposed amendment to the Lacey Act that would prohibit interstate shipments within the United States; authorize the Department of the Interior to issue an emergency designation prohibiting the importation of a species that would pose an imminent threat to human beings, to the interests of agriculture, horticulture, forestry, or to wildlife, or to the wildlife resources of the United States; and presumptively prohibit importing any non-native vertebrate, mollusk, or crustacean species, or the eggs of any such species. The prohibition would be lifted if the Department of the Interior determined that the species does not pose a significant risk of invasiveness to the United States. The amendment was opposed by many enthusiastic keepers of non-native animals as well as commercial interests such as pet dealers, and it failed to be enacted.

1. E.g., Fowler et al., "Failure of the Lacey Act to Protect US Ecosystems."
2. Jewell, "Century of Injurious Wildlife Listing."
3. Jewell, "Century of Injurious Wildlife Listing"; Jewell and Fuller, "Unsung Success of Injurious Wildlife Listing."

The argument that actions to prevent biological invasions, at least in the United States, are simply displaced xenophobia associated with heightened opposition to immigration has resonated with other authors. For instance, feminist biologist Banu Subramaniam, professor of English John Leland, anthropologist Hugh Raffles, and historian Garland Allen, along with the aforementioned writer Michael Pollan, have linked the greatly increased concern with invasions beginning in the 1980s to increasing anti-immigrant sentiment in the United States in the late twentieth and early twenty-first centuries.[44]

Another thread of the argument about a relationship between invasion science and xenophobia, nativism, and racism is actually part of a larger, persistent claim that environmentalism and ecological conservation are similarly infected.[45] Detailed treatment of this claim is well beyond the scope of this book, but historian Peter Coates provided an introduction to the entire claim emphasizing its intersection with matters related to non-native species.[46] The gist of the overall argument is that many prominent late nineteenth- and early twentieth-century American advocates for conservation were associated with anti-immigrant activities or the eugenics movement, or both. To a greater or lesser extent, several of these people also deplored non-native species, sometimes explicitly linking the two perceived problems.

Best known is Madison Grant, member of the board of trustees of the American Museum of Natural History, eugenicist, and an explicit racist who greatly influenced public attitudes and policy, including United States immigration laws.[47] His 1916 book, *The Passing of the Great Race; or, The Racial Basis of European History*, was a panegyric to the supremacy of "Nordic" peoples and a lament for their replacement in the United States by immigrants from southern and eastern Europe. Grant was also a titan of early twentieth-century conservation, so the fact that this book was lauded by Adolf Hitler and served as a defense document in the Nuremberg trial of a prominent Nazi has occasionally been adduced to charge that the entire conservation movement is infected with racism. Grant was a major figure (with William T. Hornaday; more on him shortly) in the successful attempt to rescue bison from extinction and an important influence, through his writings and close ties to important political figures, in tightening hunting laws to prevent a hecatomb of American birds and mammals.

One of three themes pervading Grant's many writings on natural history was that invasive species could mate with native species or even drive them to extinction.[48] This concern was one component of Grant's fervent support of wildlife refuges, arguing that it must be addressed

> if we wish to continue to enjoy the spectacle of animated nature around us. To bring about such a change in public opinion is a gigantic undertaking, and it may be necessary in many places to go through, in our characteristic national way, the process of complete destruction of the animals we have, and the restocking of the country with new and perhaps in many cases with foreign and less attractive forms. To avoid this last misfortune, the continuation of the native wild stock through the medium of game refuges is absolutely essential.[49]

It is no surprise that Grant, who deplored human race-mixing (which he termed "mongrelization"), was also revulsed by the prospect of native species hybridizing with introduced ones, arguing, with respect to the breeding of members of the dwindling bison population with cattle, that "a half breed is an abomination" and that "it is of the utmost importance to preserve all remnants of the American bison without any cross-breeding."[50]

Grant—at least initially—did not ostensibly link his concern with non-native species to nativism but rather to the impact of non-native species on native species:

> Efforts to introduce European game, instead of protecting the native American animals, are constantly cropping out. Why anyone should prefer a chamois to the far finer native animal is somewhat of a mystery. Nature has provided for every portion of our country, mammals, birds and fish well adapted to the needs of the locality, and the introduction of foreign animals simply means, in case they survive, the crowding out of some native form.[51]

However, one can doubt that his view that native mountain goats are "far finer" than chamois or that foreign "forms" are "less attractive" was a purely aesthetic judgment.

Henry Fairfield Osborn was a collaborator of Grant's, a paleontologist, president of the American Museum of Natural History from 1908 through 1933 and of the New York Zoological Society from 1909 through 1925, and a prominent conservationist. He was also a leading academic supporter of the eugenics movement and a racist. Osborn was a friend of the publisher Charles Scribner, who had published Osborn's book, *Men of the Old Stone Age*, in 1915.[52] Osborn introduced Grant to Scribner and facilitated publication of Grant's *The Passing of the Great Race* the very next year.[53] Osborn also endorsed, wrote, and then supplemented a preface to subsequent editions of the book.[54]

Grant's "comrade on the firing-line and esteemed friend" William T. Hornaday, zoologist and director of what is now known as the Bronx Zoo, publicized massive losses of wildlife, especially to hunting, and deplored "the slaughter of song, insectivorous and all other birds by Italians and other aliens from southern Europe."[55] Terming Italian immigrants "a human mongoose," he was also no admirer of "southern negroes and poor whites" and felt impelled to deprecate them even though he expected to be pilloried for it.[56] He wrote:

> Whenever the people of a particular race make a specialty of some particular type of wrong-doing, anyone who pointedly rebukes the faulty members of that race is immediately accused of "race prejudice." On account of the facts I am now setting forth about the doings of Italian and negro bird-killers, I expect to be accused along that line.[57]

Hornaday was well aware of Palmer's widely bruited warning about the small Indian mongoose (discussed in chap. 6, sec. 6.1), and he devoted a chapter of his 1913 book *Our Vanishing Wildlife* to the ills perpetrated by non-native species, inveighing against the spongy moth (discussed in chap. 4, sec. 4.4), the rabbit in Australia and New Zealand (discussed in chap. 4, sec. 4.1), and the English sparrow.[58] The English sparrow, an extremely common non-native species in North America and the target of much controversy, earned Hornaday's particular scorn: "The English sparrow is a nuisance and a pest, and if it could be returned to the land of its nativity we would gain much."[59] As was Grant, Hornaday was a major advocate of wildlife refuges and hunting restrictions in the service of conservation, distributing his 1913 book to every member of Congress. But today he is at least as well known for having exhibited a living Congolese Pygmy human in the monkey house of the Bronx Zoo as he is for his considerable contributions to conservation.

The circle of prominent New Yorkers associated during the early twentieth century with the American Museum of Natural History and the New York Zoological Society included leading scientists. A striking example is Frank M. Chapman, curator of birds at the American Museum from 1908 through 1942 and member of the United States National Academy of Sciences.[60] Chapman interacted with Grant in the Zoological Society, where they both served on the board of managers and Grant was on the executive committee, and with Osborn, who was president of the American Museum during much of Chapman's tenure there as curator of birds. Indicative of the tight-knit nature of this circle was a celebration of Osborn's sixtieth birthday at which both Grant and Chapman spoke and were signatories, along with Hornaday (and Theodore Roosevelt), of an "illuminated message of congratulations."[61] Chapman was not only a leading ornithologist but also a committed conservationist. He founded and for thirty-five years edited the journal *Bird-Lore*, which later became the *Audubon Magazine*, and he contributed many notes on conservation.[62] In 1925, he published an article on the distribution and impacts of the starling, which had been introduced from Europe to the United States in Central Park in 1890 and was spreading rapidly in North America.[63] Embedded in a description of ecological and economic impacts is a strikingly xenophobic passage:

> Having with thoughtless hospitality accorded the starling, house sparrow, San José scale, gypsy [spongy] moth, and other pests, including certain members of the genus *Homo*, free and unchallenged entry to our ports, we now ask (if to our sorrow, we have not already learned), "are they desirable?"[64]

Charles Goethe, another contemporary of Grant, was heavily influenced by him. He was the only member of this prominent group to survive into the mid-twentieth century (he died in 1966), long after their views on eugenics and racial characteristics had been widely discredited. A leading eugenicist, Goethe was also a prominent conservationist and founding member of the Sierra Club, and he substantially contributed to national parks and to saving redwoods.[65] In a book edited by Grant, Goethe linked his vigorous objection to immigrants explicitly to non-native species, analogizing Mexican immigrants to the vilified non-native house sparrow: "The peon problem, biologically, is akin to the English sparrow problem. . . . In a new favorable environment, it multiplied like the peon, with startling rapidity."[66] He was particularly concerned with "hybridization" between young female immigrants from Mexico and "Anglo-Saxons." His antipathy toward both immigrants and non-native species peaked in his 1955 book, *Garden Philosopher*, a jumble of insights on vegetation and rants against immigrants, whom he termed "hyphenateds" and "human weeds."[67] His frequent juxtaposition of immigrants with invasive non-native species included comparisons to the house sparrow, spongy moth, rabbit, Argentine ant, Oriental fruit fly, Bermuda grass, crabgrass, Johnson grass, Klamath weed, bindweed, wild radish, prickly lettuce, cheese mallow, chicory, water hyacinth, hoary cress, dandelion, milk thistle, yellow star thistle, redstem filaree, halogeton, a non-native dodder, and various nameless South African and Asian plants.

Goethe claimed similarities between particular immigrants and specific non-native species. Al Capone, son of Italian immigrants, was a favored target, depicted as sharing features with yellow star thistle and hoary cress, while Goethe's chief obsession, German physicist and Soviet spy Klaus Fuchs (immigrant to Great Britain recruited to work at the Los Alamos Laboratory), was compared to the Oriental fruit fly, bindweed, and native parasitic mistletoe. Fuchs was part of a recurring gripe, the loss of atomic secrets to the Soviet Union, which led Goethe to digress from his focus on immigrants to pillory the native-born American Jews Ethel and Julius Rosenberg and Harry Dexter White, termed "second generation" to distinguish them from "first generation" Fuchs.[68]

Some recent critics who have perceived xenophobia and racism in modern invasion science have traced a direct lineage back to the Nazis or the circle associated with Grant, or both, but there is no direct evidence for such a linkage.[69] Others have not quite claimed a "hereditary" link but have tagged invasion science as tainted by xenophobia, as exemplified by landscape geographer Kenneth Olwig: "By carefully disentangling the interwoven texts of science and nationalism, it should be possible both to improve the quality of the science and to work to divorce it from nationalism, racism and xenophobia."[70]

More frequently, twenty-first century critics have criticized the language for resembling that of xenophobia. For instance, historian T. Chris Smout argued that "the language of the conservation scientist can sometimes sound, to the impartial ear, in danger of coming close to the neo-fascist."[71] Geographer William O'Brien explained this view:

> Demand for proof of overt racist and xenophobic intent downplays the ways in which language can transport such values beyond the intent of any individual or group of speakers. The main concern among critics is not that restorationists, conservationists, and others are racists per se, but rather that the long and deep history of racism and xenophobia, particularly in the US, provides a conveniently supportive framework for expression that helps perpetuate denigrating views of foreigners. . . .
>
> And at a time when anti-immigrant activism within environmentalism is resurging, critics' suspicions that human and non-human forms of nativism are mutually reinforcing are all the more justified. Yet we do have a centuries-long track record of racism and xenophobia that does anything but ease concerns that the anti-exotics campaign informing ecological restoration in some way reproduces this history. The criticism is not about any individual's personal intent, but rather the perpetuation of a discourse with a troubling genealogy that continues to guide the way we represent the immigrant Other.[72]

It is undeniable that several persons, including biologists, who deplored invasions in the early twentieth century were motivated at least partly by racism. However, in the absence of evidence of similar motives among modern invasion scientists, it is illogical to assume that just because those were the motivations of certain people in the past, the same motivations persist today.[73] Persistent attempts by scholars in the social sciences and humanities to seek direct links or to infer indirect ones between past xenophobes and modern invasion science may reflect more about disciplinary modes of research and explanation than about motivations of invasion biologists and managers. Historians, by the nature of their research and interests, are inclined to seek connections between facts or phenomena of the past and later phenomena, including those of the present. Historian Marc Bloch in the 1940s termed this modus operandi the "idol of origins" and argued that it can become an obsession beclouding an understanding of current phenomena:

> The explanation of the very recent in terms of the remotest past, naturally attractive to men who have made of this past their chief subject of research, has sometimes dominated our studies to the point of a hypnosis.[74]

He claimed that historians frequently engage the idol of origins to present value judgments as facts: "So in many cases the demon of origins has been, perhaps, only the incarnation of that other satanic enemy of true history: the mania for making judgements."[75]

18.2 Terminology and Categories

The charge that the terminology of modern invasion science encourages xenophobia is one facet of a broader controversy over terminology. To buttress their claims that invasion scientists at least facilitate xenophobia and nativism, critics point to terminology used in both the lay and, less frequently, scientific literature to describe non-native species that resonates with xenophobic sentiment. For instance, biologist and gender studies scholar Banu Subramaniam cited headlines and titles such as "Aliens Reeking [*sic*] Havoc: The Invasion of the Woodlands Soil Snatchers," by journalist Barbara Stewart; "Creepy Strangler Climbs Oregon's Least Wanted List," by journalist Jonathan Brinckman; "U.S. Can't Handle Today's Tide of Immigrants," by Ling-Ling Yeh, representative of an organization advocating population stabilization; "Congress Threatens Wild Immigrants," by lawyer Heather Weiner, and "Invasive Species: Pathogens of Globalization," by environmental writer Christopher Bright.[76] Examples by journalists abound in the popular literature: "Invasion of the Habitat Snatchers," "Botanical Barbarians," and "territory-gobbling imperialist."[77] Scientific articles less frequently feature such sensationalistic language.[78] One can easily find examples, however—for instance, "There Goes the Neighborhood: Apparent Competition Between Invasive and Native Orchids Mediated by a Specialist Florivorous Weevil" and "Do Noxious Neighbors Spread Disease?"[79]

Recently, scientists have advocated replacing common names of high-profile invasive species in North America because of xenophobic connotations. In 2021, the Entomological Society of America removed from its list of approved common names the term "gypsy moth" for *Lymantria dispar*, recognizing it as an offensive term for people of Romani heritage; this was the first time the society had removed a common name because it is offensive to a community.[80] In 2022, the society approved "spongy moth" as the new name.[81] A previous plea was to replace the term "Asian carp," widely used for four species, partly on the grounds that it is culturally insensitive, following the prohibition by the State of Minnesota of state employee use of the term in official documents.[82] Although the American Society of Ichthyologists and Herpetologists and the American Fisheries Society have a joint committee on common names, "Asian carp" does not fall under its purview because the term refers to a group of species rather than a single species.

A related charge, leveled most prominently by sociologist Brendon Larson, is that language used to describe biological invaders in both scientific and especially popular literature is laden with militaristic metaphors, beginning with the words "invasion" and "invader" themselves.[83] Use of martial metaphors in discussions of weeds and pests has long been common, particularly with reference to non-native species. In a 1910 speech to Illinois farmers and teachers, Stephen A. Forbes, a founder of modern ecology and early president of the Ecological Society of America, constructed a detailed, explicit military analogy in calling for a war on insect pests, focusing on non-native invaders.[84] His talk was rife with martial metaphors and similes. Forbes termed a state entomologist the "private, captain, colonel, adjutant, and major general" of a "corporal's guard of trained and experienced fighters" in a "fighting squad" engaged in a "war against insects" facing "marauding enemies" that "cross our borders at frequent intervals."[85] The struggle was triggered by introduction of non-native species, he contended:

> Our rapidly growing, world-wide commerce of fruits and grains, our importation of new plants from the remotest regions of the earth, often bring their special insect enemies with them, . . . such that we seem certain to become the eventual prey of every insect species living anywhere on earth that can do us any harm.[86]

Forbes used three high-profile invaders to exemplify the enemy. First, the San Jose scale (native to Asia), from the time of its establishment on the Pacific coast,

> dispatched its armies of invaders into every state in the Union. . . . Armies of orchardists are now fighting it . . . [with] trainloads of ammunition in the form of sulfur and lime, thousands of small arms in the form of spraying

> pumps, and hundreds of pieces of artillery—the great power sprayers. . . . In order that no indifferent noncombatant may give aid and comfort to the enemy . . . it is made by law a misdemeanor . . . to give it food and lodgment on one's premises.[87]

The second invader was the browntail moth (native to Eurasia), a forest pest rapidly spreading through New England forests at the time (see chap. 4, final sentence of sec. 4.4 and its note)—Forbes's own forces had "captured and burned alive several hundred thousand of these French invaders."[88] His third target was the Hessian fly, an Asian midge that had previously been introduced to Europe and reached the United States during the American Revolution in the straw bedding of mercenary Hessian troops or grasses imported for their horses.[89] Five years later, early in the First World War, Forbes arranged for his speech to be published by the Illinois State Laboratory of Natural History, which he directed.[90]

The First World War (1914–18) also inspired Leland O. Howard, chief entomologist of the USDA and early herald of damaging impacts of non-native species (as discussed in chap. 6, sec. 6.1), to portray entomologists as an army fighting insects. Howard had depicted entomologists as important in combating economic and public health threats posed by insects, and during the First World War he had emphasized this work as aiding the war effort.[91] Beginning with his 1921 speech as retiring president of the American Association for the Advancement of Science, "On Some Presidential Addresses: The War Against Insects," he switched from portraying entomologists as assisting a war effort to describing them as engaging in a war, and this viewpoint continued in publications for the rest of his life.[92] Howard credited as inspirations a 1915 journal article by British explorer and botanist Sir Harry Johnston and a 1919 speech by Colonel W. Glen Liston of the Indian Medical Service, both entitled "The Next War: Man versus Insects," but historian Edmund Russell pointed out that the speech resembled the 1915 speech by Howard's colleague Forbes.[93] In any event, Howard's speech was widely noted in the popular press, which often adopted his and Forbes's military metaphors. For instance, in 1925, William Atherton Du Puy, who wrote many popular books on nature, published "The Insects Are Winning: A Report on the Thousand-year War" in *Harper's Monthly*, lauding Howard and his colleagues in the USDA Bureau of Entomology as "staff officers, the intelligence corps of the thousand-year war."[94] His article abounded in military metaphors and similes and cited several prominent invasions of non-native insects as key battles in the struggle, analogizing them to the sinking of the *Lusitania* in the First World War.[95]

The Second World War (1939–45) brought yet another flurry of martial metaphors. Chemist and entomologist Anthony Standen published a lay treatment of entomology titled *Insect Invaders* featuring several non-native insects in North America—the spongy moth, browntail moth, San Jose scale, Japanese beetle, and white-fringed beetle—as well as two that had originated in North America and moved elsewhere: the phylloxera (detailed in chap. 4, sec. 4.2) and the Colorado potato beetle.[96] Popular science journals ran features such as "Our Next World War—Against Insects" and "Total Insect War Urged," and *Time* contributed "War on Insects."[97]

Terms such as "war," "beachhead," "explosion," "advancing front," and "battle" remain common, and the scientific literature has provided additional phrases such as "novel weapons" and "invasional meltdown" that are occasionally then seen in lay media.[98] Larson argued that all metaphors are inherently imprecise, and his concern with militaristic metaphors to describe invasions was twofold. First, they imply that all non-native populations are "enemies," and the very conception of a war assumes a goal of the triumph of good over evil, so if the goal of invasion managers is to restore ecosystems to a pristine state, this will never happen. Second, he worried that invasion scientists will lose credibility with the public if they are seen as not being objective or as failing to achieve objectives, and he feared that this loss of credibility will in turn hinder the effectiveness of managing or preventing invasions that are actually harmful.

In Larson's view, the problem begins with the term "invasion," which "is not a neutral metaphor, but a nationalistic feedback metaphor that raises associated fears that our country will be invaded by foreigners and our bodies by disease."[99] The final point refers to the evident analogy of a biological invasion to a pathogenic disease, and Larson also deplored the frequent use of military metaphors in medicine, preferring alternatives such as metaphors in traditional Chinese medicine that imply balance and even "kindness to tumors."[100] Militaristic language is as frequently employed throughout the literatures of ecology and conservation biology in general as in invasion science.[101] Whether or not the imprecision of the term "invasion" to describe the spread of a non-native species renders its use ineffective in the short or long term, the analogy to maps of military actions is evident. Charles Darwin referred to an "intruder" having "invaded" another species' range, and Alfred Russel Wallace perceived the Norway rat "invading . . . all over the world."[102] Mark Davis's quixotic campaign to rename the field of invasion biology "SPRED ecology" (*sp*ecies *red*istribution ecology) has not yet gained traction.[103]

It is doubtful that military metaphors will disappear from the invasion or medical literatures, whatever their merits or disadvantages. They are simply too entrenched. Whether invasion science and public understanding of invasions would benefit if military metaphors were expunged from reports on research is a difficult question.[104] We cannot avoid thinking metaphorically, especially about complicated matters—and invasions are complex. However, metaphors do limit what we notice, they influence what we emphasize and deemphasize, and they are part of the mental process that we automatically use to reason with and engage to converse.[105] Writer Jamie Fahey deplored the use of military metaphors in reportage of almost everything—he cited wars on drugs, obesity, idleness, poverty, oil-price speculators, science, saturated fats, drink, Waitrose supermarkets, and Brussels (itself a metaphor for the European Union). His personal favorite was the Australian war on the introduced cane toad.[106] But the very pervasiveness of military metaphors suggests that any attempt to rid popular media of them would be futile. Metaphors with familiar referents effectively engage the public, and engagement is the goal of popular media.[107] No matter how scientists present their research in scientific publications or other arenas, popular media are unlikely to rid themselves of militaristic metaphors and terms that might connote xenophobia.[108] However, at times the scientific literature on invasions would benefit from more precise, "neutral" language.[109]

Controversies over terminology intersect with the claim that invasion science is xenophobic in arguments over the term "alien" to describe non-native species. Use of "alien" for this purpose has been criticized on the grounds that it has pejorative connotations associated with periods of intense xenophobia.[110] "Alien" is used in many nations as an official designation of human noncitizens, and it is not implausible that a reader of an article about a dangerous "alien" fish or plant would unconsciously associate human aliens with danger. Similarly, the trope in science fiction films of humans fighting extraterrestrial aliens, as in the spectacularly popular *Alien* film series beginning in 1979, likely magnifies the negative connotation of the term "alien" in many people's minds.[111] As with certain military metaphors, "alien" is certainly well embedded in the scientific invasion literature as well as policy and other government documents. The Invasive Species Specialist Group of the International Union for the Conservation of Nature (IUCN) published its newsletter *Aliens* from 1995 through 2013 (as described in chap. 10, sec. 10.1) as well as a recently updated document widely cited in both the scientific literature and popular media, *100 of the World's Worst Invasive Alien Species*.[112] The European Commission supported the production of *Handbook of Alien Species of Europe: DAISIE*, which is continually updated by the DAISIE Consortium.[113] Franz Essl and coauthors developed a formal procedure for determining the non-native status of a species based on the term "alien."[114] IPBES, the Intergovernmental Science-Policy Platform on Biodiversity and Ecosystem Services, completed an "Invasive Alien Species Assessment" in 2023.[115] Despite concerns about pejorative connotations associated with xenophobia, widespread use of the term "alien" might persist partly because of differing connotations associated with the word in different nations. Use of "alien" in invasion science literature has decreased in North America but not in Europe. Connotations in English may not be evident to authors, including most Europeans, whose first language is not English.

A related terminological controversy concerns definitions of the terms "alien" and "native" and centers on the contention that the implied dualism is at best obsolete and at worst contributes a xenophobic or militaristic cast to the entire field of invasion science. As described in chapter 3, section 3.3, the classification of species into native, non-native, and what is now termed "cryptogenic" (of unknown origin) arose in the nineteenth century, and the general tenor of those efforts dominates the field today. Controversy revolves around three issues.[116] First, "cryptogenic" species may generate controversies when certain stakeholders wish to maintain a species and claim it is native while others wish to eliminate it and argue it is introduced. That is, the controversy is not about the term "cryptogenic" but rather about the designation as "origin unknown." Inconclusive evidence on geographic origins of populations is particularly common in the marine realm and for species introduced long ago.[117] However, the accumulation of evidence, including historical records and fossils but especially molecular genetic data, continues to reduce the size of the cryptogenic category.[118] In any event, such local controversies do not challenge the fundamental notion of managing non-native species.

A second, broader controversy centers on the suggestion that certain non-native species, especially but not only long-established ones, such as dingos in Australia, should be granted a sort of "honorary" native status for purposes of classification, policy, and management. One suggested criterion for a non-native species to be considered native is that it develop "mutual dependence and mutual controls" with a community of native species.[119] Another proposal is that an introduced species that "coadapts" with the native biota and physical environment be considered native, though the criteria for coadaptation are unspecified.[120] A recent suggestion is that a species deliberately translocated to a region outside its native range in an attempt to prevent its extinction owing to anthropogenic climate change be termed a "neonative" (as discussed in chap. 14, sec. 14.5).[121] Non-native species often become embedded in local cultures, sometimes remarkably quickly. Some even become cultural icons or key elements of a local economy and are appreciated to the extent that many people do not even realize they are not native or, if they do, wish to exempt them from strictures related to non-native status.[122]

Donna Haraway, striving to erase many dualisms (e.g., living vs. nonliving, human vs. nonhuman), coined the term "natureculture" for assemblages of humans, other animals, and plants with complex, interrelated histories as part of her project to erase the ancient distinction between culture, as a construct of human actions, and nature, as an entity separate from human influence.[123] In this context, Haraway, Subramaniam, and anthropologist Anna Tsing, while not denying harmful impacts of certain non-native species, advocate for the continued maintenance of many non-native species integrated into naturecultures.[124] Some biologists have proposed that long-established introduced plants form a separate category—archeophytes—distinguished from more recent introductions, termed neophytes.[125] This convention has been widely adopted in Europe but not elsewhere. In Europe, the demarcation date is generally AD 1500, the beginning of the Columbian exchange. However, controversies arising from these various attempts to give special status to certain introduced species, or even an entire class of them, do not strike at the question of whether "native" and "alien" are valid terms.

The third controversy surrounding these terms is fundamental, as it challenges the idea that this categorization into native and alien (or non-native) is valid or meaningful in the first place. One aspect concerns cases in which a population cannot be classified in either category for reasons other than lack of knowledge. As an example, the native capercaillie disappeared from Scotland in 1775, largely owing to habitat destruction, but a new population was introduced from Scandinavia in the 1830s. Similarly, common rhododendron existed as a native in Great Britain and Ireland in the geological past but then disappeared as its range contracted owing to climate change. It was introduced to Great Britain in 1763 and to Ireland in the nineteenth century, and it is highly invasive in both places.[126] The new populations of capercaille and rhododendron are not lineally descended from the extinct ones; are they native or not? Hybridization also causes difficulties of classification. For instance, the wisent population in Europe was hunted nearly to extinction, leaving seventeen individuals in zoos by the early twentieth century. Populations today in the Białowieza Forest (Poland) and the Caucasus reflect rescue attempts that included hybridization. The Białowieza population contains genes from a bull taken from the Caucasus subspecies before the latter was extinguished, and the current Caucasus population is descended from animals from the Białowieza subspecies, which were also mated with female plains bison from North America.[127] Should either population be called "native"? Although such idiosyncratic cases are a very small fraction of all populations, they have led to calls to jettison the entire idea of classifying species as native or alien.[128]

Another challenge to the very notion of two separate categories relates to Haraway's idea of naturecultures. Anthropologist Stefan Helmreich pointed to the inherent contradiction between terming as "native" (and part of "nature") descendants of the first people to arrive in the Hawaiian Archipelago no more than two thousand years ago, while designating as non-native the species they brought with them, either deliberately, such as pigs and several plants, or inadvertently, such as Pacific rats and seven lizard species (as chronicled in chap. 2, sec. 2.1).[129] Biologist Stephen Jay Gould raised the same issue with respect to the term "Native Americans."[130] These observations relate to unease with words used to designate humans. The Mascarenes, known by Arab sailors about a thousand years ago, were first settled by the Dutch in the seventeenth century, but descendants of these and other European settlers are not seen as "native" in the same sense that native Hawaiians are termed "native." It also relates to the arbitrariness of any specific dating system, such as classifying plant species introduced to Europe during or after the Columbian exchange as neophytes. The fact that "native" and "alien" can be defined only with reference to an arbitrary date and a specific location has been another basis for arguments that the terms should no longer be used, at least in promulgating and implementing policy.[131] Conservation Trust manager Mike Townsend went so far as to contend that these terms harm conservation.[132]

18.3 Which Populations to Target: Beneficial Non-Natives and Wasteful Management

Several critics of invasion science have conceded that certain non-native species have caused substantial harm but have argued that the great majority of non-native species cause no significant impacts, that vast resources are wasted attempting to manage them, and that efforts should be focused on the problematic minority. Thus, for example, although the title of a journal article by Mark Davis and colleagues—"Don't Judge Species on Their Origins"—implies xenophobia, the main point of the article was that many non-native species are actually beneficial and few are harmful; the harmful ones are the only ones that should be managed.[133] The authors also pointed to examples where management efforts were costly, ineffective, and harmful to nontarget species. Developing a take-home message from their edited volume on attitudes, perceptions, and many histories of biological invasion, Ian D. Rotherham and Robert A. Lambert came to the same conclusion, with the added caveat that cultural concerns and the basis for determining non-native status be taken into account.[134] The three contentions of these critics are that (1) most non-native populations are innocuous, (2) many non-native populations are beneficial, and (3) management is often futile and sometimes damaging. The third contention is explored in chapter 17. The following subsections 18.3.1 and 18.3.2 address what can be said about the other two.

18.3.1 How Many Introduced Populations Are Innocuous?

Those claiming that most non-native populations are innocuous frequently cite an early rule of thumb proposed at a SCOPE conference (the topic of chap. 9, sec. 9.1), initially conceived as the "ten-ten rule."[135] Scanning data on non-native vertebrates, insects, and plants in Great Britain, ecologists Mark Williamson and Kevin C. Brown contended that about 10 percent of non-native species established themselves (at least one population was recorded as reproducing over many years), and about 10 percent of these became "pests" (any species "regarded as such by some authorities"). They viewed this as a "rough general rule," but in fact their data appeared to show that, for most data sets they examined, far more than 10 percent of non-native species established populations. Williamson subsequently clarified that "about 10%" meant between 5 percent and 20 percent.[136] Following presentation of the SCOPE paper, an audience member suggested that the definition of "pest" was anthropocentric and proposed a more "ecological" definition: any species that has a "detrimental effect on pre-existing species." The conference summary simply repeated the "ten-ten" rule and acknowledged the imprecision of the term "pest."[137]

Williamson and Alastair Fitter transformed the ten-ten rule into the widely cited "tens rule," adding a ten to the previous two tens by estimating the likelihood that a species introduced to Great Britain produces feral individuals, which may or may not establish a population.[138] Williamson and Fitter defined "pest" as an introduced species "with a negative economic impact." Thus, the frequency with which introduced species become pests was estimated as $(0.1)^3$ = one in a thousand rather than one in a hundred. Examining data from various taxa, they viewed the tens rule as satisfactory for British angiosperms, Pinaceae, pasture plants in Australia's Northern Territory, and non-native species of several taxa defined as "high-impact" by the United States Congress Office of Technology Assessment but not for edible crop plants in Great Britain, birds in Hawaii, insects used in biological control, and mammals introduced to Newfoundland and Ireland.[139] For both crop plants and biocontrol insects, Williamson and Fitter argued that natural selection or human choice before introduction probably caused higher likelihood of establishment and escape to pest (invasive) status than would otherwise have been the case. For the island mammals, they saw the high establishment rate as reflecting excellent preadaptation to the environment, and for the non-native birds in Hawaii, they noted that the high establishment rate was entirely attributable to invasion into anthropogenic lowland habitats. In summary, they found the tens rule "crude" but useful and the exceptions enlightening.

Following Williamson and colleagues' publications, many authors used various data sets to assess whether particular taxa in particular places followed the tens rule, some finding that they do and others that they do not.[140] In 2018, invasion scientists Jonathan Jeschke and Petr Pyšek expressed frustration that the tens rule continues to be cited, terming it a "zombie idea" (that is, an idea that persists despite much contrary evidence).[141] Fisheries biologists Ivan Jarić and Gorčin Cvijanović went further, contending that its persistence is dangerous because it leads some people to claim, without real justification, that the likelihood of a harmful impact of a newly introduced species is very low (0.01 or 0.001).[142] They particularly took issue with the last "ten," the proportion of established non-native species that become pests.

Certainly the estimates for each "ten" are indeed "crude" or "rough," as Williamson and his coauthors acknowledged, and one can question whether the rule is useful at all and whether it might even dangerously underestimate the risk of substantial impact. The issue of lag times must be considered when one estimates the probability of an established species spreading. Jeffrey A. Crooks and Michael Soulé described several examples in which a non-native species established a small population that remained innocuous and geographically restricted for an extended period but then rapidly exploded across the landscape to become a widespread pest.[143] Ingo Kowarik presented data suggesting that most plant introductions undergo a lag of greater or lesser extent before spreading.[144] These publications triggered searches for similar cases, explanations for lags, and estimates of their frequency.[145] This interest yielded many cases and several possible explanations (environmental change, evolution of the established population, etc.) but no evident rules of thumb for prevalence or duration of lags. Even aside from the lag phenomenon, all invasions must pass through a series of distinct stages, including being transported to the non-native site, surviving there, establishing a population, and so forth (as discussed in chap. 15, sec. 15.1). Eventually, some proceed to the stage where they are common and widespread enough to be perceived as having an impact, and different species proceed through these stages at different rates.[146] Therefore, any estimate of the number of impactful invaders will omit established non-native populations that simply have not yet reached the widespread stage but ultimately will do so. Even if a population causes an impact as soon as it establishes, the likelihood that the impact will be noticed is greater the longer the population is present and the farther it spreads. In sum, this research has shown that the status of some non-native populations that are now restricted and not seen as problematic will eventually change, and they will be perceived as impactful.[147]

In addition, there is reason to think, based on the history of invasion science, that any current estimate of proportion of non-native species causing impacts will increase, perhaps greatly. The growth of the science has led to more and more non-native species being studied more intensively and also to the detection of more and more kinds of impacts. Early suggestions of invasion impact, such as that of goats on St. Helena denuding mountainsides (described in chap. 2, sec. 2.3, and chap. 3, sec. 3.2), non-native plants dominating vast stretches of Argentinian pampas (chap. 3, sec. 3.2), and carnivores introduced to islands attacking ground-nesting birds involved dramatic, easily observed impacts.[148] Other impacts, even highly consequential ones, are far subtler, such as the ecosystem impact of a non-native nitrogen-fixing plant in an area of nutrient-poor soil lacking native nitrogen-fixers (discussed in chap. 12, sec. 12.3). Part of the growth of invasion science has consisted of the progressive discovery of more impacts of more types. Late nineteenth-century- and early twenty-first-century suggestions to the effect that if a non-native population causes no visible harm, we should learn to love it, did not account for the findings of modern invasion science.[149]

There is also the issue of what qualifies as an impact, or of an important impact. Most early observations of invasion impact focused on those affecting some human endeavor, whereas modern invasion science, arising soon after the growth of modern conservation biology and environmentalism (as chronicled in chap. 7, secs. 7.4 and 7.5), has focused more on impacts on native species and on ecosystem structure and function. Both iterations of the tens rule were based on the probability of an established species becoming a "pest."[150] As noted in the question after presentation of Williamson's SCOPE paper, a more ecological definition would be useful; the point was conceded by Williamson, with the statement, "That would be a very broad definition."[151] However, that is indeed the sort of definition that many researchers and policymakers seek and that has been the subject of research since the inception of modern invasion science (as described in chap. 16, sec. 16.1). Notably, a tendency of some critics of invasion science to count as a harmful impact only extinction of a native species has been criticized on the grounds that many other ecological and environmental impacts are perceived by both scientists and the public as important.[152]

18.3.2 What Proportion of Non-Native Species Are Beneficial?

"Beneficial" is a relative term: beneficial to whom or what? Human stakeholders may view a non-native species as beneficial if it provides food, hunting or fishing opportunities, aesthetic pleasure, or simply economic profit. Mark Davis and colleagues, whose article exemplified and stimulated much of the controversy surrounding invasion science and management, called for new "guiding principles" of the field highlighting beneficial impacts of invaders:

> Specifically, policy and management decisions must take into account the positive effects of many invaders. During the 1990s, the US Department of Agriculture (USDA) declared several species of introduced honeysuckles to be alien (harmful), and banned their sale in more than 25 states. Ironically, from the 1960s to the 1980s, the USDA had introduced many of these same species in land reclamation projects, and to improve bird habitats. Recent data suggest that the agency's initial instincts may have been appropriate. In Pennsylvania, more non-native honeysuckles mean more native bird species. Also the seed dispersal of native berry-producing plants is higher in places where non-native honeysuckles are most abundant.[153]

It is noteworthy that the claimed benefit in this case is to native plants and birds—that is, an environmental or conservation benefit, in line with interests of many invasion scientists and managers. Earlier, Richard B. King and colleagues had provided an example in which a non-native species widely heralded for harmful impacts, the round goby, appeared to benefit a threatened species, the Lake Erie water snake.[154] In their study, the goby constituted more than 92 percent of prey consumed by the snakes, which were then growing more rapidly and achieving larger body size than before the goby invasion. In 2006, Laura F. Rodriguez attempted the first synthesis of environmental benefits of non-native species, sketching out several ways in which they might facilitate native species, including those of conservation concern.[155] However, these publications, in specialized scientific journals targeting relatively narrow readerships, did not achieve nearly the visibility of the comment by Davis and colleagues in *Nature* in 2011.

The same year, Martin A. Schlaepfer and associates listed ostensible conservation benefits of non-native species, conceding that the listed species may also have other impacts that they did not discuss.[156] A trickle of examples of widely vilified invaders benefiting some native species or ecosystem of concern has ensued. For instance, non-native green crabs in New England consume the native herbivorous marsh crab, whose increase in the wake of overfishing had led to denuded marshes. Recovery of native smooth cordgrass marshes followed.[157] Experiments suggest that non-native wakame can serve as a habitat for native reef fishes whose populations have plummeted in southeastern Australia where overgrazing by sea urchins has devastated native kelps.[158]

The problem is that any non-native species, especially if it becomes common, and even if it has harmful impacts, serves as a resource and is likely to benefit some native species.[159] Thus, a calculus is required of positive and negative impacts, and some method is needed to determine whether a designated set of stakeholders view the net impact as negative or positive. As described in chapter 16, quantifying and comparing impacts has been a persistent challenge from the inception of modern invasion science, complicated by the fact that many impacts cannot be detected or measured without onerous studies. Without such a calculus, it is difficult to assess the overall significance of research pointing to particular benefits of specific non-native populations, though of course this research is relevant to policy and management decisions in the domain of the studied invasion. Suffice it to say that research on green crabs, wakame, and round goby invasions has revealed abundant detrimental impacts. The non-native honeysuckle case cited by Davis and colleagues is an instructive example.

The research to which Davis's team referred examined correlations in a diverse central Pennsylvania landscape between densities of two non-native honeysuckles (Amur honeysuckle and Morrow's honeysuckle, from Asia) and native birds.[160] They found that numbers of three species (gray catbird, American robin, American goldfinch) were positively associated with abundance of fruits of these honeysuckles, while three other species (red-bellied woodpecker, downy woodpecker, and dark-eyed junco) were negatively associated with honeysuckle fruit abundance. All six species have wide ranges, are common in the study region, and are not of conservation concern. The research also examined removal of fruits from potted plants of a single species (black nightshade) in two locations and found it was associated with abundance of the two non-native honeysuckles, thereby suggesting greater seed dispersal. Black nightshade is common and probably native in the study region and is viewed as a noxious weed in some habitats. Seed dispersal was not examined for any other plant, but the researchers noted that frugivores such as catbirds and robins could also be facilitating the invasion of the two honeysuckles and other non-native species by dispersing their fruits. Whatever benefits might be conferred by the honeysuckles on birds or plants, or both, in this area should be set against their negative impacts. Amur honeysuckle is widely viewed as one of the most invasive shrubs in eastern North American forests, with myriad harmful impacts, including serving as an ecological trap for certain songbirds by inducing nesting in a habitat suffering greater likelihood of nest predation.[161] Much less research exists on impacts of Morrow's honeysuckle, but they are believed to resemble those of Amur honeysuckle. Research in Pennsylvania on field sparrows revealed that presence of Morrow's honeysuckle in the vicinity of their nests lowered nesting success and created a population sink for the bird.[162]

The 2023 IPBES report (cited in section 18.2) should convince policymakers that, on balance, the damaging impacts of non-native species far outweigh their benefits.[163] A comprehensive report on impacts of non-native species compiled over four years by more than two hundred experts from every region of the world and approved by 143 member states established that there is a global consensus on the negative impacts of non-native species.[164] In particular, a review by those experts of over thirty thousand peer-reviewed publications and a huge amount of gray literature found that 85 percent of documented impacts were negative, these occur in every part of the globe and in all habitats, and the rate of establishment of newly introduced non-native species is increasing in every part of the world.

18.4 "Native Invaders" and Ecological Succession

Critics of invasion science have argued that, in terms of their processes and impacts, no real difference exists between native and non-native species, implying that the entire field of invasion science thus studies the wrong problem. Rather, concern and research should be focused on populations generating unwanted impacts, whether or not they are non-native.[165] Species in their native range can certainly produce impacts on other native species similar to some of those produced by non-native populations.[166]

In Scotland, native bracken and ragwort are both troublesome weeds in certain places, for instance, as are native smooth sumac, blackberry, and gray dogwood in parts of the US Midwest.[167] In this view, there is already a well-developed science, at least for plants, focusing on how species interact with and replace one another—research on ecological succession—and non-native species should be studied along with native species in the same way.[168] Research on succession traces back over a century and has been rife with its own controversies since the early twentieth century.[169]

Certain sorts of interactions and impacts of non-native species fit well into the research framework of ecological succession. In 1942, Frank Egler, an enigmatic mid-twentieth-century plant ecologist who anticipated several aspects of ecology and invasion science that became research foci much later, argued that the spread or lack thereof of many non-native plants on the Hawaiian island of Oahu could be best understood by studying them as part of the successional process of the native community.[170] His rationale was that the types of vegetational change they caused did not differ from that observed as native species replaced one another during succession. However, he also noted several invasions that did not fit comfortably into this framework, by virtue of the great difference between the biology and growth form of the non-native species and those of the natives.[171] For instance, red mangrove—native to the Americas and introduced to Hawaii in 1902—formed extensive intertidal forests where no trees had previously existed, replacing saltwort, a shrub native to coasts of the Americas that had itself been introduced in the nineteenth century and formed large monospecific stands, replacing what had probably been wetlands dominated by algae, fungi, and a few small herbs and shrubs.[172] Similarly, mesquite—a tree native to South America and introduced in 1838—formed extensive thickets behind beaches in previously treeless areas. Subsequent research revealed other plant invasions of the Hawaiian Islands that, by modifying ecological conditions over wide areas, affected entire ecosystems in ways that differed from those of successional processes of the native vegetation. One such ecosystem impact, caused by the introduction of a nitrogen-fixing plant, fire tree, from the eastern Atlantic region to areas of nitrogen-deficient volcanic soils lacking such species, was discussed in chapter 12, section 12.3.[173] Similarly, introduction to the largely native submontane woodland of bush beardgrass from South America and molasses grass from Africa converted the ecosystem to a grassland dominated by non-native species by increasing fire frequency and extent (as discussed in chap. 14, sec. 14.3); this change deviated drastically from the original succession of native plant species.[174]

Succession models also do not readily accommodate major changes produced by non-native animals at both the population and ecosystem levels, such as the conversion of southern beech forest on Tierra del Fuego to meadows following invasion by North American beavers in 1946 or the utter transformation of native vegetation by introduced goats on St. Helena noted by Charles Darwin, Joseph Hooker, and Alfred Russel Wallace (discussed in chap. 3, sec. 3.2).[175] Neither do succession models incorporate introduction of non-native plant pathogens that can devastate native species, as the Dutch elm disease fungi have done in Europe and North America, or entire ecosystems, such as in the chestnut blight invasion of North America (detailed in chap. 4, sec. 4.8).[176]

Although some native populations do generate impacts similar to those of some non-native invaders, several authors responded to the contention that this fact invalidated the entire field of invasion science by pointing to great disparities between either the numbers or impacts of native and non-native species. For instance, non-native plants in the United States were forty times more likely than native plants to produce impacts perceived as harmful, and when such impacts of native plants were noted, they were invariably associated with other anthropogenic environmental changes, such as modified fire or hydrological regime.[177] In a global survey, reports of impacts by non-native predators and herbivores were greater, more frequently negative, and less frequently positive than for native predators and herbivores.[178] Aquatic non-native species in Europe and North America were found to be six times more likely than natives to be reported as pests.[179] Tim M. Blackburn and coauthors compared how frequently native and non-native species were implicated as drivers of recent species extinction in the 2017 IUCN Red List of Threatened Species.[180] They found that non-native species were believed to have contributed to 25 percent of plant extinctions, compared with less than 5 percent for native species, and to 33 percent of animal extinctions, compared with less than 3 percent for native species. In addition, in line with the finding for native versus non-native plant impacts in the United States, they found that the relatively few times native species were tagged as factors in extinction, they were associated with other anthropogenic forces, such as agriculture, whereas non-native species implicated in extinctions were believed to be the sole factor in 42 percent of extinctions in which they were implicated.

A related argument about whether differences between native and non-native species are large enough to merit a separate science dedicated to non-natives is grounded in the fact that occasionally native species disperse autonomously into new regions. Once in a while a discontinuous expansion carries a species to a new region very distant from its original range. The cattle egret, which dispersed from Africa to South America in the early twentieth century, is a prime example (as discussed in chap. 1 and depicted in fig. 1.1). Though critics of invasion science have adduced such cases as part of the argument that nothing is special about species helped by humans to arrive in new regions, the great majority of cases of long-distance dispersal are achieved with direct or indirect human assistance. This is not to say that research on cases such as that of the cattle egret, like that of "native invaders," cannot advance our understanding of impacts of non-native species. The advent of rapid anthropogenic climate change has added to the controversy, because the ranges of many species, both native and non-native, will be changing (as discussed in chap. 14, sec. 14.4), a factor in renewed suggestions that how a species gets to a new area—that is, with or without human assistance—is unimportant, thus casting into doubt the utility of a separate science focused on the non-natives.[181]

Human-assisted dispersals often differ from unaided ones generated by climate change in ways important with respect to impact as well as to management and policy.[182] Key among these with respect to impact are the amount of genetic variation in the newly occupied range and the likelihood of continued genetic interaction. Populations introduced by humans often have reduced genetic variation (although several invaders, especially recent ones, have received propagules from multiple sources and thus may have even more genetic variation than do populations in the native range).[183] By contrast, populations of a native species on the edge of an expanding range will likely retain gene flow with the population in the original range.[184] In any event, the impact may be greatly affected by genetic differences, or lack thereof, between populations that spread autonomously and those introduced by humans. (chap. 11, sec. 11.3.2). Also, populations that expand their range on their own are more likely to do so in concert with at least some coevolved species from their native range, which in turn would likely lessen the probability of substantial impact in the new part of the range.[185]

This last point suggests an answer to the question of why one might expect a pronounced tendency for non-native species to be more impactful than native ones. In 1942, well before the existence of invasion science, Aldo Leopold perceived the pattern and proposed that coevolution through vast expanses of time has adapted native species to one another and to their physical environment and that introducing a non-native species lacking this coevolutionary history has a relatively greater likelihood of disrupting some part of the ecosystem.[186] Much modern research supports this hypothesis—that is, that a differing evolutionary context of a non-native species is a major factor in differences in invasion impacts of native and non-native species.[187] The fact that impacts of native species that resemble those of non-native invaders are generally associated with one or more anthropogenic changes, such as agriculture or changed fire regime, is consistent with the hypothesis, as the circumstances of the coevolution among natives are modified.

18.5 Invasive Species Denialism

In 2017, invasion scientists James C. Russell and Tim M. Blackburn coined the term "invasive species denialism" to describe attempts to reframe, downplay, or deny various impacts of non-native invasions, especially those revealed by modern invasion science research beginning in the 1980s. They identified the features of invasive species denialism as "where evidence is disregarded, or motivations are disingenuous," and they noted frequent such attempts in media reports.[188] But they particularly lamented that scientific journal opinion pieces and books by ecologists had recently included similar efforts. Locating invasive species denialism in the general category of science denialism, they differentiated it from scientific debate, which they defined as arising from the uncertainty implicit in the scientific method and which is inherent in scientific progress. Denialism differs from debate, in their view, because it attempts to manufacture uncertainty where a scientific consensus exists. They cited, for instance, denialism associated with the causes of climate change or AIDS, risks of smoking tobacco, and evidence for evolution. Such manufactured controversy often leads journalists to present consensus and contrarian views in the name of "balance" and to conflate debates about policy with debates about the underlying science.[189]

Trends in Ecology and Evolution, which published the Russell and Blackburn article, quickly published four retorts that included among the authors five scientists cited for denialism in the original article.[190] The strongest was by Mark A. Davis and Matthew K. Chew, who accused Russell and Blackburn of having "thoroughly misrepresented our views and motivations" (in fact, Russell and Blackburn had not mentioned either author) and termed the article "invasion biology's desperate last stand."[191] Less aggressively but not less firmly, four authors specifically mentioned by Russell and Blackburn accused them of aiming to shut down valid discussion by using "unpleasant comparators to suppress dissenting voices," a tactic they suggested would ultimately discredit the scientific community.[192] They conceded that there is a scientific consensus on negative impacts of non-native species but argued that the definition of "negative" is so heavily influenced by social values that the pervasiveness and importance of negative impacts cannot be determined solely by scientific research, and publications based on or addressing this proposition included several that Russell and Blackburn had labeled as denialist. Sarah L. Crowley and coauthors recognized science denialism of consensuses surrounding climate change, vaccination, and smoking but disagreed that a similar scientific consensus exists on non-native species impacts, pointing to some of the controversies discussed in previous sections of this chapter, such as those about nativism (18.1) and terminology (18.2). They acknowledged that "some who contest specific conclusions and recommendations of invasion biologists disbelieve or disregard the evidence presented," but they did not concede that such cases amount to widespread, mounting denialism.[193] Finally, John C. Briggs disputed the idea that there is a scientific consensus that invaders substantially negatively affect biodiversity (or he could be construed as arguing simply that the consensus is wrong on this point), and he also pointed to the existence of many beneficial impacts of invasions, reprising the arguments discussed in section 18.3.[194]

Although Russell and Blackburn had not actually enumerated and listed the great majority of the publications they perceived as denialist, Anthony Ricciardi and Rachael Ryan searched through Web of Science, Google Scholar, and Google News archives to identify seventy-seven articles published between 1990 and 2016 that they classified as examples of invasive species denialism (including several by Davis and Chew), and these manifested a continuous rise in number, including a sharp upward inflection beginning around 2006.[195] The common thread they perceived was that these publications ignored or misrepresented scientific findings. Ricciardi and Ryan also noted that most appeared in forums not subjected to expert peer review and that many impugned the motivation or ethics of invasion scientists. They did not identify an underlying reason for the increase in such publications, but they observed, as had Russell and Blackburn, that science reportage for lay audiences often disseminates denialist views by presenting, in the name of balance, contrarian assertions along with statements by "mainstream ecologists."[196]

Ricciardi and Ryan's article elicited a quick, anguished response from philosopher Mark Sagoff, five of whose publications appeared in the list of seventy-seven denialism examples.[197] Sagoff's main concern was his perception that the term "denialist" is an ad hominem epithet, with a pejorative connotation associated with people who argue that climate science is a hoax. He then restated the claims he made in the five listed articles. Ricciardi and Ryan responded by specifying the points in the five papers and in Sagoff's response that warranted the term "denialism."[198] David Munro and colleagues examined the list of seventy-seven papers and found at most five that they believed satisfied criteria for denialism.[199] Demetrio Boltovskoy and coauthors rejected the significance of the claim of a rapid rise in denialism on the grounds that the trajectory of the number of the seventy-seven examples against time is no greater than that produced by the rapid increase of total number of publications on invasions over the same period. They disapproved of what appeared to them as an attempt to stifle legitimate dissent, arguing that the practice of exposing inaccurate or deceptive work should "apply not only to those that maliciously or simply erroneously dismiss the negative impacts of NIS [nonindigenous species], but also to the ones that groundlessly inflate their effects."[200]

Biologist Radu Cornel Guiaşu (author of one paper in Ricciardi and Ryan's denialism list) and philosopher Christopher W. Tindale deplored "the tone of the debate [as] not always as professional as one would like to think or even expect," characterizing it as unhealthy and potentially impeding progress in understanding impacts of non-native species.[201] However, their effort to clarify the debate and elevate the dialogue was rejected by philosopher

David M. Frank and several biologist colleagues as unsuccessful because of cherry-picked examples and data and the construction of "straw people."[202] Frank subsequently attempted a clarification and a middle ground, noting that aspects of invasion science are value-laden but that this does not mean that a scientific consensus cannot arise. He pointed to normative issues defining risk in firm scientific consensuses in medicine, public health, and climate change. Frank perceived a scientific consensus in invasion science, but one perhaps not as overwhelming as Ricciardi and Ryan had implied. He saw Sagoff's publications as clearly denialist in that they misrepresented or misled on scientific matters, but he argued that aspects of his papers that discussed ethical and philosophical concerns should not on their own be labeled denialist. He also believed that Guiaşu and Tindale used denialist tactics. In sum, although he did not evaluate each of the seventy-seven claimed denialism examples, he concluded that "some of the literature criticizing invasion science clearly merits the label 'denialist,'" while "other criticisms may be otherwise valuable insofar as they point to legitimate ethical concerns."[203] Finally, he predicted that charges of denialism "are not likely to improve the quality or tenor of ethical debate, even when denialism is a real concern."[204] It is unlikely that a call for restraint will quell this controversy, however, as witness several recently published papers.[205]

18.6 Animal Rights, Animal Welfare, and Compassionate Conservation

Animal rights and animal welfare advocates have often fought attempts to control or eradicate non-native vertebrate populations, especially mammals.[206] A lawsuit based on animal rights stymied a campaign to eradicate a restricted population of the North American gray squirrel in Italy that is now spreading widely and is expected to invade neighboring nations, threatening the native red squirrel and damaging native trees (see box 18.3).[207] In the Hawaiian Islands, rooting by non-native pigs facilitates reproduction by introduced mosquitoes that vector avian malaria, a contributor to the extinction and endangerment of many Hawaiian bird species, and the pigs also spread seeds of non-native plants, fire tree and strawberry guava, that transform entire ecosystems.[208] Snaring is believed to be the only currently available means to control pigs in certain mountainous areas, and use of this painful method has elicited massive opposition from animal rights advocates.[209] Fencing by federal and state agencies and nongovernmental organizations angers hunters, including native Hawaiians whose ancestors brought pigs to the archipelago and who have hunted them ever since, and they have allied with animal rights advocates such as People for the Ethical Treatment of Animals that have traditionally deplored hunting in opposing local management and eradication campaigns against pig populations.[210] The upshot is a situation in which pigs are fenced from some areas, while in others hunting is permitted, but only enough to maintain the population at certain levels and with bag limits to prevent total eradication.[211] Similarly, an alliance between an animal rights organization (the Fund for Animals) and hunters stalled a project by the US National Park Service to remove non-native mountain goats from Olympic National Park in Washington state. Although the campaign ultimately proceeded and includes shooting goats, it is far from completed.[212] Animal rights groups also delayed but did not ultimately stymie a campaign involving shooting to eradicate feral domestic goats from Santa Catalina Island, California.[213]

BOX 18.3 Gray Squirrels in Italy

The gray squirrel population in Italy was founded in 1948 by two pairs brought to a private park in Turin, in the northern Piedmont region. Until about 1970, the population remained in nearby woodlands, after which it spread rapidly into the surrounding area. The first warnings that its spread would have substantial conservation impacts were issued in the early 1980s, based on extensive research of the earlier spread of the same species in Great Britain, which led to a drastic decline in numbers of the native red squirrel and to substantial debarking of some tree species.[1] In 1997, the National Wildlife Institute and the University of Turin formulated a plan to eradicate the gray squirrel from Italy. An April meeting between researchers and representatives of animal rights groups failed to garner support for the plan from the animal rights groups, although several conservation organizations supported it. In May the researchers initiated a trial eradication of a small population in Turin to evaluate the techniques. Although the trial indicated eradication was feasible, animal rights groups took the researchers to court in June, halting the project. In December 1999, the court ruled in favor of the plaintiffs, and although an appeals court reversed the verdict, the delay in the project led to a great expansion of the squirrel's range and rendered the proposed eradication approach unworkable. A second population was established in Umbria, in central Italy, from seven squirrels placed in outdoor cages in a zoo—from which they escaped in the first decade of the twenty-first century and spread to over 50 square kilometers (about 19 square miles) in a nearby natural area.[2] Genetic analysis showed that this population originated from individuals from the Piedmont population.[3] Although an Italian law of 2013 bans possession, trade, and breeding of the gray squirrel, several other individuals and small groups of gray squirrels have been found in other areas of Italy not contiguous with the Piedmont and Umbria populations.[4] The gray squirrel is now spreading from Italy toward France, Austria, Switzerland, and Slovenia, replacing red squirrel as it establishes populations (fig. B3).[5]

1. Genovesi and Bertolino, "Human Dimension Aspects in Invasive Alien Species Issues"; Bertolino and Genovesi, "Spread and Attempted Eradication of the Grey Squirrel"; Gurnell et al., "Ecosystem Impacts of an Alien Invader."

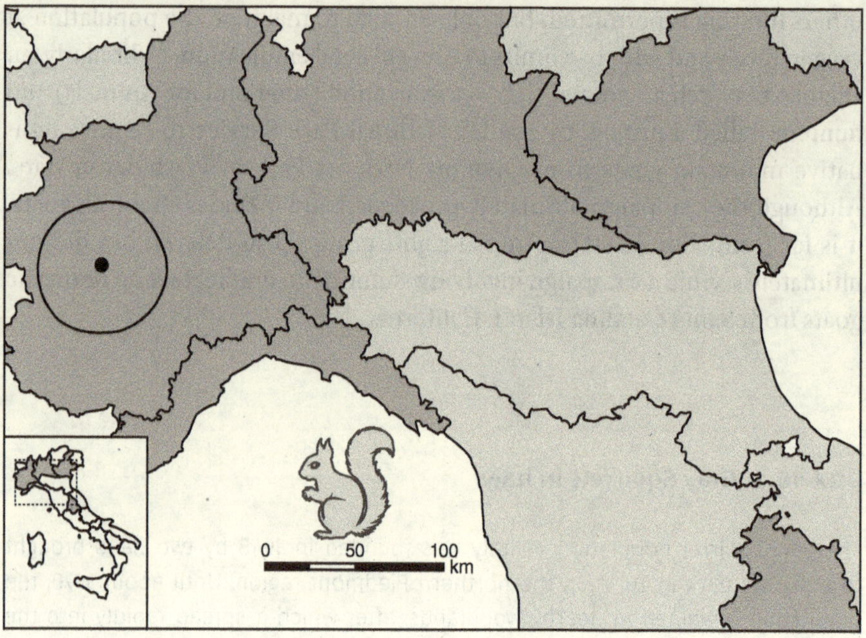

Fig. B3 Range of North American gray squirrel in Italy (dark gray) following initial introduction in 1948. Black dot: location of introduction. Circle: approximate range by 2010. Credit: By permission from Piero Genovesi.

2. Maranesi et al., "Reproductive Traits of an Invasive Alien Population of Grey Squirrel."
3. Signorile et al., "Grey Squirrels in Central Italy."
4. Mori et al., "Alien Shades of Grey."
5. Bertolino et al., "Grey Future for Europe."

One might not expect killing invasive black rats to arouse heated opposition on animal rights grounds, but an unsuccessful lawsuit by the Fund for Animals delayed an ultimately successful campaign to eradicate them from Anacapa Island (California), where they threatened the existence of Xantus's murrelet, a native bird of conservation concern, and also an endemic deer mouse subspecies.[214] A local animal rights organization, Channel Islands Animal Protection Association, actually mounted an unsuccessful sabotage operation in which vitamin K pellets were distributed as an antidote to the poison deployed to eradicate the rats. Retired Channel Islands National Park superintendent T. J. Setnicka observed, "We didn't think we would have much problem in the media with this project. Who could love a rat? As it turned out, lots of people."[215] Campaigns to eradicate invasive nonnative bird populations have similarly run afoul of animal rights advocates, as has happened repeatedly with the monk parakeet, native to South America, in the United States and Europe. The species nests on electric utility structures, causing a major economic impact, and is also of concern because of the possibility that it will inflict agricultural damage.[216] But plans to eradicate urban populations established from cage escapes have been rebuffed in London, New York, Chicago, Connecticut, Florida, and elsewhere.[217]

Controversies about animal rights and invasive populations have increasingly entailed discussions of ethics and morality as the locus of such controversies shifted beginning in 2015 into the realm of "compassionate conservation," discussed further a bit later in this section. To clarify the nature of these discussions, it is useful to digress by explaining certain terms and ideas prominent in the literature of philosophy and especially in the subfield of environmental ethics. As these relate to biological invasions, they were summarized succinctly by philosopher Clare Palmer.[218] One issue is to determine which organisms other than human beings have "moral status." Such organisms' interests should be borne in mind when humans decide whether and how to deal with them, although it does not automatically follow that they all have equal moral status. Palmer noted the existence of two "fault lines" or gradients among environmental ethicists as regards moral status. The first is the "anthropocentric/nonanthropocentric fault line." Anthropocentrists argue that humans have greater moral status than other species. Nonanthropocentrists disagree. Animal rights advocates (and plant rights advocates) are squarely on the nonanthropocentric side.[219] Most invasion scientists and managers would also consider themselves nonanthropocentrist or at most "weakly anthropocentrist," because they generally are endeavoring to establish or maintain a semblance of a conception of nature in which human influence is minimized or absent (see the discussion earlier in this chapter about the nature/culture dichotomy).[220]

The second fault line is an "individualist/holistic fault line" that has engendered particularly intense controversy with respect to invasions. A strict individualist sees moral status residing only in individual organisms, while a holist allows collective entities such as populations or species to have moral status.[221] A holist perceiving collectives in addition to individuals as having moral status would likely advocate killing introduced rats or cats threatening native bird populations, particularly if a native species were threatened with extinction, while an individualist would object. The strength of the advocacy or rejection would depend on the criteria for attributing moral status and the calculus for assigning different degrees of moral status to different individuals and populations or species.

Criteria for conferring moral status are contested. Several philosophers contend that sentience is the key or even sole criterion, although the precise definition of "sentience" is debated.[222] Peter Singer saw the ability to suffer as the relevant indicator of sentience and argued that all sentient, self-conscious organisms have equal moral status.[223] Research on animal cognition emerged as a distinct field in the 1980s, and the hallmark of its rapid growth is the recognition of cognitive abilities in an ever-increasing range of species, including invertebrates.[224] Research on sentience, however defined, is closely linked to animal cognition research and has also grown, though more slowly, and sentience has been claimed in a growing number of species.[225] Sentience as one criterion of moral status raises issues with many efforts to manage invasive populations. Sentience as the sole criterion would also imply that individuals lacking a nervous system, such as plants, would lack moral status. Philosopher Paul W. Taylor contested this view, perceiving individuals of all species as having moral status.[226] A criterion of sentience also excludes collective entities as having moral status, whereas some philosophers, such as Holmes Rolston III, saw certain collectives as having greater moral status than individuals.[227] Aldo Leopold's land ethic assigned greater moral status to the biological community than to any individual within it, and this relationship was reflected in his evolving views on non-native species.[228] Ethical frameworks founded on the notion of the intrinsic value of all living things can also ascribe moral status to collectives such as species. For instance, the biocentric ethics of philosopher Nicholas Agar included a complex calculus for measuring the relative moral status of individuals of different species and of collectives such as populations, species, and ecosystems.[229]

Even if there were agreement about how to assign and measure moral status, distinct ethical frameworks could lead stakeholders to differ on what actually to do about a non-native population. Three such frameworks have dominated the discussion.[230]

1. *Consequentialist ethics* assert that one should act to maximize the balance of good outcomes over bad outcomes, averaged over all entities with moral status that are affected by the action. Utilitarianism is based on consequentialist ethics.
2. *Virtue ethics* demands that an individual act with moral motives, based on assumed constituents of morality such as honesty, generosity, integrity, and, notably, compassion and empathy.
3. *Deontological ethics*, or *deontology*, requires that an action should adhere to a series of rules, rather than to motives or predicted consequences.

The "compassionate conservation" movement evolved from a series of papers in the first decade of the twenty-first century by biologist Marc Bekoff, who argued for more emphasis on ethics in conservation, a plea echoed by ecologist John A. Vucetich and philosopher Michael P. Nelson.[231] From the beginning, Bekoff advocated an ethical stance in the category of virtue ethics, including specifically that one should act with compassion. Vucetich and Nelson, by contrast, stressed elements of both consequentialist and virtue ethics. These and similar writings elicited much interest. In 2010, the first international meeting devoted to compassionate conservation (Compassionate Conservation 2010) took place at the University of Oxford, followed by a special session at the 2011 meeting in Chengdu of Asia for Animals, and by 2013, Bekoff could claim that "compassionate conservation" had become a "buzz phrase."[232] By then, most adherents advocated a virtue ethics framework, as most explicitly developed by Vucetich and Nelson, who called for replacing the consequentialist ethics they saw as underlying much of conservation.[233] Daniel Ramp and Bekoff stressed the key role of compassion in an ethical framework for conservation and rejected deontology as an appropriate basis.[234]

A publication in 2015 in *Conservation Biology* established a new leader among adherents of compassionate conservation, Arian D. Wallach, a wildlife biologist based at the Centre for Compassionate Conservation, established in 2013 at the University of Technology Sydney.[235] This publication turned the focus of compassionate conservation, which had been broad, directly toward managing and eradicating non-native species, and it inspired a strong rebuttal followed by a series of increasingly impassioned papers, many published in *Conservation Biology*, which became the main locus of scholarship on the controversy.[236] Among non-native species whose management Wallach and colleagues deplored were cane toads, boar, and red foxes in Australia.

The gist of the rebuttal, by invasion scientist James C. Russell and fifteen others who had worked on managing invasive populations in the name of conservation, was captured by its title: "Importance of Lethal Control of Invasive Predators for Island Conservation."[237] They provided many examples of inimical effects of non-native predators on islands and of salubrious outcomes of lethal control (usually as part of an eradication campaign). They also appeared to be attempting a conciliatory response to Wallach and her coauthors, acknowledging Wallach for "constructive feedback" on the manuscript and supporting the call of Wallach and coauthors for a compassionate approach to conservation. However, they raised two issues that increasingly generated disagreements. First, they argued that "doing nothing" is itself an action and that, depending on one's value system, it may not be the most compassionate course. Second, they stressed the threat of non-native predators on islands to collective entities—populations, species, and ecosystems.

In 2018, Wallach and collaborators contested both of these points in a broadly ranging essay emphasizing non-native species. With respect to doing nothing, they stressed a "virtue" adapted from medical ethics, emphasized from the outset by Bekoff, "first, do no harm," adding that "it may be better not to do something or to do nothing than to risk causing more harm than good."[238] However, harm and good are in the eye of the beholder. As for the moral status of collective entities, they acknowledged that such entities have intrinsic value, but from the perspective of a threat to the intrinsic value of individuals,

> the principle that the lives of individuals matter acknowledges the intrinsic value of wildlife individuals and resists the tendency to reduce them or their value solely to their position as members of collectives. Inclusivity acknowledges the intrinsic value of all wildlife individuals and collectives.[239]

They attacked "collectivism," defined as "a singular focus on the protection of wildlife collectives," equating this to a version of what the philosopher Tom Regan termed "environmental fascism."[240] They also criticized "nativism," invoking a series of ideas that are themselves controversial: inveighing against the militaristic language of invasion science (discussed previously in sec. 18.2), arguing that most non-native populations have not contributed to extinctions (as discussed in sec. 18.3), ascribing xenophobic motives to much of the activity aimed at managing non-native populations (discussed in sec. 18.1), and endorsing novel ecosystems as a useful approach to dealing with non-native species.[241] They explicitly based compassionate conservation on virtue ethics and rejected deontology as well as utilitarianism, a form of consequentialist ethics. Finally, expanding a theme developed by Bekoff, they assailed the New Zealand campaign to eradicate invasive predators, Predator Free 2050 (mentioned in chap. 17, sec. 17.2), on grounds that it is tainted by nativism and collectivism and envisions use of chemicals that induce great pain.[242]

This paper by Wallach and coauthors in turn elicited responses claiming that compassionate conservation harms conservation. Perhaps the most pointed response was by Don A. Driscoll and Maggie J. Watson, who viewed the paper as an example of science denialism (discussed in sec. 18.5) because it downplayed observed benefits of lethal culling of non-native populations and ignored the many cases in which no other method that could actually have been implemented would have protected a native population or species. They also dismissed the charge of nativism and what they saw as adherence to a value system that prioritized individual animal lives over existence of collectives.[243] Equally alarmed was an international group of thirty-six conservation biologists whose response was entitled "Envisioning the Future with 'Compassionate Conservation': An Ominous Projection for Native Wildlife and Biodiversity." With respect to controlling non-native species, these authors criticized Wallach's group for advocating untested solutions to perceived problems and ignoring examples in which lethal control had successfully protected native populations and species. They also suggested

Fig. 18.1 On Gough Island, house mice devouring carcass of MacGillivray's prion chick. Credit: Roelf J. Daling.

that the specific proposals advanced in the name of compassionate conservation to date (i.e., June 2019) "may be extreme views within the movement and urge[d] less radical 'Compassionate Conservationists' to clarify their position."[244] To that end, Emma Marris and Yasha Rohwer established that, in fact, Wallach would countenance no killing of non-native individuals to prevent suffering of native individuals or extinction of native populations and species, including in the prominent case of non-native house mice that threaten the existence of the Tristan albatross on Gough Island and inflict horrific suffering on its individuals (fig. 18.1).[245] The mice are the target of a planned eradication project by the Royal Society for the Protection of Birds.[246] Wallach's coauthor Chelsea Batavia was uncertain about whether she would permit killing the mice but asserted that individuals who plan to undertake eradicating them should mourn killing the mice "as a tragedy and a loss" and suffer moral angst once they have performed the act.[247]

Jordan O. Hampton and colleagues criticized the portrayal of ethical frameworks and their consequences by Wallach's team in 2018 on the grounds that they downplayed many harms that a virtue ethics stance would inflict on individual animals, and they advocated a consequentialist ethical stance instead.[248] Matt W. Hayward and thirty-four coauthors, focusing largely on the treatment of non-native populations in the compassionate conservation literature, rejected the frequent invocation of ethics as muddled and tagged the Centre for Compassionate Conservation as "an animal rights group posing as a scientific conservation organization."[249] The main danger they perceived is one that compassionate conservationists themselves have warned about—loss of public support for conservation.[250] However, their perspective on the cause of this threat was different. Whereas the compassionate conservationists suggested that the public will be revulsed by the details of management and eradication campaigns that entail killing animals, Hayward and colleagues feared that

> tried and tested conservation practices that have allowed threatened species to persist (Hoffmann et al. 2010; Hoffmann et al. 2011) could lose political and financial support through uninformed and ill-directed emotion and subsequent public pressure. The compassionate conservation arguments could sway public opinion by appearing as a viable alternative to existing conservation methods, yet science shows this is not the case.[251]

Andrea S. Griffin and colleagues, asserting that "Compassionate Conservation, if adopted, could be more harmful for native biodiversity than any other conservation action implemented thus far, while also causing more net harm to individuals than it aims to stop," argued against establishing a "radical form" of compassionate conservation as a legally enshrined principle of conservation activities.[252] They focused on a large literature, primarily in psychology, showing that empathy and compassion—hallmark virtues advocated by Wallach and coauthors—are based on emotion and subject to major biases, and they argued that stringent adherence to moral rules tends to lead to a "do nothing" approach. Batavia and colleagues responded that emotion contributes novel, essential insights to ethics and that compassion, as an emotional experience of interdependence and shared vulnerability, should be a core conservation virtue.[253] Meera Anna Oommen and colleagues indicted compassionate conservation as described by Wallach's team for its focus above all on the welfare of individual animals, pointing in particular to the harm this would inflict on humans in various cultures who benefit materially from lethal control of non-native animals.[254] Wallach and coauthors replied by attacking the notion of "human exceptionalism," associating it with racism, sexism, and homophobia and calling for granting "personhood" to all sentient animals, advancing a view adumbrated by Bekoff.[255]

Paul J. Johnson and thirty coauthors attempted a middle ground in the increasingly contentious debate, adducing several examples of impacts of non-native species and pointing to the variation in statements by advocates of compassionate conservation about exactly what such a stance entails, ranging from Wallach's refusal to countenance any killing of sentient individuals for the sake of a collective entity, noted previously, to the statement by her coauthors Ramp and Bekoff that "that is not to say that individual welfare supersedes species or ecosystem welfare, only that we cannot continue to ignore individuals in conservation practice."[256] Johnson and colleagues interpreted this statement as permitting lethal control if some sort of consequentialist calculation deemed the harm of lethal control the lesser evil. They also raised an issue stressed by Oommen and coauthors of potential harm to humans of an extreme version of compassionate conservation. They suggested that such harm would counteract the possibility that traditional conservation employing lethal measures risks alienating the public, as is frequently asserted by compassionate conservationists.[257] Christopher Bobier and Benjamin Allen focused on variation among compassionate conservationists with respect to killing animals, observing that in cases such as those of non-native predators harming native animals, a virtuous person might determine that killing the predators is the best available option.[258]

In the first review of the metastasizing controversy, in 2022, Simon Coghlan and Adam P. A. Cardilini analyzed the criticisms in eleven responses to the 2018 paper by Wallach and colleagues, assigning them to four broad categories: (1) ethical foundations of compassionate conservation, (2) definitions of "compassionate," (3) scientific credentials of compassionate conservation, and (4) clarification of the application of compassionate conservation, identifying within these categories many recurring specific criticisms.[259] They found that some critiques assumed the truth of questionable views or were simplistic but that certain criticisms raised important ongoing philosophical issues as well as empirical questions, and they called for an interdisciplinary dialogue about ethics, values, and conservation. It seems certain that such a dialogue will ensue, as well as continued polemics. The ultimate impact of compassionate conservation on perception and management of non-native species remains to be seen. In particular, the predicted or feared loss of public support for killing individuals of non-native populations, including sentient individuals, is hypothetical so far. The substantial fraction of the controversy displayed in the pages of one journal is striking. From 2015 through mid-2021, at least twenty contributions appeared in *Conservation Biology*, the great majority treating the general issue of non-native species or specific invasions. Biology is not really the major part, or in some cases any part, of these papers, which increasingly focus on ethics and values.

: : :

Controversies have beset modern invasion science from its inception. The charge of xenophobia or nativism, the most consistently caustic criticism, has largely emanated from scholars in the social sciences and humanities.

Several critics of biological invasions in the nineteenth and twentieth centuries, including Nazis, were clearly nativists and even analogized human immigrants to biological invaders, but simply because these critics held those views in the past does not mean researchers and managers of invasions today are similarly motivated. Most critics coming from this direction downplay or even ignore the impacts that currently spur study and management of invasions. Rather, their conception of invasion science is a social construction, in which social factors and the psychology of scientists propel the field rather than observations in nature or the laboratory.[260]

The controversy over invasive species denialism has largely played out within the community of ecologists and conservation biologists. As originally defined—"where evidence is disregarded, or motivations are disingenuous"—it invited controversy, because ascribing motivation was bound to elicit objections from at least some authors accused of having disingenuous motives. The other part of the definition—disregarded evidence—is amenable to normal scientific discourse with particular assertions being more or less persuasive. Attacks on invasive species management in the name of "compassionate conservation" and rebuttals to these attacks are unlikely to lead to many changed opinions or a resolution. They reflect value judgments that resist, at least in the short term, empirical findings and scientific data. Probably some terminology will evolve—the quite widespread replacement of "exotic" by "non-native" is an example. But military metaphors are too thoroughly embedded in the field (as in medicine) to disappear, and aspects of the analogy of biological invasions to military invasions are too obvious for writers of lay media to forgo using terms such as "explosion," "beachhead," and "war" that they believe will resonate with readers.

The fact that some—perhaps many—non-native populations benefit some coexisting native species or human stakeholder, whatever other impacts they have, has never been doubted by modern invasion scientists and, in fact, has often been explicitly noted. How to weigh the different impacts—negative and positive—with respect to policy and management decisions is the focus of the recent EICAT, SEICAT, and EICAT+ frameworks (described in chap. 16, sec. 16.3) and will doubtless be an ongoing research focus. Similarly, invasion scientists have long observed that some native species have negative impacts similar to those of some non-native species. Nevertheless, these two well-known facts continue to be repeated by critics of invasion science, and surely this repetition will continue. This is part of a frequent feature of much literature surrounding controversial aspects of invasion science: the agglomeration of unrelated charges in a critique largely based on one concern. Thus, for instance, a report on beneficial impacts of non-native species may state or hint at a xenophobic motive associated with managing them, or a plea for compassionate conservation may question claimed negative impacts of a non-native species or argue that management is futile.

The controversies have occasionally impeded management efforts (e.g., the aforementioned failure to eradicate a spreading gray squirrel population in Italy). However, they have not led to a broadscale reassessment of management because negative impacts of so many non-native species are apparent to at least some stakeholders and often the wider public (e.g., the Burmese python in Florida). Neither have they yet slowed the rapid growth of the modern field of invasion science documented in this book.

19

The Near Future of Invasion Science

Biological invasions and especially efforts to manage them have long attracted scholars' attention in disciplines other than biology. Invasion science pioneer Charles Elton (profiled in chap. 7, sec. 7.1), focusing heavily on ecology, entitled his 1958 seminal work *The Ecology of Invasions by Animals and Plants*. The term "invasion ecology" thus came to be used frequently to embrace research on invasions that often delved into areas that were not fundamentally ecological. Indeed, a comprehensive book published in 2011 summarizing progress in invasion science since Elton's book was entitled *Fifty Years of Invasion Ecology: The Legacy of Charles Elton*, even though it included substantial treatment of evolution.[1] As late as 2013, the second edition of the textbook by Lockwood and colleagues (discussed in chap. 10, sec. 10.2.3.2) was entitled *Invasion Ecology*, despite treating economics of invasions and having an entire chapter on evolution. However, by then the field was more frequently characterized as "invasion biology," as in Mark Davis's 2009 text, *Invasion Biology* (described in chap. 10, sec. 10.2.3.3). Particularly with regard to managing invasions (as discussed in chaps. 17 and 18), appeals to engage social scientists proliferated.[2] Christophe Kueffer and Gertrude Hirsch Hadorn viewed the study of invasions as "transdisciplinary" and simply called the gamut of research addressing biological invasions "invasion research."[3] However, the term that caught on widely was "invasion science," coined by David M. Richardson, who pointed to the many facets of invasion research relevant to biology, but also to several fields with extensive literatures independent of biology, such as sociology, economics, policy science, philosophy, and climatology.[4]

Richardson also speculated about future areas of growth in invasion science, beyond increasing interdisciplinarity both within biological sciences and between biological sciences and other disciplines. Among other accurate predictions, he foresaw growing attention to networks (discussed in chap. 13), evolution and genetics (chap. 11), propagule pressure (chap. 15), climate change (chap. 14), and quantifying and ranking impacts (chap. 16). Having traced the trajectory of invasion science from antiquity through the present, I now offer, in the eight sections that follow, some thoughts on likely developments in the near future.

19.1 Collaborations

The decline in single-authored scholarly publications and concomitant rise in multiauthored ones was first depicted for chemical sciences and subsequently in many fields, including economics, social sciences and humanities, medical sciences, natural sciences, and ecology and environmental sciences.[5] Invasion science has seen a similar trend—among 2,535 publications in 2020, fewer than 4 percent were single-authored, and the mean number of authors increased almost linearly from 1980 through 2020 from 1.33 to 5.26.[6] Beginning in the mid-1990s, the maximum number of authors increased rapidly, approximately exponentially. In 2020, 48.3 percent of all publications had 5 or more authors; one had 642![7]

Several factors drive these trends.[8] Research on invasions encompasses an increasing range of questions involving multiple disciplines (e.g., sociology, economics, philosophy). Additionally, even within disciplines, technological advances have spurred increasing collaboration. In biology, for instance, the rapid development of molecular genetic methods and software to analyze output of these methods (detailed in chap. 11, sec. 11.2) has led to many publications coauthored by specialists in the ecology, evolution, or systematics of a taxon and experts in the use and interpretation of the genetic data. Such collaborations are already prominent in invasion science and will likely increase.

A structural development that arose throughout sciences in the 1990s facilitated collaborations that would have occurred more slowly otherwise and certainly has fostered many new ones: synthesis centers that assemble researchers with diverse interests and expertise in working groups that meet once or several times to attempt to achieve new insights and to pose interesting new questions.[9] In invasion science, the Scientific Committee on Problems of the Environment (SCOPE) working groups of the 1980s that established the modern field (the topic of chap. 9) are a prime example. In the United States, the National Center for Ecological Analysis and Synthesis (NCEAS), begun in 1995 at the University of California, Santa Barbara and supported for its first seventeen years by the US National Science

Foundation, has convened ten working group projects specifically on aspects of biological invasions and several others on topics that included invasions. The ten main projects alone engaged two hundred participants from fields as diverse as economics, policy science, and climatology and yielded dozens of publications. For instance, the first invasion working group (one of the first working group projects of NCEAS, "Invasion Biology: Toward a Theory of Impacts," 1998) produced several influential publications, including the foundational paper on quantifying impacts (discussed in chap. 16, sec. 16.1) and the contribution naming and pointing to the existence of invasional meltdown (as described in chap. 15, sec. 15.6).[10] An invasion-oriented synthesis center is the South African Centre for Invasion Biology (CIB), established in 2004, whose scope encompasses many fields. Through 2018, CIB produced 1,745 peer-reviewed publications featuring collaborations among 4,237 authors from 110 nations.[11] Several other invasion-oriented collaborative networks exist; most are international.[12] These all generate increasing research collaborations, many of which are multidisciplinary.

The rise of email and the advent of the World Wide Web spurred increased collaboration.[13] Email arose from an initial effort to aid military research in the United States under the sponsorship of the Advanced Research Projects Agency (ARPA), established in 1958. ARPA research through the 1960s led to the network ARPANET to facilitate communication among ARPA facilities, with the first transmission in 1969. Several other networks arose over the next decade, including BITNET in 1981, for communication among university researchers in 1981. My first use of email, including seeking information on biological invasions, was through BITNET in the mid-1980s. BITNET as a major conduit of research communication began to decline in the early 1990s as networks for various user communities proliferated, leading to the emergence of a global internet.

Initial steps toward the creation of what became the World Wide Web occurred in the 1980s, but the World Wide Web as it now exists was invented by British computer scientist Timothy J. Berners-Lee, working in 1990 as a contractor for CERN, the European Organization for Nuclear Research (which kept the abbreviation from its original French name, Conseil Européen pour la Recherche Nucléaire). The early 1990s saw rapid development of browsers for the World Wide Web, and by the new millennium it was routine for collaborating researchers to exchange comments and drafts of manuscripts by email. Videoconferencing arose by the early 1960s, initially used primarily for large corporate meetings. Although quickly heralded as beginning a new era of conferencing and communication, the technology languished until the 1990s.[14] However, technological developments, rising cost of travel, and increasing familiarity with videoconferences led to increased usage, and the COVID-19 pandemic beginning in 2019 precipitated rapid adoption of videoconferencing with Microsoft Teams and especially

Zoom by entire conferences, working groups, and researchers engaged in less formal interactions. The approximate and fortuitous synchrony of the rise of email and the internet with the rapid development of invasion science (described in chap. 9, sec. 9.1) suggests the field was preadapted to substantial collaboration at the outset.

The increase in multidisciplinary research in invasion science, though pronounced, has been uneven. Ana S. Vaz and colleagues extracted records of invasion science publications from the Web of Science between 1950 and 2014; they classified about half as "monodisciplinary" and half as multidisciplinary.[15] During the late 1980s and the 1990s, as the modern field of invasion science rapidly developed, the frequency of multidisciplinary research and the number of disciplines involved in particular research publications increased, but 92.4 percent of the multidisciplinary publications focused on ecological questions (which included ecology, evolution, and environmental sciences in their classification), 4.4 percent on social questions (first arising in the 1990s), and only 3.2 percent on what the authors termed "social-ecological questions" (first arising in the first decade of the twenty-first century). These authors ascribed the growing interest in invasions from a sociological perspective to three themes: human influences on the invasion process, impacts of invasions on humans, and practical aspects of management. They perceived a "deficit of social-ecological approaches in invasion research" and advocated reducing the "ecological-environmental focus of invasion science" and increasing interaction with the social sciences and humanities by reframing research questions as a social-ecological challenge rather than a purely ecological matter. They went so far as to say it would be beneficial "to promote debates that target the social construction of invasive species based on scientific facts or cultural norms."[16]

Although the growth of invasion science literature will surely include growth in the number of publications on social or social-ecological questions, as well as a continuing increase in collaboration in general, it is not clear that percentages of multidisciplinary publications on social or social-ecological questions will increase and that of ecological questions will decrease, at least for the next two or three decades. For the great majority of non-native species, little if anything is known about their impacts, and research in the realm of ecology, evolution, and environmental science in this area attracts increasing attention. Most of the major hypotheses proposed to "explain" invasion patterns (presented in chap. 15) remain attractive research topics, both in general and for particular species. Some, such as the role of meltdown and related facilitatory processes and the intersection of climate change with invasions, are expanding because they are particularly timely or explore important but previously relatively neglected phenomena, or both. Collaborative research on topics such as these, and on impacts generally, may well be multidisciplinary but need not involve interactions with social scientists. At least until the basic biological aspects of many more invasions are quite well understood, I doubt that the number of publications on social questions or social-ecological questions related to invasions will increase relative to those in the "ecological" category.

The advent of SEICAT (Socio-Economic Impact Classification of Alien Taxa) will certainly encourage research on social questions and social-ecological questions; probably much of this research will include interactions across the ecology-sociology divide. However, the advent of EICAT (Environmental Impact Classification for Alien Taxa) and EICAT+ will encourage research that does not automatically entail interaction with sociologists. (EICAT, EICAT+, and SEICAT are all discussed in chap. 16, sec. 16.3.) As for whether it would in fact be beneficial to promote debates that target the social construction of invasive species, this would depend on the precise definition of "beneficial"—who benefits?—and also a consideration of who might be harmed. Managing an invasive species often falls in the category of a "wicked problem"—one that is complex and has no definitive solution.[17] Rather, diverse stakeholders with different views assess various possible actions and solutions differently. In the case of invasion science, and particularly policy and management of invasive populations, often the key cause of "wickedness" is different values, rather than inadequate science.[18] Philosopher David Frank suggested that disparate values held by individuals concerned with non-native species may be useful at least insofar as they may improve considerations of ethics in research, and possibly even in problem solving if ethical commitments among stakeholders substantially overlap.[19] However, engagement among stakeholders holding disparate values about controversial matters does not always lead to some fruitful middle ground and way forward, although disparate values can nevertheless generate an ongoing literature.[20] In invasion science, the controversy surrounding compassionate conservation (detailed in chap. 18, sec. 18.6), appears to be a persistent stalemate, with some participants vetoing any currently available approaches to lessening impacts of particular invasions and others arguing that, as has often been observed with respect to managing non-native species, taking no management action is itself an action.[21]

19.2 Social Media

Social media such as Twitter, Facebook, and TikTok first became prominent in the early twenty-first century and are already used for activities associated with invasion research and management. The role of social media in invasion science will surely increase at least in the near term as today's younger generation that more fully embraces social media than my generation does increasingly dominates and directs the field.[22] Among such activities, new records of the spread of non-native species will be prominent—an early example featured the use of Twitter to record the first appearance of the white-shouldered fly from Japan on the Atlantic coast of North America.[23] Posting of publications or their citations on social media can lead to discussions that inspire research. An example is the posting on Twitter of a publication on media framing of spiders that precipitated collaborations through social media and email resulting in a global survey of misinformation on spiders.[24] Social media are also used to search for research partners, as in a Twitter call for collaborators on a project to describe and explain variation in results among studies of similar phenomena.[25]

Social media also affect research in less direct but nonetheless important ways. In my generation, faculty and graduate students developed their knowledge base and sense of the direction of a field primarily by reading publications and through in-person discussions with colleagues. Today, insights from the large and still-increasing number of discussions and arguments on social media about scientific matters greatly outnumber those that might be gleaned in person. In addition, the casual nature of social media minimizes the power structure inherent in potential interactions at in-person events such as conferences, workshops, and seminars, thus surely facilitating interactions between early-career researchers and more senior scientists.

19.3 Citizen Science

The term "citizen science" has come to encompass a range of projects involving collaborations between nonprofessionals and professional scientists in addressing scientific questions, especially those that engage the public in gathering massive amounts of data.[26] Public, nonprofessional participation in ecological research goes back centuries, before what is sometimes viewed as the professionalization of science in the late nineteenth century.[27] Examples include a spate of recent papers on biological changes over the past 150 years in the flora of Thoreau's Woods (Concord, Massachusetts) focused on the impacts of climate change, including how climate change influences the spread of non-native species at the site.[28] This research was enabled by many data recorded by Henry David Thoreau and later observers. The advent of the internet and smartphones has dramatically increased the scope of citizen science through websites that facilitate data transfer (including photographs) from observers to database managers who attempt to validate the identifications and coordinate and systematize records into a form scientists can access and use (chap. 17, sec. 17.1).[29]

Citizen science was quickly perceived as a partial solution to a persistent problem in invasive species management: finding an invasion early. It has long been a mantra of invasion scientists and managers that "an ounce of prevention is worth a pound of cure," leading to the widespread implementation of early detection, rapid response (EDRR) programs. Possibly the first EDRR program in the United States began in the 1960s—the Witchweed Eradication Program, targeting an introduced root parasite of agricultural crops in the Carolinas.[30] Several programs trained volunteers to monitor specific areas for invasive plants (chap. 17, sec. 17.1).

The use of citizen science for both early detection of invasions and refined and updated maps of the ranges of non-native species—for example, for use in species distribution models (SDMs, as described in chap. 14, sec. 14.4)—will surely increase.[31] Smartphones were used by 66 percent of the global population in 2020, and many less-developed nations achieved high smartphone usage rates as quickly as wealthy nations did.[32] Use of such data for early detection of non-natives is already common, and other uses are occasionally reported, such as integration of such data with SDMs.[33] Enthusiastic proposals for engaging citizen scientists beyond early detection suggest increasing employment of both smartphone records and other public activities in research and management.[34] Quality and verification have been consistent concerns for data reported by citizen scientists.[35] Advances will surely feature in the near future, aided by rapid developments in artificial intelligence.

19.4 The Global Scope of the Science

The ambit of invasion science should certainly be global, because so many invasions, and such a high proportion of impactful ones, are intercontinental. However, research published in international journals is skewed heavily toward certain continents and nations. The most intensive analysis, focusing on submissions to the journal *Biological Invasions* (discussed in chap. 10, sec. 10.1.2.1) and articles published there (each submission assigned to affiliation of senior author), determined that North America (1,466), Europe (946), and Australasia (783) published the most articles from the first issue in 1999 through 2020, while South America (245), Asia (232), and Africa (93) published far fewer.[36] Submissions from North America, Europe, and Australasia also had the highest acceptance rates, and these continents also provided the most reviewers. If population size is taken into account, New Zealand, Australia, and Switzerland had the most publications per capita. The number of nations represented among published articles (76 total) has increased every year, and the relative proportion from North America (and especially the United States) has decreased, but authors from a few nations and continents still publish the great majority of papers.

That this disparity does not serve the science well seems obvious. We can dispense with the notion that the number of publications is roughly proportional to the number of non-native species in a continent or region. For continents, for instance, the numbers of *known* non-native plant species are greatest for North America, closely followed by Europe, while Africa has only about 60 percent as many recorded non-native species.[37] However, most of Africa is believed to have very incomplete national inventories of non-native plant species, while North America and Europe are believed to have very complete inventories.[38] Similar shortcomings in data completeness characterize databases on invasions by many other taxa.[39]

A substantial part of the disparity can surely be attributed to the growing use of English as an international language of science and the associated fact that the most prestigious journals are published in English.[40] Although facilitating research in ecology in general and invasion science in particular, the dominance of English in scientific literature has increasingly been recognized in invasion science and related fields as hindering access of researchers, managers, and policymakers to important literature and publication outlets.[41] This hindrance slows the advance of the science and limits effective responses to invasions in general and to impacts of specific invasions.[42] Solutions to problems posed by English dominance are not obvious. Steps that invasion science as a discipline can take that may ameliorate some of the impediments posed by the dominance of English include publishing abstracts in other languages, diversifying editorial boards and manuscript reviewers, and providing enhanced copyediting assistance.[43] Other more fundamental changes, such as increasing access to English language training, transcend invasion science.

An important aspect of the dominance of English is that web searches with tools such as Google Scholar and Web of Science omit a substantial fraction of relevant literature, especially but not only if search terms are all in English, and this problem pertains particularly but not only to gray literature.[44] Thus, reviews and meta-analyses based on such searches may yield inaccurate conclusions. Elena Angulo and colleagues provided an interesting example in invasion science. They assessed the economic costs of invasions, comparing data gathered from English documents with equivalent data from documents in fifteen non-English languages. The non-English sources added costs of 249 species and 15 nations, and these additions raised their estimate of the global economic cost of invasions by $214 billion (16.6 percent), as calculated in US dollars.[45] As this problem with searches limited to English keywords becomes more widely recognized, one might expect that authors of publications based on such surveys will attempt to widen their scope or acknowledge potential shortcomings in interpreting their results, or both.

19.5 New Disciplines and Underrepresented Research Areas

Chapter 11 recounts the rapid development of molecular genetic technologies and concomitant increase in invasion studies in the areas of evolution and genetics, describing such research as a "leading edge" of invasion science. I doubt that this status will wane in the next two or three decades. The relevant technologies continue to advance, the corps of researchers that use them is growing rapidly, and the number of non-native species that can be studied from this perspective is enormous. Similarly, chapter 14, section 14.5, outlines the increase in research on the interaction of global climate change with non-native species, spurred by heightened general concern about climate change and by new statistical tools to depict how a species' present or potential range relate to climatic and other factors and to generate predictions of survival and spread of non-native species. Concern with climate change will not abate in the near future, and an endless pool of non-native species is available for such statistical analysis. However, chapter 14, section 14.4.3, also describes a different, less frequently applied approach to such predictions—termed "mechanistic"—that relies on detailed understanding of how a species' physiological processes relate to climatic factors such as temperature and humidity.

19.5.1 Physiology

In fact, physiologists have argued that physiological research is underused in invasion science and management and that this deficit must be redressed.[46] Physiology has recently been depicted as a declining or even dying discipline, eliciting spirited defenses of its relevance.[47] Sixty years ago, when I was a graduate student, biology departments had faculty who studied the physiology of whole organisms—both animals and plants—or at least entire organs, such as the kidney, or systems, such as the digestive tract. As biology underwent a "molecular revolution" in the 1950s and 1960s, research programs became increasingly reductionist, focused at the cellular and molecular levels, and dramatic findings at those levels led to the view in some quarters that whole organism physiology is passé.[48] Biology departments hired ever fewer individuals studying physiology at the whole organism level, new departments of cell biology and molecular biology arose, and the science simply termed "physiology" underwent a persistent crisis of confidence.

One response to this angst was a manifesto proclaiming a new subfield of conservation biology, termed "conservation physiology," with examples of how research on physiological mechanisms underlying declines in species of conservation concern, usually at the whole organism level, can help redress the declines.[49] Despite the close relationship of invasions to many conservation issues, this early discussion of conservation physiology did not directly address non-native species. A subsequent proposed framework for conservation physiology continued to focus on redressing declines of threatened species but included non-native species as one factor causing declines.[50] It also discussed the mechanistic basis of thermal tolerances for the non-native emerald ash borer with respect to the importance of such research to an understanding of population range limits in the face of climate change. The journal *Conservation Physiology*, founded in 2013, has published articles on non-native species from its inception.[51] A 2021 article listing one hundred key questions for conservation physiologists included seven specifically on non-native species.[52]

The first explicit call for expanded physiological research in invasions found that 6 percent of articles in the journal *Biological Invasions* from its inception in 1999 through 2013 incorporated physiological research, with the majority focused on predicting survival and spread of non-native species.[53] However, the authors saw much room for increased integration of physiological research with invasion science, pointing to three major areas: (1) the potential use of physiological stress indicators to assess adaptation of non-native species to their new environment, (2) thermal physiology studies to improve predictions of survival and spread of non-native species under climate change, and (3) the possible use of physiological information in designing measures to prevent establishment and spread. Seven years later, Leigh Boardman and colleagues found increasing mention of physiology in invasion papers but argued that physiology is "still underrepresented in invasion science."[54]

What would the "correct representation" be? Answering this question would require specifying the goals of invasion science, as well as which stakeholders advance which goals and what mix of research questions and approaches would best address these goals. Such a project is surely too prescriptive. That much more research involving physiology relevant to invasion science could be undertaken is obvious, but the same could be said for many other disciplines, given that invasion science is a relatively new field and most non-native populations have barely or never been studied. As with the increased attention of evolutionists and geneticists to invasion science, to some extent the integration of physiology with invasion science will depend on physiologists' recognizing that invasion science is important and that they can contribute major insights by conducting relevant research that also interests them. To that end, Boardman and colleagues pitched their contribution to physiologists as much as to invasion scientists, providing suggestions on how physiologists could readily adapt their research methods to yield important insights into invasions.

19.5.2 Behavior

As with the study of physiology, research on animal behavior underwent a crisis in confidence in the mid- and late twentieth century.[55] The discipline termed "ethology"—the study of animal behavior—arose early in that century in Europe as a science and was seemingly crowned in 1973 when the Nobel Prize in Physiology or Medicine was awarded to Karl von Frisch, Konrad Lorenz, and Nikolaas Tinbergen, recognized as founding ethology, "for their discoveries concerning organization and elicitation of individual and social behaviour patterns."[56] However, by the 1970s, ethology was already widely seen as in decline and the Nobel Prize as honoring research of the past. Ethology was largely focused at the individual level, often dealing with animal behavior viewed as innate or instinctive, and not strongly tied to genetic and evolutionary advances that were then exciting ecologists as well as evolutionists. By the early 1980s, many students of behavior saw themselves as part of a successor field of "behavioral ecology," with ecological context, evolution, and population-level research as prominent features.[57]

As with physiology, behavior researchers sought to show that their findings were relevant to the new field of modern conservation biology. Thus Eberhard Curio could claim in 1996 that "the time is ripe for looking at the potential benefits [that] ethology and behavioural ecology (its intellectual offspring) can offer to the field of conservation biology."[58] In 1998, William J. Sutherland lamented the dearth of conservation-related papers in the behavior literature and corresponding scarcity of behavior-oriented papers in the conservation literature, imploring behavioral ecologists to involve themselves in conservation.[59] A collection of relevant readings soon followed, but these had little about invasions.[60] The first textbooks in conservation biology, in the early 1990s (discussed in chap. 10, sec. 10.2), barely mentioned behavioral research, nor did subsequent editions. In 2007, Richard Buchholz announced a new discipline of "conservation behaviour," which he saw as necessary because of "the neglect shown ethology in conservation-biology textbooks."[61] A textbook about the new field quickly appeared, although with scant attention to invasive species.[62]

The incorporation of behavioral science into invasion science thus did not begin primarily through conservation biology, but rather through scientists interested in invasion as a process and in invasion impacts, including impacts relevant to conservation. Well before the advent of both modern conservation biology (described in chap. 7, sec. 7.4) and modern invasion science with the SCOPE program (discussed in chap. 9, sec. 9.1), evolutionist Ernst Mayr in 1965 attempted to answer the first SCOPE question—what separates species that tend to succeed when introduced from those that tend to fail?[63] Examining birds that had invaded islands or had rapidly expanded to new regions in their native range, he focused on six traits, most of which have a behavioral component, such as a tendency to be social or an ability to discover and use new habitats. Within the field of invasion science, perhaps the first call for greater integration of behavioral research came in 1999 from David A. Holway and Andrew V. Suarez, specialists in ant biology.[64] They noted a role for behavior in impacts of two prominent non-native ant species—the Argentine ant and the red imported fire ant—and also research on the invasion of Hawaii by the Pacific field cricket and a North American fly (chap. 11, sec. 11.3.2) as an example of how invasions can help address questions linking behavior research and invasion science. David G. Chapple and colleagues observed that invasions consist of a series of stages (outlined in chap. 15, sec. 15.1) and that the aforementioned 1999 review had focused on invader behavior after they had established populations. They broadened the call for invasion scientists to pay more attention to behavior by noting that behavior can affect earlier states—both dispersal of a non-native species to a distant new location and initial population establishment there.[65] In 2013, Claudio Carere and Francesca Gherardi bemoaned the relative lack of interest in behavioral traits of invaders by invasion scientists compared with what they termed an overemphasis on other hypotheses such as those featuring propagule pressure and enemy release (hypotheses discussed in chap. 15, secs. 15.2 and 15.4, respectively).[66]

Behavioral ecologists and invasion scientists are now particularly interested in the concept of "personality," defined as "a suite of correlated behaviors expressed either within a given behavioral context (e.g., correlations between foraging behaviors in different habitats) or across different contexts (e.g., correlations among feeding, antipredator, mating, aggressive, and dispersal behaviors)."[67] Particular behavioral traits (e.g., aggressiveness, boldness) have been characterized in somewhat different ways for different species, but this kind of within- and between-population variation in such traits has been demonstrated for many animals. Antecedents to research on animal personalities began in the 1920s and 1930s, the concept was widely accepted by the 1970s and 1980s, and research on many species proliferated greatly by the 1990s and continues to be an exciting, prominent component of behavioral research.[68] Part of the development of this area of research was the identification of an "invasion syndrome"—a group of behaviors that facilitate invasions.[69] These behaviors would likely be associated with success at the various stages of an invasion: certain traits such as boldness and exploration tendency would enhance the likelihood that propagules disperse to a new location, other traits such as predator avoidance might aid establishment and persistence in a new location, and still other traits such as aggressiveness or cooperativeness might facilitate spread and population growth.

An online conference in 2013 led to the first book addressing the role of behavior in invasions.[70] The eighteen chapters described behavioral aspects of many non-native species that appear to contribute to success at various stages of the invasion process. For instance, guppies, native to Trinidad and adjacent regions of northern South America, have widely established globally. Typical behavioral traits of guppies have facilitated their transport, establishment, and spread.[71] For example, the fact that they eat mosquito larvae has led humans to transport and introduce them as biological control agents. Newborn guppies have innate predator-avoidance behavior that aids initial establishment, as does behavioral flexibility in foraging. A tendency to disperse and jumping behavior facilitate spread after establishment, as does the presence of individuals with a bold, exploratory personality. Lynn B. Martin and colleagues linked physiology with behavior in determining invasiveness and invasion impact via hormonal regulation of phenotypic plasticity (as discussed in chap. 11, sec. 11.3.2).[72] They traced how hormones can affect behavior at different invasion stages—for example, in birds, glucocorticoids affect boldness, response to novelty, and exploratory tendencies. Such behavioral plasticity can be important at various points in the invasion process.[73]

Behavioral studies will doubtless increasingly inform invasion science. The vast majority of non-native species have not been studied substantially, especially from this perspective, and the field of behavioral ecology is flourishing. Surely its practitioners will recognize a largely unexploited opportunity, particularly as enlightening examples appear. Behavioral plasticity is already of great interest to students of behavior—both the mechanistic (often hormonal) underpinning and the ecological and evolutionary context. This interest is part of a growing research program in invasion science on the relative importance of phenotypic plasticity and evolution in successful traverses of the invasion stages. Animal cognition studies, which arose as a distinct field in the 1980s (as mentioned in chap. 18, sec. 18.6), explore the basis for behavioral responses. The field is growing rapidly, and this important linkage is just beginning to be explored.[74]

19.6 Context Dependence

Many authors have noted examples in which species known to be invasive and impactful when introduced to certain locations have either failed to survive upon introduction to other sites or failed to generate a substantial impact. Rafael D. Zenni and Martín A. Nuñez listed seventy-six such species, including many types of plants and animals and a variety of locations and kinds of ecosystems.[75] Most striking are cases in which a species is a widespread, prominent invader in one or more locations and has failed even to establish a persistent population when introduced elsewhere. For instance, the Chinese mitten crab, native to the Pacific coast of China and Korea, is widely established in North America and Europe, where its impacts are major and deplored.[76] However, despite a recorded introduction to Japan and strong evidence that it has arrived there often, this crab has failed to establish a population. Zhixin Zhang and colleagues proposed that the failure results from the fact that no Japanese introduction site to date has both a suitable freshwater habitat for the adults and a suitable marine habitat for the larvae.[77] In some instances, the failure of a notable invader even to establish a population when introduced to a particular location can be plausibly ascribed either to the stochastic outcome of a small propagule or to an obviously unsuitable environment—it would not be surprising if a single pair of some species did not establish a population in a new area even with a suitable physical environment (chap. 15, sec. 15.2.2), nor would one expect elands, native to Africa, to survive in Ireland (chap. 15, sec. 15.2.1). Other times, as with the Chinese mitten crab in Japan, the explanation is far from obvious.

A related and growing literature describes situations in which a species is invasive in a new location but declining and of conservation concern in its native range.[78] Again, in some cases, an explanation is apparent. In its native range in Vietnam and China, wattle-necked soft-shelled turtle populations have plummeted because these turtles are hunted for food and as a source of traditional medicine.[79] Meanwhile, on Kauai in the Hawaiian Islands, this introduced species is thriving and altering freshwater ecosystems, posing an additional threat to native species already often imperiled by other forces.[80] Other times, the reason for the disparate trends is more obscure. Populations of the European rabbit, legendarily invasive in Australia (as detailed in chap. 4, sec. 4.1, and in chap. 11, box 11.1), have declined in the Iberian Peninsula, its native range. Although introduced myxoma virus and calici virus, hunting, and changes in land use are all believed to have contributed to the decline, the exact roles of these and other possible causes have not been determined.[81]

These varying results for the same species have long perplexed invasion scientists, both as challenging our understanding of particular invasions and how to manage them and as devaluing the entire science as devoid of useful generalizations or rules. Several authors have suggested that the field is basically a catalog of anecdotes without evident general principles.[82] To an extent, ecology as a whole has been subjected to similar criticism, especially at the community level of organization that has been a frequent locus of invasion research (as discussed in chap. 12, sec. 12.2).[83]

These disparate fates and impacts of introductions of the same species have come to be termed "context dependence," and, under that term, Jane Catford and coauthors found a rapidly accelerating increase in the number of papers addressing this phenomenon in invasion science beginning around 1995.[84] They suggested several factors that could result in apparent context dependence in invasions. For two of these factors, straightforward adjustments in methods might show that much of the apparent variation is illusory. In some cases, different researchers used different statistical approaches or interpreted statistical measures differently, while in other instances researchers observed or measured different variables, or the same variable but in different ways. Another factor entails more interesting research that is, in fact, attracting increasing scientific attention and surely will continue to do so over the next two or three decades because relatively few invasions have been studied in detail and apparent context dependence presents a striking challenge. In some instances, the relationship between a non-native species and a native species or ecosystem differs because some biological or physical variables (e.g., another species, a climatic variable) fundamentally change the relationship. Determining what these factors might be is a start. Observations and statistical analysis can suggest why the relationship differs, but pinning down the mechanism requires detailed and often difficult empirical research—similar to the challenge of achieving a mechanistic explanation for a species distribution (as discussed in chap. 14, sec. 14.4.3).

Research from Australia on impacts of the dingo, red fox, and house cat exemplify how complicated interactions can generate context dependence and the types of research that can help explain it. Dingoes arrived in Australia from Asia, where they had been domesticated, between three thousand and five thousand years ago; Asian seafarers probably intentionally transported them.[85] The red fox was deliberately introduced to Australia in the 1870s and has been a scourge to native small vertebrates as well as to poultry and livestock.[86] However, its impact on native species varies from site to site, and at least part of this variation results from dingo predation on foxes and competition with them.[87] Although dingoes prey on foxes, the relationship of that predation to fox population density and thus to fox impact on their prey is not straightforward, with some reports of substantial impact on fox populations and others of no impact.[88] The house cat arrived in Australia with the First Fleet in 1788 or possibly even earlier, quickly went feral, spread throughout the continent, and plagues small vertebrates, just as the fox does.[89] Dingoes affect cat populations most when cat prey are scarce and least or not at all when prey are abundant.[90] Cats prefer to prey on small mammals but also eat birds, reptiles, and invertebrates.[91] Prey densities, however, are largely governed by weather. The relationship between dingoes and cats therefore depends on a complex set of factors.

19.7 Over- and Underemphasized Research Foci

As noted in previous sections, various scientists and practitioners have argued that invasion science has overemphasized research in certain areas (e.g., the role of propagule pressure in the establishment of non-native populations or the role of enemy release in their spread and impact) and underemphasized other areas (e.g., the physiology or behavior of established or potential non-native species, the sociological context of perception or management of non-native species). These observations are invariably intended to encourage a change in research approaches or goals. Such exhortations seem problematic for both fundamental and practical reasons.

Fundamentally, to say that some sort of research is "overemphasized" or "underemphasized" entails a value judgment because it requires a conception of an ideal or at least a better science than now exists. But such a conception varies from person to person. Some scientists view the main goal of invasion science to be a mechanistic understanding of the trajectory and impacts of every invasion, while others see the key objective as the ability to predict, based on statistical methods, the trajectory of impacts. These are not the same thing, as witness the vitriolic debate between Robert H. Peters and his critics over whether the primary goal of ecology should be prediction pure and simple or understanding the mechanisms that allow an accurate prediction (discussed in chap. 14, sec. 14.4). Yet other stakeholders might see the main purpose of invasion science as developing means to minimize negative impacts of invasions or to reduce the flow of non-native species.

In practical terms, I doubt that calling an approach or a research area over- or underemphasized will inspire many scientists to modify their research, at least in the short term. In a field growing as rapidly as invasion science is, and with such a small proportion of non-native populations intensively studied, even research of the sort primarily undertaken at the beginning of the modern field—impacts of particular invaders on particular native species—is likely to yield interesting results. The incentive to switch gears for a scientist successfully performing such research is thus not great. As new technologies arise, such as the molecular genetic approaches described in chapter 11, section 11.2, they may encourage experts in those technologies to apply them to non-native species or to collaborate with invasion scientists, but that does not mean use of such tools had been previously underemphasized. Scientists' motivations may include wanting a science to be "better" (however "better" is defined) but surely can include other factors. Some may simply be fascinated by the challenges of a certain research direction, independently of whether advances in that direction somehow improve the science. Others may be driven to want to answer some question or understand some phenomenon. The aesthetic aspects of a research problem or technique may seem attractive. These are intrinsic psychological factors. Other factors may be extrinsic: a desire for respect, career advancement, better research funding, monetary gain. It is unlikely that saying some research direction is overemphasized will steer practitioners or even students away from it. Saying a research direction is underemphasized is problematic because of the value judgment it entails. Perhaps simply pointing to opportunities not yet exploited would lead a scientist to consider this factor along with other motivating factors.

19.8 Will the Invasion Juggernaut Be Slowed?

It is a commonplace that invasions keep multiplying and even accelerating, with concomitant increasing impacts.[92] Even before modern invasion science, many scientists who recognized the scope of impacts called for measures to stop or at least slow the phenomenon (as detailed in chaps. 6 and 7). As concerns increased rapidly in the early years of modern invasion science, such calls were reflected in the 1992 Earth Summit at Rio de Janeiro, which produced the Convention on Biodiversity (CBD), implemented in 1993 and ratified by 196 parties (but not the United States).[93] Article 8h of the CBD, "In-situ Conservation," stated, "Each contracting Party shall, as far as possible and as appropriate, prevent the introduction of, control or eradicate those alien species which threaten ecosystems, habitats, or species."[94] The 6th Conference of Parties of the CBD in 2002 in The Hague adopted decision VI/23, guiding principles for implementing Article 8h, noting that "full and effective implementation of Article 8(h) is a priority."[95] The decision recognized international instruments, such as the International Plant Protection Convention, and international organizations that contribute to implementing Article 8h, including the Office International des Epizooties, the Food and Agriculture Organization of the United Nations, the International Maritime Organization, and the World Health Organization. However, it noted the key role of national biodiversity strategies and action plans and urged nations to develop these in line with Article 8h.

The 10th Conference of Parties, in 2010 in Nagoya, Japan, acknowledged the global failure to reach previous biodiversity protection goals. The United Nations declared 2010 the International Year of Biodiversity, and the Conference of Parties developed a strategic biodiversity plan for 2011 through 2020. The strategy included twenty goals, the Aichi Biodiversity targets, including Target 9: "Invasive alien species and pathways are identified and prioritized, priority species are controlled or eradicated, and measures are in place to manage pathways to prevent their introduction and establishment."[96] According to the "Global Biodiversity Outlook 5" report issued by the United Nations in 2020, no target was achieved globally, but six were partly achieved, including Target 9.[97] The report described individual eradication successes but noted that these pertained to a small proportion of all invasive populations. It found no evidence that the number of new non-native species introductions had slowed. Only 10 percent of reporting parties had national targets similar to Target 9 and were substantially on track to meet them.

The 15th Conference of Parties, in 2022 in Kunming, China, and Montreal, Canada, adopted four goals and twenty-three targets, including Target 6: "Eliminate, minimize, reduce and/or mitigate the impacts of invasive alien species on biodiversity and ecosystem services by identifying and managing pathways of the introduction of alien species, preventing the introduction and establishment of priority invasive alien species, reducing the rates of introduction and establishment of other known or potential invasive alien species by at least 50 per cent, by 2030, eradicating or controlling invasive alien species especially in priority sites, such as islands."[98] Target 6 primarily reiterated and strengthened Aichi Target 9. The Conference of Parties acknowledged progress toward meeting Aichi Target 9, but overall failure and no reduction in impacts of non-native species on biodiversity, ecosystems, or society owing largely to inability of nations to establish or implement policies.[99]

One could cynically view the trajectory of CBD targets as a fruitless exercise, kicking the can down the road every ten years. However, surely these targets are just that—inspirational targets—and the authors did not expect them to be achieved. Progress *has* occurred, though the overall problem has worsened. The CBD reports and the 2023 report of the Intergovernmental Science-Policy Platform on Biodiversity and Ecosystem Services make it clear that the major impediment is sociopolitical rather than scientific.[100] Perhaps the growing research on sociopolitical aspects of invasions and the increasing recognition of the need to increase global participation in invasion science will lead to greater progress toward these targets.

: : :

In 2017, ecologists Mark Davis and Matthew Chew, responding to claims of "invasive species denialism" (detailed in chap. 18, sec. 18.5), predicted the imminent demise of invasion science:

> [James] Russell and [Tim] Blackburn's call to action against "invasive species denialism" appears to be invasion biology's desperate last stand. Invasion biologists' insistence that species introductions constitute a unified ecological crisis is losing support from both the scientific community and the informed public. It is being replaced by a more equivocal notion of ecological novelty, which can accommodate recognizing and responding to a spectrum of phenomena ranging from great good to great harm, with a good deal of relative insignificance in between.[101]

This prediction followed Davis's argument that the entire project of studying invasions by non-native species is not quite legitimate, that such research would have been more productive had it been incorporated into research on ecological succession (a topic covered in chap. 18, sec. 18.4), and that it should be absorbed into a science dealing more generally with species geographic distributions and range changes that Davis termed "SPRED ecology" (*s*pecies *red*istribution ecology).[102]

The prediction that invasion science would or should end as a distinct field was forcefully rejected by several of its practitioners, and its imminent death seems unlikely. The dramatic growth of invasion science described in this book shows no indication of slowing, and surely so many researchers—especially graduate students and postdoctoral scholars—would not enter the field if they did not perceive it as interesting, dynamic, productive, and important.[103] The science has already evolved greatly. This book has described its evolution as well as its expansion, which was particularly rapid in the twenty-first century. Invasion science will continue to evolve, partly spurred by new technologies, such as those in molecular genetics, geographic information systems, and stable isotope analysis, and partly by exciting research that suggests new questions or new analytic methods, or both.[104] It will also continue to evolve because problems generated by non-native species will not disappear, and their existence will spur research on preventing and managing invasions.

It would be foolish to forecast what research on non-native species will look like a century from now, but the sort of rapid paradigm shift perceived by Davis and Chew is not underway now, and it is not likely to occur over the next two or three decades. In Thomas Kuhn's conception, a paradigm shift occurs when an existing science senses it is in crisis because its practitioners are unable to answer certain questions or explain perceived anomalies under the reigning principles and with available methods.[105] Few workers sense such a crisis, hence the enthusiastic activity and rapid expansion and evolution of the field. Most challenges that appear unconquerable as they are being addressed today are not scientific but sociopolitical. They concern policy and management to alleviate problems caused by invasion impacts, not inability to learn enough about particular invasions or invasion as a general phenomenon. The scientific challenge is context dependence, along with continuing unease with being seen as a science of case studies rather than of overarching general rules—hence the proliferation of hypotheses and attempts to consolidate them or subsume them into a more general framework (as discussed in chap. 15). However, as in ecology, such an attempt can take research on invasions only so far because invasions are, in fact, idiosyncratic owing to the multiplicity of potential influences on their trajectories and impacts. So a revolution is not going to come from that direction—a new paradigm encompassing invasions that eliminates or greatly reduces context dependence. Scientists studying invasions two or three decades from now will be conducting similar research to that now underway, and the view of non-native species as frequently problematic will still prevail. However, one can hope that, a century from now, this research and other studies using methods not yet imagined will have helped generate a more promising response to the sociopolitical inertia that many invasion scientists see as the real crisis.

Notes

Chapter 1

1. Dueñas et al., "Role Played by Invasive Species."

2. Perrings, "Invasion Economics"; Diagne, Leroy, Vaissière, et al., "High and Rising Economic Costs of Biological Invasions Worldwide."

3. Sikka et al., "Emergence of Zika Virus"; Rawtani et al., *COVID-19 in the Environment.*

4. Crosby, *Columbian Exchange*; Crosby, *Ecological Imperialism.*

5. Crosby, *Ecological Imperialism.*

6. Rangan, Carney, et al., "Environmental History of Botanical Exchanges"; di Castri, "History of Biological Invasions"; Pascal et al., *Invasions biologiques et extinctions.*

7. Lindroth, *Faunal Connections*; Sailer, "Our Immigrant Insect Fauna"; Sailer, "History of Insect Introductions."

8. Egerton, *Roots of Ecology*; Egerton, "History of Ecological Sciences, Part 44"; Haeckel, *Generelle Morphologie.*

9. Elton, *Ecology of Invasions by Animals and Plants,* 15.

10. Simberloff, "Charles Elton—Neither Founder nor Siren."

11. Simberloff, "SCOPE Project."

12. Data derived from searches of Nexis Uni, Factiva, and ProQuest. Search terms: "alien specie*" OR "non-native specie*" OR "nonnative specie*" OR "biological invasion*" OR "invasive specie*" OR "introduced specie*".

13. J. A. Lockwood, *Six-Legged Soldiers.*

14. Knopf, "Changing Landscapes."

15. Diller et al., "Demographic Response of Northern Spotted Owls"; Livezey, "Range Expansion of Barred Owls."

16. Romagosa and Labisky, "Establishment and Dispersal of the Asian Collared-Dove."

17. Botkin, "Naturalness of Biological Invasions."

18. De Queiroz, *Monkey's Voyage.*

19. For critics of invasion science, see Theodoropoulos, *Invasion Biology: Critique of a Pseudoscience*; K. Thompson, *Where Do Camels Belong?*

20. M. Davis, *Invasion Biology.*

21. Ricciardi, "Are Modern Biological Invasions an Unprecedented Form of Global Change?"

22. Reichard and White, "Horticulture as a Pathway of Invasive Plant Introductions."

23. Fosberg, "Derivation of the Flora of the Hawaiian Islands."

24. Cordell, "Regional Summaries: Hawaii and US-Affiliated Pacific Islands."

25. Ricciardi, "Are Modern Biological Invasions an Unprecedented Form of Global Change?"

26. Pereyra, "Revisiting the Use of the Invasive Species Concept."

27. Pereyra, "Rethinking the Native Range Concept."

28. Essl, Bacher, et al., "Which Taxa Are Alien?"; Essl, Dullinger, et al., "Conceptual Framework for Range-Expanding Species."

29. Essl, Bacher, et al., "Which Taxa Are Alien?"

30. Pereyra, "Rethinking the Native Range Concept."

31. Colautti and MacIsaac, "A Neutral Terminology to Define 'Invasive' Species."

32. Richardson, Pyšek, et al., "Naturalization and Invasion of Alien Plants."

33. Chew and Hamilton, "Rise and Fall of Biotic Nativeness."

34. Richardson, Pyšek, et al., "Naturalization and Invasion of Alien Plants."

35. Exec. Order No. 13112, 64 C. F. R. 6183 (1999).

36. Williamson and Brown, "Analysis and Modelling of British Invasions"; Williamson and Fitter, "Varying Success of Invaders."

37. "Invasive Alien Species (IAS) are animals and plants that are introduced accidentally or deliberately into a natural environment where they are not normally found, with serious negative consequences for their new environment. They represent a major threat to native plants and animals in Europe and are one of the five major causes of biodiversity loss. They can also cause significant adverse impacts on the economy (their economic impact in the EU was estimated at around EUR 12 billion per year) as well as human health, such as severe allergies and burns." European Commission, "Invasive Alien Species."

38. Fuller et al., *Nonindigenous Fishes.*

39. D. Strong and Ayres, "Ecological and Evolutionary Misadventures of *Spartina.*"

40. Vitule, Bornatowski, et al., "Extralimital Introductions of *Salminus brasiliensis.*"

41. Richardson, Pyšek, et al., "Naturalization and Invasion of Alien Plants."

42. Kowarik, "Time Lags in Biological Invasions"; Crooks, "Lag Times."

43. B. Larson, "War of the Roses"; Simberloff, "Invasional Meltdown 6 Years Later."

44. Mukherjee, "Invasion Equation."

45. Wang et al., "Stable Isotope Ratios."

46. Connell and Slatyer, "Mechanisms of Succession."

47. M. Davis, Pergel, et al., "Vegetation Change."

48. On similarities, cf. Egler, "Indigene versus Alien in the Development of Arid Hawaiian Vegetation." On differences, see Simberloff, "Invasions of Plant Communities."

49. Simberloff, "Native Invaders"; Simberloff, Souza, et al., "Natives Are Restless."

50. Inderjit et al., "Framework for Understanding Human-Driven Vegetation Change."

51. Zenni, Ziller, et al., "Invasion Science in the Developing World"; Nuñez, Chiuffo, Pauchard, et al., "Making Ecology Really Global."

Chapter 2

1. Wilmé et al., "Human Translocation as an Alternative Hypothesis to Explain the Presence of Giant Tortoises"; Migaud, "First Approach to Links Between Animals and Life on Board Sailing Vessels."

2. Hofman and Rick, "Ancient Biological Invasions and Island Ecosystems."

3. Matisoo-Smith, "Ancient DNA and the Human Settlement of the Pacific"; K. West et al., "The Pacific Rat Race to Easter Island."

4. Steadman, *Extinction and Biogeography of Tropical Pacific Birds*; Hunt, "Rethinking Easter Island's Ecological Catastrophe."

5. Bahn and Flenley, *Easter Island, Earth Island*; J. Diamond, "Easter's End."

6. Athens, "*Rattus exulans*."

7. Matisoo-Smith, "Ancient DNA and the Human Settlement of the Pacific."

8. Greig et al., "Complex History of Dog (*Canis familiaris*) Origins."

9. Four-clawed gecko, Indo-Pacific gecko, Indo-Pacific tree gecko, mourning gecko, mottled snake-eyed skink, blue-tailed striped skink, moth skink (Stejneger, "Land Reptiles of the Hawaiian Islands"; Lever, *Naturalized Reptiles and Amphibians of the World*; Kraus, *Alien Reptiles and Amphibians*).

10. Lord et al., "Complete Mitogenomes of Ancient Caribbean Guinea Pigs."

11. Giovas et al., "Sr and Pb Isotopic Investigation of Mammal Introductions."

12. Giovas et al., "Sr and Pb Isotopic Investigation of Mammal Introductions."

13. E. Wing and S. Wing, "Introduction of Animals as an Adaptation to Colonization of Islands."

14. Grouard, "Modes de vie des précolombiens des Antilles françaises"; Giovas et al., "Sr and Pb Isotopic Investigation of Mammal Introductions."

15. Fairnell and Barrett, "Fur-bearing Species and Scottish Islands."

16. Berry and Rose, "Islands and the Evolution of *Microtus arvalis* (Microtinae)"; Haynes et al., "Phylogeography of the Common Vole."

17. Stanton et al., "Colonization of the Scottish Islands."

18. Fairnell and Barrett, "Fur-Bearing Species and Scottish Islands."

19. Carden et al., "Phylogeographic, Ancient DNA, Fossil and Morphometric Analyses."

20. di Castri, "History of Biological Invasions"; Pascal et al., *Invasions biologiques et extinctions*.

21. Bonhomme et al., "Genetic Differentiation of the House Mouse."

22. Vigne, "Large 'True' Mediterranean Islands"; Vigne, "Origins of Mammals on the Mediterranean Islands"; Vigne, Zazzo, et al., "Pre-Neolithic Wild Boar Management."

23. Vigne, "Large 'True' Mediterranean Islands."

24. C. Driscoll et al., "Taming of the Cat"; Ottoni et al., "Palaeogenetics of Cat Dispersal."

25. Doherty, Glen, et al., "Invasive Predators and Global Biodiversity Loss"; Lepczyk et al., "Science-Based Policy for Managing Free-Roaming Cats."

26. Detry et al., "Did the Romans Introduce the Egyptian Mongoose?"

27. Pons et al., "Vingt cinq ans de recherches en analyse pollinique."

28. Newsom and Wing, *On Land and Sea*.

29. Rangan, Carney, et al., "Environmental History of Botanical Exchanges"; Vigne, Briois, et al., "First Wave of Cultivators Spread to Cyprus at Least 10,600 y Ago."

30. Roullier et al., "Historical Collections Reveal Patterns of Diffusion"; Tromp and Dudgeon, "Differentiating Dietary and Non-Dietary Microfossils."

31. Ladefoged et al., "Introduction of Sweet Potato in Polynesia."

32. Athens et al., "Palaeoenvironmental and Archaeological Model-Based Estimate for the Colonisation of Hawai'i."

33. Chang et al., "Holistic Picture of Austronesian Migrations."

34. Rangan, Bell, et al., "New Genetic and Linguistic Analyses."

35. Prebble and Wilmshurst, "Detecting the Initial Impact of Humans and Introduced Species on Island Environments"; Hofman and Rick, "Ancient Biological Invasions and Island Ecosystems."

36. Panagiotakopulu and Buckland, "Thousand Bites"; Panagiotakopulu and Buckland, "Early Invaders."

37. Dawson, "Life History Strategy and Evolutionary History of *Tribolium* Flour Beetles."

38. Panagiotakopulu and Buckland, "Thousand Bites"; Panagiotakopulu and Buckland, "Early Invaders."

39. Buckland and Sadler, "Biogeography of the Human Flea"; Panagiotakopulu and Buckland, "Thousand Bites"; Panagiotakopulu and Buckland, "Early Invaders."

40. Wodzicki, *Introduced Mammals of New Zealand*; Matisoo-Smith, "Ancient DNA and the Human Settlement of the Pacific."

41. Carrier and Bort, "History of Kentucky Bluegrass"; Global Invasive Species Database, "Species Profile: *Poa pratensis*."

42. Carlton, "Global Framework of Introduced Species"; Carlton, "Biological Invasions and Cryptogenic Species."

43. Gomoiu and Skolka, "Changements récents dans biodiversité de la Mer Noire."

44. Gomoiu et al., "Black Sea."

45. Petersen et al., "Clams Before Columbus?"

46. Rakov and Brodianski, "Ancient Oyster Farming"; Sahrage and Lundbeck, *History of Fishing*; Bartley and Subasinghe, "Historical Aspects of International Movement of Living Aquatic Species."

47. Bostock and Riley, *Natural History of Pliny*, 400–401; Kaster, *Saturnalia of Macrobius, Books 3–5*, 113–14.

48. Pliny quoted in Bostock and Riley, *Natural History of Pliny*, 400–401.

49. Marzano, *Harvesting the Sea*.

50. On the beginnings of aquaculture, see Beveridge and Little, "History of Aquaculture in Traditional Societies."

51. Spitzy, "Prehistoric Man as a Possible Crayfish Transplantor."

52. Balon, "Origin and Domestication of the Wild Carp."

53. Gosse, *St. Helena 1502–1938*; Grove, *Green Imperialism*.

54. Gosse, *St. Helena 1502–1938*.

55. On introduction of black rats, see J. L. Long, *Introduced Mammals of the World*. On their recognition as abundant and a pest of food stocks, see Gosse, *St. Helena 1502–1938*.

56. Gosse, *St. Helena 1502–1938*.

57. Von Mandelslo, *Voyages and Travels of Ambassadors*.

58. Grove, *Green Imperialism*.

59. Beeckman, *Voyage to and from the Island of Borneo*, 193.

60. De Burgh-Edwardes, *History of Mauritius 1507–1914*.

61. De Burgh-Edwardes, *History of Mauritius 1507–1914*; J. L. Long, *Introduced Mammals of the World*. A. S. Cheke argued that goats and pigs were not introduced until 1606, by the Dutch (Cheke, "Ecological History of the Mascarene Islands"). Just as the Portuguese, although not establishing a settlement, introduced goats and pigs

to Mauritius for future travelers, so a British ship in 1629 landed briefly on nearby Île de la Réunion and did not stay but left goats and pigs "that by a happy multiplication and encrease the future Passenger might be releeved and blesse the Planter" (Herbert, *Some Yeares Travels*, 351).

62. On the sixteenth-century introduction of black rats, see J. L. Long, *Introduced Mammals of the World*. For the 1606 Dutch report, see Cheke, "Ecological History of the Mascarene Islands."

63. On rusa deer, see Cheke, "Ecological History of the Mascarene Islands." On control of their harassers, see Grove, *Green Imperialism*.

64. Cheke, "Ecological History of the Mascarene Islands."

65. Leguat, *Aventures aux Mascareignes*; Pasfield-Oliver, *Voyage of François Le Guat*.

66. For Leguat's report on house mice, see Cheke, "Ecological History of the Mascarene Islands." On the rats, see Leguat, *Aventures aux Mascareignes*. Geoffrey Atkinson argued that Leguat was a fictional character and his account of the voyage and sojourns on Rodrigues and Mauritius a novel (Atkinson, "French Desert Island Novel"; Atkinson, *Extraordinary Voyage in French Literature*). Literary scholars then repeated this assertion. For instance, Defoe scholar Arthur W. Secord stated, citing Atkinson ("French Desert Island Novel"), that Leguat was a fictional character ("imaginary hero") and his 1707 book a "desert island fiction" that inspired some incidents in *Robinson Crusoe* (Secord, "Studies in the Narrative Method of Defoe," 92). Atkinson defended his claim in various publications until his death in 1962, despite overwhelming evidence for both Leguat's existence and the veracity of his account (North-Coombes, *Vindication of François Leguat*). The travel writer Maximilien Misson edited small, discrete portions of Leguat's account (Pasfield-Oliver, *Voyage of François Le Guat*; North-Coombes, *Vindication of François Leguat*).

67. This 1761 manuscript was finally published in 2004. See Pingré, "Voyage à Rodrigue."

68. Bernardin de Saint-Pierre, *Voyage à l'Île de France*.

69. Egerton, "Changing Concepts in the Balance of Nature."

70. Bernardin de Saint-Pierre, *Études de la nature*.

71. Bernardin de Saint-Pierre, *Studies of Nature*, 1:244.

72. On the "enemy release" hypothesis as the explanation for the explosive growth of non-native populations, see Keane and Crawley, "Exotic Plant Invasions and the Enemy Release Hypothesis."

73. Bernardin de Saint-Pierre, *Studies of Nature*, 2:249.

74. Grove, *Green Imperialism*.

75. "C'est un malheur que nous n'ayons pas icy plus d'oiseaux destructeurs de la gent insectière. Cette jsle seule offer le spectacle de grandes forêts sans aucun pique-bois. C'est là le grand ennemi des cariats, des fourmis, des petites et grosses chenilles. Quel service ne rendroit-on pas à cette colonie si on pouvoit y introduire des espèces de pies grièches, de dominicains, de tyrans, de gobe-mouches, de pics, d'arniers, et autres mangeurs d'insectes, qui n'attaquent jamais les grains; de petits faucons, de bouchers, d'oiseaux de nuits, pour balancer la multiplication des oisillons granivores; comme aussy de serpens innocens pour détruire les rats; il n'est pas jusqu'à des grenouilles qu'il ne fût utile d'apporter pour purger les eaux croupissantes de la prodigieuse quantité de larves de cousins qui y fourmille." Quoted in Cap, *Philibert Commerson*, 142, my translation.

76. Wood, *New England's Prospect*; Mack, "Plant Naturalizations and Invasions."

77. Josselyn, *New-Englands Rarities*.

78. Josselyn, *New-Englands Rarities*, 86, 130.

79. Mack, "Plant Naturalizations and Invasions."

80. On the black rat, see Godin, *Wild Mammals of New England*. On the honeybee, see Josselyn, *Account of Two Voyages to New England*.

81. Fulling, "Plant Life and the Law of Man."

82. These laws included remarkable enforcement provisions. Thus, for example, according to the Massachusetts statute, after giving a property owner three months to remove barberry, anyone could enter the property, remove the plants, and charge the owner for the labor. Landowners were also obliged to reimburse tenants for expenses and labor involved in removing the bushes, and tenants could withhold rent in the absence of reimbursement. Municipalities were also fined if they did not eradicate barberry bushes along roads. The Massachusetts law expired in 1764, but the Connecticut and Rhode Island laws persisted in modified form through statehood in the new nation of the United States into the early nineteenth century. As barberry spread west in the nineteenth century and reached the great wheat belt of the northern Great Plains, as did the rust, it was perhaps inevitable that barberry would again come under attack. Between 1917 and 1923, eleven Midwestern and Western states passed laws mandating destruction of barberry, and sometimes other plants, such as species of the related genus *Mahonia* (Fulling, "Plant Life and the Law of Man").

83. Darlington, *Memorials of John Bartram and Humphry Marshall*.

84. Wulf, *Brother Gardeners*.

85. P. Meyer, "John Bartram: Botanist & Horticulturist"; Egerton, "History of the Ecological Sciences, Part 25."

86. Reichard and White, "Horticulture as a Pathway of Invasive Plant Introductions."

87. On Scotch broom, see Kalm, *Peter Kalm's Travels in North America*.

88. Darlington, *Memorials of John Bartram and Humphry Marshall*, 381–82; D. Nowak and Rowntree, "History and Range of Norway Maple."

89. Darlington, *Memorials of John Bartram and Humphry Marshall*, 342–43.

90. On invasiveness of non-native bullfrogs, see Kraus, *Alien Reptiles and Amphibians*. Bartram undoubtedly introduced many plants to Great Britain, probably about three hundred species, primarily through shipments to Collinson and to Philip Miller (P. Meyer, "John Bartram: Botanist & Horticulturist"). No systematic analysis of which of these became invasive has been published.

91. Darlington, *Memorials of John Bartram and Humphry Marshall*, 383.

92. Kalm, *Peter Kalm's Travels in North America*.

93. Kalm, *Peter Kalm's Travels in North America*, 1:152.

94. On the pea weevil, see Kergoat et al., "Phylogeny and Host-Specificity of European Seed Beetles." In 1751, Kalm published a small pamphlet in Swedish summarizing his observations on American plants that he felt merited introduction to Sweden and offering to provide seeds for these (Larsen, "Peter Kalm's Short Account"). I have no idea what plantings this publication triggered, but several of the species on offer subsequently became invasive in Europe, including common milkweed, black cherry, honey locust, and black locust.

95. Dean, *A Botânica e a Política Imperial*; Dean, *A Ferro e Fogo*.

96. On the seventeenth-century introduction of jackfruit, see Dean, *A Botânica e a Política Imperial*. On its twentieth-century invasion, see Freitas et al., "Invasion Impact of *Artocarpus heterophyllus* Lam. (Moraceae)"; Assunção et al., "Invasion of *Artocarpus heterophyllus*, Jackfruit."

97. Diogo et al., "Portuguese Naturalist Correia da Serra."

98. Geographers James J. Parsons and Hilgard O. Sternberg traced this history and consequences of the massive introduction of African grasses to South America, as will be discussed in detail in chapter 8.

99. D'Antonio and Vitousek, "Biological Invasions by Exotic Grasses"; Da Silva Alves et al., "Biological Invasion by *Cenchrus ciliaris* L."

100. Egerton, *Roots of Ecology*.

Chapter 3

1. Egerton, *Roots of Ecology*.

2. A. Price, *Explorations of Captain James Cook*.

3. The "when" and "how" parts of biogeography are still being debated (e.g., de Queiroz, *Monkey's Voyage*), especially in light of the recognition of continental drift and the ability to date taxonomic events by examining DNA of current species or recent fossils. These developments have led to continuing arguments about the roles of dispersal and continental drift (or other geological events) in establishing current distributions of particular species and to statistical methods to attempt to resolve disagreements (e.g., Matzke, "Model Selection in Historical Biogeography").

4. See De Candolle, *Essai élémentaire de géographie botanique*; A. Wallace, *Geographical Distribution of Animals*. For antecedents, see Nelson, "From Candolle to Croizat."

5. von Humboldt and Bonpland, *Essai sur la géographie des plantes*.

6. Buffon, *Histoire naturelle générale et particulière*.

7. von Humboldt, "Sur les lois que l'on observe dans la distribution des formes végétales"; Latreille, "Introduction à la géographie générale des arachnides et des insectes." On "Buffon's law" and the definitive comprehensive statements formulated by de Candolle and Wallace, see Nelson, "From Candolle to Croizat."

8. Buffon, *Histoire naturelle générale et particulière*; Fabricius, *Philosophia Entomologica*; Latreille, "Introduction à la géographie générale des arachnides et des insectes"; Prichard, *Researches into the Physical History of Mankind*.

9. De Candolle, *Essai élémentaire de géographie botanique*; A. Wallace, *Geographical Distribution of Animals*; Sclater, "On the General Geographical Distribution of the Members of the Class Aves."

10. "Quelques plantes, qui font l'objet du jardinage et de l'agriculture depuis les temps les plus reculés, ont accompagné l'homme d'un bout du globe à l'autre" (von Humboldt and Bonpland, *Essai sur la géographie des plantes*, 25). On von Humboldt's explorations, see Botting, *Humboldt and the Cosmos*.

11. Von Haller quoted in W. Watson, "Account of a Treatise," 377.

12. Schweinitz, "Remarks on the Plants of Europe Which Have Become Naturalized."

13. Chew, "Ending with Elton."

14. Willdenow, *The Principles of Botany and of Vegetable Physiology*, 389–90.

15. Lyell, *Principles of Geology*.

16. Lyell, *Principles of Geology*, 83. On this point, Lyell cites Linnaeus, "Essay on the Habitable Earth."

17. Bentham, "De Candolle's 'Geographical Botany.'"

18. Darwin, *Journal of Researches* [. . .] *from 1832 to 1836*, 138.

19. Darwin, *Journal of Researches* [. . .] *from 1832 to 1836*, 172.

20. Darwin, *Journal of Researches* [. . .] *from 1832 to 1836*, 460.

21. Darwin, *Journal of Researches* [. . .] *Under the Command of Capt. Fitz Roy*, 378. In fact, the rat on James Island (now Santiago) is truly an endemic Galápagos rice rat,

Nesoryzomys swarthi (Dowler et al., "Rediscovery of Rodents"). This fact does not negate Darwin's attempt to infer evolution from what he suspected to be a recently introduced species. Of four endemic Galápagos rodent species, only *Nesoryzomys swarthi* coexists with non-native ship rats, house mice, and cats. The other three species persist only on Fernandina and Santa Fe, which lack these non-native competitors and predators. Two other endemic species are believed to be extinct, likely owing to these introductions. One, *Oryzomys galapagoensis,* existed on San Cristóbal (formerly Chatham) and was the mouse referred to by Darwin and described by Waterhouse as *Mus galapagoensis* (Dowler et al., "Rediscovery of Rodents (Genus *Nesoryzomys*) Considered Extinct in the Galápagos Islands").

22. Darwin, *Journal of Researches* [. . .] *from 1832 to 1836*, 580.

23. See Beatson, *Tracts Relative to the Island of St. Helena*, 1–2.

24. Beatson, *Tracts Relative to the Island of St. Helena*, 295–326.

25. Journalist Fred Pearce argued that the largely non-native plant community on Green Mountain shows that groups of species introduced from diverse places with no history of coevolution can constitute a biodiverse ecosystem functioning similarly to highly coevolved native communities (Pearce, "On a Remote Island"). This notion traces to a suggestion by evolutionary ecologist Daniel H. Janzen that "ecological fitting" among species newly brought together can produce communities difficult to distinguish from coevolved communities and that it would be important for ecologists to study how they may differ (Janzen, "On Ecological Fitting"). Pearce's description of the Green Mountain ecosystem contains many inaccuracies (Simberloff and Strong, "Counterpoint: Scientists Offer a Dissenting View on Ascension Island"), but Pearce and others occasionally cite "ecological fitting" in support of the controversial notion that "novel ecosystems" including non-native species are suitable goals for ecological restoration (Pearce, "On a Remote Island").

26. Darwin, *Journal of Researches* [. . .] *Under the Command of Capt. Fitz Roy*, 492–93.

27. Darwin, *On the Origin of Species*.

28. Darwin, *On the Origin of Species*, 337.

29. Darwin, *On the Origin of Species*, 337.

30. Darwin, *On the Origin of Species*, 115, in reference to de Candolle, *Géographie botanique raisonnée*.

31. Daehler, "Darwin's Naturalization Hypothesis Revisited."

32. A. Wallace, *Island Life*.

33. A. Wallace, *Island Life*, 286.

34. For the 1639 observations, see von Mandelslo, *Voyages and Travels of Ambassadors*.

35. A. Wallace, *Island Life*, 252.

36. Chew, "Ending with Elton"; Chew and Hamilton, "Rise and Fall of Biotic Nativeness."

37. Godman, *Natural History of the Azores*.

38. A. Wallace, *Darwinism*.

39. A. Wallace, *Darwinism*, 15–34.

40. Shurin, "Top-Down and Bottom-Up Regulation of Communities."

41. The lecture was published as "Insular Floras" in an 1867 issue of *Gardeners' Chronicle*, reprinted in 1984 in Williamson, "Sir Joseph Hooker's Lecture on Insular Floras."

42. Williamson, "Sir Joseph Hooker's Lecture on Insular Floras," 63.

43. Hooker, *Botany of the Antarctic Voyage* [. . .] *Part III; Flora Tasmaniæ*, cv.

44. C. M. Smith, "Changed and Changing Vegetation."

45. Beatson, *Papers Relating to the Devastation by Goats on the Island of St. Helena*, 12–20.

46. Williamson, "Sir Joseph Hooker's Lecture on Insular Floras," 67.

47. Chew and Hamilton, "Rise and Fall of Biotic Nativeness"; J. Henslow, "Observations Concerning the Indigenousness and Distinctness of Certain Species."

48. J. Henslow, "Observations Concerning the Indigenousness and Distinctness of Certain Species," 3. For the referent Watson publications, see H. Watson, "Outlines of the Geographical Distribution of British Plants"; H. Watson, "On the Altitude of the Habitats of Plants in Cumberland."

49. Egerton, "Hewett C. Watson"; Winch, *Essay on the Geographical Distribution of Plants*, 25–36.

50. Winch, *Essay on the Geographical Distribution of Plants*, 25–36.

51. J. Henslow, "Observations Concerning the Indigenousness and Distinctness of Certain Species," 85, in reference to early papers by Watson.

52. H. Watson, *Cybele Brittanica*, 63–64.

53. Carlton, "Biological Invasions and Cryptogenic Species."

54. G. Marsh, *Man and Nature; or, Physical Geography as Modified by Human Action*; G. Marsh, *Earth as Modified by Human Action*; G. Marsh, *Man and Nature.*

55. Leopold, *Sand County Almanac*; Carson, *Silent Spring.*

56. Lowenthal, *George Perkins Marsh, Prophet of Conservation*; Lowenthal, foreword to *Man and Nature.*

57. Lowenthal, foreword to *Man and Nature.*

58. G. Marsh, *Man and Nature; or, Physical Geography as Modified by Human Action*, 57.

59. G. Marsh, *Man and Nature; or, Physical Geography as Modified by Human Action*, 58.

60. G. Marsh, *Man and Nature; or, Physical Geography as Modified by Human Action*, 59.

61. G. Marsh, *Camel*; G. Marsh, *Report* [. . .] *on the Artificial Propagation of Fish.*

62. Darwin, *Journal of Researches* [. . .] *from 1832 to 1836.*

63. Simberloff and Gibbons, "Now You See Them, Now You Don't!"

64. G. Marsh, *Man and Nature; or, Physical Geography as Modified by Human Action*, 105.

65. DeBach and Rosen, *Biological Control by Natural Enemies.*

66. G. Marsh, *Man and Nature; or, Physical Geography as Modified by Human Action*, 520–21.

67. Por, *Lessepsian Migration.*

68. Galil and Zenetos, "Sea Change"; Golani and Appelbaum-Golani, *Fish Invasions of the Mediterranean Sea.*

69. G. Marsh, *Man and Nature; or, Physical Geography as Modified by Human Action*, 441.

70. G. Marsh, *Earth as Modified by Human Action*, 87.

71. G. Marsh, *Man and Nature; or, Physical Geography as Modified by Human Action.*

72. G. Marsh, *Man and Nature; or, Physical Geography as Modified by Human Action*, 111–12. In the late nineteenth and early twentieth centuries, the house sparrow was widely reviled in the United States as a non-native pest (Coates, *Strangers on the Land*).

73. G. Marsh, *Man and Nature; or, Physical Geography as Modified by Human Action*, 104.

74. Lowenthal, *George Perkins Marsh, Prophet of Conservation*, 174.

75. Lowenthal, "George Perkins Marsh and the American Geographical Tradition," 212.

76. For Marsh's influence on Pinchot, see Lowenthal, *Forest Stewardship*. On "wise use," see Pinchot, *Breaking New Ground*, 505. The term "wise use" has become the watchword in the United States of advocates of unlimited private property rights and reduced regulation of public property (cf. Helvarg, *War Against the Greens*).

77. Pinchot, *Fight for Conservation*, 43.

78. A. Wallace, *Island Life*; A. Wallace, *Darwinism*; G. Marsh, *Man and Nature; or, Physical Geography as Modified by Human Action*, 117.

79. Thomson, *Naturalisation of Animals and Plants in New Zealand*; Elton, *Animal Ecology*.

80. Simberloff, "Integrity, Stability, and Beauty."

81. Elton, *Ecology of Invasions by Animals and Plants*.

82. W. Thomas, *Man's Role in Changing the Face of the Earth*; E. Anderson, "Man as a Maker of New Plants"; M. Bates, "Man as an Agent in the Spread of Organisms"; Clark, "Impact of Exotic Invasions."

83. Ludsin and Wolfe, "Biological Invasion Theory."

Chapter 4

1. For the first plant ecology book, see Cowles, "Ecological Relations of the Vegetation on the Sand Dunes of Lake Michigan." For early twentieth-century animal ecology works, see Shelford, *Animal Communities in Temperate America*; Elton, *Animal Ecology*. On the incipient science of ecology, see Egerton, *Roots of Ecology*.

2. Egerton, "History of Ecological Sciences, Part 44."

3. Flux, "World Distribution."

4. H. Thompson and King, *European Rabbit*.

5. Rolls, *They All Ran Wild: The Story of Pests on the Land in Australia*; K. Myers et al., "Rabbit in Australia."

6. Rolls, *They All Ran Wild: The Story of Pests on the Land in Australia*; K. Myers et al., "Rabbit in Australia."

7. J. M. Alves and colleagues employed molecular genetic techniques to show that, despite the other introductions of rabbits, including that by the First Fleet, rabbits in Australia today are overwhelmingly descended from those few wild individuals that Austin released (Alves et al., "Single Introduction of Wild Rabbits"). See also chapter 11, box 11.1.

8. Rolls, *They All Ran Wild: The Story of Pests on the Land in Australia*; Harley, "Rabbits and the Rise of Australian Nativism."

9. K. Myers et al., "Rabbit in Australia."

10. Rolls, *They All Ran Wild: The Story of Pests on the Land in Australia*.

11. Rolls, *They All Ran Wild: The Story of Pests on the Land in Australia*.

12. Rolls, *They All Ran Wild: The Story of Pests on the Land in Australia*.

13. Risbey et al., "Impact of Cats and Foxes"; K. Myers et al., "Rabbit in Australia."

14. C. King, *Immigrant Killers*.

15. B. Cooke, *Australia's War Against Rabbits*.

16. [J. S. G.], "Rabbit in Australia."

17. Rodier, *Rabbit Pest in Australia*; Rolls, *They All Ran Wild: The Story of Pests on the Land in Australia*.

18. On Rodier's endorsement, see Sclater, "Rabbit Pest." On the impracticality of Rodier's plan, see Rolls, *They All Ran Wild: The Story of Pests on the Land in Australia*.

19. Rolls, *They All Ran Wild: The Story of Pests on the Land in Australia.*

20. Rolls, *They All Ran Wild: The Story of Pests on the Land in Australia*; Fenner and Ross, "Myxomatosis."

21. Fenner and Ross, "Myxomatosis"; Fenner and Fantini, *Biological Control of Vertebrate Pests*; B. Cooke, *Australia's War Against Rabbits.*

22. Fenner and Fantini, *Biological Control of Vertebrate Pests*; B. Cooke, *Australia's War Against Rabbits.*

23. Sobey and Conolly, "Myxomatosis."

24. Sobey and Conolly, "Myxomatosis"; Fenner and Ross, "Myxomatosis"; Fenner and Fantini, *Biological Control of Vertebrate Pests.*

25. Rolls, *They All Ran Wild: The Story of Pests on the Land in Australia.*

26. Sobey and Conolly, "Myxomatosis."

27. B. Cooke, *Australia's War Against Rabbits.*

28. B. Cooke, *Australia's War Against Rabbits.*

29. Stevenson, "Diffusion of Disaster"; Pouget, *Histoire de la lutte contre le phylloxéra.*

30. Pouget, *Histoire de la lutte contre le phylloxéra.*

31. Pouget, *Histoire de la lutte contre le phylloxéra.*

32. Goodell et al., "Biological Impacts of Species Invasions."

33. Granett et al., "Biology and Management of Grape Phylloxera."

34. Pouget, *Histoire de la lutte contre le phylloxéra.*

35. J. Sanders, "Save Us from Invading Pests."

36. Gossard, *Cottony Cushion Scale*; DeBach and Rosen, *Biological Control by Natural Enemies.*

37. Gossard, *Cottony Cushion Scale.*

38. Caltagirone and Doutt, "History of the Vedalia Beetle Importation."

39. Caltagirone and Doutt, "History of the Vedalia Beetle Importation," 3.

40. DeBach and Rosen, *Biological Control by Natural Enemies*, 142.

41. Gossard, *The Cottony Cushion Scale*; Caltagirone and Doutt, "History of the Vedalia Beetle Importation"; DeBach and Rosen, *Biological Control by Natural Enemies.*

42. DeBach and Rosen, *Biological Control by Natural Enemies.*

43. Caltagirone and Doutt, "History of the Vedalia Beetle Importation."

44. On the vedalia beetles attacking or competing with nontarget insects, see Pell et al., "Intraguild Predation Involving *Harmonia axyridis.*" On the beetles massing in people's homes, see Koch and Galvan, "Bad Side of a Good Beetle." On their fouling wine, see Sloggett et al., "Chemical Ecology of *Harmonia axyridis.*"

45. On the flies' arrival by another route, see DeBach and Rosen, *Biological Control by Natural Enemies.*

46. DeBach and Rosen, *Biological Control by Natural Enemies.*

47. Caltagirone and Doutt, "History of the Vedalia Beetle Importation."

48. Gossard, *Cottony Cushion Scale.*

49. Stansly and Rogers, "Soft-Bodied Insects Attacking Foliage and Fruit"; Hamon and Fosulo, "Cottony Cushion Scale."

50. Hoddle et al., "Post Release Evaluation of *Rodolia cardinalis.*"

51. Forbush and Fernald, *Gypsy Moth.*

52. Forbush and Fernald, *Gypsy Moth*; T. Dunlap, "Gypsy Moth"; Spear, *Great Gypsy Moth War.*

53. Dreistadt and Weber, "Gypsy Moth in the Northeast."

54. Spear, *Great Gypsy Moth War*; T. Dunlap, "Gypsy Moth."

55. Rutgers School of Environmental and Biological Sciences, Department of Entomology, "Gypsy Moth."

56. Liebhold, Halverson, et al., "Quantitative Analysis of the Invasion of Gypsy Moth."

57. R. Morin et al., *Mapping Host-Species Abundance.*

58. Sharov et al., "'Slow the Spread.'"

59. Hajek and Tobin, "Micro-Managing Arthropod Invasions."

60. Liebhold, Elkinton, et al., "What Causes Outbreaks of the Gypsy Moth in North America?"

61. Liebhold and McManus, "Evolving Use of Insecticides."

62. Howard and Fiske, *Importation into the United States of the Parasites of the Gipsy Moth*; Hoy, "Establishment of Gypsy Moth Parasitoids in North America"; T. Dunlap, "Gypsy Moth."

63. Elkinton and Liebhold, "Population Dynamics of Gypsy Moth in North America."

64. Hajek, Humber, and Elkinton, "Mysterious Origins of *Entomophaga maimaiga*"; Hajek et al., "Allozyme and Restriction Fragment Length Polymorphism Analyses."

65. Hoy, "Establishment of Gypsy Moth Parasitoids in North America"; Elkinton and Liebhold, "Population Dynamics of Gypsy Moth in North America"; DeBach and Rosen, *Biological Control by Natural Enemies.*

66. Elkinton et al., "Implicating an Introduced Generalist Parasitoid." Beginning in 1989, the browntail moth underwent a series of outbreaks in Maine, and the population increased dramatically. It is suspected that this rather abrupt change after decades of absence except on two small islands resulted from ongoing climatic changes (Boyd et al., "Factors Influencing the Population Fluctuations of *Euproctis chrysorrhoea*"). Also, the parasitic fly *Compsilura concinnata* has become the major source of mortality for giant silk moths of conservation concern in the northeastern United States (Elkinton and Boettner, "Benefits and Harm Caused by the Introduced Generalist Tachinid").

67. Hoagland et al., "Biogeography and Population Biology of the Mongoose in the West Indies."

68. Espeut, "On the Acclimatization of the Indian Mungoos in Jamaica."

69. Hoagland et al., "Biogeography and Population Biology of the Mongoose in the West Indies."

70. Bryan, "Much Maligned Mongoose."

71. Simberloff, Dayan, et al., "Character Displacement and Release in the Small Indian Mongoose"; Hays, "Small Indian Mongoose"; Barun, Simberloff, et al., "Possible Character Displacement of an Introduced Mongoose."

72. Nellis and Everard, "Biology of the Mongoose in the Caribbean."

73. Feilden, "Notes on the Terrestrial Mammals of Barbados."

74. Urich, "Mongoose in Trinidad and Methods of Destroying It."

75. G. M. Allen, "Mammals of the West Indies"; Burdon, *Handbook of St. Kitts–Nevis.*

76. Macmillan, *Mauritius Illustrated.*

77. Carié, *L'acclimatation à l'Île Maurice.*

78. Bryan, "Much Maligned Mongoose."

79. "Mongoose," *Proceedings of the Agricultural Society of Trinidad and Tobago.*

80. Nellis and Everard, "Biology of the Mongoose in the Caribbean."

81. J. G. Myers, "Mongoose: The Present Position"; Urich, "Mongoose in Trinidad."

82. Carié, *L'acclimatation à l'Île Maurice.*

83. On Puerto Rico and the efforts to find an effective poison and suitable bait, see Pimentel, "Control of the Mongoose in Puerto Rico." On Grenada, see Nellis and

Everard, "Biology of the Mongoose in the Caribbean." On recent research and management, see Barun, Campbell, et al., "Review of Small Indian Mongoose Management and Eradications on Islands."

84. Watari et al., "Final Stage of the Mongoose-Eradication Project"; "Japan Close to Eradicating Mongoose"; Ministry of the Environment, Government of Japan, "Declaration of the Eradication of the Small Indian Mongoose."

85. Schmitz et al., "Exotic Aquatic Plants in Florida"; Schmitz, "Historical Introductions of Water Hyacinth and Hydrilla."

86. Penfound and Earle, "Biology of the Water Hyacinth."

87. Schmitz et al., "Exotic Aquatic Plants in Florida."

88. Buker, "Engineers vs. Florida's Green Menace."

89. Schmitz et al., "Exotic Aquatic Plants in Florida."

90. Penfound and Earle, "Biology of the Water Hyacinth."

91. Schardt, "Maintenance Control."

92. Schardt, "Maintenance Control."

93. Buker, "Engineers vs. Florida's Green Menace."

94. Buker, "Engineers vs. Florida's Green Menace."

95. Schardt, "Maintenance Control."

96. Buker, "Engineers vs. Florida's Green Menace"; Schardt, "Maintenance Control."

97. Schardt, "Maintenance Control."

98. Buker, "Engineers vs. Florida's Green Menace"; Schardt, "Maintenance Control."

99. Schardt, "Maintenance Control."

100. Center, Frank, et al., "Biological Control."

101. A. Williams et al., "Water Hyacinth in Lake Victoria"; J. Wilson, Ajuonu, et al., "Decline of Water Hyacinth on Lake Victoria."

102. Center, Dray, et al., "Biological Control of Water Hyacinth."

103. Hummer, "History of the Origin and Dispersal of White Pine Blister Rust."

104. Maloy, "White Pine Blister Rust."

105. Benedict, *History of White Pine Blister Rust Control.*

106. Maloy, "White Pine Blister Rust Control in North America."

107. Maloy, "White Pine Blister Rust Control in North America"; Hummer, "History of the Origin and Dispersal of White Pine Blister Rust."

108. Benedict, *History of White Pine Blister Rust Control.*

109. Maloy, "White Pine Blister Rust Control in North America."

110. Benedict, *History of White Pine Blister Rust Control.*

111. Maloy, "White Pine Blister Rust Control in North America."

112. Benedict, *History of White Pine Blister Rust Control*; Maloy, "White Pine Blister Rust Control in North America."

113. Maloy, "White Pine Blister Rust Control in North America."

114. Maloy, "White Pine Blister Rust Control in North America."

115. Maloy, "White Pine Blister Rust Control in North America."

116. P. Peterson, "Barberry Eradication Program in Minnesota."

117. Moss, "Antibiotics for Control of Blister Rust"; Maloy, "White Pine Blister Rust Control in North America."

118. E.g., McDonald et al., "Breeding Rust-Resistant Five-Needle Pines."

119. Maloy, "White Pine Blister Rust Control in North America."

120. Freinkel, *American Chestnut.*

121. Murrill, "New Chestnut Disease."

122. I. Williams, "New Chestnut Bark Disease"; Freinkel, *American Chestnut.*

123. Freinkel, *American Chestnut.*

124. Freinkel, *American Chestnut.*

125. Anagnostakis, "Chestnut Blight."

126. Dutech et al., "Chestnut Blight Fungus World Tour."

127. Robin and Heiniger, "Chestnut Blight in Europe"; Milgroom and Cortesi, "Biological Control of Chestnut Blight with Hypovirulence"; Anagnostakis, "Chestnut Blight"; Dutech et al., "Chestnut Blight Fungus World Tour."

128. Dutech et al., "Chestnut Blight Fungus World Tour."

129. D. Zhang and Nuss, "Engineering Super Mycovirus Donor Strains of Chestnut Blight Fungus."

130. A. Graves, "Future of the Chestnut Tree"; Freinkel, *American Chestnut.*

131. Diskin et al., "Recovery of American Chestnut Characteristics."

132. Newhouse et al., "Transgenic American Chestnuts Show Enhanced Blight Resistance"; Grandoni, "Genetic Engineering Was Meant to Save Chestnut Trees."

133. Ricker, "Effects of Compensatory Mortality on Populations."

134. Nicholson, "Balance of Animal Populations."

135. Fenner and Fantini, *Biological Control of Vertebrate Pests.*

136. DeBach and Rosen, *Biological Control by Natural Enemies.*

137. Buker, "Engineers vs. Florida's Green Menace."

138. On the cane toad in Australia, see Lever, *Cane Toad.* On mustelids in New Zealand, see C. King, *Immigrant Killers.* On the myna in Hawaii, see Funasaki et al., "Review of Biological Control Introductions in Hawaii."

139. R. D. Van Driesche, Pratt, et al., "Cases of Biological Control Restoring Natural Systems."

140. Howard and Fiske, *Importation into the United States of the Parasites of the Gipsy Moth.*

141. DeBach and Rosen, *Biological Control by Natural Enemies.*

142. Howard and Fiske, *Importation into the United States of the Parasites of the Gipsy Moth.*

Chapter 5

1. Jenkins, *Noah's Ark Syndrome*; Lever, *They Dined on Eland*; McDowall, *Gamekeepers for the Nation.*

2. Lever, *They Dined on Eland.*

3. On the Acclimatisation Society of the United Kingdom, see Lever, *They Dined on Eland.* On the acclimatization societies of Australia, see Jenkins, *Noah's Ark Syndrome*; Lever, *They Dined on Eland.*

4. McDowall, *Gamekeepers for the Nation.*

5. Lever, *They Dined on Eland.*

6. Lever, *They Dined on Eland.*

7. Simberloff and Boecklen, "Patterns of Extinction in the Introduced Hawaiian Avifauna"; Lewis, *Belonging on an Island.*

8. Lever, *They Dined on Eland.*

9. Lampman, *Coming of the Pond Fishes.*

10. Rolls, *They All Ran Wild: The Story of Pests on the Land in Australia*, 230.

11. Fendley, "McCoy, Sir Frederick."

12. Millet, *Les Merveilles des fleuves et des ruisseaux.*

13. Rolls, *They All Ran Wild: The Story of Pests on the Land in Australia.*

14. On starlings introduced to North America, see Lever, *They Dined on Eland.* On wels catfish in the UK, see Hickley and Chare, "Fisheries for Non-Native Species

in England and Wales." On European rabbits in Australia, see Jenkins, *Noah's Ark Syndrome*. On house sparrows, starlings, and blackbirds in Australia, see Rolls, *They All Ran Wild: The Story of Pests on the Land in Australia*; Jenkins, *Noah's Ark Syndrome*. On possums, hedgehogs, house sparrows, owls, and mallards in New Zealand, see McDowall, *Gamekeepers for the Nation*.

15. C. King, *Immigrant Killers*; P. Wells, "'Enemy of the Rabbit.'"

16. See, e.g., C. King, *Immigrant Killers*, and McDowall, *Gamekeepers for the Nation*, for New Zealand.

17. Rolls, *They All Ran Wild: The Story of Pests on the Land in Australia*.

18. Rolls, *They All Ran Wild: The Story of Pests on the Land in Australia*, 227.

19. "Ornithology of New Zealand," 219.

20. P. Wells, "'Enemy of the Rabbit.'"

21. C. King, *Immigrant Killers*; P. Wells, "'Enemy of the Rabbit.'"

22. On legal releases, see P. Wells, "'Enemy of the Rabbit.'" On private releases, see C. King, *Immigrant Killers*. On the Rabbit Nuisance Act, see P. Wells, "'Enemy of the Rabbit.'"

23. Druett, *Exotic Intruders*, 114.

24. Jenkins, *Noah's Ark Syndrome*; Lever, *They Dined on Eland*.

25. McDowall, *Gamekeepers for the Nation*.

26. A. Wallace, *Darwinism*.

27. McDowall, *Gamekeepers for the Nation*; D. Lewis, *Belonging on an Island*.

28. Rolls, *They All Ran Wild: The Story of Pests on the Land in Australia*; McDowall, *Gamekeepers for the Nation*.

29. Moffitt et al., "Historical Perspectives on Inland Fisheries Management." In 1903, the US Fish Commission, previously independent, became the United States Bureau of Fisheries, part of the new US Department of Commerce and Labor. In 1913, the Department of Commerce and Labor was divided into its two components, and the bureau became part of the Department of Commerce. In 1939, the Bureau of Fisheries was transferred to the US Department of the Interior (USDI), and in 1940, it was merged with the Bureau of Biological Survey of the USDI to form the US Fish and Wildlife Service within the USDI.

30. Hedgepeth, "Founders of Fish Culture"; Gooley, "Livingston Stone."

31. H. M. Smith, "Review of the History and Results of the Attempts to Acclimatize Fish." Description of a US Fish Commission aquarium car, from *Oregon State Journal*, February 3, 1883: "Young carp were shipped from Washington [DC] on the night of January 28, for Texas, New Mexico, Arizona, California, Nevada, Washington, Oregon, Idaho, and Utah. The shipment goes a large part of the way over the Pennsylvania Railroad, which has established special rates for the purpose. It is attached to one fast train. The car is especially fitted up for the purpose of carrying young fish of all kinds and is provided with refrigerating apparatus of most improved construction. The greater part of the weight is placed at the bottom of the car so as to avoid top-heaviness. Five messengers accompany the shipment as the journey is very fatiguing, and it is necessary that two be on hand night and day to watch the fish. The present equipment is somewhat of an experiment, as there were certain new features in this car that have never been tried. If these fish get through in safety, German [common] carp will be colonized in every state and territory of the Union" (quoted in Lampman, *Coming of the Pond Fishes*, 30–31).

32. H. M. Smith, "Review of the History and Results of the Attempts to Acclimatize Fish." Two of the Old World species—tench and brown trout—had been brought from Europe by the US Fish Commission (Fuller et al., *Nonindigenous Fishes*).

33. The formal name of the California Fish Commission was the California Board of Fish Commissioners, established in 1870. In 1909 this board became part of the new California Fish and Game Commission.

34. In addition to surviving foreign fish listed in the text, the US Fish Commission introduced several European fish that did not establish persistent populations, such as the powan to Michigan in 1877 and the vendace to Maine in the 1880s (Fuller et al., *Nonindigenous Fishes*).

35. Vitule, "Introdução de peixes em ecossistemas continentais brasileiros." The Bureau of Fisheries, successor to the Fish Commission, continued to be active in aiding intercontinental introductions. For instance, in 1904, the bureau shipped eggs of Atlantic salmon, brook trout, lake trout, and lake whitefish to Argentina (Tulian, "Acclimatization of American Fishes in Argentina"), and from 1901 through 1907, it shipped eggs of chinook salmon and lake whitefish to New Zealand (Ayson, "Introduction of American Fishes into New Zealand").

36. H. M. Smith, "Review of the History and Results of the Attempts to Acclimatize Fish."

37. Fuller et al., *Nonindigenous Fishes.*

38. Problems with common carp foretold in H. M. Smith, "Review of the History and Results of the Attempts to Acclimatize Fish."

39. H. M. Smith, "Review of the History and Results of the Attempts to Acclimatize Fish."

40. Lampman, *Coming of the Pond Fishes.*

41. Courtenay and Robins, "Fish Introductions."

42. Lachner et al., "Exotic Fishes and Other Aquatic Organisms Introduced into North America."

43. On ohrid trout, see Fuller et al., *Nonindigenous Fishes.* On black carp, see Nico et al., *Black Carp.*

44. Sellars, *Preserving Nature in the National Parks*; Simberloff, "Biological Invasions in the National Parks."

45. Sellars, *Preserving Nature in the National Parks.*

46. National Park Service, *Administrative Policies for Natural Areas*, 17, 56. For Hubbs's objections, see Hubbs, "Fishes from the Big Bend Region of Texas"; Hubbs and Wallis, "Native Fish Fauna of Yosemite National Park"; Hubbs and Lagler, "Fishes of Isle Royale, Lake Superior, Michigan."

47. Courtenay and Robins, "Fish Introductions."

48. Van Oosten, "Dispersal of Smelt"; Fuller et al., *Nonindigenous Fishes.*

49. Fuller et al., *Nonindigenous Fishes.*

50. Courtenay and Robins, "Fish Introductions"; Kurdila, "Introduction of Exotic Species into the United States."

51. Courtenay and Robins, "Fish Introductions."

52. Courtenay, "Biological Pollution Through Fish Introductions."

53. Lever, *Naturalized Fishes of the World*; Fuller et al., *Nonindigenous Fishes.*

54. C. Mitchell, "Liberator."

55. Dolin, *Snakehead.*

56. "Pike Reappear"; Elmendorf et al., *Fear and Fishing in Lake Davis.*

57. The entity often called US Biological Survey began in the United States Department of Agriculture in 1885, originally called the Section of Ornithology. In 1886, the name was changed to Division of Economic Ornithology and Mammalogy; in 1896, it became the Division of Biological Survey; and in 1905, it became the Bureau of Biological Survey. In 1939, it was transferred to the Department of the Interior, where,

in 1940, it was merged with the Bureau of Fisheries to create the United States Fish and Wildlife Service (A. Gardner, "United States Biological Survey"). Throughout this chapter it will be noted as "US Biological Survey."

58. A. Gardner, "United States Biological Survey."

59. J. C. Phillips, *Wild Birds Introduced or Transplanted in North America.*

60. The non-native birds introduced to Sapelo Island included great tinamou, great curassow, plain chachalaca, and ocellated turkey.

61. Chalmers, *Ecology of Sapelo Island.*

62. J. C. Phillips, *Wild Birds Introduced or Transplanted in North America.*

63. McAtee, *Game Birds Suitable for Naturalizing in the United States.*

64. On Palmer's criticism, see Palmer, "Danger of Introducing Noxious Animals and Birds." On permitted introductions, see J. C. Phillips, *Wild Birds Introduced or Transplanted in North America.*

65. McAtee, *Game Birds Suitable for Naturalizing in the United States*, 3.

66. McAtee, "Introduction upon Introduction," 160.

67. United States Fish and Wildlife Service, "Dr. Gardiner Bump Joins Research Staff of FWS."

68. Bump, "Introduction and Transplantation of Game Birds and Mammals into the State of New York."

69. Bump's publication was part of a symposium on non-native game species held in 1940 at the Fifth North American Wildlife Conference. Of nine presentations, only that by O. J. Murie on game introduction in Alaska raised any concern about ecological impacts (Murie, "Wildlife Introductions in Alaska"). The majority focused on how to reduce the high rate of failure or attempted to explain the reasons for observed successful and failed introductions. Lawrence E. Hicks of the Ohio Cooperative Wildlife Management Research Unit quite aggressively supported game introductions, subject only to precautions against inadvertently introducing disease. Referring to the Ohio Valley, he averred that "it is inconceivable that any introduced game species could become numerous enough to acquire pest status, since our enormous hunting pressures guarantee adequate control" (Hicks, "Role of Exotics in the Ohio Valley," 422). Ironically, chairing the symposium was Theodore S. Palmer, retired from the US Biological Survey, who had so strongly urged great caution in introductions because of the high risk of unintended consequences. The conference transactions do not record discussions of the presentations.

70. United States Fish and Wildlife Service, "Dr. Gardiner Bump Joins Research Staff of FWS."

71. Bump, "Game Introductions," 316; R. T. King, "Is It Wise Policy to Introduce Exotic Game Birds?"

72. Bump, "Game Introductions," 321.

73. Bump, "Introduction and Transplantation of Game Birds and Mammals into the State of New York."

74. Levi, "Evaluation of Wildlife Importations."

75. De Vos et al., "Introduced Mammals and Their Influence on Native Biota."

76. Bump, "Foreign Game Investigation," 2.

77. Bump, "Foreign Game Investigation," 13.

78. Bump, "Foreign Game Investigation," 13.

79. E.g., Gullion, "Critique Concerning Foreign Game-Bird Introductions."

80. The species in Hawaii are black francolin, gray francolin, Erckel's francolin, chestnut-bellied sandgrouse, and kalij pheasant (Banks, *Summary of Foreign Game Bird Liberations*). The North American species are Himalayan snowcock and green pheasant

(Banks, *Summary of Foreign Game Bird Liberations*; Condon et al., "Red Junglefowl Introductions").

81. C. Guild, *Rainbow in Tahiti*; Peaker, "Tahiti—Introduced Birds"; Peaker, "Tahiti—Introduced Birds—More Information."

82. E. Guild, "Tahitian Aviculture," 8; E. Guild, "Western Bluebirds in Tahiti."

83. Peaker, "Tahiti—Introduced Birds"; Peaker, "Tahiti—Introduced Birds—More Information."

84. Spotswood et al., "Invasive Tree."

85. C. Guild, *Rainbow in Tahiti*.

86. Haydock, "Survey of the Birds of St. Helena Island."

87. Dora became vice president of the Hui Manu, the Hawaiian acclimatization society, and also introduced several non-native plants to Kauai (Soboleski, "Island History for Friday, November 13, 2009") as well as non-native wasps for biological control (Fullaway, "Report of Entomologist, November, 1924").

88. Soboleski, "Bird Lover, Amateur Ornithologist Alexander H. Isenberg."

89. Isenberg, "Transplanting Foreign Birds to the Hawaiian Islands."

90. T. Palmer, "Danger of Introducing Noxious Animals and Birds"; McAtee, "Introduction upon Introduction"; Grinnell, "Risks Incurred in the Introduction of Alien Game Birds."

91. On Isenberg's rejection of concerns, see Isenberg, "Exotic Birds Not Detrimental to Native Birds in the Hawaiian Islands." On his being taken to task, see H. Fisher, "Question of Avian Introductions in Hawaii."

92. Isenberg, "Transplanting Foreign Birds to the Hawaiian Islands."

93. Hawaii Invasive Species Council, "Red-Whiskered Bulbul."

94. Locey, "Introduced Game Birds of Hawaii," 28.

95. Locey, "Introduced Game Birds of Hawaii," 30.

96. Locey, "Introduced Game Birds of Hawaii," 6.

97. Fitter, *Ark in Our Midst*.

98. Lever, *Naturalized Mammals of the World*.

99. Bump, "Introduction and Transplantation of Game Birds and Mammals into the State of New York."

100. Bump, "Game Introductions," 320.

101. Thompson, "Wild European Rabbit," 8.

102. De Vos et al., "Introduced Mammals and Their Influence on Native Biota," 173, 177–78; cf. chapter 6.

103. Lever, *Naturalized Mammals of the World*.

104. Bump, "Foreign Game Investigation," 2.

105. Mungall and Sheffield, *Exotics on the Range*.

106. Boar (1930), fallow deer (1930), axis deer (1932), blackbuck antelope (1932), eland (1932), sambar deer (1932), sika deer (1932), red deer (1930s), barasingha (1939), mouflon (1946), Barbary sheep (1950), beisa oryx (1960).

107. Mungall and Sheffield, *Exotics on the Range*.

108. Mungall and Sheffield, *Exotics on the Range*.

109. Schreiner, "Uses of Exotic Animals in a Commercial Hunting Program"; Elbein, "Texotics."

110. Craighead and Dasmann. *Exotic Big Game on Public Lands*.

111. Mungall and Sheffield, *Exotics on the Range*; Edgington, "Safari of the Southwest."

112. D. Brown, *Bringing Back the Game*; D. Brown et al., "Charles Taylor Vorhies."

113. In particular, Hibben discovered human remains and artifacts in New Mexico reflecting a culture occupying the Sandia Cave twenty-five thousand years ago, far older than previously accepted human occupancy of the Americas, and flint arrowheads at Cook Inlet, Alaska, similar to those of the Folsom people who lived in New Mexico ten thousand years ago. Both of these claims were largely disputed by many archeologists, who charged Hibben with offenses ranging from sloppily mixing samples to "salting" sites with artifacts that were discovered later (D. Preston, "Mystery of Sandia Cave"; Dalton, "University Buildings Named on Shaky Ground"). However, recent dating of footprints at White Sands National Park, about 300 kilometers (186 miles) from the cave, places humans as present at least twenty-one thousand years ago (Pigati et al., "Independent Age Estimates").

114. For examples of his books, see Hibben, *Hunting in Africa*; Hibben, *Under the African Sun*.

115. D. Preston, "Mystery of Sandia Cave."

116. Edgington, "Safari of the Southwest."

117. On introduction of Barbary sheep, see Hibben, "Role of Exotic Animals in Game Management"; J. L. Long, *Introduced Mammals of the World*.

118. J. L. Long, *Introduced Mammals of the World*; Edgington, "Safari of the Southwest."

119. Rowley, "Graceful Gazelle Becomes a Pest."

120. Huey quoted in Edgington, "Safari of the Southwest." On his minimization of concerns, see W. Huey, "New New Mexicans."

121. Rowley, "Graceful Gazelle Becomes a Pest"; Andreoni et al., "Exotic Oryx Interact with Shrub Encroachment."

122. Edgington, "Safari of the Southwest."

123. Hibben, "Role of Exotic Animals in Game Management."

124. Hibben, "Role of Exotic Animals in Game Management." See also McAtee, *Game Birds Suitable for Naturalizing in the United States*; Bump, "Game Introductions."

125. Craighead and Dasmann, *Exotic Big Game on Public Lands*.

126. Hibben, "Role of Exotic Animals in Game Management."

127. Renault, "Texas Can't Quit the Aoudad."

128. On the ranchers importing non-native mammals, see, e.g., see Schreiner, "Uses of Exotic Animals in a Commercial Hunting Program."

129. D. Weiner, "Community Ecology in Stalin's Russia."

130. Medvedev, *Rise and Fall of T. D. Lysenko*; D. Weiner, *Little Corner of Freedom*.

131. D. Weiner, *Little Corner of Freedom*. Stanchinskii's treatment by Stalin is discussed further in chapter 6, section 6.4.

132. For Zhitkov's monograph, see Zhitkov, *Akklimatizatsiya zhivotnykh* [Acclimatization of animals]. On introductions of nutria, see Pavlov, "Results of the Acclimatization of Nutria in the U.S.S.R."; J. L. Long, *Introduced Mammals of the World*.

133. Khlyap et al., "Biological Invasions on Russian Territory: Mammals."

134. Yanushevich, *Acclimatization of Animals in the U.S.S.R.*; Khlyap et al., "Biological Invasions on Russian Territory: Mammals."

135. Rukovskii, "Acclimatization of the Teleut Squirrel."

136. Khlyap et al., "Biological Invasions on Russian Territory: Mammals." Introduction of non-native species to provide food for non-native and native species was not restricted to small mammals. For example, in the Soviet Union beginning in the 1930s, many aquatic invertebrate species were introduced to facilitate fishes in inland waters, such as those that had been introduced into reservoirs (Ioffe, "Present State and Prospects of the Acclimatization of Invertebrates").

137. Sludskii, "Theory and Practice of Acclimatization of Game Animals."

138. Kleinenberg quoted in D. Weiner, *Little Corner of Freedom*, 229.

139. Skalon quoted in D. Weiner, *Little Corner of Freedom*, 380.

140. Nasimovich, "Some General Problems and Results of Acclimatization of Terrestrial Animals," 967. For the critics he cited, see, e.g., Calahane, "Some Effects of Exotics on Nature"; M. Bates, "Man as an Agent in the Spread of Organisms"; Elton, *Ecology of Invasions by Animals and Plants*.

141. For his conference presentation, see Nasimovich, "Theoretical Basis of the Acclimatization of Land Mammals." On other mentions of possible harmful ecological impacts, see, e.g., Eliseev, "Results and Prospects of Acclimatization of Mammals and Birds in the RSFSR"; Elkin, "Results of Acclimatization of Game and Fur Animals in Mountain Altai"; Samusenko, "Acclimatization of Fur Animals in Belorussia."

142. E.g., Lavov, "Acclimatization of Mammals on Sakhalin Island"; Tatarinov, "Problems of Enrichment of the Carpathian Fauna."

143. Geptner, "Kakovy zhe puti obogashcheniia fauny." The name Geptner is sometimes transliterated as "Heptner."

144. Fitter, *Ark in Our Midst*.

145. Fitter, *Ark in Our Midst*; Lever, *Naturalized Animals of the British Isles*.

146. Despott, "Reptiles of the Maltese Islands"; Savona-Ventura, "Herpetofauna of the Maltese Islands."

147. Despott, "I nostri rettili."

148. Kraus, *Alien Reptiles and Amphibians*.

149. Kraus, *Alien Reptiles and Amphibians*.

150. Lever, *Naturalized Reptiles and Amphibians of the World*.

151. Lever, *Cane Toad*.

152. On the number of established populations, see Krysko et al., "New Verified Nonindigenous Amphibians and Reptiles in Florida." On the species mentioned, see Meshaka et al., *Exotic Amphibians and Reptiles of Florida*.

153. W. King and Krakauer, "Exotic Herpetofauna of Southern Florida"; L. Wilson and Porras, *Ecological Impact of Man on the South Florida Herpetofauna*; Meshaka et al., *Exotic Amphibians and Reptiles of Florida*.

154. Kraus, *Alien Reptiles and Amphibians*.

155. Kraus, "New Records of Alien Reptiles."

156. Lever, *Naturalized Reptiles and Amphibians of the World*.

157. Kraus, *Alien Reptiles and Amphibians*.

158. Kraus, *Alien Reptiles and Amphibians*.

159. Kraus, *Alien Reptiles and Amphibians*.

160. Lever, *Naturalized Reptiles and Amphibians of the World*; Kraus, *Alien Reptiles and Amphibians*.

161. Stuart, "Insularity and Quaternary Vertebrate Faunas in Britain and Ireland."

162. Frazer, "Introduced Species of Amphibians and Reptiles in Mainland Britain."

163. Frazer, "Introduced Species of Amphibians and Reptiles in Mainland Britain"; Kraus, *Alien Reptiles and Amphibians*.

164. Fitter, "Man's Addition to the British Fauna," 58.

165. Kirk, "Displacement of Species in New Zealand"; Thomson, *Naturalisation of Animals and Plants in New Zealand*.

166. Marlatt, *Danger of General Spread of the Gipsy and Brown-Tail Moths*; Jefferson and Fusonie, *Japanese Flowering Cherry Trees of Washington, D.C.*

167. For the 1941 articles, see Leopold, "Fifth Column of the Fencerow"; Leopold, "Cheat Takes Over." For the reprint of the latter, see Leopold, *Sand County Almanac*, 154–58.

168. Egler, "Indigene Versus Alien in the Development of arid Hawaiian Vegetation"; Egler, "Arid Southeast Oahu Vegetation"; Allan, "Indigene Versus Alien in the New Zealand Plant World."

169. Bates, "Man as an Agent in the Spread of Organisms."

170. Oosting, *Study of Plant Communities*; Oosting, *Study of Plant Communities*, 2nd ed.

171. Oosting, *Study of Plant Communities*; Oosting, *Study of Plant Communities*, 2nd ed.; Clark, *Invasion of New Zealand*; Clark, "Impact of Exotic Invasions"; Aubert de la Rüe, "Man's Influence on Tropical Vegetation."

172. Regarding the BBC broadcasts, see Elton, "Balance and Barrier."

173. Levi, "Evaluation of Wildlife Importations."

174. Calahane, "Some Effects of Exotics on Nature."

175. On Adams, see Griesbach, *150 Years of Research*. On agencies, see Klose, *America's Crop Heritage*.

176. In 1994, the Soil Conservation Service was renamed the Natural Resources Conservation Service.

177. Graham, *Natural Principles of Land Use*.

178. Graham, *Land and Wildlife*.

179. Amrine and Stasny, "Biocontrol of Multiflora Rose."

180. Griesbach, *150 Years of Research*.

181. A. Harris, *David Fairchild and America's Plant Explorers*. The Office of Foreign Seed and Plant Introduction became part of the Division of Botany, one of several research entities merged to form the Bureau of Plant Industry when the Division of Botany was established in 1901 within the United States Department of Agriculture. It was renamed the Office of Foreign Plant Introduction around 1924. In 1934, the Division of Botany, including the Office of Foreign Plant Introduction, was merged into a newly created Division of Plant Exploration and Introduction. After other various mergers, renamings, and reorganizations, the unit ended up as the Plant Introduction Section of the Horticultural Research Branch of the Agricultural Research Service, which was created in 1953 (National Archives, "Records of the Bureau of Plant Industry").

182. Kimmelman, "Progressive Era Discipline"; Pauly, "Beauty and Menace of the Japanese Cherry Trees."

183. Fairchild, *World Was My Garden*.

184. On the catalog, see Fairchild, "Annual Catalogue of Plant Immigrants." In urging speedy requests for seeds, Fairchild wrote: "Application for seeds or plants listed in Plant Immigrants may be sent at any time, however, and will be filed in the order of their receipt. When material is ready for distribution, these requests will be given first attention; if their number is sufficient to exhaust the available supply of a given species, it will not be included in the Annual Catalogue" (Fairchild, "Note").

185. "Note." Commercial nurseries complained bitterly about competition from both federal and state government nurseries, and their own importation of non-native plant species was greatly reduced by the federal Plant Quarantine Act of 1912, which authorized the USDA to inspect agricultural products, to organize border quarantines, and to restrict entry of infested agricultural goods, particularly Quarantine No. 37, effective in 1919 (Beattie, "Operation of Quarantine No. 37"; R. White, *Century of Service*). Individuals and amateur horticulture groups were restricted but not prohibited from importing limited quantities of plants, subject to inspection and quarantine.

186. "*Elaeagnus angustifolia*: Oleander."

187. G. Katz and Shafroth, "Biology, Ecology, and Management of *Elaeagnus angustifolia* L."

188. Fairchild, *World Was My Garden*, 164.

189. Winter et al., "Improved Pasture Plants"; A. Cameron, *Review of Pasture Plant Introduction.*

190. Winter et al., "Improved Pasture Plants."

191. Lonsdale, "Inviting Trouble."

192. Mott, "Planned Invasions of Australian Tropical Savannas."

Chapter 6

1. The entity known as the US Biological Survey went through a number of name changes between its inception in 1885 and today; for the details of those changes, see note 57 to chapter 5. Throughout this chapter it will be referred to as the US Biological Survey in all instances.

2. T. Palmer, "Danger of Introducing Noxious Animals and Birds."

3. Howard, "Danger of Importing Insect Pests."

4. Sailer, "Our Immigrant Insect Fauna"; Sailer, "History of Insect Introductions."

5. Howard, *Fighting the Insects*; L. Russell, "Leland Ossian Howard."

6. Marlatt, *Danger of General Spread of the Gipsy and Brown-Tail Moths.*

7. Marlatt, "Losses Caused by Imported Tree and Plant Pests"; Liebhold and Griffin, "Legacy of Charles Marlatt." On the phylloxera invasion of Europe, see J. Sanders, "Save Us from Invading Pests."

8. Marlatt, "Insect Control in California."

9. Jefferson and Fusonie, *Japanese Flowering Cherry Trees of Washington, D.C.*; Pauly, "Beauty and Menace of the Japanese Cherry Trees"; Liebhold and Griffin, "Legacy of Charles Marlatt." Marlatt's early work on periodical cicadas (Marlatt, *Periodical Cicada*) included meticulous mapping of the locations of the distinct broods in the United States, which he named with Roman numerals. His work received renewed public recognition with the simultaneous emergence of two broods in 2024.

10. Pauly, "Beauty and Menace of the Japanese Cherry Trees"; Pauly, reply to Simberloff letter to the editor, *Isis*. See chapter 18 and Simberloff, letter to the editor, *Isis*.

11. J. C. Phillips, *Wild Birds Introduced or Transplanted in North America.*

12. J. C. Phillips, *Wild Birds Introduced or Transplanted in North America*, 2.

13. J. C. Phillips, *Wild Birds Introduced or Transplanted in North America*, 5.

14. Grinnell, "Risks Incurred in the Introduction of Alien Game Birds."

15. Phillips cited T. Palmer, "Danger of Introducing Noxious Animals and Birds."

16. Graham, *Natural Principles of Land Use.* See also Leopold, "Cheat Takes Over"; Elton, *Animal Ecology*; J. C. Phillips, *Wild Birds Introduced or Transplanted in North America.*

17. H. Zimmermann et al., "Renowned Cactus Moth *Cactoblastis cactorum*: Its Natural History."

18. Graham, "Statement to Select Committee"; Graham, "Statement to Subcommittee No. 3 ."

19. Graham, *Land and Wildlife.*

20. Graham, "Nature Protection as Part of Land Development."

21. Elton, *Ecology of Invasions by Animals and Plants.*

22. Graham, *Natural Principles of Land Use.*

23. Crooks and Soulé, "Lag Times in Population Explosions of Invasive Species."

24. Amrine and Stasny, "Biocontrol of Multiflora Rose."

25. R. M. Anderson, "Effect of the Introduction of Exotic Animal Forms."

26. R. M. Anderson, "Effect of the Introduction of Exotic Animal Forms," 769. See also Ritchie, *Influence of Man on Animal Life in Scotland.*

27. J. C. Phillips, *Wild Birds Introduced or Transplanted in North America.*

28. R. M. Anderson, "Faunas of Canada."

29. E.g., Leopold, "Mixing Trout in Western Waters"; Leopold, "Forestry and Game Conservation."

30. E.g., Leopold, "Chukaremia"; Leopold, "Fifth Column of the Fencerow"; Leopold, "Bluebirds Welcome"; Leopold, "Cheat Takes Over" (1941).

31. E.g., Leopold, "Biotic View of the Land."

32. Cf. Elton, *Animal Ecology*, Elton, *Ecology of Invasions by Animals and Plants.* On Leopold's 1931 meeting with Elton, see Simberloff, "Integrity, Stability, and Beauty."

33. Simberloff, "Integrity, Stability, and Beauty"; J. Warren, *Aldo Leopold's Odyssey.*

34. Sargent, "Editorial"; cf. Simberloff, "Confronting Introduced Species."

35. Leopold, "Cheat Takes Over" (1949).

36. Leopold, "Land Ethic." Cf. Leopold, "Land-Health Concept and Conservation."

37. Cf. Callicott, "Leopold's Land Aesthetic"; Simberloff, "Integrity, Stability, and Beauty."

38. R. T. King, "Is It Wise Policy to Introduce Exotic Game Birds?" The New York State College of Forestry at Syracuse University subsequently became part of the State University of New York.

39. Hammerstrom, "Exotic Game Birds."

40. Leopold, "Mixing Trout in Western Waters."

41. Simberloff, "Integrity, Stability, and Beauty."

42. Leopold, *Game and Fish Handbook*, 95.

43. Leopold, "Mixing Trout in Western Waters," 102.

44. E.g., Bartlett, "Discussion on Carp."

45. Talbott, "Invasion of the Potomac," 168, 171, 173.

46. See, e.g., Henshall, "Indiscriminate and Inconsiderate Planting of Fish"; Griffiths, "Considerations of the Introduction and Distribution of Exotic Fishes."

47. Rivero, "Introduction of Largemouth Bass."

48. Madsen, "Protection of Native Fishes in the National Parks," 395.

49. Sellars, "The Rise and Decline of Ecological Attitudes."

50. Simberloff, "Integrity, Stability, and Beauty."

51. Leopold, "Cheat Takes Over" (1941); Leopold, "Conservation: In Whole or in Part?"

52. Leopold, "Fifth Column of the Fencerow."

53. [G. Myers], "Use of Predatory Vertebrate Animals."

54. [G. Myers], "Use of Predatory Vertebrate Animals," 323–24.

55. G. Myers, "Foreign Introductions of North American Fishes."

56. Pringle, "Nile Perch."

57. Chamberlain, "There Is No Universal Pond Fish."

58. G. Myers, "*Gambusia*, the Fish Destroyer."

59. Innes, *Exotic Aquarium Fishes.*

60. Pyke, "Plague Minnow or Mosquito Fish?"

61. On the distributional confusion engendered by escaped aquarium fishes, see Myers, "American Cyprinodont Fish."

62. Dritschilo, *Magnificent Failure.*

63. Egler, "Indigene Versus Alien in the Development of Arid Hawaiian Vegetation."

64. Egler, "Arid Southeast Oahu Vegetation."

65. Egler, *Nature of Naturalization II.*

66. On Egler's experimentation, see, e.g., Egler, "Herbicide Effects in Connecticut Vegetation"; Egler, "Vegetation Management for Right-of-Ways and Roadsides." For an example of his papers on nontarget impacts, see Egler, "Pointed Perspectives."

67. Dritschilo, *Magnificent Failure.* See also Carson, *Silent Spring.*

68. E. Anderson, *Plants, Man, and Life.*

69. Harrison, "Plant History."

70. E. Anderson, *Plants, Man, and Life.*

71. E. Anderson, *Plants, Man, and Life*, 28.

72. E. Anderson, *Plants, Man, and Life.*

73. Thomas, *Man's Role in Changing the Face of the Earth.*

74. Oosting, *Study of Plant Communities.*

75. Odum, *Fundamentals of Ecology*; Odum, *Fundamentals of Ecology*, 2nd ed.

76. Oosting, *Study of Plant Communities*, 2nd ed.

77. Levi, "Evaluation of Wildlife Importations."

78. For those Levi cited, see Elton, *Animal Ecology*; Grinnell, "Risks Incurred in the Introduction of Alien Game Birds"; J. C. Phillips, *Wild Birds Introduced or Transplanted in North America*; Wodzicki, *Introduced Mammals of New Zealand.*

79. Calahane, "Some Effects of Exotics on Nature."

80. Calahane, "Some Effects of Exotics on Nature," 177–78.

81. De Vos et al., "Introduced Mammals and Their Influence on Native Biota."

82. Lever, *Naturalized Mammals of the World*; J. L. Long, *Introduced Mammals of the World.*

83. De Vos et al., "Introduced Mammals and Their Influence on Native Biota," 185.

84. W. Thomas, *Man's Role in Changing the Face of the Earth.*

85. E. Anderson, "Man as a Maker of New Plants."

86. Clark, "Impact of Exotic Invasions."

87. Bates, "Man as an Agent in the Spread of Organisms."

88. Allan, "Indigene Versus Alien in the New Zealand Plant World"; Egler, "Indigene Versus Alien in the Development of Arid Hawaiian Vegetation."

89. Metcalf and Flint, *Destructive and Useful Insects.*

90. Bates, "Man as an Agent in the Spread of Organisms," 797.

91. On overall efficacy of biological control, see, e.g., M. B. Thomas and Reid, "Are Exotic Natural Enemies an Effective Way of Controlling Invasive Plants?"; Goldson et al., "If and When Successful Classical Biological Control Fails." On failed invasions being informative about the invasion process, see Zenni and Nuñez, "Elephant in the Room."

92. Thomson, *Naturalisation of Animals and Plants in New Zealand*; R. R. Miller and Alcorn, "Introduced Fishes of Nevada."

93. Myers, "*Gambusia*, the Fish Destroyer"; Pyke, "Plague Minnow or Mosquito Fish?"

94. J. C. Phillips, *Wild Birds Introduced or Transplanted in North America.*

95. Carter and Leonard, "Review of the Literature."

96. Bates, "Man as an Agent in the Spread of Organisms," 799.

97. E.g., Weis and Sol, *Biological Invasions and Animal Behaviour.*

98. Ritchie, *The Influence of Man on Animal Life in Scotland.*

99. Chew, "Ending with Elton."

100. On reasons for failed invasions, cf. Zenni and Nuñez, "Elephant in the Room." On the role of habitat disturbance in facilitating invasions, see Elton, *Ecology of Invasions by Animals and Plants*; MacDougall and Turkington, "Are Invasive Species the Drivers or Passengers of Change?"

101. Elton, *Animal Ecology*.

102. Ritchie, *Birds and Beasts as Farm Pests*; Ritchie, "Changing Face of Nature, I"; Ritchie, "Changing Face of Nature, II"; Ritchie, "Changing Face of Nature, III"; Ritchie, "Changing Face of Nature, IV"; Ritchie, "Changing Face of Nature, V"; Ritchie, "Changing Face of Nature, VI."

103. Elton, "Plagues of Animals."

104. Lindroth, *Faunal Connections*.

105. E.g., Niemelä and Mattson, "Invasion of North American Forests by European Phytophagous Insects."

106. Kirk, "Displacement of Species in New Zealand."

107. Thomson, *Naturalisation of Animals and Plants in New Zealand*.

108. Darwin, *On the Origin of Species*, 337.

109. Thomson, *Naturalisation of Animals and Plants in New Zealand*, 509.

110. Hooker, *Botany of the Antarctic Voyage* [. . .] *Part I; Flowering Plants*; Wallace, *Island Life*.

111. MacDougall and Turkington, "Are Invasive Species the Drivers or Passengers of Change in Degraded Ecosystems?"

112. Thomson, *Naturalisation of Animals and Plants in New Zealand*, 527–28.

113. Veltman et al., "Correlates of Introduction Success"; R. Duncan, "Role of Competition and Introduction Effort"; Green, "Influence of Numbers Released."

114. E.g., Moulton, Cropper, and Avery, "Reassessment of the Role of Propagule Pressure."

115. Guthrie-Smith, *Tutira*.

116. Absalom, preface to *Tutira*, 4th ed.

117. Cf. Simberloff and Gibbons, "Now You See Them, Now You Don't!"

118. Guthrie-Smith, *Tutira*, 355.

119. Guthrie-Smith, *Tutira*, xxiii.

120. Cf. Druett, *Exotic Intruders*; McDowall, *Gamekeepers for the Nation*; D. Young, *Our Islands, Our Selves*.

121. On Norway rats, see Buller, "Further Notes on the Ornithology of New Zealand." On non-native mustelids, see Buller, "On the Proposed Introduction of the Polecat into New Zealand"; Buller, "Some Curiosities of Birdlife."

122. Oliver, *New Zealand Birds*.

123. Allan, "Indigene Versus Alien in the New Zealand Plant World."

124. Allan, "Indigene Versus Alien in the New Zealand Plant World." Cf. Cockayne, "Observations Concerning Evolution"; Cockayne, *Vegetation of New Zealand*.

125. Wodzicki, *Introduced Mammals of New Zealand*.

126. A later work (Wodzicki, "The Status of Some Exotic Vertebrates in the Ecology of New Zealand") developed the history of bird introductions more systematically.

127. Wodzicki, "Ecology and Management of Introduced Ungulates in New Zealand"; Wodzicki, "Introduced Mammals in New Zealand Forests"; Wodzicki, foreword to C. King, *Handbook of New Zealand Mammals*.

128. Clark, *Invasion of New Zealand*.

129. Crosby, *Ecological Imperialism*.

130. Clark, "Impact of Exotic Invasions."

131. D. Weiner, "Community Ecology in Stalin's Russia"; D. Weiner, *Little Corner of Freedom*.

132. Stanchinskii was one of several Soviet scientists who either anticipated ecological advances widely attributed to Western ecologists or developed the same ideas in parallel, yet are relegated for the most part to footnotes or passing mention

in discussion of the evolution of these ideas. In the 1930s, he had developed the idea of *trophodynamics*—the flow of energy through food webs—primarily associated with Raymond Lindeman (see Lindeman, "Trophic-Dynamic Aspect of Ecology") and the pyramid of biomass associated with Charles Elton's pyramid of numbers (Elton, *Animal Ecology*), which F. S. Bodenheimer expanded to a pyramid of biomass (Bodenheimer, *Problems of Animal Ecology*). See D. Weiner, "Community Ecology in Stalin's Russia."

133. D. Weiner, "Historical Origins of Soviet Environmentalism"; D. Weiner, *Little Corner of Freedom.*

134. D. Weiner, *Little Corner of Freedom*; Foster, "Late Soviet Ecology and the Planetary Crisis."

135. On birds, see Grinnell, "Risks Incurred in the Introduction of Alien Game Birds"; Oliver, *New Zealand Birds*; R. T. King, "Is It Wise Policy to Introduce Exotic Game Birds?" On mammals, see Wodzicki, *Introduced Mammals of New Zealand*; De Vos et al., "Introduced Mammals and Their Influence on Native Biota." On birds and mammals, see T. Palmer, "Danger of Introducing Noxious Animals and Birds." On fishes, see Myers, "Use of Predatory Vertebrate Animals"; Myers, "Foreign Introductions of North American Fishes"; Myers, "*Gambusia*, the Fish Destroyer." On insects, see Howard, "Danger of Importing Insect Pests"; Howard, *Fighting the Insects*; Marlatt, "Insect Control in California"; Marlatt, *Danger of General Spread of the Gipsy and Brown-Tail Moths*; Lindroth, *Faunal Connections.* On plants, see Cockayne, "Observations Concerning Evolution"; Cockayne, *Vegetation of New Zealand*; Allan, "Indigene Versus Alien in the New Zealand Plant World"; Egler, "Indigene Versus Alien in the Development of arid Hawaiian Vegetation."

136. R. T. King, "Is It Wise Policy to Introduce Exotic Game Birds?"; Myers, "Use of Predatory Vertebrate Animals," Myers, "*Gambusia*, the Fish Destroyer."

137. Leopold, "Cheat Takes Over."

138. Ritchie, *Influence of Man on Animal Life in Scotland*, Ritchie, *Birds and Beasts as Farm Pests.*

139. R. M. Anderson, "Effect of the Introduction of Exotic Animal Forms."

140. Wodzicki, *Introduced Mammals of New Zealand.*

141. Graham, *Natural Principles of Land Use.* On water hyacinth, see Zierer, "Brisbane—River Metropolis of Queensland." On prickly pear, see A. Dodd, *Biological Campaign Against Prickly-Pear.*

142. Leopold, *Game Management*; Leopold, "Chukaremia."

143. Ritchie, *Influence of Man on Animal Life in Scotland*; Grinnell, "Risks Incurred in the Introduction of Alien Game Birds"; R. T. King, "Is It Wise Policy to Introduce Exotic Game Birds?"; Graham, *Natural Principles of Land Use*; Levi, "Evaluation of Wildlife Importations"; M. Bates, "Man as an Agent in the Spread of Organisms"; De Vos et al., "Introduced Mammals and Their Influence on Native Biota."

144. For such citations, see, e.g., T. Palmer, "Danger of Introducing Noxious Animals and Birds"; Ritchie, *Influence of Man on Animal Life in Scotland*; Guthrie-Smith, *Tutira*; Thomson, *Naturalisation of Animals and Plants in New Zealand*; R. M. Anderson, "Effect of the Introduction of Exotic Animal Forms"; Leopold, "Cheat Takes Over"; R. T. King, "Is It Wise Policy to Introduce Exotic Game Birds?"; Wodzicki, *Introduced Mammals of New Zealand.*

145. E.g., Cockayne, "Observations Concerning Evolution"; Cockayne, *Vegetation of New Zealand*; Ritchie, *Influence of Man on Animal Life in Scotland.*; Thomson, *Naturalisation of Animals and Plants in New Zealand*; Allan, "Indigene Versus Alien in the New Zealand Plant World"; Egler, "Indigene Versus Alien in the Development of

arid Hawaiian Vegetation"; Graham, *Natural Principles of Land Use*; M. Bates, "Man as an Agent in the Spread of Organisms."

146. Cockayne, "Observations Concerning Evolution," 32–33. Italics in the original.

147. MacDougall and Turkington, "Are Invasive Species the Drivers or Passengers of Change?"

148. Clark, *Invasion of New Zealand*; Crooks and Soulé, "Lag Times in Population Explosions of Invasive Species."

149. Leopold, "Cheat Takes Over"; Graham, *Natural Principles of Land Use*; Clark, *Invasion of New Zealand*; Levi, "Evaluation of Wildlife Importations." On "invasion meltdown," see Simberloff and Von Holle, "Positive Interactions of Nonindigenous Species."

150. Rhymer and Simberloff, "Extinction by Hybridization and Introgression."

151. Simberloff and Gibbons, "Now You See Them, Now You Don't!"

152. On the question of why some plant species become weedy and related ones do not, see H. Baker, "Characteristics and Modes of Origin of Weeds."

153. Leopold, "Biotic View of the Land"; Leopold, "Land-Health Concept and Conservation"; Leopold, "Land Ethic"; Vitousek, "Biological Invasions and Ecosystem Properties"; Ehrenfeld, "Ecosystem Consequences of Biological Invasions"; Simberloff, "How Common Are Invasion-Induced Ecosystem Impacts?"

154. For diagrammed food web, see, e.g., Elton, *Animal Ecology*.

155. Tansley, "Use and Abuse of Vegetational Concepts and Terms"; Odum, *Fundamentals of Ecology*.

156. Golley, *History of the Ecosystem Concept in Ecology*, esp. fig. 4.4.

157. Leopold, "Land Ethic."

158. On late twentieth-century legislation, see M. Miller and Fabian, *Harmful Invasive Species: Legal Responses*; Genovesi, Carbonera, et al., "EU Adopts Innovative Legislation on Invasive Species"; R. Johnson, *Invasive Species: Major Laws*; Lukey and Hall, "Biological Invasion Policy." On the persistence of inadequate regulation, see Pyšek, Hulme, et al., "Scientists' Warning on Invasive Alien Species"; H. Roy et al., *Summary for Policymakers*.

Chapter 7

1. Elton, *Ecology of Invasions by Animals and Plants*. See, e.g., Rejmánek et al., "Biological Invasions: Politics"; Ricciardi and MacIsaac, "Book That Began Invasion Biology"; Richardson and Pyšek, "Classics in Physical Geography Revisited."

2. Kitching, "World of Thought."

3. Kitching, "World of Thought"; Simberloff, "Charles Elton: Pioneer Conservation Biologist."

4. Elton, "Balance and Barrier"; Elton, "Plagues of Animals"; Elton, "Nature Sanctuaries."

5. Ritchie, *Influence of Man on Animal Life in Scotland*; Bates, "Man as an Agent in the Spread of Organisms."

6. On its status as a milestone, see Richardson, introduction to *Fifty Years of Invasion Ecology*. On its true influence, see Simberloff, "Charles Elton—Neither Founder nor Siren."

7. Elton, "Small Adventures."

8. Elton, *Animal Ecology*.

9. Elton, *Animal Ecology*, 20.

10. Elton, "Alien Invaders." On the Chinese mitten crab, see Elton, "New Invader," quotation at 192.

11. Elton, *Ecology of Animals*; Elton, *Exploring the Animal World*.

12. Elton, *Exploring the Animal World*; Elton, "Plagues of Animals."

13. Crowcroft, *Elton's Ecologists*.

14. Warwick, "Distribution of the Muskrat"; Warwick, "Contribution to the Ecology of the Musk-Rat."

15. Warwick, "Some Escapes of Coypus"; L. Gosling and Baker, "Eradication of Muskrats and Coypus from Britain."

16. Middleton, "Ecology of the American Grey Squirrel"; Middleton, "Grey Squirrel (*Sciurus carolinensis*) in the British Isles"; Middleton, "Distribution of the Grey Squirrel."

17. C. Shuttleworth et al., *Grey Squirrel*.

18. Chitty and Southern, *Control of Rats and Mice*.

19. Paviour-Smith, "Ecology of the Fauna Associated with Macrofungi"; Paviour-Smith, "Invasion of Britain by *Cis bilamellatus* Fowler"; Paviour-Smith, "Population Study of *Cis bilamellatus* Wood."

20. Elton, *Pattern of Animal Communities*.

21. Elton, "First 30 Years of the Bureau of Animal Population."

22. Orledge et al., "Non-Pest Australasian Fungivore"; Lopes-Andrade and Lüer, "Invasive Beetle."

23. Crowcroft, *Elton's Ecologists*. Another woman, Monica Shorten, was affiliated with the Bureau of Animal Population from 1945 through 1954 and quickly became the leading squirrel researcher in the United Kingdom (see Coates, *Squirrel Nation*), beginning with a 1944–45 survey of the distributions of both the native red squirrel and the non-native gray squirrel (Shorten, "Survey of the Distribution"). In 1954 she moved to the Infestation Control Division of the Ministry of Agriculture, and in both the bureau and the ministry she published both academic and more lay-oriented works on the spread of the gray squirrel and its impacts, especially on its native congener (e.g., Barkalow and Shorten, *World of the Gray Squirrel*; Shorten, "Inheritance of Melanism in Grey Squirrels"; Shorten, *Squirrels*; Shorten, "Damage Caused by Squirrels"; Shorten, "Squirrels in England, Wales and Scotland"; Shorten and Courtier, "Population Study of the Grey Squirrel"; Shorten and Elton, "Some Aspects of the Biology of the Grey Squirrel"; K. Taylor et al., "Movements of the Grey Squirrel"). Early in her career she married Arturo D. Vizoso, specialist in wildlife diseases, and published papers as "Monica Vizoso" or as "Monica Shorten (Mrs. Vizoso)." She conducted research on squirrel parasites with Vizoso (Vizoso et al., "Isolation of a Virus [. . .] from a Red Squirrel"). During her tenure at the bureau she also studied non-native rats (e.g., Chitty and Shorten, "Techniques for the Study of the Norway Rat"), and she later worked for the Game Conservancy on the native Eurasian woodcock (e.g., Morgan and Shorten, "Breeding of the Woodcock"). Peter Crowcroft, Elton's student and biographer, suggests a hesitancy on Elton's part to employ women, citing a comment by Crowcroft's wife Gillian Godfrey (who subsequently studied voles at Wytham Wood for her doctorate) to the effect that, when she approached Elton about working at the bureau around 1950, he told her he "didn't care to have women in the Bureau just yet" (Crowcroft, *Elton's Ecologists*, 58; cf. Montgomery, "'Never So at Home'"). It is difficult to reconcile this suggestion with Shorten's prominence in the bureau at that time.

24. On the sycamore invasion, see Paviour-Smith, "How Far Has *Cryptostroma corticale* Spread in Britain?"

25. Simberloff, "Charles Elton—Neither Founder nor Siren."

26. Murdoch, "Population Ecology of Certain Carabid Beetles."

27. W. W. Murdoch, email message to the author, 2008.

28. Murdoch, "Diversity, Complexity, Stability and Pest Control."

29. E.g., Murdoch et al., "Biological Control in Theory and Practice."

30. Murdoch, email message.

31. On his interaction with Paviour-Smith, see Kitching, *Food Webs and Container Habitats*. On his interaction with Elton, see Kitching, "World of Thought." For his dissertation, see Kitching, "Fauna of Tree-Holes."

32. R. L. Kitching, email message to the author, 2008; Kitching, "World of Thought."

33. Kitching, email message.

34. Kitching and Jones, *Ecology of Pests*.

35. Kitching, *Ecology of Exotic Animals and Plants*.

36. Kitching, "Exotics in Australia."

37. R. L. Kitching, email message.

38. Pimentel, email message to the author, 2008.

39. Pimentel, "Biology of the Indian Mongoose"; Pimentel, "Control of the Mongoose in Puerto Rico."

40. Pimentel, Glenister, et al., "Environmental Risks of Biological Pest Controls."

41. Pimentel, "Animal Population Regulation."

42. Elton, *Pattern of Animal Communities*.

43. Pimentel, Lach, et al., "Environmental and Economic Costs of Nonindigenous Species in the United States"; Pimentel, *Biological Invasions: Economic and Environmental Costs*, 2nd ed.; United States Congress, Office of Technology Assessment, *Harmful Non-Indigenous Species in the United States*; Pimentel, email message.

44. Elton, *Ecology of Invasions by Animals and Plants*, 2nd ed.

45. Simberloff, foreword to *Ecology of Invasions by Animals and Plants*; Elton, *Ecology of Invasions by Animals and Plants*, 2nd ed.

46. Ricciardi and MacIsaac, "Book That Began Invasion Biology"; Richardson and Pyšek, "Fifty Years of Invasion Ecology."

47. Simberloff, "Charles Elton—Neither Founder nor Siren."

48. Baker and Stebbins, *Genetics of Colonizing Species*.

49. Woodwell and Smith, *Diversity and Stability in Ecological Systems*.

50. Simberloff, "Integrity, Stability, and Beauty"; Elton, *Ecology of Invasions by Animals and Plants*.

51. Alan Gray, Crawley, and Edwards, *Colonization, Succession and Stability*.

52. S. Barrett, "Foundations of Invasion Genetics" S. Barrett et al., *Invasion Genetics*.

53. Leopold, "Mixing Trout in Western Waters"; Leopold, "Forestry and Game Conservation."

54. Leopold, "Forestry and Game Conservation."

55. Callicott, "Leopold's Land Aesthetic"; Simberloff, "Integrity, Stability, and Beauty."

56. Simberloff, "Integrity, Stability, and Beauty."

57. Leopold, "Land Ethic"; Simberloff, "Integrity, Stability, and Beauty."

58. Leopold, "Cheat Takes Over" (1949).

59. Leopold, *Sand County Almanac*; Carson, *Silent Spring*; S. Duffy, "*Silent Spring* and *A Sand County Almanac*."

60. R. Carson to E. O. Wilson, 6 November 1958, in Correspondence and Papers of Charles Sutherland Elton (hereafter cited as Elton papers).

61. R. Carson to C. Elton, 24 September 1962, Elton papers.

62. Carson, *Silent Spring*, 233–34. See Elton, *Ecology of Invasions by Animals and Plants*, 141.

63. Elton, *Ecology of Invasions by Animals and Plants*, 2nd ed. For the three articles he cited, see J. George, "Program to Eradicate the Imported Fire Ant"; E. Wilson, "Fire Ant"; E. Wilson and Brown, "Recent Changes in the Introduced Population of the Fire Ant."

64. Sukopp, "Neophyten in natürlichen Pflanzengesellschaften Mitteleuropas." For use of "neophytes" to distinguish them from species introduced in antiquity, see Pyšek, Sádlo, and Mandák, "Catalogue of Alien Plants of the Czech Republic"; Pyšek, Richardson, et al., "Alien Plants in Checklists and Floras."

65. E.g., Kowarik, "Time Lags in Biological Invasions."

66. Lachner et al., "Exotic Fishes and Other Aquatic Organisms Introduced into North America."

67. Courtenay and Robins, "Exotic Aquatic Organisms in Florida"; Courtenay et al., "Exotic Fishes in Fresh and Brackish Waters of Florida."

68. Courtenay and Robins, "Exotic Organisms."

69. Moyle, "Fish Introductions into North America"; Ewel, "Invasibility: Lessons from South Florida."

70. Whitfield et al., "Biological Invasion of the Indo-Pacific lionfish"; Courtenay and Williams, *Snakeheads (Pisces: Channidae)*; Kobell, "Florida's Dr. Snakehead."

71. Dasmann, *No Further Retreat.*

72. Dasmann, *Destruction of California.* Craighead and Dasmann, *Exotic Big Game on Public Lands.*

73. Ewel, "Invasibility: Lessons from South Florida"; H. Mooney, Hamburg, and Drake, "Invasion of Plants and Animals into California," respectively. For the volume as a whole, see H. Mooney and Drake, *Ecology of Biological Invasions of North America and Hawaii.*

74. Austin, "Exotic Plants and Their Effects in Southeastern Florida"; Morton, "Pestiferous Spread of Many Ornamental and Fruit Species."

75. Ewel, "Invasibility: Lessons from South Florida."

76. Simberloff, Schmitz, and Brown, *Strangers in Paradise.*

77. Hardy, "Feral Exotic Birds in Southern California"; Bull, "Exotic Birds in the New York City Area"; Owre, "Consideration of the Exotic Avifauna of Southeastern Florida."

78. For the 1980s SCOPE book citation, see Ewel, "Invasibility: Lessons from South Florida." For a later citation from the 1990s, see e.g., James, "Nonindigenous Birds."

79. G. Williams, "Dispersal from New Zealand and Australia of Some Introduced European Passerines."

80. Milne and Milne, *Balance of Nature.*

81. Laycock, *Alien Animals.*

82. M. Davis, "Invasion Biology 1958–2005."

83. Roots, *Animal Invaders.*

84. Rolls, *They All Ran Wild: The Story of Pests on the Land in Australia.*

85. Lever, *Naturalized Animals of the British Isles.*

86. Lever later published several books on introduced species in particular taxa, as well as one on acclimatization societies: Lever, *Naturalized Mammals of the World*; Lever, *Naturalized Birds of the World*; Lever, *They Dined on Eland*; Lever, *Naturalized Fishes of the World*; Lever, *Cane Toad.*

87. R. Fitter, *Ark in Our Midst.*

88. R. Fitter, "Man's Addition to the British Fauna"; R. Fitter and Leigh-Pemberton, *Britain's Wildlife.*

89. Lever, *Naturalized Animals of the British Isles*. Fitter's son Alastair Fitter became a prominent ecologist specializing in plant ecology and plant interactions with mycorrhizal fungi. Father and son jointly published an early paper on flowering time changes induced by climate warming (Fitter and Fitter, "Rapid Changes in Flowering Time in British Plants") as well as several nature guides. Alastair Fitter coauthored with Mark Williamson two widely cited papers in the early post-SCOPE invasions literature, one estimating the fraction of invasions that lead to substantial impacts (Williamson and Fitter, "Varying Success of Invaders") and the other seeking features that distinguish successful invaders from unsuccessful ones (Williamson and Fitter, "Characters of Successful Invaders").

90. R. Fitter, *Ark in Our Midst*; Laycock, *Alien Animals*; Roots, *Animal Invaders*; Lever, *Naturalized Animals of the British Isles*.

91. Rolls, *They All Ran Wild: The Story of Pests on the Land in Australia*, xv.

92. Rolls, *They All Ran Wild: The Animals and Plants That Plague Australia*, xv.

93. Low, *Feral Future*, 295.

94. Elton, *Ecology of Invasions by Animals and Plants*, 2nd ed.

95. On walking catfish, see Courtney and Ogilvie, "Species Pollution." On pike killifish, see Kallman, "Pike Killifish." On giant African snail, see "Snails as Big as Your Fist." On barnacle, see Kawahara, "Invasion into Japanese Waters of the European Barnacle." On pink salmon, see W. B. Scott to C. Elton, 27 May 1964, Elton papers. On Japanese wireweed, see Hawkes, "These Triffids Have Had Their Day"; Hawkes, "Seaweed That Won't Stop Growing"; John Cross, "Simply Science." On Texas mud crab, see Naylor, "North American Xanthoid Crab New to Britain."

96. On monk parakeet, see Freeland, "Some Food Preferences and Aggressive Behavior by Monk Parakeets"; Neidermyer and Hickey, "Monk Parakeet in the United States." On raccoon dog, see Röben, "Zur Ausbreitung des Waschbären." On Sirex wood wasp, see Irvine, "Forest and Timber Insects in Victoria." On giant African snail, see "Miami Snail Control"; Sturgeon, "*Achatina fulica* Infestation in North Miami"; Mead, "Ecological Malacology." On hydrilla, see W. McLane, "Aquatic Plant Business"; Bob Norton, "Hydrilla Menace."

97. Zaret and Paine, "Species Introduction in a Tropical Lake." Zaret and Paine referred to the peacock bass in Gatun Lake as *Cichla ocellaris*. However, subsequent systematic and biogeographic research (Kullander and Ferreira, "Review of the South American Cichlid Genus *Cichla*"; S. Willis et al., "Systematics, Biogeography, and Evolution of the Neotropical Peacock Basses") and detailed knowledge of the initial introduction demonstrate that the species introduced was actually *C. monoculus* (D. M. T. Sharpe, email message to the author, May 12, 2020).

98. Zaret, "Studies on Predator-Prey Interaction"; Zaret, "Predator-Prey Interaction."

99. Vitousek, "Biological Invasions and Ecosystem Properties."

100. C. Spencer et al., "Shrimp Stocking, Salmon Collapse, and Eagle Displacement"; Ellis et al., "Long-Term Effects of a Trophic Cascade."

101. Ligtvoet and Witte, "Perturbation Through Predator Introduction Effect on the Food Web"; Moreau, "Analysis of Species Changes in Lake Victoria."

102. Paine and Zaret, "Ecological Gambling."

103. Welcomme, *International Introductions of Inland Aquatic Species*, 22, 71.

104. Shafland, *Proposal for Introducing Peacock Bass*, 10. For Zaret's observation, see Zaret, "Predation in Freshwater Fish Communities."

105. Sharpe et al., "Tropical Fish Community." For Zaret's later conjecture, see Zaret, "Stability/Diversity Controversy."

106. Valverde et al., "Trophic Shifts in a Native Predator."

107. Zaret, "Ecology of Introductions."

108. Worthington, "Ecology of Introductions"; Pringle, "Nile Perch."

109. Zaret, "Predation in Freshwater Fish Communities"; Zaret, "Stability/Diversity Controversy."

110. Zaret, "Life History and Growth Relationships of *Cichla ocellaris*."

111. Swartzmann and Zaret, "Modeling Fish Species Introduction and Prey Extermination."

112. Zaret, "Stability/Diversity Controversy." Cf. Simberloff, "Integrity, Stability, and Beauty"; Elton, *Ecology of Invasions by Animals and Plants*.

113. Zaret, *Predation and Freshwater Communities*.

114. On keystone species, see Paine, "Note on Trophic Complexity." On top-down predation in maintaining biodiversity, see Paine, "Food Web Complexity and Species Diversity." On trophic cascades, see Paine, "Food Webs."

115. Paine, "Conversation on Refining the Concept of Keystone Species."

116. On the 1998 study of cases, see Paine et al., "Compounded Perturbations Yield Ecological Surprises." On the flood plus the Asian overbite clam, see Nichols et al., "Remarkable Invasion of San Francisco Bay." On the volcanic eruption plus the fire tree invasion, see Vitousek and Walker, "Biological Invasion by *Myrica faya* in Hawai'i."

117. For the 2002 commentary, see Paine and Schindler, "Ecological Pork." For the 2011 publication, see Estes et al., "Trophic Downgrading of Planet Earth."

118. Carson, *Silent Spring*; R. Dunlap and Mertig, "Evolution of the U.S. Environmental Movement"; Sale, *Green Revolution*.

119. Marsh, *Man and Nature; or, Physical Geography as Modified by Human Action*; Marsh, *Earth as Modified by Human Action*.

120. W. Thomas, *Man's Role in Changing the Face of the Earth*.

121. Detwyler, *Man's Impact on Environment*.

122. United States National Research Council, *Global Change: Ecosystems Research*. Interestingly, Marsh discussed only local climatic impacts of humans (Marsh, *Man and Nature; or, Physical Geography as Modified by Human Action*; Marsh, *Earth as Modified by Human Action*), while the books edited by Thomas and Detwyler each had only one chapter on this issue (W. Thomas, *Man's Role in Changing the Face of the Earth*; Detwyler, *Man's Impact on Environment*), whereas anthropogenic global climate change now dominates most discussions of planetwide human impacts.

123. Elton, *Pattern of Animal Communities*.

124. Elton, *Pattern of Animal Communities*; Simberloff, "Nature, Natives, Nativism, and Management."

125. Walter G. Rosen first coined the term "biodiversity" in the mid-1980s in the lead-up to a major conservation conference, the National Forum on BioDiversity, held in 1986 and cosponsored by the US National Academy of Sciences and the Smithsonian Institution (Takacs, *Idea of Biodiversity*). The book *Biodiversity* (E. Wilson, *Biodiversity*), drawn from the conference, drew an enormous readership, and the term quickly became shorthand for all aspects of the diversity of life, but especially for the number of species at a site (Takacs, *Idea of Biodiversity*).

126. Simberloff, "Nature, Natives, Nativism, and Management."

127. Elton, *Exploring the Animal World*. Cf. Elton, "Three Hundred Years of Conservation."

128. Simberloff, "Contribution of Population and Community Ecology to Conservation Science."

129. MacArthur and Wilson, *Theory of Island Biogeography*.

130. Terborgh, "Preservation of Natural Diversity"; Terborgh, "Faunal Equilibria and the Design of Wildlife Preserves"; J. Diamond, "Island Dilemma"; E. Wilson and Willis, "Applied Biogeography."

131. N. Moore, "Heaths of Dorset and Their Conservation"; M. Hooper, "Size and Surroundings of Nature Reserves."

132. Frankel, "Variation, the Essence of Life"; Frankel, "Genetic Conservation"; Berry, "Conservation Aspects of the Genetical Constitution of Populations."

133. On avoidance of small populations, see I. Franklin, "Evolutionary Change in Small Populations"; Senner, "Inbreeding Depression and the Survival of Zoo Populations"; Soulé, "Thresholds for Survival." On minimum viable population size, see, e.g., Shaffer, "Minimum Population Sizes for Species Conservation."

134. Caughley, "Directions in Conservation Biology."

135. Simberloff, "Community Effects of Introduced Species."

136. Simberloff, "Contribution of Population and Community Ecology to Conservation Science."

137. "Editorial Statement," *Biological Conservation*.

138. E.g., B. Reid, "Survival Status of the Takahe"; Weber, "Pinta, Galápagos: Une île à sauver."

139. Heddle, "South African Daisy in the National Parks of South Australia."

140. M. Bates, "Man as an Agent in the Spread of Organisms."

Chapter 8

1. Entrikin and Brunn, *Reflections on Richard Hartshorne's "The Nature of Geography"*; Koelsch, "Academic Geography, American Style."

2. Simberloff, "Biogeography."

3. Hehn, *Kulturpflanzen und Hausthiere in ihrem Übergang*; in English as Hehn, *Wanderings of Plants and Animals*.

4. Cierjacks et al., "Biological Flora of the British Isles: *Robinia pseudoacacia*."

5. Hehn, *Wanderings of Plants and Animals*, 398.

6. Wulff, *Введение в историческую географию растений*; Wulff, *Introduction to Historical Plant Geography*.

7. In his focus on the role of disturbance in permitting introduced propagules to establish a population and the population to spread, Wulff resembled contemporary plant ecologists such as two discussed in chapter 6, Harry H. Allan (Allan, "Indigene Versus Alien in the New Zealand Plant World") and Frank Egler (Egler, "Indigene Versus Alien in the Development of arid Hawaiian Vegetation"; Egler, "Arid Southeast Oahu Vegetation, Hawaii"). The role of disturbance, especially anthropogenic disturbance, later became a major hypothesis in modern invasion science (explored further in chap. 15).

8. Wulff, *Introduction to Historical Plant Geography*, 112, and citing Allan, "Indigene Versus Alien in the New Zealand Plant World."

9. C. Sauer, "Morphology of Landscape."

10. Koelsch, "Academic Geography, American Style."

11. E.g., C. Sauer, "Geographic Sketch of Early Man in America."

12. C. Sauer, "Early Relations of Man to Plants," 23. For the source cited in the quotation, see C. Sauer, "Geographic Sketch of Early Man in America."

13. C. Sauer, "Agency of Man on the Earth." For the volume as a whole, see W. Thomas, *Man's Role in Changing the Face of the Earth*.

14. C. Sauer, "Agency of Man on the Earth," 61.

15. For his symposium contribution, see Clark, "Impact of Exotic Invasions." For his major earlier work, see Clark, *Invasion of New Zealand.*

16. C. Sauer, "Man in the Ecology of Tropical America," 109.

17. Aubert de la Rüe, "Man's Influence on Tropical Vegetation."

18. Parsons, "Spread of African Pasture Grasses"; Parsons, "Forest to Pasture."

19. Parsons, "Spread of African Pasture Grasses," 17.

20. Sternberg, "Man and Environmental Change in South America."

21. Sternberg accurately predicted that the forestry experiments with non-native trees were "laying the groundwork for far-reaching landscape transformations in the future" (Sternberg, "Man and Environmental Change in South America," 424). When he wrote, he described 250,000 hectares of Monterey pine in Chile and 3,000 hectares of southern blue gum in Peru. The area planted to these two species in Chile alone increased from 400,000 hectares in 1974 to around 3 million hectares in 2019 (Carte et al., "Expanding Exotic Forest Plantations"), and Monterey pine has invaded other forest types (Bustamante et al., "Forest Fragmentation"). Sternberg's son, Leonel Sternberg, is a botanist specializing in hydrological processes and the use of stable isotope analysis, with occasional research on non-native plants (e.g., Ewe and Sternberg, "Water Uptake Patterns of an Invasive Exotic Plant").

22. Sternberg, "Man and Environmental Change in South America," 428–29.

23. Brothers et al., "Jonathan D. Sauer."

24. J. Sauer, *Plant Migration.*

25. Brothers et al., "Jonathan D. Sauer."

26. J. Sauer, "Oceanic Islands and Biogeographical Theory"; J. Sauer, "Biogeographical Theory and Cultural Analogies."

27. Simberloff, "Equilibrium Theory of Island Biogeography and Ecology"; Simberloff, "Biogeographic Models."

28. Simberloff, "Elton, Charles S."

29. Merrill, *Plant Life of the Pacific World.*

30. Merrill, *Plant Life of the Pacific World*, 123.

31. Meinig, *Great Columbia Plain.*

32. Mack, "Invasion of *Bromus tectorum* L."

33. Mack, "Invasion of *Bromus tectorum* L."

34. Mack, "Cheatgrass"; Leopold, "Cheat Takes Over" (1941).

35. Clark, "South Island of New Zealand."

36. Clark, *Invasion of New Zealand.*

37. Clark, "Impact of Exotic Invasions."

38. Clark, *Three Centuries and the Island*, v.

39. Clark, *Acadia.*

40. Clark, *Three Centuries and the Island*, 41.

41. Clark, *Acadia.*

42. Clark, *Acadia*, 379.

43. Clark, *Acadia*, 86; presumably the black rat, as the Norway rat did not arrive in North America until the eighteenth century (Guiry et al., "Ratting of North America").

44. Roche, "A. H. Clark's Framing of Geographical Change."

45. Clark, *Acadia*, 379.

46. Wynn, "Museums, Laboratories, Showcases."

47. On introductions of game animals, see Curley, "Wildlife Matters." On mention of non-native marine plants, see Novaczek, "Mermaid's Tresses."

48. Erskine, *French Period in Nova Scotia*; Hill and Blaney, "Exotic and Invasive Vascular Plants of the Atlantic Maritime Ecozone."

49. On invasions becoming a concern in Nova Scotia and Prince Edward Island, see, e.g., Catling and Carbyn, "Invasive Scots Pine." On marine invaders, see, e.g., Watanabe et al., "Contrasting Patterns of Spread."

50. Meinig, "Prologue: Andrew Hill Clark."

51. Meinig, "Prologue: Andrew Hill Clark."

52. Clark, *Acadia*, 371.

53. Clark, "Impact of Exotic Invasions," 755.

54. Meinig, "Prologue: Andrew Hill Clark."

55. Simberloff, "Charles Elton: Pioneer Conservation Biologist"; Simberloff, "Charles Elton."

56. Elton, *Pattern of Animal Communities*; Simberloff, "Charles Elton."

57. Simberloff, "Community Ecology."

58. A. Price, *White Settlers in the Tropics*; A. Price, *White Settlers and Native Peoples.*

59. A. Price, *Importance of Disease in History.*

60. A. Price, *Western Invasions of the Pacific*, 177.

61. A. Price, *Western Invasions of the Pacific*, 201.

62. A. Price, *Western Invasions of the Pacific*, 1.

63. A. Price, *Western Invasions of the Pacific*, 200.

64. Clark, review of *Western Invasions of the Pacific.*

65. Crosby, *Columbian Exchange*; Crosby, *Ecological Imperialism.*

66. Crosby, *Germs, Seeds, and Animals*, x.

67. W. McNeill, *Rise of the West.*

68. Crosby, *Ecological Imperialism.*

69. Crosby, *Ecological Imperialism*, 287–88, 293.

70. H. Mooney and Drake, *Ecology of Biological Invasions of North America and Hawaii.*

71. Clark, "Impact of Exotic Invasions."

72. Clark, *Invasion of New Zealand.*

73. Mann, *1493.*

74. W. McNeill, *Plagues and Peoples.* For the earlier book that influenced Crosby's *Columbian Exchange*, see W. McNeill, *Rise of the West.*

75. E.g., Vitousek, Antonio, Loope, et al., "Introduced Species"; Mack, Simberloff, et al., "Biotic Invasions."

76. W. McNeill, *Rise of the West*, 572; Crosby, *Germs, Seeds, and Animals*, xii; A. Price, *Western Invasions of the Pacific*, 145. For the Ashburn work they all cited, see Ashburn and Ashburn, *Ranks of Death.*

77. Crosby, *Germs, Seeds, and Animals*, xiii.

78. Niemelä and Mattson, "Invasion of North American Forests by European Phytophagous Insects."

79. P. Martin, "Prehistoric Overkill."

80. Drake et al., *Biological Invasions*; Simberloff, "SCOPE Project."

81. Di Castri, "History of Biological Invasions."

82. Di Castri, "History of Biological Invasions," 17. See also Braudel, *Civilisation Matérielle.*

83. Simberloff and Von Holle, "Positive Interactions of Nonindigenous Species."

84. Simberloff and Von Holle, "Positive Interactions of Nonindigenous Species," 22.

85. Simberloff, "Invasional Meltdown 6 Years Later."

86. Crosby, *Germs, Seeds, and Animals.*

87. Di Castri et al., *Biological Invasions in Europe and the Mediterranean Basin.*

88. J. Diamond, *Guns, Germs, and Steel.*

89. J. Diamond, "Overview of Recent Extinctions."

90. J. Diamond, "Historic Extinctions."

91. E.g., J. Diamond, "Avifaunas of Rennell and Bellona Islands"; Mayr and Diamond, *Birds of Northern Melanesia.*

92. However, the data underpinning the rule—presence/absence distributions of Melanesian bird species on various islands—were static: which species were where at one particular time. No invasions or attempted invasions were actually observed. The assembly rules generated an enormous amount of controversy among ecologists (Gotelli, "Research Frontiers in Null Model Analysis.").

93. E.g., W. McNeill, "History Upside Down"; W. McNeill, "Guns, Germs, and Steel (Reply)"; J. McNeill, "World According to Jared Diamond."

94. Alderman, "Channing Cope and the Making of a Miracle Vine," 173. See also Alderman, "Changing South"; Alderman and Alderman, "Kudzu."

95. *Geographical Review* 94, no. 2 (2004).

96. On the beaver as symbol of identity, see C. Anderson, Pizarro, et al., "Reconceiving the Biological Invasion of North American Beavers." On the cartoon version, see Haider and Jax, "Application of Environmental Ethics in Biological Conservation."

97. Subalusky et al., "Potential Ecological and Socio-Economic Effects of a Novel Megaherbivore Introduction."

98. Bukowski, "Using the Commons."

99. Halverson, "Relatedness Effects on Wood Frog Ecology and Evolution"; Halverson, *Entirely Synthetic Fish.* In the book, Halverson focused almost exclusively on spread and impacts in the United States, although rainbow trout have been widely introduced in many parts of the world with various ecological impacts (e.g., Crowl et al., "Impact of Introduced Brown and Rainbow Trout on Native Fish"; Shelton et al., "Predatory Impact of Non-Native Rainbow Trout on Endemic Fish Populations"; Stanković, Crivelli, and Snoj, "Rainbow Trout in Europe").

100. On the call for invasion scientists to pay more attention to the full scope of human interactions with invasive species, see Schüttler et al., "Towards a Societal Discourse on Invasive Species Management"; Heger et al., "Conceptual Frameworks and Methods for Advancing Invasion Ecology."

101. Cf. Simberloff, "Confronting Introduced Species"; Simberloff, "Nature, Natives, Nativism, and Management."

102. L. Head, "Decentring 1788"; L. Head, "Social Dimensions of Invasive Plants"; L. Head and Muir, "Nativeness, Invasiveness, and Nation in Australian Plants"; L. Head et al., "Living with Invasive Plants in the Anthropocene"; Warren, "Perspectives on the 'Alien' Versus 'Native' Species Debate"; Warren, "Nativeness and Nationhood."

103. Robbins, "Comparing Invasive Networks."

104. Kowarik, "Time Lags in Biological Invasions"; Crooks and Soulé, "Lag Times in Population Explosions of Invasive Species."

Chapter 9

1. H. Mooney, *Globalization of Ecological Thought*; van Wilgen, "Brief, Selective History of Research"; H. A. Mooney, email message to the author, 2008.

2. Simberloff, "SCOPE Project"; Simberloff, "Charles Elton—Neither Founder nor Siren."

3. Kruger, *Description of the Fynbos Biome Project*, 4. See also di Castri and Mooney, *Mediterranean Type Ecosystems.*

4. Simberloff, " Charles Elton—Neither Founder nor Siren." Before the SCOPE program, Mooney and Slatyer had collaborated on a plant physiological ecology paper

(H. Mooney, Ferrar, and Slatyer, "Photosynthetic Capacity and Carbon Allocation Patterns"). Slatyer worked primarily on the ecological effects of fire and on plant ecophysiology of water use, and he had no particular interest in invasion impacts (Australian Academy of Sciences, "Professor Ralph Slatyer"). However, living in Australia, he must have been exposed to publicity and research on introduced species. In 1975, he published a paper on features and management of ecological reserves that was almost completely in line with the "small population paradigm" focused wholly on population size, as discussed in chapter 7 (Slatyer, "Ecological Reserves"). However, the final three paragraphs began, "Finally, I would like to refer briefly to the other aspect of population management, that of exotic species, both plant and animal" (Slatyer, "Ecological Reserves," 35). His main points were that non-native species should be kept out of reserves except where they played some particularly important ecological role and that population reduction and eradication measures had not been "particularly successful." Mooney, a physiological ecologist, also had not previously focused on biological invasions; his research especially addressed photosynthesis, the carbon cycle, and the function and ecology of roots, although in 1973 he coedited a book on Mediterranean type ecosystems in which many chapters discussed non-native species (di Castri and Mooney, *Mediterranean Type Ecosystems*). However, he continued to play a key role in developing invasion science (Simberloff, Meyerson, et al., "Honoring Harold A. Mooney"). Although only 17 of his 280 papers dealt with invasions, several influenced the field, and he also coedited two of the SCOPE invasion program volumes (H. Mooney and Drake, *Ecology of Biological Invasions of North America and Hawaii*; Drake et al., *Biological Invasions: A Global Perspective*) and coedited three subsequent books on invasions (H. Mooney and Hobbs, *Invasive Species in a Changing World*; H. Mooney, Mack, et al., *Invasive Alien Species*; Perrings, Mooney, and Williamson, *Bioinvasions and Globalization*). Interestingly, in a 1993 interview reflecting on his experiences as president of SCOPE, Slatyer did not comment on the program in biological invasions, which played a key role in establishing a research field. Rather, he discussed as a high point a report on the potential consequences of a nuclear war. Mooney, by contrast, in a broadly ranging reflection on his many contributions to the globalization of ecological thought, highlighted the SCOPE invasions program as one of his most important accomplishments (H. Mooney, *Globalization of Ecological Thought*).

5. "Ecology of Biological Invasions," *SCOPE Newsletter*.

6. "Reports of 11 SCOPE Working Groups," in SCOPE archives.

7. On the questions posed in the two volumes emerging from the SCOPE program, see Vitousek, "Biological Invasions and Ecosystem Properties"; Ramakrishnan and Vitousek, "Ecosystem-Level Processes."

8. The six books: Groves and Burdon, *Ecology of Biological Invasions*; I. Macdonald, Kruger, and Ferrar, *Ecology and Management of Biological Invasions in Southern Africa*; H. Mooney and Drake, *Ecology of Biological Invasions of North America and Hawaii*; Drake et al., *Biological Invasions: A Global Perspective*; di Castri, Hansen, and Debussche, *Biological Invasions in Europe and the Mediterranean Basin*; Ramakrishnan, *Ecology of Biological Invasion in the Tropics*. The two journal special issues: Kornberg and Williamson, "Quantitative Aspects of the Ecology of Biological Invasions"; Joenje et al., "Ecology of Biological Invasions."

9. Joenje et al., "Ecology of Biological Invasions."

10. Joenje, "SCOPE Programme on the Ecology of Biological Invasions."

11. Simberloff, "SCOPE Project." The two most widely cited books: H. Mooney and Drake, *Ecology of Biological Invasions of North America and Hawaii*; Drake et al., *Biological Invasions: A Global Perspective*.

12. On Australia, see Groves and Burdon, *Ecology of Biological Invasions*. On South Africa, see I. Macdonald, Kruger, and Ferrar, *Ecology and Management of Biological Invasions in Southern Africa*. On UK, see Kornberg and Williamson, "Quantitative Aspects of the Ecology of Biological Invasions" On US, see H. Mooney and Drake, *Ecology of Biological Invasions of North America and Hawaii*. On Europe, see di Castri, Hansen, and Debussche, *Biological Invasions in Europe and the Mediterranean Basin*..

13. Joenje, "SCOPE Programme on the Ecology of Biological Invasions."

14. Sandlund, Schei, and Viken, *Proceedings of the Norway/UN Conference on Alien Species*.

15. "Legal Component of the SCOPE/IUCN Global Invasive Species Programme."

16. Simberloff, "SCOPE Project"; Richard H. Groves, email message to the author, 2009.

17. Thomson, *Naturalisation of Animals and Plants in New Zealand*; Elton, *Animal Ecology*; Elton, *Ecology of Invasions by Animals and Plants*.

18. E.g., R. Allen and Lee, *Biological Invasions in New Zealand*; Nuñez, Chiuffo, Seebens, et al., "Two Decades of Data"; cf. Simberloff, "New Zealand as a Leader."

19. Brockie et al., "Biological Invasions of Nature Reserves"; Usher et al., "Ecology of Biological Invasions into Nature Reserves."

20. Drake et al., *Biological Invasions: A Global Perspective*.

21. Duffey and Usher, "Biological Invasions of Nature Reserves."

22. Kornberg and Williamson, "Quantitative Aspects of the Ecology of Biological Invasions."

23. Di Castri, Hansen, and Debussche, *Biological Invasions in Europe and the Mediterranean Basin*; Groves and di Castri, *Biogeography of Mediterranean Invasions*.

24. Ramakrishnan, preface to *Ecology of Biological Invasion in the Tropics*.

25. Ramakrishnan, "Biological Invasion in the Tropics."

26. On invasions in nature reserves, see I. Macdonald and Frame, "Invasion of Introduced Species"; I. Macdonald, Loope, et al., "Wildlife Conservation and the Invasion of Nature Reserves"; I. Macdonald, Graber, et al., "Introduced Species in Nature Reserves." On mathematical models, see Drake and Williamson, "Invasions of Natural Communities."

27. Holdgate, "Summary and Conclusions," 232.

28. Macdonald, Kruger, and Ferrar, *Ecology and Management of Biological Invasions in Southern Africa*.

29. H. Mooney, "Global Invasive Species Program (GISP)," 97.

30. H. Mooney, *Globalization of Ecological Thought*; Sandlund et al., *Proceedings of the Norway/UN Conference on Alien Species*; Mack, Neville, and Stewart, preface to *Invasive Alien Species*.

31. H. Mooney, "Global Invasive Species Program (GISP)"; Mack, Neville, and Stewart, preface to *Invasive Alien Species*.

32. H. Mooney, "Global Invasive Species Program (GISP)," 97.

33. H. Mooney, "Global Invasive Species Program (GISP)."

34. H. Mooney and Hobbs, *Invasive Species in a Changing World*.

35. Perrings, Williamson, and Dalmazzone, eds. *Economics of Biological Invasions*, 3.

36. On progress of GISP projects, see Mack, Neville, and Stewart, preface to *Invasive Alien Species*, the synthesis volume edited by H. Mooney et al. For the ten-point global strategy, see McNeely et al., "Global Strategy on Invasive Alien Species." For the earlier GISP document, see McNeely et al., *Global Strategy on Invasive Alien Species*.

37. Wittenberg and Cock, *Invasive Alien Species: A Toolkit of Best Prevention and Management Practices*.

38. H. Mooney and Bernardi, *Introduction of Genetically Modified Organisms*; M. Levin and Strauss, *Risk Assessment in Genetic Engineering*; P. Regal, "True Meaning of 'Exotic Species' as a Model."

39. P. Regal, "Models of Genetically Engineered Organisms."

40. Sharples, "Spread of Organisms with Novel Genotypes." Cf. Sharples, "Ecological Aspects of Hazard Identification"; J. H. Brown, Colwell, et al., "Report on Workshop."

41. E.g., S. Levin and Harwell, "Potential Ecological Consequences of Genetically Engineered Organisms."

42. For the contestation by genetic engineers, see, e.g., B. Davis, "Scientist's Opinion"; Brill, "Safety Concerns." For the defense by ecologists, see Colwell et al., "Genetic Engineering in Agriculture." For the rejection of that defense, see Brill, "Reply."

43. P. Regal, "Models of Genetically Engineered Organisms"; P. Regal, "True Meaning of 'Exotic Species.'"

44. S. Levin and Harwell, "Potential Ecological Consequences of Genetically Engineered Organisms."

45. E.g., Tiedje et al., "Planned Introduction of Genetically Engineered Organisms."

46. National Research Council, *Introduction of Recombinant DNA-Engineered Organisms.*

47. Whelan, "Purpose of the Meeting."

48. H. Mooney and Drake, "Release of Genetically Designed Organisms." For the book as a whole, see Mooney and Bernardi, *Introduction of Genetically Modified Organisms.*

49. S. Levin, "Ecological Issues Related to the Release of Genetically Modified Organisms."

50. H. Mooney and Bernardi, "*Introduction of Genetically Modified Organisms,*" xviii. A workshop sponsored by COGENE in Basel in 1988 covered similar ground to the SCOPE-COGENE workshop and came to an identical conclusion about regulating products of genetic engineering rather than the process of genetic engineering itself (Stearns et al., "COGENE 88 at Basel.").

51. P. Regal, "True Meaning of 'Exotic Species' as a Model."

52. E.g., Esvelt et al., "Concerning RNA-Guided Gene Drives." See also a multinational project to eradicate invasive rodents from islands: GBIRd, "Genetic Biocontrol of Invasive Rodents."

53. Friends of the Earth, "Genetic 'Extinction' Technology Rejected by International Groups."

54. E.g., Esvelt and Gemmell, "Conservation Demands Safe Gene Drive."

55. National Academies of Sciences, Engineering, and Medicine, *Gene Drives on the Horizon*; National Research Council, *Field Testing Genetically Modified Organisms*; E. Callaway, "UN Treaty Agrees to Limit Gene Drives."

56. Simberloff, preface to *Invasions biologiques et extinctions.*

57. H. Mooney, *The Globalization of Ecological Thought.*

58. Nuñez and Pauchard, "Biological Invasions in Developing and Developed Countries"; Angulo et al., "Non-English Languages Enrich Scientific Knowledge"; Nuñez, Ciuffo, Pauchard, et al., "Making Ecology Really Global."

59. Angulo et al., "Non-English Languages Enrich Scientific Knowledge."

60. Nuñez, Chiuffo, Pauchard, et al., "Making Ecology Really Global"; Nuñez, Chiuffo, Seebens, et al., "Two Decades of Data"; Kuebbing et al., "Self-Study of Editorial Board Diversity."

61. H. Roy et al., *Summary for Policymakers.*

62. Amano and Sutherland, "Four Barriers to the Global Understanding of Biodiversity Conservation"; Amano, González-Varo, and Sutherland, "Languages Are Still a Major Barrier to Global Science"; Amano, Ramírez-Castañeda, et al., "Manifold Costs of Being a Non-native English Speaker in Science"; Nuñez and Amano, "Monolingual Searches Can Limit and Bias Results."

63. On the global nature of the discipline, see, e.g., Van Kleunen, Dawson, Essl, et al., "Global Exchange and Accumulation of Non-Native Plants."

64. Anker, *Imperial Ecology.*

65. Peretti, "Nativism and Nature."

Chapter 10

1. E.g., the article by Thomas M. Zaret and Robert T. Paine in *Science* (Zaret and Paine, "Species Introduction in a Tropical Lake"); cf. chapter 12.

2. D. Campbell, "Some Botanical and Environmental Aspects of Hawaii."

3. On the European corn borer: Barber, "Study of the Cause"; Barber, "Some Factors Responsible"; Babcock, "European Corn Borer *Pyrausta nubilalis* Hübn. I"; Babcock, "European Corn Borer, *Pyrausta nubilalis* Hübn. II." On the Mexican bean beetle: Transeau, "Vegetation Types and Insect Devastations"; Sweetman, "Precipitation and Irrigation as Factors in the Distribution of the Mexican Bean Beetle."

4. Kern, "Observations of the Dissemination of the Barberry"; W. Cooper, "Ecological Life History of Certain Species of *Ribes.*"

5. Allan, "Remarkable New Zealand Scrub Association"; Gleason, "Tutira (Book Review)."

6. R. C. Miller, "Ecological Relations of Marine Wood-Boring Organisms." Miller believed that two of the four resident wood-borers (the feathery shipworm and a shipworm he called the "San Diego shipworm," *Teredo diegensis*) were native, one (an isopod he called *Limnoria lignorum*) was cryptogenic, and one (an isopod he called *Sphaeroma pentadon*) perhaps introduced. In fact, *Teredo diegensis* turned out to be the blacktip shipworm, introduced from the Indo-Pacific region, and *Limnoria lignorum* was found to be two species (*Limnoria tripunctata* [tripunctate gribble] and *Limnoria quadripunctata*), both introduced, probably from the Indo-West Pacific. *Sphaeroma pentadon* was, in fact, *Sphaeroma quoianum*, native to New Zealand and Australia. Miller's prediction that all four (in fact, five) would survive the invasion of the naval shipworm was correct (J. Carlton, email message to the author, n.d.). He also correctly predicted, based on their tolerance of cold water, that the naval shipworm would spread to Puget Sound but that the species he called the San Diego shipworm would not (R. C. Miller, "Marine Borer Problem on the Pacific Coast").

7. Kingsland, "Battling Botanist."

8. MacDougal, "Reactions of Plants to New Habitats," 2.

9. E. Phillips, "Honey Crop from Introduced Plants," 182.

10. Mackin et al., "Rapid Evolution of a Floral Trait."

11. Cahn, "Effect of Carp on a Small Lake."

12. Data from a search of Clarivate Analytics Web of Science Core Collections topics between May and October 2021 for the following search terms: "non-native" OR "nonnative" OR "exotic speci*" OR "alien" OR "invas* speci*" OR "biolog* invas*" (details in S. Campbell and Simberloff, "Forty Years of Invasion Research").

13. E.g., Franceschet and Costantini, "Effect of Scholar Collaboration."

14. S. Campbell and Simberloff, "Forty Years of Invasion Research."

15. The one with 642 authors: Rosa et al., "Neotropical Alien Mammals." The one with 126 authors: Katsanevakis et al., "Unpublished Mediterranean Records."

16. S. Campbell and Simberloff, "Forty Years of Invasion Research."

17. Lynch et al., "Transdisciplinary Synthesis"; Baron et al., "Synthesis Centers as Critical Research Infrastructure."

18. I. Parker et al., "Impact."

19. S. Campbell and Simberloff, "Productivity Puzzle in Invasion Science."

20. Ladd Johnson, "Zebra Mussel."

21. Paul Focazio, email message to the author, October 25, 2021. James Carlton, an authority on aquatic invasions, has copies of several other, more ephemeral newsletters on zebra mussels from the 1990s, including *Zebra Mussel Informer*, *Mussel Morsels*, *Zebra Mussel Watch/Moules Zébrées*, *Zebra Mussel Newsletter* (two independent publications with the same name), *Zebra Mussel Newsletter Southern Region*, and *Zebra Mussel Update* (three independent publications with the same name), as well as other newsletters reporting aquatic invasions more generally but including zebra mussels. In 2011, the California Department of Fish and Game published the first issue of an online newsletter, *Eye on Invasives*, intended as a quarterly effort on invasions in general, with the initial issue on the quagga and zebra mussels. It has since been published sporadically, most recently (volume 6) in 2020.

22. Jubinsky, "Editorial," 13. Jubinsky was chair of FL-EPPC.

23. Cleary, "Editorial."

24. R. L. Cleary, email message to the author, October 23, 2021.

25. Lowe et al., *100 of the World's Worst Invasive Alien Species*.

26. Michael Day, email message to the author, October 27, 2021.

27. *Biosecurity*, no. 78 (September 2007): 2.

28. *Biosecurity*, no. 102 (April 2011): 27.

29. *Biosecurity* was not the only informal, lay-oriented publication on invasions in New Zealand during its existence. In 2000, the National Institute of Water and Atmospheric Research, a government-owned, stand-alone company, through its National Centre for Aquatic Biodiversity and Biosecurity, initiated a quarterly, glossy print and online newsletter, *Biodiversity Update*, that ran for five issues through 2002. It published articles about freshwater and especially marine plants and animals, with a few features on non-native species during its short existence. In 2002, it was replaced by *Aquatic Biodiversity & Biosecurity*, similarly glossy, both in print and online, and planned as an eight-page quarterly. This newsletter strongly emphasized invasions, with detailed, beautifully illustrated pieces on particular established non-native species, newly detected incursions, and warnings for the public to be on the lookout for species from elsewhere that were not (yet) in New Zealand waters. The first number alone featured six of the latter (the Mediterranean fanworm, green crab, Chinese mitten crab, northern Pacific sea star, Asian clam, and killer alga), three recent detections (the Asian paddle crab, striped sandgoby, and a snail, the red-rimmed melania), and several well-established non-natives. After seven issues, this newsletter was superseded in 2004 by the *National Centre for Aquatic Biodiversity & Biosecurity Update*, with a similar format and focus on invasions but only two pages in each issue. Initially this newsletter appeared every two months, which lengthened to every three months by the end of its existence (number 28, in 2008). Two species on the watchlist published in the first issue of *Aquatic Biodiversity & Biosecurity* have since arrived in New Zealand: the clam and the fanworm.

30. Lowe, "From the Editor," 2.

31. Piero Genovesi, email message to the author, October 26, 2021. As this book goes to press, a biannual, digital successor to *Aliens* has just been launched: *Invasions Newsletter* (Novoa et al., "Announcing the Invasions Newsletter").

32. In 1998, Kluwer acquired the British publisher Chapman & Hall, which in the 1930s had merged with Methuen, which published Elton's 1958 monograph. The paperback reprint of the monograph in 1972 was thus under the imprint of Chapman & Hall. In 2002, Chapman & Hall was acquired by venture capitalists, who also acquired the German publishing company Bertelsmann Springer, and in 2004, they merged the two companies to form Springer Science+Business Media. In 2015, Springer Science+Business Media merged with the Holtzbrinck Publishing Group/Nature Publishing Group to form Springer Nature, based in Switzerland and henceforth the publisher of *Biological Invasions*.

33. James Carlton, email message and in-person discussion with the author, May 12, 2021.

34. Carlton, email and in-person discussion.

35. Nuñez, Chiuffo, Pauchard, et al., "Making Ecology Really Global."

36. Kuebbing et al., "Self-Study of Editorial Board Diversity."

37. Nuñez, Chiuffo, Pauchard, et al., "Making Ecology Really Global."

38. Nuñez, Chiuffo, Pauchard, et al., "Making Ecology Really Global"; Kuebbing et al., "Self-Study of Editorial Board Diversity."

39. Nuñez and Amano, "Monolingual Searches Can Limit and Bias Results"; Nuñez, Chiuffo, Pauchard, et al., "Making Ecology Really Global"; Amano, Ramírez-Castañeda, et al., "Manifold Costs of Being a Non-Native English Speaker in Science."

40. Kowarik and Starfinger, "Neobiota."

41. Kowarik, "Biologische Invasionen in Deutschland"; Kowarik and Starfinger, "Neobiota," 23.

42. Kowarik and Starfinger ("Neobiota") described the fragmentation of research on invasions in Germany "as in other European countries" and reported that public awareness of invasions there began to grow in 1992, after the SCOPE project. It is noteworthy that German scientists were not heavily involved in the SCOPE project, and an anticipated contribution from Germany did not appear during the project. In 1999, a workshop in Bonn was convened on the "Legal Component of the SCOPE/IUCN Global Invasive Species Programme." At the Norway/UN Conference on Alien Species convened in 1996 in collaboration with SCOPE (Sandlund, Schei, and Viken, *Proceedings of the Norway/UN Conference on Alien Species*), at which a draft of a Global Strategy for Alien Species was developed under the auspices of SCOPE, sixty-four nations were represented by 116 delegates (not counting those of host nation Norway). Of these, twenty-two nations were in Europe, and of the 38 European delegates not based in Norway, only 1 was from Germany, compared to 5 each from France and the United Kingdom. That delegate was Lyle Glowka, a biologist and lawyer based in Bonn. The SCOPE project is the topic of chapter 9.

43. Kowarik and Starfinger, "Introduction."

44. Kühn et al., "Open Minded and Open Access."

45. Penev et al., "Fast, Linked, and Open."

46. The *NeoBiota* editorial board from 2011 has consistently listed as managing editor Lyubomir Penev, the founder and chief executive officer of Pensoft. He is not included in the tallies reported here.

47. Kowarik and Starfinger, "Neobiota."

48. Richardson, "Editorial," 1.

49. Richardson, "Editorial," 1.

50. J. Franklin, "*Diversity and Distributions* Is (Still) a Journal of Conservation Biogeography."

51. Burns et al., "Building a Better Future."

52. Panov and Gollasch, "Aquatic Invasions," 1.

53. Zimdahl, *History of Weed Science.*

54. Asa Gray, "Predominance and Pertinacity of Weeds," 161–62.

55. Zimdahl, *History of Weed Science.*

56. Brenchley, *Weeds of Farm Land,* 3–4.

57. Zimdahl, *History of Weed Science.*

58. DiTomaso, "WSSA Enters a New Era," 2. Despite the advent of the new journal, a group of weed scientists and invasion scientists in 2014 suggested that weed science still focuses too narrowly on management and would benefit from greater engagement with invasion science (S. Ward et al., "Agricultural Weed Research").

59. Dana and Graham. "Management of Biological Invasions."

60. Lucy and Panov, "BioInvasions Records."

61. Heimpel and Mills, *Biological Control.*

62. Shields et al., "History, Current Situation and Challenges."

63. The American Association of Economic Entomologists was initiated in 1889 as the Association of Economic Entomologists, and volume 1 of the *Journal of Economic Entomology,* in 1908, was listed on its first page as the "Official Organ of the Association of Economic Entomologists." Beginning with volume 2 in 1909, it was listed as "Official Organ American Association of Economic Entomologists." This heading remained through 1952 (volume 45). In 1953, the American Association of Economic Entomologists became part of the Entomological Society of America, an organization begun in 1906 (E. Smith, "Entomological Society of America"), and volume 46 in 1953 had this heading: "ENTOMOLOGICAL SOCIETY OF AMERICA. Formerly Official Organ American Association of Economic Entomologists."

64. For papers reporting on new insect arrivals, see, e.g., Felt, "Recent Observations upon European Insects in America"; Marlatt, "Some Recent New Importations."

65. E.g., Fiske, "Work with Parasites of the Gipsy and Brown-Tail Moths"; W. Thompson, "Notes on the Pupation and Hibernation of Tachinid Parasites."

66. Hokkanen, "Demonstrating the Safety of Biocontrol"; Van Lenteren et al., "Environmental Risk Assessment of Exotic Natural Enemies."

67. H. Roy and Wajnberg. "From Biological Control to Invasion."

68. Harmon and Andow, "Indirect Effects Between Shared Prey."

69. E.g., Barton, "Predictability of Pathogen Host Range"; Elkinton and Boettner, "Benefits and Harm Caused by the Introduced Generalist Tachinid"; Pratt and Center, "Biocontrol Without Borders"; Simberloff, "Risks of Biological Control for Conservation Purposes."

70. Andow et al., "Exotic Generalist Arthropod Biological Control Agents."

71. On analysis of interaction networks, see J. Todd et al., "Using Qualitative Food Webs to Predict Species at Risk." On what species traits and phylogeny can reveal about invasion impacts, see McGrath et al., "Framework for Predicting Competition." On the enemy release hypothesis, see Paula et al., "Scrutinizing the Enemy Release Hypothesis."

72. Barratt, Moran, et al.," Status of Biological Control."

73. Although the founding editors of *Biological Control* rationalized the need for the journal by claiming, "At present, no journal offers the broad coverage and the accent on current research on biological control envisioned for our journal" (Charudattan et al., editorial, 1), there has been no substantial difference in the breadth of subjects covered from that in *BioControl.*

74. Ehler, "Invasion Biology and Biological Control."

75. Heimpel and Mills, *Biological Control.*

76. R. G. Van Driesche et al., “Classical Biological Control.”

77. Hoy, “Biological Control of Arthropods.”

78. E.g., Carruthers and D’Antonio, introduction to “Science and Decision Making in Biological Control of Weeds”; Louda, Rand, et al., “Assessment of Ecological Risks in Weed Biocontrol”; M. Wright et al., “Evaluating Risks of Biological Control Introductions”; Pearson and Callaway, “Indirect Nontarget Effects of Host-Specific Biological Control Agents.”

79. Barratt, Howarth, et al., “Progress in Risk Assessment.”

80. V. Moran and Hoffmann, “Fourteen International Symposia on Biological Control of Weeds”; Barratt, Moran, et al., “Status of Biological Control.”

81. E.g., J. Frank and McCoy, “Risk of Classical Biological Control in Florida”; Marlin et al., “Successful Biocontrol Agent in the USA.”

82. Elton, *Animal Ecology*.

83. Clements and Shelford, *Bio-Ecology*; Shantz, “Frederic Edward Clements”; Tjossem, “Victor Shelford.”

84. Clements and Shelford, *Bio-Ecology*, 120.

85. Cahn, “Effect of Carp on a Small Lake.” In addition to anticipating by many years later developments in invasion science, Cahn made significant scientific contributions in other fields as he passed from job to job (Callomon, “Alvin Cahn”). He received his doctorate in 1924 from the University of Illinois under the direction of Victor Shelford, one of the leading figures in early twentieth-century ecology (more on him in chapter 12); Cahn’s publication on the impact of carp arose from research for his dissertation. He then taught at the University of Illinois. However, in the 1930s, he ran clandestine wrestling rings and recruited high school and college wrestlers, an activity that led to his forced resignation. The agreement ending his Illinois position permitted him to present a résumé with no indication of the reason for his departure, which allowed him to obtain a position in 1936 with the Tennessee Valley Authority (TVA). By 1939, he headed the new TVA Biological Readjustment Division, charged with developing and implementing conservation strategies for TVA lands. It was during this period that he explored the likely fate of game birds introduced to the region based on climate data (discussed further in chapter 14). However, his continued involvement in wrestling rings again led to his dismissal.

Cahn had enlisted in the US Medical Reserve Corps and served in France in 1918–19. In 1942, as part of the great buildup of US forces and supporting staff in the Aleutian Islands during World War II, Cahn, who was a Naval Reserve officer, was called to active duty (McCartney, “Alvin R. Cahn”). He attended US Navy training school and was then posted to the Dutch Harbor Naval Operating Base on Amaknak Island, shortly after the Japanese attack on the base. He discovered numerous foxholes, for protection against future attacks, excavated shell middens, and collected many artifacts from excavations into the middens. He shipped these to the Field Museum and later to the American Museum of Natural History. Cahn also made drawings and diagrams associated with the excavations. These collections and drawings were the basis of important publications on Aleutian archeology, although lack of provenience data for artifacts lessened their value (McCartney, “Alvin R. Cahn”).

After the war, Cahn returned to his home in Florida, collecting shells and seeking employment. His search led to an offer from the US War Department of a post in the occupation administration in Japan, and he moved to Tokyo in 1947, working in the Fisheries Division of the Natural Resources Section (Callomon, “Alvin Cahn”). He spent the rest of his life in Japan. Cahn corresponded with William J. Clench, curator of mollusks at the Museum of Comparative Zoology (Harvard), and Clench suggested

Japanese malacologists with whom he might be able to trade shells. This activity led to collecting trips and to interactions with several leading figures that contributed substantially to the development of Japanese malacology. In addition, Cahn wrote major reports on pearl, oyster, and clam culture in Japan.

However, beginning in 1947, Cahn began managing boxers and promoting bouts, and this activity became his full-time occupation by 1952 (Callomon, "Alvin Cahn"). He trained and managed Yoshio Shirai, who in 1952 became world flyweight champion, the first Japanese boxer to achieve world champion status. Cahn used inherited wealth to invest in Tokyo real estate and subsequently built a house for Shirai's family; he eventually moved in with them, and they cared for him as he declined into dementia, dying in 1971 (Shōji, "Man Who Trained Japan's First World Boxing Champion").

86. Oosting, *Study of Plant Communities*; Oosting, *Study of Plant Communities*, 2nd ed.

87. R. M. Anderson, "Effect of the Introduction of Exotic Animal Forms."

88. Odum, *Fundamentals of Ecology*.

89. Craige, *Eugene Odum*.

90. G. Barrett and Mabry, "Twentieth-Century Classic Books." The next three ranked books all played significant roles in the history of invasion science: Rachel Carson's *Silent Spring* (published in 1962; see chap. 7), Charles Darwin's *On the Origin of Species* (published in 1859; see chap. 3), and Aldo Leopold's *A Sand County Almanac* (published in 1949; see chap. 6).

91. The second edition was written with an acknowledgment of Eugene Odum's younger brother Howard Odum, a longtime collaborator and himself a key figure in the development of ecosystem ecology. The fourth edition had a different title, *Basic Ecology*, and was a revision of the third edition but omitted the lengthy section (part 2) of the third edition describing the various ecosystems typical of different habitats. The fifth edition, published posthumously, was coauthored by Gary W. Barrett, an ecosystem ecologist, the Eugene P. Odum Professor of Ecology at the University of Georgia, whose 1967 doctorate had been advised by Odum. Odum died in 2002, having submitted a draft of the fifth edition shortly before his death (Odum and Barrett, *Fundamentals of Ecology*, 5th ed., xvii).

92. Odum, *Fundamentals of Ecology*, 328.

93. Odum, *Fundamentals of Ecology*, 177.

94. Odum, *Fundamentals of Ecology*, 3rd ed. In 1963, between the second and third editions of *Fundamentals of Ecology*, Odum published a 150-page book, *Ecology*, in the Modern Biology series of Holt, Rinehart and Winston. This series targeted "the beginning student in college biology—as well as the gifted high school student and all interested readers" (Odum, *Ecology*, i). Odum thanked W. B. Saunders "for permission to adopt certain features used in . . . *Fundamentals of Ecology*" and emphasized the "different level of readers" (vi). He cited Elton's 1958 first edition of *The Ecology of Invasions by Animals and Plants* and, although omitting major examples from his own second edition, such as the chestnut blight and eucalypts, he included a few cases featured by Elton (e.g., the huge invasion impacts in New Zealand and the Hawaiian Islands). In fact, Odum began his book by saying that "the new orders of magnitude emerging with the atomic age necessitate new orders of thinking" and, as a first example, stated that "animals carelessly introduced to new environments may be hard to get rid of after they become pests. The introduction into North America of the English sparrow and starling . . . come to mind in this connection" (2); Elton had detailed these two invasions. Strikingly, Odum also claimed that "we have all heard about the introduction of the mongoose, a small weasel-like predator, on islands; usually it was brought in for the purpose of controlling rats, but instead it often wreaks havoc with ground-nesting

birds" (105). One wonders who the "all" is. Elton had not mentioned the mongoose. Though its impacts were certainly known to many scientists by 1963 (as discussed in chap. 4), scientists were not Odum's target audience.

95. Odum, *Fundamentals of Ecology*, 3rd ed. Following Elton, Odum termed the invading hybrid *Spartina townsendii.* It was subsequently determined that the initial hybridization produced a sterile, noninvasive diploid, now termed *S. townsendii*, and that the invading species, *S. anglica*, arose from a chromosomal mutation, a spontaneous duplication of chromosomes in an individual of *S. townsendii* (Elton, *Ecology of Invasions by Animals and Plants*, 2nd ed.).

96. Brooks and Dodson, "Predation, Body Size, and Composition of Plankton."

97. Odum and Barrett, *Fundamentals of Ecology*, 5th ed.

98. Orians, "Diversity of Textbooks"; Hagen, "Teaching Ecology."

99. Krebs, *Ecology*; Krebs, *Ecology*, 2nd ed.; Krebs, *Ecology*, 3rd ed.; Krebs, *Ecology*, 4th ed.; Krebs, *Ecology*, 5th ed.; Krebs, *Ecology*, 6th ed.

100. Lowe et al., *100 of the World's Worst Invasive Alien Species.*

101. Lever, *Naturalized Birds of the World*; Bissessur and Florens, "Predation of the Mauritius Endemic."

102. Howarth, "Environmental Impacts of Classical Biological Control"; Simberloff and Stiling, "How Risky Is Biological Control?"

103. Krebs, *Ecology*, 5th ed. See Simberloff and Stiling, "How Risky Is Biological Control?"; Civeyrel and Simberloff, "Tale of Two Snails."

104. Krebs, *Ecology*, 6th ed. See Howarth, "Environmental Impacts of Classical Biological Control"; Koch et al., "Predicted Impact of an Exotic Generalist Predator on Monarch Butterfly."

105. Horppila et al., "Top-Down or Bottom-Up Effects by Fish."

106. Ricklefs, *Ecology.*

107. Ricklefs and Miller, *Ecology*, 4th ed.

108. Ricklefs, *Economy of Nature.*

109. Ricklefs and Relyea, *Ecology: The Economy of Nature*, 7th ed.; Relyea and Ricklefs, *Ecology: The Economy of Nature*, 8th ed.

110. Relyea, *Ecology: The Economy of Nature*, 9th ed.

111. Ricklefs, *Ecology*, 2nd ed.

112. Ricklefs, *Ecology*, 3rd ed.

113. Ricklefs and Miller, *Ecology*, 4th ed.

114. Ricklefs, *Economy of Nature*; cf. Ricklefs, *Ecology*, 2nd ed.

115. Zaret and Paine, "Species Introduction in a Tropical Lake."

116. Ricklefs, *Economy of Nature*; cf. Ricklefs, *Ecology*, 2nd ed.

117. Ricklefs, *The Economy of Nature*, 6th ed; see also Zuk, Rotenberry, and Tinghitella, "Silent Night."

118. Ricklefs and Relyea, *Ecology: The Economy of Nature*, 7th ed.

119. Herms and McCullough, "Emerald Ash Borer Invasion of North America"; Springer, "Eradication of Invasive Species on Macquarie Island."

120. Relyea, *Ecology: The Economy of Nature*, 9th ed.

121. Pianka, *Evolutionary Ecology*, 154.

122. Pianka, *Evolutionary Ecology*, 2nd ed. Cf. Krebs, *Ecology*.

123. Pianka, *Evolutionary Ecology*, 3rd ed. Cf. Krebs, *Ecology*; Ricklefs, *Ecology*.

124. Pianka, *Evolutionary Ecology*, 5th ed. Cf. Ricklefs, *Ecology*.

125. Pianka, *Evolutionary Ecology*, 6th ed.; Pianka, *Evolutionary Ecology*, 7th ed.

126. Pianka, *Evolutionary Ecology*; Pianka, *Evolutionary Ecology*, 2nd ed.; Pianka, *Evolutionary Ecology*, 3rd ed.; Pianka, *Evolutionary Ecology*, 4th ed.; Pianka,

Evolutionary Ecology, 5th ed.; Pianka, *Evolutionary Ecology*, 6th ed.; Pianka, *Evolutionary Ecology*, 7th ed. Cf. Odum, *Fundamentals of Ecology*, 2nd ed.

127. Begon, Harper, and Townsend, *Ecology: Individuals, Populations, and Communities*; Begon, Harper, and Townsend, *Ecology: Individuals, Populations, and Communities*, 2nd ed.; Begon, Harper, and Townsend, *Ecology: Individuals, Populations, and Communities*, 3rd ed.; Begon, Townsend, and Harper, *Ecology: From Individuals to Ecosystems*, 4th ed.; Begon and Townsend, *Ecology: From Individuals to Ecosystems*, 5th ed. (The fourth edition permuted authorship to Begon, Townsend, and Harper and was retitled *Ecology: From Individuals to Ecosystems*. The foreword noted Harper's retirement, and he died in 2009. Begon and Townsend wrote the fifth edition.)

128. E.g., Crowl et al., "Impact of Introduced Brown and Rainbow Trout on Native Fish"; Flecker and Townsend, "Community-Wide Consequences of Trout Introduction"; C. Townsend, "Invasion Biology and Ecological Impacts of Brown Trout"; C. Townsend, "Individual, Population, Community, and Ecosystem Consequences"; Simon and Townsend, "Impacts of Freshwater Invaders."

129. Begon, Harper, and Townsend, *Ecology: Individuals, Populations, and Communities*. For instance, two studies of plants in California were cited to exemplify interspecific competition, but in neither case (compact brome and ripgut brome; oat and slender wild oat) was it noted that both species are non-native.

130. Begon, Harper, and Townsend, *Ecology: Individuals, Populations, and Communities*, 2nd ed.

131. Begon, Harper, and Townsend, *Ecology: Individuals, Populations, and Communities*, 3rd ed.

132. Begon, Townsend, and Harper, *Ecology: From Individuals to Ecosystems*, 4th ed. See also Louda, Arnett, et al., "Invasiveness of Some Biological Control Insects."

133. Begon and Townsend, *Ecology: From Individuals to Ecosystems*, 5th ed. Other applications related to invasions included 6.4—"Invasive Fragmenting Aquatic Weeds," 6.8—"Species Distribution Modeling for (Re)introductions and Invasions," 7.3—"The Fast-Slow Continuum, Invasion and Conservation," 8.3—"Apparent Competition Threatens an Endangered Plant," 8.4—"Red and Grey Squirrels and Squirrelpox Virus," 10.1—"Generalist, Switching Predators as Effective Biocontrol Agents?," 10.3—"What's Required of a Good Biological Control Agent?," 11.2—"The Value of Dung Beetles to Agriculture," 12.3—"Myxomatosis," 16.1—"A Parasitic Threat to Darwin's Finches," 17.1—"Mesopredator Release."

134. Cain, Bowman, and Hacker, *Ecology*: Cain, Bowman, and Hacker, *Ecology*, 2nd ed.; Cain, Bowman, and Hacker, *Ecology*, 3rd ed.; Bowman, Hacker, and Cain, *Ecology*, 4th ed., 525; Bowman and Hacker, *Ecology*, 5th ed., Bowman and Hacker, *Ecology*, 6th ed. (The fourth edition permuted the authors' names to Bowman, Hacker, and Cain; and Cain did not participate in the fifth or sixth edition.)

135. The chapter on conservation biology in the first edition was written by Loraine Kohorn, although similar chapters in subsequent editions did not mention a separate author.

136. Enloe et al., "Impacts of Prescribed Fire, Glyphosate, and Seeding."

137. Carroll, Dingle, and Klassen, "Genetic Differentiation of Fitness-Associated Traits."

138. Cain, Bowman, and Hacker, *Ecology*, 2nd ed.

139. Cain, Bowman, and Hacker, *Ecology*, 3rd ed. See also Vredenburg, "Reversing Introduced Species Effects."

140. Livdahl and Willey, "Prospects for an Invasion."

141. Schlaepfer et al., "Potential Conservation Value of Non-Native Species"; cf. Vitule, Freire, et al., "Revisiting the Potential Conservation Value of Non-Native Species."

142. Bowman, Hacker, and Cain, *Ecology*, 4th ed., 525.

143. Bowman, Hacker, and Cain, *Ecology*, 4th ed.

144. Bowman, Hacker, and Cain, *Ecology*, 5th ed.

145. Simberloff, "Contribution of Population and Community Ecology to Conservation Science."

146. Primack, *Essentials of Conservation Biology*; Primack, *Essentials of Conservation Biology*, 2nd ed.; Primack, *Essentials of Conservation Biology*, 3rd ed.; Primack, *Essentials of Conservation Biology*, 4th ed.; Primack, *Essentials of Conservation Biology*, 5th ed.; Primack, *Essentials of Conservation Biology*, 6th ed.

147. Primack, *Essentials of Conservation Biology*.

148. Primack, *Essentials of Conservation Biology*, 2nd ed.; Simberloff, Schmitz, and Brown, *Strangers in Paradise*.

149. Primack, *Essentials of Conservation Biology*, 3rd ed. For the paper, see Boettner et al., "Effects of a Biological Control Introduction."

150. Primack, *Essentials of Conservation Biology*, 4th ed. See also Louda, Kendall, et al., "Ecological Effects of an Insect Introduced for the Biological Control of Weeds."

151. Primack, *Essentials of Conservation Biology*, 5th ed.

152. Primack, *Essentials of Conservation Biology*, 6th ed.

153. Meffe and Carroll, *Principles of Conservation Biology*.

154. Meffe and Carroll, *Principles of Conservation Biology*, 218.

155. Meffe and Carroll, *Principles of Conservation Biology*, 2nd ed.

156. Groom, Meffe, and Carroll, *Principles of Conservation Biology*, 3rd ed.

157. Groom, Meffe, and Carroll, *Principles of Conservation Biology*, 3rd ed., 295.

158. Kareiva and Marvier, *Conservation Science*.

159. E.g., Kareiva et al., *Risk Analysis as a Tool for Making Decisions About the Introduction of Non-Indigenous Species*; Ruesink et al., "Reducing the Risks of Non-Indigenous Species Introductions"; Marvier et al., "How Do the Design of Monitoring and Control Strategies Affect the Chance of Detecting and Containing Transgenic Weeds?" At the University of Washington, Kareiva was Groom's doctoral advisor, Marvier was his postdoctoral fellow, and the doctoral dissertations of both Wonham and Parker acknowledged him for assistance and encouragement.

160. For the manifesto, see Kareiva and Marvier, "What Is Conservation Science?"; cf. the earlier Kareiva and Marvier, "Conserving Biodiversity Coldspots."

161. Voosen, "Myth-Busting Scientist"; Kuebbing and Simberloff, "Missing the Bandwagon."

162. Kareiva quoted in Lalasz, "Fighting Invasive Species," 8.

163. Kareiva and Marvier, *Conservation Science*. For Marris's suggestion, see Marris, "Ragamuffin Earth." On the salt cedar invasion, see, e.g. Stromberg, Chew, et al., "Changing Perceptions of Change."

164. Kareiva and Marvier, *Conservation Science*, 2nd ed.

165. Kareiva and Marvier, *Conservation Science*, 2nd ed., 559.

166. Kareiva, Marvier, and Silliman, *Effective Conservation Science*.

167. Cf. Hobbs, Higgs, and Hall, *Novel Ecosystems*; Murcia et al., "Critique of the 'Novel Ecosystem' Concept."

168. Silliman et al., "Business as Usual Leads to Underperformance in Coastal Restoration." For the research, see, e.g., Grosholz, "Recent Biological Invasion May Hasten Invasional Meltdown"; Grosholz et al., "Modeling the Impacts of the European Green Crab."

169. Among the books focusing on invasion by particular species were Nesler and Bergersen, *Mysids in Fisheries*; D. Williams, *Exotic Ants*; Goldschmidt, *Darwin's*

Dreampond (about the invasion of Lake Victoria by Nile perch, a translation of Goldschmidt, *Darwin's Hofvijver*); Jaffe, *And No Birds Sing* (about the invasion of Guam by the brown tree snake); Fuller et al., *Nonindigenous Fishes*; Montague, *Brushtail Possum*; Dolin, *Snakehead* (about the invasion by the northern snakehead of the Washington, DC, region); and Nico et al., *Black Carp*. Examples of books discussing particular aspects of invasions include Dahlsten and Garcia, *Eradication of Exotic Pests*; Rosenfield and Mann, *Dispersal of Living Organisms into Aquatic Ecosystems*; Low, *Feral Future* (about invasions into Australia); Pederson, *Marine Bioinvasions*; H. Mooney and Hobbs, *Invasive Species in a Changing World*; Pimentel, "Introduction: Non-Native Species in the World"; J. H. Myers and Bazely, *Ecology and Control of Introduced Plants*; and Ruiz and Carlton, *Invasive Species: Vectors and Management Strategies*. Shigesada and Kawasaki, *Biological Invasions: Theory and Practice*, aimed to be a textbook for undergraduate and graduate students but also to find an audience among field researchers and scholars. However, coverage of most topics was cursory and for some (e.g., evolution) was nonexistent. The strong suit was mathematical models of various processes that occur during invasions, such as spread (four of the ten chapters), competition, and predation, and the book was in fact based on an earlier book in Japanese by Shigesada that was wholly about modeling. Thus, this book could not really serve as a textbook for a university course on invasion science as it existed in 1997, but rather for a mathematics course of modeling various ecological processes. Noteworthy is that the early invasion science textbooks by Mark Williamson (M. Williamson, *Biological Invasions*) and Julie Lockwood and colleagues (Lockwood, Hoopes, and Marchetti, *Invasion Ecology*) discussed mathematical models of certain processes in some detail, particularly reaction-diffusion equations for spread of an invasion from its initial location in the new range.

170. M. Williamson, *Biological Invasions*.

171. M. Williamson, *Biological Invasions*, 4.

172. J. L. Lockwood, Hoopes, and Marchetti, *Invasion Ecology*.

173. J. L. Lockwood, Hoopes, and Marchetti, *Invasion Ecology*, 2nd ed.

174. The first edition cited 607 references and the second edition 881.

175. M. Davis, *Invasion Biology*.

176. M. Davis, *Invasion Biology*, 165. For the later publication, see M. Davis, Chew, et al., "Don't Judge Species on Their Origins."

177. Bright, *Life Out of Bounds*; Devine, *Alien Invasion*; J. Van Driesche and Van Driesche, *Nature Out of Place*; Baskin, *Plague of Rats and Rubbervines*; Burdick, *Out of Eden*; Lefeuvre, *Les invasions biologiques*; van der Weijden et al., *Biological Globalisation*; Anthony, *Aliens Among Us*.

178. On cane toad, see S. Lewis, *Cane Toads: An Unnatural History*. Stephanie Lewis based her book on her son Mark Lewis's 1988 cult classic documentary film of the same name. Mark Lewis directed a sequel, *Cane Toads: The Conquest*, released in 2011. On Nile perch, see Goldschmidt, *Darwin's Hofvijver*, Goldschmidt, *Darwin's Dreampond*. In 2004 Hubert Sauper directed a full-length feature film and Academy Award best documentary nominee, *Darwin's Nightmare*, on the invasion of Lake Victoria by the Nile perch. On killer alga, see Meinesz, *Le roman noir de l'algue "tueuse"*; Meinesz, *Killer Algae*. On brown tree snake, see Jaffe, *And No Birds Sing*. On northern snakehead, see Dolin, *Snakehead*. On chestnut blight, see Freinkel, *American Chestnut*. On common wasp, see Lester, *Vulgar Wasp*.

179. On Australia, see Low, *Feral Future*. On Patagonia, see de la Vega, *Invasión en Patagonia*. On the Laurentian Great Lakes, see Rapai, *Lake Invaders*; Dan Egan, *Death and Life of the Great Lakes*.

180. C. King, *Immigrant Killers*, C. King, *Invasive Predators in New Zealand.*; K. Todd, *Tinkering with Eden.*

181. Stolzenburg, *Rat Island*; Muñoz, *Caulerpa Conquest.*

182. Buhs, *Fire Ant Wars*; Spear, *Great Gypsy Moth War*; Marché, *Green Menace.*

183. Seibert, *Toad Overload*; Gleitzman, *Toad Rage*; Genovesi and Natalini, *Per un pugno di ghiande.*

184. G. Hamilton, *Super Species*; Marris, *Rambunctious Garden*; K. Thompson, *Where Do Camels Belong?*; Tassin, *La grande invasion*; Pearce, *New Wild*; Molnar, *Playing God in the Meadow.*

185. Orion, *Beyond the War on Invasive Species*; Shah, *Next Great Migration.*

186. Essig, "Sketch History of Entomology."

187. Carlton, "Journal of Biological Invasions"; Kühn et al., "Open Minded and Open Access."

Chapter 11

1. Darwin, *On the Origin of Species.*

2. Volpe, *Understanding Evolution*, 21.

3. Detmer, "Zum Problem der Vererbung."

4. Bordage, "À propos de l'hérédité des caractères acquis." Peach is native to Asia but was introduced to Europe at least as early as 300 BC.

5. Grinnell, "English Sparrow Has Arrived in Death Valley," 469.

6. Calhoun, "Role of Temperature and Natural Selection." Calhoun buttressed his conclusions with measurements that H. C. Bumpus had published in 1898 for house sparrows killed in Rhode Island during a storm, a study that became famous for demonstrating the action of "stabilizing" natural selection—that is, selection against individuals with extreme phenotypes and in favor of those near the mean phenotype (Bumpus, "Elimination of the Unfit"). Ironically, Calhoun demonstrated that, in fact, at least for size, selection in these birds was "directional" rather than stabilizing—that is, selection was against individuals of mean phenotype and instead favored individuals whose phenotype differed in one direction from the mean in a particular trait. The surviving birds were larger.

7. R. Johnston and Selander, "House Sparrows"; R. Johnston and Selander, "Evolution in the House Sparrow II"; R. Johnston and Selander, "Evolution in the House Sparrow III"; Selander and Johnston, "Evolution in the House Sparrow: Intrapopulation Variation." Selander pioneered the use of molecular genetic approaches to study population genetic structure, beginning soon after the advent in the mid-1960s of electrophoretic determination of isoenzyme distributions. By 1969, he had applied this technique to house mouse population relationships, by 1972 to those of blind cave fish, and in the 1980s to bacteria. His work on the house sparrow relied wholly on morphology, although later others studied differences and relationships of native and non-native house sparrows with more recently developed molecular techniques, such as Schrey and colleagues with microsatellites (Schrey, Grispo, et al., "Broad-Scale Latitudinal Patterns of Genetic Diversity") and with amplified fragment length polymorphisms (Schrey, Coon, et al., "Epigenetic Variation May Compensate for Decreased Genetic Variation with Introductions").

8. R. Johnston and Selander, "House Sparrows," 550.

9. R. Johnston and Selander, "Evolution in the House Sparrow II," 26.

10. A genetic *bottleneck* arises when few individuals from a larger population establish a new population—the very scenario assumed to have occurred in many invasions. By chance alone, one would expect less genetic variation in this founding population,

just as a few marbles chosen randomly from a large jar with many differently colored marbles would be expected to contain fewer colors. A *founder effect* is a type of bottleneck that occurs when the reduced genetic variation in the small founding population happens to have a genetic constitution that is far from the sort of genetic constitution that would usually arise if the founders were randomly drawn. To return to the marble analogy: if the large jar had one hundred marbles, ten each of ten colors, one could in principle randomly draw a sample of five marbles all of the same color, but that would be a very unusual result (in fact, if one performed this exercise a million times, this result would be expected to occur only thirty-three times). *Genetic drift* is the random change in genetic constitution that arises in any population because of random aspects of which individuals with which genotypes happen to live, die, and reproduce in each generation. In large populations, the effect of genetic drift pales in comparison to that of natural selection. However, in small populations, such as many of those introduced to new regions, genetic drift can be much more substantial.

11. E. Ashton and Zuckerman, "Influence of Geographic Isolation on the Skull of the Green Monkey (*Cercopithecus aethiops sabaeus*) I"; E. Ashton and Zuckerman, "Influence of Geographic Isolation on the Skull of the Green Monkey (*Cercopithecus aethiops sabaeus*) II"; E. Ashton and Zuckerman, "Influence of Geographic Isolation on the Skull of the Green Monkey (*Cercopithecus aethiops sabaeus*) III"; E. Ashton and Zuckerman, "Influence of Geographic Isolation on the Skull of the Green Monkey (*Cercopithecus aethiops sabaeus*) IV"; E. Ashton, "Influence of Geographic Isolation on the Skull of the Green Monkey (*Cercopithecus aethiops sabaeus*) V."

12. Ashton and Zuckerman, "Influence of Geographic Isolation on the Skull of the Green Monkey (*Cercopithecus aethiops sabaeus*) IV," 354.

13. Andrewartha and Birch, *Distribution and Abundance of Animals*; Birch, "Evolutionary Opportunity for Insects and Mammals in Australia."

14. Vinal, "European Corn Borer."

15. Dopman et al., "Components of Reproductive Isolation."

16. Andrewartha and Birch, *Distribution and Abundance of Animals.*

17. Zimmerman, *Insects of Hawaii*; Zimmerman, "Possible Evidence of Rapid Evolution in Hawaiian Moths."

18. Wilmshurst et al., "High Precision Radiocarbon Dating Shows Recent and Rapid Initial Human Colonization of East Polynesia."

19. Berry, "Evolution of an Island Population of the House Mouse"; Berry and Murphy, "Biochemical Genetics of an Island Population of the House Mouse."

20. On founder effect, see Berry, "Evolution of an Island Population of the House Mouse." On quick-acting natural selection, see Berry and Murphy, "Biochemical Genetics of an Island Population of the House Mouse"; Berry, "Diversity and Differentiation."

21. Berry et al., "House Mice of the Faroe Islands."

22. Clarke, "On Some Forms of *Mus musculus*, Linn," 163.

23. E.g., Huxley, *Evolution, the Modern Synthesis*; L. Matthews, *British Mammals.*

24. Berry, "Diversity and Differentiation."

25. Berry et al., "House Mice of the Faroe Islands."

26. Berry, "Diversity and Differentiation."

27. K. Myers, "Rabbit in Australia."

28. K. Myers, "Rabbit in Australia"; C. K. Williams and Moore, "Phenotypic Adaptation and Natural Selection in the Wild Rabbit."

29. C. K. Williams and Moore, "Phenotypic Adaptation and Natural Selection in the Wild Rabbit."

30. Elton, *Animal Ecology and Evolution.*

31. Pimentel, "Animal Population Regulation."

32. Baker and Stebbins, preface to Baker and Stebbins, *Genetics of Colonizing Species*, vii.

33. Simberloff, "Charles Elton—Neither Founder nor Siren."

34. For the book arising from the Asilomar workshop, see Baker and Stebbins, *Genetics of Colonizing Species.*

35. R. Hunter and Markert, "Histochemical Demonstration of Enzymes."

36. H. Harris, "Enzyme Polymorphisms in Man"; Hubby and Lewontin, "Molecular Approach to the Study of Genic Heterozygosity in Natural Populations, I"; Lewontin and Hubby, "A Molecular Approach to the Study of Genic Heterozygosity in Natural Populations, II."

37. Lewontin, "Electrophoresis in the Development of Evolutionary Genetics."

38. Clegg and Allard, "Patterns of Genetic Differentiation in the Slender Wild Oat."

39. Estoup and Guillemaud, "Reconstructing Routes of Invasion Using Genetic Data"; B. Fitzpatrick, Fordyce, et al., "What Can DNA Tell Us About Biological Invasions?"; Cristescu, "Genetic Reconstructions of Invasion History."

40. Selander and Kaufman, "Self-Fertilization and Genetic Population Structure in a Colonizing Land Snail."

41. Brussard, "Geographic Variation in North American Colonies of *Cepaea nemoralis.*"

42. Klitz, "Empirical Population Genetics."

43. R. Johnston and Klitz, "Variation and Evolution in a Granivorous Bird."

44. Parkin and Cole, "Genetic Differentiation and Rates of Evolution."

45. Dlugosch and Parker, "Founding Events in Species Invasions."

46. Ross, "Genetic Differentiation of Starling (*Sturnus vulgaris*: Aves) Populations"; A. Baker and Moeed, "Rapid Genetic Differentiation and Founder Effect in Colonizing Populations of Common Mynas"; A. Baker, "Genetic and Morphometric Divergence in Ancestral European and Descendant New Zealand Populations of Chaffinches."

47. Lavergne and Molofsky, "Increased Genetic Variation and Evolutionary Potential Drive the Success of an Invasive Grass."

48. Botstein et al., "Construction of a Genetic Map."

49. Mullis et al., "Specific Enzymatic Amplification of DNA In vitro."

50. Weathersbee et al., "Differentiation of *Diaprepes abbreviatus* and *Pachnaeus litus.*"

51. J. G. Williams et al., "DNA Polymorphisms Amplified by Arbitrary Primers."

52. C. L. Williams et al., "Geographical Origin of an Introduced Insect Pest."

53. Avise, *Molecular Markers, Natural History, and Evolution.*

54. Vos et al., "AFLP."

55. On AFLP's technical advantages over RFLP and RAPD, see Meudt and Clarke, "Almost Forgotten or Latest Practice?" On AFLP primarily revealing only presence of a locus, see Avise, *Molecular Markers, Natural History, and Evolution.*

56. Amsellem et al., "Comparison of Genetic Diversity of the Invasive Weed *Rubus alceifolius.*"

57. On microsatellites' discovery, see Weller et al., "Organization of the Human Myoglobin Gene." On their uses, see Avise, *Molecular Markers, Natural History, and Evolution.*

58. Tsutsui et al., "Reduced Genetic Variation and the Success of an Invasive Species."

59. Giraud et al., "Evolution of Supercolonies."

60. Reusch et al., "Microsatellites Reveal Origin and Genetic Diversity of Eurasian Invasions."

61. Schatz et al., "Deoxyribonucleic Acid Associated with Yeast Mitochondria"; Nass et al., "General Occurrence of Mitochondrial DNA."

62. Galtier et al., "Mitochondrial DNA as a Marker of Molecular Diversity."

63. Hebert, Cywinska, et al., "Biological Identifications Through DNA Barcodes"; Hebert, Hollingsworth, et al., "From Writing to Reading the Encyclopedia of Life."

64. Kolbe, Glor, Schettino, et al., "Genetic Variation Increases During Biological Invasion by a Cuban Lizard."

65. On the "genetic paradox of invasions," see Allendorf and Lundquist, "Introduction: Population Biology." See also Roman and Darling, "Paradox Lost."

66. Collins et al., "Variations on a Theme."

67. Gupta et al., "Single Nucleotide Polymorphisms"; Brumfield et al., "Utility of Single Nucleotide Polymorphisms."

68. B. Fitzpatrick, Johnson, et al., "Rapid Fixation of Non-Native Alleles Revealed."

69. Vidal et al., "SNP Diversity in Introduced Populations."

70. Vandepitte et al., "Rapid Genetic Adaptation."

71. E.g., M. Gardner et al., "Rise of the Machines."

72. Luikart et al., "The Power and Promise of Population Genomics."

73. Carvalho, "Molecular Ecology"; Avise, *Molecular Markers, Natural History, and Evolution.*

74. S. Barrett, "Foundations of Invasion Genetics."

75. Sax et al., *Species Invasions.*

76. R. Huey, Gilchrist, and Hendry, "Using Invasive Species to Study Evolution"; R. Huey, Gilchrist, et al., "Rapid Evolution of a Geographic Cline."

77. R. Huey, Gilchrist, et al., "Rapid Evolution of a Geographic Cline."

78. For the 1975 observation, see R. David and Bocquet, "Evolution in a Cosmopolitan Species."

79. Ricklefs, "Taxon Cycles." See also E. Wilson, "Nature of the Taxon Cycle."

80. S. Novak and Mack, "Genetic Bottlenecks in Alien Plant Species."

81. Wares et al., "Mechanisms That Drive Evolutionary Change."

82. Holt et al., "Theories of Niche Conservatism and Evolution."

83. Rice and Sax, "Testing Fundamental Evolutionary Questions."

84. E. Moran and Alexander, "Evolutionary Responses to Global Change."

85. Sexton et al., "Plasticity and Genetic Diversity May Allow Saltcedar to Invade Cold Climates"; Dybdahl and Kane, "Adaptation vs. Phenotypic Plasticity"; Nyamukondiwa et al., "Phenotypic Plasticity of Thermal Tolerance Contributes to the Invasion Potential of Mediterranean Fruit Flies"; Riis et al., "Invasion Strategies in Clonal Aquatic Plants"; Godoy et al., "Forests Are Not Immune to Plant Invasions"; Hahn et al., "Increased Phenotypic Plasticity to Climate May Have Boosted the Invasion Success of Polyploid *Centaurea stoebe*"; Kolbe et al., "Physiological Variation Among Invasive Populations of the Brown Anole"; Sultan and Matesanz, "Ideal Weed."

86. E.g., Ghalambor et al., "Adaptive Versus Non-adaptive Phenotypic Plasticity"; T. Wright et al., "Behavioral Flexibility and Species Invasions"; Davidson et al., "Do Invasive Species Show Higher Phenotypic Plasticity than Native Species?"; Hendry, *Eco-Evolutionary Dynamics.*

87. Krysan et al., "Diapause in *Diabrotica virgifera*"; M. E. Gray et al., "Adaptation and Invasiveness of Western Corn Rootworm."

88. E. Levine et al., "Adaptation of the Western Corn Rootworm to Crop Rotation."

89. M. E. Gray et al., "Adaptation and Invasiveness of Western Corn Rootworm."

90. Chu et al., "Gut Bacteria Facilitate Adaptation to Crop Rotation."

91. S. Barrett et al., *Invasion Genetics*; Bock et al., "What We Still Don't Know About Invasion Genetics." This special issue is *Molecular Ecology* 24, no. 9 (2015).

92. Cox, *Alien Species and Evolution.*

93. Estoup et al., "Is There a Genetic Paradox of Biological Invasion?"

94. Kolbe, Glor, Rodriguez-Schettino, et al., "Multiple Sources, Admixture, and Genetic Variation."

95. Keller et al., "Recent Admixture Generates Heterozygosity-Fitness Correlations."

96. Roman, "Diluting the Founder Effect."

97. On hybridization between individuals of non-native species and individuals of related native species, see Rhymer and Simberloff, "Extinction by Hybridization and Introgression." On hybridization between two different non-native species, see, e.g., Gaskin and Kazmer, "Introgression Between Invasive Saltcedars."

98. Ellstrand and Schierenbeck, "Hybridization as a Stimulus for the Evolution of Invasiveness in Plants?"

99. Ayers et al., "Introduction to 'Ecological and Evolutionary Consequences After Invaders Hybridize.'" For the cordgrass example, see Ainouche et al., "Hybridization, Polyploidy and Invasion."

100. Gaskin and Kazmer, "Introgression Between Invasive Saltcedars."

101. Gutekinst et al., "Clonal Genome Evolution."

102. Thomson, *Naturalisation of Animals and Plants in New Zealand.*

103. Gillespie, "Hybridization, Introgression, and Morphometric Differentiation."

104. Rhymer et al., "Mitochondrial Analysis of Gene Flow."

105. Muñoz-Fuentes et al., "Hybridization Between White-Headed Ducks and Introduced Ruddy Ducks."

106. Robertson et al., "Towards the European Eradication of the North American Ruddy Duck."

107. These changes in gene expression may be heritable or simply persist for a long period in the cells in which they occur (Gibney and Nolan, "Epigenetics and Gene Expression").

108. Richards et al., "Invasion of Diverse Habitats by Few Japanese Knotweed Genotypes."

109. Marin et al., "Biological Invasion."

110. Stapley et al., "Transposable Elements as Agents of Rapid Adaptation"; Marin et al., "Biological Invasion."

111. Schrader et al., "Transposable Element Islands."

112. Morais and Reichard, "Cryptic Invasions."

113. Saltonstall, "Cryptic Invasion by Non-Native Genotypes of the Common Reed."

114. Saltonstall and Meyerson, "*Phragmites australis.*"

115. Mabuchi et al., "Mitochondrial DNA Analysis Reveals Cryptic Large-Scale Invasion."

116. Geller, Carlton, and Powers, "PCR-Based Detection of mtDNA Haplotypes"; Geller, "Decline of a Native Mussel."

117. Esterhuizen et al., "Genetic Identification of Members of the *Bemisia tabaci* Cryptic Species Complex."

118. J. K. Brown et al., "Sweetpotato or Silverleaf Whiteflies"; Yu et al., "Different Thermal Tolerance and HSP Gene Expression."

119. Geller, Walton, et al., "Cryptic Invasions of the Crab *Carcinus.*"

120. W. Meyer et al., "Two for One."

121. Bush, "Sympatric Host Race Formation and Speciation in Frugivorous Flies."

122. For differences in allozymes, see Feder, Chilcote, and Bush, "Genetic Differentiation Between Sympatric Host Races." For chromosomal inversion, see Feder, Berlocher, et al., "Allopatric Genetic Origins for Sympatric Host-Plant Shifts"; Feder, Roethele, et al., "Evidence for Inversion Polymorphism." For mtDNA and nuclear gene sequences, see Xie et al., "Radiation and Divergence in the *Rhagoletis pomonella* Species Complex." For SNPs, see S. Egan et al., "Experimental Evidence of Genome-Wide Impact of Ecological Selection."

123. S. Egan et al., "Experimental Evidence of Genome-Wide Impact of Ecological Selection"; R. Taylor and Friesen, "Role of Allochrony in Speciation."

124. Ehrlich and Hanski, *On the Wings of Checkerspots*.

125. M. Singer, "Evolution of Food-Plant Preferences in the Butterfly, *Euphydryas Editha*."

126. M. Singer, "Evolution of Food-Plant Preferences in the Butterfly, *Euphydryas Editha*"; C. Thomas et al., "Incorporation of a European Weed into the Diet of a North American Herbivore."

127. C. Thomas et al., "Incorporation of a European Weed into the Diet of a North American Herbivore"; M. Singer, Thomas, and Parmesan, "Rapid Human-Induced Evolution of Insect-Host Associations."

128. C. Thomas et al., "Incorporation of a European Weed into the Diet of a North American Herbivore."

129. Radtkey and Singer, "Repeated Reversals of Host-Preference Evolution"; Wee, "Effects of Geographic Distance, Landscape Features and Host Association on Genetic Differentiation"; Mikheyev et al., "Rapid Microsatellite Isolation from a Butterfly"; M. Singer and McBride, "Multitrait, Host-Associated Divergence Among Sets of Butterfly Populations."

130. Strauss, Lau, and Carroll, "Evolutionary Responses of Natives to Introduced Species."

131. Carroll, Dingle, and Klassen, "Genetic Differentiation of Fitness-Associated Traits."

132. Carroll and Boyd, "Host Race Radiation in the Soapberry Bug"; Carroll, Dingle, et al., "Genetic Architecture of Adaptive Differentiation"; Carroll, Dingle, and Famula, "Rapid Appearance of Epistasis During Adaptive Divergence."

133. Carroll, Loye, et al., "And the Beak Shall Inherit."

134. Cox, *Alien Species and Evolution*; Strauss, Lau, and Carroll, "Evolutionary Responses of Natives to Introduced Species." For butterflies, the observation that many species use non-native host plants, and for some these are now the predominant hosts (S. Graves and Shapiro, "Exotics as Host Plants"), has even led to the suggestion that suppression of non-native plants is misguided (Pearce, *New Wild*).

135. Strauss, Lau, and Carroll, "Evolutionary Responses of Natives to Introduced Species."

136. Lever, *Cane Toad*; Shine, "Invasive Species as Drivers of Evolutionary Change."

137. B. Phillips and Shine, "Adapting to an Invasive Species."

138. B. Phillips and Shine, "Invasive Species Induces Rapid Adaptive Change in a Native Predator."

139. B. Phillips et al., "Predator Behaviour and Morphology Mediates the Impact of an Invasive Species."

140. Zuk, Simmons, and Cupp, "Calling Characteristics." The Pacific field cricket could have arrived in the Hawaiian Islands with the Polynesians over a thousand years

ago or might have been introduced from ships in the nineteenth century (Zuk, Rotenberry, and Simmons, "Calling Songs of Field Crickets").

141. Silent males, known as "satellite males," are found in several otherwise singing cricket species and mate by intercepting females attracted to calling males. Such behavior may be adaptive in particularly dense populations or in certain other circumstances (Tinghitella et al., "Preexisting Behavior Renders a Mutation Adaptive").

142. Zuk, Rotenberry, and Simmons, "Calling Songs of Field Crickets."

143. Tinghitella et al., "Preexisting Behavior Renders a Mutation Adaptive."

144. Tinghitella, "Rapid Evolutionary Change in a Sexual Signal"; Tinghitella et al., "Preexisting Behavior Renders a Mutation Adaptive."

145. Strauss, "Ecological and Evolutionary Responses [. . .] Eco-Evolutionary Feedbacks"; Hendry, *Eco-Evolutionary Dynamics*.

146. Schluter and McPhail, "Ecological Character Displacement."

147. On disrupting the mating behavior of limnetic males, see Velema, Rosenfeld, and Taylor, "Effects of Invasive American Signal Crayfish." On population dominance of an intermediate form, see Rudman and Schluter, "Ecological Impacts of Reverse Speciation."

148. Combrink et al., "Parallel Shifts in Trout Feeding Morphology."

149. L. Boyle, "Relative Effects of Fish and Nutrient Introduction."

150. On Darwin's writings relating to invasions, see Ludsin and Wolfe, "Biological Invasion Theory." On the naturalization hypothesis, see Darwin, *On the Origin of Species*; Daehler, "Darwin's Naturalization Hypothesis Revisited."

151. De Candolle, *Géographie botanique raisonnée*.

152. Darwin, *On the Origin of Species*, 86.

153. Daehler, "Darwin's Naturalization Hypothesis Revisited."

154. For the earlier studies, see Mack, "Biotic Barriers to Plant Naturalization"; Rejmánek, "Theory of Seed Plant Invasiveness"; Rejmánek, "Invasive Plant Species and Invadible Ecosystems."

155. Strauss, Webb, and Salamin, "Exotic Taxa Less Related to Native Species Are More Invasive."

156. Blossey and Nötzold, "Evolution of Increased Competitive Ability in Invasive Non-Indigenous Plants."

157. Ridley, *Dispersal of Plants Throughout the World*, 636.

158. Thebaud and Simberloff, "Are Plants Really Bigger in Their Introduced Ranges?"

159. Honor and Colautti, "EICA 2.0."

160. Felker-Quinn et al., "Meta-Analysis Reveals Evolution in Invasive Plant Species but Little Support for Evolution of Increased Competitive Ability"; Rotter and Holeski, "Meta-Analysis of the Evolution of Increased Competitive Ability Hypothesis."

161. Honor and Colautti, "EICA 2.0."

Chapter 12

1. Millet, *Les Merveilles des Fleuves et des Ruisseaux*; Newton cited by C. King, *Immigrant Killers*; P. Wells, "Enemy of the Rabbit."

2. T. Palmer, "Danger of Introducing Noxious Animals and Birds."

3. Chamberlain, "There Is No Universal Pond Fish"; J. Deacon et al., "Some Effects of Introduced Fishes."

4. Elton, *Ecology of Invasions by Animals and Plants*.

5. Nicholson, "Balance of Animal Populations"; Errington, "Predation and Vertebrate Populations."

6. Savidge, "Extinction of an Island Forest Avifauna by an Introduced Snake"; Dorcas et al., "Severe Mammal Declines Coincide with Proliferation of Burmese Pythons."

7. E.g., Fields, "Invasion of the Snakeheads."

8. Grinnell, "Risks Incurred in the Introduction of Alien Game Birds."

9. Levi, "Evaluation of Wildlife Importations."

10. Elton, *Animal Ecology*, 28.

11. Elton, *Ecology of Invasions by Animals and Plants*, 123.

12. Middleton, *Grey Squirrel*; Shorten, *Squirrels*; Edwards, "Red Squirrel Disease."

13. Reynolds, "Details of the Geographic Replacement of the Red Squirrel," 161.

14. Dale and Chantrey, "Subtle Endemic Virus in Grey Squirrels."

15. Pyšek, Blackburn, et al., "Displacement and Local Extinction of Native and Endemic Species."

16. Orwig and Foster, "Forest Response to the Introduced Hemlock Woolly Adelgid"; Herms and McCullough, "Emerald Ash Borer Invasion of North America"; Daniels et al., "Thousand Cankers Disease Complex."

17. Brulliard, "Highly Contagious and Fatal Virus"; Dugovich and Ruder, "Rabbit Hemorrhagic Disease Virus 2 Reported."

18. Simberloff, "Risks of Biological Control for Conservation Purposes."

19. Williamson, "Sir Joseph Hooker's Lecture on Insular Floras," 68.

20. For antecedents, cf. A. McFadyen, *Animal Ecology*; Major, "Historical Development of the Ecosystem Concept." For arguments over whether plants and animals constitute separate communities, see J. V. Phillips, "Biotic Community"; Shelford, "Some Concepts of Bioecology." On the "superorganism" concept, see A. McFadyen, *Animal Ecology*; McIntosh, *Background of Ecology*.

21. E.g., Shelford, *Animal Communities in Temperate America*.

22. Shelford, *Laboratory and Field Ecology*, 2.

23. E.g., Stephenson," Validity of the Community Concept in Marine Biology"; Lawton, "Are There General Laws in Ecology?"; Lawton, *Community Ecology in a Changing World*; Simberloff, "Community Ecology"; Ricklefs, "Disintegration of the Ecological Community"; Vellend, "Conceptual Synthesis in Community Ecology."

24. Cragg, review of *Natural Communities*, 3. Perhaps the most thoroughgoing defense of broad use of the term was by Robert MacArthur, one of the most influential twentieth-century ecologists, who saw his classic study of a group of coexisting warblers (MacArthur, "Population Ecology of Some Warblers") as testing a key tenet of community ecology. In 1971, he announced that "irrespective of how other people use the term 'community'—and there are almost as many uses as there are ecologists—I use it here to mean any set of organisms currently living near each other and about which it is interesting to talk. Thus, we may speak about the community of grassland birds in a particular field or about the community of seed-eating birds that is a subset of the first community" (MacArthur, "Patterns of Terrestrial Bird Communities," 189–90).

25. Leopold, "Cheat Takes Over" (1941); Leopold, *Sand County Almanac*.

26. Breytenbach, "Impacts of Alien Organisms on Terrestrial Communities"; H. Mooney, Hamburg, and Drake, "Invasion of Plants and Animals into California."

27. Mack, "Temperate Grasslands Vulnerable to Plant Invasion."

28. Tansley, "Use and Abuse of Vegetational Concepts and Terms," 299.

29. On the groundwork, see McFadyen, *Animal Ecology*; Major, "Historical Development of the Ecosystem Concept." On the conflicts, see Tansley, "Use and Abuse of Vegetational Concepts and Terms"; McIntosh, *Background of Ecology*.

30. Vilà, Espinar, et al., "Ecological Impacts of Invasive Alien Plants"; E. Cameron et al., "Global Meta-Analysis of the Impacts of Terrestrial Invertebrate Invaders"; Vilà and Hulme, "Non-Native Species, Ecosystem Services, and Human Well-Being."

31. Vitousek, "Biological Invasions and Ecosystem Properties"; Ramakrishnan and Vitousek, "Ecosystem-Level Processes."

32. Vitousek, "Biological Invasions and Ecosystem Properties."

33. Vitousek credits growing up in Hawaii, with its wealth of non-native species, with playing a major role in his interest in invasions. He read Elton's 1958 invasion monograph in an English class on the literature of science, and it resonated with his Hawaiian experience (P. Vitousek, email message to the author, February 11, 2023).

34. H. Mooney, Hamburg, and Drake, "Invasion of Plants and Animals into California"; Vitousek, Loope, and Stone, "Introduced Species in Hawaii."

35. Vitousek, Loope, and Stone, "Introduced Species in Hawaii."

36. D'Antonio and Vitousek, "Biological Invasions by Exotic Grasses."

37. Vitousek, "Biological Invasions and Ecosystem Processes," 9.

38. Williamson, *Biological Invasions*.

39. McIntosh, *Background of Ecology*; Worster, *Nature's Economy*.

40. I. Macdonald, Loope, et al., "Wildlife Conservation."

41. Vitousek, D'Antonio, and Asner, "Invasions and Ecosystems"; Fei et al., "Biogeomorphic Impacts of Invasive Species."

42. Simberloff, "Keystone Species and Community Effects." For more on "keystone species," see Paine, "Note on Trophic Complexity."

43. Power et al., "Challenges in the Quest for Keystones."

44. J. Chapman et al., "Premature Refutation of a Human-Mediated Marine Species Introduction"; Blakeslee et al., "Solving Cryptogenic Histories."

45. On changes to high tide pools, see Lubchenco, "Plant Species Diversity in a Marine Intertidal Community." On changes to coastal areas, see Bertness, "Habitat and Community Modification."

46. Dayton, "Toward an Understanding of Community Resilience," 84. See also Dayton, "Experimental Evaluation of Ecological Dominance."

47. M. Wells et al., "History of Introduction of Invasive Alien Plants to Southern Africa."

48. Jones et al., "Organisms as Ecosystem Engineer"; Jones et al., "Positive and Negative Effects of Organisms as Physical Ecosystem Engineers"; Cuddington et al., *Ecosystem Engineers*.

49. Crooks, "Characterizing Ecosystem-Level Consequences."

50. Vitousek, Loope, and Stone, "Introduced Species in Hawaii," 226.

51. Ehrenfeld, "Ecosystem Consequences of Biological Invasions"; Simberloff, "How Common Are Invasion-Induced Ecosystem Impacts?"

52. Judith Fisher et al., "Enhanced Soil and Leaf Nutrient Status."

53. Lubchenco, "Plant Species Diversity in a Marine Intertidal Community"; Bertness, "Habitat and Community Modification."

54. On cordgrasses, see Alan Gray, Raybould, and Brown, "Environmental Impact of *Spartina anglica*." On paperbark, see Serbesoff-King, "Melaleuca in Florida."

55. On the sour fig, see Manning and Paterson-Jones, *Field Guide to Fynbos*. On its impact in California, see Molinari et al., "*Carpobrotus* as a Case Study." On its effect in Portugal, see Maltez-Mouro, Maestre, and Freitas, "Weak Effects of the Exotic Invasive *Carpobrotus edulis*."

56. Savidge, "Extinction of an Island Forest Avifauna by an Introduced Snake."

57. Mortensen et al., "Snake in Paradise."

58. Rogers et al., "'Natural Experiment' Demonstrates Top-Down Control of Spiders."

59. Fritts and Rodda, "Role of Introduced Species in the Degradation of Island Ecosystems."

60. Cf. Ehrlich and Ehrlich, *Extinction*; Ehrlich and Mooney, "Extinction, Substitution, and Ecosystem Services."

61. Daily, "Introduction: What Are Ecosystem Services?," 3.

62. H. Mooney and Ehrlich, "Ecosystem Services." For the 1970 report, see SCEP [Study of Critical Ecosystem Problems], *Man's Impact on the Global Environment*. For the 1997 volume, see Daily, *Nature's Services*.

63. Millennium Ecosystem Assessment, *Ecosystems and Human Well-Being*.

64. Millennium Ecosystem Assessment, *Ecosystems and Human Well-Being*, fig. 4.3.

65. Subsequent research casts doubt on whether impacts on several of the ecosystem types cited as unchanging are in fact not increasing. For instance, climate change and increasing human access to high-elevation sites are increasing plant invasions in mountains (Pauchard et al., "Ain't No Mountain High Enough"; Alexander, Lembrechts, et al., "Plant Invasions into Mountains and Alpine Ecosystems"), while several authors suggest that the same forces are increasing invasions in both polar regions (Ruiz and Hewitt, "Latitudinal Patterns of Biological Invasions in Marine Ecosystems"; K. Hughes, Pertierra, et al., "Biological Invasions in Terrestrial Antarctica"; K. Hughes, Pescott, et al., "Invasive Nonnative Species Likely to Threaten Biodiversity"; McGeoch et al., "Monitoring Biological Invasion Across the Broader Antarctic"; G. Duffy et al., "Barriers to Globally Invasive Species Are Weakening Across the Antarctic"; Ricciardi et al., "Invasion Science"; C. Fraser et al., "Antarctica's Ecological Isolation Will Be Broken"; Cardenas et al., "First Mussel Settlement Observed in Antarctica").

66. Charles and Dukes, "Impacts of Invasive Species on Ecosystem Services."

67. Pejchar and Mooney, "Invasive Species, Ecosystem Services and Human Well-Being."

68. Vilà, Basnou, et al., "How Well Do We Understand the Impacts of Alien Species on Ecosystem Services?"

69. Vilà and Hulme, *Impact of Biological Invasions on Ecosystem Services*.

70. Bennett, review of *Impact of Biological Invasions on Ecosystem Services*, by Vilà and Hulme.

71. Hulme and Vilà, "Integrating the Impacts of Non-Native Species on Ecosystem Services into Environmental Policy," 327.

72. Kueffer and Kull, "Non-Native Species and the Aesthetics of Nature."

73. Castro-Díez et al., "Global Effects of Non-Native Tree Species."

74. Schlaepfer et al., "Potential Conservation Value of Non-Native Species."

75. Leopold, "Biotic Land-Use"; Leopold, "Conservation: In Whole or in Part?"; Simberloff, "Integrity, Stability, and Beauty."

76. McCann, "Diversity-Stability Debate."

77. For the argument, see, e.g., Wardle et al., "Introduced Browsing Mammals in New Zealand Natural Forests"; D. Hooper et al., "Effects of Biodiversity on Ecosystem Functioning"; Srivastava and Vellend, "Biodiversity-Ecosystem Function Research"; Tilman et al., "Biodiversity and Ecosystem Functioning." For the research, see, e.g., Bellwood et al., "Limited Functional Redundancy in High Diversity Systems"; Kareiva and Levin, *Importance of Species*.

78. Vitousek, "Biological Invasions and Ecosystem Properties."

79. Wardle, Bardgett, et al., "Ecological Linkages."

80. On plantation trees, see Egunjobi, "Impact of Some Exotic Plant Species." For previous research, see Lundgren, *Soil Conditions and Nutrient Cycling*.

81. *Science* 304, no. 5677 (2004).

82. Van der Putten et al., "Linking Above- and Belowground Multitrophic Interactions"; Bardgett and Wardle, "Herbivore-Mediated Linkages."

83. Wardle, Bardgett, et al., "Ecological Linkages."

84. Wardle, Barker, et al., "Introduced Browsing Mammals in New Zealand Natural Forests"; Klironomos, "Feedback with Soil Biota."

85. Wardle, *Communities and Ecosystems*; Bardgett and Wardle, *Aboveground–Belowground Linkages*.

86. E.g., Crous et al., "Identifying and Naming Plant Pathogenic Fungi."

87. For early reviews focused on changes in climate and land use, see, e.g., Wardle, Verhoef, and Clarholm, "Trophic Relationships"; Wolters et al., "Effects of Global Changes." For later ones focused on invasions, see, e.g., Klironomos, "Feedback with Soil Biota"; Wardle, Bardgett, et al., "Ecological Linkages."

88. Baas-Becking, *Geobiologie of Inleiding tot de Milieukunde*; cf. O'Malley, "'Everything Is Everywhere.'"

89. Martiny et al., "Microbial Biogeography."

90. Thakur et al., "Microbial Invasions in Terrestrial Ecosystems."

Chapter 13

1. For the dissertation, see Vitousek, "Regulation of Element Concentrations." For early career work, see, e.g., Vitousek, "Nutrient Cycling and Nutrient Use Efficiency."

2. E.g., Vitousek, Van Cleve, et al., "Soil Development and Nitrogen Turnover"; Vitousek, "Biological Invasions and Ecosystem Properties."

3. Glass, "Primary, Secondary, and Meta-Analysis of Research," 3.

4. Glass was at pains to argue for the superiority of his approach to the method that was then most commonly used, and is still frequently used, to examine groups of studies: vote counting. In this method, each individual study is construed as a test of the same hypothesis, and the numbers of studies supporting and rejecting the hypothesis are tallied. The single outcome is determined by this tally: the hypothesis is rejected or not. Glass particularly criticized the inability of this approach to incorporate the magnitude of the result of each individual study, so that, for example, a close rejection counts exactly as much as an overwhelming rejection. Although meta-analysts routinely deplore vote counting on this and other grounds, it continues to be used in ecology (cf. G. Stewart, "Meta-Analysis in Applied Ecology").

5. McGaw and Glass, "Choice of the Metric."

6. Rosenberg et al., "Effect Sizes." Several authors have called for a more restrictive definition of the term "meta-analysis" in ecology, in particular excluding two approaches Glass included, vote counting and tests simply combining probabilities of a number of independent tests of hypotheses (though he criticized vote counting). D. Vetter and colleagues and J. Koricheva and J. Gurevitch do not accept such studies as "proper" meta-analyses (Vetter et al., "Meta-Analysis: A Need for Well-Defined Usage"; Koricheva and Gurevitch, "Uses and Misuses of Meta-Analysis"), though both approaches are still occasionally employed in invasion science (e.g., Beric and MacIsaac, "Determinants of Rapid Response Success"), as are others not embraced by these critics.

7. On "formal meta-analyses," D. Vetter et al., "Meta-Analysis: A Need for Well-Defined Usage"; Koricheva and Gurevitch, "Uses and Misuses of Meta-Analysis." On "broad-sense meta-analyses," see Boltovskoy et al., "Traits and Impacts of Introduced Species."

8. Lau et al., "History and Progress of Meta-Analysis."

9. For textbooks pitched at social scientists, see, e.g., Glass et al., *Meta-Analysis of Social Research*; J. Hunter et al., *Meta-Analysis: Cumulating Research Findings*; R. Rosenthal, *Meta-Analytic Procedures for Social Research*. For one aimed at medical researchers, see Jenicek, *Meta-analyse en médecine*.

10. Lau et al., "History and Progress of Meta-Analysis."

11. Gurevitch et al., "Meta-Analysis in Ecology." They identified the first two papers as those by Järvinen on the impact of female age on egg-laying in two bird species and by Marchant and McGrew on laterality of function in apes, both published in 1991 (Järvinen, "Meta-Analytic Study of the Effects of Female Age"; Marchant and McGrew, "Laterality of Function in Apes"). Marchant and McGrew did not actually perform a statistical meta-analysis of the sort advocated by social scientists in the 1980s but rather discussed how the variability of the fifty-eight apparently relevant studies was too great to test the hypothesis that asymmetry of cerebral structure is correlated with laterality of function (e.g., hand use).

12. On the 2010 articles, see Lau et al., "History and Progress of Meta-Analysis." For the 2013 textbook, see Koricheva et al., *Handbook of Meta-Analysis in Ecology and Evolution*.

13. Lau et al., "History and Progress of Meta-Analysis."

14. Borenstein et al., *Introduction to Meta-Analysis*.

15. R. Rosenthal, "File Drawer Problem."

16. E.g., Borenstein et al., *Introduction to Meta-Analysis*.

17. Wonham et al., "Ecological and Geographic Patterns of Impact in Invaded Communities."

18. J. Levine et al., "Meta-Analysis of Biotic Resistance."

19. The Web of Science search employed the keywords (("non-native" OR "non-native" OR "exotic speci*" OR "alien" OR "invas* speci*" OR "biolog* Invas*") AND ("network*" OR "meta-anal*" OR "metaanal*")).

20. Boltovskoy et al., "Traits and Impacts of Introduced Species."

21. See note 4 for references on "vote counting."

22. Vilà, Williamson, and Lonsdale, "Competition Experiments on Alien Weeds with Crops."

23. Davidson et al., "Do Invasive Species Show Higher Phenotypic Plasticity than Native Species?," 421.

24. Gaertner et al., "Impacts of Alien Plant Invasions."

25. Gaertner et al., "Impacts of Alien Plant Invasions," 319.

26. Cavaleri and Sack, "Comparative Water Use."

27. Vilà, Espinar, et al., "Ecological Impacts of Invasive Alien Plants," 702.

28. Read et al., "Accounting for the Nested Nature of Genetic Variation."

29. G. J. Vermeij, for instance, proposed an agenda for invasion science in order to "move beyond the anecdotal stage in which it currently resides" (Vermeij, "Agenda for Invasion Biology," 5). This charge has frequently been leveled against ecology as a whole and especially against community ecology, the ecological subfield most closely associated with invasion science (e.g., Lawton, "Are There General Laws in Ecology?").

30. Hendry, "Key Questions on the Role of Phenotypic Plasticity," 32. For the source cited by Hendry, see Davidson et al., "Do Invasive Species Show Higher Phenotypic Plasticity than Native Species?"

31. Oduor et al., "Exotic Vertebrate and Invertebrate Herbivores," 407; Bird et al., "Damage Caused by Low-Density Exotic Herbivore Populations," 743. For the other

source cited by Bird, see J. Parker et al., "Opposing Effects of Native and Exotic Herbivores."

32. Koricheva et al., *Handbook of Meta-Analysis in Ecology and Evolution.*

33. Barabási, *Network Science*; Proulx et al., "Network Thinking in Ecology and Evolution"; Borrett et al., "Rise of Network Ecology."

34. A "generalized community matrix" is a matrix in which each species is represented by both a row and a column, and the value at the intersection of a row and column is a term representing the interaction strength between the species represented by the row and the species represented by the column. This term is often the coefficient of Lotka-Volterra equations, but other values have been used (M. Novak et al., "Characterizing Species Interactions").

35. Dehling, "Structure of Ecological Networks."

36. Egerton, "Understanding Food Chains and Food Webs."

37. R. Bradley, *New Improvements of Planting and Gardening.*

38. Camerano, "Dell'equilibrio dei vivente mercé la reciproca distruzione"; Joel Cohen, "Lorenzo Camerano's Contribution to Early Food Web Theory"; Summerhayes and Elton, "Bear Island." Elton, *Animal Ecology.*

39. Eklöf and Allesina, "Networks, Ecological."

40. Levins, *Evolution in Changing Environments.* On the Lotka-Volterra competition equations, see Gotelli, *Primer of Ecology.*

41. Barabási, *Network Science.*

42. Barabási, *Network Science*; Pavlopoulos et al., "Bipartite Graphs." Interactions between individuals within a single-species population have also been depicted as networks of various types (Croft et al., "Hypothesis Testing in Animal Social Networks"; Hobson et al., "Analytical Framework").

43. Olesen et al., "Invasion of Pollination Networks"; Lopezaraiza-Mikel et al., "Impact of an Alien Plant."

44. May, *Stability and Complexity in Model Ecosystems*; cf. M. Novak et al., "Characterizing Species Interactions." For research that focused on one sort of interaction at a time, see, e.g., Ings et al., "Ecological Networks—Beyond Food Webs"; Sauve, Fontaine, and Thébault, "Structure–Stability Relationships."

45. Ringel et al., "Stability and Persistence of Mutualisms"; Jang, "Dynamics of Herbivore-Plant-Pollinator Models."

46. Grimm and Wissel, "Babel, or the Ecological Stability Discussions."

47. Dunne et al., "Network Structure and Biodiversity Loss in Food Webs."

48. This version of stability has sometimes been termed "tolerance" (Memmott, Waser, and Price, "Tolerance of Pollination Networks to Species Extinctions.") or "robustness" (Dunne et al., "Network Structure and Biodiversity Loss in Food Webs").

49. Blüthgen, "Why Network Analysis Is Often Disconnected from Community Ecology."

50. Goldstein and Zych, "What If We Lose a Hub?"

51. Yadav and Babu, "NEXCADE: Perturbation Analysis."

52. E.g., Goodman, "Theory of Diversity-Stability Relationships in Ecology"; Murdoch, "Diversity, Complexity, Stability and Pest Control."

53. Elton's view of the role of diversity in conferring stability was perhaps inspired by earlier writings by his friend Aldo Leopold (Simberloff, "Integrity, Stability, and Beauty").

54. May, "Will a Large Complex System Be Stable?"; May, *Stability and Complexity in Model Ecosystems.*

55. Pimm, *Food Webs.*

56. Case, "Invasion Resistance"; Mithen and Lawton, "Food-Web Models."

57. Proulx et al., "Network Thinking in Ecology and Evolution"; Borrett et al., "Rise of Network Ecology"; Poisot et al., "Structure of Probabilistic Networks"; Ings and Hawes, "History of Ecological Networks."

58. Borrett et al., "Rise of Network Ecology."

59. The Web of Science search employed the keywords (("non-native" OR "non-native" OR "exotic speci*" OR "alien" OR "invas* speci*" OR "biolog* Invas*") AND ("network*")). For the 2002 publication, see Oleson et al., "Invasion of Pollination Networks."

60. Knight et al., "Reflections on, and Visions for, the Changing Field of Pollination Ecology."

61. Kaiser-Bunbury et al., "Community Structure of Pollination Webs."

62. D. Larson et al., "Exotic Plant Infestation."

63. Memmott, Waser, and Price, "Tolerance of Pollination Networks to Species Extinctions."

64. Valdovinos, Ramos-Jiliberto, et al., "Structure and Dynamics of Pollination Networks."

65. Lopezaraiza-Mikel et al., "Impact of an Alien Plant."

66. Russo, Memmott, et al., "Patterns of Introduced Species Interactions Affect Multiple Aspects of Network Structure."

67. Russo, Nichol, and Shea, "Pollinator Floral Provisioning by a Plant Invader"; Russo, Albert, et al., "Experimental Species Introduction Shapes Network Interactions."

68. Valdovinos, Berlow, et al., "Species Traits and Network Structure."

69. Knight et al., "Reflections on, and Visions for, the Changing Field of Pollination Ecology."

70. J. Strong and Leroux, "Impact of Non-Native Terrestrial Mammals."

71. Salvaterra et al., "Impacts of the Invasive Alga *Sargassum muticum*."

72. Galiana et al., "Invasions Cause Biodiversity Loss."

73. Romanuk et al., "Predicting Invasion Success."

74. Lurgi et al., "Network Complexity and Species Traits Mediate the Effects."

75. Emer et al., "Impact of the Invasive Alien Plant."

76. On the network of ants and plants, see Sugiura, "Species Interactions—Area Relationships." For the other two network studies, see Thomsen and South, "Communities and Attachment Networks." For the lake network of fish and parasites, see Amundsen et al., "New Parasites and Predators."

77. Vizentin-Bugoni et al., "Structure, Spatial Dynamics, and Stability of Novel Seed Dispersal Mutualistic Networks."

78. This suggestion supports a prominent and controversial contention, known as "ecological fitting," that coevolution is less important than previously suggested and that very recently formed communities, such as those including many non-native species, function similarly to those that have long coevolved (Janzen, "On Ecological Fitting"; Simberloff and Strong, "Counterpoint"). Importantly, the implications of the Hawaiian study with respect to stability of the community are unclear because the claim of similar network "robustness" to those calculated for largely native communities rests on simulated removals of species in the network and hypothesized subsequent extinctions for both the Hawaiian network and the largely native ones.

79. Borrett et al., "Rise of Network Ecology."

80. Vanderhoeven et al., "Linking Concepts."

81. Abrahams et al., "Exploring the Dynamics of Research Collaborations."

82. P. David et al., "Impacts of Invasive Species on Food Webs"; Knight et al., "Reflections on, and Visions for, the Changing Field of Pollination Ecology"; Bohan et al., *Networks of Invasion: A Synthesis of Concepts*.

83. Knight et al., "Reflections on, and Visions for, the Changing Field of Pollination Ecology."

84. Among software packages available for calculation of network statistics, see Bastian et al.,, "Gephi"; Dormann, Gruber, and Fründ, "Introducing the Bipartite Package"; Dorman, Fründ, et al., "Indices, Graphs and Null Models"; Almeida-Neto and Ulrich, "Straightforward Computational Approach for Quantifying Nestedness"; Poisot, "ESM"; Marquitti et al., "MODULAR: Software." R. Antoniazzi Jr. and colleagues provide a guide to many software packages, including which ones calculate which network statistics (Antoniazzi et al., "Useful Guide of Main Indices and Software").

85. Frost et al., "Using Network Theory."

86. E.g., Ings et al., "Ecological Networks—Beyond Food Webs."

87. Knight et al., "Reflections on, and Visions for, the Changing Field of Pollination Ecology."

88. Kéfi, Berlow, Wieters, Navarrete, et al., "More Than a Meal"; Kéfi, Berlow, Wieters, Joppa, et al., "Network Structure Beyond Food Webs"; Kéfi, Miele, et al., "How Structured Is the Entangled Bank?"

89. P. David et al., "Impacts of Invasive Species on Food Webs."

90. Sauve et al., "Structure–Stability Relationships."

91. Most elements exist in nature in two or more forms, or "isotopes." Different isotopes of the same element have the same number of protons but different numbers of neutrons, so their masses differ. Stable isotope analysis consists of determining the relative fractions of different isotopes of elements (usually carbon or nitrogen) in organismal tissues or other substances found in nature (Michener and Lajtha, *Stable Isotopes in Ecology and Environmental Science*).

92. On beavers, see C. Anderson et al., "Effects of Invasive North American Beavers on Riparian Plant Communities." On zebra mussels, see De Vanna et al., "Alternative Hypothesis to Invasional Meltdown."

93. Milo et al., "Network Motifs"; L. Stone et al., "Network Motifs and Their Origins."

94. L. Stone et al., "Network Motifs and Their Origins."

95. Ings et al., "Ecological Networks—Beyond Food Webs"; Knight et al., "Reflections on, and Visions for, the Changing Field of Pollination Ecology."

96. Ings et al., "Ecological Networks—Beyond Food Webs."

97. P. David et al., "Impacts of Invasive Species on Food Webs."

98. On unintended effects, see Zavaleta et al., "Viewing Invasive Species Removal in a Whole-Ecosystem Context"; Zavaleta, "It's Often Better to Eradicate, but Can We Eradicate Better?" On examining the entire network, see P. David et al., "Impacts of Invasive Species on Food Webs."

99. Emer et al., "Impact of the Invasive Alien Plant."

100. On individual-based networks, see Dupont et al., "Spatial Structure of an Individual-Based Plant-Pollinator Network"; Arroyo-Correa et al., "Individual-Based Plant–Pollinator Networks." On multilevel networks, see M. C. Hutchinson et al., "Seeing the Forest for the Trees."

101. Cagua et al., "Keystoneness, Centrality, and the Structural Controllability of Ecological Networks"; C. Campbell, Yang, et al., "Network Model for Plant–Pollinator Community Assembly"; C. Campbell, Russo, et al., "Whole Community Invasions."

Chapter 14

1. United States National Research Council, *Global Change: Ecosystems Research.*

2. Marsh, *Man and Nature; or, Physical Geography as Modified by Human Action*, Marsh, *Earth as Modified by Human Action*; W. Thomas, *Man's Role in Changing the Face of the Earth.*

3. Carson, *Silent Spring.*

4. Detwyler, *Man's Impact on Environment.*

5. Weart, *Discovery of Global Warming.*

6. Gray quoted in Lever, *They Dined on Eland.*

7. Nebeker, *Calculating the Weather.*

8. Fourier, "Remarques générales sur les températures du globe terrestre et des espaces planétaire."

9. Tyndall, "Bakerian Lecture."

10. Foote, "Circumstances Affecting the Heat of the Sun's Rays"; R. Jackson, "Eunice Foote, John Tyndall and a Question of Priority."

11. Arrhenius, "On the Influence of Carbonic Acid."

12. Keeling, "Concentration and Isotopic Abundances of Carbon Dioxide."

13. Pales and Keeling, "Concentration of Atmospheric Carbon Dioxide in Hawaii."

14. E.g., Bolin and Keeling, "Large-Scale Atmospheric Mixing"; Keeling et al., "Atmospheric Carbon Dioxide Variations."

15. The first of the two mid-twentieth-century books on human-induced changes to the earth (W. Thomas, *Man's Role in Changing the Face of the Earth*) had no reference to greenhouse gases or global climate change. In the second (Detwyler, *Man's Impact on Environment*), one chapter made a glancing reference to the possibility of CO_2-driven climate change in the closing section, "Future Problems": "The possible 'greenhouse' effect of carbon dioxide is not known precisely, yet extrapolation of present measurement indicates a global increase of this constituent of about 40% by the turn of the century" (Pack, "Meteorology of Air Pollution," 111). In a 1990 volume (Turner et al., *Earth as Transformed by Human Action*) that saw itself as a sequel to Marsh (*Man and Nature; or, Physical Geography as Modified by Human Action*) and Thomas, in addition to several passing references to greenhouse gases or climate change, two chapters confronted the issue directly. In a chapter on carbon, Houghton and Skole sounded an alarm: "There is an urgent need to stabilize the atmospheric concentration of carbon dioxide and to prevent the politically destabilizing effects of climate change" (Houghton and Skole, "Carbon," 406). The authors of the chapter on climate were considerably more cautious: "On the global scale, it is possible that changes in atmospheric composition as a result of human activities are changing the climate, but again, there is as yet no unequivocal proof" (Jäger and Barry, "Climate," 349).

16. Gore, *Earth in the Balance*; Guggenheim, *Inconvenient Truth.*

17. United States National Research Council, *Energy and Climate*, viii.

18. United States National Research Council, *Carbon Dioxide and Climate.*

19. Hansen et al., "Climate Impact of Increasing Atmospheric Carbon Dioxide."

20. Bolin et al., *Greenhouse Effect.*

21. IPCC, *Climate Change 2022.*

22. Woodwell, "Carbon Dioxide Question"; Woodwell et al., "Biota and the World Carbon Budget."

23. E.g., N. Myers, *Primary Source.*

24. Weltzin, Belote, and Sanders, "Biological Invaders in a Greenhouse World."

25. R. Nowak et al., "Functional Responses of Plants"; Norby and Zak, "Ecological Lessons."

26. Belote et al., "Response of an Understory Plant Community."

27. L. Hughes and Bazzaz, "Effects of Elevated CO_2 on Five Plant-Aphid Interactions."

28. Shelford, "Physiological Animal Geography"; Shelford, *Animal Communities in Temperate America*. Victor Shelford, an American animal ecologist and one of the leading figures in the development of the new science of ecology (discussed in chap. 12), intended his law of tolerance to apply to animals. The English botanist Ronald Good, apparently independently, described a virtually identical principle, which he termed "the theory of tolerance," for plants (Good, "Theory of Plant Geography").

29. Controlled experiments have determined thermal limits in plants (e.g., Saavedra et al., "Changes in Flowering and Abundance of *Delphinium nuttallianum* (Ranunculaceae)"; Jentsch et al., "New Generation of Climate-Change Experiments"; Zinn et al., "Temperature Stress and Plant Sexual Reproduction"; Y. Liu et al., "Global Warming Reduces Plant Reproductive Output"; Capovilla et al., "Control of Flowering by Ambient Temperature"), vertebrates (e.g., Angilletta et al., "Thermal Effects on the Energetics of Lizard Embryos"; Deutsch et al., "Impacts of Climate Warming on Terrestrial Ectotherms"; Eliason et al., "Differences in Thermal Tolerance"; Levy et al., "Resolving the Life Cycle"; Cortes et al., "Thermal Ecological Physiology"; N. Tremblay et al., "Cold Temperature Tolerance"), and invertebrates (e.g., Deutsch et al., "Impacts of Climate Warming on Terrestrial Ectotherms"; Terblanche et al., "Ecologically Relevant Measures of Tolerance"; Boardman, Grout, and Terblanche, "False Codling Moth"; Boardman, Sørensen, et al., "Cold Tolerance Is Unaffected by Oxygen Availability"; Boardman and Terblanche, "Oxygen Safety Margins Set Thermal Limits"; Verberk et al., "Does Oxygen Limit Thermal Tolerance in Arthropods?"; Jørgensen et al., "How to Assess *Drosophila* Heat Tolerance").

30. Sorte et al., "Poised to Prosper?"

31. A. Bates et al., "Geographical Range, Heat Tolerance and Invasion Success."

32. E. McKnight et al., "Non-Native Species Outperform Natives."

33. E.g., Neilson and Wullstein, "Biogeography of Two Southwest American Oaks"; Prince et al., "Geographical Distribution of Prickly Lettuce."

34. E.g., Parmesan, "Ecological and Evolutionary Responses to Recent Climate Change"; C. Thomas, "Climate, Climate Change and Range Boundaries"; Sheldon, "Climate Change in the Tropics."

35. Krehenwinkel and Tautz, "Northern Range Expansion."

36. Hulme, "Climate Change and Biological Invasions."

37. For Great Britain, see Shardlow and Taylor, "Southern Green Shield Bug"; Salisbury et al., "Current Status of the Southern Green Shield Bug." For Japan, see Kiritani, "Impacts of Global Warming on *Nezara viridula*." For Hungary, see Rédei and Torma, "Occurrence of the Southern Green Stink Bug." For Germany, see Werner, "*Nezara viridula* (Linnaeus, 1758) in Köln."

38. Musolin, "Surviving Winter."

39. Purse et al., "Climate Change and the Recent Emergence of Bluetongue"; Mlot, "Spread of Tropical Livestock Virus"; Hendrickx, "Spread of Blue Tongue in Europe."

40. Y. Thomas et al., "Global Change and Climate-Driven Invasion of the Pacific Oyster."

41. Stachowicz et al., "Linking Climate Change and Biological Invasions."

42. Giejsztowt et al., "Climate Change and Invasion May Synergistically Affect Native Plant Reproduction."

43. Kim, "How Swaths of Invasive Grass Made Maui's Fires So Devastating"; D'Antonio and Vitousek, "Biological Invasions by Exotic Grasses"; D'Antonio, Tunison, and Loh, "Variation in the Impact of Exotic Grasses."

44. Kim, "How Swaths of Invasive Grass Made Maui's Fires So Devastating"; Rains, "Maui's Deadly Blazes"; Piper et al., "Maui's Neglected Grasslands."

45. IPCC, *Climate Change 2022.*

46. Diez et al., "Will Extreme Climatic Events Facilitate Biological Invasions?"; Grosholz, "Climate Change and Invasive Species."

47. Sikora et al., "Coordinated Effort to Manage Soybean Rust."

48. Diez et al., "Will Extreme Climatic Events Facilitate Biological Invasions?"

49. Carlton et al., "Tsunami-Driven Rafting."

50. On dispersal of marine organisms, see Grosholz, "Climate Change and Invasive Species." On Xantus' swimming crab, see Sadowski et al., "Temperature-Induced Range Expansion of a Subtropical Crab."

51. B. Bradley, "Regional Analysis of the Impacts of Climate Change"; B. Bradley, Oppenheimer, and Wilcove, "Climate Change and Plant Invasions"; B. Bradley, Wilcove, and Oppenheimer, "Climate Change Increases Risk of Plant Invasion."

52. J. M. Allen and Bradley, "Out of the Weeds?"

53. Nori et al., "Climate Change and American Bullfrog Invasion"; Barbet-Massin et al., "Climate Change Increases the Risk of Invasion by the Yellow-Legged Hornet."

54. Lowe et al., *100 of the World's Worst Invasive Alien Species.*

55. Bellard et al., "Will Climate Change Promote Future Invasions?"

56. Sanderson, "Influence of Minimum Temperatures."

57. Elkinton, Parry, and Boettner, "Implicating an Introduced Generalist Parasitoid."

58. The browntail moth was a major pest whose population mysteriously began to decline in 1914, soon after Sanderson's research, and was eventually restricted to small, isolated populations on Cape Cod and islands and peninsulas of Casco Bay, Maine. J. S. Elkinton and colleagues convincingly argued that the decline resulted from the introduction of a parasitic tachinid fly, *Compsilura concinnata*, released in a failed attempt to control the spongy moth, that survived and attacked native moths and the browntail moth (Elkinton, Parry, and Boettner, "Implicating an Introduced Generalist Parasitoid"). Beginning in 1989, the browntail moth began to outbreak in several areas of Maine. This change is suspected to be caused by climate change (Boyd et al., "Factors Influencing the Population Fluctuations of *Euproctis chrysorrhoea*").

59. J. Ball, "Climatological Diagrams."

60. T. Taylor, *Control of Settlement by Humidity and Temperature.*

61. An early example, from 1916, is the work of W. Dwight Pierce, who applied climographs to the spread of the boll weevil northward into the United States (Pierce, "New Interpretation").

62. Cook, "Distribution of the Alfalfa Weevil." For his earlier use of climographs to understand aspects of native species' ecology, see Cook, "Studies in Physical Ecology of the Noctuidae"; Cook, "Distribution of the Pale Western Cutworm."

63. On the Maryland invasion, see Radcliffe and Flanders, "Biological Control of Alfalfa Weevil." In 1925, John E. Graf similarly applied climographs to a non-native insect in the United States, the Mexican bean beetle (Graf, "Climate in Relation to Mexican Bean Beetle Distribution"). Native to southern Mexico, this lady beetle was introduced to the US Southwest in the late nineteenth century and separately to the East and Midwest in the twentieth century. Comparing climographs in the recently invaded Midwest, long-invaded Southwest, and native southern Mexico to several climographs

in the East, Graf predicted (correctly, it turned out) that climate would not prevent invasion of the eastern states.

64. Cook, "Notes on Predicting the Probable Future Distribution of Introduced Insects."

65. Gjullin, "Probable Distribution of the Mediterranean Fruit Fly."

66. Despite its common name, the Mediterranean fruit fly is native to tropical Africa. Its presence in South Africa dates to sometime before the end of the nineteenth century (Karsten et al., "Population Genetics of *Ceratitis capitata*").

67. Walters et al., "Eradication: Strategies and Tactics."

68. Quayle, "Possible Future Distribution of the Mediterranean Fruit Fly."

69. A. Hopkins, *Bioclimatics.* In 1958, Friedrich S. Bodenheimer constructed climographs (which he sometimes termed "ecoclimographs") for many species in their native ranges (and some also in their non-native ranges) to try to understand geographic variation in density (Bodenheimer, *Animal Ecology To-Day*). He combined climographs for the medfly in many sites in its native and non-native ranges with information on its number of generations per year. He also presented information on variation in the climograph between years for several sites, all in an attempt to understand the potential for the medfly to damage agriculture.

70. Twomey, "Climographic Studies of Certain Introduced and Migratory Birds."

71. Furthermore, because climographs depict weather each month, Twomey was able to argue that weather in the months of May through September, when eggs are laid and the young hatched and reared, is particularly important.

72. Much later, conversion of agricultural land on Vancouver Island led to a skylark population crash and by the 1990s to great range restriction (Toochin and Meredith, *Status and Occurrence of Eurasian Skylark*).

73. Cahn, "Climographic Analysis."

74. Moulton, Cropper, et al., "Patterns of Success in Game Bird Introductions."

75. For example, in 1946, R. A. McCabe and A. S. Hawkins, exploring the status of the grey partridge in the American Midwest and parts of Canada, reprinted one of Twomey's climographs, constructed new climographs for five sites in their region of interest plus a sixth in Texas, and added a third dimension to temperature and rainfall: hours of sunshine (McCabe and Hawkins, "Hungarian Partridge in Wisconsin"). They observed that the degree of overlap of the climographs of the six sites of introduction with the climograph of the native range generally correlated with the fate and status of the partridge at the six sites, and they used the climographs to generate hypotheses about the direct causes of partridge success or failure. They then gathered observational data on survival and reproduction in the field to test the hypotheses. In 1974, two researchers in Canada similarly examined climographs of temperature and daily photoperiod in the native and introduced ranges of two non-native bird species in British Columbia, the crested myna and the starling, and then used artificial nest boxes to study reproduction, growth, and survival in the field to determine the specific stages of the myna life cycle challenged by the physical environment in the Vancouver area (C. Johnson and Cowan, "Thermal Adaptation as a Factor Affecting Colonizing Success").

76. Grinnell, "Niche-Relationships of the California Thrasher"; G. Hutchinson, "Concluding Remarks"; cf. Soberón, "Grinnellian and Eltonian Niches."

77. American ecologist Victor E. Shelford in 1929 constructed climographs from around the world to depict optimal and suboptimal regions for sheep production in North America. Shelford also expanded the range of variables used in climographs. Thus, for example, in 1954, he depicted suitable conditions for the ring-necked pheasant in Ohio by means of a "heliohydrogram," a climograph of two dimensions: April

sunshine and April rainfall. He chose these dimensions "because of general knowledge of the reproduction of birds" (Shelford, "Experimental Approach to the Study of Bird Populations," 254).

78. Gjullin produced a map indirectly, first by constructing a two-dimensional space of temperature and rainfall, then by locating in that space all the points in climographs of locations where the medfly had survived, and then, on a map of the United States, marking all the areas whose temperature and rainfall were within one or more of the climographs (Gjullin, "Probable Distribution of the Mediterranean Fruit Fly"). Cook had previously produced a map of the distribution of the native pale western cutworm by similar means (Cook, "Distribution of the Pale Western Cutworm").

79. McCabe and Hawkins, "Hungarian Partridge in Wisconsin."

80. A. Twomey, "Climographic Studies of Certain Introduced and Migratory Birds."

81. Colinet et al., "Insects in Fluctuating Thermal Environments."

82. Tyrwhitt, "Surveys for Planning"; Steinitz et al., "Hand-Drawn Overlays."

83. McHarg, *Design with Nature.*

84. M. Wing and Bettinger, *Geographic Information Systems.*

85. WorldClim, "Maps, Graphs, Tables, and Data of the Global Climate"; Earth Resources Observation and Science (EROS) Center, "USGS EROS Archive—Digital Elevation—HYDRO1K."

86. J. Franklin, *Mapping Species Distributions.* An early effort at such interpolation was by E. Fitzpatrick and Nix, "Climatic Factor."

87. Wahba and Wendelberger, "Some New Mathematical Methods"; M. F. Hutchinson, "Interpolating Mean Rainfall"; Booth et al., "BIOCLIM."

88. Dewey, Price, and Ramsey, "Satellite Remote Sensing."

89. Diez et al., "Will Extreme Climatic Events Facilitate Biological Invasions?"

90. Messenger and Flitters, "Bioclimatic Studies of Three Species of Fruit Flies."

91. Hopkins, *Bioclimatics.*

92. A. Hopkins, *Bioclimatics*; Messenger, "Bioclimatic Studies with Insects."

93. Porter and Gates, "Thermodynamic Equilibria of Animals with Environment."

94. Porter et al., "Behavioral Implications of Mechanistic Ecology." On the cane toad, see Kearney et al., "Modelling Species Distributions."

95. Kearney and Porter, "NicheMapR."

96. Tracy, "Biophysical Modelling in Reptilian Physiology and Ecology."

97. S. McLane and Aitken, "Whitebark pine (*Pinus albicaulis*) Assisted Migration Potential."

98. On famine weed, see Doley, "Parthenium Weed (*Parthenium hysterophorus* L.)." On musk thistle, see Medd and Smith. "Prediction of the Potential Distribution of *Carduus nutans* (Nodding Thistle)." On ragweed, see D. Chapman et al., "Phenology Predicts the Native and Invasive Range Limits of Common Ragweed." On cane toad, see Mittan and Zamudio, "Rapid Adaptation to Cold in the Invasive Cane Toad."

99. Nix et al., "Potential Areas of Production"; Booth et al., "BIOCLIM."

100. Sutherst and Maywald, "Computerised System for Matching Climates."

101. Worner, "Ecoclimatic Assessment of Potential Establishment of Exotic Pests."

102. Poutsma et al., "Predicting the Potential Geographical Distribution of the Harlequin Ladybird"; He et al., "Modeling the Potential Global Distribution of Light Brown Apple Moth."

103. E.g., Pattison and Mack, "Potential Distribution of the Invasive Tree *Triadica sebifera*."

104. Kearney et al., "Modelling Species Distributions."

105. Chown, "Trait-Based Approaches to Conservation Physiology"; Lennox et al., "Improving Science-Based Invasive Species Management."

106. On number of factors and complexity of their interaction, see e.g., Tracy, "Biophysical Modelling in Reptilian Physiology and Ecology." On scarcity of comprehensive data, see Sutherst, Maywald, and Skarratt, "Predicting Insect Distributions in a Changed Climate"; Urban, Bocedi, et al., "Improving the Forecast for Biodiversity."

107. On thermal performance curves, see Sinclair et al., "Can We Predict Ectotherm Responses to Climate Change?" On knockdown times, see Jørgensen et al., "How to Assess *Drosophila* Heat Tolerance."

108. For the basic version, see, e.g., A. T. Peterson, "Predicting the Geography of Species' Invasions." GBIF, established in 2001, is an international network and database that provides open access to data about all species, accumulating and coordinating information from many sources, ranging from museum specimens to DNA barcodes and smartphone photographs. GBIF arose from a suggestion offered in 1999 by the Megascience Forum of the Organization for Economic Cooperation and Development (GBIF, "What Is GBIF?"). Exactly which of many possible climatic and other variables to select for use in statistical models is not a trivial matter. B. Petitpierre and colleagues suggested that, to predict where non-native plant species will survive, one should use either variables strongly believed a priori to be important for a particular plant species or a statistical procedure to select a reduced set among all possible variables (Petitpierre et al., "Selecting Predictors to Maximize the Transferability of Species Distribution Models").

109. Cf. A. T. Peterson, Soberón, et al., *Ecological Niches and Geographic Distributions.*

110. Cf. J. Franklin, *Mapping Species Distributions.*

111. On development of BIOCLIM, see Nix, "Biogeographic Analysis of Australian Elapid Snakes"; Busby, "BIOCLIM—A Bioclimate Analysis and Prediction System"; Booth and Williams, "Developing Biodiverse Plantings"; Booth et al., "BIOCLIM." For first publication, see Prendergast and Hattersley, "Distribution and Cytology of Australian *Neurachne*"; Booth, "New Method for Assisting Species Selection."

112. E.g., Xu and Hutchinson, *ANUCLIM Version 6.1 User Guide.*

113. For New Zealand, see Panetta and Mitchell, "Bioclimatic Prediction of the Potential Distribution of Some Weed Species." For Australia, see Sindel and Michael, "Spread and Potential Distribution of *Senecio madagascariensis* Poir. (Fireweed)." For South Africa, see Honig et al., "Invasive Potential of Australian Banksias."

114. On the Mahalanobis distance method, see Farber and Kadmon, "Assessment of Alternative Approaches." On the other frameworks, see J. Franklin, *Mapping Species Distributions*; Booth and Williams, "Developing Biodiverse Plantings."

115. Stockwell and Noble, "Induction of Sets of Rules from Animal Distribution Data"; Stockwell and Peters, "GARP Modelling System"; J. Franklin, *Mapping Species Distributions*; A. T. Peterson, Papeş, and Eaton, "Transferability and Model Evaluation in Ecological Niche Modeling."

116. A. T. Peterson and Vieglais, "Predicting Species Invasions."

117. A. T. Peterson and Robins, "Using Ecological-Niche Modeling to Predict Barred Owl Invasions." On the earlier arrival of the barred owl, see Knopf, "Faunal Mixing, Fauna Integrity"; Knopf, "Avian Assemblages on Altered Grasslands"; Livezey, "Range Expansion of Barred Owls."

118. Northern spotted owl populations have continued to decline in the Pacific Northwest as barred owl populations increase, and mounting evidence implicates competition and aggression from the barred owl in the decline. This invasion will likely end in extinction of the northern spotted owl in much if not all of its current range

unless intervention is undertaken, such as shooting barred owls and possibly modifying habitat (L. Long and Wolfe, "Review of the Effects of Barred Owls on Spotted Owls"; Yackulic et al., "Past and Future Roles of Competition and Habitat"). Hybridization between the two species also occurs and could increase as an additional threat to the northern spotted owl's persistence as competition causes its population to dwindle.

119. S. Phillips et al., "Maximum Entropy Modelling"; S. Phillips and Dudík, "Modeling of Species Distributions with Maxent"; Elith, Phillips, et al., "Statistical Explanation for MaxEnt."

120. For criticism, see, e.g., Royle et al., "Likelihood Analysis of Species Occurrence Probability"; M. Fitzpatrick et al., "MaxEnt Versus MaxLike." On accuracy, see, e.g., Elith, Graham, et al., "Novel Methods Improve Prediction of Species' Distributions."

121. Ficetola, Thuiller, and Miaud, "Prediction and Validation." Importantly, bullfrogs, unlike many anurans, must have permanent ponds or lakes because their tadpoles overwinter in water. Climatic data alone do not include the presence or absence of such permanent waterbodies. Ficetola and colleagues attempted to approximate whether or not a site includes such water bodies by including two precipitation variables (total precipitation between December and February and total precipitation between June and August) that together capture water availability during both summer and winter.

122. Ward, "Modelling the Potential Geographic Distribution of Invasive Ant Species."

123. Broennimann and Guisan, "Predicting Current and Future Biological Invasions"; Elith, Kearney, and Phillips, "Art of Modelling Range-Shifting Species."

124. Beaumont et al., "Different Climatic Envelopes Among Invasive Populations."

125. E.g., Engler and Guisan, "MIGCLIM"; Engler et al., "MIGCLIM R Package."

126. Smolik et al., "Integrating Species Distribution Models."

127. Urban, Phillips, et al., "Cane Toad's (*Chaunus* [*Bufo*] *Marinus*) Increasing Ability to Invade."

128. McKey and Kaufmann, "Naturalization of Exotic *Ficus* Species"; Nadel et al., "Escapees and Accomplices."

129. C. Shuttleworth et al., *Grey Squirrel.*

130. Elkinton, Parry, and Boettner, "Implicating an Introduced Generalist Parasitoid"; Boyd et al., "Factors Influencing the Population Fluctuations of *Euproctis chrysorrhoea.*"

131. G. Hutchinson, "Concluding Remarks."

132. For instance, many plant species cannot occupy sites in which levels of some nutrient are below a particular threshold value. However, the presence of mutualist mycorrhizal fungi able to access nutrients at lower levels than the plant can permits many species to thrive in sites from which they would otherwise be excluded—an expansion of the realized niche beyond the fundamental niche (Peay, "Mutualistic Niche").

133. E.g., Guisan et al., "Unifying Niche Shift Studies."

134. E.g., Griggs, "Observations on the Behavior of Species."

135. Araújo and Rozenfeld, "Geographic Scaling of Biotic Interactions."

136. E.g., Wisz et al., "Role of Biotic Interactions in Shaping Distributions"; R. P. Anderson, "When and How Should Biotic Interactions Be Considered?"

137. Alexander, Diez, et al., "When Climate Reshuffles Competitors."

138. Araújo and Luoto, "Importance of Biotic Interactions."

139. Buckingham, "Biological Control of Alligatorweed."

140. Hellmann et al., "Five Potential Consequences of Climate Change"; Schooler, "*Alternanthera philoxeroides* (Martius) Grisebach (Alligator Weed)."

141. On CLIMEX effort, see Julien et al., "Potential Geographical Distribution of Alligator Weed." On MaxEnt, see N. Young et al., *MaxEnt Model*; D. Liu et al., "Predicting Plant Invasions"; Yan et al., "Prediction of the Spatial Distribution of *Alternanthera philoxeroides*."

142. N. Young et al., *MaxEnt Model*; Koncki and Aronson, "Invasion Risk in a Warmer World"; Shabani et al., "Invasive Weed Species' Threats to Global Biodiversity."

143. Julien et al., "Potential Geographical Distribution of Alligator Weed."

144. On experimental warming, see Wu et al., "Global Warming Increases the Interspecific Competitiveness." On adaptation through phenotypic plasticity, see Luo et al., "Comparative Transcriptome Analysis."

145. On direct effects, e.g., increased reproductive rate and decreased generation time, see J. Fu et al., "Demography and Population Projection of Flea Beetle". On indirect effects, e.g., increases in both fecundity and net rate of population, see Shi et al., "Indirect Effect of Elevated CO_2."

146. E.g., competition might be assumed to vary along a productivity gradient (Wisz et al., "Role of Biotic Interactions").

147. Sutherst, Maywald, and Bottomley, "From CLIMEX to PESKY."

148. Maguire, Nieto-Lugilde, J. L. Blois, et al., "Controlled Comparison of Species- and Community-Level Models."

149. Beerling et al., "Climate and the Distribution of *Fallopia japonica*"; Rafoss and Sæthre, "Spatial and Temporal Distribution of Bioclimatic Potential."

150. B. Bradley, Oppenheimer, and Wilcove, "Climate Change and Plant Invasions"; Bradley, Wilcove, and Oppenheimer, "Climate Change Increases Risk of Plant Invasion."

151. Thuiller, "Patterns and Uncertainties of Species' Range Shifts Under Climate Change"; Araújo and New, "Ensemble Forecasting of Species Distributions."

152. Thuiller et al., "BIOMOD."

153. Marmion et al., "Evaluation of Consensus Methods."

154. On olive, see Crossman and Bass, "Application of Common Predictive Habitat Techniques." On Argentine ant, see Roura-Pascual et al., "Consensual Predictions of Potential Distributional Areas." On invasive plant species in US, see Stohlgren, Ma, et al., "Ensemble Habitat Mapping."

155. Poulos et al., "Ensemble Forecasting of Potential Habitat."

156. Gama et al., "Ensemble Forecasting of *Corbicula fluminea* Worldwide Distribution"; Bertelsmeier et al., "Worldwide Ant Invasions Under Climate Change."

157. B. Bradley, Wilcove, and Oppenheimer, "Climate Change Increases Risk of Plant Invasion."

158. Buczkowski and Bertelsmeier, "Invasive Termites in a Changing Climate."

159. L. Buckley et al., "Can Mechanism Inform Species' Distribution Models?"

160. X. Morin and Thuiller, "Comparing Niche-and Process-Based Models."

161. A. T. Peterson, "Predicting the Geography of Species' Invasions"; A. T. Peterson, "Ecological Niche Conservatism."

162. Hendry, *Eco-Evolutionary Dynamics*.

163. Lankau et al., "Evolutionary Limits"; Lankau, "Coevolution Between Invasive and Native Plants."

164. Ainouche et al., "Hybridization, Polyploidy and Invasion."

165. Skelly et al., "Evolutionary Responses to Climate Change."

166. R. Barrett et al., "Rapid Evolution of Cold Tolerance in Stickleback"; Geerts et al., "Rapid Evolution of Thermal Tolerance in the Water Flea"; S. Diamond et al., "Rapid Evolution of Ant Thermal Tolerance."

167. Balanyá et al., "Global Genetic Change."

168. Skelly et al., "Evolutionary Responses to Climate Change."

169. Bradshaw and Holzapfel, "Evolutionary Response to Rapid Climate Change."

170. Mittan and Zamudio, "Rapid Adaptation to Cold in the Invasive Cane Toad."

171. J. W. Williams and Jackson, "Novel Climates, No-Analog Communities, and Ecological Surprises."

172. Guisan et al., "Unifying Niche Shift Studies."

173. Boiffin et al., "Species Distribution Models May Misdirect Assisted Migration."

174. Guisan et al., "Unifying Niche Shift Studies."

175. Veloz et al., "No-Analog Climates"; Maguire, Nieto-Lugilde, Fitzpatrick, et al., "Modeling Species and Community Responses."

176. Smolik et al., "Integrating Species Distribution Models."

177. Skellam, "Random Dispersal in Theoretical Populations."

178. For those who fit the model to invasions of particular species, see, e.g., Okubo, *Diffusion and Ecological Problems*; Lubina and Levin, "Spread of a Reinvading Species"; Hengeveld, *Dynamics of Biological Invasions*. For those who elaborated and expanded the model, see, e.g., E. Holmes et al., "Partial Differential Equations in Ecology"; Shigesada and Kawasaki, *Biological Invasions: Theory and Practice*.

179. Moody and Mack, "Controlling the Spread of Plant Invasions."

180. Sutherst, Maywald, and Skarratt, "Predicting Insect Distributions in a Changed Climate."

181. Kearney and Porter, "Mechanistic Niche Modelling."

182. L. Buckley et al., "Does Including Physiology Improve Species Distribution Model Predictions?"

183. Feng and Papeş, "Can Incomplete Knowledge of Species' Physiology Facilitate Ecological Niche Modelling?"; Feng and Papeş, "Physiological Limits in an Ecological Niche Modeling Framework."

184. Urban, Bocedi, et al., "Improving the Forecast."

185. R. H. Peters, *Critique for Ecology*, 145.

186. Quotations from, respectively, Lawton, "Predictable Plots," 444; Scheiner, "Additional Thoughts on 'A Critique for Ecology,'" 180; J. Weiner, "On the Practice of Ecology," 154.

187. Such arguments over what constitutes an adequate scientific explanation have a long history. In the second edition of his *Philosophiæ Naturalis Principia Mathematica*, Isaac Newton in 1713 was impelled to concede that he had not actually explained *why* gravity acted as it did, but he argued that the fact that his equations correctly described the resultant movement of objects sufficed: "Hitherto we have explained the phaenomena of heavens and of our sea by the power of gravity, but have not yet assigned the cause of this power. . . . I have not been able to discover the cause of those properties of gravity from phaenomena, and I frame no hypotheses; for whatever is not deduced from phaenomena is to be called an hypothesis; and hypotheses, whether metaphysical or physical, whether of occult qualities or mechanical, have no place in experimental philosophy. . . . And to us it is enough that gravity does really exist, and act according to the laws which we have explained, and abundantly serves to account for all the motions of the celestial bodies, and of our sea" (Newton, *Newton's Principia*, 506–7).

188. Diez et al., "Will Extreme Climatic Events Facilitate Biological Invasions?"

189. Grosholz, "Climate Change and Invasive Species."

190. Gerlach and Rice, "Testing Life History Correlates of Invasiveness."

191. Fridley, "Extended Leaf Phenology."

192. Pyšek and Richardson, "Traits Associated with Invasiveness in Alien Plants"; Pearson, Ortega, and Sears, "Darwin's Naturalization Hypothesis Up-Close."

193. Zettlemoyer et al., "Limited Evidence for Phenological Differences."

194. Polgar et al., "Drivers of Leaf-Out Phenology."

195. Wolkovich and Cleland, "Phenology of Plant Invasions"; Wolkovich et al., "Temperature-Dependent Shifts in Phenology."

196. McClure et al., "Projecting the Compound Effects."

197. J. M. Allen and Bradley, "Out of the Weeds?"

198. B. Bradley, Oppenheimer, and Wilcove, "Climate Change and Plant Invasions."

199. B. Bradley, "Regional Analysis of the Impacts of Climate Change."

200. B. Bradley, Oppenheimer, and Wilcove, "Climate Change and Plant Invasions."

201. Gerlach and Rice, "Testing Life History Correlates of Invasiveness."

202. Salo, "Red Brome."

203. Essl, Dullinger, et al., "Conceptual Framework for Range-Expanding Species."

204. E.g., Coope, "Response of Insect Faunas"; Kullman, "Non-Analogous Tree Flora."

205. E.g., M. Warren et al., "Rapid Responses of British Butterflies."

206. First proposed by R. L. Peters and Darling, "Greenhouse Effect and Nature Reserves." For the 2014 search, see Hällfors et al., "Coming to Terms with the Concept of Moving Species Threatened by Climate Change."

207. Pedlar et al., "Implementation of Assisted Migration in Canadian Forests"; Klenk and Larson, "Assisted Migration of Western Larch."

208. E.g., Ricciardi and Simberloff, "Assisted Colonization Is Not a Viable Conservation Strategy"; Ricciardi and Simberloff, "Assisted Colonization: Good Intentions and Dubious Risk Assessment."

209. Fazey and Fischer, "Assisted Colonization Is a Techno-Fix."

210. Among those who contended that the concern was overblown, for example, M. L. Hunter ("Climate Change and Moving Species") and O. Hoegh-Guldberg and colleagues ("Assisted Colonization and Rapid Climate Change") argued that most impactful invasions entailed intercontinental or continent-to-island movements, while proposals for assisted migration aim to move populations to other parts of the same broad biogeographic region sharing similar organisms with the native range. For those who argued that careful planning could avoid such events, see J. McLachlan et al., "Framework for Debate of Assisted Migration"; Richardson, Hellmann, et al., "Multidimensional Evaluation of Managed Relocation"; C. Palmer and Larson, "Should We Move the Whitebark Pine?"

211. Mueller and Hellmann, "Assessment of Invasion Risk from Assisted Migration."

212. E.g., Galloway et al., "Framework for Assessing the Feasibility of Native Fish Conservation Translocations."

213. E.g., Sandler, "Value of Species"; Albrecht et al., "Ethics of Assisted Colonization"; Maier and Simberloff, "Assisted Migration in Normative and Scientific Context"; C. Palmer and Larson, "Should We Move the Whitebark Pine?"

214. Bucharova, "Assisted Migration Within Species Range Ignores Biotic Interactions," 16.

215. Torreya Guardians, "Assisted Migration."

216. Downs et al., *Glacier National Park Fisheries Monitoring and Management Report*; Galloway et al., "Framework for Assessing the Feasibility of Native Fish Conservation Translocations."

217. United States Fish and Wildlife Service, "Endangered and Threatened Wildlife and Plants."

218. United States Department of the Interior, "Department of the Interior Proposes Expanding Conservation Technique."

219. Brodie et al., "Global Policy for Assisted Colonization of Species."

220. Convention on Biological Diversity, *Decision Adopted by the Conference.*

221. On the term "neonative," see Essl, Dullinger, et al., "Conceptual Framework for Range-Expanding Species."

222. Ricciardi and Simberloff, "Assisted Colonization Risk Assessment." J. R. Wilson deplored the neologism "neonative" on several grounds, including that C. A. Stace and M. J. Crawley had already coined the term "neonative" to mean something totally different from either of the meanings given by Essl and colleagues or Brodie and colleagues: "A plant that has originated in the area without direct human involvement but has arisen as the result of hybridization either between a native and an alien taxon or between two alien taxa or as a result of evolution from an alien or neonative taxon" (Stace and Crawley, *Alien Plants*, 7). See J. Wilson, "Definitions Can Confuse"; Essl, Dullinger, et al., "Conceptual Framework for Range-Expanding Species"; Brodie et al., "Global Policy for Assisted Colonization of Species."

Chapter 15

1. Windelband, "History and Natural Science."

2. Joel Cohen, "Mathematics as Metaphor"; Simberloff, "Community Ecology."

3. For example, an attempt to explain many ecological patterns by a simple mechanistic explanation (G. West, "General Model for the Origin of Allometric Scaling Laws"; J. H. Brown et al., "Toward a Metabolic Theory of Ecology") was heralded, despite its authors' disclaimers, as "a metabolic theory for everything" (B. Grant, "Powers That Might Be") and lambasted as widely overextended (e.g., O'Connor et al., "Reconsidering the Mechanistic Basis of the Metabolic Theory"; B. Hawkins et al., "Global Evaluation of Metabolic Theory").

4. Catford, Jansson, and Nilsson, "Reducing Redundancy in Invasion Ecology."

5. Enders, Hütt, and Jeschke, "Drawing a Map of Invasion Biology"; Enders, Havemann, and Jeschke, "Citation-Based Map of Concepts in Invasion Biology"; Enders, Havemann, et al., "Conceptual Map of Invasion Biology."

6. Catford, Jansson, and Nilsson, "Reducing Redundancy in Invasion Ecology."

7. P. Ashton and Mitchell, "Aquatic Plants"; Bazzaz, "Life History of Colonizing Plants"; Arthington and Mitchell, "Aquatic Invasive Species"; Groves, "Plant Invasion in Australia"; Groves, "Invasion of Mediterranean Ecosystems by Weeds."

8. Vermeij, "Agenda for Invasion Biology."

9. M. Williamson, *Biological Invasions.*

10. E.g., M. Williamson, "Explaining and Predicting the Success of Invading Species"; Theoharides and Dukes, "Plant Invasion Across Space and Time."

11. Blackburn, Pyšek, et al., "Proposed Unified Framework for Biological Invasions."

12. Lever, *They Dined on Eland.*

13. E.g., H. Baker, "Patterns in Plant Invasion in North America"; Ewel, "Invasibility: Lessons from South Florida"; Groves, "Plant Invasion in Australia"; Mack, "Alien Plant Invasion into the Intermountain West"; Nix and Wapshere, "Biogeographic Origin of Invading Species"; Swincer, "Physical Characteristics of Sites in Relation to Invasions"; Crawley, "Chance and Timing in Biological Invasions"; Kruger et al., "Characteristics of Invaded Mediterranean-Climate Regions."

14. Baker, "Patterns in Plant Invasion in North America," 54.

15. Crawley, "Chance and Timing in Biological Invasions," 417. For the 1986 work referenced in the quotation, see Crawley, "Population Biology of Invaders."

16. On forestry, see Booth, "New Method for Assisting Species Selection." On biological control, see Messenger, "Bioclimatic Inputs to Biological Control and Pest Management Programs"; Goeden, "Critique and Revision of Harris' Scoring System."

17. M. Williamson, *Biological Invasions.*

18. M. Williamson, *Biological Invasions.* Propagule pressure incorporates two factors: (1) the number of individuals introduced during a single event, such as a shipping container carrying several individuals of a non-native insect, and (2) the number of such events, such as a continuing flow of occasional containers carrying several individuals of the same species.

19. Beirne, "Biological Control Attempts."

20. For SCOPE participants who considered the role of propagule pressure, see, e.g., Newsome and Noble, "Ecological and Physiological Characters of Invading Species"; Simberloff, "Which Insect Introductions Succeed and Which Fail?"

21. Simberloff, "Role of Propagule Pressure in Biological Invasions."

22. For horticultural activity, see Mulvaney, "Effect of Introduction Pressure." For aquarium trade data, see Semmens et al., "Hotspot of Non-Native Aquarium Fishes." For boat and shipping traffic, see Schneider et al., "Transportation Model Assessment of Risk"; Colautti, Niimi, et al., "Spatial and Temporal Analysis of Transoceanic Shipping Vectors"; Semmens et al., "Hotspot of Non-Native Aquarium Fishes."

23. For New Zealand, see Veltman et al., "Correlates of Introduction Success"; R. Duncan, "Role of Competition and Introduction Effort"; R. Green, "Influence of Numbers Released." For Australia, see Newsome and Noble, "Ecological and Physiological Characters of Invading Species."

24. Memmott, Fowler, and Hill, "Effect of Release Size"; Grevstad, "Experimental Invasions Using Biological Control Introductions."

25. Colautti, Grigorovich, and MacIsaac, "Propagule Pressure." That increased number of individuals introduced increases the probability of persistence does not mean that major invasions cannot arise from very small propagule sizes. Perhaps the best-known examples are the invasions of Europe by the muskrat, introduced from North America, and of the West Indies by the small Indian mongoose. Three female and two male muskrats from Alaska were released in 1905 on an estate in Bohemia as an amenity, and their progeny were responsible for a major part of the spread through central Europe before being supplemented by other individuals (Mohr, "Muskrat"; J. L. Long, *Introduced Mammals of the World*). The initial spread of the mongoose through the West Indies arose from only four males and five females brought from Calcutta (present-day Kolkatta) to Jamaica in 1872 for control of rats in sugar cane plantations (Hoagland, Horst, and Kilpatrick, "Biogeography and Population Biology of the Mongoose in the West Indies"; see also chap. 4, sec. 4.5).

26. MacArthur and Wilson, *Theory of Island Biogeography.*

27. On "minimum viable population size," see Shaffer, "Minimum Viable Populations"; Lande, "Genetics and Demography in Conservation." On its use in planning for management of threatened populations, see Traill et al., "Minimum Viable Population Size: A Meta-Analysis"; B. Brook et al., "Minimum Viable Population Size: Not Magic, But Necessary."

28. E.g., Sax and Brown, "Paradox of Invasion."

29. H. Baker and Stebbins, *Genetics of Colonizing Species*; H. Baker, "Characteristics and Modes of Origin of Weeds"; Ehrendorfer, "Dispersal Mechanisms"; Mulligan, "Recent Colonization by Herbaceous Plants."

30. H. Baker, "Evolution of Weeds."

31. For Baker's earlier research, see H. Baker, "Self-Compatibility and Establishment." On the antecedents, see S. Barrett, "Foundations of Invasion Genetics."

32. G. Henslow, "On the Self-Fertilization of Plants"; Asa Gray, "Predominance and Pertinacity of Weeds."

33. E.g., Perrins et al., "Survey of Differing Views of Weed Classification."

34. I. Parker and Kareiva, "Assessing the Risks of Invasion for Genetically Engineered Plants."

35. H. Baker, "Patterns in Plant Invasion in North America."

36. On the Baker List as a rough guide at best, cf. S. Sutherland, "What Makes a Weed a Weed." For a trait list, see, e.g., Van Kleunen, Weber, and Fischer, "Meta-Analysis of Trait Differences."

37. For the Australian weed risk assessment procedure, see Pheloung et al., "Weed Risk Assessment Model." For its adaptations, see K. Lewis and Porter, "Global Approaches to Addressing Biofuel-Related Invasive Species Risks"; Barney et al., *Life-Cycle Approach to Low-Invasion Potential Bioenergy Production.*

38. Ehrlich, "Which Animal Will Invade?"; Ehrlich, "Attributes of Invaders and the Invading Processes: Vertebrates."

39. I. Noble, "Attributes of Invaders and the Invading Process: Terrestrial and Vascular Plants."

40. Roy, "In Search of the Characteristics of Plant Invaders."

41. Rejmánek and Richardson, "What Attributes Make Some Plant Species More Invasive?"

42. On statistical discriminant analysis, see G. McLachlan, *Discriminant Analysis and Statistical Pattern Recognition.*

43. Richardson and Rejmánek, "Conifers as Invasive Aliens."

44. Grotkopp et al., "Toward a Causal Explanation of Plant Invasiveness."

45. Ladd Johnson and Carlton, "Post-Establishment Spread in Large-Scale Invasions."

46. Ricciardi, "Global Range Expansion of the Asian Mussel"; Ricciardi, "Predicting the Impacts of an Introduced Species from Its Invasion History."

47. Moutinho, "Golden Menace."

48. Catford, Jansson, and Nilsson, "Reducing Redundancy in Invasion Ecology"; Enders, Hütt, and Jeschke, "Drawing a Map of Invasion Biology"; Enders, Havemann, and Jeschke, "Citation-Based Map of Concepts in Invasion Biology"; Enders, Havemann, et al., "Conceptual Map of Invasion Biology."

49. McGill et al., "Rebuilding Community Ecology from Functional Traits"; Violle et al., "Emergence and Promise of Functional Biogeography"; Chown and Gaston, "Macrophysiology."

50. Van Kleunen, Weber, and Fischer, "Meta-Analysis of Trait Differences"; Van Kleunen, Dawson, and Dostal, "Research on Invasive-Plant Traits"; Boltovskoy et al., "Traits and Impacts of Introduced Species"; Chown and Clusella-Trullas, "Functional Trait Variation Along Animal Invasion Pathways."

51. M. Solomon, "Natural Control of Animal Populations."

52. Holling, "Components of Predation"; Holling, "Some Characteristics of Simple Types of Predation"; Leather, "Ten Papers That Shook My World."

53. Holling, "Components of Predation."

54. Riechert and Lockley, "Spiders as Biological Control Agents."

55. Lester et al., "Pyrethroid Encapsulation."

56. Lester and Harmsen, "Functional and Numerical Responses."

57. Fernández-Arhex and Corley, "Functional Response of Parasitoids."

58. On multicolored lady beetle, see Lee and Kang, "Functional Response of *Harmonia axyridis*." On predatory mites, see Poletti et al., "Toxicity of Neonicotinoid Insecticides"; Carrillo and Peña, "Prey-Stage Preferences"; Filgueiras et al., "Prey Stage Preference."

59. Bollache et al., "Comparison of the Functional Responses of Invasive and Native Amphipods."

60. South et al., "Predicting Predatory Impact of Juvenile Invasive Lionfish"; Joyce et al., "Using Functional Responses and Prey Switching to Quantify Invasion Success of the Pacific Oyster."

61. Dick, Alexander, Jeschke, et al., "Advancing Impact Prediction."

62. Cuthbert et al., "Functional Response Ratio (FRR)."

63. Dick, Alexander, Ricciardi, et al., "Functional Responses Can Unify Invasion Ecology."

64. Tilman, "Resource Competition Between Plankton Algae," 347.

65. Aguilera, "Influence of Soil Community Density on Plant-Soil Feedbacks."

66. Dick, Alexander, Ricciardi, et al., "Functional Responses Can Unify Invasion Ecology." See also Catford, Jansson, and Nilsson, "Reducing Redundancy in Invasion Ecology."

67. Cf. Dick, Alexander, Jeschke, et al., "Advancing Impact Prediction."

68. Vonesh et al., "Functional Responses Can't Unify Invasion Ecology"; Landi et al., "Predicting Invasive Consumer Impact."

69. Dickey et al., "Assessing the Relative Potential Ecological Impacts."

70. The observation that islands are more susceptible than continents to invasions was one of six points raised by Elton in support of his hypothesis that greater numbers of species or increased complexity of their interactions confers increased stability on the community (Elton, *Ecology of Invasions by Animals and Plants*). He similarly ascribed the apparent frequent invasion of agricultural fields to the reduced number of species present, thus absence of biotic resistance. His 1958 hypothesis is often cited as the founding document (e.g., Goodman, "Theory of Diversity-Stability Relationships in Ecology") of a widely adopted rationale for biodiversity conservation—that greater biodiversity confers ecological stability—although the nature of the relationship of diversity to stability is controversial (McCann, "Diversity-Stability Debate"). This rationale subsequently was transmogrified into the equally controversial rationale that greater biodiversity improves ecosystem function, especially productivity (Srivastava and Vellend, "Biodiversity-Ecosystem Function Research"; Tilman et al., "Biodiversity and Ecosystem Functioning"; D. Frank, "Science and Values in the Biodiversity-Ecosystem Function Debate"). Elton's hypothesis was probably inspired by his interactions with Aldo Leopold (Simberloff, "Integrity, Stability, and Beauty").

71. MacArthur, "Fluctuations of Animal Populations"; MacArthur, "Species-Packing and Competitive Equilibrium for Many Species."

72. E.g., Simberloff, "Introduced Insects"; Simberloff, "Which Insect Introductions Succeed and Which Fail?"; J. H. Brown, "Patterns, Modes and Extents of Invasions by Vertebrates"; Ehrlich, "Attributes of Invaders and the Invading Processes: Vertebrates"; Pimm, "Theories of Predicting Success and Impact."

73. J. Levine and D'Antonio, "Elton Revisited."

74. E.g., Stohlgren, Binkley, et al., "Exotic Plant Species Invade Hot Spots of Native Plant Diversity"; Stadler et al., "Exotic Plant Species Invade Diversity Hot Spots." For plant invasions in the United States, E. M. Beaury and colleagues found the opposite correlation—fewer non-native species in communities with more native species; they

viewed this result as evidence for biotic resistance (Beaury et al., "Biotic Resistance to Invasion Is Ubiquitous").

75. Fridley et al., "Invasion Paradox."

76. Peng et al., "Correlation of Native and Exotic Species Richness."

77. J. Levine and D'Antonio, "Elton Revisited."

78. MacArthur, "Fluctuations of Animal Populations," is an exception.

79. For plants as the single trophic level, see Crawley et al., "Invasion-Resistance in Experimental Grassland Communities"; T. Kennedy et al., "Biodiversity as a Barrier to Ecological Invasion"; Emery and Gross, "Dominant Species Identity Regulates Invasibility." For aquatic invertebrates, see Dzialowski, "Experimental Effect of Consumer Identity."

80. For herbivory, see, e.g., Nuñez, Simberloff, and Relva, "Seed Predation as a Barrier to Alien Conifer Invasions." For predation, see, e.g., Henriksson et al., "Biotic Resistance in Freshwater Fish Communities."

81. Henriksson et al., "Biotic Resistance in Freshwater Fish Communities."

82. The enemy release hypothesis has most often been applied to plant invasions (e.g., Keane and Crawley, "Exotic Plant Invasions and the Enemy Release Hypothesis"), but it has also been cited as a more general hypothesis applicable to animal invasions as well (e.g., Colautti et al., "Is Invasion Success Explained by the Enemy Release Hypothesis?").

83. Elton's enthusiasm for the role of enemy release in invasion success may appear paradoxical in light of his proposal of biotic resistance as an explanation for many failed invasions. However, for biotic resistance he envisioned a role for native natural enemies in the new range, whereas for enemy release he was citing the absence in the new range of natural enemies from the native range of the invader.

84. "Die starke Ausbreitung neu eingeschleppter Pflanzen hängt meistens damit zusammen, daß nicht nur ihre natürlichen Konkurrenten, die in einer für das Gleichgewicht der Flora und Vegetation sehr förderlichen Weise das starke Überhandnehmen einer einzelnen Art verhindern, in dem neuen Gebiete fehlen, sondern häufig auch gewisse Feinde (Parasiten u. dgl. vgl. dazu den später zu besprechenden Fall der Lantana Camara auf den Hawai-Inseln)." Thellung, "Pflanzenwanderungen unter dem Einfluss des Menschen," 62. (My translation: "The rapid spread of newly introduced plants is mostly due to the fact that not only are their natural competitors, who prevent a single species from gaining the upper hand in a way that is very beneficial to the balance of flora and vegetation, missing in the new area, but also of certain enemies [parasites and the like as in the case of the Lantana Camara on the Hawaiian Islands to be discussed later])." Cf. Kowarik and Pyšek, "First Steps Towards Unifying Concepts in Invasion Ecology"; Heger and Jeschke, "Enemy Release Hypothesis as a Hierarchy of Hypotheses."

85. Elton, *Animal Ecology.*

86. Elton, *Ecology of Invasions by Animals and Plants,* 2nd ed.

87. For 1930s, see Sweetman, *Biological Control of Insects.* For 1940s–50s, see Huffaker, "Fundamentals of Biological Control of Weeds."

88. D. Strong et al., *Insects on Plants.*

89. E.g., Heimpel and Mills, *Biological Control.*

90. E.g., Williamson, *Biological Invasions*; J. L. Lockwood, Hoopes, and Marchetti, *Invasion Ecology.*

91. Keane and Crawley, "Exotic Plant Invasions and the Enemy Release Hypothesis"; Colautti, Ricciardi, et al., "Is Invasion Success Explained by the Enemy Release Hypothesis?"; Heger and Jeschke, "Enemy Release Hypothesis as a Hierarchy of Hypotheses."

92. Heger and Jeschke, "Enemy Release Hypothesis as a Hierarchy of Hypotheses."

93. D. Strong, "Density-Vague Population Change"; Krebs, *Ecology*; Begon and Townsend, *Ecology: Individuals, Populations, and Communities*.

94. Keane and Crawley, "Exotic Plant Invasions and the Enemy Release Hypothesis"; Heger and Jeschke, "Enemy Release Hypothesis as a Hierarchy of Hypotheses."

95. Huffaker, "Fundamentals of Biological Control of Weeds," 125.

96. Keane and Crawley, "Exotic Plant Invasions and the Enemy Release Hypothesis."

97. Colautti, Ricciardi, et al., "Is Invasion Success Explained by the Enemy Release Hypothesis?"

98. Karban, "Ecology and Evolution of Induced Resistance Against Herbivores"; Yactayo-Chang et al., "Plant Defense Chemicals Against Insect Pests"; Strauss and Agrawal, "Ecology and Evolution of Plant Tolerance to Herbivory."

99. Blossey and Nötzold, "Evolution of Increased Competitive Ability in Invasive Non-Indigenous Plants."

100. Thebaud and Simberloff, "Are Plants Really Bigger in Their Introduced Ranges?"

101. Felker-Quinn et al., "Meta-Analysis Reveals Evolution in Invasive Plant Species but Little Support for Evolution of Increased Competitive Ability"; Rotter and Holeski, "Meta-Analysis of the Evolution of Increased Competitive Ability Hypothesis"; Honor and Colautti, "EICA 2.0."

102. Asa Gray, "Predominance and Pertinacity of Weeds," 161.

103. Lozon and MacIsaac, "Biological Invasions: Are They Dependent on Disturbance?"

104. Elton, *Ecology of Invasions by Animals and Plants*, 117, 147.

105. Egerton, "History of Ecological Sciences, Part 54."

106. Cowles, "Ecological Relations of the Vegetation on the Sand Dunes of Lake Michigan"; Clements, *Plant Succession*; Gleason, "Individualistic Concept of the Plant Association."

107. Egler, "Indigene Versus Alien in the Development of Arid Hawaiian Vegetation"; Egler, "Arid Southeast Oahu Vegetation"; Noble and Slatyer, "Effect of Disturbance on Plant Succession"; M. Davis, Thompson, and Grime, "Charles S. Elton and the Dissociation of Invasion Ecology"; M. Davis, Pergl, et al., "Vegetation Change."

108. Therriault and Kolasa, "Explicit Links."

109. Odum, *Ecology*, 32.

110. Grime, "Competitive Exclusion in Herbaceous Vegetation"; Grime, "Control of Species Density in Herbaceous Vegetation."

111. Connell, "Diversity in Tropical Rain Forests and Coral Reefs"; Wilkinson, "Disturbing History of Intermediate Disturbance."

112. Catford, Daehler, et al. "Intermediate Disturbance Hypothesis."

113. Wilkinson, "Ecology Before Ecology."

114. P. White and Pickett, "Natural Disturbance and Patch Dynamics," 7. Similarly, W. P. Sousa defined disturbance as "a discrete, punctuated killing, displacement, or damaging of one or more individuals that directly or indirectly creates an opportunity for new individuals (or colonies) to become established" (Sousa, "Role of Disturbance in Natural Communities," 356).

115. Grime, *Plant Strategies and Vegetation Processes*, 39.

116. Amor and Piggin, "Factors Influencing the Establishment and Success of Exotic Plants in Australia."

117. Christensen and Burrows, "Fire: An Old Tool with New Uses."

118. Swincer, "Physical Characteristics of Sites in Relation to Invasions," 69.

119. H. Baker, "Patterns in Plant Invasion in North America."

120. Moyle, "Fish Introductions into North America."

121. Mack, "Alien Plant Invasion into the Intermountain West"; Mooney, Hamburg, and Drake, "Invasion of Plants and Animals into California"; Ewel, "Invasibility: Lessons from South Florida."

122. Joenje, Bakker, and Vlijm, "Ecology of Biological Invasions."

123. Di Castri, Hansen, and Debussche, *Biological Invasions in Europe and the Mediterranean Basin*; Drake et al., *Biological Invasions: A Global Perspective.*

124. Fox and Fox, "Susceptibility of Natural Communities to Invasion," 65.

125. Orians, "Site Characteristics Favoring Invasions." Orians introduced his chapter by pointing to Elton's conclusion that "invaders were more likely to establish viable populations in cultivated or otherwise disturbed and, usually, simplified communities" (133) and investigated whether subsequent evidence supported Elton's view. In fact, as noted earlier, Elton restricted this statement to plants and focused far more attention on animals and biotic resistance.

126. Kornaś, "Man's Impact upon the Flora and Vegetation in Central Europe"; Kornaś, "Plant Invasions in Central Europe."

127. Kruger et al., "Characteristics of Invaded Mediterranean-Climate Regions."

128. Hobbs, "Nature and Effects of Disturbance Relative to Invasions."

129. Hämet-Ahti, "Human Impact on Closed Boreal Forest (Taiga)"; D. Moore, "Human Impact on Island Vegetation."

130. Hobbs and Huenneke, "Disturbance, Diversity, and Invasion."

131. On the "fluctuating resource hypothesis," see M. Davis, Grime, and Thompson, "Fluctuating Resources in Plant Communities."

132. Lozon and MacIsaac, "Biological Invasions: Are They Dependent on Disturbance?" On BIOSIS, see H. Kennedy, "BIOSIS to the Rescue."

133. D'Antonio, Dudley, and Mack, "Disturbance and Biological Invasions." See also Orians, "Site Characteristics Favoring Invasions"; Hobbs, "Nature and Effects of Disturbance Relative to Invasions."

134. J. L. Lockwood, Hoopes, and Marchetti, *Invasion Ecology*, 76.

135. Moles et al., "Invasions."

136. Catford, Daehler, et al., "Intermediate Disturbance Hypothesis."

137. Jauni et al., "Non-Native Plant Species Benefit from Disturbance."

138. E.g., Lembrechts et al., "Disturbance Is the Key"; Lear et al., "Disentangling the Mechanisms Underpinning Disturbance-Mediated Invasion."

139. MacDougall and Turkington, "Are Invasive Species the Drivers or Passengers of Change?"

140. E.g., Didham et al., "Are Invasive Species the Drivers of Ecological Change?"; Stromberg, Lite, et al., "Altered Stream-Flow Regimes."

141. Bauer, "Invasive Species: 'Back-seat Drivers' of Ecosystem Change?"

142. Clewell and Aronson, *Ecological Restoration*; Bauer and Reynolds, "Restoring Native Understory"; Catford, "Using Management to Determine Drivers of Alien Plant Invasion"; Marks and Van Driesche, "Designing Restoration Programs"; Catterall, "Fauna as Passengers and Drivers in Vegetation Restoration."

143. M. Reid et al., "Does Invasive Plant Management Aid the Restoration of Natural Ecosystems?"

144. M. B. Thomas and Reid, "Are Exotic Natural Enemies an Effective Way of Controlling Invasive Plants?"

145. For critics of invasion science and management, see, e.g., Stromberg, Chew, et al., "Changing Perceptions of Change."

146. M. Davis, Grime, and Thompson, "Fluctuating Resources in Plant Communities."

147. Grime, "Vegetation Classification by Reference to Strategies"; Grime, "C-S-R Model of Primary Plant Strategies."

148. M. Davis, Wrage, and Reich, "Competition Between Tree Seedlings and Herbaceous Vegetation."

149. M. Davis, Grime, and Thompson, "Fluctuating Resources in Plant Communities."

150. M. Davis, Thompson, and Grime, "Charles S. Elton and the Dissociation of Invasion Ecology"; M. Davis, Pergl, et al., "Vegetation Change"; M. Davis, *Invasion Biology*.

151. E.g., Li and Stevens, "Fluctuating Resource Availability Increases Invasibility."

152. See references in M. Davis, *Invasion Biology*.

153. Grotkopp et al., "Toward a Causal Explanation of Plant Invasiveness."

154. Walker et al., "Does Fluctuating Resource Availability Increase Invasibility?"

155. Pearson, Ortega, Villarreal, et al., "Fluctuating Resource Hypothesis."

156. Yang et al., "What Can We Learn from Resource Pulses?"

157. M. Davis, Grime, and Thompson, "Fluctuating Resources in Plant Communities."

158. Holdgate, "Summary and Conclusions"; Simberloff, "Which Insect Introductions Succeed and Which Fail?"

159. Catford, Jansson, and Nilsson, "Reducing Redundancy in Invasion Ecology"; Enders, Hütt, and Jeschke, "Drawing a Map of Invasion Biology"; Enders, Havemann, and Jeschke, "Citation-Based Map of Concepts in Invasion Biology"; Enders, Havemann, et al., "Conceptual Map of Invasion Biology."

160. Gibson-Reinemer, "Vacant Niche." See also Grinnell, "Niche-Relationships of the California Thrasher"; Elton, *Animal Ecology*; G. Hutchinson, "Concluding Remarks."

161. Chase and Leibold, *Ecological Niches*.

162. G. Hutchinson, "Concluding Remarks," 424.

163. Elton, *Ecology of Invasions by Animals and Plants*, 135–36.

164. Verkaik, "Muskrat in the Netherlands." Other SCOPE participants citing a vacant niche as part of an explanation for an invasion outcome include Loope and Mueller-Dombois ("Characteristics of Invaded Islands") for bird species in upland Hawaii, Rejmánek ("Invasibility of Plant Communities") for trees in the South African fynbos region and Japanese climbing fern in bottomland hardwood forests of the southeastern United States, and Drent ("Biological Invasions: Birds and Mammals") for the invasion of the gray squirrel from North America into Great Britain. For the last example, subsequent research (C. Shuttleworth et al., *Grey Squirrel*) demonstrated that the vacant niche explanation was incorrect.

165. On invasions ascribed to the presence of a vacant niche in the invaded ecosystem, see, e.g., Ladd Johnson and Carlton, "Post-Establishment Spread in Large-Scale Invasions"; Southward et al., "Invasion of Hawaiian Shores by an Atlantic Barnacle"; Cizek et al., "Vacant Niche in Alpine Habitat"; Hierro et al., "Biogeographical Approach to Plant Invasions"; Ivanova and Symes, "Invasion of *Psittacula krameri* in Gauteng, South Africa." On the evidence supporting equally well the hypothesis that some sort of disturbance has changed the habitat to favor a non-native species, see Herbold and Moyle, "Introduced Species and Vacant Niches."

166. Simberloff and Von Holle, "Positive Interactions of Nonindigenous Species"; Simberloff, "Invasional Meltdown 6 Years Later."

167. Crosby, *Ecological Imperialism.*

168. Howarth, "Impacts of Alien Land Arthropods."

169. McKey and Kaufmann, "Naturalization of Exotic *Ficus* Species"; Nadel et al., "Escapees and Accomplices."

170. DeBach et al., "Biological Check Method."

171. Simberloff and Von Holle, "Positive Interactions of Nonindigenous Species"; Richardson, Pyšek, et al., "Naturalization and Invasion of Alien Plants."

172. Gurevitch, "Commentary on Simberloff (2006)," the referent article being Simberloff, "Invasional Meltdown 6 Years Later."

173. Jeschke et al., "Support for Major Hypotheses."

174. Braga et al., "Structuring Evidence for Invasional Meltdown"; Braga et al., "Invasional Meltdown Hypothesis."

175. Ricciardi, "Facilitative Interactions Among Aquatic Invaders."

176. P. Green et al., "Invasional Meltdown: Invader–Invader Mutualism."

177. Gullan, "Relationships with Ants."

178. Lake and O'Dowd, "Red Crabs in Rain Forest."

179. Colautti, Ricciardi, et al., "Is Invasion Success Explained by the Enemy Release Hypothesis?"; Eppinga et al., "Accumulation of Local Pathogens."

180. I. Ball, "Nature and Formulation of Biogeographical Hypotheses."

181. G. Henslow, "On the Self-Fertilization of Plants"; Asa Gray, "Predominance and Pertinacity of Weeds"; Elton, *Ecology of Invasions by Animals and Plants.*

182. J. Platt, "Strong Inference"; Popper, *Conjectures and Refutations*, Popper, *Logic of Scientific Discovery*; Lakatos, "Falsification and the Methodology of Scientific Research Programs." I. R. Ball termed this stage the "analytic" stage (I. Ball, "Nature and Formulation of Biogeographical Hypotheses").

183. Ruse, *Philosophy of Biology*; Hull, *Philosophy of Biological Science*; Roughgarden, "Competition and Theory in Community Ecology"; Simberloff, "Competition Theory."

184. Joel Cohen, "Mathematics as Metaphor," 675.

185. On the focusing and fine-tuning of quite general hypotheses, cf. the hierarchy-of-hypotheses approach (Jeschke et al., "Support for Major Hypotheses"; Heger et al., "Conceptual Frameworks and Methods for Advancing Invasion Ecology"; Heger and Jeschke, "Enemy Release Hypothesis as a Hierarchy of Hypotheses").

Chapter 16

1. Beever et al., "Social–Ecological Mismatches Create Conservation Challenges."

2. Office of Technology Assessment, *Harmful Non-Indigenous Species in the United States.*

3. Perrings, Williamson, and Dalmazzone, *Economics of Biological Invasions.*

4. E.g., Eiswerth and Johnson, "Managing Nonindigenous Invasive Species"; Olson and Roy, "Economics of Controlling a Stochastic Biological Invasion"; Born et al., "Economic Evaluation of Biological Invasions."

5. E.g., Perrings, Williamson, and Dalmazzone, *Bioinvasions and Globalization.*

6. Vilà et al., "How Well Do We Understand the Impacts of Alien Species on Ecosystem Services?"

7. Office of Technology Assessment, *Harmful Non-Indigenous Species in the United States*, 68.

8. Pimentel, Lach, et al., "Environmental and Economic Costs of Nonindigenous Species in the United States."

9. Pimentel, Lach, et al., "Environmental and Economic Costs Associated with Non-indigenous Species in the United States"; Pimentel, "Introduction: Non-Native Species in the World." The six nations were Australia, Brazil, British Isles, India, South Africa, and the United States.

10. Pimentel, *Biological Invasions: Economic and Environmental Costs*, 2nd ed.

11. Diagne, Leroy, Gozlan, et al., "InvaCost."

12. Diagne, Leroy, Vaissière, et al., "High and Rising Economic Costs of Biological Invasions Worldwide."

13. For the *NeoBiota* special issue, see Zenni et al., "Economic Costs of Biological Invasions Around the World." For *Biological Invasions*, see Richardson, "Special Section: Economic Costs."

14. On biodiversity as crucial to aspects of ecosystem function, see Wardle, Barker, et al., "Introduced Browsing Mammals"; D. Hooper et al., "Effects of Biodiversity on Ecosystem Functioning"; Srivastava and Vellend, "Biodiversity-Ecosystem Function Research"; Tilman et al., "Biodiversity and Ecosystem Functioning."

15. Carrasco et al., "Economic Valuation of Ecosystem Services."

16. Charles and Dukes, "Impacts of Invasive Species on Ecosystem Services."

17. R. Mitchell and Carson, *Using Surveys to Value Public Goods*; Stanley, "Local Perception of Public Goods"; Lew, "Willingness to Pay for Threatened and Endangered Marine Species."

18. P. Taylor, "Ethics of Respect for Nature"; P. Taylor, *Respect for Nature: A Theory of Environmental Ethics*; Agar, *Life's Intrinsic Value*.

19. E.g., Kanya et al., "Criterion Validity of Willingness to Pay Methods."

20. Case, "Invasion Resistance Arises"; Case, "Invasion Resistance"; Case, "Surprising Behavior from a Familiar Model"; Mithen and Lawton, "Food-Web Models"; R. Morton and Law, "Regional Species Pools."

21. Sagoff, "Do Non-Native Species Threaten the Natural Environment?"

22. I Parker et al., "Impact."

23. Parker explained the rationale for the equation in a box in a conservation biology textbook (Groom, Meffe, and Carroll, *Principles of Conservation Biology*, 3rd ed.; see also chapter 10 in this book).

24. Ricciardi, "Predicting the Impacts of an Introduced Species from Its Invasion History."

25. Thiele et al., "Impact Assessment Revisited."

26. Latombe et al., "GIRAE."

27. The IUCN is an international organization based in Gland, Switzerland. Founded in 1948 with the encouragement of UNESCO, it has influenced environmental and conservation activities of the United Nations, individual nations, and many nongovernmental organizations.

28. Betts et al., "Framework for Evaluating the Impact of the IUCN Red List."

29. Hilton-Taylor, "History of the IUCN Red List."

30. Mace and Lande, "Assessing Extinction Threats." See also Boyce, "Population Viability Analysis."

31. IUCN Species Survival Commission, *IUCN Red List Categories*.

32. Regarding the alterations, for instance, the category *critically endangered* (CR) was defined by any of the following five features: (1) an observed, estimated, inferred, or suspected population reduction of at least 80 percent over the last years or three generations, whichever is longer; (2) extent of occurrence estimated to be less than 100 km^2 or area of occupancy estimated to be less than 10 km^2, and any two of the following—(a) severely fragmented or known to exist at only a single location;

(b) continuing decline, observed, inferred, or projected, in extent or area of occupancy, extent or quality of habitat, number of locations or subpopulations, or number of mature individuals; (c) extreme fluctuations in extent of occurrence, area of occupancy, number of locations or subpopulations, or number of mature individuals; (3) population estimated to be less than 250 mature individuals and either an estimated decline of at least 25 percent within three years or one generation or a continuing decline in numbers of mature individuals and population structure; (4) population estimated to be less than 50 mature individuals; (5) quantitative analysis showing that the probability of extinction in the wild is at 50 percent within ten years or three generations, whichever is longer. See IUCN Species Survival Commission, *IUCN Red List Categories.*

33. Collar, Crosby, and Stattersfield, *Birds to Watch 2*; IUCN Conservation Monitoring Centre, World Conservation Monitoring Centre, IUCN Species Survival Commission, International Council for Bird Preservation and Birdlife International, *IUCN Red List of Threatened Animals*; Oldfield et al., *World List of Threatened Trees.*

34. IUCN Species Survival Commission, *IUCN Red List Categories and Criteria.*

35. IUCN Standards and Petitions Committee, *Guidelines for Using the IUCN Red List Categories and Criteria.*

36. On the disputes, see, e.g., Campbell, "Seeing Red."

37. Nentwig et al., "Generic Impact-Scoring System."

38. The Delphi method, named after the Oracle of Delphi, was developed for military purposes by the Rand Corporation in the 1950s and 1960s. It is a form of expert opinion that rests on the principle that a judgment by a group of experts is likely to be more accurate than the judgment of a single expert (Adler and Ziglio, *Gazing into the Oracle*; Rescher, *Predicting the Future*).

39. Kumschick and Nentwig, "Some Alien Birds Have as Severe an Impact as the Most Effectual Alien Mammals"; Kumschick, Bacher, and Blackburn, "What Determines the Impact of Alien Birds and Mammals in Europe?"

40. Kumschick, Bacher, Dawson, et al., "Conceptual Framework for Prioritization of Invasive Alien Species."

41. Cullen and Delfosse, "*Echium plantagineum* in Australia"; P. Harris, "Environmental Impact of Weed-Control Insects."

42. Schlaepfer et al., "Potential Conservation Value of Non-Native Species"; Serbesoff-King, "Melaleuca in Florida."

43. Blackburn, Essl, et al., "Unified Classification of Alien Species." Charlotte Hawkins and colleagues proposed the name of the method and the acronym EICAT (C. Hawkins et al., "Framework and Guidelines").

44. C. Hawkins et al., "Framework and Guidelines."

45. Kumschick, Bacher, S. Bertolino, et al., "Appropriate Uses of EICAT Protocol."

46. Volery et al., "Improving the Environmental Impact Classification for Alien Taxa."

47. IUCN, "IUCN Standard to Support Global Action."

48. Vimercati, Probert, et al., "EICAT+ Framework Enables Classification of Positive Impacts"; following up on suggestions in Kumschick, Bacher, Dawson, et al., "Conceptual Framework for Prioritization of Invasive Alien Species"; Vimercati, Kumschick, et al., "Importance of Assessing Positive and Beneficial Impacts of Alien Species."

49. Ten types of positive impacts, most of which are further divided into sets of mechanisms that can produce the impacts.

50. Vitule, Freire, et al., "Revising the Potential Conservation Value of Non-Native Species."

51. Bacher et al., "Socio-Economic Impact Classification of Alien Taxa (SEICAT)."

52. Millennium Ecosystem Assessment, *Ecosystems and Human Well-Being.*

53. Bacher et al., "Socio-Economic Impact Classification of Alien Taxa (SEICAT)." Quotations are from supporting information available in the online version of this journal article, pages 2 and 1, respectively.

54. Galanidi et al., "Assessing the Socio-Economic Impacts."

55. Much of the formal risk analysis literature distinguishes between *risk assessment*, the estimation of risk, especially quantitative estimation, and *risk analysis*, a process that includes risk assessment and risk management (Orr, "Generic Nonindigenous Aquatic Organisms Risk Analysis Review Process"). Often this distinction is blurred in general discussions of risk. B. Leung and colleagues have argued that, from an economic perspective, it is illogical to conduct a risk assessment without explicitly considering costs of management (Leung et al., "Ounce of Prevention or a Pound of Cure"). Without quantifying such management costs, several subsequent risk assessment tools implicitly consider management costs in certain questions asked of experts in determining levels of risk.

56. Food and Agricultural Organization, *International Standards for Phytosanitary Measures*; Secretariat of the International Plant Protection Convention, *International Plant Protection Convention*; Hayes, "Biosecurity and the Role of Risk Assessment."

57. G. Gray et al., "Principles for Conduct of Pest Risk Analyses."

58. Andersen et al., "Risk Assessment for Invasive Species."

59. E.g., logs of Siberian larch from Siberia and the Soviet Far East, Monterey pine and Douglas fir from New Zealand, and Monterey pine, coihue, and tepa from Chile (United States Department of Agriculture Forest Service, *Pest Risk Assessment of the Importation of Larch*; United States Department of Agriculture Forest Service, *Pest Risk Assessment of the Importation of Pinus radiata and Douglas-fir Logs*; United States Department of Agriculture Forest Service, *Pest Risk Assessment of the Importation of Pinus radiata, Nothofagus dombeyi, and Laurelia philippiana Logs*).

60. Nico and Williams, *Risk Assessment on Black Carp.*

61. Environmental Protection Agency, *Framework for Ecological Risk Assessment.*

62. Simberloff and Alexander, "Issue Paper on Biological Stressors"; cf. Simberloff and Alexander, "Assessing Risks to Ecological Systems from Biological Introductions."

63. Environmental Protection Agency, *Guidelines for Ecological Risk Assessment*, Appendix A, 3.

64. United States Department of Agriculture Forest Service, *Pest Risk Assessment of the Importation of Pinus radiata, Nothofagus dombeyi, and Laurelia philippiana Logs.*

65. National Invasive Species Council, *Meeting the Invasive Species Challenge.*

66. Office of Technology Assessment, *Harmful Non-Indigenous Species in the United States*, 113.

67. Tucker and Richardson, "Expert System."

68. Hayes, *Review of Ecological Risk Assessment Methodologies.*

69. Pheloung, *Determining the Weed Potential of New Plant Introductions*; Pheloung et al., "Weed Risk Assessment Model"; Lewis and Porter, "Global Approaches to Addressing Biofuel-Related Invasive Species Risks."

70. Koop et al., "Development and Validation of a Weed Screening Tool for the United States."

71. Mumford et al. "Invasive Non-Native Species Risk Assessment in Great Britain."

72. Cousens, "Risk Assessment of Biofuel Species."

73. Orr, "Generic Nonindigenous Aquatic Organisms Risk Analysis Review Process."

74. Simberloff, "Politics of Assessing Risk for Biological Invasions"; Simberloff, "Risk Assessments, Blacklists, and White Lists for Introduced Species."

75. Lodge et al., "Risk Analysis and Bioeconomics of Invasive Species."

76. C. S. Smith et al., "When to Ignore Advice."

Chapter 17

1. M. Davis et al., "Don't Judge Species on Their Origins," 154.

2. Vince, "Embracing Invasives."

3. Spear, *Great Gypsy Moth War* (gypsy moth is the former name of spongy moth); Maloy, "White Pine Blister Rust Control in North America"; Buhs, *Fire Ant Wars.*

4. Brody, "Agriculture Department to Abandon Campaign Against the Fire Ant."

5. Matzek et al., "Closing the Knowing–Doing Gap in Invasive Plant Management"; Lavoie and Brisson, "Training Environmental Managers to Control Invasive Plants"; Foxcroft et al., "Knowing-Doing Continuum or Knowing-Doing Gap?"

6. "Workshop: Role Model Interventions—US."

7. On the successful eradication of the giant African snail, see Mead, "Ecological Malacology."

8. McNeely et al., "Global Strategy on Invasive Alien Species"; Wittenberg and Cock, *Invasive Alien Species: A Toolkit of Best Prevention and Management Practices.*

9. Westbrooks and Eplee, "Early Detection and Rapid Response."

10. United States General Accounting Office (GAO), *Invasive Species: Obstacles Hinder Federal Rapid Response.*

11. Westbrooks, "New Approaches for Early Detection and Rapid Response."

12. Meyerson and Simberloff, "Early Detection and Rapid Response."

13. Reaser et al., "Early Detection of and Rapid Response (EDRR) to Invasive Species."

14. Genovesi and Shine, *European Strategy on Invasive Alien Species.*

15. Genovesi, Scalera, et al., *Towards an Early Warning and Information System for Invasive Alien Species.*

16. Bois et al., "Invasive Plant Atlas of New England"; McInerney and Smith, "Victorian Weed Alert Program"; Munakamwe et al., "Victorian Weed Spotters."

17. EDDMapS, "EDDMapS: Find, Map, Track"; R. Wallace et al., "IveGot1."

18. See note 108 to chapter 14.

19. Simberloff, Keitt, et al., "Yes We Can!"

20. Lapeyssonie, *La Médecine Coloniale* (on tsetse fly); Hoelmer and Grace, "Citrus Blackfly"; Baumhover et al., "Screw-Worm Control;" J. Watson et al., "Seychelles Magpie Robin"; Ashmole et al., "Seabird Conservation and Feral Cats."

21. Soper and Wilson, *Anopheles gambiae in Brazil 1930–1940*; Elton, "Biological Cost of Modern Transport."

22. During World War II, Elton focused the entire Bureau of Animal Population on aiding the British war effort by studying the biology and possible control mechanisms for four non-native species in Britain: Norway rat, black rat, house mouse, and European rabbit (Crowcroft, *Elton's Ecologists*; Southwood and Clarke, "Charles Sutherland Elton").

23. Elton, *Ecology of Invasions by Animals and Plants*, 154.

24. Dahlsten and Garcia, *Eradication of Exotic Pests.*

25. Veitch and Clout, *Turning the Tide*; Veitch, Clout, and Towns, *Island Invasives: Eradication and Management*; Veitch, Clout, Martin, et al., *Island Invasives: Scaling Up.*

26. Keitt et al., "Global Islands Invasive Vertebrate Eradication Database."

27. N. Holmes et al., "Tracking Invasive Species Eradications."

28. Simberloff, Keitt, et al., "Yes We Can!"

29. Veitch et al., *Island Invasives: Scaling Up*.

30. Bioethics Panel, *Predator-free New Zealand*; Peltzer et al., "Scale and Complexity Implications."

31. Kean et al., "Global Eradication and Response Database."

32. Suckling et al., "Trends in Arthropod Eradication Programmes."

33. Woldendorp and Bomford, *Weed Eradication Strategies*.

34. Simberloff, Keitt, et al., "Yes We Can!"

35. On invasion scientists' and managers' skepticism, see Rejmánek and Pitcairn, "When Is Eradication of Exotic Pest Plants a Realistic Goal?"; Gardener et al., "Eradications and People." On persistence of soil seedbanks, see Downey and Richardson, "Alien Plant Invasions and Native Plant Extinctions."

36. Simberloff, Keitt, et al., "Yes We Can!"

37. E.g., Cory and Knapp, "Program to Eradicate Twenty-Four Nonnative Invasive Plant Species"; R. Rodriguez et al., "Enhancing Invasive Species Control."

38. Simberloff, "Eradication and Maintenance Management of Established Aquatic Invaders."

39. Rytwinski et al., "Effectiveness of Non-Native Fish Removal Techniques in Freshwater Ecosystems."

40. Simberloff, "Eradication and Maintenance Management of Established Aquatic Invaders."

41. On Caribbean black-striped mussel, see Bax et al., "Man-made Marinas." On "killer alga," see L. Anderson, "California's Reaction to *Caulerpa taxifolia*"; Muñoz, *Caulerpa Conquest*.

42. Pickart, "Mechanical Control."

43. Bourne, *History of Rat Control in Alberta*.

44. Pickart, "Dune Restoration over Two Decades."

45. Vats, "Herbicides."

46. Timmons, "History of Weed Control."

47. Timmons, "History of Weed Control."

48. On use of sulfer dust, see P. Russell, "Century of Fungicide Evolution." On "Bordeaux mixture," see G. Matthews, *History of Pesticides*.

49. P. Russell, "Century of Fungicide Evolution."

50. G. Matthews, *History of Pesticides*.

51. G. Matthews, *History of Pesticides*.

52. On Carson's work, see F. Davis, *Banned*; G. Matthews, *History of Pesticides*.

53. T. Williams, "Killer Weeds."

54. Simberloff, "Biological Invasions: What's Worth Fighting and What Can Be Won?"

55. Puri et al., "Stability of Fluridone-Resistant Hydrilla."

56. O'Dowd et al., "Invasional 'Meltdown' on an Oceanic Island"; Boland et al., "Heli-Baiting Using Low Concentration Fipronil."

57. On brodifacoum, see Whisson, "Pesticides (Mammal)." On rotenone, see Rytwinski et al., "Effectiveness of Non-Native Fish Removal Techniques in Freshwater Ecosystems."

58. On fipronil, see P. Stewart et al., "Alternative Baiting Method of Yellow Crazy Ants." On brodifacoum, see Oetting, "Controversy Grows over Recipe for Rodent Problem."

59. Dalu et al., "Assessment of the Effect of Rotenone."

60. Georghiou and Saito, *Pest Resistance to Pesticides.*

61. For the American Association of Economic Entomologists 1940 presidential address, see H. S. Smith, "Racial Segregation in Insect Populations."

62. Michel et al., "Somatic Mutation-Mediated Evolution of Herbicide Resistance"; Puri et al., "Stability of Fluridone-Resistant Hydrilla."

63. Twigg et al., "Evidence of Pesticide Resistance in Medium-Sized Mammalian Pests."

64. J. Mooney et al., "VKORC1 Sequence Variants Associated with Resistance"; Ruiz-López et al., "Widespread Resistance to Anticoagulant Rodenticides"; Carromeu-Santos et al., "Widespread Distribution of Rodenticide Resistance-Conferring Mutations"; Stöck et al., "Potential of VKORC1 Polymorphisms in Mustelidae."

65. Talekar and Shelton, "Biology, Ecology, and Management of the Diamondback Moth"; Furlong et al., "Diamondback Moth Ecology and Management."

66. Simberloff, "Biological Invasions: What's Worth Fighting and What Can Be Won?"

67. D. Cooper et al., "1080 and Wildlife."

68. E.g., Littin et al., "Behaviour, Coagulopathy and Pathology of Brushtail Possums."

69. Whisson, "Pesticides (Mammal)"; Eason et al., "Toxicology and Ecotoxicology of Para-Aminopropiophenone (PAPP)."

70. Clausen, "Insect Parasitism and Biological Control"; Sweetman, *Principles of Biological Control.*

71. V. Moran and Zimmermann, "Biological Control of Cactus Weeds."

72. V. Moran and Zimmermann, "Biological Control of Cactus Weeds"; Stiling, "Worm That Turned"; H. Zimmermann et al., "Renowned Cactus Moth, *Cactoblastis cactorum* (Lepidoptera: Pyralidae)."

73. H. Zimmermann et al., "Renowned Cactus Moth, *Cactoblastis cactorum* (Lepidoptera: Pyralidae)."

74. V. Moran and Zimmermann. "Biological Control of Cactus Weeds." Biological control is occasionally employed against native species perceived as pests, although such uses do not fall in the category of classical biological control. For instance, the tarnished plant bug is native to North America, where it damages several crop plants, especially alfalfa (which is itself native to Asia). To attempt to control this insect, the United States Department of Agriculture in 1983 introduced a European parasitic wasp, *Peristenus digoneutis,* in New Jersey, from which it spread both autonomously and with assistance from agricultural agents to eight northeastern states (Day, Hedlund, et al., "Establishment of *Peristenus digoneutis*"; Day, Tilmon, et al., "Recent Range Expansions of *Peristenus digoneutis*").

75. J. Moore, "Use of Fishes for Control of Mosquitos."

76. Pyke, "Plague Minnow or Mosquito Fish?"

77. Clausen, "Insect Parasitism and Biological Control."

78. Clausen, "Insect Parasitism and Biological Control"; Sweetman, *Principles of Biological Control.*

79. R. D. Van Driesche, Simberloff, et al., *Integrating Biological Control into Conservation Practice.*

80. Buckingham, "Biological Control of Alligatorweed."

81. Schooler, "*Alternanthera philoxeroides* (Martius) Grisebach (Alligator Weed)."

82. Statistic from Cock et al., "Trends in the Classical Biological Control of Insect Pests by Insects." On criteria for success, see Syrett et al., "Success in Biological Control."

83. Hinz et al., "Global Review of Target Impact."

84. Cock et al., "Trends in the Classical Biological Control of Insect Pests by Insects."

85. In fact, *Aphytis chrysomphali*, which was introduced to California in the early twentieth century, turned out not to control the red scale adequately (Compere, "Red Scale and Its Insect Enemies"). In the late 1940s, *Aphytis lingnanensis* was imported from China (Rosen and DeBach, *Introduced Parasites and Predators*), and by 1958, it had displaced *A. chrysomphali* except in some coastal sites (DeBach and Sundby, "Competitive Displacement Between Ecological Homologues"). However, *A. lingnanensis* did not sufficiently control the scale in warm, dry areas of interior southern California. In 1958, *Aphytis melinus* from northern India and Pakistan was introduced and quickly displaced *A. lingnanensis* in southern California except along coastal areas, where *A. lingnanensis* was dominant or at least common (DeBach, "Competitive Displacement and Coexistence Principles"). The entire sequence not only exemplifies a successful series of introductions ultimately conferring control but also the context-dependency of biological invasions with respect to survival and impact.

86. Elton, *Ecology of Invasions by Animals and Plants*, 2nd ed.; Nicholson, "Progress in the Control of *Hypericum* by Insects."

87. Elton, *Ecology of Invasions by Animals and Plants*, 2nd ed.

88. Howarth, "Environmental Impacts of Classical Biological Control"; Simberloff and Stiling, "How Risky Is Biological Control?"

89. C. King, *Invasive Predators in New Zealand.*

90. Cowie, "Invertebrate Invasions on Pacific Islands"; Cowie, "Snails and Slugs." What had previously been believed to be a single species, the rosy wolf snail, has recently been determined to be two similar species, both native at least to Florida in the southeastern United States and both introduced as biological control agents to most islands on which the "rosy wolf snail" was released (W. Meyer et al., "Two for One"). See also chapter 11, section 11.3.

91. Muniappan et al., "Giant African Snail Control"; Muniappan, "Biological Control of the Giant African Snail."

92. Lowe et al., *100 of the World's Worst Invasive Alien Species.*

93. Lever, *Cane Toad*; J. L. Long, *Introduced Birds of the World.*

94. Boettner et al., "Effects of a Biological Control Introduction"; Elkinton and Boettner, "Benefits and Harm Caused by the Introduced Generalist Tachinid."

95. Elkinton, Parry, and Boettner, "Implicating an Introduced Generalist Parasitoid"; Elkinton and Boettner, "Benefits and Harm Caused by the Introduced Generalist Tachinid." See note 57 to chapter 14.

96. On Hawaii, see Raine et al., "Impact of an introduced Avian Predator the Barn Owl." On the Seychelles, see A. Diamond, "Seabirds of the Seychelles." On the earlier failed introductions, see J. L. Long, *Introduced Birds of the World.*

97. On taking nontarget impacts seriously, see Heimpel and Mills, *Biological Control.* Whereas modern biological control practitioners acknowledge the early disastrous introductions, several defenses of biological control have employed counterfactuals to contend that nontarget impacts would have been worse had a biological control agent not been introduced. For instance, F. J. Simmonds and colleagues suggested that, despite contributions to extinctions of native species by the small Indian mongoose, the degree to which it lessened rat populations in the West Indies contributed more to conservation of native species than whatever losses it caused (Simmonds et al., "History of Biological Control").

98. Simmonds and Bennett, "Biological Control of *Opuntia* spp."

99. Pemberton, "*Cactoblastis cactorum* (Lepidoptera: Pyralidae) in the United States."

100. Zimmermann et al., "Renowned Cactus Moth, *Cactoblastis cactorum* (Lepidoptera: Pyralidae)."

101. Louda, Kendall, et al., "Ecological Effects of an Insect Introduced for the Biological Control of Weeds"; Simberloff, "Risks of Biological Control for Conservation Purposes."

102. E.g., Heimpel and Mills, *Biological Control.*

103. E.g., R. McFadyen, "Biological Control of Weeds"; Culliney, "Benefits of Classical Biological Control for Managing Invasive Plants."

104. Zwölfer and Harris, "Biology and Host Specificity."

105. Suckling and Sforza, "What Magnitude Are Observed Non-Target Impacts from Weed Biocontrol?"

106. E.g., V. Moran and Hoffmann, "Fourteen International Symposia on Biological Control of Weeds"; Suckling and Sforza, "What Magnitude Are Observed Non-Target Impacts from Weed Biocontrol?"; DiTomaso et al., "Addressing the Needs for Improving Classical Biological Control Programs"; Barratt, Moran, et al., "Status of Biological Control"; Hinz, Winston, and Schwarzländer, "Global Review of Target Impact."

107. E.g., Simberloff, "Risks of Biological Control for Conservation Purposes"; Havens et al., "Risks and Rewards."

108. Simberloff, "Risks of Biological Control for Conservation Purposes"; Heimpel and Mills, *Biological Control.*

109. Pearson and Callaway, "Indirect Effects of Host-Specific Biological Control Agents"; Ortega et al., "Effects of Biological Control Agents."

110. Holt and Hochberg, "Indirect Interactions, Community Modules and Biological Control."

111. Heimpel and Mills, *Biological Control.*

112. Knipling, "Possibilities in Insect Control"; Baumhover, "Personal Account of Developing the Sterile Insect Technique."

113. Klassen et al., "History of the Sterile Insect Technique."

114. E.g., Shelly et al., "Chilling and Flight Ability."

115. On beetles and moths, see Klassen et al., "History of the Sterile Insect Technique." On crayfish, see N. Green et al., "Dominance, Reproductive Behaviours and Female Mate Choice."

116. Simberloff, "Eradication and Maintenance Management of Established Aquatic Invaders."

117. James (Jim) R. Miller and Gut, "Mating Disruption for the 21st Century."

118. Simberloff, "Eradication and Maintenance Management of Established Aquatic Invaders."

119. James (Jim) R. Miller and Gut, "Mating Disruption for the 21st Century"; Lance et al., "Mating Disruption as a Suppression Tactic."

120. Lance et al., "Mating Disruption as a Suppression Tactic."

121. Aldridge et al., "Microencapsulated BioBullets."

122. Martinez et al., "Technology Innovation."

123. M. D. Thomas, "Radio Telemetry System for Animal Tracking."

124. D. Taylor and Katahira, "Radio Telemetry as an Aid in Eradicating Remnant Feral Goats."

125. K. Campbell, "Advances in Judas Goat Methodology"; K. Campbell and Donlan, "Feral Goat Eradications on Islands"; K. Campbell et al., "Increasing the Efficacy of Judas Goats"; K. Campbell et al., "Development of a Prolonged Estrus Effect."

126. Carrion et al., "Archipelago-Wide Island Restoration."

127. McIlroy, "New Techniques for an Old Problem"; Ditchkoff and Bodenchuk, "Management of Wild Pig."; P. Spencer et al., "Genetic Relationships Within Social Groups"; Morrison et al., "Irruption Interrupted."

128. Ramsey et al., "Quantifying the Probability of Detection."

129. Eamens et al., "RNA Silencing in Plants"; Mahmood-ur-Rahman et al., "RNA Interference."

130. E.g., Baum et al., "Control of Coleopteran Insect Pests."

131. E.g., San Miguel and Scott, "The Next Generation of Insecticides"; Zotti et al., "RNAi Technology in Crop Protection."

132. Bachman et al., "Characterization of the Spectrum of Insecticidal Activity of a Double-Stranded RNA"; S. Zhang, "EPA Quietly Approved Monsanto's New Genetic-Engineering Technology"; Zotti et al., "RNAi Technology in Crop Protection."

133. McGonigle et al., *Compositions and Methods for Insecticidal Control of Stinkbugs.*

134. Leonard et al., "Engineered Symbionts."

135. On common reed, see Martinez et al., "Technology Innovation." On crustaceans, see Savaya-Alkalay et al., "Size-Selective Predation by All-Male Prawns."

136. W. Hamilton, "Extraordinary Sex Ratios."

137. For the 2006 proposal, see Gutierrez and Teem, "Model Describing the Effect of Sex-Reversed YY Fish"; Teem and Gutierrez, "Theoretical Strategy for Eradication of Asian Carps." On the technology that had been developed in the 1970s, see, e.g., K. Hopkins et al., "Estrogen Sex-Reversal of *Tilapia aurea.*"

138. Schill et al., "Production of a YY Male Brook Trout Broodstock"; Schill, Meyer, and Hansen, "Simulated Effects of YY-Male Stocking."

139. P. Kennedy et al., "Survival and Reproduction Success of Hatchery YY Male Brook Trout."

140. Jiang et al., "Production of YY Supermale and XY Physiological Female Common Carp."

141. On discussion of transgenes to control invasive species, see F. Gould, "Broadening the Application."

142. G. Fu et al., "Female-Specific Insect Lethality."

143. Angulo and Gilna, "International Law Should Govern Release of GM Mosquitoes"; Enserink, "GM Mosquito Trial Alarms Opponents." An alternative approach to controlling diseases transmitted by *Aedes aegypti* that does not entail manipulating the mosquito's genotype uses a bacterium, *Wolbachia*, that lives inside cells of many insect species, including some mosquitoes, but was not found in *A. aegypti. Wolbachia* are passed from one generation to the next through an insect's eggs. A team led by Scott O'Neill managed to infect *A. aegypti* with *Wolbachia*, and in these mosquitoes *Wolbachia* competes with viruses such as dengue, Zika, chikungunya and yellow fever, causing viral reproduction to plummet, so the mosquitoes are much less likely to transmit the viruses from person to person (S. O'Neill, "Dengue Stopper"; Innis, "Trial to See if 'Good' Mosquitoes Can Edge Out Bad"). Field trials were conducted in Australia and Vietnam (Nguyen et al., "Field Evaluation"), and *Wolbachia*-infected *A. aegypti* were released in both Malaysia (Nazni et al., "Establishment of *Wolbachia* Strain") and northern Queensland, Australia (P. Ryan et al., "Establishment of wMel *Wolbachia*"), persisted, and were associated with a decline of dengue fever.

144. Rutkin, "GM Mosquitoes Approved"; Servick, "Winged Warriors."

145. Evans et al., "Transgenic Aedes Aegypti Mosquitoes."

146. Evans et al., "Editorial Expression of Concern."

147. Oxitec, "Our Technology."

148. Lander, "Heroes of CRISPR"; Jon Cohen, "Cut Above."

149. Esvelt et al., "Concerning RNA-Guided Gene Drives"; Harvey-Samuel et al., "Towards the Genetic Control of Invasive Species"; Moro et al., "Identifying Knowledge Gaps."

150. Target Malaria, "Who We Are"; GBIRd, "Genetic Biocontrol of Invasive Rodents."

151. On mosquitos, see V. Gantz et al., "Highly Efficient Cas9-mediated Gene Drive"; Hammond et al., "CRISPR-Cas9 Gene Drive System"; Kyrou et al. "CRISPR–Cas9 Gene Drive Targeting *Doublesex*." On mice, see Grunwald et al., "Super-Mendelian Inheritance"; Pfitzner et al. "Progress Toward Zygotic and Germline Gene Drives in Mice."

152. Suhr and Senut, *Eradication of Invasive Quagga and Zebra Mussels.*; US Bureau of Reclamation, "Bureau of Reclamation Selects Three Submissions to Receive Prizes."

153. Friends of the Earth, "Genetic 'Extinction' Technology Rejected by International Groups."

154. National Academies of Sciences, Engineering, and Medicine, *Gene Drives on the Horizon*; E. Callaway, "UN Treaty Agrees to Limit Gene Drives but Rejects a Moratorium."

155. E.g., Unckless et al., "Evolution of Resistance Against CRISPR/Cas9 Gene Drive."

156. C. Noble et al., "Current CRISPR Gene Drive Systems Are Likely to Be Highly Invasive."

157. Esvelt and Gemmell, "Conservation Demands Safe Gene Drive."

158. Jon Cohen, "Reporter Does CRISPR"; Baumgaertner, "Gene Lab Next Door."

159. Díaz-Ferguson and Moyer, "History, Applications, Methodological Issues and Perspectives"; Cristescu and Hebert, "Uses and Misuses of Environmental DNA"; Beng and Corlett, "Applications of Environmental DNA"; Darling, "How to Learn to Stop Worrying."

160. Ibabe et al., "Environmental DNA from Plastic and Textile Marine Litter."

161. On marine sediments, see Ogram et al., "Extraction and Purification of Microbial DNA from Sediments." On phytoplankton blooms, see, e.g., Bailiff and Karl, "Dissolved and Particulate DNA Dynamics During a Spring Bloom."

162. Ficetola, Miaud, et al., "Species Detection Using Environmental DNA."

163. Jerde et al., "'Sight-Unseen' Detection of Rare Aquatic Species."

164. Díaz-Ferguson and Moyer, "History, Applications, Methodological Issues and Perspectives."

165. Cristescu and Hebert, "Uses and Misuses of Environmental DNA."

166. Beng and Corlett, "Applications of Environmental DNA (eDNA)."

167. On aquatic plants, see Scriver et al., "Development of Species-Specific Environmental DNA (eDNA) Markers." On hydrilla, see Matsuhashi et al., "Evaluation of the Environmental DNA Method"; C. Gantz et al., "Environmental DNA Detection of Aquatic Invasive Plants." On Burmese python, see Piaggio et al., "Detecting an Elusive Invasive Species." On New Zealand mud snail, see Goldberg et al., "Environmental DNA as a New Method for Early Detection." On Brazilian waterweed, see Fujiwara et al., "Use of Environmental DNA"; Matsuhashi et al., "Evaluation of the Environmental DNA Method." On rusty crayfish, see Dougherty et al., "Environmental DNA (eDNA) Detects the Invasive Rusty Crayfish." On red swamp crayfish, spiny-cheek crayfish, and signal crayfish, see Mauvisseau et al., "Environmental DNA as an Efficient Tool for Detecting Invasive Crayfishes." On feral pigs, see K. Williams et al., "Detection and Persistence of Environmental DNA."

168. Valentin et al., "Early Detection of Invasive Exotic Insect Infestations Using eDNA."

169. É. Tremblay et al., "High-Resolution Biomonitoring."

170. Stokstad, "DNA Plucked from Air Identifies Nearby Animals."

171. Cristescu and Hebert, "Uses and Misuses of Environmental DNA."

172. Cristescu and Hebert, "Uses and Misuses of Environmental DNA"; Beng and Corlett, "Applications of Environmental DNA"; Darling, "How to Learn to Stop Worrying."

173. Simberloff, "Rats Are Not the Only Introduced Rodents Producing Ecosystem Impacts."

Chapter 18

1. Simberloff, "Nature, Natives, Nativism, and Management."

2. M. Davis et al., "Don't Judge Species on Their Origins."

3. For the argument to abandon invasion science, see M. Davis, *Invasion Biology*; Valéry et al., "Another Call for the End of Invasion Biology"; B. Hoffmann and Courchamp, "Biological Invasions and Natural Colonisations." For the rebuttal, see Pyšek and Hulme, "Invasion Biology Is a Discipline That's too Young to Die"; Simberloff and Vitule, "Call for an End to Calls for the End of Invasion Biology"; Richardson and Ricciardi, "Misleading Criticisms of Invasion Science"; J. Wilson, García-Díaz, et al., "Biological Invasions and Natural Colonisations Are Different."

4. Gbedomon et al., "Diverse Views"; Shackleton et al., "Consensus and Controversy."

5. Simberloff, "Confronting Introduced Species."

6. Quotation from Gröning and Wolschke-Bulmahn, "Some Notes on the Mania for Native Plants in Germany," 123, citing a 1939 work by Reinhold Tüxen, a botanist who headed the Reich Central Office for Vegetation Mapping. Tüxen performed many services to advance the ideology and war effort of the Reich, yet he became a respected scientific figure in postwar Germany. The journal *Tuexenia* of the Floristisch-soziologische Arbeitsgemeinschaft, a German-speaking association of experts and enthusiasts of floristics and plant sociology, is named in his honor. For Gröning and Wolschke-Bulmahn's series of papers, see Gröning and Wolschke-Bulmahn, "Some Notes on the Mania for Native Plants in Germany"; Gröning and Wolschke-Bulmahn, "Native Plant Enthusiasm"; Gröning and Wolschke-Bulmahn, "Response: If the Shoe Fits, Wear It!"; Wolschke-Bulmahn and Gröning, "Ideology of the Nature Garden"; Wolschke-Bulmahn, " 'Wild Garden' and 'Nature Garden' "; Wolschke-Bulmahn, review of *Jens Jensen*, by R. E. Grese; Wolschke-Bulmahn, introduction to Wolschke-Bulmahn, *Nature and Ideology*; Wolschke-Bulmahn, "Nationalization of Nature."

7. Gröning and Wolschke-Bulmahn, "Native Plant Enthusiasm"; Gröning and Wolschke-Bulmahn, "Response: If the Shoe Fits, Wear It!"

8. Peretti, "Nativism and Nature"; Head and Muir, "Nativeness, Invasiveness, and Nation in Australian Plants"; C. Warren, "Perspectives on the 'Alien' Versus 'Native' Species Debate"; Fall, "Biosecurity and Ecology"; Trigger and Head, "Restored Nature, Familiar Culture"; I. Brook, "Making Here Like There"; E. Katz, "Nazi Comparison in the Debate over Restoration"; Tassin, *La grande invasion*.

9. Pollan, "Against Nativism," 54, 55.

10. S. Gould, "Evolutionary Perspective," 12. This piece was published in the volume edited by Wolschke-Bulmahn, *Nature and Ideology*. Gould published it two other times: (1) in the journal *Arnoldia* in 1998, and (2) in his own book *I Have Landed*

in 2012 (S. Gould, "Evolutionary Perspective," 2–10; S. Gould, *I Have Landed*, 335–46). Social scientist Tomaz Mastnak and colleagues defended native plant enthusiasts against the charges by both Gould and Pollan, instead casting invasive non-native plants, with particular reference to California, as part of the colonization of much of the world by European nations (Mastnak et al., "Botanical Decolonization"). From this perspective, native plant societies are undertaking a botanical decolonization, and Pollan and Gould were defending an imperialistic legacy. Mastnak and colleagues refer to the conception by the geographer Alfred Crosby of introduction of non-native species as part of the imperialistic project of European nations (discussed in chap. 8, sec. 8.3).

11. Haraway and Harvey, "Nature, Politics, and Possibilities."

12. Theodoropoulos, *Invasion Biology: Critique of a Pseudoscience.*

13. Comaroff and Comaroff, "Naturing the Nation"; Tassin, *La grande invasion.*

14. Peretti, "Nativism and Nature," 188–89. For the source cited in the quotation, see I. Macdonald, Kruger, and Farrar, *Ecology and Management of Biological Invasions in Southern Africa.*

15. E. Katz, "Nazi Comparison in the Debate over Restoration." The matter of potential Nazi heritage for modern invasion science is part of a larger controversy about the potential relationship of Nazi interest in conservation and the environment to modern-day environmentalism and conservation, which is well beyond the scope of this book. Frank Uekötter provided an introduction to the issue (Uekötter, *Green and the Brown*).

16. E.g., "Germany has been chosen to lend its name to this style in the history of gardens and to become once again an 'improver of the world'" (Lange, *Gartenbilder*, cited by Wolschke-Bulmahn, "'Wild Garden' and 'Nature Garden'"). On scant evidence of xenophobia or racism, see Simberloff, "Confronting Introduced Species."

17. Wolschke-Bulmahn, "'Wild Garden' and 'Nature Garden,'" 198.

18. von Humboldt, *Ideen zu einer Physiognomik der Gewächse.*

19. E.g., Orians, "Ecological and Evolutionary Approach."

20. Wolschke-Bulmahn, "'Peculiar Garden'"; Wolschke-Bulmahn, review of *Jens Jensen*; Wolschke-Bulmahn, introduction to *Nature and Ideology.*

21. Jensen, "Die Lichtung (The Clearing)," 177.

22. Dave Egan and Tishler, "Jens Jensen, Native Plants, and the Concept of Nordic Superiority"; McKinney and Lockwood, "Biotic Homogenization"; Lockwood and McKinney, *Biotic Homogenization.*

23. E.g., Gould, "Evolutionary Perspective," *Arnoldia*; Hettinger, "Exotic Species, Naturalisation, and Biological Nativism." On "McDonaldization," see Gould, "Evolutionary Perspective," *Arnoldia*, 9: "I would certainly be horrified to watch the botanical equivalent of McDonald's uniform architecture and cuisine wiping out every local diner in America. Cherishing native plants does allow us to defend and preserve a maximal amount of local variety."

24. Jamieson, "Ecosystem Health."

25. On the desire to retain intact native assemblages seeming analogous to attempts to retain distinct human cultures, see Council of Europe Committee of Ministers, *Declaration on Cultural Diversity*; UNESCO, "UNESCO Universal Declaration on Cultural Diversity"; Simberloff, Martin, et al., "Impacts of Biological Invasions." On the idea that it need not be linked to nativism, see Flores, "Environmentalism and Multiculturalism." W. O'Brien addressed, with a degree of skepticism, the attempt by invasion scientists and restoration ecologists to deflect the charge of nativism or xenophobia by invoking cultural McDonaldization (O'Brien, "Exotic Invasions, Nativism, and Ecological Restoration").

26. Olmsted, "Foreign Plants and American Scenery"; Sargent, "Editorial"; Zaitzevsky, *Frederick Law Olmsted and the Boston Park System.*

27. Zaitzevsky, *Frederick Law Olmsted and the Boston Park System.*

28. Sargent, "Editorial," 266. Sargent consistently advocated for untrammeled nature, subsequently deploring Gifford Pinchot's model for conserving United States forests partially through sustainable timber harvests, instead urging that they be maintained in a state of wilderness (McGeary, *Gifford Pinchot*). He engaged in a long effort to remove all non-native trees planted at Mount Vernon since George Washington's time (Fairchild, *World Was My Garden*, 426).

29. Olmsted, "Foreign Plants and American Scenery," 418.

30. Blanchan at times appealed directly to the notion of aesthetic appeal in terms similar to those of Sargent:

> Do not try to collect a museum of plants; avoid freaks of variegated foliage, . . . strange exotics that look out of place in our American landscape. (Blanchan, *American Flower Garden*, 21)

Other times, she seemed at least as concerned with the likelihood that a non-native species would thrive as that it be beautiful:

> To the purist it may seem an impertinence to transplant the flora of other lands to any of those parterres of nature's garden we are pleased to call "ours" when so many of our native wild flowers offer delightful possibilities as yet little realised by American gardeners. But let him remember that the commonest wild flowers we have, for example, the daisy that now whitens the fields throughout the United States and Canada, was unknown on this continent until it smuggled its passage across the Atlantic in the hay used for packing the early Colonists' china. Very many other so-called weeds—the exquisite Queen Anne's lace or wild carrot, the dusty white yarrow, the buttercup that spangles our meadows, and "succory to match the sky"—to name only a few among many—are merely naturalised foreigners, not natives, that thrive far better here, however, than they did at home, just as the Irish and Italian immigrants do. When nature does not fix sectional limitations, why should we? (Blanchan, *American Flower Garden*, 81)

> If the yew and holly are the best evergreens for England because, being native, they thrive there to perfection, so our spruce, hemlock, arborvite, pine and junipers are best for us to use as a basis for other planting. On the solid foundation of our native trees we may build the lighter superstructure and embellish it, according to fancy, with details from the ends of the earth; but let us not forget the enormous sums of American money wasted on European evergreens—on English yews alone. After exhausting the possibilities of our beautiful native trees, our hope lies in those from lands with climatic conditions, similar to our own, notably Siberia, China and Japan. (Blanchan, *American Flower Garden*, 138)

31. Dave Egan and Tishler, "Jens Jensen, Native Plants, and the Concept of Nordic Superiority."

32. I. Brook, "Making Here Like There."

33. A. Helmreich, "Re-presenting Nature." Wallace quotation in Raby, *Alfred Russel Wallace*, 159. Such effusive expressions of appreciation for native plants may

have been quite common in nineteenth-century England. The Romantic poet William Wordsworth worried that importing plants "from every portion of the globe" would eliminate his local favorites, expressing particular concern in 1837 for hollyhocks, fretting that "the Dahlia, and other foreigners recently introduced should have done so much to drive this imperial flower out of notice" (quoted in Noyes, *Wordsworth and the Art of Landscape*, 100). Ironically, hollyhocks are native to Asia but were introduced to the United Kingdom at least by the fifteenth century (Gardens Trust, "Hollyhocks"); dahlias, native to Mexico and Central America did not arrive until about 1800 (Hogg, *Dahlia*). Wordsworth's 1835 *Guide Through the District of the Lakes* inveighed against the changes new residents in the Lake District had wrought on the landscape, with non-native plants the leading edge of the aesthetic disaster. For instance, regarding trees, he argued that "trees have been introduced within these last fifty years, such as beeches, larches, limes, &c. and plantations of firs, seldom with advantage, and often with great injury to the appearance of the country" (Wordsworth, *Guide Through the District*, 27). He particularly deplored the larch, native to central Europe and introduced in the seventeenth century to Great Britain (Jansen and Geburek, "Historic Translocations of European Larch"). Wordsworth himself cast his interest in landscapes—in paintings, gardens, and nature—as an aesthetic matter, but A. Hazucha depicted his antipathy to non-native plants as increasingly related to xenophobia, terming him an "eco-nationalist" afflicted by "ecoxenophobia" (Hazucha, "Neither Deep nor Shallow but National," 61, 69).

34. Raby, *Alfred Russel Wallace*.

35. Simberloff, "Confronting Introduced Species."

36. Pauly, "Beauty and Menace of the Japanese Cherry Trees"; Pauly, reply to Simberloff letter to the editor, *Isis*.

37. Pauly, "Beauty and Menace of the Japanese Cherry Trees," 54, 70.

38. On the absence of direct evidence of nativist tendencies, see Simberloff, letter to the editor, *Isis*.

39. Pauly, reply to Simberloff letter to the editor, *Isis*, 677.

40. Cf. Kraut, *Silent Travelers*; Tomes, *Gospel of Germs*.

41. Simberloff, letter to the editor, *Isis*.

42. Seftel, "Government Regulation and the Rise of the California Fruit Industry."

43. Pauly, "Beauty and Menace of the Japanese Cherry Trees."

44. Subramaniam, "Aliens Have Landed!"; Subramaniam, "Aliens in Our Midst"; Subramaniam, *Botany of Empire*; Leland, *Aliens in the Backyard*; Raffles, "Mother Nature's Melting Pot"; G. E. Allen, " 'Culling the Herd': Eugenics and the Conservation Movement."

45. E.g., Darnovsky, "Stories Less Told"; Budiansky, *Nature's Keepers*. Budiansky captured this claim succinctly in a chapter title in *Nature's Keepers*: "Nazis, Planners, Eugenicists, and Other Ecologists."

46. Coates, *Strangers on the Land*.

47. Spiro, *Defending the Master Race*.

48. Spiro, "Puritan Racist"; Spiro, *Defending the Master Race*.

49. M. Grant, "Future of Our Fauna," 504.

50. Spiro, *Defending the Master Race*, 63.

51. M. Grant, *Rocky Mountain Goat*, 34.

52. Osborn, *Men of the Old Stone Age*.

53. B. Regal, *Henry Fairfield Osborn*.

54. M. Grant, *Passing of the Great Race*, 4th ed.

55. Hornaday, *Our Vanishing Wild Life*, quotations at x, 95.

56. Hornaday, *Our Vanishing Wild Life*, quotations at 101, 105.

57. Hornaday, *Our Vanishing Wild Life*, 105.

58. Hornaday, *Our Vanishing Wild Life*.

59. Hornaday, *Our Vanishing Wild Life*, 334. On the English sparrow, see also Coates, *Strangers on the Land*.

60. Gregory, "Biographical Memoir of Frank Michler Chapman"; Vuilleumier, "Dean of American Ornithologists."

61. "Celebration in Honor of Dr. Henry Fairfield Osborn."

62. Vuilleumier, "Dean of American Ornithologists."

63. Several previous introductions of the starling to North America are believed to have failed to establish persistent populations (J. L. Long, *Introduced Birds of the World*).

64. F. Chapman "European Starling," 482.

65. Spiro, *Defending the Master Race*.

66. Goethe, "Immigration from Mexico," 140.

67. Goethe, *Garden Philosopher*.

68. Goethe acquired great wealth through stock investments, real estate, and land development. As a "founding father" (T. Platt, "Engaging the Past") and frequent benefactor (including in his will) of Sacramento State College (subsequently California State University, Sacramento), Goethe was honored when the university arboretum was named after him. His local star began to wane in the 1960s, when the university named a new science building for him and was forced by protests to rename it two years later (T. Platt, "Engaging the Past"). The arboretum was finally renamed in 2005. Soon thereafter, a Sacramento public school and Sacramento County park named after Goethe were also renamed (G. E. Allen, "'Culling the Herd': Eugenics and the Conservation Movement").

69. Among the critics, see, e.g., Pollan, "Against Nativism"; Theodoropoulos, *Invasion Biology: Critique of a Pseudoscience*. On lack of evidence for linkage with Grant's circle or Nazism, see Simberloff, "Confronting Introduced Species"; Coates, *Strangers on the Land*.

70. Olwig, "Natives and Aliens in the National Landscape," 63.

71. Smout, "Alien Species in 20th-Century Britain," 17.

72. O'Brien, "Exotic Invasions, Nativism, and Ecological Restoration," 66–67, 75.

73. Simberloff, "Confronting Introduced Species"; Coates, *Strangers on the Land*.

74. Bloch, *Historian's Craft*, 24.

75. Bloch, *Historian's Craft*, 26. Bloch wrote *The Historian's Craft* in 1941–42 but did not complete it, as he joined the French Resistance and was captured and executed by firing squad in 1944. It was first published in France in 1949 as *Apologie pour l'histoire ou Métier d'historien* and was later translated into English.

76. Subramaniam, "Aliens Have Landed!"; Subramaniam, *Botany of Empire*; B. Stewart, "Invasion of the Soil Snatchers"; Brinckman, "Creepy Strangler Climbs Oregon's Least-Wanted List"; Yeh, "U.S. Can't Handle Today's Tide of Immigrants"; H. Weiner, "Congress Threatens Wild Immigrants"; Bright, "Invasive Species: Pathogens of Globalization." The headline for Stewart's article in the online archive of the *New York Times*, as listed in the references, differs from that quoted by Subramaniam. Several times I have copied *New York Times* articles as they first appeared and found that headlines or photographs had later been changed.

77. "Invasion of the Habitat Snatchers" (in *Grist*, no byline); Lemonick, "Invasion of the Habitat Snatchers"; McKibben, "Invasion of the Habitat Snatchers"; O. Marsh, "Invasion of the Habitat Snatchers," among others; Devine, "Botanical Barbarians," the text of which contains the phrase "territory-gobbling imperialist," 53.

78. Subramaniam, "Aliens Have Landed!"

79. Recart et al., "There Goes the Neighborhood"; Weinhold, "Do Noxious Neighbors Spread Disease?"

80. Imbler, "This Moth's Name Is a Slur."

81. Imbler, "Introducing Spongy Moth."

82. Kočovský et al., "'Asian Carp' Is Societally and Scientifically Problematic." The four species were bighead carp, silver carp, black carp, and grass carp.

83. B. Larson, "War of the Roses"; B. Larson, *Metaphors for Environmental Sustainability*.

84. On Forbes as a founder of modern ecology, see his 1887 scientific journal article: Forbes, "Lake as a Microcosm."

85. Forbes, *Insect* [. . .] *and the State*, 7–9.

86. Forbes, *Insect* [. . .] *and the State*, 3.

87. Forbes, *Insect* [. . .] *and the State*, 12–13.

88. Forbes, *Insect* [. . .] *and the State*, 13.

89. Pauly, *Fruits and Plains*.

90. Forbes, *Insect* [. . .] *and the State*.

91. Howard, "Entomology and the War."

92. Howard, "On Some Presidential Addresses"; E. Russell, "L. O. Howard Promoted War Metaphors." Historian Edmund P. Russell III depicts the role of metaphors in both entomology and popular portrayal of wartime activities during the period encompassing the First and Second World Wars, including the use of entomological metaphors for human combatants and military metaphors for campaigns against insects (E. Russell, "'Speaking of Annihilation': Mobilizing for War"; E. Russell, *War and Nature*).

93. H. Johnston, "Next War"; Liston, "'Next War'"; Russell, "L. O. Howard Promoted War Metaphors."

94. Du Puy, "Insects Are Winning," 440.

95. Military metaphors and similes included, e.g., "invasions are a part of the thousand-year war," "marauders," "piratic fleets," "strategists in a great campaign." Insect invaders cited were cottony cushion scale, Japanese beetle, gypsy moth (spongy moth), and San Jose scale.

96. Standen, *Insect Invaders*. See chapter 17, box 17.1.

97. "Our Next World War—Against Insects"; "Total Insect War Urged"; "War on Insects." (None of these articles had a bylined author.)

98. R. Callaway and Ridenour, "Novel Weapons"; Simberloff and Von Holle, "Positive Interactions of Nonindigenous Species: Invasional Meltdown?"

99. B. Larson, *Metaphors for Environmental Sustainability*, 162–63.

100. Interestingly, oncologist Siddartha Mukherjee believes that an approach modeled on invasion science, with its emphasis on context dependence, would be more effective than the current medical research paradigm for cancer (Mukherjee, "Invasion Equation").

101. Janovsky and Larson, "Does Invasive Species Research Use More Militaristic Language?"

102. Darwin, *On the Origin of Species*, 2nd ed., 259, 263; A. Wallace, *Darwinism*, 34.

103. M. Davis, *Invasion Biology*.

104. Simberloff, "Invasional Meltdown 6 Years Later."

105. On metaphors used to reason with, see Lakoff, "Metaphor and War." On metaphors engaged to converse, see Deutscher, *Unfolding of Language*.

106. Fahey, "Let's Declare War on These Tired Military Metaphors."

107. Lakoff and Johnson, *Philosophy in the Flesh*.

108. Simberloff, "Invasional Meltdown 6 Years Later."

109. Colautti and MacIsaac, "A Neutral Terminology to Define 'Invasive' Species."

110. Coates, "Editorial Postscript"; Lepczyk, "Time to Retire 'Alien' from the Invasion Ecology Lexicon."

111. S. George, "Women in Film Time."

112. Lowe et al., *100 of the World's Worst Invasive Alien Species*; Luque et al., "100th of the World's Worst Invasive Alien Species."

113. Hulme et al., *Handbook of Alien Species in Europe: DAISIE*. DAISIE is an acronym for Delivering Alien Invasive Species Inventory for Europe.

114. Essl, Bacher, et al., "Which Taxa Are Alien?"

115. H. Roy et al., *Summary for Policymakers*.

116. Simberloff, "Nature, Natives, Nativism, and Management."

117. On inconclusive evidence in the marine realm, see Carlton, "Biological Invasions and Cryptogenic Species."

118. Hofman and Rick, "Ancient Biological Invasions and Island Ecosystems."

119. Rodman, "Restoring Nature: Natives and Exotics," 150.

120. Hettinger, "Exotic Species, Naturalisation, and Biological Nativism."

121. Brodie et al., "Global Policy for Assisted Colonization of Species." Franz Essl and colleagues have proposed the term "neonative" for native species that, owing to climate change, expand their ranges without human assistance to previously unoccupied regions (Essl, Dullinger, et al. "Conceptual Framework for Range-Expanding Species.").

122. Nuñez and Simberloff, "Invasive Species and the Cultural Keystone Concept"; Nuñez, Dimarco, and Simberloff, "Why Some Exotic Species Are Deeply Integrated."

123. Haraway, *How Like a Leaf*.

124. Haraway, *When Species Meet*; Subramaniam, "Aliens in Our Midst"; Subramaniam, *Botany of Empire*; Tsing, *Friction*; Simberloff, "Nature, Culture, and Natureculture." The terms "social-natural landscape" (Tsing, *Friction*) and "bioculture" (Rozzi, "Biocultural Ethics") have been used to describe constructs very similar to natureculture.

125. Pyšek, Sádlo, and Mandák, "Catalogue of Alien Plants of the Czech Republic"; Pyšek, Richardson, et al., "Alien Plants in Checklists and Floras."

126. John Ridley Cross, "Biological Flora of the British Isles"; John Ridley Cross, "Establishment of *Rhododendron ponticum*"; J. Sauer, *Plant Migration*; Simberloff, "Nature, Natives, Nativism, and Management."

127. Slatis, "Analysis of Inbreeding in the European Bison"; James Fisher, Simon, and Vincent, *Wildlife in Danger*.

128. E.g., Kendle and Rose, "Aliens Have Landed!"; M. Davis, *Invasion Biology*; C. Warren, "Nativeness and Nationhood."

129. S. Helmreich, "How Scientists Think About 'Natives.'"

130. S. Gould, "Evolutionary Perspective."

131. C. Warren, "Nativeness and Nationhood."

132. M. Townsend, "Is the Social Construction of Native Species a Threat to Biodiversity?"

133. M. Davis, Chew, et al., "Don't Judge Species on Their Origins."

134. Rotherham and Lambert, "Good Science, Good History, and Pragmatism." For the volume as a whole, see Rotherham and Lambert, *Invasive and Introduced Plants and Animals*.

135. Williamson and Brown, "Analysis and Modelling of British Invasions"; Williamson, "Measuring the Impact of Plant Invaders in Britain."

136. Williamson, "Environmental Risks from the Release of Genetically Modified Organisms"; Williamson, "Invaders, Weeds and the Risk from Genetically Modified Organisms"; Williamson, "Measuring the Impact of Plant Invaders in Britain."

137. Holdgate, "Summary and Conclusions."

138. Williamson and Fitter, "Varying Success of Invaders." This journal article had been cited 1,788 times, according to a Google Scholar search conducted on June 19, 2021. Jeschke and Pyšek ("Tens Rule") provided many examples of the popularity and influence of Williamson and Brown ("Analysis and Modelling of British Invasions"), including in popular literature and exhibitions. On estimating the likelihood that an introduced species produces feral individuals, see Williamson, "Invaders, Weeds and the Risk from Genetically Modified Organisms."

139. Office of Technology Assessment, US Congress, *Harmful Non-Indigenous Species in the United States.*

140. For findings that did follow the tens rule, see, e.g., J. L. Lockwood et al., "How Many, and Which, Plants Will Invade Natural Areas?" For those that did not, see, e.g. Jeschke and Strayer, "Invasion Success of Vertebrates."

141. Jeschke and Pyšek, "Tens Rule."

142. Jarić and Cvijanović, "Tens Rule in Invasion Biology."

143. Crooks and Soulé, "Lag Times in Population Explosions of Invasive Species."

144. Kowarik, "Time Lags in Biological Invasions."

145. Aagaard and Lockwood, "Exotic Birds Show Lags in Population Growth."

146. Richardson and Pyšek, "Naturalization of Introduced Plants."

147. Occasionally the population of a non-native species that had spread and become invasive declines, often quite quickly, to become restricted and innocuous. Reasons for these apparently spontaneous population crashes can sometimes be deduced by careful research (e.g., they may be due to habitat change) and other times they remain mysterious. Along with my student Leah Gibbons, I have depicted many examples (Simberloff and Gibbons, "Now You See Them, Now You Don't!"). In any event, such cases are far fewer than situations in which a formerly restricted and relatively harmless population of a non-native species rapidly increases to become highly invasive.

148. Lesson, "Île de l'Ascension"; Black, *Atka*; Bailey, *Introduction of Foxes to Alaskan Islands.*

149. Sagoff, "What's Wrong with Alien Species?"; Slobodkin, "Good, the Bad and the Reified."

150. Williamson and Brown, "Analysis and Modelling of British Invasions"; Williamson and Fitter, "Characters of Successful Invaders"; Williamson and Fitter, "Varying Success of Invaders."

151. Williamson and Brown, "Analysis and Modelling of British Invasions," 522.

152. Sagoff, "Do Non-Native Species Threaten the Natural Environment?"; Simberloff, "Non-Native Species Do Threaten the Natural Environment!"

153. M. Davis, Chew, et al., "Don't Judge Species on Their Origins," 154.

154. R. B. King et al., "Gorging on Gobies."

155. L. Rodriguez, "Can Invasive Species Facilitate Native Species?"

156. Schlaepfer et al., "Potential Conservation Value of Non-Native Species."

157. Bertness and Coverdale, "Invasive Species Facilitates the Recovery of Salt Marsh Ecosystems."

158. L. Barrett et al., "Nonnative Habitat-Former Mitigates Native Habitat Losses."

159. Vitule, Freire, et al., "Revising the Potential Conservation Value of Non-Native Species."

160. Gleditsch and Carlo, "Fruit Quantity of Invasive Shrubs."

161. McNeish and McEwan, "Review on the Invasion Ecology of Amur Honeysuckle"; Rodewald et al., "Exotic Shrubs as Ephemeral Ecological Traps for Nesting Birds."

162. McChesney and Anderson, "Reproductive Success of Field Sparrows."

163. H. Roy et al., *Summary for Policymakers.*

164. Schwindt et al., "Overwhelming Evidence."

165. K. Thompson et al., "Native and Alien Invasive Plants"; Sagoff, "What's Wrong with Alien Species?"; Sagoff, "Do Non-Native Species Threaten the Natural Environment?"; M. Davis, Grime, and Thompson, "Fluctuating Resources in Plant Communities"; M. Davis, Thompson, and Grime, "Charles S. Elton and the Dissociation of Invasion Ecology"; M. Davis, Pergl, et al., "Vegetation Change"; M. Davis, Chew, et al., "Don't Judge Species on Their Origins"; Hoffmann and Courchamp, "Biological Invasions and Natural Colonisations."

166. Carey et al., "Native Invaders."

167. On Scotland, see C. Warren, "Perspectives on the 'Alien' versus 'Native' Species Debate"; C. Warren, "Nativeness and Nationhood." On US, see Allison, "Paradox of Invasive Species in Ecological Restoration."

168. M. Davis, Thompson, and Grime, "Charles S. Elton and the Dissociation of Invasion Ecology."

169. McCook, "Understanding Ecological Community Succession."

170. On later research foci that Egler anticipated, see Dritschilo, *Magnificent Failure.* For his Oahu study, see Egler, "Indigene versus Alien in the Development of Arid Hawaiian Vegetation."

171. Egler, "Arid Southeast Oahu Vegetation."

172. J. A. Allen, "Mangroves as Alien Species."

173. Vitousek, "Biological Invasions and Ecosystem Properties."

174. D'Antonio and Vitousek, "Biological Invasions by Exotic Grasses."

175. Simberloff, "Invasions of Plant Communities." On Tierra de Fuego, see Lizarralde et al., "Invader Species in Argentina"; C. Anderson et al., "Effects of Invasive North American Beavers on Riparian Plant Communities."

176. On Dutch elm disease, see Brasier, "Intercontinental Spread."

177. Simberloff, Souza, et al., "Natives Are Restless."

178. Paolucci et al., "Origin Matters."

179. Hassan and Ricciardi, "Are Non-Native Species More Likely to Become Pests?"

180. Blackburn, Bellard, and Ricciardi, "Alien Versus Native Species as Drivers of Recent Extinction."

181. Webber and Scott, "Rapid Global Change"; B. Hoffmann and Courchamp, "Biological Invasions and Natural Colonisations."

182. J. Wilson, García-Díaz, et al., "Biological Invasions and Natural Colonisations Are Different"; Essl, Dullinger, et al., "Conceptual Framework for Range-Expanding Species."

183. J. Wilson, Dormontt, et al., "Something in the Way You Move."

184. Inderjit et al., "Framework for Understanding Human-Driven Vegetation Change."

185. J. Wilson, Dormontt, et al., "Something in the Way You Move"; Essl, Dullinger, et al., "Conceptual Framework for Range-Expanding Species."

186. Leopold, "Biotic Land-Use"; Leopold, "Land-Use and Democracy."

187. Ricciardi, "Invasive Species"; Y. Buckley and Catford, "Does the Biogeographic Origin of Species Matter?"

188. J. Russell and Blackburn, "Rise of Invasive Species Denialism," 4.

189. Oreskes and Conway, *Merchants of Doubt*; Ceccarelli, "Manufactured Scientific Controversy."

190. The five scientists cited by Russell and Blackburn are Steve M. Redpath, Jacques Tassin, Ken Thompson, Scott P. Carroll, and Chris Thomas.

191. M. Davis and Chew, "'Denialists Are Coming! Well, Not Exactly,'" quotations at 230, 229.

192. Quotation from Tassin et al., "Determining Whether the Impacts of Introduced Species Are Negative Cannot Be Based Solely on Science," 231.

193. S. Crowley et al., "Disagreement About Invasive Species Does Not Equate to Denialism," 228.

194. Briggs, "Rise of Invasive Species Denialism?"

195. Ricciardi and Ryan, "Exponential Growth of Invasive Species Denialism."

196. Ricciardi and Ryan, "Exponential Growth of Invasive Species Denialism," 552. A similar issue has pervaded discussion of anthropogenic climate change (Boykoff and Boykoff, "Balance as Bias"; Sinatra and Hofer, *Science Denial*).

197. Sagoff, "Invasive Species Denialism: A Reply to Ricciardi and Ryan."

198. Ricciardi and Ryan, "Invasive Species Denialism Revisited."

199. Munro et al., "On Allegations of Invasive Species Denialism."

200. Boltovskoy, Sylvester, and Paolucci, "Invasive Species Denialism: Sorting Out Facts, Beliefs, and Definitions," 11195.

201. Guiaşu and Tindale, "Logical Fallacies and Invasion Biology." 2.

202. Frank et al., "Logical Fallacies and Reasonable Debates."

203. Frank, "Disagreement or Denialism?," S6099.

204. Frank, "Disagreement or Denialism?," S6109.

205. Guerin, "Invoking Denialism Does Not Strengthen Invasion Science"; M. Davis, "Let's Welcome a Variety of Voices to Invasion Biology"; Sagoff, "Fact and Value in Invasion Biology."

206. Animal rights and animal welfare, though often treated as synonyms in controversies over managing non-native species, are not the same (Bekoff, "Introduction: What Does It Feel Like to Be a Fox?"). The difference is that a commitment to animal welfare does not forbid causing pain to animals but obliges one not to cause excessive or unnecessary pain. The definitions of "excessive" and "unnecessary" vary. Animal rights adherents believe it is wrong to cause animals to suffer, including in the limit to keep them in zoos, to eat them, or to conduct certain sorts of research on them. In their view, animals' lives have intrinsic value that endows them with rights, including the right not to suffer and the right to life itself. Animal welfarists, by contrast, do not grant animals rights, beyond minimizing suffering.

207. Gurnell et al., "Ecosystem Impacts of an Alien Invader."

208. LaPointe et al., "Ecology and Conservation Biology of Avian Malaria."

209. S. Anderson and Stone, "Snaring to Control Feral Pigs"; Knickerbocker, "Snaring of Hawaii's Feral Pigs Angers Animal Activists."

210. Burdick, *Out of Eden*. Pigs in Hawaii now comprise hybrids of animals brought by the Polynesian colonists and others brought much later by European colonists (Linderholm et al., "Novel MC1R Allele").

211. Ikagawa, "Invasive Ungulate Policy."

212. Houston, Schreiner, and Moorhead, *Mountain Goats in Olympic National Park*; National Park Service, "Mountain Goat Removal Complete for 2020."

213. Rainey, "Catalina Goats Being Shot"; Schuyler, Garcelon, and Escover, "Control of Feral Goats."

214. Howald, Faulkner, et al., "Eradication of Black Rat from Anacapa Island"; Howald, Donlan, et al., "Eradication of Black Rats *Rattus rattus*."

215. Animal People. "Channel Islands National Park Ex-Chief Hits Cruelty of Killing 'Invasive Species.'" One saboteur pled guilty and was fined, while the other was ultimately exonerated on the grounds that his case was moot by the time it was heard because the rats had been eradicated. The eradication operation and attempted sabotage were sufficiently publicized that they led to a striking roman à clef about the case (T. Boyle, *When the Killing's Done*) that was reviewed in *Nature* (Marris, "Reconstituted Edens"), *Biological Invasions* (Simberloff, review of *When the Killing's Done*), and *Marine Ornithology* (Rauzon, review of *When the Killing's Done*) with emphasis on the clash in the novel between animal rights and conservation, while a writer for *Ecozon@* criticized the ethical consistency of protagonists of both sides (Leikham, "Environmental Imaginations of the California Channel Islands").

216. Pruett-Jones et al., "Population Viability Analysis of Monk Parakeets"; Burger and Gochfeld, "Exotic Monk Parakeets"; Calzada Preston et al., "Monk Parakeets as a Globally Naturalized Species."

217. Whalen, "In Britain, a Game of Cat and Mouse Breaks Out Over a Parakeet"; Welch, "Save the Parakeets"; Gorman, "Putting Nature on the Pill"; Salzman, "Eradication of Parakeets Draws Protests"; Brotman, "Parrot Troopers Defend Their Feathered Friends."; Browne, "Why Act Now?"

218. Palmer, "Contested Frameworks in Environmental Ethics"; Simberloff, "Nature, Natives, Nativism, and Management."

219. On plant rights advocates, see Simberloff, "Nature, Natives, Nativism, and Management."

220. Bryan Norton, "Environmental Ethics and Weak Anthropocentrism."

221. For the individualist's perspective, see Varner, *In Nature's Interests?*; Sandler and Crane, "On the Moral Considerability of *Homo sapiens*." For the holist's perspective, see Rolston, *Environmental Ethics*; Rolston, *Conserving Natural Value*; Lawrence Johnson, "Toward the Moral Considerability of Species"; Lawrence Johnson, *Morally Deep World*; C. Stone, "Toward Ethical Treatment of Animals."

222. On sentience as key criterion, see Singer, *Animal Liberation*, Singer, *Practical Ethics*; Regan, *Case for Animal Rights*. On definitions of "sentience," see D. Rosenthal, "How Many Kinds of Consciousness?"; I. Duncan, "Changing Concept of Animal Sentience."

223. Singer, *Animal Liberation*, Singer, *Practical Ethics*.

224. S. Shuttleworth, *Cognition, Evolution, and Behavior*; Wasserman and Zentall, *Comparative Cognition*; Healy, "Face of Animal Cognition."

225. Proctor et al., "Searching for Animal Sentience."

226. P. Taylor, "Ethics of Respect for Nature"; P. Taylor, *Respect for Nature*.

227. Rolston, *Conserving Natural Value*.

228. Callicott, "Conceptual Foundations of the Land Ethic"; Simberloff, "Integrity, Stability, and Beauty."

229. Agar, *Life's Intrinsic Value*.

230. C. Palmer, "Contested Frameworks in Environmental Ethics."

231. E.g., Bekoff, "The Importance of Ethics in Conservation Biology"; Vucetich and Nelson, "What Are 60 Warblers Worth?"

232. Bekoff, "Preface: Who Lives, Who Dies, and Why."

233. Vucetich and Nelson, "Infirm Ethical Foundations of Conservation."

234. Ramp and Bekoff, "Compassion as a Practical and Evolved Ethic for Conservation."

235. Wallach, Bekoff, et al., "Promoting Predators and Compassionate Conservation."

236. From the outset, advocates of compassionate conservation have addressed many areas, including culling native predators in an effort to support threatened native prey species, killing animals that damage agriculture or raise concerns in an urban environment, trophy hunting, and killing surplus zoo animals (cf. Ben-Ami, "Compassionate Conservation"; Wallach, Bekoff, et al., "Summoning Compassion").

237. J. Russell et al., "Importance of Lethal Control."

238. Wallach, Bekoff, et al., "Summoning Compassion," 1260.

239. Wallach, Bekoff, et al., "Summoning Compassion," 1260.

240. Wallach, Bekoff, et al., "Summoning Compassion," 1262; Regan, *Case for Animal Rights*.

241. The concept of a "novel ecosystem," first proposed by F. S. Chapin and A. M. Starfield in 1997, was initially focused wholly on how climate change and associated factors, such as altered fire regimes, would produce new collections of species embedded in new ecosystems with new processes and flows of nutrients and energy (Chapin and Starfield, "Time Lags and Novel Ecosystems"). However, it attracted much more attention as well as controversy in 2006 with a publication heralding a "new world order" caused by climate change, non-native species, and anthropogenic land use changes (Hobbs, Arico, et al., "Novel Ecosystems: Theoretical and Management Aspects"). Although definitions are varied (Murcia et al., "Critique of the 'Novel Ecosystem' Concept"), the fundamental elements of a novel ecosystem are that it differs from any that prevailed previously, tends to self-organize, and may retain its novelty without human assistance (Hobbs, Higgs, and Hall, "Defining Novel Ecosystems"). Novel ecosystematists advocate such systems as the correct goal of restoration, and also attempts to shape novel ecosystems to serve human ends (e.g., by producing food or impeding floods), so the concept has been highly controversial because restoration ecology has traditionally aimed at restoring historical communities and trajectories, with removal of non-native species often the leading edge. The burgeoning debate about the role of novel ecosystems in ecological restoration has thus often focused on non-native species, with advocates and critics delving into other controversies, such as the long-term feasibility of control (discussed in chap. 16) and beneficial versus harmful impacts of non-native populations (discussed in section 18.3.2). And critics of conservation biology and management based on other factors occasionally cite novel ecosystems as the wave of the future to support their views. The controversy shows no sign of imminent decline, with many adherents (Hobbs, Higgs, and Harris, "Novel Ecosystems: Implications"; Hobbs, Higgs, and Hall, *Novel Ecosystems*; James Robert Miller and Bestelmeyer, "What's Wrong with Novel Ecosystems, Really?"; James Robert Miller and Bestelmeyer, "What the Novel Ecosystem Concept Provides"; Higgs et al., "Keep Ecological Restoration Open") and critics (Aronson, Murcia, et al., "Road to Confusion"; Aronson, Simberloff, et al., "Restoration Science Does Not Need Redefinition"; Murcia et al., "Critique of the 'Novel Ecosystem' Concept"; Kattan et al., "Does the Novel Ecosystem Concept Provide a Framework for Practical Applications?").

242. Bekoff, " 'Possum Stomp' vs. Compassionate Conservation."

243. Driscoll and Watson, "Science Denialism and Compassionate Conservation."

244. Callen et al., "Envisioning the Future with 'Compassionate Conservation,' " 9.

245. Marris, "When Conservationists Kill Lots (and Lots) of Animals"; Rohwer and Marris, "Clarifying Compassionate Conservation with Hypotheticals"; Caravaggi et al., "Impacts of Introduced House Mice." Sealers inadvertently introduced house mice to the remote, uninhabited island of Gough in the South Atlantic in the nineteenth

century (M. M. Gray, Wegmann, et al., "Demographic History of a Recent Invasion of House Mice"), and, in the absence of rats, mice thrived on the abundant seabird colonies and evolved to be twice the size of continental house mice (M. M. Gray, Parmenter, et al., "Genetics of Rapid and Extreme Size Evolution"). Their devastating impact on the seabird populations has inspired the Royal Society for the Protection of Birds to plan an ambitious eradication program (Royal Society for the Protection of Birds, "Saving Species from Extinction") based on baits with brodifacoum, which causes a painful death (Littin et al., "Behaviour, Coagulopathy and Pathology of Brushtail Possums"). The plan, in turn, has triggered strenuous opposition from animal rights advocates (Barkham, "Should We Cull One Species to Save Another?").

246. Royal Society for the Protection of Birds, "Saving Species from Extinction."

247. Marris, "When Conservationists Kill Lots (and Lots) of Animals"; Batavia, Nelson, and Wallach, "Moral Residue of Conservation."

248. Hampton et al., "Compassionate Versus Consequentialist Conservation."

249. Hayward et al., "Deconstructing Compassionate Conservation," 765. Cf. Fleming, "Compassionate Conservation or Misplaced Conservation?"

250. Wallach, Bekoff, et al., "Promoting Predators and Compassionate Conservation"; Wallach, Bekoff, et al., "Summoning Compassion"; Bruskotter et al., "Animal Rights and Wildlife Conservation"; van Eeden et al., "Shifting Public Values."

251. Hayward et al., "Deconstructing Compassionate Conservation," 766. For the sources cited by Hayward, see M. Hoffmann, Hilton-Taylor, et al., "Impact of Conservation"; M. Hoffmann, Belant, et al., "Changing Fates of the World's Mammals."

252. Griffin, Callen, et al., "Compassionate Conservation Clashes with Conservation Biology," 1.

253. Batavia et al., "Emotion as a Source of Moral Understanding in Conservation."

254. Oommen et al., "Fatal Flaws of Compassionate Conservation."

255. Wallach, Batavia, et al., "Recognizing Animal Personhood"; Bekoff, "Introduction: What Does It Feel Like to Be a Fox?"

256. P. Johnson et al., "Consequences Matter"; Ramp and Bekoff, "Compassion as a Practical and Evolved Ethic for Conservation," 325.

257. Wallach, Bekoff, et al., "Promoting Predators and Compassionate Conservation"; Wallach, Bekoff, et al., "Summoning Compassion"; Bruskotter et al., "Animal Rights and Wildlife Conservation"; van Eeden et al., "Shifting Public Values."

258. Bobier and Allen, "Virtue of Compassion in Compassionate Conservation."

259. Coghlan and Cardilini, "A Critical Review of the Compassionate Conservation Debate."

260. Hacking and Hacking, *The Social Construction of What?*; J. R. Brown, *Who Rules in Science?*

Chapter 19

1. Richardson, *Fifty Years of Invasion Ecology.*

2. E.g., Kueffer, "Integrating Natural and Social Sciences."

3. Kueffer and Hadorn, "How to Achieve Effectiveness in Problem-Oriented Landscape Research."

4. Richardson, "Invasion Science."

5. For economics, see Hudson, "Trends in Multi-Authored Papers in Economics." For chemical sciences, see D. Price, *Little Science, Big Science.* For other fields, see, e.g., Regalado, "Multiauthor Papers on the Rise"; Khan et al., "Controlled Analysis of Authorship Trends"; Weltzin et al., "Authorship in Ecology"; Gorham and Kelly,

"Multiauthorship"; Ossenblok et al., "Coauthorship of Journal Articles and Book Chapters"; Barlow et al., "On the Extinction of the Single-Authored Paper."

6. S. Campbell and Simberloff, "Forty Years of Invasion Research."

7. Rosa et al., "Neotropical Alien Mammals."

8. S. Campbell and Simberloff, "Forty Years of Invasion Research."

9. Lynch et al., "Transdisciplinary Synthesis"; Baron et al., "Synthesis Centers."

10. I. Parker et al., "Impact"; Simberloff and Von Holle, "Positive Interactions of Nonindigenous Species."

11. Richardson, Abrahams, et al., "South Africa's Centre for Invasion Biology."

12. S. Campbell and Simberloff, "Forty Years of Invasion Research."

13. Abbate, *Inventing the Internet*; J. Ryan, *History of the Internet and the Digital Future*.

14. C. Egido described videoconferencing as having "failed to become anything more than a revolutionary concept on the brink of success" (Egido, "Video Conferencing as a Technology to Support Group Work," 13).

15. Vaz et al., "Progress of Interdisciplinarity in Invasion Science."

16. Vaz et al., "Progress of Interdisciplinarity in Invasion Science," 437.

17. Rittel and Webber, "Dilemmas in a General Theory of Planning"; B. Head, "Wicked Problems."

18. Estévez et al., "Clarifying Values, Risk Perceptions, and Attitudes"; D. Frank, "Disagreement or Denialism?"

19. D. Frank, "Disagreement or Denialism?"

20. Estévez et al., "Clarifying Values, Risk Perceptions, and Attitudes."

21. Russell et al., "Importance of Lethal Control."

22. Ortiz-Ospina, "Rise of Social Media." References are to Twitter before it was acquired by Elon Musk in October 2022 and renamed "X" in July 2023. This shift has led to much experimentation with substitute media (Roose, "Bluesky, Smiling at Me"). Many scientists in particular, including invasion scientists, have switched to Bluesky. Whether this migration from X grows and whether Bluesky continues to be the main beneficiary remains to be seen.

23. M. Jackson et al., "New Record of the Non-Native Long-Legged Fly."

24. Mammola, Nanni, et al., "Media Framing of Spiders"; Mammola, Malumbres-Olarte, et al., "Global Spread of Misinformation on Spiders"; A. Chuang, email message to the author, June 19, 2023.

25. H. Fraser et al, "Same Data, Different Analysts."

26. The widely used term "citizen science" has been assailed as being exclusionary, and "community science" has been the most frequently suggested alternative. Other terms such as "participatory science" and "public science" have also been advanced (Bonney, "Expanding the Impact of Citizen Science"; C. Cooper et al., "Inclusion in Citizen Science"; Elwood et al., "Citizen Science Needs a Name Change"). This attack has in turn elicited skeptical responses (e.g., D. Hunter et al., "What's In a Name?") and triggered a heated, ongoing exchange through Twitter and other internet platforms. The term "citizen science" came into use in both the United States and Europe in 1995 (Bonney, "Expanding the Impact of Citizen Science") and appears in the titles, abstracts, and texts of thousands of publications, including many in invasion science. I will use it in this book simply in the interest of clarity (other connotations of "community science" exist [Bonney, "Expanding the Impact of Citizen Science"; C. Cooper et al., "Inclusion in Citizen Science"]), acknowledging that its use may be superseded in the future.

27. Miller-Rushing et al., "History of Public Participation in Ecological Research."

28. E.g., Miller-Rushing and Primack, "Global Warming and Flowering Times in Thoreau's Concord"; C. Willis et al., "Favorable Climate Change Response Explains Non-Native Species' Success in Thoreau's Woods."

29. Bonney, "Expanding the Impact of Citizen Science."

30. Westbrooks and Eplee, "Early Detection and Rapid Response."

31. E. Larson et al., "From eDNA to Citizen Science."

32. Statistic from Jay, "Number of Smartphone and Mobile Phone Users Worldwide." On smartphone usage in less-developed versus more-developed nations, see Horst and Miller, *Cell Phone*.

33. On use of such data for early detection of non-natives, see, e.g., Antúnez-Fonseca et al., "New Records and Range Expansion"; Werenkraut et al., "Citizen Science Reveals the Distribution of the Invasive Harlequin Ladybird"; Angelidou et al., "Establishment and Spread of the Invasive Ladybird"; Gaspar et al., "Exponential Outspread of *Halyomorpha halys*"; Pemberton and Escalona, "Spread and Distribution of the Naturalized Orchid Bee." On other reported uses, see, e.g., Serniak et al., "Predicting Habitat Suitability."

34. For such proposals, see, e.g., Encarnação et al., "Citizen Science and Biological Invasions"; Hulburt et al., "Citizen Science Can Enhance Strategies."

35. Bonney, "Expanding the Impact of Citizen Science"; Encarnação et al., "Citizen Science and Biological Invasions."

36. Nuñez, Chiuffo, Seebens, et al., "Two Decades of Data."

37. Van Kleunen, Dawson, et al., "Global Exchange and Accumulation of Non-Native Plants."

38. Van Kleunen, Pyšek, et al., "Global Naturalized Alien Flora (GloNAF) Database."

39. Turbelin et al., "Mapping the Global State of Invasive Alien Species"; A. Hughes et al., "Sampling Biases."

40. O'Neil, "English as the Lingua Franca"; Bortolus, "Running Like Alice"; Amano, González-Varo, and Sutherland, "Languages Are Still a Major Barrier to Global Science."

41. On English facilitating research, see H. Mooney, *Globalization of Ecological Thought*.

42. Nuñez, Barlow, et al., "Assessing the Uneven Global Distribution of Readership, Submissions and Publications"; Nuñez, Chiuffo, Seebens, et al., "Two Decades of Data."

43. Regarding publishing abstracts in languages other than English, in 2020, in its instructions to authors, *Biological Invasions* invited authors to submit an abstract in an additional language, though very few have availed themselves of the opportunity. *NeoBiota* does not explicitly invite such abstracts but occasionally publishes them (I. Kühn, email message to the author, March 13, 2023). One such paper (Haubrock et al., "Economic Costs of Invasive Alien Species") had abstracts in eleven languages. To my knowledge, no published research assesses use of non-English abstracts in English-language journals. On diversifying editorial boards and manuscript reviewers and providing enhanced copyediting assistance, see Kuebbing et al., "Self-Study of Editorial Board Diversity"; Nuñez, Chiuffo, Seebens, et al., "Two Decades of Data."

44. Amano, González-Varo, and Sutherland, "Languages Are Still a Major Barrier to Global Science"; Nuñez and Amano, "Monolingual Searches Can Limit and Bias Results."

45. Angulo et al., "Non-English Languages Enrich Scientific Knowledge."

46. Lennox et al., "Improving Science-Based Invasive Species Management"; Boardman, Lockwood, et al., "Future of Invasion Science Needs Physiology."

47. Cf. Pohl, "Physiology Without Borders"; Barman et al., "Reports of Physiology's Demise Have Been Greatly Exaggerated"; Lemoine and Pradeu, "Dissecting the Meanings of 'Physiology.'"

48. On the "molecular revolution," see Morange, *History of Molecular Biology*.

49. Wikelski and Cooke, "Conservation Physiology."

50. Coristine, "Conceptual Framework for the Emerging Discipline of Conservation Physiology."

51. For instance, the first volume contained articles on invasive plants (Funk, "Physiology of Invasive Plants") and fishes (Liss et al., "Spatial and Temporal Influences").

52. S. Cooke, Bergman, et al., "One Hundred Research Questions."

53. Lennox et al., "Improving Science-Based Invasive Species Management."

54. Boardman, Lockwood, et al., "Future of Invasion Science Needs Physiology."

55. Stuhrmann, "'It Felt More Like a Revolution.'"

56. Nobel Prize, "1973 Nobel Prize in Physiology or Medicine."

57. Stuhrmann, "'It Felt More Like a Revolution.'"

58. Curio, "Conservation Needs Ethology," 261.

59. Sutherland, "Importance of Behavioural Studies in Conservation Biology."

60. L. Gosling and Sutherland, *Behaviour and Conservation*. Of nineteen chapters in Gosling and Sutherland's collection, only one mentioned non-native species.

61. Buchholz, "Should Animal Behaviorists Teach Conservation?," 3. For his 2007 announcement of the new discipline, see Buchholz, "Behavioural Biology." The Animal Behavior Society established a standing Conservation Committee in 1997 to encourage its members to conduct research relevant to conservation. In 2003, the committee established a biannual electronic newsletter—*The Conservation Behaviorist*—to inform society members about research and scientific news at the interface of conservation and behavior. After appearing regularly through 2014, it disappeared, as have numerous newsletters in invasion science (as noted in chapter 10), because the volunteer effort required to produce them became too burdensome (B. Schulte, email message to the author, March 6, 2023).

62. Blumstein and Fernández-Juricic, *Primer of Conservation Behavior*. Blumstein and Fernández-Juricic in one short paragraph observed that climate change and invasive species will generate new interactions between species and that the behavioral aspect of those interactions will be worth studying.

63. Mayr, "Nature of Colonizations in Birds."

64. Holway and Suarez, "Animal Behavior."

65. Chapple et al., "Can Behavioral and Personality Traits Influence the Success of Unintentional Species Introductions?"

66. Carere and Gherardi, "Animal Personalities Matter."

67. Sih et al., "Behavioral Syndromes: Integrative Overview," 242. Several authors (e.g., Sih, Bell, and Johnson, "Behavioral Syndromes: Ecological and Evolutionary Overview"; Sih et al., "Behavioral Syndromes: Integrative Overview") reserved the term "personality" for human behaviors and used other terms, such as "temperament" and "behavioral syndrome," for an analogous suite of correlated behaviors in other animals. As research on personalities of other animals increased (Laskowski et al., "Consistent Individual Behavioral Variation"), the term "personality" came to be used universally.

68. S. Gosling, "From Mice to Men"; Laskowski et al., "Consistent Individual Behavioral Variation."

69. Wolf and Weissing, "Animal Personalities"; Carere and Gherardi, "Animal Personalities Matter."

70. Weis and Sol, *Biological Invasions and Animal Behaviour.*

71. A. Deacon and Magurran, "How Behaviour Contributes."

72. L. Martin et al., "Invader Endocrinology."

73. S. J. Cooke and colleagues called for integrating the fields of conservation physiology and conservation behavior, pointing to several weaknesses in each approach that can be compensated for by the other and to several examples in which such integration advanced conservation (S. Cooke et al., "Physiology, Behavior, and Conservation"). Although none of these examples referred to conservation challenges posed by invasions, the advances often involved an understanding of how hormones—especially glucocorticoids—affect behavior.

74. Griffin, Guez, et al., "Invading New Environments."

75. Zenni and Nuñez, "Elephant in the Room."

76. Charles Elton had warned about the impact of this crab after its initial introduction to Great Britain, in a review of a German book about its biology and impact on the European mainland (Elton, "New Invader"). He closed this review deploring a long list of the invaders already present in Great Britain.

77. Zhang et al., "To Invade or Not to Invade?"

78. Marchetti and Engstrom, "Conservation Paradox of Endangered and Invasive Species."

79. The wattle-necked soft-shelled turtle is listed as endangered by the IUCN; in Hawaii, it is now reported on Oahu and Maui in addition to Kauai (Marchetti and Engstrom, "Conservation Paradox of Endangered and Invasive Species").

80. Marchetti and Engstrom, "Conservation Paradox of Endangered and Invasive Species."

81. Lees and Bell, "Conservation Paradox for the 21st Century."

82. E.g., "But despite the exponentially growing literature, invasion biology remains a field characterized by a wealth of anecdotal studies with few general principles" (Chisholm, "Invasion Biology," 314).

83. Lawton, "Are There General Laws in Ecology?"; Simberloff, "Community Ecology."

84. Catford and colleagues conducted separate literature searches for context dependence in invasion science and invasion ecology (Catford, Wilson, et al, "Addressing Context Dependence"). Both showed a rapid acceleration in number of papers beginning around 1995, with almost half of all references through 2019 published from 2015 onward.

85. Corbett, *Dingo in Australia and Asia.*

86. Saunders et al., "Impacts and Management of Foxes."

87. Cupples et al., "Dietary Overlap and Prey Selectivity."

88. Feit et al., "Apex Predators Decouple Population Dynamics."

89. Dickman, "Impact of Exotic Generalist Predators"; Abbott, "Spread of the Cat"; Doherty et al., "Impacts and Management of Feral Cats."

90. Abbott, "Spread of the Cat"; Doherty et al., "Impacts and Management of Feral Cats."

91. Dickman, "Impact of Exotic Generalist Predators."

92. H. Roy et al., *Summary for Policymakers.*

93. For information available from the CBD, see Convention on Biodiversity, "Information Centre." The conventional and largely correct explanation for the failure of the United States to ratify the CBD is that Republican President George H. W. Bush refused to sign it when it was proposed in 1992, and that after Democratic President Bill Clinton signed it in 1993, a substantial number of Republican senators prevented

the Senate from ratifying it, a situation that has persisted until the present. However, this unnuanced depiction omits several contributing factors (Blomquist, "Ratification Resisted").

94. Convention on Biodiversity, "Article 8: In-situ Conservation."

95. Convention on Biodiversity, "COP 6Decision VI/23."

96. Convention on Biodiversity, "Aichi Target 9."

97. Convention on Biodiversity, "Global Biodiversity Outlook 5."

98. Convention on Biodiversity, "Global Biodiversity Framework Target 6."

99. Convention on Biodiversity, "Global Biodiversity Outlook 5."

100. For the 2023 IPBES report, see H. Roy et al., *Summary for Policymakers.*

101. M. Davis and Chew, "'Denialists Are Coming!' Well, Not Exactly," 229, in response to J. Russell and Blackburn, "Rise of Invasive Species Denialism."

102. M. Davis, *Invasion Biology.*

103. Simberloff, "Invasions of Plant Communities."

104. On molecular genetics, see chap. 11, sec. 11.2. On GIS, see chap. 14. On stable isotope analysis, see Fry, *Stable Isotope Ecology*; Shiels et al., "Dietary Niche Differentiation"; McCue et al., "Using Stable Isotope Analysis"; Polito et al., "Population Dynamics and Resource Availability."

105. Kuhn, *Structure of Scientific Revolutions.*

Bibliography

Aagaard, K., and J. Lockwood. "Exotic Birds Show Lags in Population Growth." *Diversity and Distributions* 20 (2014): 547–54.

Abbate, J. *Inventing the Internet.* Cambridge, MA: MIT Press, 1999.

Abbott, I. "The Spread of the Cat, *Felis catus,* in Australia: Re-examination of the Current Conceptual Model with Additional Information." *Conservation Science Western Australia* 7 (2008): 1–17.

Abrahams, B., N. Sitas, and K. J. Esler. "Exploring the Dynamics of Research Collaborations by Mapping Social Networks in Invasion Science." *Journal of Environmental Management* 229 (2019): 27–37.

Absalom, B. M. Preface to *Tutira: The Story of a New Zealand Sheep Station,* 4th ed., edited by H. Guthrie-Smith. Seattle: University of Washington Press, 1969.

Adler, M., and E. Ziglio. *Gazing into the Oracle: The Delphi Method and Its Application to Social Policy and Public Health.* London: Jessica Kingsley, 1996.

Agar, N. *Life's Intrinsic Value: Science, Ethics, and Nature.* New York: Columbia University Press, 2001.

Aguilera, A. G. "The Influence of Soil Community Density on Plant-Soil Feedbacks: An Important Unknown in Plant Invasion." *Ecological Modelling* 222 (2011): 3413–20.

Ainouche, M. L., P. M. Fortune, A. Salmon, C. Parisod, M.-A. Grandbastien, et al. "Hybridization, Polyploidy and Invasion: Lessons from *Spartina* (Poaceae)." *Biological Invasions* 11 (2009): 1159–73.

Aizen, M. A., C. L. Morales, and J. M. Morales. "Invasive Mutualists Erode Native Pollination Webs." *PLoS Biology* 6, no. 2 (2008): e31.

Albrecht, G. A., C. Brooke, D. H. Bennett, and S. T. Garnett. "The Ethics of Assisted Colonization in the Age of Anthropogenic Climate Change." *Journal of Agricultural and Environmental Ethics* 26 (2013): 827–45.

Alderman, D. H. "The Changing South: A Vine for Postmodern Times: An Update on Kudzu at the Close of the Twentieth Century." *Southeastern Geographer* 38 (1998): 167–79.

Alderman, D. H. "Channing Cope and the Making of a Miracle Vine." *Geographical Review* 94 (2004): 157–77.

Alderman, D. H., and D. G. Alderman. "Kudzu: A Tale of Two Vines." *Southern Cultures* 7, no. 3 (2001): 49–64.

Aldridge, D. C., P. Elliott, and G. D. Moggridge. "Microencapsulated BioBullets for the Control of Biofouling Zebra Mussels." *Environmental Science and Technology* 40 (2006): 975–79.

Alexander, J. M., J. M. Diez, S. P. Hart, and J. M. Levine. "When Climate Reshuffles Competitors: A Call for Experimental Macroecology." *Trends in Ecology and Evolution* 31 (2016): 831–41.

Alexander, J. M., J. J. Lembrechts, L. A. Cavieres, C. C. Daehler, S. Haider, et al. "Plant Invasions into Mountains and Alpine Ecosystems: Current Status and Future Challenges." *Alpine Botany* 126 (2016): 89–103.

Allan, H. H. "Indigene Versus Alien in the New Zealand Plant World." *Ecology* 17 (1936): 187–93.

Allan, H. H. "A Remarkable New Zealand Scrub Association." *Ecology* 7 (1926): 72–76.

Allen, G. E. "'Culling the Herd': Eugenics and the Conservation Movement in the United States, 1900–1940." *Journal of the History of Biology* 46 (2013): 31–72.

Allen, G. M. "Mammals of the West Indies." *Bulletin of the Museum of Comparative Zoology, Harvard College* 54 (1911): 175–263.

Allen, J. A. "Mangroves as Alien Species: The Case of Hawaii." *Global Ecology and Biogeography Letters* 7 (1998): 61–71.

Allen, J. M., and B. A. Bradley. "Out of the Weeds? Reduced Plant Invasion Risk with Climate Change in the Continental United States." *Biological Conservation* 203 (2016): 306–12.

Allen, R. B., and W. G. Lee, eds. *Biological Invasions in New Zealand*. Berlin: Springer, 2006.

Allendorf, F. W., and L. L. Lundquist. "Introduction: Population Biology, Evolution, and Control of Invasive Species." *Conservation Biology* 17 (2003): 24–30.

Allison, S. K. "The Paradox of Invasive Species in Ecological Restoration: Do Restorationists Worry about Them Too Much or Too Little?" In Rotherham and Lambert, *Invasive and Introduced Plants and Animals*, 265–76.

Almeida-Neto, M., and W. Ulrich. "A Straightforward Computational Approach for Quantifying Nestedness Using Quantitative Matrices." *Environmental Modelling and Software* 26 (2011): 173–78.

Alves, J. M., M. Carneiro, J. P. Day, J. J. Welch, J. A. Duckworth, et al. "A Single Introduction of Wild Rabbits Triggered the Biological Invasion of Australia." *Proceedings of the National Academy of Sciences of the United States of America* 119, no. 35 (2022): e2122734119.

Amano, T., J. P. González-Varo, and W. J. Sutherland. "Languages Are Still a Major Barrier to Global Science." *PLoS Biology* 14 (2016): e2000933.

Amano, T., V. Ramírez-Castañeda, V. Berdejo-Espinola, I. Borokini, S. Chowdhury, et al. "The Manifold Costs of Being a Non-Native English Speaker in Science." *PLoS Biology* 21 (2023): e3002184.

Amano, T., and W. J. Sutherland. "Four Barriers to the Global Understanding of Biodiversity Conservation: Wealth, Language, Geographical Location and Security." *Proceedings of the Royal Society B* 280 (2013): 20122649.

Amor, R. L., and C. M. Piggin. "Factors Influencing the Establishment and Success of Exotic Plants in Australia." In D. Anderson, *Exotic Species in Australia—Their Establishment and Success*, 15–26.

Amrine, J. W., Jr., and T. A. Stasny. "Biocontrol of Multiflora Rose." In B. McKnight, *Biological Pollution*, 9–21.

Amsellem, L., J. L. Noyer, T. Le Bourgeois, and M. Hossaert-McKey. "Comparison of Genetic Diversity of the Invasive Weed *Rubus alceifolius* Poir. (Rosaceae) in Its Native Range and in Areas of Introduction, Using Amplified Fragment Length Polymorphism (AFLP) Markers." *Molecular Ecology* 9 (2000): 443–55.

Amundsen, P. A., K. D. Lafferty, R. Knudsen, R. Primicerio, R. Kristoffersen, et al. "New Parasites and Predators Follow the Introduction of Two Fish Species to a Subarctic Lake: Implications for Food-Web Structure and Functioning." *Oecologia* 171 (2013): 993–1002.

Anagnostakis, S. L. "Chestnut Blight—The Classical Problem of an Introduced Pathogen." *Mycologia* 79 (1987): 23–37.

Andersen, M. C., H. Adams, B. Hope, and M. Powell. "Risk Assessment for Invasive Species." *Risk Analysis* 24 (2004): 787–93.

Anderson, C. B., C. R. Griffith, A. D. Rosemond, R. Rozzi, and O. Dollenz. "The Effects of Invasive North American Beavers on Riparian Plant Communities in Cape Horn, Chile: Do Exotic Beavers Engineer Differently in Sub-Antarctic Ecosystems?" *Biological Conservation* 128 (2006): 467–74.

Anderson, C. B., J. C. Pizarro, A. E. Valenzuela, N. Ader, S. Ballari, et al. "Reconceiving the Biological Invasion of North American Beavers (*Castor canadensis*) in Southern Patagonia as a Socio-Ecological Problem: Implications and Opportunities for Research and Management." In *Biological Invasions in the South American Anthropocene*, edited by F. M. Jaksic and S. A. Castro, 231–53. Cham, Switz.: Springer, 2021.

Anderson, D., ed. *Exotic Species in Australia—Their Establishment and Success*. Vol. 10, *Proceedings of the Ecological Society of Australia*. Canberra: Ecological Society of Australia, 1977.

Anderson, E. "Man as a Maker of New Plants and New Plant Communities." In Thomas, *Man's Role in Changing the Face of the Earth*, 763–77.

Anderson, E. *Plants, Man, and Life*. Berkeley: University of California Press, 1952.

Anderson, L. W. J. "California's Reaction to *Caulerpa taxifolia*: A Model for Invasive Species Rapid Response." *Biological Invasions* 7 (2005): 1003–16.

Anderson, R. M. "Effect of the Introduction of Exotic Animal Forms." In *Proceedings of the Fifth Pacific Science Congress*, vol. 1, edited by General Secretary, 769–78. Toronto: University of Toronto Press, 1934.

Anderson, R. M. "Faunas of Canada." In *The Canada Year Book 1937*, 29–52. Ottawa: J. O. Patenaude, 1937.

Anderson, R. P. "When and How Should Biotic Interactions Be Considered in Models of Species Niches and Distributions?" *Journal of Biogeography* 44 (2017): 8–17.

Anderson, S. J., and C. P. Stone. "Snaring to Control Feral Pigs *Sus scrofa* in a Remote Hawaiian Rain Forest." *Biological Conservation* 63 (1993): 195–201.

Andow, D. A., B. I. P. Barratt, R. S. Pfannenstiel, and D. P. Paula. "Exotic Generalist Arthropod Biological Control Agents: Need to Improve Environmental Risk Assessment to Ensure Safe Use." *BioControl* 66 (2021): 1–8.

Andreoni, K. J., C. J. Wagnon, B. T. Bestelmeyer, and R. L. Schooley. "Exotic Oryx Interact with Shrub Encroachment in the Chihuahuan Desert." *Journal of Arid Environments* 184 (2021): 104302.

Andrewartha, H. G., and L. C. Birch. *The Distribution and Abundance of Animals*. Chicago: University of Chicago Press, 1954.

Angelidou, I., J. Demetriou, M. Christou, E. Koutsoukos, C. Kazilas, et al. "Establishment and Spread of the Invasive Ladybird *Harmonia axyridis* (Coleoptera:

Coccinellidae) in Greece: Based on Contributions from Citizen Scientists." *Biological Invasions* 25 (2023): 889–900.

Angilletta, M. J., Jr., S. R. Winters, and A. E. Dunham. "Thermal Effects on the Energetics of Lizard Embryos: Implications for Hatching Phenotypes." *Ecology* 81 (2000): 2957–68.

Angulo, E., C. Diagne, L. Ballesteros-Mejia, T. Adamjy, D. A. Ahmed, et al. "Non-English Languages Enrich Scientific Knowledge: The Example of Economic Costs of Biological Invasions." *Science of the Total Environment* 775 (2021): 144441.

Angulo, E., and B. Gilna. "International Law Should Govern Release of GM Mosquitoes." *Nature* 454 (2008): 158.

Animal People. "Channel Islands National Park Ex-Chief Hits Cruelty of Killing 'Invasive Species.'" April 2005. http://www.animalpeoplenews.org/05/4/tsg.channelIslands4.05.htm (content no longer available). Image of web page viewable via Internet Archive Wayback Machine, https://web.archive.org/web/20160303235143/http://www.animalpeoplenews.org/05/4/tsg.channelIslands4.05.htm.

Anker, P. *Imperial Ecology: Environmental Order in the British Empire, 1895–1945.* Cambridge, MA: Harvard University Press, 2001.

Anthony, L. *The Aliens Among Us: How Invasive Species Are Transforming the Planet—and Ourselves.* New Haven, CT: Yale University Press, 2017.

Antoniazzi, R., Jr., W. Dáttilo, and V. Rico-Gray. "A Useful Guide of Main Indices and Software Used for Ecological Network Studies." In Dáttilo and Rico-Gray, *Ecological Networks in the Tropics*, 185–96.

Antúnez-Fonseca, C. A., E. P. Hofmann, A. A. Reyes-Barahona, D. I. Ordoñez-Mazier, F. J. Dubón, et al. "New Records and Range Expansion of *Norops sagrei* (Squamata: Dactyloidae) in Honduras Highlight the Importance of Citizen Science in Documenting Non-Native Species." *Caldasia* 44 (2022): 325–31.

Araújo, M. B., and M. Luoto. "The Importance of Biotic Interactions for Modelling Species Distributions Under Climate Change." *Global Ecology and Biogeography* 16 (2007): 743–53.

Araújo, M. B., and M. New. "Ensemble Forecasting of Species Distributions." *Trends in Ecology and Evolution* 22 (2007): 42–47.

Araújo, M. B., and A. Rozenfeld. "The Geographic Scaling of Biotic Interactions." *Ecography* 37 (2014): 406–15.

Aronson, J., C. Murcia, G. H. Kattan, D. Moreno-Mateosa, K. Dixon, and D. Simberloff. "The Road to Confusion Is Paved with Novel Ecosystem Labels: A Reply to Hobbs et al." *Trends in Ecology and Evolution* 29 (2014): 646–47.

Aronson, J., D. Simberloff, A. Ricciardi, and N. Goodwin. "Restoration Science Does Not Need Redefinition." *Nature Ecology and Evolution* 2 (2018): 916.

Arrhenius, S. "On the Influence of Carbonic Acid in the Air upon the Temperature of the Ground." *London, Edinburgh, and Dublin Philosophical Magazine and Journal of Science* 41, no. 251 (1896): 237–76.

Arroyo-Correa, B., I. Bartomeus, and P. Jordano. "Individual-Based Plant–Pollinator Networks Are Structured by Phenotypic and Microsite Plant Traits." *Journal of Ecology* 109 (2021): 2832–44.

Arthington, A. H., and D. S. Mitchell. "Aquatic Invasive Species." In Groves and Burdon, *Ecology of Biological Invasions*, 34–53.

Ashburn, P. M., and F. D. Ashburn. *The Ranks of Death: A Medical History of the Conquest of America.* New York: Coward-McCann, 1947.

Ashmole, N. P., M. J. Ashmole, and K. E. L. Simmons. "Seabird Conservation and Feral Cats on Ascension Island, South Atlantic." In Nettleship, Burger, and Gochfeld, *Seabirds on Islands*, 94–121.

Ashton, E. H. "The Influence of Geographic Isolation on the Skull of the Green Monkey (*Cercopithecus aethiops sabaeus*) V: The Degree and Pattern of Differentiation in the Cranial Dimensions in the St. Kitts Green Monkey." *Proceedings of the Royal Society of London B* 151 (1959): 563–83.

Ashton, E. H., and S. Zuckerman. "The Influence of Geographic Isolation on the Skull of the Green Monkey (*Cercopithecus aethiops sabaeus*) I: A Comparison Between the Teeth of the St. Kitts and African Green Monkey." *Proceedings of the Royal Society of London B* 137 (1950): 212–38.

Ashton, E. H., and S. Zuckerman. "The Influence of Geographic Isolation on the Skull of the Green Monkey (*Cercopithecus aethiops sabaeus*) II: The Cranial Dimension of the St. Kitts and African Green Monkey." *Proceedings of the Royal Society of London B* 138 (1951): 204–13.

Ashton, E. H., and S. Zuckerman. "The Influence of Geographic Isolation on the Skull of the Green Monkey (*Cercopithecus aethiops sabaeus*) III: The Developmental Stability of the Skulls and Teeth of the St. Kitts and the African Green Monkey." *Proceedings of the Royal Society of London B* 138 (1951): 213–18.

Ashton, E. H., and S. Zuckerman. "The Influence of Geographic Isolation on the Skull of the Green Monkey (*Cercopithecus aethiops sabaeus*) IV: The Degree and Speed of Dental Differentiation in the St. Kitts Green Monkey." *Proceedings of the Royal Society of London B* 138 (1951): 354–74.

Ashton, P. J., and D. S. Mitchell. "Aquatic Plants: Patterns and Modes of Invasion, Attributes of Invading Species and Assessment of Control Programs." In Drake et al., *Biological Invasions: A Global Perspective*, 111–54.

Assunção, A. C. R., R. V. Alexandrino, A. N. Caiafa, and G. de Oliveira. "The Invasion of *Artocarpus heterophyllus*, Jackfruit, in Protected Areas Under Climate Change and Across Scales: From Atlantic Forest to a Natural Heritage Private Reserve." *Biological Invasions* 21 (2019): 481–92.

Athens, J. S. "*Rattus exulans* and the Catastrophic Disappearance of Hawai'i's Native Lowland Forest." *Biological Invasions* 11 (2009): 1489–501.

Athens, J. S., T. M. Rieth, and T. S. Dye. "A Palaeoenvironmental and Archaeological Model-Based Estimate for the Colonisation of Hawai'i." *American Antiquity* 79 (2014): 144–55.

Atkinson, G. "A French Desert Island Novel." *Publications of the Modern Language Association of America* 36 (1921): 509–28.

Atkinson, G. *The Extraordinary Voyage in French Literature from 1700 to 1720.* Paris: Librairie Ancienne Honoré Champion, 1922.

Aubert de la Rüe, E. "Man's Influence on Tropical Vegetation." In *Proceedings of the Ninth Pacific Science Congress of the Pacific Science Association (1957)*, vol. 20, *Special Symposium on Climate, Vegetation, and Rational Land Utilization in the Humid Tropics*, 81–94. Bangkok: Secretariat, Ninth Pacific Science Congress, Department of Science, 1958.

Austin, D. F. "Exotic Plants and Their Effects in Southeastern Florida." *Environmental Conservation* 5 (1978): 25–34.

Australian Academy of Sciences. "Professor Ralph Slatyer (1929–2012), Ecologist." Accessed November 1, 2024. https://www.science.org.au/learning/general-audience/history/interviews-australian-scientists/professor-ralph-slatyer-1929-2012#17.

Avise, J. C. *Molecular Markers, Natural History, and Evolution*. 2nd ed. Sunderland, MA: Sinauer, 2004.

Ayers, D., K. Schierenbeck, and N. Ellstrand. "Introduction to 'Ecological and Evolutionary Consequences After Invaders Hybridize.'" *Biological Invasions* 11 (2009): 1087–88.

Ayson, L. F. "Introduction of American Fishes into New Zealand." *Bulletin of the Bureau of Fisheries* 28 (1908): 967–75.

Baas-Becking, L. G. M. *Geobiologie of Inleiding tot de Milieukunde*. The Hague: Van Stockum & Zoon, 1934.

Babcock, K. W. "The European Corn Borer *Pyrausta nubilalis* Hübn. I: A Discussion of Its Dormant Period." *Ecology* 8 (1927): 45–59.

Babcock, K. W. "The European Corn Borer, *Pyrausta nubilalis* Hübn. II: A Discussion of Its Seasonal History in Relation to Various Climates." *Ecology* 8 (1927): 177–93.

Bacher, S., T. M. Blackburn, F. Essl, P. Genovesi, J. Heikkilä, et al. "Socio-Economic Impact Classification of Alien Taxa (SEICAT)." *Methods in Ecology and Evolution* 9 (2018): 159–68. https://doi.org/10.1111/2041-210X.12844.

Bachman, P. M., R. Bolognesi, W. J. Moar, G. M. Mueller, M. S. Paradise, et al. "Characterization of the Spectrum of Insecticidal Activity of a Double-Stranded RNA with Targeted Activity Against Western Corn Rootworm (*Diabrotica virgifera virgifera* LeConte)." *Transgenic Research* 22 (2013): 1207–22.

Bahn, P. G., and J. Flenley. *Easter Island, Earth Island*. New York: Thames and Hudson, 1992.

Bailey, E. P. *Introduction of Foxes to Alaskan Islands—History, Effects on Avifauna, and Eradication*. Fish and Wildlife Service Resource Publication 193. Washington, DC: United States Department of the Interior, 1993.

Bailiff, M., and D. Karl. "Dissolved and Particulate DNA Dynamics During a Spring Bloom in the Antarctic Peninsula Region, 1986–1987." *Deep Sea Research Part 1: Oceanographic Research Papers* 38 (1991): 1077–95.

Baker, A. J. "Genetic and Morphometric Divergence in Ancestral European and Descendant New Zealand Populations of Chaffinches (*Fringilla coelebs*)." *Evolution* 46 (1992): 1784–800.

Baker, A. J., and A. Moeed. "Rapid Genetic Differentiation and Founder Effect in Colonizing Populations of Common Mynas (*Acridotheres tristis*)." *Evolution* 41 (1987): 525–38.

Baker, H. G. "Characteristics and Modes of Origin of Weeds." In Baker and Stebbins, *Genetics of Colonizing Species*, 147–72.

Baker, H. G. "The Evolution of Weeds." *Annual Review of Ecology and Systematics* 5 (1974): 1–24.

Baker, H. G. "Patterns in Plant Invasion in North America." In Mooney and Drake, *Ecology of Biological Invasions of North America and Hawaii*, 44–57.

Baker, H. G. "Self-Compatibility and Establishment After 'Long-distance' Dispersal." *Evolution* 9 (1955): 347–49.

Baker, H. G., and G. L. Stebbins, eds. *The Genetics of Colonizing Species*. New York: Academic Press, 1965.

Baker, H. G., and G. L. Stebbins. Preface to Baker and Stebbins, *Genetics of Colonizing Species*, vii–viii.

Balanyá, J., J. M. Oller, R. B. Huey, G. W. Gilchrist, and L. Serra. "Global Genetic Change Tracks Global Climate Warming in *Drosophila subobscura*." *Science* 313 (2006): 1773–75.

Ball, I. R. "Nature and Formulation of Biogeographical Hypotheses." *Systematic Biology* 24 (1975): 407–30.

Ball, J. "Climatological Diagrams." *Cairo Scientific Journal* 4 (1910): 280–81.

Balon, E. K. "Origin and Domestication of the Wild Carp, *Cyprinus carpio*: From Roman Gourmets to the Swimming Flowers." *Aquaculture* 129 (1995): 3–48.

Banks, R. C. *Summary of Foreign Game Bird Liberations, 1969–78*. Special Scientific Report, Wildlife No. 239. Washington, DC: US Fish and Wildlife Service, Bureau of Sport Fisheries and Wildlife, Division of Wildlife Research, 1981.

Barabási, A.-L. *Network Science*. Cambridge: Cambridge University Press, 2016.

Barber, G. W. "Some Factors Responsible for the Decrease of the European Corn Borer in New England During 1923 and 1924." *Ecology* 7 (1926): 143–62.

Barber, G. W. "A Study of the Cause of the Decrease in the Infestation of the European Corn Borer (*Pyrausta nubilalis*, Hubn.) in the New England Area During 1923." *Ecology* 6 (1925): 39–47.

Barbet-Massin, M., Q. Rome, F. Muller, A. Perrard, C. Villemant, and F. Jiguet. "Climate Change Increases the Risk of Invasion by the Yellow-Legged Hornet." *Biological Conservation* 157 (2013): 4–10.

Bardgett, R. D., and D. A. Wardle. *Aboveground–Belowground Linkages: Biotic Interactions, Ecosystem Processes, and Global Change*. New York: Oxford University Press, 2010.

Bardgett, R. D., and D. A. Wardle. "Herbivore-Mediated Linkages Between Aboveground and Belowground Communities." *Ecology* 84 (2003): 2258–68.

Barkalow, F. S., and M. Shorten. *The World of the Gray Squirrel*. Philadelphia: J. B. Lippincott, 1973.

Barkham, P. "Should We Cull One Species to Save Another?" *Guardian* (US edition), June 28, 2020. https://www.theguardian.com/environment/2020/jun/28/should-we-cull-one-species-to-save-another-huge-mice-killing-birds-gough-island.

Barlow, J., P. A. Stephens, M. Bode, M. W. Cadotte, K. Lucas, et al. "On the Extinction of the Single-Authored Paper: The Causes and Consequences of Increasingly Collaborative Applied Ecological Research." *Journal of Applied Ecology* 55 (2018): 1–4.

Barman, S. M., K. E. Barrett, and D. Pollock. "Reports of Physiology's Demise Have Been Greatly Exaggerated." *Physiology* 28 (2013): 360–62.

Barney, J., A. Davis, R. Porter, and D. Simberloff. *A Life-Cycle Approach to Low-Invasion Potential Bioenergy Production*. CAST Commentary QTA 2016-1. Ames, IA: Council for Agricultural Science and Technology, 2016.

Baron, J. S., A. Specht, E. Garnier, P. Bishop, C. A. Campbell, et al. "Synthesis Centers as Critical Research Infrastructure." *BioScience* 67 (2017): 750–59.

Barratt, B. I. P., F. G. Howarth, T. M. Withers, J. M. Kean, and G. S. Ridley. "Progress in Risk Assessment for Classical Biological Control." *Biological Control* 52 (2010): 245–54.

Barratt, B. I. P., V. C. Moran, F. Bigler, and J. C. van Lenteren. "The Status of Biological Control and Recommendations for Improving Uptake for the Future." *BioControl* 63 (2018): 155–67.

Barrett, G. W., and K. E. Mabry. "Twentieth-Century Classic Books and Benchmark Publications in Biology." *BioScience* 52 (2002): 282–85.

Barrett, L. T., T. Dempster, and S. E. Swearer. "A Nonnative Habitat-Former Mitigates Native Habitat Loss for Endemic Reef Fishes." *Ecological Applications* 29 (2019): 1–12.

Barrett, R. D. H., A. Paccard, T. M. Healy, S. Bergek, P. M. Schulte, et al. "Rapid Evolution of Cold Tolerance in Stickleback." *Proceedings of the Royal Society B* 278 (2011): 233–38.

Barrett, S. C. H. "Foundations of Invasion Genetics: The Baker and Stebbins Legacy." *Molecular Ecology* 24 (2015): 1927–41.

Barrett, S. C. H., R. I. Colautti, K. M. Dlugosch, and L. H. Rieseberg, eds. *Invasion Genetics: The Baker and Stebbins Legacy*. Chichester: Wiley, 2017.

Bartlett, S. P. "Discussion on Carp." *Transactions of the American Fisheries Society* 30 (1901): 114–32.

Bartley, D. M., and R. P. Subasinghe. "Historical Aspects of International Movement of Living Aquatic Species." *Revue scientifique et technique de l'Office International des Épizooties* 15 (1996): 387–400.

Barton, J. "Predictability of Pathogen Host Range in Classical Biological Control of Weeds: An Update." *BioControl* 57 (2012): 289–305.

Barun, A., K. C. Campbell, C. C. Hanson, and D. Simberloff. "A Review of Small Indian Mongoose Management and Eradications on Islands." In Veitch, Clout, and Towns, *Island Invasives: Eradication and Management*, 17–25.

Barun, A., D. Simberloff, S. Meiri, N. Tvrtković, and Z. Tadić. "Possible Character Displacement of an Introduced Mongoose and Native Marten on Adriatic Islands, Croatia." *Journal of Biogeography* 42 (2015): 2257–69.

Baskin, Y. *A Plague of Rats and Rubbervines: The Growing Threat of Species Invasions*. Washington, DC: Island Press, 2002.

Bastian, M., S. Heymann, and M. Jacomy. "Gephi: An Open Source Software for Exploring and Manipulating Networks." In *International AAAI Conference on Weblogs and Social Media*, 361–62. San Jose, CA: Association for the Advancement of Artificial Intelligence, 2009.

Batavia, C., M. P. Nelson, J. T. Bruskotter, M. S. Jones, E. Yanco, et al. "Emotion as a Source of Moral Understanding in Conservation." *Conservation Biology* 35 (2021): 1380–87.

Batavia, C., M. P. Nelson, and A. D. Wallach. "The Moral Residue of Conservation." *Conservation Biology* 34 (2020): 1114–21.

Bates, A. E., C. M. McKelvie, C. J. Sorte, S. A. Morley, N. A. Jones, et al. "Geographical Range, Heat Tolerance and Invasion Success in Aquatic Species." *Proceedings of the Royal Society B: Biological Sciences* 280 (2013): 20131958.

Bates, M. "Man as an Agent in the Spread of Organisms." In Thomas, *Man's Role in Changing the Face of the Earth*, 788–804.

Bauer, J. T. "Invasive Species: 'Back-seat Drivers' of Ecosystem Change?" *Biological Invasions* 14 (2012): 1295–304.

Bauer, J. T., and H. L. Reynolds. "Restoring Native Understory to a Woodland Invaded by *Euonymus fortunei*: Multiple Factors Affect Success." *Restoration Ecology* 24 (2016): 45–52.

Baum, J., T. Bogaert, W. Clinton, G. R. Heck, P. Feldmann, et al. "Control of Coleopteran Insect Pests through RNA Interference." *Nature Biotechnology* 25 (2007): 1322–26.

Baumgaertner, E. "The Gene Lab Next Door." *New York Times*, May 15, 2018, D1.

Baumhover, A. H. "A Personal Account of Developing the Sterile Insect Technique to Eradicate the Screwworm from Curacao, Florida and the Southeastern United States." *Florida Entomologist* 85 (2002): 666–73.

Baumhover, A. H., A. J. Graham, B. A. Bitter, D. E. Hopkins, W. D. New, et al. "Screw-Worm Control Through Release of Sterilized Flies." *Journal of Economic Entomology* 48 (1955): 462–66.

Bax, N., K. Hayes, A. Marshall, D. Parry, and R. Thresher. "Man-made Marinas as Sheltered Islands for Alien Marine Organisms: Establishment and Eradication of an Alien Invasive Marine Species." In Veitch and Clout, *Turning the Tide: The Eradication of Invasive Species*, 26–39.

Bayliss, S. L. J., M. Papeş, J. A. Schweitzer, and J. K. Bailey. "Aggregate Population-Level Models Informed by Genetics Predict More Suitable Habitat than Traditional Species-Level Model Across the Range of a Widespread Riparian Tree." *PLoS One* 17 (2022): e0274892.

Bazzaz, F. A. "Life History of Colonizing Plants: Some Demographic, Genetic, and Physiological Features." In Mooney and Drake, *Ecology of Biological Invasions of North America and Hawaii*, 96–110.

Beatson, A. *Papers Relating to the Devastation by Goats on the Island of St. Helena.* St. Helena: S. Solomon, 1810.

Beatson, A. *Tracts Relative to the Island of St. Helena.* London: W. Bulmer, 1816.

Beattie, K. R. "The Operation of Quarantine No. 37." *Journal of Economic Entomology* 14 (1921): 201–5.

Beaumont, L. J., R. V. Gallagher, W. Thuiller, P. O. Downey, M. Leishman, and L. Hughes. "Different Climatic Envelopes Among Invasive Populations May Lead to Underestimation of Current and Future Biological Invasions." *Diversity and Distributions* 15 (2009): 409–20.

Beaury, E. M., J. T. Finn, J. D. Corbin, V. Barr, and B. A. Bradley. "Biotic Resistance to Invasion Is Ubiquitous Across Ecosystems of the United States." *Ecology Letters* 23 (2020): 476–82.

Beeckman, D. *A Voyage to and from the Island of Borneo.* London: Printed for T. Warner at the Black Boy, and J. Batley at the Dow, in Pater-Noster-Row, 1718.

Beerling, D. J., B. Huntley, and J. P. Bailey. "Climate and the Distribution of *Fallopia japonica*: Use of an Introduced Species to Test the Predictive Capacity of Response Surfaces." *Journal of Vegetation Science* 6 (1995): 269–82.

Beever, E. A., D. Simberloff, S. L. Crowley, R. Al-Chokhachy, H. A. Jackson, and S. L. Petersen. "Social–Ecological Mismatches Create Conservation Challenges in Introduced Species Management." *Frontiers in Ecology and the Environment* 17 (2019): 117–25.

Begon, M., J. L. Harper, and C. R. Townsend. *Ecology: Individuals, Populations, and Communities.* Sunderland, MA: Sinauer Associates, 1986.

Begon, M., J. L. Harper, and C. R. Townsend. *Ecology: Individuals, Populations, and Communities.* 2nd ed. Cambridge, MA: Blackwell Scientific, 1990.

Begon, M., J. L. Harper, and C. R. Townsend. *Ecology: Individuals, Populations, and Communities.* 3rd ed. Oxford: Blackwell Science, 1996.

Begon, M., and C. R. Townsend. *Ecology: From Individuals to Ecosystems.* 5th ed. Oxford: Wiley, 2021.

Begon, M., C. R. Townsend, and J. L. Harper. *Ecology: From Individuals to Ecosystems.* 4th ed. Malden, MA: Blackwell, 2006.

Beirne, B. P. "Biological Control Attempts by Introductions Against Pest Insects in the Field in Canada." *Canadian Entomologist* 107 (1975): 225–36.

Bekoff, M., ed. *Animal Passions and Beastly Virtues: Reflections on Redecorating Nature.* Philadelphia: Temple University Press, 2005.

Bekoff, M., ed. *Ignoring Nature No More: The Case for Compassionate Conservation.* Chicago: University of Chicago Press, 2013.

Bekoff, M. "The Importance of Ethics in Conservation Biology: Let's Be Ethicists and Not Ostriches." *Endangered Species Update* 19, no. 2 (2002): 23–26. Reprinted in Bekoff, *Animal Passions and Beastly Virtues*, 225–31.

Bekoff, M. "Introduction: What Does It Feel Like to Be a Fox?" In Bekoff, *Animal Passions and Beastly Virtues*, 1–21.

Bekoff, M. "The 'Possum Stomp' vs. Compassionate Conservation and Ethics." *Ethics and Morality* (blog), *Psychology Today*, October 21, 2017. https://www.psychology

today.com/us/blog/animal-emotions/201710/the-possum-stomp-vs-compassionate-conservation-and-ethics.

Bekoff, M. Preface to Bekoff, *Ignoring Nature No More*, xiii–xxvii.

Bellard, C., W. Thuiller, B. Leroy, P. Genovesi, M. Bakkenes, and F. Courchamp. "Will Climate Change Promote Future Invasions?" *Global Change Biology* 19 (2013): 3740–48.

Bellwood, D. R., A. S. Hoey, and J. H. Choat. "Limited Functional Redundancy in High Diversity Systems: Resilience and Ecosystem Function on Coral Reefs." *Ecology Letters* 6 (2003): 281–85.

Belote, R. T., J. F. Weltzin, and R. J. Norby. "Response of an Understory Plant Community to Elevated [CO_2] Depends on Differential Responses of Dominant Invasive Species and Is Mediated by Soil Water Availability." *New Phytologist* 161 (2003): 827–35.

Ben-Ami, D. "Compassionate Conservation: Where to from Here?" *Israel Journal of Ecology and Evolution* 63, no. 3–4 (2017): 1–4.

Benedict, W. V. *History of White Pine Blister Rust Control—A Personal Account*. Washington, DC: USDA Forest Service, 1981.

Beng, K. C., and R. T. Corlett. "Applications of Environmental DNA (eDNA) in Ecology and Conservation: Opportunities, Challenges and Prospects." *Biodiversity and Conservation* 29 (2020): 2089–2121.

Bennett, E. M. Review of *Impact of Biological Invasions on Ecosystem Services*, by Montserrat Vilà and Philip E. Hulme, eds. *Biological Invasions* 20 (2018): 813–14.

Bentham, G. "De Candolle's 'Geographical Botany.'" *Edinburgh Review or Critical Journal* 104 (1856): 490–518.

Beric, B., and H. J. MacIsaac. "Determinants of Rapid Response Success for Alien Invasive Species in Aquatic Ecosystems." *Biological Invasions* 17 (2015): 3327–35.

Bernardin de Saint-Pierre, J.-H. *Études de la nature*. Paris: Jean-Pierre Didot, 1784.

Bernardin de Saint-Pierre, J.-H. *Studies of Nature*. 2 vols. Translated by H. Hunter. Boston, 1797.

Bernardin de Saint-Pierre, J.-H. *Voyage à l'Île de France, à l'Île Bourbon et au Cap de Bonne-Espérance*. Paris: Merlin, 1773.

Berry, R. J. "Conservation Aspects of the Genetical Constitution of Populations." In Duffey and Watt, *Scientific Management of Animal and Plant Communities for Conservation*, 177–206.

Berry, R. J. "Diversity and Differentiation: The Importance of Island Biology for General Theory." *Oikos* 41 (1983): 523–29.

Berry, R. J. "The Evolution of an Island Population of the House Mouse." *Evolution* 18 (1964): 468–83.

Berry, R. J., M. E. Jakobson, and J. Peters. "The House Mice of the Faroe Islands: A Study in Microdifferentiation." *Journal of Zoology* 185 (1978): 73–92.

Berry, R. J., and H. M. Murphy. "The Biochemical Genetics of an Island Population of the House Mouse." *Proceedings of the Royal Society of London B* 175 (1970): 255–67.

Berry, R. J., and F. E. N. Rose. "Islands and the Evolution of *Microtus arvalis* (Microtinae)." *Journal of Zoology* 177 (1975): 395–409.

Bertelsmeier, C., G. M. Luque, B. D. Hoffmann, and F. Courchamp. "Worldwide Ant Invasions Under Climate Change." *Biodiversity and Conservation* 24 (2015): 117–28.

Bertness, M. D. "Habitat and Community Modification by an Introduced Herbivorous Snail." *Ecology* 65 (1984): 370–81.

Bertness, M. D., and T. C. Coverdale. "An Invasive Species Facilitates the Recovery of Salt Marsh Ecosystems on Cape Cod." *Ecology* 94 (2013): 1937–43.

Bertolino, S., N. Cordero di Montezemolo, D. G. Preatoni, L. A. Wauters, and A. Martinoli. "A Grey Future for Europe: *Sciurus carolinensisis* Replacing Native Red Squirrels in Italy." *Biological Invasions* 16 (2014): 53–62.

Bertolino, S., and P. Genovesi. "Spread and Attempted Eradication of the Grey Squirrel (*Sciurus carolinensis*) in Italy, and Consequences for the Red Squirrel (*Sciurus vulgaris*) in Eurasia." *Biological Conservation* 109 (2003): 351–58.

Betts, J., R. P. Young, C. Hilton-Taylor, M. Hoffmann, J. P. Rodríguez, et al. "A Framework for Evaluating the Impact of the IUCN Red List of Threatened Species." *Conservation Biology* 34 (2020): 632–43.

Beveridge, M. C. M., and D. C. Little. "History of Aquaculture in Traditional Societies." In *Ecological Aquaculture*, edited by B. A. Costa-Pierce, 3–29. Oxford: Blackwell Science, 2002.

Bioethics Panel. *Predator-free New Zealand: Social, Cultural, and Ethical Challenges.* Lincoln: BioHeritage Challenge, 2019.

Birch, L. C. "Evolutionary Opportunity for Insects and Mammals in Australia." In Baker and Stebbins, *Genetics of Colonizing Species*, 197–213.

Bird, P., G. Mutze, D. Peacock, and S. Jennings. "Damage Caused by Low-Density Exotic Herbivore Populations: The Impact of Introduced European Rabbits on Marsupial Herbivores and *Allocasuarina* and *Bursaria* Seedling Survival in Australian Coastal Shrubland." *Biological Invasions* 14 (2012): 743–55.

Bissessur, P., and F. V. Florens. "Predation of the Mauritius Endemic Macchabé Skink, *Gongylomorphus bojerii fontenayi*, by the Common Myna, *Acridotheres tristis*." *Bulletin Phaethon* 48 (2018): 91–93.

Black, L. T. *Atka: An Ethnohistory of the Western Aleutians*. Kingston, ON: Limestone Press, 1984.

Blackburn, T. M., C. Bellard, and A. Ricciardi. "Alien Versus Native Species as Drivers of Recent Extinction." *Frontiers in Ecology and the Environment* 17 (2019): 203–7.

Blackburn, T. M., F. Essl, T. Evans, P. E. Hulme, J. M. Jeschke, et al. "A Unified Classification of Alien Species Based on the Magnitude of Their Environmental Impacts." *PLoS Biology* 12 (2014): e1001850.

Blackburn, T. M., P. Pyšek, S. Bacher, J. T. Carlton, R. P. Duncan, et al. "A Proposed Unified Framework for Biological Invasions." *Trends in Ecology and Evolution* 26 (2011): 333–39.

Blakeslee, A. M. H., J. E. Byers, and M. P. Lesser. "Solving Cryptogenic Histories Using Host and Parasite Molecular Genetics: The Resolution of *Littorina littorea*'s North American Origin." *Molecular Ecology* 17 (2008): 3684–96.

Blanchan, N. *The American Flower Garden*. New York: Doubleday, Page, 1909.

Bloch, M. *The Historian's Craft*. Manchester: Manchester University Press, 1954.

Blomquist, R. F. "Ratification Resisted: Understanding America's Response to the Convention on Biological Diversity, 1989–2002." *Golden Gate University Law Review* 32 (2002): 493–586.

Blossey, B., and R. Nötzold. "Evolution of Increased Competitive Ability in Invasive Non-Indigenous Plants: A Hypothesis." *Journal of Ecology* 83 (1995): 887–89.

Blumstein, D. T., and E. Fernández-Juricic. *A Primer of Conservation Behavior*. Sunderland, MA: Sinauer Associates, 2010.

Blüthgen, N. "Why Network Analysis Is Often Disconnected from Community Ecology: A Critique and an Ecologist's Guide." *Basic Applied Ecology* 11 (2010): 185–95.

Boardman, L., T. G. Grout, and J. S. Terblanche. "False Codling Moth *Thaumatotibia leucotreta* (Lepidoptera, Tortricidae) Larvae Are Chill-Susceptible." *Insect Science* 19 (2012): 315–28.

Boardman, L., J. L. Lockwood, M. J. Angilletta Jr., J. S. Krause, J. A. Lau, et al. "The Future of Invasion Science Needs Physiology." *BioScience* 72 (2022): 1204–19.

Boardman, L., J. G. Sørensen, V. Koštál, P. Šimek, and J. S. Terblanche. "Cold Tolerance Is Unaffected by Oxygen Availability Despite Changes in Anaerobic Metabolism." *Scientific Reports* 6 (2016): 32856.

Boardman, L., and J. S. Terblanche. "Oxygen Safety Margins Set Thermal Limits in an Insect Model System." *Journal of Experimental Biology* 218 (2015): 1677–85.

Bobier, C., and B. Allen. "The Virtue of Compassion in Compassionate Conservation." *Conservation Biology* 36 (2022): e13776.

Bock, D. G., C. Caseys, R. D. Cousens, M. A. Hahn, S. M. Heredia, et al. "What We Still Don't Know About Invasion Genetics." *Molecular Ecology* 24 (2015): 2277–97.

Bodenheimer, F. S. *Animal Ecology To-Day*. The Hague: W. Junk, 1958.

Bodenheimer, F. S. *Problems of Animal Ecology*. London: Oxford University Press, 1938.

Boettner, G. H., J. S. Elkinton, and C. J. Boettner. "Effects of a Biological Control Introduction on Three Nontarget Native Species of Saturniid Moths." *Conservation Biology* 14 (2000): 1798–1806.

Bohan, D., A. Dumbrell, and F. Massol, eds. *Networks of Invasion: A Synthesis of Concepts*. London: Academic Press, 2017.

Boiffin, J., V. Badeau, and N. Bréda. "Species Distribution Models May Misdirect Assisted Migration: Insights from the Introduction of Douglas-fir to Europe." *Ecological Applications* 27 (2017): 446–57.

Bois, S. T., J. A. Silander Jr., and L. J. Mehrhoff. "Invasive Plant Atlas of New England: The Role of Citizens in the Science of Invasive Alien Species Detection." *BioScience* 61 (2011): 763–70.

Boland, C. R. J., M. J. Smith, B. Maple, R. Tiernan, R. Barr, et al. "Heli-Baiting Using Low Concentration Fipronil to Control Invasive Yellow Crazy Ant Supercolonies on Christmas Island, Indian Ocean." In Veitch, Clout, and Towns, *Island Invasives: Eradication and Management*, 152–56.

Bolin, B., B. R. Döös, J. Jäger, and R. A. Warrick, eds. *The Greenhouse Effect, Climatic Change, and Ecosystems*. SCOPE 29. Chichester: J. Wiley, 1986.

Bolin, B., and C. D. Keeling. "Large-Scale Atmospheric Mixing as Deduced from the Seasonal and Meridional Variations of Carbon Dioxide." *Journal of Geophysical Research* 68 (1963): 3899–3920.

Bollache, L., J. T. A. Dick, K. D. Farnsworth, and W. I. Montgomery. "Comparison of the Functional Responses of Invasive and Native Amphipods." *Biology Letters* 4 (2008): 166–69.

Boltovskoy, D., N. M Correa, L. E. Burlakova, A. Y. Karatayev, E. V. Thuesen, et al. "Traits and Impacts of Introduced Species: A Quantitative Review of Meta-Analyses." *Hydrobiologia* 848 (2021): 2225–58.

Boltovskoy, D., F. Sylvester, and E. M. Paolucci. "Invasive Species Denialism: Sorting Out Facts, Beliefs, and Definitions." *Ecology and Evolution* 8 (2018): 11190–98.

Bonhomme, F., A. Orth, T. Cucchi, H. Rajabi-Maham, J. Catalan, et al. "Genetic Differentiation of the House Mouse Around the Mediterranean Basin: Matrilineal Footprints of Early and Late Colonization." *Proceedings of the Royal Society B* 278 (2011): 1034–43.

Bonney, R. "Expanding the Impact of Citizen Science." *BioScience* 71 (2021): 448–51.

Booth, T. H. "A New Method for Assisting Species Selection." *Commonwealth Forestry Review* 64 (1985): 241–50.

Booth, T. H., H. A. Nix, J. R. Busby, and M. F. Hutchinson. "BIOCLIM: The First Species Distribution Modelling Package, its Early Applications and Relevance to Most Current MAXENT Studies." *Diversity and Distributions* 20 (2014): 1–9.

Booth, T. H., and K. J. Williams. "Developing Biodiverse Plantings Suitable for Changing Climatic Conditions 1: Underpinning Scientific Methods." *Ecological Management and Restoration* 13 (2012): 267–73.

Bordage, E. "À propos de l'hérédité des caractères acquis." *Bullétin biologique de la France et de la Belgique* 44 (1910): 51–88.

Borenstein, M., L. V. Hedges, J. P. T. Higgins, and H. R. Rothstein. *Introduction to Meta-Analysis*. Chichester: Wiley, 2009.

Born, W., F. Rauschmayer, and I. Bräuer. "Economic Evaluation of Biological Invasions—A Survey." *Ecological Economics* 55 (2005): 321–36.

Borrett, S. R., J. Moody, and A. Edelmann. "The Rise of Network Ecology: Maps of the Topic Diversity and Scientific Collaboration." *Ecological Modelling* 10 (2014): 111–27.

Bortolus, A. "Running Like Alice and Losing Good Ideas: On the Quasi-Compulsive Use of English by Non-Native English Speaking Scientists." *Ambio* 41 (2012): 769–72.

Bostock, J., and H. G. Riley, eds. and trans. *The Natural History of Pliny*. Vol. 2. London: George Bell and Sons, 1900.

Botkin, D. B. "The Naturalness of Biological Invasions." *Western North American Naturalist* 61 (2001): 261–66.

Botstein, D., R. L. White, M. Skolnick, and R. W. Davis. "Construction of a Genetic Map in Man Using Restriction Fragment Length Polymorphisms." *American Journal of Human Genetics* 32 (1980): 314–31.

Botting, D. *Humboldt and the Cosmos*. London: Sphere Books, 1973.

Bourne, J. *A History of Rat Control in Alberta*. Edmonton: Alberta Agriculture, Food and Rural Development, 2000.

Bowman, W. D., and S. D. Hacker. *Ecology*. 5th ed. New York: Sinauer Associates, 2021.

Bowman, W. D., and S. D. Hacker. *Ecology*. 6th ed. New York: Sinauer Associates, 2021.

Bowman, W. D., S. D. Hacker, and M. L. Cain. *Ecology*. 4th ed. Sunderland, MA: Sinauer Associates, 2017.

Boyce, M. S. "Population Viability Analysis." *Annual Review of Ecology and Systematics* 23 (1992): 481–506.

Boyd, K. S., F. Drummond, C. Donahue, and E. Groden. "Factors Influencing the Population Fluctuations of *Euproctis chrysorrhoea* (Lepidoptera: Erebidae) in Maine." *Environmental Entomology* 50 (2021): 1203–16.

Boykoff, M. T., and J. M. Boykoff. "Balance as Bias: Global Warming and the US Prestige Press." *Global Environmental Change* 14 (2004): 125–36.

Boyle, L. J. "The Relative Effects of Fish and Nutrient Introduction on Zooplankton Assemblages in Alpine Lakes." Master's thesis, University of Wyoming, 2021.

Boyle, T. C. *When the Killing's Done*. New York: Viking, 2011.

Bradley, B. A. "Regional Analysis of the Impacts of Climate Change on Cheatgrass Invasion Shows Potential Risk and Opportunity." *Global Change Biology* 15 (2009): 196–208.

Bradley, B. A., M. Oppenheimer, and D. S. Wilcove. "Climate Change and Plant Invasions: Restoration Opportunities Ahead?" *Global Change Biology* 15 (2009): 1511–21.

Bradley, B. A., D. S. Wilcove, and M. Oppenheimer. "Climate Change Increases Risk of Plant Invasion in the Eastern United States." *Biological Invasions* 12 (2010): 1855–72.

Bradley, R. *New Improvements of Planting and Gardening, Both Philosophical and Practical.* 2nd ed. London: W. Mears, 1718.

Bradshaw, W. E., and C. M. Holzapfel. "Evolutionary Response to Rapid Climate Change." *Science* 312 (2006): 1477–78.

Braga, R. R., L. Gómez-Aparicio, T. Heger, J. R. S. Vitule, and J. M. Jeschke. "Structuring Evidence for Invasional Meltdown: Broad Support but with Biases and Gaps." *Biological Invasions* 20 (2018): 923–36.

Braga, R. R., L. Gómez-Aparicio, T. Heger, J. R. S. Vitule, and J. M. Jeschke. "Invasional Meltdown Hypothesis." In Jeschke and Heger, *Invasion Biology: Hypotheses and Evidence*, 79–91.

Brasier, C. M. "Intercontinental Spread and Continuing Evolution of the Dutch Elm Disease Pathogens." In *The Elms*, edited by C. P. Dunn, 61–72. Boston: Springer, 2000.

Braudel, F. *Civilisation Matérielle, Économie et Capitalisme, XVe–XVIIIe siècle.* 3 vols. Paris: Armand Colin, 1979.

Brenchley, W. *Weeds of Farm Land.* London: Longman Green, 1920.

Breytenbach, G. J. "Impacts of Alien Organisms on Terrestrial Communities with Emphasis on Communities of the South-Western Cape." In Macdonald, Kruger, and Ferrar, *Ecology and Management of Biological Invasions in Southern Africa*, 229–38.

Briggs, J. C. "Rise of Invasive Species Denialism? A Response to Russell and Blackburn." *Trends in Ecology and Evolution* 32 (2017): 231–32.

Bright, C. "Invasive Species: Pathogens of Globalization." *Foreign Policy* 116 (1999): 50–64.

Bright, C. *Life Out of Bounds: Bioinvasion in a Borderless World.* New York: W. W. Norton, 1998.

Brill, W. J. "Reply." *Science* 229 (1985): 115–17. [Response to Colwell et al., "Genetic Engineering in Agriculture."]

Brill, W. J. "Safety Concerns and Genetic Engineering in Agriculture." *Science* 227 (1985): 381–84.

Brinckman, J. "Creepy Strangler Climbs Oregon's Least-Wanted List." *Oregonian* (Portland), February 28, 2001, A1, A14.

Brockie, R. E., L. L. Loope, M. B. Usher, and O. Hamann. "Biological Invasions of Nature Reserves." *Biological Conservation* 44 (1988): 9–36.

Brodie, J. F., S. Lieberman, A. Moehrenschlager, K. H. Redford, J. P. Rodríguez, and M. Schwartz. "Global Policy for Assisted Colonization of Species." *Science* 372 (2021): 456–58.

Brody, J. E. "Agriculture Department to Abandon Campaign Against the Fire Ant." *New York Times*, April 20, 1975, 46.

Broennimann, O., and A. Guisan. "Predicting Current and Future Biological Invasions: Both Native and Invaded Ranges Matter." *Biology Letters* 4 (2008): 585–89.

Brook, B. W., C. J. Bradshaw, L. W. Traill, and R. Frankham. "Minimum Viable Population Size: Not Magic, But Necessary." *Trends in Ecology and Evolution* 26 (2011): 619–20.

Brook, I. "Making Here Like There: Place Attachment, Displacement and the Urge to Garden." *Ethics, Place and Environment* 6 (2003): 227–34.

Brooks, J. L., and S. I. Dodson. "Predation, Body Size, and Composition of Plankton." *Science* 150 (1965): 28–35.

Brothers, T. S., B. Fredrich, D. W. Gade, and C. T. Kimber. "Jonathan D. Sauer (1918–2008): Perspectives on His Life and Work in Latin America and Beyond." *Journal of Latin American Geography* 8 (2009): 165–80.

Brotman, B. "Parrot Troopers Defend Their Feathered Friends." *Chicago Tribune*, April 19, 1988, A1.

Brown, D. E. *Bringing Back the Game: Arizona Wildlife Management, 1912–1962*. Phoenix: Arizona Game and Fish Department, 2012.

Brown, D. E., R. D. Babb, and R. R. Johnson. "Charles Taylor Vorhies (1879–1949): Arizona's Premier Nongame Biologist." *Journal of the Arizona-Nevada Academy of Science* 46 (2015): 18–23.

Brown, J. H. "Patterns, Modes and Extents of Invasions by Vertebrates." In Drake et al., *Biological Invasions: A Global Perspective*, 85–109.

Brown, J. H., R. K. Colwell, R. E. Lenski, B. R. Levin, M. Lloyd, et al. "Report on Workshop on Possible Ecological and Evolutionary Impacts of Bioengineered Organisms Released into the Environment." *Bulletin of the Ecological Society of America* 65 (1984): 436–38.

Brown, J. H., J. E. Gillooly, A. P. Allen, V. M. Savage, and G. B. West. "Toward a Metabolic Theory of Ecology." *Ecology* 85 (2004): 1771–89.

Brown, J. K., D. R. Frohlich, and R. C. Rosell. "The Sweetpotato or Silverleaf Whiteflies: Biotypes of *Bemisia tabaci* or a Species Complex?" *Annual Review of Entomology* 40 (1995): 511–34.

Brown, J. R. *Who Rules in Science?* Cambridge, MA: Harvard University Press, 2001.

Browne, A. "Why Act Now?" *Chicago Tribune*, November 18, 1997, SW16.

Brulliard, K. "A Highly Contagious and Fatal Virus Is Spreading in Several States: Its Victims Are Rabbits." *Washington Post*, May 18, 2020.

Brumfield, R. T., P. Beerli, D. A. Nickerson, and S. V. Edwards. "The Utility of Single Nucleotide Polymorphisms in Inference of Population History." *Trends in Ecology and Evolution* 18 (2003): 249–56.

Bruskotter, J. T., J. A. Vucetich, and M. P. Nelson. "Animal Rights and Wildlife Conservation: Conflicting or Compatible." *Wildlife Professional*, July/August 2017, 40–43.

Brussard, P. F. "Geographic Variation in North American Colonies of *Cepaea nemoralis*." *Evolution* 29 (1975): 402–10.

Bryan, E. H., Jr. "The Much Maligned Mongoose." *Paradise of the Pacific* 50, no. 4 (1938): 32–34.

Bucharova, A. "Assisted Migration Within Species Range Ignores Biotic Interactions and Lacks Evidence." *Restoration Ecology* 25 (2017): 14–18.

Buchholz, R. "Behavioural Biology: An Effective and Relevant Conservation Tool." *Trends in Ecology and Evolution* 22 (2007): 401–7.

Buchholz, R. "Should Animal Behaviorists Teach Conservation?" *Conservation Behaviorist* 4, no. 1 (2006): 3–4.

Buckingham, G. R. "Biological Control of Alligatorweed, *Alternanthera philoxeroides*, the World's First Aquatic Weed Success Story." *Castanea* 61 (1996): 232–43.

Buckland, P. C., and J. Sadler. "A Biogeography of the Human Flea, *Pulex irritans* L. (Siphonaptera: Pulicidae)." *Journal of Biogeography* 16 (1989): 115–20.

Buckley, L. B., M. C. Urban, M. J. Angilletta, L. G. Crozier, L. J. Rissler, and M. W. Sears. "Can Mechanism Inform Species' Distribution Models?" *Ecology Letters* 13 (2010): 1041–54.

Buckley, L. B., S. Waaser, H. MacLean, and R. Fox. "Does Including Physiology Improve Species Distribution Model Predictions of Responses to Recent Climate Change?" *Ecology* 92 (2011): 2214–21.

Buckley, Y. M., and J. Catford. "Does the Biogeographic Origin of Species Matter? Ecological Effects of Native and Non-Native Species and the Use of Origin to Guide Management." *Journal of Ecology* 104 (2016): 4–17.

Buczkowski, G., and C. Bertelsmeier. "Invasive Termites in a Changing Climate: A Global Perspective." *Ecology and Evolution* 7, no. 3 (2017): 974–85.

Budiansky, S. *Nature's Keepers: The New Science of Nature Management*. New York: Free Press, 1995.

Buffon, G.-L., Comte de. *Histoire naturelle générale et particulière*. Vol. 32, Supplement 3, *Servant de suite à l'histoire des animaux quadrupèdes*. Paris: L'Imprimerie royale, 1776.

Buhs, J. B. *The Fire Ant Wars: Nature, Science, and Public Policy in Twentieth-Century America*. Chicago: University of Chicago Press, 2004.

Buker, G. E. "Engineers vs. Florida's Green Menace." *Florida Historical Quarterly* 60, no. 4 (1982): 413–27.

Bukowski, E. "Using the Commons to Understand the Dutch Elm Disease in Syracuse, NY." *Geographical Review* 109 (2019): 180–98.

Bull, J. "Exotic Birds in the New York City Area." *Wilson Bulletin* 85 (1973): 501–5.

Buller, W. L. "Further Notes on the Ornithology of New Zealand." *Transactions of the New Zealand Institute* 3 (1870): 37–56.

Buller, W. L. "On the Proposed Introduction of the Polecat into New Zealand." *Transactions and Proceedings of the New Zealand Institute* 9 (1877): 634–35.

Buller, W. L. "Some Curiosities of Birdlife." *Transactions and Proceedings of the New Zealand Institute* 27 (1894): 134–42.

Bump, G. "Foreign Game Investigation: A Federal-State Cooperative Program." Resource Publication No 49. Washington, DC: US Department of the Interior, US Fish and Wildlife Service, Bureau of Sport Fisheries and Wildlife, 1968.

Bump, G. "Game Introductions—When, Where, and How." In *Transactions of the 16th North American Wildlife Conference*, 316–25. Washington, DC: Wildlife Management Institute, 1951.

Bump, G. "The Introduction and Transplantation of Game Birds and Mammals into the State of New York." In *Transactions of the 5th North American Wildlife Conference*, 409–20. Washington, DC: Wildlife Management Institute, 1941.

Bumpus, H. C. "The Elimination of the Unfit as Illustrated by the Introduced Sparrow, *Passer domesticus*." *Biological Lectures: Woods Hole Marine Biological Laboratory* (1898): 209–26.

Burdick, A. *Out of Eden: An Odyssey of Ecological Invasion*. New York: Farrar, Straus and Giroux, 2005.

Burdon, K. J. *A Handbook of St. Kitts–Nevis*. London: West Indies Committee, 1920.

Burger, J., and M. Gochfeld. "Exotic Monk Parakeets (*Myiopsitta monachus*) in New Jersey: Nest Site Selection, Rebuilding Following Removal, and Their Urban Wildlife Appeal." *Urban Ecosystems* 12 (2009): 185–96.

Burns, K. C., L. Santini, and A. Zhan. "Building a Better Future for Diversity & Distributions." *Diversity and Distributions* 25, no. 7 (2019): 1010–11.

Busby, J. R. "BIOCLIM—A Bioclimate Analysis and Prediction System." *Plant Protection Quarterly* 6 (1991): 8–9.

Bush, G. L. "Sympatric Host Race Formation and Speciation in Frugivorous Flies of the Genus *Rhagoletis* (Diptera, Tephritidae)." *Evolution* 23 (1969): 237–51.

Bustamante, R. O., I. A. Serey, and S. T. A. Pickett. "Forest Fragmentation, Plant Regeneration and Invasion Processes Across Edges in Central Chile." In *How Landscapes Change: Human Disturbance and Ecosystem Fragmentation in the Americas*, edited by G. A. Bradshaw and P. A. Marquet, 145–60. Berlin: Springer, 2003.

Cagua, E. F., K. L. Wootton, and D. B. Stouffer. "Keystoneness, Centrality, and the Structural Controllability of Ecological Networks." *Journal of Ecology* 107 (2019): 1779–90.

Cahn, A. R. "A Climographic Analysis of the Problem of Introducing Three Exotic Game Birds into the Tennessee Valley and Vicinity." In *Transactions of the Third North American Wildlife Conference*, 807–17. Washington, DC: American Wildlife Institute, 1938.

Cahn, A. R. "The Effect of Carp on a Small Lake: The Carp as a Dominant." *Ecology* 10 (1929): 271–74.

Cain, M. L., W. D. Bowman, and S. D. Hacker. *Ecology*. Sunderland, MA: Sinauer Associates, 2008.

Cain, M. L., W. D. Bowman, and S. D. Hacker. *Ecology*. 2nd ed. Sunderland, MA: Sinauer Associates, 2011.

Cain, M. L., W. D. Bowman, and S. D. Hacker. *Ecology*. 3rd ed. Sunderland, MA: Sinauer Associates, 2014.

Calahane, V. H. "Some Effects of Exotics on Nature." *Atlantic Naturalist* 10, no. 4 (1955): 176–85.

Calhoun, J. B. "The Role of Temperature and Natural Selection in Relation to the Variations in the Size of the English Sparrow in the United States." *American Naturalist* 81 (1947): 203–28.

Callaway, E. "UN Treaty Agrees to Limit Gene Drives but Rejects a Moratorium." *Nature*, November 30, 2018. https://doi.org/10.1038/d41586-018-07600-w.

Callaway, R. M., and W. M. Ridenour. "Novel Weapons: Invasive Success and the Evolution of Increased Competitive Ability." *Frontiers in Ecology and Evolution* 2 (2004): 436–43.

Callen, A., M. W. Hayward, K. Klop-Toker, B. L. Allen, G. Ballard, et al. "Envisioning the Future with 'Compassionate Conservation': An Ominous Projection for Native Wildlife and Biodiversity." *Biological Conservation* 241 (2020): 108365.

Callicott, J. B. "The Conceptual Foundations of the Land Ethic." In *The Environmental Ethics and Policy Book: Philosophy, Ecology, Economics*, edited by D. Van DeVeer and C. Pierce, 224–38. Belmont, CA: Wadsworth, 1994.

Callicott, J. B. "Leopold's Land Aesthetic." In *Nature, Aesthetics, and Environmentalism: From Beauty to Duty*, edited by A. Carlson and S. Lintott, 105–18. New York: Columbia University Press, 2008.

Callicott, J. B., and E. T. Freyfogle, eds. *Aldo Leopold: For the Health of the Land; Previously Unpublished Essays and Other Writings*. Washington, DC: Island Press, 1999.

Callomon, P. "Alvin Cahn: A Man of Science in Post-War Japan." *Proceedings of the Academy of Natural Sciences of Philadelphia* 167 (2019): 27–47.

Caltagirone, L. E., and R. L. Doutt. "The History of the Vedalia Beetle Importation to California and Its Impact on the Development of Biological Control." *Annual Review of Entomology* 34 (1989): 1–16.

Calzada Preston, C. E., S. Pruett-Jones, and J. R. Eberhard. "Monk Parakeets as a Globally Naturalized Species." In *Naturalized Parrots of the World*, edited by S. Pruett-Jones, 173–92. Princeton, NJ: Princeton University Press, 2021.

Cameron, A. G. *A Review of Pasture Plant Introduction in the 600–1500 mm Rainfall Zone of the Northern Territory 1981–1985*. Technical Bulletin No. 144. Darwin: Northern Territory Department of Primary Industry and Fisheries, 1989.

Cameron, E. K., M. Vilà, and M. Cabeza. "Global Meta-Analysis of the Impacts of Terrestrial Invertebrate Invaders on Species, Communities and Ecosystems." *Global Ecology and Biogeography* 25 (2016): 596–606.

Camerano, L. "Dell' equilibrio dei viventi mercé la reciproca distruzione." *Atti della Reale Accademia delle Scienze di Torino* 15 (1880): 393–414.

Campbell, C., L. Russo, R. Albert, A. Buckling, and K. Shea. "Whole Community Invasions and the Integration of Novel Ecosystems." *PLoS Computational Biology* 18, no. 6 (2022): e1010151.

Campbell, C., S. Yang, R. Albert, and K. Shea. "A Network Model for Plant-Pollinator Community Assembly." *Proceedings of the National Academy of Sciences of the United States of America* 108 (2011): 197–202.

Campbell, D. H. "Some Botanical and Environmental Aspects of Hawaii." *Ecology* 1 (1920): 257–69.

Campbell, K. "Advances in Judas Goat Methodology in the Galápagos Islands: Manipulating the Animals." In *Judas Workshop 2002*, edited by J. Gregory, B. Kyle, and M. Simmons, 70–77. Dunedin: New Zealand Department of Conservation, Otago Conservancy, 2002.

Campbell, K., G. Baxter, P. Murray, B. E. Coblentz, and C. J. Donlan. "Development of a Prolonged Estrus Effect for Use in Judas Goats." *Applied Animal Behaviour Science* 102 (2007): 12–23.

Campbell, K., G. Baxter, P. Murray, B. E. Coblentz, C. J. Donlan, et al. "Increasing the Efficacy of Judas Goats by Sterilisation and Pregnancy Termination." *Wildlife Research* 32 (2005): 737–43.

Campbell, K., and C. J. Donlan. "Feral Goat Eradications on Islands." *Conservation Biology* 19 (2005): 1362–74.

Campbell, L. M. "Seeing Red: Inside the Science and Politics of the IUCN Red List." *Conservation and Society* 10 (2012): 367–80.

Campbell, S. E., and D. Simberloff. "Forty Years of Invasion Research: More Papers, More Collaboration . . . Bigger Impact?" *NeoBiota* 75 (2022): 57–77.

Campbell, S. E., and D. Simberloff. "The Productivity Puzzle in Invasion Science: Declining but Persisting Gender Imbalances in Research Performance." *BioScience* 72 (2022): 1220–29.

Cap, P. A. *Philibert Commerson: Naturaliste voyageur*. Paris: Victor Masson et fils, 1861.

Capovilla, G., M. Schmid, and D. Posé. "Control of Flowering by Ambient Temperature." *Journal of Experimental Botany* 66 (2015): 59–69.

Caravaggi, A., R. J. Cuthbert, P. G. Ryan, J. Cooper, and A. L. Bond. "The Impacts of Introduced House Mice on the Breeding Success of Nesting Seabirds on Gough Island." *Ibis* 161 (2019): 648–61.

Carden, R. F., A. D. McDevitt, F. E. Zachos, P. C. Woodman, P. O'Toole, et al. "Phylogeographic, Ancient DNA, Fossil and Morphometric Analyses Reveal Ancient and Modern Introductions of a Large Mammal: The Complex Case of Red Deer (*Cervus elaphus*) in Ireland." *Quaternary Science Reviews* 42 (2012): 74–84.

Cardenas, L., J.-C. Leclerc, P. Brunning, I. Garrido, C. Detree, et al. "First Mussel Settlement Observed in Antarctica Reveals the Potential for Future Invasions." *Scientific Reports* 10 (2020): 5552.

Carere, C., and F. Gherardi. "Animal Personalities Matter for Biological Invasions." *Trends in Ecology and Evolution* 28 (2013): 5–6.

Carey, M. P., B. L. Sanderson, K. A. Barnas, and J. D. Olden. "Native Invaders—Challenges for Science, Management, Policy, and Society." *Frontiers in Ecology and the Environment* 10 (2012): 373–81.

Carié, P. *L'acclimatation à l'Île Maurice*. Paris: Société National d'Acclimatation de France, 1916.

Carlton, J. T. "Biological Invasions and Cryptogenic Species." *Ecology* 77 (1996): 1653–55.

Carlton, J. T. "A Global Framework of Introduced Species (abstract)." Paper presented at Western Society of Naturalists annual meeting, Long Beach, CA, December 27–30, 1982.

Carlton, J. T. "A Journal of Biological Invasions." *Biological Invasions* 1 (1999): 1.

Carlton, J. T., J. W. Chapman, J. B. Geller, J. A. Miller, D. A. Carlton, et al. "Tsunami-Driven Rafting: Transoceanic Species Dispersal and Implications for Marine Biogeography." *Science* 29 (2017): 1402–6.

Carrasco, L. R., T. P. L. Nghiem, T. Sunderland, and L. P. Koh. "Economic Valuation of Ecosystem Services Fails to Capture Biodiversity Value of Tropical Forests." *Biological Conservation* 178 (2014): 163–70.

Carrier, L., and K. S. Bort. "The History of Kentucky Bluegrass and White Clover in the United States." *Journal of the American Society of Agronomy* 8 (1916): 256–66.

Carrillo, D., and J. E. Peña. "Prey-Stage Preferences and Functional and Numerical Responses of *Amblyseius largoensis* (Acari: Phytoseiidae) to *Raoiella indica* (Acari: Tenuipalpidae)." *Experimental and Applied Acarology* 57 (2012): 361–72.

Carrion, V., C. J. Donlan, K. Campbell, C. Lavoie, and F. Cruz. "Archipelago-wide Island Restoration in the Galápagos Islands: Reducing Costs of Invasive Mammal Eradication Programs and Reinvasion Risk." *PLoS One* 6, no. 5 (2011): e18835.

Carroll, S. P., and C. Boyd. "Host Race Radiation in the Soapberry Bug: Natural History, with the History." *Evolution* 46 (1992): 1052–69.

Carroll, S. P., H. Dingle, and T. R. Famula. "Rapid Appearance of Epistasis During Adaptive Divergence Following Colonization." Supplement, *Proceedings of the Royal Society of London B* 270 (2003): S80–S83.

Carroll, S. P., H. Dingle, T. R. Famula, and C. W. Fox. "Genetic Architecture of Adaptive Differentiation in Evolving Host Races of the Soapberry Bug (*Jadera haematoloma*)." *Genetica* 112–13 (2001): 257–72.

Carroll, S. P., H. Dingle, and S. P. Klassen. "Genetic Differentiation of Fitness-Associated Traits Among Rapidly Evolving Populations of the Soapberry Bug." *Evolution* 51 (1997): 1182–88.

Carroll, S. P., J. E. Loye, H. Dingle, M. Mathieson, T. R. Famula, and M. Zalucki. "And the Beak Shall Inherit—Evolution in Response to Invasion." *Ecology Letters* 8 (2005): 944–51.

Carromeu-Santos, A., M. L. Mathias, and S. I. Gabriel. "Widespread Distribution of Rodenticide Resistance-Conferring Mutations in the Vkorc1 Gene Among House Mouse Populations in Portuguese Macaronesian Islands and Iberian Atlantic Areas." *Science of the Total Environment* 900 (November 20, 2023): article 166290. https://doi.org/10.1016/j.scitotenv.2023.166290.

Carruthers, R. I., and C. M. D'Antonio. Introduction to "Science and Decision Making in Biological Control of Weeds: Benefits and Risks of Biological Control," special issue, *Biological Control* 35 (2005): 181–82.

Carson, R. *Silent Spring*. Boston: Houghton Mifflin, 1962.

Carte, L., Á. Hofflinger, and M. H. Polk. "Expanding Exotic Forest Plantations and Declining Rural Populations in La Araucanía, Chile." *Land* 10 (2021): 283.

Carter, J., and B. P. Leonard. "A Review of the Literature on the Worldwide Distribution, Spread of, and Efforts to Eradicate the Coypu (*Myocastor coypus*)." *Wildlife Society Bulletin* 30 (2002): 162–75.

Carvalho, G. R. "Molecular Ecology: Origins and Approach." In *Advances in Molecular Ecology*, edited by G. R. Carvalho, 1–23. Amsterdam: IOS Press, 1998.

Case, T. J. "Invasion Resistance Arises in Strongly Interacting Species-Rich Model Competition Communities." *Proceedings of the National Academy of Sciences of the United States of America* 87 (1990): 9610–14.

Case, T. J. "Invasion Resistance, Species Build-Up and Community Collapse in Metapopulation Models with Interspecies Competition." *Biological Journal of the Linnean Society* 42 (1991): 239–66.

Case, T. J. "Surprising Behavior from a Familiar Model and Implications for Competition Theory." *American Naturalist* 146 (1995): 961–66.

Castro-Díez, P., A. S. Vaz, J. S. Silva, M. Loo, Á. Alonso, et al. "Global Effects of Non-Native Tree Species on Multiple Ecosystem Services." *Biological Reviews* 94 (2019): 1477–1501.

Catford, J. A. "Using Management to Determine Drivers of Alien Plant Invasion and Limits to Native Restoration." *Applied Vegetation Science* 19 (2016): 5–6.

Catford, J. A., C. C. Daehler, H. T. Murphy, A. W. Sheppard, B. D. Hardesty, et al. "The Intermediate Disturbance Hypothesis and Plant Invasions: Implications for Species Richness and Management." *Perspectives in Plant Ecology, Evolution and Systematics* 14 (2012): 231–41.

Catford, J. A., R. Jansson, and C. Nilsson. "Reducing Redundancy in Invasion Ecology by Integrating Hypotheses into a Single Theoretical Framework." *Diversity and Distributions* 15 (2009): 22–40.

Catford, J. A., J. R. Wilson, P. Pyšek, P. E. Hulme, and R. P. Duncan. "Addressing Context Dependence in Ecology." *Trends in Ecology and Evolution* 37 (2022): 158–70.

Catling, P. M., and S. Carbyn. "Invasive Scots Pine, *Pinus sylvestris*, Replacing Corema, *Corema conradii*, Heathland in the Annapolis Valley, Nova Scotia." *Canadian Field Naturalist* 119 (2005): 237–44.

Catterall, C. P. "Fauna as Passengers and Drivers in Vegetation Restoration: A Synthesis of Processes and Evidence." *Ecological Management and Restoration* 19 (2018): 54–62.

Caughley, G. "Directions in Conservation Biology." *Journal of Animal Ecology* 63 (1994): 215–44.

Cavaleri, M. A., and L. Sack. "Comparative Water Use of Native and Invasive Plants at Multiple Scales: A Global Meta-Analysis." *Ecology* 91 (2010): 2705–15.

Ceccarelli, L. "Manufactured Scientific Controversy: Science, Rhetoric, and Public Debate." *Rhetoric and Public Affairs* 14 (2011): 195–228.

"Celebration in Honor of Dr. Henry Fairfield Osborn." *Science* 46 (1917): 477.

Center, T. D., F. A. Dray Jr., G. P. Jubinsky, and M. J. Grodowitz. "Biological Control of Water Hyacinth Under Conditions of Maintenance Management: Can Herbicides and Insects Be Integrated?" *Environmental Management* 23 (1999): 241–56.

Center, T. D., J. H. Frank, and F. A. Dray Jr. "Biological Control." In Simberloff, Schmitz, and Brown, *Strangers in Paradise: Impact and Management of Nonindigenous Species in Florida*, 245–73.

Chalmers, A. G., ed. *The Ecology of Sapelo Island National Estuarine Research Reserve*. Washington, DC: Office of Coastal Resource Management, National Oceanic and Atmospheric Administration, 1997.

Chamberlain, T. K. "There Is No Universal Pond Fish." *Progressive Fish-Culturist* 9 (1947): 78–81.

Chang, C.-S., H. L. Liu, X. Moncada, A. Seelenfreund, D. Seelenfreund, and K.-F. Chung. "A Holistic Picture of Austronesian Migrations Revealed by Phylogeography of Pacific Paper Mulberry." *Proceedings of the National Academy of Sciences of the United States of America* 112 (2015): 13537–42.

Chapin, F. S., and A. M. Starfield. "Time Lags and Novel Ecosystems in Response to Transient Climatic Change in Alaska." *Climate Change* 35 (1997): 449–61.

Chapman, D., T. Haynes, S. Beal, F. Essl, and J. M. Bullock. "Phenology Predicts the Native and Invasive Range Limits of Common Ragweed." *Global Change Biology* 20 (2014): 192–202.

Chapman, F. M. "The European Starling as an American Citizen." *Natural History* 25, no. 5 (1925): 480–85.

Chapman, J. W., J. T. Carlton, M. R. Bellinger, and A. M. Blakeslee. "Premature Refutation of a Human-Mediated Marine Species Introduction: The Case History of the Marine Snail *Littorina littorea* in the Northwest Atlantic." *Biological Invasions* 9 (2007): 737–50.

Chapple, D. G., S. M. Simmonds, and B. B. M. Wong. "Can Behavioral and Personality Traits Influence the Success of Unintentional Species Introductions?" *Trends in Ecology and Evolution* 27 (2012): 57–64.

Charles, H., and J. S. Dukes. "Impacts of Invasive Species on Ecosystem Services." In *Biological Invasions*, vol. 193, *Ecological Studies*, edited by W. Nentwig, 217–37. Heidelberg: Springer-Verlag, 2007.

Charudattan, R., W. J. Lewis, H. K. Kaya, and C. E. Rogers. Editorial. *Biological Control* 1 (1991): 1.

Chase, J. M., and M. A. Leibold. *Ecological Niches: Linking Classical and Contemporary Approaches*. Chicago: University of Chicago Press, 2003.

Cheke, A. S. "An Ecological History of the Mascarene Islands, with Particular Reference to Extinctions and Introductions of Land Vertebrates." In *Studies of Mascarene Island Birds*, edited by A. W. Diamond, 5–89. Cambridge: Cambridge University Press, 1987.

Chew, M. K. "Ending with Elton: Preludes to Invasion Biology." PhD diss., Arizona State University, 2006.

Chew, M. K., and A. Hamilton. "The Rise and Fall of Biotic Nativeness: A Historical Perspective." In Richardson, *Fifty Years of Invasion Ecology*, 35–47.

Chisholm, R. A. "Invasion Biology." *BioScience* 60 (2010): 314–15.

Chitty, D., and M. Shorten. "Techniques for the Study of the Norway Rat (*Rattus norvegicus*)." *Journal of Mammalogy* 27 (1946): 63–78.

Chitty, D., and H. N. Southern, eds. *Control of Rats and Mice*. Oxford: Oxford University Press, 1954.

Chown, S. L. "Trait-Based Approaches to Conservation Physiology: Forecasting Environmental Change Risks from the Bottom Up." *Philosophical Transactions of the Royal Society B* 367 (2012): 1615–27.

Chown, S. L., and S. Clusella-Trullas. "Functional Trait Variation Along Animal Invasion Pathways." *Annual Review of Ecology, Evolution, and Systematics* 54 (2023): 151–70.

Chown, S. L., and K. J. Gaston. "Macrophysiology—Progress and Prospects." *Functional Ecology* 30 (2016): 330–44.

Christensen, P. E., and N. D. Burrows. "Fire: An Old Tool with New Uses." In Groves and Burdon, *Ecology of Biological Invasions*, 97–105.

Chu, C. C., J. L. Spencer, M. J. Curzi, J. A. Zavala, and M. J. Seufferheld. "Gut Bacteria Facilitate Adaptation to Crop Rotation in the Western Corn Rootworm." *Proceedings of the National Academy of Sciences of the United States of America* 110 (2013): 11917–22.

Cierjacks, A., I. Kowarik, J. Joshi, S. Hempel, M. Ristow, et al. "Biological Flora of the British Isles: *Robinia pseudoacacia*." *Journal of Ecology* 101 (2013): 1623–40.

Civeyrel, L., and D. Simberloff. "A Tale of Two Snails: Is the Cure Worse than the Disease?" *Biodiversity and Conservation* 5 (1996): 1231–52.

Cizek, O, A. Bakesová, T. Kuras, J. Benes, and M. Konvicka. "Vacant Niche in Alpine Habitat: The Case of an Introduced Population of the Butterfly *Erebia epiphron* in the Krkonoše Mountains." *Acta Oecologica* 24 (2003): 15–23.

Clark, A. H. *Acadia: The Geography of Early Nova Scotia to 1760*. Madison: University of Wisconsin Press, 1968.

Clark, A. H. "The Impact of Exotic Invasions on the Remaining New World Mid-Latitude Grasslands." In Thomas, *Man's Role in Changing the Face of the Earth*, 737–62.

Clark, A. H. *The Invasion of New Zealand by People, Plants, and Animals: The South Island*. New Brunswick, NJ: Rutgers University Press, 1949.

Clark, A. H. Review of *The Western Invasions of the Pacific and Its Continents*, by Archibald Grenfell Price. *Geographical Reviews* 54 (1964): 291–93.

Clark, A. H. "The South Island of New Zealand: A Geographic Study of the Introduction and Modification of British Rural Patterns and Practices Associated with the Exotic Plants and Animals of the Island." PhD diss., University of California, Berkeley, 1944.

Clark, A. H. *Three Centuries and the Island: A Historical Geography of Settlement and Agriculture in Prince Edward Island, Canada*. Toronto: University of Toronto Press, 1959.

Clarke, W. E. "On Some Forms of *Mus musculus*, Linn., with Description of a New Subspecies from Færoe Islands." *Proceedings of the Royal Physical Society of Edinburgh* 15 (1904): 160–67.

Clausen, C. P. "Insect Parasitism and Biological Control." *Annals of the Entomological Society of America* 29 (1936): 201–23.

Cleary, R. L. "Editorial." *Wildland Weeds* 17 (2015–16): 1.

Clegg, M. T., and R. W. Allard. "Patterns of Genetic Differentiation in the Slender Wild Oat Species *Avena barbata*." *Proceedings of the National Academy of Sciences of the United States of America* 69 (1972): 1820–24.

Clements, F. E. *Plant Succession: An Analysis of the Development of Vegetation*. Washington, DC: Carnegie Institution of Washington, 1916.

Clements, F. E., and V. E. Shelford. *Bio-Ecology*. New York: John Wiley and Sons, 1939.

Clewell, A. F., and J. Aronson. *Ecological Restoration: Principles, Values, and Structure of an Emerging Profession*. 2nd ed. Washington, DC: Island Press, 2013.

Coates, P. "Editorial Postscript: The Naming of Strangers in the Landscape." *Landscape Research* 28 (2003): 131–37.

Coates, P. *Squirrel Nation: Reds, Greys, and the Meaning of Home*. London: Reaktion Books, 2023.

Coates, P. *Strangers on the Land: American Perceptions of Immigrant and Invasive Species*. Berkeley: University of California Press, 2006.

Cock, M. J. W., S. T. Murphy, M. T. K. Kairo, E. Thompson, R. J. Murphy, and A. W. Francis. "Trends in the Classical Biological Control of Insect Pests by Insects: An Update of the BIOCAT Database." *BioControl* 61 (2016): 349–63.

Cockayne, L. "Observations Concerning Evolution, Derived from Ecological Studies in New Zealand." *Transactions of the New Zealand Institute* 44 (1911): 1–50.

Cockayne, L. *The Vegetation of New Zealand*. 2nd ed. Leipzig: Verlag von Wilhelm Engelmann, 1928.

Cody, M. L., and J. M. Diamond, eds. *Ecology and Evolution of Communities*. Cambridge, MA: Belknap Press of Harvard University Press, 1975.

Coghlan, S., and A. P. A. Cardilini. "A Critical Review of the Compassionate Conservation Debate." *Conservation Biology* 36 (2022): e13760.

Cohen, Joel. "Mathematics as Metaphor." *Science* 172 (1971): 674–75.

Cohen, Joel. "Lorenzo Camerano's Contribution to Early Food Web Theory." In *Frontiers in Mathematical Biology*, edited by S. A. Levin, 351–59. Berlin: Springer, 1994.

Cohen, Jon. "A Cut Above: Pair That Developed CRISPR Earns Historic Award." *Science* 370 (2020): 271–72.

Cohen, Jon. "A Reporter Does CRISPR." *Science* 354 (2016): 541.

Colautti, R. I., I. A. Grigorovich, and H. J. MacIsaac. "Propagule Pressure: A Null Model for Biological Invasions." *Biological Invasions* 8 (2006): 1023–37.

Colautti, R. I., and H. J. MacIsaac. "A Neutral Terminology to Define 'Invasive' Species." *Diversity and Distributions* 10 (2004): 135–41.

Colautti, R. I., A. J. C. Niimi, D. A. van Overdijk, E. L. Mills, K. Holeck, and H. J. MacIsaac. "Spatial and Temporal Analysis of Transoceanic Shipping Vectors to the Great Lakes." In Ruiz and Carlton, *Invasive Species: Vectors and Management Strategies*, 227–46.

Colautti, R. I., A. Ricciardi, I. A. Grigorovich, and H. J. MacIsaac. "Is Invasion Success Explained by the Enemy Release Hypothesis?" *Ecology Letters* 7 (2004): 721–33.

Colinet, H., B. J. Sinclair, P. Vernon, and D. Renault. "Insects in Fluctuating Thermal Environments." *Annual Review of Entomology* 60 (2015): 123–40.

Collar, N. J., M. J. Crosby, and A. J. Stattersfield. *Birds to Watch 2: The World List of Threatened Birds.* BirdLife Conservation Series No. 4. Cambridge: BirdLife International, 1994.

Collins, F. S., M. S. Guyer, and A. Chakravarti. "Variations on a Theme: Cataloging Human DNA Sequence Variation." *Science* 278 (1997): 1580–81.

Colwell, R. K., E. A. Norse, D. Pimentel, F. E. Sharples, and D. Simberloff. "Genetic Engineering in Agriculture." *Science* 229 (1985): 111–12.

Comaroff, J., and J. L. Comaroff. "Naturing the Nation: Aliens, Apocalypse and the Postcolonial State." *Journal of Southern African Studies* 27 (2001): 627–51.

Combrink, L. L., W. C. Rosenthal, L. J. Boyle, J. A. Rick, E. G. Mandeville, et al. "Parallel Shifts in Trout Feeding Morphology Suggest Rapid Adaptation to Alpine Lake Environments." *Evolution* 77 (2023): 1522–38.

Compere, H. "The Red Scale and Its Insect Enemies." *Hilgardia* 31 (1961): 173–278.

Condon, T., I. L. Brisbin Jr., and C. R. Chandler. "Red Junglefowl Introductions in the Southeastern United States: History and Research Legacy." *Southeastern Naturalist* 18 (2019): 37–52.

Connell, J. H. "Diversity in Tropical Rain Forests and Coral Reefs: High Diversity of Trees and Corals Is Maintained Only in a Nonequilibrium State." *Science* 199 (1978): 1302–10.

Connell, J. H., and R. O. Slatyer. "Mechanisms of Succession in Natural Communities and Their Role in Community Stability and Organization." *American Naturalist* 111 (1977): 1119–44.

Convention on Biological Diversity. "Aichi Target 9." Accessed December 18, 2024. https://www.cbd.int/aichi-targets/target/9.

Convention on Biological Diversity. "Article 8: In-situ Conservation." March 30, 2007. https://www.cbd.int/convention/articles/default.shtml?a=cbd-08.

Convention on Biological Diversity. "COP 6Decision VI/23: Alien Species That Threaten Ecosystems, Habitats, or Species." April 2002. https://www.cbd.int/decision/cop/default.shtml?id=7197.

Convention on Biological Diversity. *Decision Adopted by the Conference of the Parties to the Convention on Biological Diversity at Its Tenth Meeting: X/33; Biodiversity and Climate Change.* October 29, 2010. https://www.cbd.int/doc/decisions/cop-10/cop-10-dec-33-en.pdf.

Convention on Biological Diversity. "Global Biodiversity Outlook 5." August 18, 2020. https://www.cbd.int/gbo5.

Convention on Biological Diversity. "Global Biodiversity Framework Target 6: 'Reduce the Introduction of Invasive Alien Species by 50% and Minimize Their Impact.'" Accessed December 18, 2024. https://www.cbd.int/gbf/targets/6.

Convention on Biological Diversity. "Information Centre." February 13, 2020. https://www.cbd.int/information.

Cook, W. C. "The Distribution of the Alfalfa Weevil (*Phytonomus posticus* Gyll.): A Study in Physical Ecology." *Journal of Agricultural Research* 30 (1925): 479–91.

Cook, W. C. "The Distribution of the Pale Western Cutworm, *Porosagrotis orthogonia* Morr.: A Study in Physical Ecology." *Ecology* 5 (1924): 60–69.

Cook, W. C. "Notes on Predicting the Probable Future Distribution of Introduced Insects." *Ecology* 12 (1931): 245–47.

Cook, W. C. 1923. "Studies in Physical Ecology of the Noctuidae." In *Minnesota Experimental Station*, Technical Bulletin 12.

Cooke, B. D. *Australia's War Against Rabbits: The Story of Rabbit Haemorrhagic Disease.* Collingwood: CSIRO Publishing, 2014.

Cooke, S. J., J. N. Bergman, C. L. Madliger, R. L. Cramp, J. Beardall, et al. "One Hundred Research Questions in Conservation Physiology for Generating Actionable Evidence to Inform Conservation Policy and Practice." *Conservation Physiology* 9, no. 1 (2021): coab009. https://doi.org/10.1093/conphys/coab009.

Cooke, S. J., D. T. Blumstein, R. Buchholz, T. Caro, E. Fernández-Juricic, et al. "Physiology, Behavior, and Conservation." *Physiological and Biochemical Zoology* 87 (2014): 1–14.

Coope, G. R. "The Response of Insect Faunas to Glacial-Interglacial Climatic Fluctuations." *Philosophical Transactions of the Royal Society of London. Series B: Biological Sciences* 344 (1994): 19–26.

Cooper, C. B., C. L. Hawn, L. R. Larson, J. K. Parrish, G. Bowser, et al. "Inclusion in Citizen Science: The Conundrum of Rebranding." *Science* 372 (2021): 1386–88.

Cooper, D., E. Larsen, and J. Shields. "1080 and Wildlife: Scientific and Ethical Issues Raised By Its Use on Australian Mammals." In *Pest or Guest: The Zoology of Overabundance*, edited by D. Lunney, P. Eby, P. Hutchings, and S. Burgin, 229–32. Mosman: Royal Society of New South Wales, 2007.

Cooper, W. S. "The Ecological Life History of Certain Species of *Ribes* and Its Application to the Control of the White Pine Blister Rust." *Ecology* 3 (1922): 7–16.

Corbett, L. *The Dingo in Australia and Asia.* Sydney: University of New South Wales Press, 1995.

Cordell, S. "Regional Summaries: Hawaii and US-Affiliated Pacific Islands." In *Invasive Species in Forests and Rangelands of the United States: A Comprehensive Science Synthesis for the United States Forest Sector*, edited by T. M. Poland, T. Patel-Weynand, D. M. Finch, C. F. Miniat, D. C. Hayes, and V. M. Lopez, 343–52. Heidelberg: Springer International, 2021.

Correspondence and Papers of Charles Sutherland Elton, FRS (1900–1991). Oxford, Bodleian Libraries.

Coristine, L. E., C. M. Robillard, J. T. Kerr, C. M. O'Connor, D. Lapointe, and S. J. Cooke. "A Conceptual Framework for the Emerging Discipline of Conservation Physiology." *Conservation Physiology* 2, no. 1 (2014): cou033. https://doi.org/10.1093/conphys/cou033.

Cortes, P. A., H. Puschel, P. Acuña, J. L. Bartheld, and F. Bozinovic. "Thermal Ecological Physiology of Native and Invasive Frog Species: Do Invaders Perform Better?" *Conservation Physiology* 4, no. 1 (2016): cow056. https://doi.org/10.1093/conphys/cow056.

Cory, C., and J. J. Knapp. "A Program to Eradicate Twenty-Four Nonnative Invasive Plant Species from Santa Cruz Island." *Monographs of the Western North American Naturalist* 7 (2014): 455–64.

Council of Europe Committee of Ministers. *Declaration on Cultural Diversity.* December 7, 2000. https://rm.coe.int/16804bfc0b.

Courtenay, W. R., Jr. "Biological Pollution Through Fish Introductions." In B. McKnight, *Biological Pollution*, 35–61.

Courtenay, W. R., Jr. "Florida's Walking Catfish." *Wards Natural Science Bulletin* 10 (1970): 1, 6.

Courtenay, W. R., Jr. "Nonindigenous Fishes." In Simberloff, Schmitz, and Brown, *Strangers in Paradise: Impact and Management of Nonindigenous Species in Florida*, 109–22.

Courtenay, W. R., Jr. *Pathways and Consequences of the Introduction of Non-Indigenous Fishes in the United States—August 1991.* Office of Technology Assessment Contractor's Report, 1991.

Courtenay, W. R., Jr., and P. B. Moyle. "Crimes Against Biodiversity: The Lasting Legacy of Fish Introductions." *Transactions of the North American Wildlife and Natural Resources Conference* 57 (1992): 365–72.

Courtenay, W. R., Jr., and V. E. Ogilvie. "Species Pollution." *Animal Kingdom* 74, no. 2 (1971): 22–28.

Courtenay, W. R., Jr., and C. R. Robins. "Exotic Aquatic Organisms in Florida with Emphasis on Fishes: A Review and Recommendations." *Transactions of the American Fisheries Society* 102 (1973): 1–12.

Courtenay, W. R., Jr., and C. R. Robins. "Exotic Organisms: An Unsolved, Complex Problem." *BioScience* 25 (1975): 306–13.

Courtenay, W. R., Jr., and C. R. Robins. "Fish Introductions: Good Management, Mismanagement, or No Management?" *CRC Critical Reviews in Aquatic Sciences* 1 (1989): 159–72.

Courtenay, W. R., Jr., H. F. Sahlman, W. W. Miley II, and D. J. Herrema. "Exotic Fishes in Fresh and Brackish Waters of Florida." *Biological Conservation* 6 (1974): 292–302.

Courtenay, W. R., Jr., and J. D. Williams. *Snakeheads (Pisces: Channidae)—A Biological Synopsis and Risk Assessment.* Denver: US Geological Survey, 2004.

Cousens, R. "Risk Assessment of Biofuel Species: An Application for Trait-Based Models for Predicting Weediness?" *Weed Science* 56 (2008): 873–82.

Cowie, R. H. "Invertebrate Invasions on Pacific Islands and the Replacement of Unique Native Faunas: A Synthesis of the Land and Freshwater Snails." *Biological Invasions* 3 (2002): 119–36.

Cowie, R. H. "Snails and Slugs." In Simberloff and Rejmánek, *Encyclopedia of Biological Invasions*, 634–43.

Cowles, H. C. "The Ecological Relations of the Vegetation on the Sand Dunes of Lake Michigan." *Botanical Gazette* 27 (1899): 95–117, 167–202, 281–308, 361–91.

Cox, G. W. *Alien Species and Evolution.* Washington, DC: Island Press, 2004.

Cragg, J. B. Review of *Natural Communities* by L. R. Dice. *Bulletin of the Institute of Biology* 1 (1953): 3.

Craige, B. J. *Eugene Odum: Ecosystem Ecologist and Environmentalist.* Athens: University of Georgia Press, 2001.

Craighead, F. C., and R. F. Dasmann. *Exotic Big Game on Public Lands.* Washington, DC: United States Department of the Interior, Bureau Land Management, 1966.

Crawley, M. J. "Chance and Timing in Biological Invasions." In Drake et al., *Biological Invasions: A Global Perspective*, 407–23.

Crawley, M. J. "The Population Biology of Invaders." *Philosophical Transactions of the Royal Society of London B* 314 (1986): 711–31.

Crawley, M. J., S. L. Brown, M. S. Heard, and G. R. Edwards. "Invasion-Resistance in Experimental Grassland Communities: Species Richness or Species Identity?" *Ecology Letters* 2 (1999): 140–48.

Cristescu, M. E. "Genetic Reconstructions of Invasion History." *Molecular Ecology* 24 (2015): 2212–25.

Cristescu, M. E., and P. D. N. Hebert. "Uses and Misuses of Environmental DNA in Biodiversity Science and Conservation." *Annual Review of Ecology, Evolution, and Systematics* 49 (2018): 209–30.

Croft, D. P., J. R. Madden, D. W. Franks, and R. James. "Hypothesis Testing in Animal Social Networks." *Trends in Ecology and Evolution* 26 (2011): 502–7.

Crooks, J. A. "Characterizing Ecosystem-Level Consequences of Biological Invasions: The Role of Ecosystem Engineers." *Oikos* 97 (2002): 153–66.

Crooks, J. A. "Lag Times." In Simberloff and Rejmánek, *Encyclopedia of Biological Invasions*, 404–10.

Crooks, J. A., and M. E. Soulé. "Lag Times in Population Explosions of Invasive Species: Causes and Implications." In Sandlund et al., *Proceedings of the Norway/UN Conference on Alien Species*, 39–46.

Crosby, A. W. *The Columbian Exchange: Biological and Cultural Consequences of 1492*. Westport, CT: Greenwood Press, 1972.

Crosby, A. W. *Ecological Imperialism: The Biological Expansion of Europe, 900–1900*. Cambridge: Cambridge University Press, 1986.

Crosby, A. W. "Ecological Imperialism: The Overseas Migration of Western Europeans as a Biological Phenomenon." *Texas Quarterly* 21, no. 1 (1978): 10–22.

Crosby, A. W. *Germs, Seeds, and Animals*. Armonk, NY: M. E. Sharpe, 1994.

Cross, John. "Simply Science." *Observer Magazine*, October 27, 1974, 88.

Cross, John Ridley. "Biological Flora of the British Isles: *Rhododendron ponticum*." *Journal of Ecology* 63 (1975): 345–64.

Cross, John Ridley. "The Establishment of Rhododendron Ponticum in the Killarney Oakwoods, Southwest Ireland." *Journal of Ecology* 69 (1981): 807–24.

Crossman, N. D., and D. A. Bass. "Application of Common Predictive Habitat Techniques for Post-Border Weed Risk Management." *Diversity and Distributions* 14 (2008): 213–24.

Crous, P. W., D. L. Hawksworth, and M. J. Wingfield. "Identifying and Naming Plant Pathogenic Fungi: Past, Present and Future." *Annual Review of Phytopathology* 53 (2015): 247–67.

Crowcroft, P. *Elton's Ecologists: A History of the Bureau of Animal Population*. Chicago: University of Chicago Press, 1991.

Crowl, T. A., C. R. Townsend, and A. R. McIntosh. "The Impact of Introduced Brown and Rainbow Trout on Native Fish: The Case of Australasia." *Reviews in Fish Biology and Fisheries* 2 (1992): 217–41.

Crowley, S. L., S. Hinchliffe, S. M. Redpath, and R. A. McDonald. "Disagreement About Invasive Species Does Not Equate to Denialism: A Response to Russell and Blackburn." *Trends in Ecology and Evolution* 32 (2017): 228–29.

Cuddington, K., J. E. Byers, W. G. Wilson, and A. Hastings, eds. *Ecosystem Engineers: Plants to Protists*. San Diego, CA: Academic/Elsevier, 2007.

Cullen, J. M., and E. S. Delfosse. "*Echium plantagineum* in Australia: Catalyst for Conflict and Change." In *Proceedings of the Sixth International Symposium on Biological Control of Weeds*, edited by E. Delfosse, 249–92. Ottawa: Agriculture Canada, 1985.

Culliney, T. W. "Benefits of Classical Biological Control for Managing Invasive Plants." *Critical Reviews in Plant Sciences* 24 (2005): 131–50.

Cupples, J. B., M. S. Crowther, G. Story, and M. Letnic. "Dietary Overlap and Prey Selectivity Among Sympatric Carnivores: Could Dingoes Suppress Foxes Through Competition for Prey?" *Journal of Mammalogy* 92 (2011): 590–600.

Curio, E. "Conservation Needs Ethology." *Trends in Ecology and Evolution* 11 (1996): 260–63.

Curley, R. "Wildlife Matters: A Historical Overview of Public Consciousness of Habitat and Wildlife Loss on Prince Edward Island." In E. MacDonald et al., *Time and Place: An Environmental History of Prince Edward Island*, 109–39.

Cuthbert, R. N., J. W. Dickey, N. E. Coughlan, P. W. Joyce, and J. T. A. Dick. "The Functional Response Ratio (FRR): Advancing Comparative Metrics for Predicting the Ecological Impacts of Invasive Alien Species." *Biological Invasions* 21 (2019): 2543–47.

Daehler, C. C. "Darwin's Naturalization Hypothesis Revisited." *American Naturalist* 158 (2001): 324–30.

Dahlsten, D. L., and R. Garcia, eds. *Eradication of Exotic Pests*. New Haven, CT: Yale University Press, 1989.

Daily, G. C. "Introduction: What Are Ecosystem Services?" In Dailey, *Nature's Services*, 1–10.

Daily, G. C., ed. *Nature's Services: Societal Dependence on Natural Ecosystems*. Washington, DC: Island Press, 1997.

Dale, T. D., and J. Chantrey. "A Subtle Endemic Virus in Grey Squirrels: Squirrelpox Virus." In Shuttleworth et al., *The Grey Squirrel: Ecology and Management of an Invasive Species in Europe*, 211–34.

Dalton, R. "University Buildings Named on Shaky Ground." *Nature* 426 (2003): 374.

Dalu, T., R. J. Wasserman, M. Jordaan, W. P. Froneman, and O. L. F. Weyl. "An Assessment of the Effect of Rotenone on Selected Non-Target Aquatic Fauna." *PLoS One* 10, no. 11 (2015): e0142140.

Dana, E. D., and J. Graham. "Management of Biological Invasions: Opening the Journal." *Management of Biological Invasions* 1 (2010): ii–iii.

Daniels, D. A., K. A. Nix, P. Wadl, L. M. Vito, G. J. Wiggins, et al. "Thousand Cankers Disease Complex: A Forest Health Issue that Threatens *Juglans* Species Across the U.S." *Forests* 7, no. 11 (2016): 260. https://doi.org/10.3390/f7110260.

D'Antonio, C. M., T. L. Dudley, and M. Mack. "Disturbance and Biological Invasions: Direct Effects and Feedbacks." In *Ecosystems of the World*, edited by L. R. Walker, 413–52. Amsterdam: Elsevier, 1999.

D'Antonio, C. M., J. T. Tunison, and R. K. Loh. "Variation in the Impact of Exotic Grasses on Native Plant Composition in Relation to Fire Across an Elevation Gradient in Hawaii." *Austral Ecology* 25 (2000): 507–22.

D'Antonio, C. M., and P. M. Vitousek. "Biological Invasions by Exotic Grasses, the Grass/Fire Cycle and Global Change." *Annual Review of Ecology and Systematics* 23 (1992): 63–87.

Darling, J. A. "How to Learn to Stop Worrying and Love Environmental DNA Monitoring." *Aquatic Ecosystem Health and Management* 22 (2020): 440–51.

Darlington, W. *Memorials of John Bartram and Humphry Marshall*. Philadelphia: Lindsay and Blakiston, 1849.

Darnovsky, M. "Stories Less Told: Histories of US Environmentalism." *Socialist Review* 22, no. 4 (1992): 12–54.

Darwin, C. *Journal of Researches into the Geology and Natural History of the Various Countries Visited by H.M.S. Beagle Under the Command of Captain Fitzroy, R.N. from 1832 to 1836*. London: Henry Colburn, 1839.

Darwin, C. *Journal of Researches into the Natural History and Geology of the Countries Visited during the Voyage of H.M.S. Beagle Round the World Under the Command of Capt. Fitz Roy*. 2nd ed., corrected, with additions. London: John Murray, 1845.

Darwin, C. *On the Origin of Species by Means of Natural Selection, or the Preservation of Favoured Races in the Struggle for Life*. London: John Murray, 1859.

Darwin, C. *On the Origin of Species by Means of Natural Selection, or the Preservation of Favoured Races in the Struggle for life*. 2nd ed. London: John Murray, 1860.

da Silva Alves, J., J. R. Fabricante, L. B. de Oliveira Reis, G. S. de Moraes, and E. K. Silva. "Biological Invasion by *Cenchrus ciliaris* L.: Is There an Impact on Caatinga Composition and Diversity of Herbaceous Stratum?" *Revista de biologia neotropical / Journal of Neotropical Biology* 14 (2017): 101–10.

Dasmann, R. F. *The Destruction of California*. New York: Macmillan, 1965.

Dasmann, R. F. *No Further Retreat: The Fight to Save Florida*. New York: Macmillan, 1971.

Dáttilo, W., and V. Rico-Gray, eds. *Ecological Networks in the Tropics*. Cham, Switz.: Springer, 2018.

David, J. R., and C. Bocquet. "Evolution in a Cosmopolitan Species: Genetic Latitudinal Clines in *Drosophila melanogaster* Wild Populations." *Experientia* 31 (1975): 164–66.

David, P., E. Thébault, O. Anneville, P.-F. Duyck, E. Chapuis, and N. Loeuille. "Impacts of Invasive Species on Food Webs: A Review of Empirical Data." *Advances in Ecological Research* 56 (2017): 1–60.

Davidson, A. M., M. Jennions, and A. B. Nicotra. "Do Invasive Species Show Higher Phenotypic Plasticity than Native Species and, If So, Is It Adaptive? A Meta-Analysis." *Ecology Letters* 14 (2011): 419–31.

Davis, B. D. "A Scientist's Opinion: Science, Fanaticism, and the Law." *Discover* 5, no. 8 (1984): 24–25.

Davis, F. R. *Banned: A History of Pesticides and the Science of Toxicology*. New Haven, CT: Yale University Press, 2014.

Davis, M. A. *Invasion Biology*. Oxford: Oxford University Press, 2009.

Davis, M. A. "Invasion Biology 1958–2005: The Pursuit of Science and Conservation." In *Conceptual Ecology and Invasion Biology: Reciprocal Approaches to Nature*, edited by M. W. Cadotte, S. M. McMahon, and T. Fukami, 35–64. Dordrecht, Neth.: Springer, 2006.

Davis, M. A. "Let's Welcome a Variety of Voices to Invasion Biology." *Conservation Biology* 34 (2020): 1329–30.

Davis, M. A., and M. K. Chew. "'The Denialists Are Coming!' Well, Not Exactly: A Response to Russell and Blackburn." *Trends in Ecology and Evolution* 32 (2017): 229–30.

Davis, M. A., M. K. Chew, R. J. Hobbs, A. E. Lugo, J. J. Ewel, et al. "Don't Judge Species on Their Origins." *Nature* 474 (2011): 153–54.

Davis, M. A., J. P. Grime, and K. Thompson. "Fluctuating Resources in Plant Communities: A General Theory of Invasibility." *Journal of Ecology* 88 (2000): 528–34.

Davis, M. A., J. Pergl, A. M. Truscott, J. Kollmann, J. P. Bakker, et al. "Vegetation Change: A Reunifying Concept in Plant Ecology." *Perspectives in Plant Ecology, Evolution and Systematics* 7 (2005): 69–76.

Davis, M. A., K. Thompson, and J. P. Grime. "Charles S. Elton and the Dissociation of Invasion Ecology from the Rest of Ecology." *Diversity and Distributions* 7 (2001): 97–102.

Davis, M. A., K. J. Wrage, and P. B. Reich. "Competition Between Tree Seedlings and Herbaceous Vegetation: Support for a Theory of Resource Supply and Demand." *Journal of Ecology* 86 (1998): 652–61.

Dawson, P. J. "Life History Strategy and Evolutionary History of *Tribolium* Flour Beetles." *Evolution* 31 (1977): 227–29.

Day, W. H., R. C. Hedlund, L. B. Saunders, and D. Coutinot. "Establishment of *Peristenus digoneutis* (Hymenoptera: Braconidae), a Parasite of the Tarnished Plant Bug (Hemiptera: Miridae), in the United States." *Environmental Entomology* 19 (1990): 1528–33.

Day, W. H., K. J. Tilmon, R. F. Romig, A. T. Eaton, and K. D. Murray. "Recent Range Expansions of *Peristenus digoneutis* (Hymenoptera: Braconidae), a Parasite of the Tarnished Plant Bug (Hemiptera: Miridae), and High Temperatures Limiting Its Geographic Distribution in North America." *Journal of the New York Entomological Society* 108 (2000): 326–31.

Dayton, P. K. "Experimental Evaluation of Ecological Dominance in a Rocky Intertidal Algal Community." *Ecological Monographs* 45 (1975): 137–59.

Dayton, P. K. "Toward an Understanding of Community Resilience and the Potential Effects of Enrichments to the Benthos at McMurdo Sound, Antarctica." In *Proceedings of the Colloquium on Conservation Problems in Antarctica*, edited by B. C. Parker, 81–96. Lawrence, KS: Allen Press, 1972.

Deacon, A. E., and A. E. Magurran. "How Behaviour Contributes to the Success of an Invasive Poeciliid Fish: The Trinidadian Guppy (*Poecilia reticulata*) as a Model Species." In Weis and Sol, *Biological Invasions and Animal Behaviour*, 266–90.

Deacon, J. E., C. Hubbs, and B. J. Zahuranec. "Some Effects of Introduced Fishes on the Native Fish Fauna of Southern Nevada." *Copeia* 1964 (1964): 384–88.

Dean, W. *A Botânica e a política imperial: Introdução e adaptação de plantas no Brasil colonial e imperial.* São Paulo: Universidade de São Paulo, Instituto de Estudos Avançados, 1992.

Dean, W. *A Ferro e fogo: A História e a devastação da mata atlântica brasileira.* São Paulo: Companhia das Letras, 1996.

DeBach, P. "The Competitive Displacement and Coexistence Principles." *Annual Review of Entomology* 11 (1966): 183–212.

DeBach, P., C. A. Fleschner, and E. J. Dietrick. "A Biological Check Method for Evaluating the Effectiveness of Entomophagous Insects." *Journal of Economic Entomology* 44 (1951): 763–66.

DeBach, P., and D. Rosen. *Biological Control by Natural Enemies.* 2nd ed. Cambridge: Cambridge University Press, 1991.

DeBach, P., and R. A. Sundby. "Competitive Displacement Between Ecological Homologues." *Hilgardia* 34 (1963): 105–66.

de Burgh-Edwardes, S. B. *History of Mauritius 1507–1914.* London: East and West, 1921.

De Candolle, A. P. *Essai élémentaire de géographie botanique.* Strasbourg: F. G. Levrault, 1820.

De Candolle, A. L. *Géographie botanique raisonnée.* Paris: Victor Masson; Geneva: J. Kessmann, 1855.

Dehling, D. M. "The Structure of Ecological Networks." In Dáttilo and Rico-Gray, *Ecological Networks in the Tropics*, 29–42.

de la Vega, S. G. *Invasión en Patagonia.* Buenos Aires: Silvestre, 2005.

de Queiroz, A. *The Monkey's Voyage: How Improbable Journeys Shaped the History of Life.* New York: Basic Books, 2014.

Despott, G. "I nostri rettili." *Archivium Melitense* 2 (1913): 93–96.

Despott, G. "The Reptiles of the Maltese Islands." *Zoologist* (London) 19, no. 891 (1915): 321–27.

Detmer, W. "Zum Problem der Vererbung." *Archiv für die gesamte Physiologie des Menschen und der Tiere* 41, no. 1 (1887): 203–15.

Detry, C., J. L. Cardoso, J. H. Mora, M. Bustamante-Álvarez, A. M. Silva, et al. "Did the Romans Introduce the Egyptian Mongoose (*Herpestes ichneumon*) into the Iberian Peninsula?" *Science of Nature* 105 (2018): 63.

Detwyler, T. R., ed. *Man's Impact on Environment.* New York: McGraw-Hill, 1971.

Deutsch, C. A., J. J. Tewksbury, R. B. Huey, K. S. Sheldon, C. K. Ghalambor, et al. "Impacts of Climate Warming on Terrestrial Ectotherms across Latitude." *Proceedings of the National Academy of Sciences of the United States of America* 105 (2008): 6668–72.

Deutscher, G. *The Unfolding of Language: An Evolutionary Tour of Mankind's Greatest Invention.* New York: Metropolitan Books, 2005.

De Vanna, K. M., B. L. Bodamer, C. G. Wellington, E. Hammer, C. M. Mayer, and J. M. Bossenbroek. "An Alternative Hypothesis to Invasional Meltdown in the Laurentian Great Lakes Region: General Facilitation by *Dreissena.*" *Journal of Great Lakes Research* 37 (2011): 632–41.

Devine, R. S. "Botanical Barbarians." *Sierra* 79, no. 1 (1994): 50–57, 71.

Devine, R. S. *Alien Invasion: America's Battle with Non-Native Animals and Plants.* Washington, DC: National Geographic Society, 1998.

De Vos, A., R. H. Manville, and R. G. Van Gelder. "Introduced Mammals and Their Influence on Native Biota." *Zoologica (Contributions of the New York Zoological Society)* 41 (1956): 163–94.

Dewey, S. A., K. P. Price, and D. Ramsey. "Satellite Remote Sensing to Predict Potential Distribution of Dyers Woad (*Isatis tinctoria*)." *Weed Technology* 5 (1991): 479–84.

Diagne, C., B. Leroy, R. E. Gozlan, A. C. Vaissière, C. Assailly, et al. "InvaCost, a Public Database of the Economic Costs of Biological Invasions Worldwide." *Scientific Data* 7, no. 1 (2020): 1–12.

Diagne, C., B. Leroy, A. C. Vaissière, R. E. Gozlan, D. Roiz, et al. "High and Rising Economic Costs of Biological Invasions Worldwide." *Nature* 592 (2021): 571–76.

Diamond, A. "Seabirds of the Seychelles, Indian Ocean." In Nettleship, Burger, and Gochfeld, *Seabirds on Islands*, 258–67.

Diamond, J. M. "The Avifaunas of Rennell and Bellona Islands." *Natural History of Rennell Island, British Solomon Islands* 8 (1984): 127–68.

Diamond, J. M. "Easter's End." *Discover* 9 (1995): 62–69.

Diamond, J. M. *Guns, Germs, and Steel: The Fates of Human Societies.* New York: W. W. Norton, 1997.

Diamond, J. M. "Historic Extinctions: A Rosetta Stone for Understanding Prehistoric Extinctions." In Martin and Klein, *Quaternary Extinctions*, 824–62.

Diamond, J. M. "The Island Dilemma: Lessons of Modern Biogeographic Studies for the Design of Natural Reserves." *Biological Conservation* 7 (1975): 129–46.

Diamond, J. M. "Overview of Recent Extinctions." In *Conservation in the Twenty-First Century*, edited by D. Western and M. C. Pearl, 37–69. New York: Oxford University Press, 1984.

Diamond, S. E., L. Chick, A. Perez, S. A. Strickler, and R. A. Martin. "Rapid Evolution of Ant Thermal Tolerance Across an Urban–Rural Temperature Cline." *Biological Journal of the Linnean Society* 121 (2017): 248–57.

Díaz-Ferguson, E. E., and G. R. Moyer. "History, Applications, Methodological Issues and Perspectives for the Use Environmental DNA (eDNA) in Marine and Freshwater Environments." *Revista de Biología Tropical* 62 (2014): 1273–84.

di Castri, F. "History of Biological Invasions with Emphasis on the Old World." In Drake et al., *Biological Invasions: A Global Perspective*, 1–30.

di Castri, F., A. J. Hansen, and M. Debussche, eds. *Biological Invasions in Europe and the Mediterranean Basin*. Dordrecht, Neth.: Kluwer, 1990.

di Castri, F., and H. A. Mooney, eds. *Mediterranean Type Ecosystems: Origin and Structure*. Berlin: Springer, 1973.

Dick, J. T. A., M. E. Alexander, J. M. Jeschke, A. Ricciardi, H. J. MacIsaac, et al. "Advancing Impact Prediction and Hypothesis Testing in Invasion Ecology Using a Comparative Functional Response Approach." *Biological Invasions* 16 (2014): 735–53.

Dick, J. T. A., M. E. Alexander, A. Ricciardi, C. Laverty, P. O. Downey, et al. "Functional Responses Can Unify Invasion Ecology." *Biological Invasions* 19 (2017): 1667–72.

Dickey, J. W., R. N. Cuthbert, M. Rea, C. Laverty, K. Crane, et al. "Assessing the Relative Potential Ecological Impacts and Invasion Risks of Emerging and Future Invasive Alien Species." *NeoBiota* 40 (2018): 1–24.

Dickman, C. R. "Impact of Exotic Generalist Predators on the Native Fauna of Australia." *Wildlife Biology* 2 (1996): 185–95.

Didham, R. K., J. M. Tylianakis, M. A. Hutchison, R. M. Ewers, and N. J. Gemmell. "Are Invasive Species the Drivers of Ecological Change?" *Trends in Ecology and Evolution* 20 (2005): 470–74.

Diez, J. M., C. M. D'Antonio, J. S. Dukes, E. D. Grosholz, J. D. Olden, et al. "Will Extreme Climatic Events Facilitate Biological Invasions?" *Frontiers in Ecology and the Environment* 10 (2012): 249–57.

Diller, L. V., K. A. Hamm, D. A. Early, D. W. Lamphear, K. M. Dugger, et al. "Demographic Response of Northern Spotted Owls to Barred Owl Removal." *Journal of Wildlife Management* 80 (2016): 691–707.

Diogo, M. P., A. Carneiro, and A. Simões. "The Portuguese Naturalist Correia da Serra (1751–1823) and His Impact on Early Nineteenth-Century Botany." *Journal of the History of Biology* 34 (2001): 353–93.

Diskin, M., K. C. Steiner, and F. V. Hebard. "Recovery of American Chestnut Characteristics Following Hybridization and Backcross Breeding to Restore Blight-Ravaged *Castanea dentata*." *Forest Ecology and Management* 223 (2006): 439–47.

Ditchkoff, S. S., and M. J. Bodenchuk. "Management of Wild Pigs." In *Invasive Wild Pigs in North America: Ecology, Impacts, and Management*, edited by K. C. VerCauteren, J. C. Beasley, S. S. Ditchkoff, J. J. Mayer, G. J. Roloff, and B. K. Strickland, 175–98. Boca Raton, FL: CRC Press, 2020.

DiTomaso, J. M. "WSSA Enters a New Era of Invasive Weed Communication." *Invasive Plant Science and Management* 1 (2008): 1–2.

DiTomaso, J. M., R. A. van Steenwyk, R. M. Nowierski, L. A. Meyerson, O. C. Doering, et al. "Addressing the Needs for Improving Classical Biological Control Programs in the USA." *Biological Control* 106 (2017): 35–39.

Dlugosch, K. M., and I. M. Parker. "Founding Events in Species Invasions: Genetic Variation, Adaptive Evolution, and the Role of Multiple Introductions." *Molecular Ecology* 17 (2008): 431–49.

Dodd, A. P. *The Biological Campaign Against Prickly-Pear*. Brisbane: Commonwealth Prickly Pear Board, 1940.

Dodd, J. A., J. T. A. Dick, M. E. Alexander, C. MacNeil, A. M. Dunn, and D. C. Aldridge. "Predicting the Ecological Impacts of a New Freshwater Invader: Functional Responses and Prey Selectivity of the 'Killer Shrimp,' *Dikerogammarus villosus*, Compared to the Native *Gammarus pulex*." *Freshwater Biology* 59 (2014): 337–52.

Doherty, T. S., C. R. Dickman, C. N. Johnson, S. M. Legge, E. G. Ritchie, and J. C. Woinarski. "Impacts and Management of Feral Cats *Felis catus* in Australia." *Mammal Review* 47 (2017): 83–97.

Doherty, T. S., A. S. Glen, D. G. Nimmo, E. G. Ritchie, and C. R. Dickman. "Invasive Predators and Global Biodiversity Loss." *Proceedings of the National Academy of Sciences of the United States of America* 113 (2016): 11261–65.

Doley, D. "Parthenium Weed (*Parthenium hysterophorus* L.): Gas Exchange Characteristics as a Basis for Prediction of Its Geographical Distribution." *Australian Journal of Agricultural Research* 28 (1977): 449–60.

Dolin, E. J. *Snakehead: A Fish Out of Water*. Washington, DC: Smithsonian Institution, 2003.

Dopman, E. B., P. S. Robbins, and A. Seaman. "Components of Reproductive Isolation Between North American Pheromone Strains of the European Corn Borer." *Evolution* 64 (2010): 881–902.

Dorcas, M. E., J. D. Willson, R. N. Reed, R. W. Snow, M. R. Rochford, et al. "Severe Mammal Declines Coincide with Proliferation of Burmese Pythons in Everglades National Park." *Proceedings of the National Academy of Sciences of the United States of America* 109 (2012): 2418–22.

Dormann, C., J. Fründ, N. Blüthgen, and B. Gruber. "Indices, Graphs and Null Models: Analyzing Bipartite Ecological Networks." *Open Ecology Journal* 2 (2009): 7–24.

Dormann, C., B. Gruber, and J. Fründ. "Introducing the Bipartite Package: Analysing Ecological Networks." *R News* 8 (2008): 8–11.

Dougherty, M. M., E. R. Larson, M. A. Renshaw, C. A. Gantz, S. P. Egan, et al. "Environmental DNA (eDNA) Detects the Invasive Rusty Crayfish *Orconectes rusticus* at Low Abundances." *Journal of Applied Ecology* 53 (2016): 722–32.

Dowler, R. C., D. S. Carroll, and C. W. Edwards. "Rediscovery of Rodents (Genus *Nesoryzomys*) Considered Extinct in the Galápagos Islands." *Oryx* 34 (2000): 109–17.

Downey, P. O., and D. M. Richardson. "Alien Plant Invasions and Native Plant Extinctions: A Six-Threshold Framework." *AoB Plants* 8 (2016): plw047. https://doi.org/10.1093/aobpla/plw047.

Downs, C. C., C. R. Fredenberg, V. S. D'Angelo, C. C. Muhlfeld, and B. McKeon. *Glacier National Park Fisheries Monitoring and Management Report, 2014*. West Glacier, MT: National Park Service, 2015.

Drake, J. A., H. A. Mooney, F. di Castri, R. H. Groves, F. J. Kruger, et al., eds. *Biological Invasions: A Global Perspective*. Chichester: Wiley, 1989.

Drake, J. A., and M. Williamson. "Invasions of Natural Communities." *Nature* 319 (1986): 718–19.

Dreistadt, S. H., and D. C. Weber. "Gypsy Moth in the Northeast and Great Lakes States." In Dahlsten and Garcia, *Eradication of Exotic Pests*, 229–56.

Drent, R. H. "Biological Invasions: Birds and Mammals." *Proceedings of the Koninklijke Nederlandse Akademie van Wetenschappen, Series C: Biological and Medical Sciences* 90 (1987): 39–44.

Driscoll, C. A., J. Clutton-Brock, A. C. Kitchener, and S. J. O'Brien. "The Taming of the Cat." *Scientific American* 300 (2009): 68–75.

Driscoll, D. A., and M. J. Watson. "Science Denialism and Compassionate Conservation: Response to Wallach et al. 2018." *Conservation Biology* 33 (2019): 777–80.

Dritschilo, W. *Magnificent Failure: Frank Egler and the Greening of American Ecology*. Proctor, VT: Samizdat, 2016.

Druett, J. *Exotic Intruders: The Introduction of Plants and Animals into New Zealand.* Auckland: Heinemann, 1983.

Dueñas, M.-A., H. J. Ruffhead, N. H. Wakefield, P. D. Roberts, D. J. Hemming, and H. Diaz-Soltero. "The Role Played by Invasive Species in Interactions with Endangered and Threatened Species in the United States: A Systematic Review." *Biodiversity and Conservation* 27 (2018): 3171–83.

Duffey, E., and M. B. Usher, eds. "Biological Invasions of Nature Reserves." Special issue, *Biological Conservation* 44 (1988): 1–135.

Duffey, E., and A. S. Watt. *The Scientific Management of Animal and Plant Communities for Conservation.* Oxford: Blackwell, 1971.

Duffy, G. A., B. W. T. Coetzee, G. Latombe, A. H. Akerman, M. A. McGeoch, et al. "Barriers to Globally Invasive Species Are Weakening Across the Antarctic." *Diversity and Distributions* 23 (2017): 982–96.

Duffy, S. "*Silent Spring* and *A Sand County Almanac*: The Two Most Significant Environmental Books of the 20th Century." *Nature Study* 44, no. 2–3 (1991): 6–8.

Dugovich, B., and M. Ruder. "Rabbit Hemorrhagic Disease Virus 2 Reported in Wild Lagomorphs." *SCWDS Briefs* 36, no. 1 (2020): 5–6.

Duncan, I. "The Changing Concept of Animal Sentience." *Applied Animal Behavior Science* 100 (2006): 11–19.

Duncan, R. P. "The Role of Competition and Introduction Effort in the Success of Passeriform Birds Introduced to New Zealand." *American Naturalist* 149 (1997): 903–15.

Dunlap, R. E., and A. G. Mertig. "The Evolution of the U.S. Environmental Movement from 1970 to 1990: An Overview." In *American Environmentalism: The U.S. Environmental Movement, 1970–1990,* edited by R. E. Dunlap and A. G. Mertig, 1–10. Washington, DC: Taylor and Francis, 1992.

Dunlap, T. R. "The Gypsy Moth: A Study in Science and Public Policy." *Journal of Forest History* 24 (1980): 116–26.

Dunne, J. A., R. J. Williams, and N. D. Martinez. "Network Structure and Biodiversity Loss in Food Webs: Robustness Increases with Connectance." *Ecology Letters* 5 (2002): 558–67.

Dupont, Y. L., K. Trøjelsgaard, M. Hagen, M. V. Henriksen, J. M. Olesen, et al. "Spatial Structure of an Individual-Based Plant-Pollinator Network." *Oikos* 123 (2014): 1301–10.

Du Puy, W. A. "The Insects Are Winning." *Harper's Monthly Magazine* 150 (1925): 435–40.

Dutech, C., B. Barres, J. Bridier, C. Robin, M. G. Milgroom, and V. Ravigné. "The Chestnut Blight Fungus World Tour: Successive Introduction Events from Diverse Origins in an Invasive Plant Fungal Pathogen." *Molecular Ecology* 21 (2012): 3931–46.

Dybdahl, M. F., and S. L. Kane. "Adaptation vs. Phenotypic Plasticity in the Success of a Clonal Invader." *Ecology* 86 (2005): 1592–601.

Dzialowski, A. R. "Experimental Effect of Consumer Identity on the Invasion Success of a Non-Native Cladoceran." *Hydrobiologia* 652 (2010): 139–48.

Eamens, A., M. B. Wang, N. A. Smith, and P. M. Waterhouse. "RNA Silencing in Plants: Yesterday, Today, and Tomorrow." *Plant Physiology* 147 (2008): 456–68.

Earth Resources Observation and Science (EROS) Center. "USGS EROS Archive—Digital Elevation—HYDRO1K." US Geological Survey. July 11, 2018. https://www.usgs.gov/centers/eros/science/usgs-eros-archive-digital-elevation-hydro1k?qt-science_center_objects=0#qt-science_center_objects.

Eason, C. T., A. Miller, D. B. MacMorran, and E. C. Murphy. "Toxicology and Ecotoxicology of Para-Aminopropiophenone (PAPP)—A New Predator Control Tool for Stoats and Feral Cats in New Zealand." *New Zealand Journal of Ecology* 38 (2014): 177–88.

"Ecology of Biological Invasions." *SCOPE Newsletter* 23 (1985): 1–5.

EDDMapS. "EDDMapS: Find, Map, Track." University of Georgia Center for Invasive Species and Ecosystem Health. Accessed December 11, 2024. https://www.eddmaps.org/.

Edgington, R. "The Safari of the Southwest: Hunting, Science, and the African Oryx on White Sands Missile Range, New Mexico, 1969–2006." *Western Historical Quarterly* 40 (2009): 469–91.

"Editorial Statement." *Biological Conservation* 1 (1968): 1–3.

Edwards, F. B. "Red Squirrel Disease." *Veterinary Record* 74 (1962): 739–41.

Egan, Dan. *The Death and Life of the Great Lakes.* New York: W. W. Norton, 2017.

Egan, Dave, and W. H. Tishler. "Jens Jensen, Native Plants, and the Concept of Nordic Superiority." *Landscape Journal* 18 (1999): 11–29.

Egan, S. P., G. J. Ragland, L. Assour, T. H. Q. Powell, G. R. Hood, S. Emrich, et al. "Experimental Evidence of Genome-Wide Impact of Ecological Selection During Early Stages of Speciation-with-Gene-Flow." *Ecology Letters* 18 (2015): 817–25.

Egerton, F. N. "Changing Concepts in the Balance of Nature." *Quarterly Review of Biology* 48 (1973): 322–50.

Egerton, F. N. "Hewett C. Watson, Great Britain's First Phytogeographer." *Huntia* 3 (1979): 87–102.

Egerton, F. N. "A History of the Ecological Sciences, Part 25: American Naturalists Explore Eastern North America; John and William Bartram." *Bulletin of the Ecological Society of America* 83 (2007): 253–68.

Egerton, F. N. "History of Ecological Sciences, Part 44: Phytopathology During the 1800s." *Bulletin of the Ecological Society of America* 93 (2012): 303–39.

Egerton, F. N. "History of Ecological Sciences, Part 54: Succession, Community, and Continuum." *Bulletin of the Ecological Society of America* 96 (2015): 426–74.

Egerton, F. N. *Roots of Ecology: Antiquity to Haeckel.* Berkeley: University of California Press, 2012.

Egerton, F. N. "Understanding Food Chains and Food Webs, 1700–1970." *Bulletin of the Ecological Society of America* 88 (2007): 50–69.

Egido, C. "Video Conferencing as a Technology to Support Group Work: A Review of Its Failures." *Proceedings of the 1988 ACM Conference on Computer-Supported Cooperative Work,* January 1988, 13–24.

Egler, F. E. "Arid Southeast Oahu Vegetation, Hawaii." *Ecological Monographs* 17 (1947): 383–435.

Egler, F. E. "Herbicide Effects in Connecticut Vegetation, 1949." *Botanical Gazette* 112 (1950): 76–85.

Egler, F. E. "Indigene Versus Alien in the Development of Arid Hawaiian Vegetation." *Ecology* 23 (1942): 14–23.

Egler, F. E. *The Nature of Naturalization II: The Introduced Flora of Aton Forest, Connecticut.* Claude E. Phillips Herbarium Publication no. 6. Dover, DE: Delaware State College, 1983.

Egler, F. E. "Pointed Perspectives: Pesticides in Our Ecosystem." *Ecology* 47 (1966): 1077–84.

Egler, F. E. "Vegetation Management for Right-of-Ways and Roadsides." *Annual Report of the Smithsonian Institution 1953* (1954): 299–322.

Egunjobi, J. K. "Impact of Some Exotic Plant Species on Soil Characteristics in Tropical Africa." In Ramakrishnan, *Ecology of Biological Invasion in the Tropics*, 75–86.

Ehler, L. E. "Invasion Biology and Biological Control." *Biological Control* 13 (1998): 127–33.

Ehrendorfer, F. "Dispersal Mechanisms, Genetic Systems, and Colonizing Abilities in Some Flowering Plant Families." In Baker and Stebbins, *Genetics of Colonizing Species*, 331–51.

Ehrenfeld, J. G. "Ecosystem Consequences of Biological Invasions." *Annual Review of Ecology, Evolution, and Systematics* 41 (2011): 59–80.

Ehrlich, P. R. "Attributes of Invaders and the Invading Processes: Vertebrates." In Drake et al., *Biological Invasions: A Global Perspective*, 315–28.

Ehrlich, P. R. "Which Animal Will Invade?" In Mooney and Drake, *Ecology of Biological Invasions of North America and Hawaii*, 79–95.

Ehrlich, P. R., and A. H. Ehrlich. *Extinction: The Causes and Consequences of the Disappearance of Species*. New York: Random House, 1981.

Ehrlich, P. R., and I. Hanski, eds. *On the Wings of Checkerspots: A Model System for Population Biology*. Oxford: Oxford University Press, 2004.

Ehrlich, P. R., and H. A. Mooney. "Extinction, Substitution, and Ecosystem Services." *BioScience* 33 (1983): 248–54.

Eiswerth, M. E., and W. S. Johnson. "Managing Nonindigenous Invasive Species: Insights from Dynamic Analysis." *Environmental and Resource Economics* 23 (2002): 319–42.

Eklöf, A., and S. Allesina. "Networks, Ecological." In *Encyclopedia of Theoretical Ecology*, edited by A. Hastings and L. Gross, 470–78. Berkeley: University of California Press, 2012.

"*Elaeagnus angustifolia*: Oleander." *Bulletin of Foreign Plant Introductions*, no. 68 (1911): 487.

Elbein, A. "The Texotics." *Texas Observer*, June 22, 2020. www.texasobserver.org/the-texotics/.

Eliason, E. J., T. D. Clark, M. J. Hague, L. M. Hanson, Z. S. Gallagher, et al. "Differences in Thermal Tolerance Among Sockeye Salmon Populations." *Science* 332 (2011): 109–12.

Eliseev, N. V. "Results and Prospects of Acclimatization of Mammals and Birds in the RSFSR." In Yanushevich, *Acclimatization of Animals in the U.S.S.R.*, 58–60.

Elith, J., C. H. Graham, R. P. Anderson, M. Dudík, S. Ferrier, et al. "Novel Methods Improve Prediction of Species' Distributions from Occurrence Data." *Ecography* 29 (2006): 129–51.

Elith, J., M. Kearney, and S. J. Phillips. "The Art of Modelling Range-Shifting Species." *Methods in Ecology and Evolution* 1 (2010): 330–42.

Elith, J., S. J. Phillips, T. Hastie, M. Dudík, Y. E. Chee, and C. J. Yates. "A Statistical Explanation for MaxEnt for Ecologists." *Diversity and Distributions* 17 (2011): 43–57.

Elkin, K. F. "Results of Acclimatization of Game and Fur Animals in Mountain Altai." In Yanushevich, *Acclimatization of Animals in the U.S.S.R.*, 61–62.

Elkinton, J. S., and G. H. Boettner. "Benefits and Harm Caused by the Introduced Generalist Tachinid, *Compsilura concinnata*, in North America." *Biocontrol* 57 (2012): 277–88.

Elkinton, J. S., and A. M. Liebhold. "Population Dynamics of Gypsy Moth in North America." *Annual Review of Entomology* 35 (1990): 571–96.

Elkinton, J. S., D. Parry, and G. H. Boettner. "Implicating an Introduced Generalist Parasitoid in the Invasive Browntail Moth's Enigmatic Demise." *Ecology* 87 (2006): 2664–72.

Ellis, B. K., J. A. Stanford, D. Goodman, C. P. Stafford, D. L. Gustafson, et al. "Long-Term Effects of a Trophic Cascade in a Large Lake Ecosystem." *Proceedings of the National Academy of Sciences of the United States of America* 108 (2011): 1070–75.

Ellstrand, N. C., and K. Schierenbeck. "Hybridization as a Stimulus for the Evolution of Invasiveness in Plants?" *Proceedings of the National Academy of Sciences of the United States of America* 97 (2000): 7043–50.

Elmendorf, S., J. Byrnes, A. Wright, S. Olyarnik, R. Fisher, and L. Chamberlin, dirs. *Fear and Fishing in Lake Davis*. Davis, CA: Flag in the Ground Productions, 2005. DVD.

Elton, C. S. "Alien Invaders: An Animal Census in Britain." Letter to the editor, *Times* (London), May 6, 1933.

Elton, C. S. *Animal Ecology*. New York: Macmillan, 1927.

Elton, C. S. *Animal Ecology and Evolution*. Oxford: Clarendon Press, 1930.

Elton, C. S. "Balance and Barrier." *Listener* 57 (1957): 514–15, 556–57, 596–97, 600.

Elton, C. S. "The Biological Cost of Modern Transport." *Journal of Animal Ecology* 13 (1944): 87–88.

Elton, C. S. *The Ecology of Animals*. London: Methuen, 1933.

Elton, C. S. *The Ecology of Invasions by Animals and Plants*. London: Methuen, 1958.

Elton, C. S. *The Ecology of Invasions by Animals and Plants*. 2nd ed. Cham, Switz.: Springer Nature, 2020.

Elton, C. S. *Exploring the Animal World*. London: Allen & Unwin, 1933.

Elton, C. S. "The First 30 Years of the Bureau of Animal Population." Unpublished talk, transcribed by C. Pond, 2014. Charles Sutherland Elton Papers, Folder A. 32, MS.Eng.c3326 A47, Weston Library, Oxford University, Oxford, 1962.

Elton, C. S. "Nature Sanctuaries." *Listener* 9 (1933): 779–80.

Elton, C. S. "A New Invader." *Journal of Animal Ecology* 5 (1936): 188–92.

Elton, C. S. *The Pattern of Animal Communities*. London: Methuen, 1966.

Elton, C. S. "Plagues of Animals." *Listener* 9 (1933): 828–30.

Elton, C. S. "Small Adventures: A Private Record of the Early Life of Charles Sutherland Elton." Charles Sutherland Elton Papers, Folder A. 32, Ms. Eng. C. 3326, Weston Library, Oxford University, Oxford, 1955.

Elton, C. S. "Three Hundred Years of Conservation." *Journal of Animal Ecology* 7 (1938): 402–3.

Elwood, E. R., G. B. Paul, J. Ahn, K. Golembiewski, L. M. Higgins, et al. "Citizen Science Needs a Name Change." *Trends in Ecology and Evolution* 38 (2023): 485–89.

Emer, C., I. P. Vaughan, S. Hiscock, and J. Memmott. "The Impact of the Invasive Alien Plant, *Impatiens glandulifera*, on Pollen Transfer Networks." *PLoS One* 10 (2015): 16.

Emery, S. M., and K. L. Gross. "Dominant Species Identity Regulates Invasibility of Old-Field Plant Communities." *Oikos* 115 (2006): 549–58.

Encarnação, J., M. A. Teodósio, and P. Morais. "Citizen Science and Biological Invasions: A Review." *Frontiers in Environmental Science* 8 (2021): 602980.

Enders, M., F. Havemann, and J. M. Jeschke. "A Citation-Based Map of Concepts in Invasion Biology." *NeoBiota* 47 (2019): 23–42.

Enders, M., F. Havemann, F. Ruland, M. Bernard-Verdier, J. A. Catford, et al. "A Conceptual Map of Invasion Biology: Integrating Hypotheses into a Consensus Network." *Global Ecology and Biogeography* 29 (2020): 978–91.

Enders, M., M. T. Hütt, and J. M. Jeschke. "Drawing a Map of Invasion Biology Based on a Network of Hypotheses." *Ecosphere* 9, no. 3 (2018): e02146.

Engler, R., and A. Guisan. "MIGCLIM: Predicting Plant Distribution and Dispersal in a Changing Climate." *Diversity and Distributions* 15 (2009): 590–601.

Engler, R., W. Hordijk, and A. Guisan. "The MIGCLIM R Package: Seamless Integration of Dispersal Constraints into Projections of Species Distribution Models." *Ecography* 35 (2012): 872–78.

Enloe, S. F., N. J. Loewenstein, D. W. Held, L. Eckhardt, and D. K. Lauer. "Impacts of Prescribed Fire, Glyphosate, and Seeding on Cogongrass, Species Richness, and Species Diversity in Longleaf Pine." *Invasive Plant Science and Management* 6 (2013): 536–44.

Enserink, M. "GM Mosquito Trial Alarms Opponents, Strains Ties in Gates-Funded Project." *Science* 330 (2010): 1030–31.

Entrikin, J. N., and S. D. Brunn, eds. *Reflections on Richard Hartshorne's "The Nature of Geography."* Washington, DC: Association of American Geographers, 1989.

Environmental Protection Agency. *Framework for Ecological Risk Assessment.* EPA 630/R-92-001. Washington, DC: Environmental Protection Agency Risk Assessment Forum, 1992.

Environmental Protection Agency. *Guidelines for Ecological Risk Assessment.* EPA/630/R-95/002F. Washington, DC: Environmental Protection Agency Risk Assessment Forum, 1998.

Eppinga, M. B., M. Rietkerk, S. C. Dekker, P. C. De Ruiter, and W. H. Van der Putten. "Accumulation of Local Pathogens: A New Hypothesis to Explain Exotic Plant Invasions." *Oikos* 114 (2006): 168–76.

Errington, P. L. "Predation and Vertebrate Populations." *Quarterly Review of Biology* 21 (1946): 144–77, 221–45.

Erskine, J. *The French Period in Nova Scotia, A.D. 1500–1758, and Present Remains: A Historical, Archaeological and Botanical Survey.* 1975. Reprint, Belle River, PEI: Reg Porter, 2019. https://regporter.com/pei/wp-content/uploads/2019/03/Erskine-Remains-of-the-French-Period-in-Nova-Scotia.pdf.

Espeut, W. B. "On the Acclimatization of the Indian Mungoos in Jamaica." *Proceedings of the Zoological Society of London* 50 (1882): 712–14.

Essig, E. O. "A Sketch History of Entomology." *Osiris* 2 (1936): 80–123.

Essl, F., S. Bacher, P. Genovesi, P. E. Hulme, J. M. Jeschke, et al. "Which Taxa Are Alien? Criteria, Applications, and Uncertainties." *BioScience* 68 (2018): 496–509.

Essl, F., S. Dullinger, P. Genovesi, P. E. Hulme, J. M. Jeschke, et al. "A Conceptual Framework for Range-Expanding Species that Track Human-Induced Environmental Change." *BioScience* 69 (2019): 908–19.

Esterhuizen, L. L., K. G. Mabasa, S. W. van Heerden, H. Czosnek, J. K. Brown, et al. "Genetic Identification of Members of the *Bemisia tabaci* Cryptic Species Complex from South Africa Reveals Native and Introduced Haplotypes." *Journal of Applied Entomology* 137 (2013): 122–35.

Estes, J. A., J. Terborgh, J. S. Brashares, M. E. Power, J. Berger, et al. "Trophic Downgrading of Planet Earth." *Science* 333 (2011): 301–6.

Estévez, R. A., C. B. Anderson, J. C. Pizarro, and M. A. Burgman. "Clarifying Values, Risk Perceptions, and Attitudes to Resolve or Avoid Social Conflicts in Invasive Species Management." *Conservation Biology* 29 (2015): 19–30.

Estoup, A., and T. Guillemaud. "Reconstructing Routes of Invasion Using Genetic Data: Why, How and So What?" *Molecular Ecology* 19 (2010): 4113–30.

Estoup, A., V. Ravigné, R. A. Hufbauer, R. Vitalis, M. Gautier, and B. Facon. "Is There a Genetic Paradox of Biological Invasion?" *Annual Review of Ecology, Evolution, and Systematics* 47 (2016): 51–72.

Esvelt, K. M., and N. J. Gemmell. "Conservation Demands Safe Gene Drive." *PLoS Biology* 15, no. 11 (2017): e2003850.

Esvelt, K. M., A. L. Smidler, F. Catteruccia, and G. M. Church. "Concerning RNA-Guided Gene Drives for the Alteration of Wild Populations." *eLife*, July 17, 2014. https://doi.org/10.7554/eLife.03401.

European Commission. "Invasive Alien Species." Accessed November 4, 2024. https://ec.europa.eu/environment/nature/invasivealien/index_en.htm.

Evans, B. R., P. Kotsakiozi, A. L. Costa-da-Silva, R. S. Ioshino, L. Garziera, et al. "Editorial Expression of Concern: Transgenic *Aedes aegypti* Mosquitoes Transfer Genes into a Natural Population." *Scientific Reports* 10 (2020): 1–2.

Evans, B. R., P. Kotsakiozi, A. L. Costa-da-Silva, R. S. Ioshino, L. Garziera, et al. "Transgenic *Aedes aegypti* Mosquitoes Transfer Genes into a Natural Population." *Scientific Reports* 9 (2019): 13047.

Ewe, S. M., and L. D. S. L. Sternberg. "Water Uptake Patterns of an Invasive Exotic Plant in Coastal Saline Habitats." *Journal of Coastal Research* 23 (2007): 255–64.

Ewel, J. J. "Invasibility: Lessons from South Florida." In Mooney and Drake, *Ecology of Biological Invasions of North America and Hawaii*, 214–30.

Fabricius, J. C. *Philosophia Entomologica*. Hamburg: Carol Ernest Bohnii, 1778.

Fahey, J. "Let's Declare War on These Tired Military Metaphors." *Guardian* (US edition), November 22, 2010. https://www.theguardian.com/media/mind-your-language/2010/nov/22/military-metaphors-mind-your-language.

Fairchild, D. "The Annual Catalogue of Plant Immigrants." *Journal of Heredity* 8 (1917): 500–508.

Fairchild, D. "9227. *Pueraria thunbergiana*. Kudzu." In *Seeds and Plants Imported during the Period from September 1900 to December 1903; Inventory No. 10; Nos. 5501–9896*, vol. 258, bulletin 66. Washington, DC: Bureau of Plant Industry, USDA, 1905.

Fairchild, D. "Note." *Plant Immigrants* 200 (1922): 1826.

Fairchild, D. *The World Was My Garden: Travels of a Plant Explorer*. New York: Charles Scribner's Sons, 1938.

Fairnell, E. H., and J. H. Barrett. "Fur-bearing Species and Scottish Islands." *Journal of Archaeological Science* 34 (2007): 463–84.

Fall, J. J. "Biosecurity and Ecology: Beyond the Nativism Debate." In *Biosecurity: The Socio-politics of Invasive Species and Infectious Disease*, edited by A. Dobson, K. Barker, and S. L. Taylor, 167–81. London: Routledge, 2013.

Farber, O., and R. Kadmon. "Assessment of Alternative Approaches for Bioclimatic Modeling with Special Emphasis on the Mahalanobis Distance." *Ecological Modelling* 160 (2003): 115–30.

Fazey, I., and J. Fischer. "Assisted Colonization Is a Techno-Fix." *Trends in Ecology and Evolution* 24 (2009): 475.

Feder, J. L., S. H. Berlocher, J. B. Roethele, H. Dambroski, J. J. Smith, et al. "Allopatric Genetic Origins for Sympatric Host-Plant Shifts and Race Formation in *Rhagoletis*." *Proceedings of the National Academy of Sciences of the United States of America* 100 (2003): 10314–19.

Feder, J. L., C. A. Chilcote, and G. L. Bush. "Genetic Differentiation Between Sympatric Host Races of *Rhagoletis pomonella*." *Nature* 336 (1988): 61–64.

Feder, J. L., J. B. Roethele, K. Filchak, J. Niedbalski, and J. Romero-Severson. "Evidence for Inversion Polymorphism Related to Sympatric Host Race Formation in the Apple Maggot Fly, *Rhagoletis pomonella*." *Genetics* 163 (2003): 939–53.

Fei, S., J. Phillips, and M. Shouse. "Biogeomorphic Impacts of Invasive Species." *Annual Review of Ecology, Evolution, and Systematics* 45 (2014): 69–87.

Feilden, H. W. "Notes on the Terrestrial Mammals of Barbados." *Zoologist* 14 (1890): 52–55.

Feit, B., A. Feit, and M. Letnic. "Apex Predators Decouple Population Dynamics Between Mesopredators and Their Prey." *Ecosystems* 22 (2019): 1606–17.

Felker-Quinn, E., J. A. Schweitzer, and J. K. Bailey. "Meta-Analysis Reveals Evolution in Invasive Plant Species but Little Support for Evolution of Increased Competitive Ability (EICA)." *Ecology and Evolution* 3 (2013): 739–51.

Felt, E. P. "Recent Observations upon European Insects in America." *Journal of Economic Entomology* 3 (1910): 341–43.

Fendley, G. C. "McCoy, Sir Frederick (1817–1899)." In *Australian Dictionary of Biography*. National Centre of Biography, Australian National University, 2006. http://adb.anu.edu.au/biography/mccoy-sir-frederick-4069/text6491.

Feng, X., and M. Papeş. "Can Incomplete Knowledge of Species' Physiology Facilitate Ecological Niche Modelling? A Case Study with Virtual Species." *Diversity and Distributions* 23 (2017): 1157–68.

Feng, X., and M. Papeş. "Physiological Limits in an Ecological Niche Modeling Framework: A Case Study of Water Temperature and Salinity Constraints of Freshwater Bivalves Invasive in USA." *Ecological Modelling* 346 (2017): 48–57.

Fenner, F., and B. Fantini. *Biological Control of Vertebrate Pests: The History of Myxomatosis—An Experiment in Evolution*. Wallingford, UK: CABI, 1999.

Fenner, F., and J. Ross. "Myxomatosis." In H. Thompson and King, *European Rabbit*, 205–39.

Fernández-Arhex, V., and J. C. Corley. "The Functional Response of Parasitoids and Its Implications for Biological Control." *Biocontrol Science and Technology* 13 (2003): 403–13.

Ficetola, G. F., C. Miaud, F. Pompanon, and P. Taberlet. "Species Detection Using Environmental DNA from Water Samples." *Biology Letters* 4 (2008): 423–25.

Ficetola, G. F., W. Thuiller, and C. Miaud. "Prediction and Validation of the Potential Global Distribution of a Problematic Alien Invasive Species—the American Bullfrog." *Diversity and Distributions* 13 (2007): 476–85.

Fields, H. "Invasion of the Snakeheads." *Smithsonian Magazine* 35, no. 11 (2005): 62–70.

Filgueiras, R. M. C., J. D. A. Mendes, F. W. B. Da Silva, E. P. D. S. Neto, and J. W. D. S. Melo. "Prey Stage Preference and Functional and Numerical Responses of *Neoseiulus barkeri* Hughes (Acari: Phytoseiidae) to Eggs of *Raoiella indica* Hirst (Acari: Tenuipalpidae)." *Systematic and Applied Acarology* 25 (2020): 1147–57.

Fisher, H. I. "The Question of Avian Introductions in Hawaii." *Pacific Science* 2 (1948): 59–64.

Fisher, James, N. Simon, and J. Vincent. *Wildlife in Danger*. New York: Viking, 1969.

Fisher, Judith L., E. J. Veneklaas, H. Lambers, and W. A. Loneragan. "Enhanced Soil and Leaf Nutrient Status of a Western Australian *Banksia* Woodland Community Invaded by *Ehrharta calycina* and *Pelargonium capitatum*." *Plant and Soil* 284 (2006): 253–64.

Fiske, W. F. "Work with Parasites of the Gipsy and Brown-Tail Moths." *Journal of Economic Entomology* 3 (1910): 179–80.

Fitter, A. H., and R. S. R. Fitter. "Rapid Changes in Flowering Time in British Plants." *Science* 296 (2002): 1689–91.

Fitter, R. S. R. *The Ark in Our Midst: The Story of the Introduced Animals of Britain—Birds, Beasts, Reptiles, Amphibians, Fishes*. London: Collins, 1959.

Fitter, R. S. R. "Man's Addition to the British Fauna." *Discovery* 11 (1950): 58–62.

Fitter, R. S. R., and J. Leigh-Pemberton. *Britain's Wildlife: Rarities and Introductions.* London: Nicholas Kay, 1966.

Fitzpatrick, B. M., J. A. Fordyce, M. L. Niemiller, and R. G. Reynolds. "What Can DNA Tell Us About Biological Invasions?" *Biological Invasions* 14 (2012): 245–53.

Fitzpatrick, B. M., J. R. Johnson, D. K. Kump, H. B. Shaffer, J. J. Smith, and S. R. Voss. "Rapid Fixation of Non-Native Alleles Revealed by Genome-Wide SNP Analysis of Hybrid Tiger Salamanders." *BMC Evolutionary Biology* 9 (2009): 176.

Fitzpatrick, E. A., and H. A. Nix. "The Climatic Factor in Australian Grassland Ecology." In *Australian Grasslands,* edited by R. M. Moore, 3–26. Canberra: Australian National University Press, 1970.

Fitzpatrick, M. C., N. J. Gotelli, and A. M. Ellison. "MaxEnt Versus MaxLike: Empirical Comparisons with Ant Species Distributions." *Ecosphere* 4, no. 5 (2013): 55.

Flecker, A. S., and C. R. Townsend. "Community-Wide Consequences of Trout Introduction in New Zealand Streams." *Ecological Applications* 4 (1994): 798–807.

Fleming, P. J. S. "Compassionate Conservation or Misplaced Conservation?" *Invasive Species Council* (blog), May 23, 2018. https://invasives.org.au/blog/compassionate-conservation/.

Flores, D. "Environmentalism and Multiculturalism." In *Reopening the American West,* edited by H. K. Rothman, 24–37. Tucson: University of Arizona Press, 1998.

Flux, J. E. C. "World Distribution." In H. Thompson and King, *European Rabbit,* 8–21.

Food and Agricultural Organization. *International Standards for Phytosanitary Measures: Part 1 Import Regulations—Guidelines for Pest Risk Analysis.* Rome: Secretariat of the International Plant Protection Convention, Food and Agriculture Organization, 1996.

Foote, E. "Circumstances Affecting the Heat of the Sun's Rays." *American Journal of Science and Arts* 22, no. 1 (1856): 382–83.

Forbes, S. A. *The Insect, the Farmer, the Teacher, the Citizen, and the State.* Urbana: Illinois State Laboratory of Natural History, 1915.

Forbes, S. A. "The Lake as a Microcosm." *Bulletin of the Peoria Scientific Association,* 1887, 77–87.

Forbush, E. H., and C. H. Fernald. *The Gypsy Moth.* Boston: Wright & Potter, 1896.

Fosberg, F. R. "Derivation of the Flora of the Hawaiian Islands." In *Insects of Hawaii,* vol. 1, edited by E. C. Zimmerman, 107–19. Honolulu: University of Hawaii Press, 1948.

Foster, J. B. "Late Soviet Ecology and the Planetary Crisis." *Monthly Review* 67, no. 2 (2015): 1–20.

Fourier, J. "Remarques générales sur les températures du globe terrestre et des espaces planétaire." *Annales de chimie et de physique* 27 (1824): 135–67.

Fowler, A. J., D. M. Lodge, and J. F. Hsia. "Failure of the Lacey Act to Protect US Ecosystems Against Animal Invasions." *Frontiers in Ecology and the Environment* 5 (2007): 353–59.

Fox, M. D., and B. J. Fox. "The Susceptibility of Natural Communities to Invasion." In Groves and Burdon, *Ecology of Biological Invasions,* 57–66.

Foxcroft, L. C., B. W. van Wilgen, B. Abrahams, K. J. Esler, and A. Wannenburgh. "Knowing-Doing Continuum or Knowing-Doing Gap? Information Flow Between Researchers and Managers of Biological Invasions in South Africa." In van Wilgen et al., *Biological Invasions in South Africa,* 831–53.

Franceschet, M., and A. Costantini. "The Effect of Scholar Collaboration on Impact and Quality of Academic Papers." *Journal of Informetrics* 4 (2010): 540–53.

Frank, D. M. "Disagreement or Denialism? 'Invasive Species Denialism' and Ethical Disagreement in Science." Supplement, *Synthese* 198, no. 25 (2021): S6085–S6113.

Frank, D. M. “Science and Values in the Biodiversity-Ecosystem Function Debate.” *Biology & Philosophy* 37 (2022): 7.

Frank, D. M., D. Simberloff, J. Bush, A. Chuang, and C. Leppanen. “Logical Fallacies and Reasonable Debates in Invasion Biology: A Response to Guiaşu and Tindale.” *Biology and Philosophy* 34 (2019): 49.

Frank, J. H., and E. D. McCoy. “The Risk of Classical Biological Control in Florida.” *Biological Control* 4 (2007): 151–74.

Frankel, O. H. “Genetic Conservation: Our Evolutionary Responsibility.” *Genetics* 78 (1974): 53–65.

Frankel, O. H. “Variation, the Essence of Life: Sir William Macleay Memorial Lecture.” *Proceedings of the Linnean Society of New South Wales* 95 (1970): 158–69.

Franklin, I. R. “Evolutionary Change in Small Populations.” In Soulé and Wilcox, *Conservation Biology: An Evolutionary-Ecological Perspective*, 135–50.

Franklin, J. “*Diversity and Distributions* Is (Still) a Journal of Conservation Biogeography.” *Diversity and Distributions* 22 (2016): 1–2.

Franklin, J. *Mapping Species Distributions: Spatial Inference and Prediction*. Cambridge: Cambridge University Press, 2009.

Fraser, C. I., A. K. Morrison, A. M. Hogg, E. C. Macaya, E. van Sebille, et al. “Antarctica’s Ecological Isolation Will Be Broken by Storm-Driven Dispersal and Warming.” *Nature Climate Change* 8 (2018): 704–8.

Fraser, H. T. H. Parker, S. Nakagawa, F. Fidler, E. Gould, et al. “Same Data, Different Analysts: Variation in Effect Sizes Due to Analytical Decisions in Ecology and Evolutionary Biology.” OSF, Center for Open Science. Updated February 15, 2024. https://osf.io/mn5aj/.

Frazer, J. F. D. “Introduced Species of Amphibians and Reptiles in Mainland Britain.” *British Journal of Herpetology* 3 (1964): 145–50.

Freeland, D. B. “Some Food Preferences and Aggressive Behavior by Monk Parakeets.” *Wilson Bulletin* 85 (1973): 332–34.

Freinkel, S. *American Chestnut: The Life, Death, and Rebirth of a Perfect Tree*. Berkeley: University of California Press, 2007.

Freitas, W. K., L. M. S. Magalhães, A. S. Resende, F. C. Brasil, L. R. Vivès, et al. “Invasion Impact of *Artocarpus heterophyllus* Lam. (Moraceae) at the Edge of an Atlantic Forest Fragment in the Municipality of Rio de Janeiro, Brazil.” *Bioscience Journal* 33 (2017): 422–33.

Fridley, J. D. “Extended Leaf Phenology and the Autumn Niche in Deciduous Forest Invasions.” *Nature* 485 (2012): 359–62.

Fridley, J. D., J. J. Stachowicz, S. Naeem, D. F. Sax, E. W. Seabloom, et al. “The Invasion Paradox: Reconciling Pattern and Process in Species Invasions.” *Ecology* 88 (2007): 3–17.

Friends of the Earth. “Genetic ‘Extinction’ Technology Rejected by International Groups of Scientists, Conservationists and Environmental Advocates.” News release, September 1, 2016. https://foe.org/news/2016-08-genetic-extinction-technology-rejected-by-international-group-of-scientists/.

Fritts, T. H., and G. H. Rodda. “The Role of Introduced Species in the Degradation of Island Ecosystems: A Case History of Guam.” *Annual Review of Ecology and Systematics* 29 (1998): 113–40.

Frost, C. M., W. J. Allen, F. Courchamp, J. M. Jeschke, W. C. Saul, and D. A. Wardle. “Using Network Theory to Understand and Predict Biological Invasions.” *Trends in Ecology and Evolution* 34 (2019): 831–43.

Fry, B. *Stable Isotope Ecology*. New York: Springer, 2006.

Fu, G., K. C. Condon, M. J. Epton, P. L. Gong, L. Jin, et al. "Female-Specific Insect Lethality Engineered Using Alternative Splicing." *Nature Biotechnology* 25 (2007): 353–57.

Fu, J. W., M. Z. Shi, T. Wang, J. Y. Li, L. Z. Zheng, and G. Wu. "Demography and Population Projection of Flea Beetle, *Agasicles hygrophila* (Coleoptera: Chrysomelidae), Fed on Alligator Weed Under Elevated CO_2." *Journal of Economic Entomology* 109 (2016): 1116–24.

Fujiwara, A., S. Matsuhashi, H. Doi, S. Yamamoto, and T. Minamoto. "Use of Environmental DNA to Survey the Distribution of an Invasive Submerged Plant in Ponds." *Freshwater Science* 35 (2016): 748–54.

Fullaway, D. T. "Report of Entomologist, November, 1924." *Hawaiian Forester and Agriculturist* 21 (1924): 178–79.

Fuller, P. L., L. G. Nico, and J. D. Williams. *Nonindigenous Fishes Introduced into Inland Waters of the United States.* Bethesda, MD: American Fisheries Society, 1999.

Fulling, E. H. "Plant Life and the Law of Man, IV: Barberry, Currant and Gooseberry, and Cedar Control." *Botanical Review* 9 (1943): 483–592.

Funasaki, G. Y., P. Y. Lai, L. M. Nakahara, J. W. Beardsley, and A. K. Ota. "A Review of Biological Control Introductions in Hawaii: 1890 to 1985." *Proceedings of the Hawaiian Entomological Society* 28 (1988): 105–60.

Funk, J. L. "The Physiology of Invasive Plants in Low-Resource Environments." *Conservation Physiology* 1, no. 1 (2013): cot026. https://doi.org/10.1093/conphys/cot026.

Furlong, M. J., D. J. Wright, and L. M. Dosdall. "Diamondback Moth Ecology and Management: Problems, Progress, and Prospects." *Annual Review of Entomology* 58 (2013): 517–41.

Gaertner, M., A. D. Breeyen, C. Hui, and D. M. Richardson. "Impacts of Alien Plant Invasions on Species Richness in Mediterranean-type Ecosystems: A Meta-Analysis." *Progress in Physical Geography* 33 (2009): 319–38.

Galanidi, M., A. Zenetos, and S. Bacher. "Assessing the Socio-Economic Impacts of Priority Marine Invasive Fishes in the Mediterranean with the Newly Proposed SEICAT Methodology." *Mediterranean Marine Science* 19 (2018): 107–23.

Galiana, N., M. Lurgi, J. M. Montoya, and B. C. Lopez. "Invasions Cause Biodiversity Loss and Community Simplification in Vertebrate Food Webs." *Oikos* 123 (2014): 721–28.

Galil, B. S., and A. Zenetos. "A Sea Change: Exotics in the Eastern Mediterranean Sea." In *Invasive Aquatic Species of Europe: Distribution, Impacts and Management*, edited by E. Leppäkoski, S. Gollasch, and S. Olenin, 325–36. Dordrecht, Neth.: Kluwer, 2002.

Galloway, B. T., C. C. Muhlfeld, C. S. Guy, C. C. Downs, and W. A. Fredenberg. "A Framework for Assessing the Feasibility of Native Fish Conservation Translocations: Applications to Threatened Bull Trout." *North American Journal of Fisheries Management* 36 (2016): 754–68.

Galtier, N., B. Nabholz, S. Glémin, and G. D. D. Hurst. "Mitochondrial DNA as a Marker of Molecular Diversity: A Reappraisal." *Molecular Ecology* 18 (2009): 4541–50.

Gama, M., D. Crespo, M. Dolbeth, and P. M. Anastácio. "Ensemble Forecasting of *Corbicula fluminea* Worldwide Distribution: Projections of the Impact of Climate Change." *Aquatic Conservation: Marine and Freshwater Ecosystems* 27 (2017): 675–84.

Gantz, C. A., M. A. Renshaw, D. Erickson, D. M. Lodge, and S. P. Egan. "Environmental DNA Detection of Aquatic Invasive Plants in Lab Mesocosm and Natural Field Conditions." *Biological Invasions* 20 (2018): 2535–52.

Gantz, V. M., N. Jasinskiene, O. Tatarenkova, A. Fazekas, V. M. Macias, et al. "Highly Efficient Cas9-mediated Gene Drive for Population Modification of the Malaria Vector Mosquito *Anopheles stephensi*." *Proceedings of the National Academy of Sciences of the United States of America* 112 (2015): E6736–E6743.

Gardener, M. R., R. Atkinson, and J. L. Rentería. "Eradications and People: Lessons from the Plant Eradication Program in Galapagos." *Restoration Ecology* 18 (2010): 20–29.

Gardens Trust. "Hollyhocks." *Gardens Trust* (blog), May 9, 2020. https://thegardenstrust.blog/2020/09/05/hollyhocks/.

Gardner, A. L. "The United States Biological Survey: A Brief History 1885–1940." In *United States Biological Survey: A Compendium of Its History, Personalities, Impacts, and Conflicts*, Monograph 64, edited by D. J. Schmidly, W. E. Tydeman, and A. L. Gardner, 1–14. Lubbock: Museum of Texas Tech University, 2016.

Gardner, M. G., A. J. Fitch, T. Bertozzi, and A. J. Lowe. "Rise of the Machines—Recommendations for Ecologists When Using Next Generation Sequencing for Microsatellite Development." *Molecular Ecology Resources* 11 (2011): 1093–1101.

Gaskin, J. F., and D. J. Kazmer. "Introgression Between Invasive Saltcedars (*Tamarix chinensis* and *T. ramosissima*) in the USA." *Biological Invasions* 11 (2009): 1121–30.

Gaspar, H., S. Castro, J. M. Grosso-Silva, T. van der Heyden, and J. Loureiro. "Exponential Outspread of *Halyomorpha halys* (Stål, 1855) (Hemiptera: Pentatomidae) in Portugal." *Arquivos entomolóxicos* 26 (2023): 13–18.

Gbedomon, R. C., V. K. Salako, and M. A. Schlaepfer. "Diverse Views Among Scientists on Non-Native Species." *NeoBiota* 54 (2020): 49–69.

GBIRd. "Genetic Biocontrol of Invasive Rodents." Accessed November 19, 2024. https://www.geneticbiocontrol.org/.

Geerts, A. N., J. Vanoverbeke, B. Vanschoenwinkel, W. Van Doorslaer, H. Feuchtmayr, et al. "Rapid Evolution of Thermal Tolerance in the Water Flea *Daphnia*." *Nature Climate Change* 5 (2015): 665–69.

Geller, J. B. "Decline of a Native Mussel Masked by Sibling Species Invasion." *Conservation Biology* 13 (1999): 661–64.

Geller, J. B., J. T. Carlton, and D. A. Powers. "PCR-Based Detection of mtDNA Haplotypes of Native and Invading Mussels on the Northeastern Pacific Coast: Latitudinal Pattern of Invasion." *Marine Biology* 119 (1994): 243–49.

Geller, J. B., E. D. Walton, E. D. Grosholz, and G. M. Ruiz. "Cryptic Invasions of the Crab *Carcinus* Detected by Molecular Phylogeography." *Molecular Ecology* 6 (1997): 901–6.

Genovesi, P., and S. Bertolino. "Human Dimension Aspects in Invasive Alien Species Issues: The Case of the Failure of the Grey Squirrel Eradication Project in Italy." In *The Great Reshuffling: Human Dimensions of Invasive Alien Species*, edited by J. A. McNeely, 113–19. Gland, Switz.: International Union for Conservation of Nature, 2001.

Genovesi, P., C. Carboneras, M. Vilà, and P. Weldon. "EU Adopts Innovative Legislation on Invasive Species: A Step Towards a Global Response to Biological Invasions?" *Biological Invasions* 17 (2015): 1307–11.

Genovesi, P., and S. Natalini. *Per un pugno di ghiande: Le specie invasive che assaltano la Terra*. Florence: Editoriale Scienza, 2018.

Genovesi, P., R. Scalera, S. Brunel, D. Roy, and W. Solarz. *Towards an Early Warning and Information System for Invasive Alien Species (IAS) Threatening Biodiversity in Europe*. Copenhagen: European Environment Agency, 2010.

Genovesi, P., and C. Shine. *European Strategy on Invasive Alien Species: Convention on the Conservation of European Wildlife and Habitats* (Bern Convention). No. 18-137. Strasbourg: Council of Europe, 2004.

George, J. L. "The Program to Eradicate the Imported Fire Ant: Report to the Conservation Foundation and the New York Zoological Society." Unpublished manuscript, 1958.

George, S. "Women in Film Time: Forty Years of the Alien Series (1979–2019)." *IAFOR Journal of Arts & Humanities* 6, no. 2 (2019): 59–68.

Georghiou, G. P., and T. Saito. *Pest Resistance to Pesticides*. New York: Plenum, 1983.

Geptner, V. G. "Kakovy zhe puti obogashcheniia fauny." *Okhota i Okhotnich'e khoziaistvo* 2 (1963): 21–26.

Gerlach, J. D., Jr., and K. J. Rice. "Testing Life History Correlates of Invasiveness Using Congeneric Plant Species." *Ecological Applications* 13 (2003): 167–79.

Ghalambor, C., J. McKay, S. Carroll, and D. Reznick. "Adaptive versus Non-Adaptive Phenotypic Plasticity and the Potential for Contemporary Adaptation in New Environments." *Functional Ecology* 21 (2007): 394–407.

Gibney, E. R., and C. M. Nolan. "Epigenetics and Gene Expression." *Heredity* 105 (2010): 4–13.

Gibson-Reinemer, D. K. "A Vacant Niche: How a Central Ecological Concept Emerged in the 19th Century." *Bulletin of the Ecological Society of America* 96 (2015): 324–35.

Giejsztowt, J., A. T. Classen, and J. R. Deslippe. "Climate Change and Invasion May Synergistically Affect Native Plant Reproduction." *Ecology* 101 (2020): e02913.

Gillespie, G. D. "Hybridization, Introgression, and Morphometric Differentiation Between Mallard (*Anas platyrhynchos*) and Grey Ducks (*Anas superciliosa*) in Otago, New Zealand." *Auk* 102 (1985): 459–69.

Giovas, C. M., G. D. Kamenov, S. M. Fitzpatrick, and J. Krigbaum. "Sr and Pb Isotopic Investigation of Mammal Introductions: Pre-Columbian Zoogeographic Records from the Lesser Antilles, West Indies." *Journal of Archaeological Science* 69 (2016): 39–53.

Giraud, T., J. S. Pedersen, and L. Keller. "Evolution of Supercolonies: The Argentine Ants of Southern Europe." *Proceedings of the National Academy of Sciences of the United States of America* 99 (2002): 6075–79.

Gjullin, C. M. "Probable Distribution of the Mediterranean Fruit Fly (*Ceratitis capitata* Weid.) in the United States." *Ecology* 12 (1931): 248–58.

Glass, G. V. "Primary, Secondary, and Meta-Analysis of Research." *Educational Researcher* 5 (1976): 3–8.

Glass, G. V., B. McGaw, and M. L. Smith. *Meta-Analysis of Social Research*. Beverly Hills, CA: Sage, 1981.

Gleason, H. A. "The Individualistic Concept of the Plant Association." *Bulletin of the Torrey Botanical Club* 53 (1926): 7–26.

Gleason, H. A. "*Tutira* (Book Review)." *Ecology* 3 (1922): 339–40.

Gleditsch, J. M., and T. A. Carlo. "Fruit Quantity of Invasive Shrubs Predicts the Abundance of Common Native Frugivores in Central Pennsylvania." *Diversity and Distributions* 17 (2011): 244–53.

Gleitzman, M. *Toad Rage*. New York: Random House, 2001.

Global Biodiversity Information Facility (GBIF). "What Is GBIF?" Accessed December 5, 2024. https://www.gbif.org/what-is-gbif.

Global Invasive Species Database. "Species Profile: *Poa pratensis*." August 23, 2010. http://www.iucngisd.org/gisd/species.php?sc=1419.

Godin, A. J. *Wild Mammals of New England*. Baltimore: Johns Hopkins University Press, 1977.

Godman, F. D. *Natural History of the Azores or Western Islands*. London: John Van Voorst, 1870.

Godoy, O., A. Saldana, N. Fuentes, F. Valladares, and E. Gianoli. "Forests Are Not Immune to Plant Invasions: Phenotypic Plasticity and Local Adaptation Allow *Prunella vulgaris* to Colonize a Temperate Evergreen Rainforest." *Biological Invasions* 13 (2011): 1615–25.

Goeden, R. D. "Critique and Revision of Harris' Scoring System for Selection of Insect Agents in Biological Control of Weeds." *Protection Ecology* 5 (1983): 287–301.

Goethe, C. M. *Garden Philosopher*. Sacramento: Keystone Press, 1955.

Goethe, C. M. "Immigration from Mexico." In *The Alien in Our Midst*, edited by M. Grant and C. S. Davison, 134–42. New York: Galton Publishing Company, 1930.

Golani, D., and B. Appelbaum-Golani, eds. *Fish Invasions of the Mediterranean Sea: Change and Renewal*. Sofia, Bulgaria: Pensoft, 2010.

Goldberg, C. S., A. Sepulveda, A. Ray, J. Baumgardt, and L. P. Waits. "Environmental DNA as a New Method for Early Detection of New Zealand Mudsnails (*Potamopyrgus antipodarum*)." *Freshwater Science* 32 (2013): 792–800.

Goldschmidt, T. *Darwin's Dreampond*. Cambridge, MA: MIT Press, 1996.

Goldschmidt, T. *Darwin's Hofvijver: Een Drama in Het Victoriameer*. Amsterdam: Prometheus, 1994.

Goldson, S. L., S. Wratten, C. Ferguson, P. J. Gerard, B. I. P. Barratt, et al. "If and When Successful Classical Biological Control Fails." *Biological Control* 72 (2014): 76–79.

Goldstein, J., and M. Zych. "What If We Lose a Hub? Experimental Testing of Pollination Network Resilience to Removal of Keystone Floral Resources." *Arthropod Plant Interactions* 10 (2016): 263–71.

Golley, F. B. *A History of the Ecosystem Concept in Ecology: More than the Sum of the Parts*. New Haven, CT: Yale University Press, 1993.

Gomoiu, M.-T., B. Alexandrov, N. Shadrin, and Y. Zaitsev. "The Black Sea—A Recipient, Donor and Transit Area for Alien Species." In *Invasive Aquatic Species of Europe: Distribution, Impact and Management*, edited by E. Leppäkoski, S. Gollasch, and S. Olenin, 341–50. Dordrecht, Neth.: Kluwer, 2002.

Gomoiu, M.-T., and M. Skolka. "Changements récents dans biodiversité de la Mer Noire dûs aux immigrants." *Geo-Eco-Marina* 1 (1996): 34–47.

Good, R. "A Theory of Plant Geography." *New Phytologist* 30, no. 3 (1931): 149–71.

Goodell, K., I. M. Parker, and G. S. Gilbert. "Biological Impacts of Species Invasions: Implications for Policy Makers." In *Incorporating Biological, Natural, and Social Sciences in Sanitary and Phytosanitary Standards in International Trade*, edited by J. Caswell, 87–117. Washington, DC: National Academies Press, 2000.

Goodman, D. "The Theory of Diversity-Stability Relationships in Ecology." *Quarterly Review of Biology* 50 (1975): 237–66.

Gooley, L. P. "Livingston Stone: Leading 19th Century Fisheries Expert." Adirondack Almanack. February 23, 2016. https://www.adirondackalmanack.com/2016/02/livingston-stone-leading-19th-century-fisheries-expert.html.

Gore, A. *Earth in the Balance: Ecology and the Human Spirit*. Boston: Houghton Mifflin, 1992.

Gorham, E., and J. Kelly. "Multiauthorship, an Indicator of the Trend Toward Team Research in Ecology." *Bulletin of the Ecological Society of America* 95 (2014): 243–49.

Gorman, J. "Putting Nature on the Pill." *New York Times*, August 31, 2004, F1.

Gosling, L. M., and S. J. Baker. "The Eradication of Muskrats and Coypus from Britain." *Biological Journal of the Linnean Society* 38 (1989): 39–51.

Gosling, L. M., and W. J. Sutherland, eds. *Behaviour and Conservation*. Cambridge: Cambridge University Press, 2000.

Gosling, S. D. "From Mice to Men: What Can We Learn About Personality from Animal Research?" *Psychological Bulletin* 127 (2001): 45–86.

Gossard, H. A. *The Cottony Cushion Scale*. Bulletin No. 56. Lake City, FL: Florida Agricultural Experiment Station, 1901.

Gosse, P. *St. Helena 1502–1938*. London: Cassell, 1938.

Gotelli, N. J. *A Primer of Ecology*. Sunderland, MA: Sinauer Associates, 1995.

Gotelli, N. J. "Research Frontiers in Null Model Analysis." *Global Ecology and Biogeography* 10 (2001): 337–43.

Gould, F. "Broadening the Application of Evolutionarily Based Genetic Pest Management." *Evolution* 62 (2008): 500–510.

Gould, S. J. "An Evolutionary Perspective on Strengths, Fallacies, and Confusions in the Concept of Native Plants." In Wolschke-Bulmahn, *Nature and Ideology*, 11–19.

Gould, S. J. "An Evolutionary Perspective on Strengths, Fallacies, and Confusions in the Concept of Native Plants." *Arnoldia* 58, no. 1 (1998): 2–10.

Gould, S. J. *I Have Landed*. Cambridge, MA: Harvard University Press, 2012.

Government of British Columbia. "Assisted Migration Adaptation Trial." Updated May 8, 2024. https://www2.gov.bc.ca/gov/content/industry/forestry/managing-our-forest-resources/tree-seed/forest-genetics/seed-transfer-climate-change/assisted-migration-adaptation-trial.

Government of British Columbia, "Chief Forester's Standards for Seed Use." Updated March 13, 2024. https://www2.gov.bc.ca/gov/content/industry/forestry/managing-our-forest-resources/tree-seed/legislation-standards/chief-forester-s-standards-for-seed-use.

Graf, J. E. "Climate in Relation to Mexican Bean Beetle Distribution." *Journal of Economic Entomology* 18 (1925): 116–21.

Graham, E. H. *The Land and Wildlife*. New York: Oxford University Press, 1947.

Graham, E. H. *Natural Principles of Land Use*. London: Oxford University Press, 1944.

Graham, E. H. "Nature Protection as Part of Land Development." In *Proceedings and Papers of the Sixth Technical Meeting Held in Edinburgh in June, 1956*, 194–201. London: International Union for the Conservation of Nature and Natural Resources, 1957.

Graham, E. H. "Statement to Select Committee on Conservation of Wildlife Resources." In *Conservation of Wildlife, Hearings, House of Representatives*, 217–34. Washington, DC: Government Printing Office, 1946.

Graham, E. H. "Statement to Subcommittee No. 3 of the Committee on Agriculture." In *Price Support for Honey, Hearing, House of Representatives*, 47–50. Washington, DC: Government Printing Office, 1949.

Grandoni, D. "Genetic Engineering Was Meant to Save Chestnut Trees: Then There Was a Mistake." *Washington Post*, December 24, 2023.

Granett, J., M. A. Walker, L. Kocsis, and A. D. Omer. "Biology and Management of Grape Phylloxera." *Annual Review of Entomology* 46 (2001): 387–412.

Grant, B. "The Powers That Might Be$^{-E/kT}$." *Scientist* 21, no. 3 (2007): 43–48.

Grant, M. "The Future of Our Fauna." *Zoological Society Bulletin*, no. 34 (1909): 504–5.

Grant, M. *The Passing of the Great Race; or, The Racial Basis of European History*. New York: Charles Scribner's Sons, 1916.

Grant, M. *The Passing of the Great Race; or, The Racial Basis of European History*. 4th ed. New York: Charles Scribner's Sons, 1936.

Grant, M. *The Rocky Mountain Goat*. New York: New York Zoological Society, 1905.

Graves, A. H. "The Future of the Chestnut Tree." *Popular Science Monthly* 84 (1914): 551–66.

Graves, S., and A. Shapiro. "Exotics as Host Plants of the California Butterfly Fauna." *Biological Conservation* 110 (2003): 413–33.

Gray, A. "The Predominance and Pertinacity of Weeds." *American Journal of Science and Arts* 118 (1879): 161–67.

Gray, A. J., M. J. Crawley, and P. J. Edwards, eds. *Colonization, Succession and Stability.* Oxford: Blackwell Scientific Publications, 1987.

Gray, A. J., A. F. Raybould, and S. L. Brown. "The Environmental Impact of *Spartina anglica*: Past, Present and Predicted." In *Second International Spartina Conference*, edited by K. Patten, 39–45. Olympia: Washington State University, Cooperative Extension, 1997.

Gray, G. M., J. C. Allen, D. E. Burmaster, S. H. Gage, J. K. Hammitt, et al. "Principles for Conduct of Pest Risk Analyses: Report of an Expert Workshop." *Risk Analysis* 18 (1998): 773–80.

Gray, M. E., T. W. Sappington, N. J. Miller, J. Moeser, and M. O. Bohn. "Adaptation and Invasiveness of Western Corn Rootworm: Intensifying Research on a Worsening Pest." *Annual Review of Entomology* 54 (2009): 303–21.

Gray, M. M., M. D. Parmenter, C. Hogan, I. Ford, R. J. Cuthbert, et al. "Genetics of Rapid and Extreme Size Evolution in Island Mice." *Genetics* 201 (2015): 213–28.

Gray, M. M., D. Wegmann, R. J. Haasl, M. A. White, S. I. Gabriel, et al. "Demographic History of a Recent Invasion of House Mice on the Isolated Island of Gough." *Molecular Ecology* 23 (2014): 1923–39.

Green, N., J. R. Britton, M. Bentley, P. Stebbing, and D. Andreou. "Dominance, Reproductive Behaviours and Female Mate Choice in Sterilised Versus Non-Sterilised Invasive Male Crayfish." *Aquatic Ecology* 54 (2020): 813–22.

Green, P. T., D. J. O'Dowd, K. L. Abbott, M. Jeffery, K. Retallick, and R. MacNally. "Invasional Meltdown: Invader–Invader Mutualism Facilitates a Secondary Invasion." *Ecology* 92 (2011): 1758–68.

Green, R. E. "The Influence of Numbers Released on the Outcome of Attempts to Introduce Exotic Bird Species to New Zealand." *Journal of Animal Ecology* 66 (1997): 25–35.

Gregory, W. K. "Biographical Memoir of Frank Michler Chapman, 1864–1945." *Biographical Memoirs of the National Academy of Sciences* 35 (1949): 111–45.

Greig, K., A. Gosling, C. J. Collins, J. Boocock, K. McDonald, et al. "Complex History of Dog (*Canis familiaris*) Origins and Translocations in the Pacific Revealed by Ancient Mitogenomes." *Scientific Reports* 8 (2018): 9130.

Grevstad, F. S. "Experimental Invasions Using Biological Control Introductions: The Influence of Release Size on the Chance of Population Establishment." *Biological Invasions* 1 (1999): 313–23.

Griesbach, R. J. *150 Years of Research at the United States Department of Agriculture: Plant Introduction and Breeding.* Washington, DC: Agricultural Research Service, US Department of Agriculture, 2013.

Griffin, A. S., A. Callen, K. Klop-Toker, R. J. Scanlon, and M. W. Hayward. "Compassionate Conservation Clashes with Conservation Biology: Should Empathy, Compassion, and Deontological Moral Principles Drive Conservation Practice?" *Frontiers in Psychology* 11 (2020): 1139.

Griffin, A. S., D. Guez, I. Federspiel, M. Diquelou, and F. Lermite. "Invading New Environments: A Mechanistic Framework Linking Motor Diversity and Cognition to Establishment Success." In Weis and Sol, *Biological Invasions and Animal Behaviour*, 26–46.

Griffiths, F. P. "Considerations of the Introduction and Distribution of Exotic Fishes in Oregon." *Transactions of the American Fisheries Society* 69 (1940): 240–43.

Griggs, R. F. "Observations on the Behavior of Species at the Edge of Their Range." *Bulletin of the Torrey Botanical Club* 41 (1914): 25–49.

Grime, J. P. "Competitive Exclusion in Herbaceous Vegetation." *Nature* 242 (1973): 344–47.

Grime, J. P. "Control of Species Density in Herbaceous Vegetation." *Journal of Environmental Management* 1 (1973): 151–67.

Grime, J. P. "The C-S-R Model of Primary Plant Strategies: Origins, Implications, and Tests." In *Plant Evolutionary Biology*, edited by L. D. Gottlieb and S. K. Jain, 371–93. London: Chapman & Hall, 1988.

Grime, J. P. *Plant Strategies and Vegetation Processes.* New York: Wiley, 1979.

Grime, J. P. "Vegetation Classification by Reference to Strategies." *Nature* 250 (1974): 26–31.

Grimm, V., and C. Wissel. "Babel, or the Ecological Stability Discussions: An Inventory and Analysis of Terminology and a Guide for Avoiding Confusion." *Oecologia* 109 (1997): 323–34.

Grinnell, J. "The English Sparrow Has Arrived in Death Valley: An Experiment in Nature." *American Naturalist* 53 (1919): 468–72.

Grinnell, J. "The Niche-Relationships of the California Thrasher." *Auk* 34 (1917): 427–33.

Grinnell, J. "Risks Incurred in the Introduction of Alien Game Birds." *Science* 61 (1925): 621–23.

Gröning, G., and J. Wolschke-Bulmahn. "The Native Plant Enthusiasm: Ecological Panacea or Xenophobia?" *Landscape Research* 28 (1993): 75–88.

Gröning, G., and J. Wolschke-Bulmahn. "Response: If the Shoe Fits, Wear It!" *Landscape Journal* 13 (1994): 62–63.

Gröning, G., and J. Wolschke-Bulmahn. "Some Notes on the Mania for Native Plants in Germany." *Landscape Journal* 11 (1992): 116–26.

Groom, M. J., G. K. Meffe, and C. R. Carroll. *Principles of Conservation Biology*. 3rd ed. Sunderland, MA: Sinauer Associates, 2006.

Grosholz, E. D. "Climate Change and Invasive Species." In *Climate Change and Estuaries*, edited by M. J. Kennish, H. W. Paerl, and J. R. Crosswell, 533–44. Boca Raton, FL: CRC Press, 2023.

Grosholz, E. D. "Recent Biological Invasion May Hasten Invasional Meltdown by Accelerating Historical Introductions." *Proceedings of the National Academy of Sciences of the United States of America* 102 (2005): 1088–91.

Grosholz, E. D., S. Lovell, and E. Besedin. "Modeling the Impacts of the European Green Crab on Commercial Shellfisheries." *Ecological Applications* 21 (2011): 915–24.

Grotkopp, E., M. Rejmánek, and T. L. Rost. "Toward a Causal Explanation of Plant Invasiveness: Seedling Growth and Life-History Strategies of 29 Pine (*Pinus*) Species." *American Naturalist* 159 (2002): 396–419.

Grouard, S. "Modes de vie des précolombiens des Antilles françaises: Synthèse des données archéozoologiques." *Les Nouvelles de l'archéologie* 108–9 (2007): 91–101.

Grove, R. H. *Green Imperialism.* Cambridge: Cambridge University Press, 1995.

Groves, R. H. "Invasion of Mediterranean Ecosystems by Weeds." In *Resilience in Mediterranean-Type Ecosystems*, edited by B. Dell, A. J. M. Hopkins, and B. B. Lamont, 129–45. Dordrecht, Neth.: Dr. W. Junk, 1986.

Groves, R. H. "Plant Invasion in Australia: An Overview." In Groves and Burdon, *Ecology of Biological Invasions*, 137–47.

Groves, R. H., and J. J. Burdon, eds. *Ecology of Biological Invasions.* Cambridge: Cambridge University Press, 1986.

Groves, R. H., and F. di Castri, eds. *Biogeography of Mediterranean Invasions.* Cambridge: Cambridge University Press, 1991.

Grunwald, H. A., V. M. Gantz, G. Poplawski, X. S. Xu, E. Bier, and K. L. Cooper. "Super-Mendelian Inheritance Mediated by CRISPR-Cas9 in the Female Mouse Germline." *Nature* 566 (2019): 105–9.

Guerin, G. R. "Invoking Denialism Does Not Strengthen Invasion Science." *Biodiversity and Conservation* 28 (2019): 1939–41.

Guggenheim, D., dir. *An Inconvenient Truth.* 2006; Los Angeles: Paramount Pictures Home Entertainment, 2017. DVD, 97 min.

Guiaşu, R. C., and C. W. Tindale. "Logical Fallacies and Invasion Biology." *Biology and Philosophy* 33 (2018): 34.

Guild, C. *Rainbow in Tahiti.* Garden City, NY: Doubleday, 1948.

Guild, E. "Tahitian Aviculture: Acclimatization of Foreign Birds." *Aviculture Magazine* 3 (1938): 8–11.

Guild, E. "Western Bluebirds in Tahiti." *Aviculture Magazine* 5 (1940): 284–85.

Guiry, E., R. Kennedy, D. Orton, P. Armitage, J. Bratten, et al. "The Ratting of North America: A 350-Year Retrospective on *Rattus* Species Compositions and Competition." *Science Advances* 10, no. 14 (2024). https://doi.org/10.1126/sciadv.adm6755.

Guisan, A., B. Petitpierre, O. Broennimann, C. Daehler, and C. Kueffer. "Unifying Niche Shift Studies: Insights from Biological Invasions." *Trends in Ecology and Evolution* 29 (2014): 260–69.

Gullan, P. J. "Relationships with Ants." In *Soft Scale Insects: Their Biology, Natural Enemies, and Control,* edited by Y. Ben-Dov and C. Hodgson, 351–74. Amsterdam: Elsevier, 1997.

Gullion, G. W. "A Critique Concerning Foreign Game-Bird Introductions." *Wilson Bulletin* 77 (1965): 409–14.

Gupta, P. K., J. K. Roy, and M. Prasad. "Single Nucleotide Polymorphisms: A New Paradigm for Molecular Marker Technology and DNA Polymorphism Detection with Emphasis on Their Use in Plants." *Current Science* 80 (2001): 524535.

Gurevitch, J. "Commentary on Simberloff (2006): Meltdowns, Snowballs and Positive Feedbacks." *Ecology Letters* 9 (2006): 919–21.

Gurevitch, J., P. S. Curtis, and M. H. Jones. "Meta-Analysis in Ecology." *Advances in Ecological Research* 32 (2001): 199–247.

Gurnell, J., P. W. W. Lurz, and C. M. Shuttleworth. "Ecosystem Impacts of an Alien Invader in Europe, the Grey Squirrel *Sciurus carolinensis.*" In Shuttleworth et al., *The Grey Squirrel: Ecology and Management of an Invasive Species in Europe*, 307–26.

Gutekinst, J., R. Andriantsoa, C. Falckenhayn, K. Hanna, W. Stein, et al. "Clonal Genome Evolution and Rapid Invasive Spread of the Marbled Crayfish." *Nature Ecology and Evolution* 2 (2018): 567–73.

Guthrie-Smith, H. *Tutira: The Story of a New Zealand Sheep Station.* Seattle: University of Washington Press, 1999. Originally published in 1921 by Blackwood. Page references are to the 1999 edition.

Gutierrez, J. B., and J. L. Teem. "A Model Describing the Effect of Sex-Reversed YY Fish in an Established Wild Population: The Use of a Trojan Y Chromosome to Cause Extinction of an Introduced Exotic Species." *Journal of Theoretical Biology* 241 (2006): 333–41.

Hacking, I., and J. Hacking. *The Social Construction of What?* Cambridge, MA: Harvard University Press, 1999.

Haeckel, E. H. P. A. *Generelle Morphologie der Organismen Allgemeine Grundzüge der organischen Formen-Wissenschaft mechanisch begründet durch die von Charles Darwin reformirte Descendenz-Theorie.* 2 vols. Berlin: Reimer, 1866.

Hagen, J. B. "Teaching Ecology During the Environmental Age, 1965–1980." *Environmental History* 13 (2008): 704–23.

Hahn, M. A., M. van Kleunen, and H. Müller-Schärer. "Increased Phenotypic Plasticity to Climate May Have Boosted the Invasion Success of Polyploid *Centaurea stoebe*." *PLoS One* 7 (2012): e50284.

Haider, S., and K. Jax. "The Application of Environmental Ethics in Biological Conservation: A Case Study from the Southernmost Tip of the Americas." *Biodiversity and Conservation* 16 (2007): 2559–73.

Hajek, A. E., R. A. Humber, and J. S. Elkinton. "Mysterious Origins of *Entomophaga maimaiga* in North America." *American Entomologist* 41 (1995): 31–42.

Hajek, A. E., R. A. Humber, J. S. Elkinton, B. May, S. R. A. Walsh, and J. C. Silver. "Allozyme and Restriction Fragment Length Polymorphism Analyses Confirm *Entomophaga maimaiga* Responsible for 1989 Epizootics in North American Gypsy Moth Populations." *Proceedings of the National Academy of Sciences of the United States of America* 87 (1990): 6979–82.

Hajek, A. E., and P. C. Tobin. "Micro-Managing Arthropod Invasions: Eradication and Control of Invasive Arthropods with Microbes." *Biological Invasions* 12 (2010): 2895–2912.

Hällfors, M. H., E. M. Vaara, M. Hyvärinen, M. Oksanen, L. E. Schulman, et al. "Coming to Terms with the Concept of Moving Species Threatened by Climate Change—A Systematic Review of the Terminology and Definitions." *PLoS One* 9, no. 7 (2014): e102979.

Halverson, [M.] Anders. *An Entirely Synthetic Fish: How Rainbow Trout Beguiled America and Overran the World*. New Haven, CT: Yale University Press, 2010.

Halverson, M. Anders. "Relatedness Effects on Wood Frog Ecology and Evolution." PhD diss., Yale University, 2005.

Hämet-Ahti, L. "Human Impact on Closed Boreal Forest (Taiga)." In Holzner et al., *Man's Impact on Vegetation*, 201–11.

Hamilton, G. *Super Species: The Species That Will Dominate the Planet*. Boston: Firefly Books, 2010.

Hamilton, W. D. "Extraordinary Sex Ratios: A Sex-Ratio Theory for Sex Linkage and Inbreeding Has New Implications in Cytogenetics and Entomology." *Science* 156 (1967): 477–88.

Hammerstrom, F. N., Jr. "Exotic Game Birds." *Wilson Bulletin* 55 (1943): 66.

Hammond, A., R. Galizi, K. Kyrou, A. Simoni, C. Siniscalchi, et al. "A CRISPR-Cas9 Gene Drive System Targeting Female Reproduction in the Malaria Mosquito Vector *Anopheles gambiae*." *Nature Biotechnology* 34 (2016): 78–83.

Hamon, A. B., and T. R. Fasulo. "Cottony Cushion Scale." Featured Creatures, University of Florida. Revised April 2017. http://entnemdept.ufl.edu/creatures/fruit/cottony_cushion_scale.htm.

Hampton, J. O., B. Warburton, and P. Sandøe. "Compassionate Versus Consequentialist Conservation." *Conservation Biology* 33 (2019): 751–59.

Hansen, J., D. Johnson, A. Lacis, S. Lebedeff, P. Lee, et al. "Climate Impact of Increasing Atmospheric Carbon Dioxide." *Science* 213 (1981): 957–66.

Haraway, D. J. *How Like a Leaf: An Interview with Thyrza Nichols Goodeve*. New York: Routledge, 1999.

Haraway, D. J. *When Species Meet*. Minneapolis: University of Minnesota Press, 2008.

Haraway, D. J., and D. Harvey. "Nature, Politics, and Possibilities: A Debate and Discussion with David Harvey and Donna Haraway." *Environment and Planning D: Society and Space* 13 (1995): 507–27.

Hardy, J. W. "Feral Exotic Birds in Southern California." *Wilson Bulletin* 85 (1973): 506–12.

Harley, A. "Rabbits and the Rise of Australian Nativism." In *Victorian Environments: Acclimatizing to Change in British Domestic and Colonial Culture*, edited by G. Moore and M. J. Smith, 39–56. London: Palgrave Macmillan, 2018.

Harmon, J. P., and D. A. Andow. "Indirect Effects Between Shared Prey: Predictions for Biological Control." *BioControl* 49 (2004): 605–26.

Harris, A. *David Fairchild and America's Plant Explorers.* Gainesville: University Press of Florida, 2015.

Harris, H. "Enzyme Polymorphisms in Man." *Proceedings of the Royal Society of London B* 164 (1966): 298–310.

Harris, P. "Environmental Impact of Weed-Control Insects." *BioScience* 38 (1988): 542–48.

Harrison, M. "Plant History: Expanding the Horizons of a Small Garden." *Arnoldia* 53, no. 1 (1993): 8–18.

Harvey, R. G., and F. Mazzotti. *The Invasion Curve: A Tool for Understanding Invasive Species Management in South Florida.* Publication WEC 347. Gainesville: University of Florida Institute of Food and Agricultural Sciences, 2021.

Harvey-Samuel, T., T. Ant, and L. Alphey. "Towards the Genetic Control of Invasive Species." *Biological Invasions* 19 (2017): 1683–703.

Hassan, A., and A. Ricciardi. "Are Non-Native Species More Likely to Become Pests? Influence of Biogeographic Origin on the Socioeconomic Impact of Freshwater Organisms." *Frontiers in Ecology and the Environment* 12 (2014): 218–23.

Haubrock, P. J., A. J. Turbelin, R. N. Cuthbert, A. Novoa, N. G. Taylor, et al. "Economic Costs of Invasive Alien Species Across Europe." *NeoBiota* 67 (2021): 153–90.

Havens, K., C. L. Jolls, T. M. Knight, and P. Vitt. "Risks and Rewards: Assessing the Effectiveness and Safety of Classical Invasive Plant Biocontrol by Arthropods." *BioScience* 69 (2019): 247–58.

Hawaii Invasive Species Council. "Red-Whiskered Bulbul." Accessed February 13, 2025. https://dlnr.hawaii.gov/hisc/info/invasive-species-profiles/red-whiskered-bulbul/.

Hawkes, N. "The Seaweed That Won't Stop Growing." *Observer*, May 12, 1974.

Hawkes, N. "These Triffids Have Had Their Day." *Observer*, June 3, 1973.

Hawkins, B. A., F. S. Albuquerque, M. B. Araújo, J. Beck, L. M. Bini, et al. "A Global Evaluation of Metabolic Theory as an Explanation for Terrestrial Species Richness Gradients." *Ecology* 88 (2007): 1877–88.

Hawkins, C. L., S. Bacher, F. Essl, P. E. Hulme, J. M. Jeschke, et al. "Framework and Guidelines for Implementing the Proposed IUCN Environmental Impact Classification for Alien Taxa (EICAT)." *Diversity and Distributions* 21 (2015): 1360–63.

Haydock, E. L. "A Survey of the Birds of St. Helena Island." *Ostrich* 25 (1954): 62–75.

Hayes, K. R. *A Review of Ecological Risk Assessment Methodologies: Centre for Research on Introduced Marine Pests.* Technical Report 13. Hobart: CSIRO Division of Marine Research, 1997.

Hayes, K. R. "Biosecurity and the Role of Risk Assessment." In Ruiz and Carlton, *Invasive Species: Vectors and Management Strategies*, 382–414.

Haynes, S., M. Jaarola, and J. B. Searle. "Phylogeography of the Common Vole (*Microtus arvalis*) with Particular Emphasis on the Colonization of the Orkney Archipelago." *Molecular Ecology* 12 (2003): 951–56.

Hays, W. "Small Indian Mongoose." In Simberloff and Rejmánek, *Encyclopedia of Biological Invasions*, 631–34.

Hayward, M. W., A. Callen, B. L. Allen, G. Ballard, F. Broekhuis, et al. "Deconstructing Compassionate Conservation." *Conservation Biology* 33 (2019): 760–68.

Hazucha, A. "Neither Deep nor Shallow but National: Eco-Nationalism in Wordsworth's *Guide to the Lakes*." *Interdisciplinary Studies in Literature and Environment* 9, no. 2 (2002): 61–73.

He, S., S. P. Worner, and T. Ikeda. "Modeling the Potential Global Distribution of Light Brown Apple Moth *Epiphyas postvittana* (Lepidoptera: Tortricidae) Using CLIMEX." *Journal of Asia-Pacific Entomology* 15 (2012): 479–85.

Head, B. W. "Wicked Problems in Public Policy." *Public Policy* 3 (2008): 101–18.

Head, L. "Decentring 1788: Beyond Biotic Nativeness." *Geographical Research* 50 (2012): 166–78.

Head, L. "The Social Dimensions of Invasive Plants." *Nature Plants* 3 (2017): 17075.

Head, L., B. M. H. Larson, R. Hobbs, J. Atchison, N. Gill, et al. "Living with Invasive Plants in the Anthropocene: The Importance of Understanding Practice and Experience." *Conservation and Society* 13 (2015): 311–18.

Head, L., and P. Muir. "Nativeness, Invasiveness, and Nation in Australian Plants." *Geographical Review* 94 (2004): 199–217.

Healy, S. D. "The Face of Animal Cognition." *Integrative Zoology* 14 (2019): 132–44.

Hebert, P. D. N., A. Cywinska, S. L. Ball, and J. R. de Waard. "Biological Identifications Through DNA Barcodes." *Proceedings of the Royal Society B* 270 (2003): 313–22.

Hebert, P. D. N., P. M. Hollingsworth, and M. Hajibabaei. "From Writing to Reading the Encyclopedia of Life." *Philosophical Transactions of the Royal Society B* 371 (2016): 20150321.

Heddle, E. M. "South African Daisy in the National Parks of South Australia." *Environmental Conservation* 1 (1974): 152.

Hedgepeth, J. "Founders of Fish Culture: Livingston Stone." *Progressive Fish-Culturist* 55 (1941): 11–14.

Heger, T., and J. M. Jeschke. "The Enemy Release Hypothesis as a Hierarchy of Hypotheses." *Oikos* 123 (2014): 741–50.

Heger, T., A. T. Pahl, Z. Botta-Dukát, F. Gherardi, C. Hoppe, et al. "Conceptual Frameworks and Methods for Advancing Invasion Ecology." *Ambio* 42 (2013): 527–40.

Hehn, V. *Kulturpflanzen und Hausthiere in ihrem Übergang aus Asien nach Griechenland und Italien sowie in das übrige Europa: Historischlinguistische Skizzen*. Berlin: Gebrüder Borntraeger, Edward Eggers, 1870.

Hehn, V. *The Wanderings of Plants and Animals from Their First Home*. Edited by J. D. Stallybrass. London: Swan Sonnenschein, 1885.

Heimpel, G. E., and N. J. Mills. *Biological Control: Ecology and Applications*. Cambridge: Cambridge University Press, 2017.

Hellmann, J. J., J. E. Byers, B. G. Bierwagen, and J. S. Dukes. "Five Potential Consequences of Climate Change for Invasive Species." *Conservation Biology* 22 (2008): 534–43.

Helmreich, A. L. "Re-presenting Nature: Ideology, Art, and Science in William Robinson's 'Wild Garden.'" In Wolschke-Bulmahn, *Nature and Ideology*, 81–111.

Helmreich, S. L. "How Scientists Think About 'Natives,' for Example: A Problem of Taxonomy Among Biologists of Alien Species in Hawaii." *Journal of the Royal Anthropological Institute* 11, no. 1 (2005): 107–28.

Helvarg, D. *The War Against the Greens: The "Wise-Use" Movement, the New Right, and the Browning of America*. Boulder, CO: Johnson Books, 2004.

Hendrickx, G. "The Spread of Blue Tongue in Europe." *Small Ruminant Research* 86, no. 1–3 (2009): 34–39.

Hendry, A. P. *Eco-Evolutionary Dynamics*. Princeton, NJ: Princeton University Press, 2017.

Hendry, A. P. "Key Questions on the Role of Phenotypic Plasticity in Eco-Evolutionary Dynamics." *Journal of Heredity* 107 (2016): 25–41.

Hengeveld, R. *Dynamics of Biological Invasions*. London: Chapman and Hall, 1989.

Henriksson, A., J. Yu, D. A. Wardle, and G. Englund. "Biotic Resistance in Freshwater Fish Communities: Species Richness, Saturation or Species Identity?" *Oikos* 124 (2015): 1058–64.

Henshall, J. A. "Indiscriminate and Inconsiderate Planting of Fish." *Transactions of the American Fisheries Society* 48 (1919): 166–69.

Henslow, G. "On the Self-Fertilization of Plants." *Transactions of the Linnean Society of London*, 2nd ser., vol. 1, no. 6 (1879): 301–422.

Henslow, J. S. "Observations Concerning the Indigenousness and Distinctness of Certain Species of Plants Included in the British Floras." *Magazine of Natural History* 8 (1835): 84–88.

Heptner, V. G., A. A. Nasimovich, and A. G. Bannikov. *Mammals of the Soviet Union*. Vol. 1, *Artiodactyla and Perissodactyla*. New Delhi: Amerind, 1980.

Herbert, T. *Some Yeares Travels into Africa and Asia the Great Especially Describing the Famous Empires of Persia and Industant as Also Divers Other Kingdoms in the Orientall Indies and I'les Adjacent*. London: Jacob Blome and Richard Bishop, 1638.

Herbold, B., and P. B. Moyle. "Introduced Species and Vacant Niches." *American Naturalist* 128 (1986): 751–60.

Herms, D. A., and D. G. McCullough. "Emerald Ash Borer Invasion of North America: History, Biology, Ecology, Impacts, and Management." *Annual Review of Entomology* 59 (2014): 13–30.

Hettinger, N. "Exotic Species, Naturalisation, and Biological Nativism." *Environmental Values* 10 (2001): 193–224.

Hibben, F. C. *Hunting in Africa*. New York: Hill and Wang, 1962.

Hibben, F. C. "The Role of Exotic Animals in Game Management." In *Proceedings of the 44th Annual Conference of Western Association of State Game and Fish Commissioners* (1964), 125–30.

Hibben, F. C. *Under the African Sun: Forty-Eight Years of Hunting the African Continent*. Huntington Beach, CA: Safari Press, 1999.

Hickley, P., and S. Chare. "Fisheries for Non-Native Species in England and Wales: Angling or the Environment?" *Fisheries Management and Ecology* 11 (2004): 203–12.

Hicks, L. E. "The Role of Exotics in the Ohio Valley." *Transactions of the Fifth North American Wildlife Conference* 5 (1941): 420–24.

Hierro, J. L., J. L. Maron, and R. M. Callaway. "A Biogeographical Approach to Plant Invasions: The Importance of Studying Exotics in Their Introduced *and* Native Range." *Journal of Ecology* 93 (2005): 5–15.

Higgs, E. S., J. A. Harris, T. Heger, R. J. Hobbs, S. D. Murphy, and K. N. Suding. "Keep Ecological Restoration Open and Flexible." *Nature Ecology and Evolution* 2 (2018): 580.

Hill, N. M, and C. S. Blaney. "Exotic and Invasive Vascular Plants of the Atlantic Maritime Ecozone." In *Assessment of Species Diversity in the Atlantic Maritime Ecozone*, edited by D. F. McAlpine and I. M. Smith, 1–18. Ottawa: NRC Research Press, 2009.

Hilton-Taylor, C. "A History of the IUCN Red List." In *The IUCN Red List: 50 Years of Conservation*, edited by J. Smart, C. Hilton-Taylor, and R. A. Mittermeier, 34–41. Washington, DC: Earth in Focus, 2014.

Hinz, H. L., R. L. Winston, and M. Schwarzländer. "A Global Review of Target Impact and Direct Nontarget Effects of Classical Weed Biological Control." *Current Opinion in Insect Science* 38 (2020): 48–54.

Hoagland, D. B., G. R. Horst, and C. W. Kilpatrick. "Biogeography and Population Biology of the Mongoose in the West Indies." In *Biogeography of the West Indies*, edited by C. A. Woods, 611–34. Gainesville, FL: Sandhill Crane Press, 1989.

Hobbs, R. J. "The Nature and Effects of Disturbance Relative to Invasions." In Drake et al., *Biological Invasions: A Global Perspective*, 389–405.

Hobbs, R. J., S. Arico, J. Aronson, J. S. Baron, P. Bridgewater, et al. "Novel Ecosystems: Theoretical and Management Aspects of the New Ecological World Order." *Global Ecology and Biogeography* 15 (2006): 1–7.

Hobbs, R. J., E. S. Higgs, and C. M. Hall. "Defining Novel Ecosystems." In Hobbs, Higgs, and Hall, *Novel Ecosystems*, 58–60.

Hobbs, R. J., E. S. Higgs, and C. M. Hall, eds. *Novel Ecosystems: Intervening in the New Ecological World Order*. Oxford: Wiley-Blackwell, 2013.

Hobbs, R. J., E. S. Higgs, and J. A. Harris. "Novel Ecosystems: Implications for Conservation and Restoration." *Trends in Ecology and Evolution* 24 (2009): 599–605.

Hobbs, R. J., and L. F. Huenneke. "Disturbance, Diversity, and Invasion: Implications for Conservation." *Conservation Biology* 6 (1992): 324–37.

Hobson, E. A., M. L. Avery, and T. F. Wright. "An Analytical Framework for Quantifying and Testing Patterns of Temporal Dynamics in Social Networks." *Animal Behaviour* 85 (2013): 83–96.

Hoddle, M. S., C. Crespo Ramirez, C. D. Hoddle, J. Loayza, M. Piedad Lincango, et al. "Post Release Evaluation of *Rodolia cardinalis* (Coleoptera: Coccinellidae) for Control of *Icerya purchasi* (Hemiptera: Monophlebidae) in the Galápagos Islands." *Biological Control* 67 (2013): 262–74.

Hoegh-Guldberg, O., L. Hughes, S. McIntyre, D. B. Lindenmayer, C. Parmesan, et al. "Assisted Colonization and Rapid Climate Change." *Science* 321 (2008): 345–46.

Hoelmer, K. A., and J. K. Grace. "Citrus Blackfly." In Dahlsten and Garcia, *Eradication of Exotic Pests*, 147–65.

Hoffmann, B. D., and F. Courchamp. "Biological Invasions and Natural Colonisations: Are They That Different?" *NeoBiota* 29 (2016): 1–14.

Hoffmann, M., J. L. Belant, J. S. Chanson, N. Cox, J. Lamoreux, et al. "The Changing Fates of the World's Mammals." *Philosophical Transactions of the Royal Society of London, Series B: Biological Sciences* 366 (2011): 2598–610.

Hoffmann, M., C. Hilton-Taylor, A. Angulo, M. Böhm, T. M. Brooks, et al. "The Impact of Conservation on the Status of the World's Vertebrates." *Science* 330 (2010): 1503–9.

Hofman, C. A., and T. C. Rick. "Ancient Biological Invasions and Island Ecosystems: Tracking Translocations of Wild Plants and Animals." *Journal of Archaeological Research* 26 (2018): 65–115.

Hogg, R. *The Dahlia: Its History and Cultivation, with Descriptions of All the Best Show Flowers*. London: Groombridge and Sons, 1853.

Hokkanen. H. M. T. "Demonstrating the Safety of Biocontrol." *BioControl* 48 (2003): 1.

Holdgate, M. W. "Summary and Conclusions: Characteristics and Consequences of Biological Invasions." *Philosophical Transactions of the Royal Society of London B* 314 (1986): 733–42.

Holling, C. S. "The Components of Predation as Revealed by a Study of Small-Mammal Predation of the European Pine Sawfly." *Canadian Entomologist* 91 (1959): 293–320.

Holling, C. S. "Some Characteristics of Simple Types of Predation and Parasitism." *Canadian Entomologist* 91 (1959): 385–98.

Holmes, E. E., M. A. Lewis, J. E. Banks, and R. R. Veit. "Partial Differential Equations in Ecology: Spatial Interactions and Population Dynamics." *Ecology* 75 (1994): 17–29.

Holmes, N. D., B. S. Keitt, D. R. Spatz, D. J. Will, S. Hein, et al. "Tracking Invasive Species Eradications on Islands at a Global Scale." In Veitch et al., *Island Invasives: Scaling Up to Meet the Challenge*, 628–32.

Holt, R. D., M. Barfield, and R. Gomulkiewicz. "Theories of Niche Conservatism and Evolution." In Sax et al., *Species Invasions: Insights into Ecology, Evolution, and Biogeography*, 259–90.

Holt, R. D., and M. E. Hochberg. "Indirect Interactions, Community Modules and Biological Control: A Theoretic Perspective." In *Evaluating Indirect Ecological Effects of Biological Control*, edited by E. Wajnberg, J. K. Scott, and P. C. Quimby, 13–37. Wallingford, UK: CABI, 2001.

Holway, D. A., and A. V. Suarez. "Animal Behavior: An Essential Component of Invasion Biology." *Trends in Ecology and Evolution* 14 (1999): 328–30.

Holzner, W., M. J. A. Werger, and I. Ikusima, eds. *Man's Impact on Vegetation*. The Hague: Dr. W. Junk, 1983.

Honig, M. A., R. M. Cowling, and D. M. Richardson. "The Invasive Potential of Australian Banksias in South African Fynbos: A Comparison of the Reproductive Potential of *Banksia ericifolia* and *Leucadendron laureolum*." *Australian Journal of Ecology* 17 (1992): 305–14.

Honor, R., and R. I. Colautti. "EICA 2.0: A General Model of Enemy Release and Defence in Plant and Animal Invasions." In *Plant Invasions: The Role of Biotic Interactions*, edited by A. Traveset and D. M. Richardson, 192–207. Wallingford, UK: CABI, 2020.

Hooker, J. D. *The Botany of the Antarctic Voyage of H.M. Discovery Ships Erebus and Terror in the Years 1839–1843: Flora Novæ-Zelandiæ, Part I; Flowering Plants*. London: Lovell Reeve, 1853.

Hooker, J. D. *The Botany of the Antarctic Voyage of H.M. Discovery Ships Erebus and Terror, in the Years 1839–1843: Part III, Flora Tasmaniæ*. Vol. 1. London: Lovell Reeve, 1860.

Hooper, D. U., F. S. Chapin III, J. J. Ewel, A. Hector, P. Inchausti, et al. "Effects of Biodiversity on Ecosystem Functioning: A Consensus of Current Knowledge." *Ecological Monographs* 75 (2005): 3–35.

Hooper, M. D. "The Size and Surroundings of Nature Reserves." In Duffey and Watt, *Scientific Management of Animal and Plant Communities for Conservation*, 555–61.

Hopkins, A. D. *Bioclimatics: A Science of Life and Climate Relations*. Miscellaneous Publication 280. Washington, DC: United States Department of Agriculture, 1938.

Hopkins, K. D., W. L. Shelton, and C. R. Engle. "Estrogen Sex-Reversal of *Tilapia aurea*." *Aquaculture* 18 (1979): 263–68.

Hornaday, W. T. *Our Vanishing Wild Life: Its Extinction and Preservation*. New York: New York Zoological Society, 1913.

Horppila, J., H. Peltonen, T. Malinen, E. Luokkanen, and T. Kairesalo. "Top-Down or Bottom-Up Effects by Fish: Issues of Concern in Biomanipulation of Lakes." *Restoration Ecology* 6 (1998): 20–28.

Horst, H. A., and D. Miller. *The Cell Phone: An Anthropology of Communication*. Oxford: Routledge, 2006.

Houghton, R. A., and D. L. Skole. "Carbon." In Turner et al., *Earth as Transformed by Human Action*, 393–408.

Houston, D. B., E. G. Schreiner, and B. B. Moorhead. *Mountain Goats in Olympic National Park: Biology and Management of an Introduced Species*. Denver: United States Department of the Interior, National Park Service, 1994.

Howald, G., C. J. Donlan, K. R. Faulkner, S. Ortega, H. Gellerman, et al. "Eradication of Black Rats *Rattus rattus* from Anacapa Island." *Oryx* 44 (2009): 3040.

Howald, G. R., K. R. Faulkner, B. Tershy, B. Keitt, H. Gellerman, et al. "Eradication of Black Rat from Anacapa Island: Biological and Social Considerations." In *Proceedings of the Sixth California Channel Islands Symposium*, edited by D. K. Garcelon and C. A. Schwemm, 299–312. Arcata, CA: Institute for Wildlife Studies, 2005.

Howard, L. O. "Danger of Importing Insect Pests." In US Department of Agriculture, *Yearbook of the Department of Agriculture, 1897*, 529–52. Washington, DC: Government Printing Office, 1898.

Howard, L. O. "Entomology and the War." *Scientific Monthly* 8, no. 2 (1919): 109–17.

Howard, L. O. *Fighting the Insects: The Story of an Entomologist*. New York: Macmillan, 1933.

Howard, L. O., and W. F. Fiske. *The Importation into the United States of the Parasites of the Gipsy Moth and the Brown-Tail Moth: A Report of Progress, with Some Consideration of Previous and Concurrent Efforts of This Kind*. US Department of Agriculture Bureau of Entomology Bulletin no. 91, July 29, 1911.

Howard, L. O. "On Some Presidential Addresses: The War Against Insects." *Science* 54 (1921): 641–51.

Howarth, F. G. "Environmental Impacts of Classical Biological Control." *Annual Review of Entomology* 36 (1991): 485–509.

Howarth, F. G. "Impacts of Alien Land Arthropods and Molluscs on Native Plants and Animals in Hawaii." In *Hawai'i's Terrestrial Ecosystems: Preservation and Management*, edited by C. P. Stone and J. M. Scott, 149–79. Honolulu: University of Hawaii, 1985.

Hoy, M. A. "Biological Control of Arthropods: Genetic Engineering and Environmental Risks." *Biological Control* 2 (1992): 166–70.

Hoy, M. A. "Establishment of Gypsy Moth Parasitoids in North America: An Evaluation of Possible Reasons for Establishment or Non-Establishment." In *Perspectives in Forest Entomology*, edited by J. F. Anderson and H. K. Haya, 215–32. New York: Academic, 1976.

Hubbs, C. L. "Fishes from the Big Bend Region of Texas." *Transactions of the Texas Academy of Science* 23 (1940): 3–12.

Hubbs, C. L., and K. F. Lagler. "Fishes of Isle Royale, Lake Superior, Michigan." *Papers of the Michigan Academy of Science, Arts, and Letters* 33 (1949): 73–133.

Hubbs, C. L., and O. L. Wallis. "The Native Fish Fauna of Yosemite National Park and Its Preservation." *Yosemite Nature Notes* 27, no. 12 (1948): 131–44.

Hubby, J. L., and R. C. Lewontin. "A Molecular Approach to the Study of Genic Heterozygosity in Natural Populations, I: The Number of Alleles at Different Loci in *Drosophila pseudoobscura*." *Genetics* 54 (1966): 577–94.

Hudson, J. "Trends in Multi-Authored Papers in Economics." *Journal of Economic Perspectives* 10, no. 3 (1996): 153–58.

Huey, R. B., G. W. Gilchrist, M. L. Carlson, D. Berrigan, and L. Serra. "Rapid Evolution of a Geographic Cline in Size in an Introduced Fly." *Science* 287 (2000): 308–9.

Huey, R. B., G. W. Gilchrist, and A. P. Hendry. "Using Invasive Species to Study Evolution: Case Studies with *Drosophila* and Salmon." In Sax et al., *Species Invasions*, 139–64.

Huey, W. H. "The New New Mexicans." *New Mexico Wildlife* 26 (September–October 1981): 25–26, 29.

Huffaker, C. "Fundamentals of Biological Control of Weeds." *Hilgardia* 27, no. 3 (1957): 101–57.

Hughes, A. C., M. C. Orr, K. Ma, M. J. Costello, J. Waller, et al. "Sampling Biases Shape Our View of the Natural World." *Ecography* 44 (2021): 1259–69.

Hughes, K. A., L. R. Pertierra, M. A. Molina-Montenegro, and P. Convey. "Biological Invasions in Terrestrial Antarctica: What Is the Current Status and Can We Respond?" *Biodiversity and Conservation* 24 (2015): 1031–35.

Hughes, K. A., O. Pescott, J. Peyton, T. Adriaens, E. J. Cotter-Cook, et al. "Invasive Nonnative Species Likely to Threaten Biodiversity and Ecosystems in the Antarctic Peninsula Region." *Global Change Biology* 26 (2020): 2702–16.

Hughes, L., and F. A. Bazzaz. "Effects of Elevated CO_2 on Five Plant-Aphid Interactions." *Entomologia Experimentalis et Applicata* 99 (2001): 87–96.

Hulbert, J. M., R. A. Hallett, H. E. Roy, and M. Cleary. "Citizen Science Can Enhance Strategies to Detect and Manage Invasive Forest Pests and Pathogens." *Frontiers in Ecology and Evolution* 11 (2023): 1113978.

Hull, D. L. *Philosophy of Biological Science*. Englewood Cliffs, NJ: Prentice-Hall, 1974.

Hulme, P. E. "Climate Change and Biological Invasions: Evidence, Expectations, and Response Options." *Biological Reviews* 92 (2017): 1297–313.

Hulme, P. E., W. Nentwig, P. Pyšek, and M. Vilà, eds. *Handbook of Alien Species in Europe: DAISIE*. Dordrecht, Neth.: Springer, 2009.

Hulme, P. E., and M. Vilà. "Integrating the Impacts of Non-Native Species on Ecosystem Services into Environmental Policy." In Vilà and Hulme, *Impact of Biological Invasions on Ecosystem Services*, 327–41.

Hume, J. P. "Undescribed Juvenile Plumages of the Laysan Rail or Crake (*Zapornia palmeri*: Frohawk, 1892) and a Detailed Chronology of Its Extinction." *Wilson Journal of Ornithology* 129 (2017): 429–45.

Hummer, K. E. "History of the Origin and Dispersal of White Pine Blister Rust." *HortTechnology* 10 (2000): 515–17.

Hunt, T. L. "Rethinking Easter Island's Ecological Catastrophe." *Journal of Archaeological Science* 34 (2007): 485–502.

Hunter, D. E. L., G. J. Newman, and M. M. Balgopal. "What's In a Name? The Paradox of Citizen Science and Community Science." *Frontiers in Ecology and the Environment* 21 (2023): 244–50.

Hunter, J. E., F. L. Schmidt, and G. B. Jackson. *Meta-Analysis: Cumulating Research Findings Across Studies*. Beverly Hills, CA: Sage, 1982.

Hunter, M. L. "Climate Change and Moving Species: Furthering the Debate on Assisted Colonization." *Conservation Biology* 21 (2007): 1356–58.

Hunter, R. L., and C. L. Markert. "Histochemical Demonstration of Enzymes Separated by Zone Electrophoresis in Starch Gels." *Science* 125 (1957): 1294–95.

Hutchinson, G. E. "Concluding Remarks." *Cold Spring Harbor Symposia on Quantitative Biology* 22 (1957): 415–27.

Hutchinson, M. C., B. Bramon Mora, S. Pilosof, A. K. Barner, S. Kéfi, et al. "Seeing the Forest for the Trees: Putting Multilayer Networks to Work for Community Ecology." *Functional Ecology* 33 (2019): 206–17.

Hutchinson, M. F. "Interpolating Mean Rainfall Using Thin Plate Smoothing Splines." *International Journal of Geographical Information Systems* 9 (1995): 385–403.

Huxley, J. S. *Evolution, the Modern Synthesis*. London: Allen and Unwin, 1943.

Ibabe, A., F. Rayon, J. L. Martinez, and E. Garcia-Vazquez. "Environmental DNA from Plastic and Textile Marine Litter Detects Exotic and Nuisance Species." *PLoS One* 15, no. 6 (2020): e0228811.

Ikagawa, M. "Invasive Ungulate Policy and Conservation in Hawaii." *Pacific Conservation Science* 19 (2013): 270–83.

Illustrations of the Ravages of the Gypsy and Brown Tail Moths. 2nd ed. Medford, MA: Medford Mercury, 1906.

Imbler, S. "Introducing Spongy Moth: For This Common Insect, a New Common Name." *New York Times,* March 8, 2022, D2.

Imbler, S. "This Moth's Name Is a Slur: Scientists Won't Use It Anymore." *New York Times,* July 10, 2021, A17.

Inderjit, J. A. Catford, S. Kalisz, D. Simberloff, and D. A. Wardle. "A Framework for Understanding Human-Driven Vegetation Change." *Oikos* 126 (2017): 1687–98.

Ings, T. C., and J. E. Hawes. "The History of Ecological Networks." In Dáttilo and Rico-Gray, *Ecological Networks in the Tropics,* 15–28.

Ings, T. C., J. M. Montoya, J. Bascompte, N. Blüthgen, L. Brown, et al. "Ecological Networks—Beyond Food Webs." *Journal of Animal Ecology* 78 (2009): 253–69.

Innes, W. T. *Exotic Aquarium Fishes.* 19th ed. Maywood, NJ: Metaframe Corporation, 1966.

Innis, M. "Trial to See if 'Good' Mosquitoes Can Edge Out Bad." *New York Times,* February 5, 2016, A11.

"Invasion of the Habitat Snatchers." *Grist,* June 11, 2003. https://grist.org/article/habitat/amp/.

Ioffe, Ts. I. "Present State and Prospects of the Acclimatization of Invertebrates in the Inland Waters of the USSR." In Yanushevich, *Acclimatization of Animals in the U.S.S.R.,* 8–9.

IPCC. *Climate Change 2022: Impacts, Adaptation and Vulnerability; Contribution of Working Group II to the Sixth Assessment Report of the Intergovernmental Panel on Climate Change.* Cambridge: Cambridge University Press, 2022.

Irvine, C. J. "Forest and Timber Insects in Victoria." *Victoria's Resources* 4 (1962): 40–43.

Isenberg, A. H. "Exotic Birds Not Detrimental to Native Birds in the Hawaiian Islands." *Aviculture* 4 (1932): 70–71.

Isenberg, A. H. "Transplanting Foreign Birds to the Hawaiian Islands, and Other Notes." *Aviculture* 3 (1931): 333–34.

IUCN. "IUCN Standard to Support Global Action on Invasive Alien Species." International Union for Conservation of Nature and Natural Resources. September 15, 2020. https://www.iucn.org/news/species/202009/iucn-standard-support-global-action-invasive-alien-species.

IUCN Conservation Monitoring Centre, World Conservation Monitoring Centre, IUCN Species Survival Commission, International Council for Bird Preservation and Birdlife International. *IUCN Red List of Threatened Animals.* Gland, Switz.: International Union for Conservation of Nature and Natural Resources, 1996.

IUCN Species Survival Commission. *IUCN Red List Categories.* Gland, Switz.: International Union for Conservation of Nature and Natural Resources, 1994.

IUCN Species Survival Commission. *IUCN Red List Categories and Criteria.* Gland, Switz.: International Union for Conservation of Nature and Natural Resources, 2001.

IUCN Standards and Petitions Committee. *Guidelines for Using the IUCN Red List Categories and Criteria.* Version 16. March 2024. https://nc.iucnredlist.org/redlist/content/attachment_files/RedListGuidelines.pdf.

Ivanova, I. M., and C. T. Symes. "Invasion of *Psittacula krameri* in Gauteng, South Africa: Are Other Birds Impacted?" *Biodiversity and Conservation* 28 (2019): 3633–56.

Jackson, M. D., N. C. Miorelli, and A. L. Bond. "New Record of the Non-Native Long-Legged Fly *Thambemyia borealis* (Takagi 1965) (Diptera: Dolichopodidae) in North America and an Example of Twitter's Utility for Natural History Documentation." *Proceedings of the Entomological Society of Washington* 119 (2017): 151–56.

Jackson, R. "Eunice Foote, John Tyndall and a Question of Priority." *Notes and Records* 74 (2020): 105–18.

Jaffe, M. *And No Birds Sing*. New York: Barricade Books, 1997.

Jäger, J., and R. B. Barry. "Climate." In Turner et al., *Earth as Transformed by Human Action*, 335–51.

James, F. C. "Nonindigenous Birds." In Simberloff, Schmitz, and Brown, *Strangers in Paradise: Impact and Management of Nonindigenous Species in Florida*, 139–56.

Jamieson, D. "Ecosystem Health: Some Preventive Medicine." *Environmental Values* 4 (1995): 333–44.

Jang, S. R. J. "Dynamics of Herbivore-Plant-Pollinator Models." *Journal of Mathematical Biology* 44 (2002): 129–49.

Janovsky, R. M., and E. R. Larson. "Does Invasive Species Research Use More Militaristic Language than Other Ecology and Conservation Biology Literature?" *NeoBiota* 44 (2019): 27–38.

Jansen, S., and T. Geburek. "Historic Translocations of European Larch (*Larix decidua* Mill.) Genetic Resources Across Europe—A Review from the 17th Until the Mid-20th Century." *Forest Ecology and Management* 379 (2016): 114–23.

Janzen, D. H. "On Ecological Fitting." *Oikos* 45 (1985): 308–10.

"Japan Close to Eradicating Mongoose from Amami-Oshima Island." *Japan Times*, May 18, 2023.

Jarić, I., and G. Cvijanović. "The Tens Rule in Invasion Biology: Measure of a True Impact or Our Lack of Knowledge and Understanding?" *Environmental Management* 50 (2012): 979–81.

Järvinen, A. "A Meta-Analytic Study of the Effects of Female Age on Laying-Date and Clutch-Size in the Great Tit *Parus major* and the Pied Flycatcher *Ficedula hypoleuca*." *Ibis* 133 (1991): 62–66.

Jauni, M., S. Gripenberg, and S. Ramula. "Non-Native Plant Species Benefit from Disturbance: A Meta-Analysis." *Oikos* 124 (2015): 122–29.

Jay, A. "Number of Smartphone and Mobile Phone Users Worldwide in 2024: Demographics, Statistics, Predictions." Modified October 24, 2024. https://financesonline.com/number-of-smartphone-users-worldwide/.

Jefferson, R. M., and A. E. Fusonie. *The Japanese Flowering Cherry Trees of Washington, D.C.* Washington, DC: USDA Agricultural Research Service, 1977.

Jenicek, M. *Meta-analyse en médecine: Évaluation et Synthèse de l'Information Clinique et Épidémiologique*. St-Hyacinthe, QC: EDISEM and Maloine, 1987.

Jenkins, C. F. H. *The Noah's Ark Syndrome*. Perth: Zoological Gardens Board of Western Australia, 1977.

Jensen, J. "Die Lichtung (The Clearing)." *Gartenkunst* 50, no. 9 (1937): 177–81.

Jentsch, A., J. Kreyling, and C. Beierkuhnlein. "A New Generation of Climate-Change Experiments: Events, Not Trends." *Frontiers in Ecology and the Environment* 5 (2007): 365–74.

Jerde, C. L., A. R. Mahon, W. L. Chadderton, and D. M. Lodge. "'Sight-unseen' Detection of Rare Aquatic Species Using Environmental DNA." *Conservation Letters* 4 (2011): 150–57.

Jeschke, J. M., L. Gómez Aparicio, S. Haider, T. Heger, C. J. Lortie, et al. "Support for Major Hypotheses in Invasion Biology Is Uneven and Declining." *NeoBiota* 14 (2012): 1–20.

Jeschke, J. M., and T. Heger. *Invasion Biology: Hypotheses and Evidence*. Wallingford, UK: CABI, 2018.

Jeschke, J. M., and P. Pyšek. "Tens Rule." In Jeschke and Heger, *Invasion Biology: Hypotheses and Evidence*, 124–32.

Jeschke, J. M., and D. L. Strayer. "Invasion Success of Vertebrates in Europe and North America." *Proceedings of the National Academy of Sciences of the United States of America* 102 (2005): 7198–202.

Jewell, S. D. "A Century of Injurious Wildlife Listing Under the Lacey Act: A History." *Management of Biological Invasions* 11 (2020): 356–71.

Jewell, S. D., and P. L. Fuller. "The Unsung Success of Injurious Wildlife Listing Under the Lacey Act." *Management of Biological Invasions* 12 (2021): 527–45.

Jiang, M., X. Wu, K. Chen, H. Luo, W. Yu, et al. "Production of YY Supermale and XY Physiological Female Common Carp for Potential Eradication of Invasive Species." *Journal of the World Aquaculture Society* 49 (2018): 315–27.

Joenje, W. "The SCOPE Programme on the Ecology of Biological Invasions: An Account of the Dutch Contribution." Special issue, *Proceedings of the Koninklijke Nederlandse Akademie van Wetenschappen, Series C: Biological and Medical Sciences* 90 (1987): 3–13.

Joenje, W., K. Bakker, and L. Vlijm, eds. "The Ecology of Biological Invasions." *Proceedings of the Koninklijke Nederlandse Akademie van Wetenschappen, Series C: Biological and Medical Sciences* 90 (1987): 1–80.

Johnson, C., and I. M. Cowan. "Thermal Adaptation as a Factor Affecting Colonizing Success of Introduced Sturnidae (Aves) in North America." *Canadian Journal of Zoology* 52 (1974): 1559–76.

Johnson, Ladd E. "Zebra Mussel." In Simberloff and Rejmánek, *Encyclopedia of Biological Invasions*, 709–13.

Johnson, Ladd E., and J. T. Carlton. "Post-Establishment Spread in Large-Scale Invasions: Dispersal Mechanisms of the Zebra Mussel *Dreissena polymorpha*." *Ecology* 77 (1996): 1686–90.

Johnson, Lawrence E. *A Morally Deep World: An Essay on Moral Significance and Environmental Ethics*. Cambridge: Cambridge University Press, 1993.

Johnson, Lawrence E. "Toward the Moral Considerability of Species and Ecosystems." *Environmental Ethics* 14 (1992): 145–57.

Johnson, P. J., V. M. Adams, D. P. Armstrong, S. E. Baker, D. Biggs, et al. "Consequences Matter: Compassion in Conservation Means Caring for Individuals, Populations, and Species." *Animals* 9 (2019): 1115.

Johnson, R., R. E. Crafton, and H. F. Upton. *Invasive Species: Major Laws, and the Role of Selected Federal Agencies*. Washington, DC: Congressional Research Service, 2017. https://crsreports.congress.gov/product/pdf/R/R43258/7.

Johnston, H. H. "The Next War: Man *Versus* Insects." *Nineteenth Century and After* 78 (1915): 151–58.

Johnston, R. F., and R. K. Selander. "House Sparrows: Rapid Evolution of Races in North America." *Science* 144 (1964): 548–50.

Johnston, R. F., and R. K. Selander. "Evolution in the House Sparrow II: Adaptive Differentiation in North American Populations." *Evolution* 25 (1971): 1–28.

Johnston, R. F., and R. K. Selander. "Evolution in the House Sparrow III: Variation in Size and Sexual Dimorphism in Europe and North and South America." *American Naturalist* 107 (1973): 373–90.

Johnston, R. F., and W. J. Klitz. "Variation and Evolution in a Granivorous Bird: The House Sparrow." In *Granivorous Birds in Ecosystems: Their Evolution, Populations,*

Energetics, Adaptations, Impact and Control, edited by J. Pinowski and S. C. Kendeigh, 15–51. Cambridge: Cambridge University Press, 1977.

Jones, C. G., J. H. Lawton, and M. Shachak. "Organisms as Ecosystem Engineers." *Oikos* 69 (1994): 373–86.

Jones, C. G., J. H. Lawton, and M. Shachak. "Positive and Negative Effects of Organisms as Physical Ecosystem Engineers." *Ecology* 78 (1997): 1946–57.

Jørgensen, L. B., H. Malte, and J. Overgaard. "How to Assess *Drosophila* Heat Tolerance: Unifying Static and Dynamic Tolerance Assays to Predict Heat Distribution Limits." *Functional Ecology* 33 (2019): 629–42.

Josselyn, J. *An Account of Two Voyages to New England*. London: Giles Widdows, 1674.

Josselyn, J. *New-Englands Rarities, Discovered in Birds, Beasts, Fishes, Serpents, and Plants of the Country*. London: G. Widdowes, 1672.

Joyce, P. W., J. W. Dickey, R. N. Cuthbert, J. T. A. Dick, and L. Kregting. "Using Functional Responses and Prey Switching to Quantify Invasion Success of the Pacific Oyster, *Crassostrea gigas*." *Marine Environmental Research* 145 (2019): 66–72.

[J. S. G.]. "The Rabbit in Australia." *Nature* 137 (1936): 806–7.

Jubinsky, G. "Editorial." *Wildland Weeds* 1, no. 1 (1997): 13.

Julien, M. H., B. Skarratt, and G. F. Maywald. "Potential Geographical Distribution of Alligator Weed and Its Biological Control by *Agasicles hygrophila*." *Journal of Aquatic Plant Management* 33 (1995): 55–60.

Kahn, E. J. "The Staffs of Life II: Man Is What He Eats." *New Yorker*, November 12, 1984, 56–106.

Kaiser-Bunbury, C. N., J. Memmott, and C. B. Müller. "Community Structure of Pollination Webs of Mauritian Heathland Habitats." *Perspectives in Plant Ecology, Evolution, and Systematics* 11 (2009): 241–54.

Kallman, K. "The Pike Killifish, Pint-Sized Predator." *Animal Kingdom* 74, no. 3 (1971): 22–24.

Kalm, P. *Peter Kalm's Travels in North America: The English Version of 1770*. 2 vols. Edited by A. B. Benson. New York: Wilson-Erickson, 1937.

Kanya, L., S. Sanghera, A. Lewin, and J. Fox-Rushby. "The Criterion Validity of Willingness to Pay Methods: A Systematic Review and Meta-Analysis of the Evidence." *Social Science and Medicine* 232 (2019): 238–61.

Karban, R. "The Ecology and Evolution of Induced Resistance Against Herbivores." *Functional Ecology* 25 (2011): 339–47.

Kareiva, P., M. Groom, I. Parker, and J. Ruesink. *Risk Analysis as a Tool for Making Decisions About the Introduction of Non-Indigenous Species into the United States*. Washington, DC: Office of Technology Assessment, 1991.

Kareiva, P., and S. Levin, eds. *The Importance of Species: Perspectives on Expendability and Triage*. Princeton, NJ: Princeton University Press, 2003.

Kareiva, P., and M. Marvier. *Conservation Science: Balancing the Needs of People and Nature*. Greenwood Village, CO: Roberts, 2011.

Kareiva, P., and M. Marvier. *Conservation Science: Balancing the Needs of People and Nature*. 2nd ed. New York: W. H. Freeman, 2017.

Kareiva, P., and M. Marvier. "Conserving Biodiversity Coldspots." *American Scientist* 91 (2003): 344–51.

Kareiva, P., and M. Marvier. "What Is Conservation Science?" *BioScience* 62 (2012): 962–69.

Kareiva, P., M. Marvier, and B. Silliman, eds. *Effective Conservation Science: Data Not Dogma*. Oxford: Oxford University Press, 2018.

Karsten, M., B. Jansen van Vuuren, A. Barnaud, and J. S. Terblanche. "Population Genetics of *Ceratitis capitata* in South Africa: Implications for Dispersal and Pest Management." *PLoS One* 8 (2013): e54281.

Kaster, R. A., ed. and trans. *Saturnalia of Macrobius, Books 3–5*. Cambridge, MA: Harvard University Press, 2011.

Katsanevakis, S., D. Poursandis, R. Hoffman, J. Rizgalla, S. B. Rothman, et al. "Unpublished Mediterranean Records of Marine Alien and Cryptogenic Species." *BioInvasions Records* 9 (2020): 165–82.

Kattan, G. H., J. Aronson, and C. Murcia. "Does the Novel Ecosystem Concept Provide a Framework for Practical Applications and a Path Forward? A Reply to Miller and Bestelmeyer." *Restoration Ecology* 24 (2016): 714–16.

Katz, E. "The Nazi Comparison in the Debate over Restoration: Nativism and Domination." *Environmental Values* 23 (2014): 377–98.

Katz, G. L., and P. B. Shafroth. "Biology, Ecology, and Management of *Elaeagnus angustifolia* L. (Russian Olive) in Western North America." *Wetlands* 23 (2003): 763–77.

Kawahara, T. "Invasion into Japanese Waters by the European Barnacle *Balanus improvisus* Darwin." *Nature* 198 (1963): 301.

Kean, J. M., D. M. Suckling, N. J. Sullivan, P. C. Tobin, L. D. Stringer, et al. "Global Eradication and Response Database." Accessed January 2, 2025. http://b3.net.nz/gerda.

Keane, R. M., and M. J. Crawley. "Exotic Plant Invasions and the Enemy Release Hypothesis." *Trends in Ecology and Evolution* 17 (2002): 164–70.

Kearney, M., B. L. Phillips, C. R. Tracy, K. A. Christian, G. Betts, and W. P. Porter. "Modelling Species Distributions Without Using Species Distributions: The Cane Toad in Australia Under Current and Future Climates." *Ecography* 31 (2008): 423–34.

Kearney, M., and W. P. Porter. "Mechanistic Niche Modelling: Combining Physiological and Spatial Data to Predict Species Ranges." *Ecology Letters* 12 (2009): 334–50.

Kearney, M. R., and W. P. Porter. "NicheMapR—an R Package for Biophysical Modelling: The Microclimate Model." *Ecography* 40 (2017): 664–74.

Keeling, C. D. "The Concentration and Isotopic Abundances of Carbon Dioxide in the Atmosphere." *Tellus* 12 (1960): 200–203.

Keeling, C. D., R. B. Bacastow, A. E. Bainbridge, C. A. Ekdahl Jr., P. R. Guenther, et al. "Atmospheric Carbon Dioxide Variations at Mauna Loa Observatory, Hawaii." *Tellus* 28 (1976): 538–51.

Kéfi, S., E. L. Berlow, E. A. Wieters, L. N. Joppa, S. A. Wood, et al. "Network Structure Beyond Food Webs: Mapping Non-Trophic and Trophic Interactions on Chilean Rocky Shores." *Ecology* 96 (2015): 291–303.

Kéfi, S., E. L. Berlow, E. A. Wieters, S. A. Navarrete, O. L. Petchey, et al. "More Than a Meal . . . Integrating Non-Feeding Interactions into Food Webs." *Ecology Letters* 15 (2012): 291–300.

Kéfi, S., V. Miele, E. A. Wieters, S. A. Navarrete, and E. L. Berlow. "How Structured Is the Entangled Bank? The Surprisingly Simple Organization of Multiplex Ecological Networks Leads to Increased Persistence and Resilience." *PLoS Biology* 14 (2016): e1002527.

Keitt, B., K. Campbell, A. Saunders, M. Clout, Y. Wang et al. "The Global Islands Invasive Vertebrate Eradication Database: A Tool to Improve and Facilitate Restoration of Island Ecosystems." In Veitch, Clout, and Towns, *Island Invasives: Eradication and Management*, 74–77.

Keller, S. R., P. D. Fields, A. E. Berardi, and D. R. Taylor. "Recent Admixture Generates Heterozygosity-Fitness Correlations During the Range Expansion of an Invading Species." *Journal of Evolutionary Biology* 27 (2014): 616–27.

Kendle, A. D., and J. E. Rose. "The Aliens Have Landed! What Are the Justifications for 'Native Only' Policies in Landscape Plantings?" *Landscape and Urban Planning* 47 (2000): 19–31.

Kennedy, H. E. "BIOSIS to the Rescue." *BioScience* 38, no. 5 (1988): 309–10.

Kennedy, P. A., K. A. Meyer, D. J. Schill, M. R. Campbell, and N. V. Vue. "Survival and Reproduction Success of Hatchery YY Male Brook Trout Stocked in Idaho Streams." *Transactions of the American Fisheries Society* 147 (2018): 419–30.

Kennedy, T. A., S. Naeem, K. M. Howe, J. M. Knops, D. Tilman, and P. Reich. "Biodiversity as a Barrier to Ecological Invasion." *Nature* 417 (2002): 636–38.

Kergoat, G. J., A. Delobel, and J-F. Silvain. "Phylogeny and Host-Specificity of European Seed Beetles (Coleoptera, Bruchidae), New Insights from Molecular and Ecological Data." *Molecular Phylogenetics and Evolution* 32 (2004): 855–65.

Kern, F. D. "Observations of the Dissemination of the Barberry." *Ecology* 2 (1921): 211–14.

Khan, K. S., C. R. Nwosu, S. F. Khan, L. S. Dwarakanath, and P. F. W. Chien. "A Controlled Analysis of Authorship Trends over Two Decades." *American Journal of Obstetrics and Gynecology* 181 (1999): 503–7.

Khlyap, L. A., V. V. Bobrov, and A. A. Warshavsky. "Biological Invasions on Russian Territory: Mammals." *Russian Journal of Biological Invasions* 1 (2010): 127–40.

Kim, S. E. "How Swaths of Invasive Grass Made Maui's Fires So Devastating." *Smithsonian Magazine*, August 15, 2023. https://www.smithsonianmag.com/smart-news/how-swaths-of-invasive-grass-made-mauis-fires-so-devastating-180982729/.

Kimmelman, B. A. "A Progressive Era Discipline: Genetics at American Agricultural Colleges and Experiment Stations, 1900–1920." PhD diss., University of Pennsylvania, 1987.

King, C. M., ed. *The Handbook of New Zealand Mammals*. Auckland: Oxford University Press, 1990.

King, C. M. *Immigrant Killers: Introduced Predators and the Conservation of Birds in New Zealand*. Auckland: Oxford University Press, 1984.

King, C. M. *Invasive Predators in New Zealand: Disaster on Four Small Paws*. Cham, Switz.: Palgrave Macmillan, 2019.

King, R. B., J. M. Ray, and K. M. Stanford. "Gorging on Gobies: Beneficial Effects of Alien Prey on a Threatened Vertebrate." *Canadian Journal of Zoology* 84 (2006): 108–15.

King, R. T. "Is It Wise Policy to Introduce Exotic Game Birds?" *Audubon Magazine* 44 (1942): 136–45, 230–36, 306–10.

King, W., and T. Krakauer. "The Exotic Herpetofauna of Southern Florida." *Quarterly Journal of Florida Academy Science* 29 (1966): 144–54.

Kingsland, S. E. "The Battling Botanist: Daniel Trembly MacDougal, Mutation Theory, and the Rise of Experimental Evolutionary Biology in America, 1900–1912." *Isis* 82 (1991): 479–509.

Kiritani, K. "Impacts of Global Warming on *Nezara viridula* and Its Native Congeneric Species." *Journal of Asia-Pacific Entomology* 14 (2011): 221–26.

Kirk, T. "The Displacement of Species in New Zealand." *Transactions of the New Zealand Institute* 28 (1895): 1–27.

Kitching, R. L., ed. *The Ecology of Exotic Animals and Plants: Some Australian Case Histories*. Brisbane: Wiley, 1986.

Kitching, R. L. "Exotics in Australia—Synopsis and Strategies." In Kitching, *Ecology of Exotic Animals and Plants: Some Australian Case Histories*, 262–69. Brisbane: Wiley, 1986.

Kitching, R. L. "The Fauna of Tree-Holes in Relation to Environmental Factors." PhD diss., Oxford University, 1969.

Kitching, R. L. *Food Webs and Container Habitats: The Natural History of Phytotelmata.* Cambridge: Cambridge University Press, 2000.

Kitching, R. L. "A World of Thought: *The Ecology of Invasions by Animals and Plants* and Charles Elton's Life's Work." In Richardson, *Fifty Years of Invasion Ecology*, 3–10.

Kitching, R. L., and R. E. Jones, eds. *The Ecology of Pests: Some Australian Case Histories.* Melbourne: CSIRO Australia, 1981.

Klassen, W., C. F. Curtis, and J. Hendrichs. "History of the Sterile Insect Technique." In *Sterile Insect Technique: Principles and Practice in Area-Wide Integrated Pest Management.* 2nd ed., edited by V. A. Dyck, J. Hendrichs, and A. S. Robinson, 1–44. Boca Raton, FL: CRC Press, 2021.

Klenk, N. L., and B. M. H. Larson. "The Assisted Migration of Western Larch in British Columbia: A Signal of Institutional Change in Forestry in Canada?" *Global Environmental Change* 31 (2015): 20–27.

Klironomos, J. N. "Feedback with Soil Biota Contributes to Plant Rarity and Invasiveness in Communities." *Nature* 417 (2002): 67–70.

Klitz, W. "Empirical Population Genetics of the North American House Sparrow." *Ornithological Monographs* 14 (1973): 34–38.

Klose, N. *America's Crop Heritage: The History of Foreign Plant Introduction by the Federal Government.* Ames: Iowa State College Press, 1950.

Knickerbocker, B. "Snaring of Hawaii's Feral Pigs Angers Animal Activists." *Christian Science Monitor*, January 6, 1994, 7.

Knight, T. M., T. L. Ashman, J. M. Bennett, J. H. Burns, S. Passonneau, and J. A. Steets. "Reflections on, and Visions for, the Changing Field of Pollination Ecology." *Ecology Letters* 21 (2018): 1282–95.

Knipling, E. F. "Possibilities in Insect Control or Eradication through the Use of Sexually Sterile Males." *Journal of Economic Entomology* 48 (1955): 459–62.

Knopf, F. L. "Avian Assemblages on Altered Grasslands." In *A Century of Avifaunal Change in Western North America*, Studies in Avian Biology 15, edited by J. R. Jehl and N. K. Johnson, 247–54. Lawrence, KS: Cooper Ornithological Society, 1994.

Knopf, F. L. "Changing Landscapes and the Cosmopolitanism of the Eastern Colorado Avifauna." *Wildlife Society Bulletin* 14 (1986): 132–42.

Knopf, F. L. "Faunal Mixing, Fauna Integrity, and the Biopolitical Template for Diversity Conservation." *Transactions of the North American Wildlife and Natural Resources Conference* 57 (1992): 330–42.

Kobell, R. "Florida's Dr. Snakehead Is Making a House Call." *Baltimore Sun*, July 19, 2002. https://www.baltimoresun.com/news/bs-xpm-2002-07-19-0207190217-story.html.

Koch, R. L., and T. L. Galvan. "Bad Side of a Good Beetle: The North American Experience with *Harmonia axyridis*." *BioControl* 53 (2008): 23–35.

Koch, R. L., R. C. Venette, and W. D. Hutchison. "Predicted Impact of an Exotic Generalist Predator on Monarch Butterfly (Lepidoptera: Nymphalidae) Populations: A Quantitative Risk Assessment." *Biological Invasions* 8 (2006): 1179–93.

Kočovský, P. M., D. C. Chapman, and S. Qian. "'Asian Carp' Is Societally and Scientifically Problematic: Let's Replace It." *Fisheries* 43 (2018): 311–16.

Koelsch, W. A. "Academic Geography, American Style: An Institutional Perspective." In *Geography: Discipline, Profession and Subject Since 1870; An International Survey*, edited by G. Dunbar, 245–79. Dordrecht, Neth.: Kluwer, 2001.

Kolbe, J. J., J. C. Ehrenberger, H. A. Moniz, and M. J. Angilletta Jr. "Physiological Variation Among Invasive Populations of the Brown Anole (*Anolis sagrei*)." *Physiological and Biochemical Zoology* 87 (2014): 92–104.

Kolbe, J. J., R. E. Glor, L. Rodríguez Schettino, A. Chamizo Lara, A. Larson, and J. B. Losos. "Multiple Sources, Admixture, and Genetic Variation in Introduced *Anolis* Lizard Populations." *Conservation Biology* 21 (2007): 1612–25.

Kolbe, J. J., R. E. Glor, L. Rodríguez Schettino, A. Chamizo Lara, A. Larson, and J. B. Losos. "Genetic Variation Increases During Biological Invasion by a Cuban Lizard." *Nature* 431 (2004): 177–81.

Koncki, N. G., and M. F. Aronson. "Invasion Risk in a Warmer World: Modeling Range Expansion and Habitat Preferences of Three Nonnative Aquatic Invasive Plants." *Invasive Plant Science and Management* 8 (2015): 436–49.

Koop, A. L., L. Fowler, L. P. Newton, and B. P. Caton. "Development and Validation of a Weed Screening Tool for the United States." *Biological Invasions* 14 (2012): 273–94.

Koricheva, J., and J. Gurevitch. "Uses and Misuses of Meta-Analysis in Plant Ecology." *Journal of Ecology* 102 (2014): 828–44.

Koricheva, J., J. Gurevitch, and K. Mengersen, eds. *Handbook of Meta-Analysis in Ecology and Evolution*. Princeton, NJ: Princeton University Press, 2013.

Kornaś, J. "Man's Impact upon the Flora and Vegetation in Central Europe." In Holzner et al., *Man's Impact on Vegetation*, 277–86.

Kornaś, J. "Plant Invasions in Central Europe: Historical and Ecological Aspects." In di Castri et al., *Biological Invasions in Europe and the Mediterranean Basin*, 19–36.

Kornberg, H., and M. H. Williamson, eds. "Quantitative Aspects of the Ecology of Biological Invasions." Special issue, *Philosophical Transactions of the Royal Society of London B* 314 (1986): 501–746.

Kowarik, I. "Biologische Invasionen in Deutschland: Zur Rolle nichteinheimischer Pflanzen." *NeoBiota* 1 (2002): 5–24.

Kowarik, I. "Time Lags in Biological Invasions with Regard to the Success and Failure of Alien Species." In *Plant Invasions: General Aspects and Special Problems*, edited by P. Pyšek, K. Prack, M. Rejmánek, and M. Wade, 15–38. Amsterdam: SPB Academic Publishing, 1995.

Kowarik, I., and P. Pyšek. "The First Steps Towards Unifying Concepts in Invasion Ecology Were Made One Hundred Years Ago: Revisiting the Work of the Swiss Botanist Albert Thellung." *Diversity and Distributions* 18 (2012): 1243–252.

Kowarik, I., and U. Starfinger. "Introduction." *Biological Invasions* 5 (2003): 279.

Kowarik, I., and U. Starfinger. "Neobiota: A European Approach." *NeoBiota* 8 (2009): 21–28.

Kraus, F. *Alien Reptiles and Amphibians: A Scientific Compendium and Analysis*. Dordrecht, Neth.: Springer, 2009.

Kraus, F. "New Records of Alien Reptiles in Hawaii." *Bishop Museum Occasional Papers* 69 (2002): 48–52.

Kraut, A. M. *Silent Travelers: Germs, Genes, and the "Immigrant Menace."* New York: Basic Books, 1994.

Krebs, C. J. *Ecology: The Experimental Analysis of Distribution and Abundance*. New York: Harper & Row, 1972.

Krebs, C. J. *Ecology: The Experimental Analysis of Distribution and Abundance*. 2nd ed. New York. Harper & Row, 1978.

Krebs, C. J. *Ecology: The Experimental Analysis of Distribution and Abundance*. 3rd ed. New York. Harper & Row, 1985.

Krebs, C. J. *Ecology: The Experimental Analysis of Distribution and Abundance*. 4th ed. New York: HarperCollins, 1994.

Krebs, C. J. *Ecology: The Experimental Analysis of Distribution and Abundance*. 5th ed. San Francisco: Benjamin Cummings, 2001.

Krebs, C. J. *Ecology: The Experimental Analysis of Distribution and Abundance*. 6th ed. San Francisco: Benjamin Cummings, 2009.

Krehenwinkel, H., and D. Tautz. "Northern Range Expansion of European Populations of the Wasp Spider *Argiope bruennichi* Is Associated with Global Warming–Correlated Genetic Admixture and Population-Specific Temperature Adaptations." *Molecular Ecology* 22 (2013): 2232–48.

Kruger, F. J. *A Description of the Fynbos Biome Project*. Pretoria: Cooperative Scientific Programmes, Council of Scientific and Industrial Research, 1978.

Kruger, F. J., G. J. Breytenbach, I. A. W. Macdonald, and D. M. Richardson. "The Characteristics of Invaded Mediterranean-Climate Regions." In Drake et al., *Biological Invasions: A Global Perspective*, 181–213.

Krysan, J. L., T. F. Branson, and G. Diaz Castro. "Diapause in *Diabrotica virgifera* (Coleoptera: Chrysomelidae): A Comparison of Eggs from Temperate and Subtropical Climates." *Entomologia Experimentalis et Applicata* 22 (1977): 81–89.

Krysko, K. L., L. A. Somma, D. C. Smith, C. R. Gillette, D. Cueva, et al. "New Verified Nonindigenous Amphibians and Reptiles in Florida, 1976 through 2015, with a Summary of Over 152 Years of Introductions." *Reptiles and Amphibians* 23 (2016): 110–43.

Kuebbing, S. E., M. A. McCary, D. Lieurance, M. A. Nuñez, M. C. Chiuffo, et al. "A Self-Study of Editorial Board Diversity at Biological Invasions." *Biological Invasions* 24 (2022): 321–32.

Kuebbing, S. E., and D. Simberloff. "Missing the Bandwagon: Nonnative Species Impacts Still Concern Managers." *NeoBiota* 25 (2015): 73–86.

Kueffer, C. "Integrating Natural and Social Sciences for Understanding and Managing Plant Invasions." In *Biodiversity and Society in the Pacific Islands*, edited by S. Larrue, 71–96. Marseille: Presses Universitaires de Provence, 2013.

Kueffer, C., and G. H. Hadorn. "How to Achieve Effectiveness in Problem-Oriented Landscape Research: The Example of Research on Biotic Invasions." *Living Reviews in Landscape Research* 2 (2008). http://www.livingreviews.org/lrlr-2008-2.

Kueffer, C., and C. A. Kull. "Non-Native Species and the Aesthetics of Nature." In Vilà and Hulme, *Impact of Biological Invasions on Ecosystem Services*, 311–24.

Kühn, I., I. Kowarik, J. Kollmann, U. Starfinger, S. Bacher, et al. "Open Minded and Open Access: Introducing *NeoBiota*, A New Peer-Reviewed Journal of Biological Invasions." *NeoBiota* 9 (2011): 1–12.

Kuhn, T. S. *The Structure of Scientific Revolutions*. Chicago: University of Chicago Press, 1962.

Kullander, S. O., and E. J. G. Ferreira. "A Review of the South American Cichlid Genus *Cichla*, with Descriptions of Nine New Species (Teleostei: Cichlidae)." *Ichthyological Exploration of Freshwaters* 17 (2006): 289–398.

Kullman, L. "Non-Analogous Tree Flora in the Scandes Mountains, Sweden, During the Early Holocene—Macrofossil Evidence of Rapid Geographic Spread and Response to Palaeoclimate." *Boreas* 27 (1998): 153–61.

Kumschick, S., S. Bacher, S. Bertolino, T. M. Blackburn, T. Evans, et al. "Appropriate Uses of EICAT Protocol, Data and Classifications." *NeoBiota* 62 (2020): 193–212.

Kumschick, S., S. Bacher, and T. M. Blackburn. "What Determines the Impact of Alien Birds and Mammals in Europe?" *Biological Invasions* 15 (2013): 785–97.

Kumschick, S., S. Bacher, W. Dawson, J. Heikkilä, A. Sendek, et al. "A Conceptual Framework for Prioritization of Invasive Alien Species for Management According to Their Impact." *NeoBiota* 15 (2012): 69–100.

Kumschick, S. and W. Nentwig. "Some Alien Birds Have as Severe an Impact as the Most Effectual Alien Mammals in Europe." *Biological Conservation* 143 (2010): 2757–62.

Kurdila, J. "The Introduction of Exotic Species into the United States: There Goes the Neighborhood." *Environmental Affairs* 16 (1995): 95–118.

Kyrou, K., A. M. Hammond, R. Galizi, N. Kranjc, A. Burt, et al. "A CRISPR–Cas9 Gene Drive Targeting *Doublesex* Causes Complete Population Suppression in Caged *Anopheles gambiae* Mosquitoes." *Nature Biotechnology* 36 (2018): 1062–66.

Lachner, E. A., C. R. Robins, and W. R. Courtenay Jr. "Exotic Fishes and Other Aquatic Organisms Introduced into North America." *Smithsonian Contributions to Zoology* 59 (1970): 1–29.

Ladefoged, T. N., M. W. Graves, and J. Coil. "The Introduction of Sweet Potato in Polynesia: Early Remains in Hawai'i." *Journal of the Polynesian Society* 114 (2005): 359–73.

Lakatos, I. "Falsification and the Methodology of Scientific Research Programs." In *Criticism and the Growth of Knowledge*, edited by I. Lakatos and A. Musgrave, 91–196. Cambridge: Cambridge University Press, 1970.

Lake, P. S., and D. J. O'Dowd. "Red Crabs in Rain Forest, Christmas Island: Biotic Resistance to Invasion by the Giant African Land Snail." *Oikos* 62 (1991): 25–29.

Lakoff, G. "Metaphor and War: The Metaphor System Used to Justify War in the Gulf." *Peace Research* 23 (2001): 25–32.

Lakoff, G., and M. Johnson. *Philosophy in the Flesh*. New York: Basic Books, 1999.

Lalasz, R. "Fighting Invasive Species: Worth the Costs? Forum with Mark Davis, Daniel Simberloff, and Peter Kareiva." *Nature Conservancy Science Chronicles*, September 2011, 2–8.

Lampman, B. H. *The Coming of the Pond Fishes*. Portland, OR: Binfords & Mort, 1946.

Lance, D. R., D. S. Leonard, V. C. Mastro, and M. L. Walters. "Mating Disruption as a Suppression Tactic in Programs Targeting Regulated Lepidopteran Pests in US." *Journal of Chemical Ecology* 42 (2016): 590–605.

Lande, R. "Genetics and Demography in Conservation." *Science* 241 (1988): 1455–60.

Lander, E. S. "The Heroes of CRISPR." *Cell* 164 (2016): 18–28.

Landi, P., M. W. McCoy, and J. R. Vonesh. "Predicting Invasive Consumer Impact via the Comparative Functional Response Approach: Linking Application to Ecological Theory." *Biological Invasions* 24 (2022): 3565–79.

Lange, W. *Gartenbilder: Mit Vorbildern aus der Natur*. Leipzig: J. J. Weber, 1922.

Lankau, R. A. "Coevolution Between Invasive and Native Plants Driven by Chemical Composition and Soil Biota." *Proceedings of the National Academy of Sciences of the United States of America* 109 (2012): 11240–45.

Lankau, R. A., V. Nuzzo, G. Spyreas, and A. S. Davis. "Evolutionary Limits Ameliorate the Negative Impact of an Invasive Plant." *Proceedings of the National Academy of Sciences of the United States of America* 106 (2009): 15362–67.

Lapeyssonie, L. *La Médecine coloniale*. Paris: Seghers, 1988.

LaPointe, D. A., C. T. Atkinson, and M. D. Samuel. "Ecology and Conservation Biology of Avian Malaria." *Annals of the New York Academy of Sciences* 1249 (2012): 211–26.

Larsen, E. L. "Peter Kalm's Short Account of the Natural Position, Use, and Care of Some Plants, of Which the Seeds Were Recently Brought Home from North America for the Service of Those Who Take Pleasure in Experimenting with the Cultivation of the Same in Our Climate." *Agricultural History* 13 (1939): 33–64.

Larson, B. M. H. *Metaphors for Environmental Sustainability: Redefining Our Relationship with Nature*. New Haven, CT: Yale University Press, 2011.

Larson, B. M. H. "The War of the Roses: Demilitarizing Invasion Biology." *Frontiers in Ecology and the Environment* 3 (2005): 495–500.

Larson, D. L., P. A. Rabie, S. Droege, J. L. Larson, and M. Haar. "Exotic Plant Infestation Is Associated with Decreased Modularity and Increased Numbers of Connectors in Mixed-Grass Prairie Pollination Networks." *PLoS One* 11, no. 5 (2016): e0155068.

Larson, E. R., B. M. Graham, R. Achury, J. J. Coon, M. K. Daniels, et al. "From eDNA to Citizen Science: Emerging Tools for the Early Detection of Invasive Species." *Frontiers in Ecology and the Environment* 18 (2020): 194–202.

Laskowski, K. L., C. Chang, K. Sheehy, and J. Aguiñaga. "Consistent Individual Behavioral Variation: What Do We Know and Where Are We Going?" *Annual Review of Ecology, Evolution, and Systematics* 53 (2022): 161–82.

Latombe, G., J. A. Catford, F. Essl, B. Lenzner, D. M. Richardson, et al. "GIRAE: A Generalised Approach for Linking the Total Impact of Invasion to Species' Range, Abundance and Per-Unit Effects." *Biological Invasions* 24 (2022): 3147–67.

Latreille, P. A. "Introduction à la géographie générale des arachnides et des insectes." *Mémoires du Muséum d'Histoire Naturelle* 3 (1817): 37–67.

Lau, J., H. R. Rothstein, and G. B. Stewart. "History and Progress of Meta-Analysis." In Koricheva et al., *Handbook of Meta-Analysis in Ecology and Evolution*, 407–19.

Lavergne, S., and J. Molofsky. "Increased Genetic Variation and Evolutionary Potential Drive the Success of an Invasive Grass." *Proceedings of the National Academy of Sciences of the United States of America* 104 (2007): 3883–88.

Lavoie, C., and J. Brisson. "Training Environmental Managers to Control Invasive Plants: Acting to Close the Knowing–Doing Gap." *Invasive Plant Science and Management* 8 (2015): 430–35.

Lavov, M. A. "Acclimatization of Mammals on Sakhalin Island." In Yanushevich, *Acclimatization of Animals in the U.S.S.R.*, 81–82.

Lawton, J. H. "Are There General Laws in Ecology?" *Oikos* 84 (1999): 177–92.

Lawton, J. H. *Community Ecology in a Changing World*. Oldendorf (Luhe): Ecology Institute, 2000.

Lawton, J. H. "Predictable Plots." *Nature* 354 (1991): 444.

Laycock, G. *The Alien Animals*. Garden City, NY: Natural History Press, 1966.

Lear, L., E. Hesse, K. Shea, and A. Buckling. "Disentangling the Mechanisms Underpinning Disturbance-Mediated Invasion." *Proceedings of the Royal Society B* 287 (2020): 20192415.

Leather, S. "Ten Papers That Shook My World—Solomon (1949)—Quantifying Predator Efficiency." *Don't Forget the Roundabouts* (blog), June 17, 2016. https://simonleather.wordpress.com/tag/maurice-solomon/.

Lee, J. H., and T. J. Kang. "Functional Response of *Harmonia axyridis* (Pallas) (Coleoptera: Coccinellidae) to *Aphis gossypii* Glover (Homoptera: Aphididae) in the Laboratory." *Biological Control* 31 (2004): 306–10.

Lees, A. C., and D. J. Bell. "A Conservation Paradox for the 21st Century: The European Wild Rabbit *Oryctolagus cuniculus*, an Invasive Alien and an Endangered Native Species." *Mammal Review* 38 (2008): 304–20.

Lefeuvre, J.-C. *Les invasions biologiques: Un danger pour la biodiversité*. Paris: Buchet Chastel, 2013.

"Legal Component of the SCOPE/IUCN Global Invasive Species Programme." Paper presented at workshop Legal and Institutional Dimensions of Invasive Alien Species Introduction and Control, Bonn, December 10–11, 1999.

Leguat, F. *Aventures aux Mascareignes*. Annotated by J.-M. Racault. Paris: Editions la Découverte, 1984.

Leguat, F. *Voyage et avantures de Francois Leguat, & et de ses compagnons en deux îles desertes des Indes Orientales*. Amsterdam: Jean Louis de Lorme, 1707.

Leikam, S. "Environmental Imaginations of the California Channel Islands and Ecological Crisis in T.C. Boyle's *When the Killing's Done*." *Ecozon@: European Journal of Literature, Culture and Environment* 5, no. 1 (2014): 136–54.

Leland, J. *Aliens in the Backyard: Plant and Animal Imports into America*. Columbia: University of South Carolina Press, 2005.

Lembrechts, J. J., A. Pauchard, J. Lenoir, M. A. Nuñez, C. Geron, et al. "Disturbance Is the Key to Plant Invasions in Cold Environments." *Proceedings of the National Academy of Sciences of the United States of America* 113, no. 49 (2016): 14061–66.

Lemoine, M., and T. Pradeu. "Dissecting the Meanings of 'Physiology' to Assess the Vitality of the Discipline." *Physiology* 33 (2018): 236–45.

Lemonick, M. D. "Invasion of the Habitat Snatchers." *Time*, September 10, 1990, 11.

Lennox, R., K. Choi, P. M. Harrison, J. E. Paterson, T. B. Peat, et al. "Improving Science-Based Invasive Species Management with Physiological Knowledge, Concepts and Tools." *Biological Invasions* 17 (2015): 2213–27.

Leonard, S. P., J. E. Powell, J. Perutka, P. Geng, L. C. Heckmann, et al. "Engineered Symbionts Activate Honey Bee Immunity and Limit Pathogens." *Science* 367 (2020): 573–76.

Leopold, A. "A Biotic View of the Land." *Journal of Forestry* 37 (1939): 727–30.

Leopold, A. "Biotic Land-Use." In Callicott and Freyfogle, *Aldo Leopold: For the Health of the Land*, 198–207. Essay written in 1942.

Leopold, A. "Bluebirds Welcome." *Wisconsin Agriculturist and Farmer* 68, no. 8 (1941): 16.

Leopold, A. "Cheat Takes Over." *The Land* 1 (1941): 310–13. Reprinted in Leopold, *Sand County Almanac*, 154–58.

Leopold, A. "Chukaremia." *Outdoor America* 3 (1938): 3.

Leopold, A. "Conservation: In Whole or in Part?" In *The River of the Mother of God and Other Essays by Aldo Leopold*, edited by S. L. Flader and J. B. Callicott, 310–19. Madison: University of Wisconsin Press, 1992.

Leopold, A. "Fifth Column of the Fencerow." *Wisconsin Agriculturist and Farmer* 68, no. 17 (1941): 11.

Leopold, A. "Forestry and Game Conservation." *Journal of Forestry* 16 (1918): 404–11.

Leopold, A. *Game and Fish Handbook*. Albuquerque: US Department of Agriculture, US Forest Service, 1915.

Leopold, A. *Game Management*. New York: Charles Scribner's Sons, 1933.

Leopold, A. "The Land-Health Concept and Conservation." In Callicott and Freyfogle, *Aldo Leopold: For the Health of the Land*, 218–26. Essay written in 1946.

Leopold, A. "The Land Ethic." In Leopold, *Sand County Almanac*, 201–26.

Leopold, A. "Land-Use and Democracy." *Audubon Magazine* 44, no. 5 (1942): 259–65.

Leopold, A. "Mixing Trout in Western Waters." *Transactions of the American Fisheries Society* 47, no. 3 (1918): 101–2.

Leopold, A. *A Sand County Almanac and Sketches Here and There*. New York: Oxford University Press, 1949.

Lepczyk, C. A. "Time to Retire 'Alien' from the Invasion Ecology Lexicon." *Frontiers in Ecology and the Environment* 20 (2022): 447.

Lepczyk, C. A., D. C. Duffy, D. M. Bird, M. Calver, D. Cherkassky, et al. "A Science-Based Policy for Managing Free-Roaming Cats." *Biological Invasions* 24 (2022): 3693–701.

Lesson, R. P. "Île de l'Ascension." In *Voyage autour du monde, exécuté par ordre du roi, sur la corvette de la majesté, la Coquille, pendant les années 1822, 1823, 1824 et 1825*, vol. 1, part 2, edited by M. L. I. DuPerrey, R. P. Lesson, and P. Garnot, 489–501. Paris: Arthus Bertrand, 1828.

Lester, P. J. *The Vulgar Wasp: The Story of a Ruthless Invader and Ingenious Predator*. Wellington, NZ: Victoria University Press, 2018.

Lester, P. J., and R. Harmsen. "Functional and Numerical Responses Do Not Always Indicate the Most Effective Predator for Biological Control: An Analysis of Two Predators in a Two-Prey System." *Journal of Applied Ecology* 39 (2002): 455–68.

Lester, P. J., D. J. Pree, H. M. A. Thistlewood, L. M. Trevisan, and R. Harmsen. "Pyrethroid Encapsulation for Conservation of Acarine Predators and Reduced Spider Mite (Acari: Tetranychidae) Outbreaks in Apple Orchards." *Environmental Entomology* 28 (1999): 72–80.

Leung, B., D. M. Lodge, D. Finnoff, J. F. Shogren, M. A. Lewis, and G. Lamberti. "An Ounce of Prevention or a Pound of Cure: Bioeconomic Risk Analysis of Invasive Species." *Proceedings of the Royal Society of London. Series B: Biological Sciences* 269 (2002): 2407–13.

Lever, C. *The Cane Toad: The History and Ecology of a Successful Colonist*. Otley, UK: Westbury, 2001.

Lever, C. *The Naturalized Animals of the British Isles*. London: Hutchinson, 1977.

Lever, C. *Naturalized Birds of the World*. Harlow, UK: Longman Scientific & Technical, 1987.

Lever, C. *Naturalized Fishes of the World*. San Diego: Academic Press, 1996.

Lever, C. *Naturalized Mammals of the World*. Harlow, UK: Longman Group, 1985.

Lever, C. *Naturalized Reptiles and Amphibians of the World*. Oxford: Oxford University Press, 2003.

Lever, C. *They Dined on Eland: The Story of the Acclimatisation Societies*. London: Quiller, 1992.

Levi, H. W. "Evaluation of Wildlife Importations." *Scientific Monthly* 74 (1952): 315–22.

Levin, M., and H. Strauss. *Risk Assessment in Genetic Engineering: Environmental Release of Organisms*. New York: McGraw-Hill, 1991.

Levin, S. A. "Ecological Issues Related to the Release of Genetically Modified Organisms into the Environment." In Mooney and Bernardi, *Introduction of Genetically Modified Organisms in the Environment*, 151–59.

Levin, S. A., and M. A. Harwell. "Potential Ecological Consequences of Genetically Engineered Organisms." *Environmental Management* 10 (1986): 495–513.

Levine, E., J. L. Spencer, S. A. Isard, D. W. Onstad, and M. E. Gray. "Adaptation of the Western Corn Rootworm to Crop Rotation: Evolution of a New Strain in Response to a Management Practice." *American Entomologist* 48 (2002): 94–107.

Levine, J. M., P. B. Adler, and S. G. Yelenik. "A Meta-Analysis of Biotic Resistance to Exotic Plant Invasions." *Ecology Letters* 7 (2004): 975–89.

Levine, J. M., and C. M. D'Antonio. "Elton Revisited: A Review of Evidence Linking Diversity and Invasibility." *Oikos* 87 (1999): 15–26.

Levins, R. *Evolution in Changing Environments: Some Theoretical Explorations*. Princeton, NJ: Princeton University Press, 1968.

Levy, O., L. B. Buckley, T. H. Keitt, C. D. Smith, K. O. Boateng, et al. "Resolving the Life Cycle Alters Expected Impacts of Climate Change." *Proceedings of the Royal Society B: Biological Sciences* 282 (2015): 20150837.

Lew, D. K. "Willingness to Pay for Threatened and Endangered Marine Species: A Review of the Literature and Prospects for Policy Use." *Frontiers in Marine Science* 2 (2015): paper 96.

Lewis, D. *Belonging on an Island: Birds, Extinction, and Evolution in Hawai'i.* New Haven, CT: Yale University Press, 2018.

Lewis, K., and R. Porter. "Global Approaches to Addressing Biofuel-Related Invasive Species Risks and Incorporation into U.S. Laws and Policies." *Ecological Monographs* 84 (2014): 171–201.

Lewis, M., dir. *Cane Toads: An Unnatural History.* Film Australia, 1988. 47 min.

Lewis, M., dir. *Cane Toads: The Conquest.* Screen Australia, Participant Media, and Discovery Studio, 2011. 85 min.

Lewis, S. *Cane Toads: An Unnatural History.* New York: Dolphin Books, 1989.

Lewontin, R. C. "Electrophoresis in the Development of Evolutionary Genetics: Milestone or Millstone?" *Genetics* 128 (1991): 657–62.

Lewontin, R. C., and J. L. Hubby. "A Molecular Approach to the Study of Genic Heterozygosity in Natural Populations, II: Amounts of Variation and Degree of Heterozygosity in Natural Populations of *Drosophila pseudoobscura.*" *Genetics* 54 (1966): 595–609.

Li, W., and M. H. Stevens. "Fluctuating Resource Availability Increases Invasibility in Microbial Microcosms." *Oikos* 121 (2012): 435–41.

Liebhold, A. M., J. Elkinton, D. Williams, and R. Muzika. "What Causes Outbreaks of the Gypsy Moth in North America?" *Population Ecology* 42 (2000): 257–66.

Liebhold, A. M., and R. L. Griffin. "The Legacy of Charles Marlatt and Efforts to Limit Plant Pest Invasions." *Bulletin of the Entomological Society of America* 62 (2016): 218–27.

Liebhold, A. M., J. Halverson, and G. Elmes. "Quantitative Analysis of the Invasion of Gypsy Moth in North America." *Journal of Biogeography* 19 (1992): 513–20.

Liebhold, A. M., and M. McManus. "The Evolving Use of Insecticides in Gypsy Moth Management." *Journal of Forestry* 97, no. 3 (1999): 20–23.

Ligtvoet, W., and F. Witte. "Perturbation Through Predator Introduction: Effect on the Food Web and Fish Yield in Lake Victoria (East Africa)." In *Proceedings of the Symposium on Perturbation and Recovery of Ecosystems, Siena, 1990,* edited by W. Ligtvoet and F. Witte, 263–68. Chichester: Ellis Horwood, 1991.

Lindeman, R. L. "The Trophic-Dynamic Aspect of Ecology." *Ecology* 23 (1942): 319–418.

Linderholm, A., D. Spencer, V. Battista, L. Frantz, R. Barnett, et al. "A Novel MC1R Allele for Black Coat Colour Reveals the Polynesian Ancestry and Hybridization Patterns of Hawaiian Feral Pigs." *Royal Society Open Science* 3 (2016): 160304.

Lindroth, C. H. *The Faunal Connections Between Europe and North America.* New York: John Wiley, 1957.

Linnaeus, C. "Essay on the Habitable Earth." *Amœnitates Academicae* 2 ([1751]): 409.

Liss, S. A., G. G. Sass, and C. D. Suski. "Spatial and Temporal Influences on the Physiological Condition of Invasive Silver Carp." *Conservation Physiology* 1, no. 1 (2013): cot017. https://doi.org/10.1093/conphys/cot017.

Liston, W. G. "'The Next War': Man *versus* Insects." *Indian Journal of Medical Research,* Special Indian Science Congress No. 6 (1919): 18–25.

Littin, K. E., C. E. O'Connor, N. G. Gregory, D. J. Mellor, and C. T. Eason. "Behaviour, Coagulopathy and Pathology of Brushtail Possums (*Trichosurus vulpecula*) Poisoned with Brodifacoum." *Wildlife Research* 29 (2002): 259–67.

Liu, D., R. Wang, D. R. Gordon, X. Sun, L. Chen, and Y. Wang. "Predicting Plant Invasions Following China's Water Diversion Project." *Environmental Science and Technology* 51 (2017): 1450–57.

Liu, Y., J. Mu, K. J. Niklas, G. Li, and S. Sun. "Global Warming Reduces Plant Reproductive Output for Temperate Multi-Inflorescence Species on the Tibetan Plateau." *New Phytologist* 195 (2012): 427–36.

Livdahl, T. P., and M. S. Willey. "Prospects for an Invasion: Competition Between *Aedes albopictus* and Native *Aedes triseriatus*." *Science* 253 (1991): 189–91.

Livezey, K. B. "Range Expansion of Barred Owls, Part II: Facilitating Ecological Changes." *American Midland Naturalist* 161 (2009): 323–49.

Lizarralde, M., J. Escobar, and G. Deferrari. "Invader Species in Argentina: A Review About the Beaver (*Castor canadensis*) Population Situation on Tierra del Fuego Ecosystem." *Interciencia* 29 (2004): 352–56.

Locey, F. H. "Introduced Game Birds of Hawaii." *Paradise of the Pacific* 49, no. 1 (1937): 5–6, 27–30.

Lockwood, J. A. *Six-Legged Soldiers*. New York: Oxford University Press, 2009.

Lockwood, J. L., M. F. Hoopes, and M. P. Marchetti. *Invasion Ecology*. Malden, MA: Blackwell, 2007.

Lockwood, J. L., M. F. Hoopes, and M. P. Marchetti. *Invasion Ecology*. 2nd ed. Chichester, UK: Wiley-Blackwell, 2013.

Lockwood, J. L., and M. L. McKinney, eds. *Biotic Homogenization*. New York: Kluwer, 2001.

Lockwood, J. L., D. Simberloff, M. McKinney, and B. Von Holle. "How Many, and Which, Plants Will Invade Natural Areas?" *Biological Invasions* 3 (2001): 1–8.

Lodge, D. M., P. W. Simonin, S. W. Burgiel, R. P. Keller, J. M. Bossenbroek, et al. "Risk Analysis and Bioeconomics of Invasive Species to Inform Policy and Management." *Annual Review of Environment and Resources* 41 (2016): 453–88.

Long, J. L. *Introduced Birds of the World*. New York: Universe Books, 1981.

Long, J. L. *Introduced Mammals of the World: Their History, Distribution and Influence*. Wallingford, UK: CABI, 2003.

Long, L. L., and J. D. Wolfe. "Review of the Effects of Barred Owls on Spotted Owls." *Journal of Wildlife Management* 83 (2019): 1281–96.

Lonsdale, W. M. "Inviting Trouble: Introduced Pasture Species in Northern Australia." *Australian Journal of Ecology* 19 (1994): 345–54.

Loope, L. L., and D. Mueller-Dombois. "Characteristics of Invaded Islands, with Special Reference to Hawaii." In Drake et al., *Biological Invasions: A Global Perspective*, 257–80.

Lopes-Andrade, A., and A. Lüer. "The Invasive Beetle *Cis bilamellatus* (Coleoptera: Ciidae) Arrives in America." *Florida Entomologist* 97 (2014): 1348–52.

Lopezaraiza-Mikel, M., R. Hayes, M. Whalley, and J. Memmott. "The Impact of an Alien Plant on a Native Plant–Pollinator Network: An Experimental Approach." *Ecology Letters* 10 (2007): 539–50.

Lord, E., C. Collins, S. deFrance, M. J. Lefebvre, and E. Matisoo-Smith. "Complete Mitogenomes of Ancient Caribbean Guinea Pigs (*Cavia porcellus*)." *Journal of Archaeological Science: Reports* 17 (2018): 678–88.

Louda, S. M., A. E. Arnett, T. Rand, and F. L. Russell. "Invasiveness of Some Biological Control Insects and Adequacy of Their Ecological Risk Assessment and Regulation." *Conservation Biology* 17 (2003): 73–82.

Louda, S. M., D. Kendall, J. Connor, and D. Simberloff. "Ecological Effects of an Insect Introduced for the Biological Control of Weeds." *Science* 277 (1997): 1088–90.

Louda, S. M., T. A. Rand, F. L. Russell, and A. E. Arnett. "Assessment of Ecological Risks in Weed Biocontrol: Input from Retrospective Ecological Analyses." *Biological Control* 35 (2005): 253–64.

Low, T. *Feral Future: The Untold Story of Australia's Exotic Invaders*. Ringwood, Australia: Viking, 1999.

Lowe, S. J. "From the Editor." *Aliens* 1 (1995): 2.

Lowe, S. J., M. Browne, and S. Boudjelas. *100 of the World's Worst Invasive Alien Species*. Auckland: IUCN/SSC Invasive Species Specialist Group (ISSG), 2000.

Lowenthal, D. *Forest Stewardship: Marsh, Pinchot, and America Today*. Milford, PA: Grey Towers Press, 2001.

Lowenthal, D. "George Perkins Marsh and the American Geographical Tradition." *Geographical Review* 43 (1953): 207–13.

Lowenthal, D. *George Perkins Marsh, Prophet of Conservation*. Seattle: University of Washington Press, 2000.

Lowenthal, D. Foreword to *Man and Nature*, by G. P. Marsh, edited by D. Lowenthal. Seattle: University of Washington Press, 2003.

Lozon, J. D., and H. J. MacIsaac. "Biological Invasions: Are They Dependent on Disturbance?" *Environmental Reviews* 5 (1997): 131–44.

Lubchenco, J. "Plant Species Diversity in a Marine Intertidal Community: Importance of Herbivore Food Preference and Algal Competitive Abilities." *American Naturalist* 112 (1978): 23–39.

Lubina, J. A., and S. A. Levin. "The Spread of a Reinvading Species: Range Expansion in the California Sea Otter." *American Naturalist* 133 (1988): 526–43.

Lucy, F. E., and V. E. Panov. "BioInvasions Records: A New International Journal on Biological Invasions." *BioInvasions Records* 1 (2012): 1–4.

Ludsin, S. A., and A. D. Wolfe. "Biological Invasion Theory: Darwin's Contributions from *The Origin of Species*." *BioScience* 51 (2001): 780–89.

Luikart, G., P. R. England, D. Tallmon, S. Jordan, and P. Taberlet. "The Power and Promise of Population Genomics: From Genotyping to Genome Typing." *Nature Reviews Genetics* 4 (2003): 981–94.

Lukey, P., and J. Hall. "Biological Invasion Policy and Legislation Development and Implementation in South Africa." In van Wilgen et al., *Biological Invasions in South Africa*, 515–52.

Lundgren, B. *Soil Conditions and Nutrient Cycling Under Natural and Plantation Forests in Tanzanian Highlands*. Uppsala: Swedish University of Agricultural Sciences (SLU), 1978.

Luo, L., X. Kong, Z. Gao, Y. Zheng, Y. Yang, et al. "Comparative Transcriptome Analysis Reveals Ecological Adaption of Cold Tolerance in Northward Invasion of *Alternanthera philoxeroides*." *BMC Genomics* 21 (2020): 1–17.

Luque, G. M., C. Bellard, C. Bertelsmeier, E. Bonnaud, P. Genovesi, et al. "The 100th of the World's Worst Invasive Alien Species." *Biological Invasions* 16 (2014): 981–85.

Lurgi, M., N. Galiana, B. C. López, L. N. Joppa, and J. M. Montoya. "Network Complexity and Species Traits Mediate the Effects of Biological Invasions on Dynamic Food Webs." *Frontiers in Ecology and Evolution* 2 (2014): 1–11.

Lyell, C. *Principles of Geology*. Vol. 2. London: John Murray, 1832.

Lynch, A. J. J., R. Thackway, A. Specht, P. J. Beggs, S. Brisbane, et al. "Transdisciplinary Synthesis for Ecosystem Science, Policy, and Management: The Australian Experience." *Science of the Total Environment* 534 (2015): 173–84.

Mabuchi, K., H. Senou, and M. Nishida. "Mitochondrial DNA Analysis Reveals Cryptic Large-Scale Invasion of Non-Native Genotypes of Common Carp (*Cyprinus carpio*) in Japan." *Molecular Ecology* 17 (2008): 796–809.

MacArthur, R. H. "Fluctuations of Animal Populations and a Measure of Community Stability." *Ecology* 36 (1955): 533–36.

MacArthur, R. H. "Patterns of Terrestrial Bird Communities." In *Avian Biology*, vol. 1, edited by D. S. Farner and J. R. King, 189–221. New York: Academic Press, 1971.

MacArthur, R. H. "Population Ecology of Some Warblers of Northeastern Coniferous Forests." *Ecology* 39 (1958): 599–619.

MacArthur, R. H. "Species-Packing and Competitive Equilibrium for Many Species." *Theoretical Population Biology* 1 (1970): 1–11.

MacArthur, R. H., and E. O. Wilson. *The Theory of Island Biogeography*. Princeton, NJ: Princeton University Press, 1967.

MacDonald, E., J. MacFadyen, and I. Novaczek, eds. *Time and Place: An Environmental History of Prince Edward Island*. Montreal: McGill-Queen's University Press, 2016.

Macdonald, I. A. W., and G. W. Frame. "The Invasion of Introduced Species into Nature Reserves in Tropical Savannas and Dry Woodlands." *Biological Conservation* 44 (1988): 67–94.

Macdonald, I. A. W., D. M. Graber, S. DeBenedetti, R. H. Groves, and E. R. Fuentes. "Introduced Species in Nature Reserves in Mediterranean-Type Climatic Regions of the World." *Biological Conservation* 44 (1988): 37–66.

Macdonald, I. A. W., F. J. Kruger, and A. A. Ferrar, eds. *The Ecology and Management of Biological Invasions in Southern Africa*. Cape Town: Oxford University Press, 1986.

Macdonald, I. A. W., L. L. Loope, M. B. Usher, and O. Hamann. "Wildlife Conservation and the Invasion of Nature Reserves by Introduced Species: A Global Perspective." In Drake et al., *Biological Invasions: A Global Perspective*, 215–55.

MacDougal, D. T. "The Reactions of Plants to New Habitats." *Ecology* 2 (1921): 1–20.

MacDougall, A. S., and R. Turkington. "Are Invasive Species the Drivers or Passengers of Change in Degraded Ecosystems?" *Ecology* 86 (2005): 42–55.

Mace, G. M., and R. Lande. "Assessing Extinction Threats: Toward a Reevaluation of IUCN Threatened Species Categories." *Conservation Biology* 5 (1991): 148–57.

Mack, R. N. "Alien Plant Invasion into the Intermountain West: A Case History." In Mooney and Drake, *Ecology of Biological Invasions of North America and Hawaii*, 191–213.

Mack, R. N. "Biotic Barriers to Plant Naturalization." In *Proceedings of the Ninth International Symposium on Biological Control of Weeds*, edited by V. C. Moran and J. H. Hoffmann, 19–26. Stellenbosch, South Africa: University of Cape Town, 1996.

Mack, R. N. "Cheatgrass." In Simberloff and Rejmánek, *Encyclopedia of Biological Invasions*, 108–13.

Mack, R. N. "Invasion of *Bromus tectorum* L. into Western North America: An Ecological Chronicle." *Agro-Ecosystems* 7 (1981): 145–65.

Mack, R. N. "Plant Naturalizations and Invasions in the Eastern United States: 1634–1860." *Annals of the Missouri Botanical Garden* 90 (2003): 77–90.

Mack, R. N. "Temperate Grasslands Vulnerable to Plant Invasion: Characteristics and Consequences." In Drake et al., *Biological Invasions: A Global Perspective*, 155–79.

Mack, R. N., L. E. Neville, and J. W. B. Stewart. Preface to Mooney et al., *Invasive Alien Species: A New Synthesis*, xiii–xvi.

Mack, R. N., D. Simberloff, M. Lonsdale, H. Evans, M. Clout, and F. A. Bazzaz. "Biotic Invasions: Causes, Epidemiology, Global Consequences, and Control." *Ecological Applications* 10 (2000): 689–710.

Mackin, C. R., J. F. Peña, M. A. Blanco, N. J. Balfour, and M. C. Castellanos. "Rapid Evolution of a Floral Trait Following Acquisition of Novel Pollinators." *Journal of Ecology* 109 (2021): 2234–46.

Macmillan, A. *Mauritius Illustrated.* London: W. H. & L. Collingridge, 1914.

Madsen, D. H. "Protection of Native Fishes in the National Parks." *Transactions of the American Fisheries Society* 66 (1937): 395–97.

Maguire, K. C., D. Nieto-Lugilde, J. L. Blois, M. C. Fitzpatrick, J. W. Williams, et al. "Controlled Comparison of Species- and Community-Level Models Across Novel Climates and Communities." *Proceedings of the Royal Society B: Biological Sciences* 283 (2016): 20152817. https://doi.org/10.1098/rspb.2015.2817.

Maguire, K. C., D. Nieto-Lugilde, M. C. Fitzpatrick, J. W. Williams, and J. L. Blois. "Modeling Species and Community Responses to Past, Present, and Future Episodes of Climatic and Ecological Change." *Annual Review of Ecology, Evolution, and Systematics* 46 (2015): 343–68.

Mahmood-ur-Rahman, I. A., T. Husnain, and S. Riazuddin. "RNA Interference: The Story of Gene Silencing in Plants and Humans." *Biotechnology Advances* 26 (2008): 202–9.

Maier, D. S., and D. Simberloff. "Assisted Migration in Normative and Scientific Context." *Journal of Agricultural and Environmental Ethics* 29 (2016): 857–82.

Major, J. "Historical Development of the Ecosystem Concept." In *The Ecosystem Concept in Natural Resource Management*, edited by G. M. Van Dyne, 9–22. New York: Academic Press, 1969.

Maloy, O. C. "White Pine Blister Rust." *Plant Health Instructor* 3 (2003). Updated 2018. https://doi.org/10.1094/PHI-I-2003-0908-01.

Maloy, O. C. "White Pine Blister Rust Control in North America: A Case History." *Annual Review of Phytopathology* 35 (1997): 87–109.

Maltez-Mouro, S., F. T. Maestre, and H. Freitas. "Weak Effects of the Exotic Invasive *Carpobrotus edulis* on the Structure and Composition of Portuguese Sand-Dune Communities." *Biological Invasions* 12 (2010): 2117–30.

Mammola, S., J. Malumbres-Olarte, V. Arabesky, D. A. Barrales-Alcalá, A. L. Barrion-Dupo, et al. "The Global Spread of Misinformation on Spiders." *Current Biology* 32 (2022): R871–R873.

Mammola, S., V. Nanni, P. Pantini, and M. Isaia. "Media Framing of Spiders May Exacerbate Arachnophobic Sentiments." *People and Nature* 2 (2020): 1145–57.

Mann, C. C. *1493: Uncovering the New World Columbus Created.* New York: Alfred A. Knopf, 2011.

Manning, J. C., and C. Paterson-Jones. *Field Guide to Fynbos.* Cape Town: Struik, 2007.

Maranesi, M., A. Bufalari, C. Dall'Aglio, D. Paoloni, G. Moretti, et al. "Reproductive Traits of an Invasive Alien Population of Grey Squirrel (*Sciurus carolinensis*) in Central Italy." *Animals* 10 (2020): 738.

Marchant, L. F., and W. C. McGrew. "Laterality of Function in Apes—A Meta-Analysis of Methods." *Journal of Human Evolution* 21 (1991): 425–38.

Marché, J. D., III. *The Green Menace: Emerald Ash Borer and the Invasive Species Problem.* New York: Oxford University Press, 2017.

Marchetti, M. P., and T. Engstrom. "The Conservation Paradox of Endangered and Invasive Species." *Conservation Biology* 30 (2016): 434–37.

Marin, P., J. Genitoni, D. Barloy, S. Maury, P. Gibert, et al. "Biological Invasion: The Influence of the Hidden Side of the (Epi) Genome." *Functional Ecology* 34 (2020): 385–400.

Marks, C. O., and R. G. Van Driesche. "Designing Restoration Programs Based on Understanding the Drivers of Ecological Change." In Van Driesche et al., *Integrating Biological Control into Conservation Practice*, 4–21.

Marlatt, C. L. *Danger of General Spread of the Gipsy and Brown-Tail Moths Through Imported Nursery Stock.* US Department of Agriculture Farmers Bulletin 433. Washington, DC: Government Printing Office, 1911.

Marlatt, C. L. "Insect Control in California." In *Yearbook of the United States Department of Agriculture, 1896,* 217–36. Washington, DC: Government Printing Office, 1897.

Marlatt, C. L. "Losses Caused by Imported Tree and Plant Pests." *American Forestry* 23 (1917): 75–80.

Marlatt, C. L. *The Periodical Cicada.* Bureau of Entomology Bulletin 71. Washington, DC: US Department of Agriculture, 1907.

Marlatt, C. L. "Some Recent New Importations." *Journal of Economic Entomology* 5 (1912): 73–77.

Marlin, D., E. R. Smit, and M. J. Byrne. "A Successful Biocontrol Agent in the USA, *Diorhabda carinulata* (Coleoptera: Chrysomelidae) on *Tamarix* spp. (Tamaricaceae), Rejected in South Africa Due to Insufficient Host Specificity." *Biological Control* 136 (2019): 104002.

Marmion, M., M. Parviainen, M. Luoto, R. K. Heikkinen, and W. Thuiller. "Evaluation of Consensus Methods in Predictive Species Distribution Modelling." *Diversity and Distributions* 15 (2009): 59–69.

Marquitti, F. M. D., P. R. Guimarães Jr., M. M. Pires, and L. F. Bittencourt. "MODULAR: Software for the Autonomous Computation of Modularity in Large Network Sets." *Ecography* 37 (2014): 221–24.

Marris, E. "Ragamuffin Earth." *Nature* 460 (2009): 450–53.

Marris, E. *Rambunctious Garden: Saving Nature in a Post-Wild World.* New York: Bloomsbury, 2011.

Marris, E. "Reconstituted Edens," review of *When the Killing's Done,* by T. C. Boyle. *Nature* 472 (2011): 414–15.

Marris, E. "When Conservationists Kill Lots (and Lots) of Animals." *Atlantic,* September 26, 2018. https://www.theatlantic.com/science/archive/2018/09/is-wildlife-conservation-too-cruel/569719/.

Marsh, G. P. *The Camel: His Organization, Habits, and Uses, Considered with Reference to His Introduction into the United States.* Boston: Gould and Lincoln, 1856.

Marsh, G. P. *The Earth as Modified by Human Action.* New York: Scribner, Armstrong, 1874.

Marsh, G. P. *Man and Nature.* Edited by D. Lowenthal. Seattle: University of Washington Press, 2003.

Marsh, G. P. *Man and Nature; or, Physical Geography as Modified by Human Action.* New York: Scribner, 1864.

Marsh, G. P. *Report Made Under Authority of the Legislature of Vermont, on the Artificial Propagation of Fish.* Burlington: Free Press Print, 1857.

Marsh, O. "Invasion of the Habitat Snatchers." *Planet,* June 16, 2019, 11–12.

Martin, L. B., A. J. Brace, H. J. Kilvitis, and S. S. Gervasi. "Invader Endocrinology: The Regulation of Behaviour in Pesky Phenotypes." In Weis and Sol, *Biological Invasions and Animal Behaviour,* 47–62.

Martin, P. S. "Prehistoric Overkill: The Global Model." In Martin and Klein, *Quaternary Extinctions,* 354–403.

Martin, P. S., and R. G. Klein, eds. *Quaternary Extinctions: A Prehistoric Revolution.* Tucson: University of Arizona Press, 1984.

Martinez, B., J. K. Reaser, A. Dehgan, B. Zamft, D. Baisch, et al. "Technology Innovation: Advancing Capacities for the Early Detection of and Rapid Response to Invasive Species." *Biological Invasions* 22 (2020): 75–100.

Martiny, J. B. H., B. J. M. Bohannan, J. H. Brown, R. K. Colwell, J. A. Furhman, et al. "Microbial Biogeography: Putting Microorganisms on the Map." *Nature Reviews Microbiology* 4 (2006): 102–12.

Marvier, M., E. Meir, and P. Kareiva. "How Do the Design of Monitoring and Control Strategies Affect the Chance of Detecting and Containing Transgenic Weeds?" In *Methods for Risk Assessment of Transgenic Plants,* vol. 3, edited by K. Ammann, Y. Jacot, V. Simonson, and G. Kjellsson, 109–22. Basel: Birkhauser Verlag Press, 1999.

Marzano, A. *Harvesting the Sea: The Exploitation of Marine Resources in the Roman Mediterranean.* Oxford: Oxford University Press, 2013.

Mastnak, T., J. Elyachar, and T. Boellstorff. "Botanical Decolonization: Rethinking Native Plants." *Environmental Planning D: Society and Space* 32 (2014): 363–80.

Matisoo-Smith, E. "Ancient DNA and the Human Settlement of the Pacific: A Review." *Journal of Human Evolution* 79 (2015): 93–104.

Matsuhashi, S., H. Doi, A. Fujiwara, S. Watanabe, and T. Minamoto. "Evaluation of the Environmental DNA Method for Estimating Distribution and Biomass of Submerged Aquatic Plants." *PLoS One* 11 (2016): e0156217.

Matthews, G. A. *A History of Pesticides.* Wallingford, UK: CABI, 2018.

Matthews, L. H. *British Mammals.* London: Collins New Naturalist, 1952.

Matzek, V., J. Covino, J. L. Funk, and M. Saunders. "Closing the Knowing–Doing Gap in Invasive Plant Management: Accessibility and Interdisciplinarity of Scientific Research." *Conservation Letters* 7 (2014): 208–15.

Matzke, N. J. "Model Selection in Historical Biogeography Reveals that Founder-Event Speciation Is a Crucial Process in Island Clades." *Systematic Biology* 63 (2014): 951–70.

Mauvisseau, Q., A. Coignet, C. Delaunay, F. Pinet, D. Bouchon, and C. Souty-Grosset. "Environmental DNA as an Efficient Tool for Detecting Invasive Crayfishes in Freshwater Ponds." *Hydrobiologia* 805 (2018): 163–75.

May, R. M. *Stability and Complexity in Model Ecosystems.* Princeton, NJ: Princeton University Press, 1973.

May, R. M. "Will a Large Complex System Be Stable?" *Nature* 238 (1972): 413–14.

Mayr, E. "The Nature of Colonizations in Birds." In Baker and Stebbins, *Genetics of Colonizing Species,* 29–47.

Mayr, E., and J. Diamond. *The Birds of Northern Melanesia: Speciation, Ecology, and Biogeography.* Oxford: Oxford University Press, 2001.

McAtee, W. L. "Introduction upon Introduction." *Auk* 42 (1925): 160.

McAtee, W. L. *Game Birds Suitable for Naturalizing in the United States.* Circular No. 96. Washington, DC: US Department of Agriculture, 1929.

McCabe, R. A., and A. S. Hawkins. "The Hungarian Partridge in Wisconsin." *American Midland Naturalist* 35 (1946): 1–76.

McCann, K. S. "The Diversity-Stability Debate." *Nature* 405 (2000): 228–33.

McCartney, A. P. "Alvin R. Cahn: World War II Aleutian Archaeologist." *Arctic Anthropology* 35 (1998): 136–46.

McChesney, H. M., and J. T. Anderson. "Reproductive Success of Field Sparrows (*Spizella pusilla*) in Response to Invasive Morrow's Honeysuckle: Does Morrow's Honeysuckle Promote Population Sinks?" *Wilson Journal of Ornithology* 127 (2015): 222–32.

McClelland, M. "Invasion of the Bio-Snatchers." *Florida Environments* 6, no. 2 (1992): 5, 13.

McClure, M. L., C. R. Hranac, C. G. Haase, S. McGinnis, B. G. Dickson, et al. "Projecting the Compound Effects of Climate Change and White-Nose Syndrome on North American Bat Species." *Climate Change Ecology* 3 (2022): 100047.

McCook, L. J. "Understanding Ecological Community Succession: Causal Models and Theories; A Review." *Vegetatio* 110 (1994): 115–47.

McCue, M. D., M. Javal, S. Clusella-Trullas, J. J. Le Roux, M. C. Jackson, et al. "Using Stable Isotope Analysis to Answer Fundamental Questions in Invasion Ecology: Progress and Prospects." *Methods in Ecology and Evolution* 11 (2020): 196–214.

McDonald, G., P. Zambino, and R. Sniezko. "Breeding Rust-Resistant Five-Needle Pines in the Western United States: Lessons from the Past and a Look to the Future." In *Breeding and Genetic Resources of Five-Needle Pines: Growth, Adaptability and Pest Resistance*, Proceedings RMRS-P-32, edited by R. A. Sniezko, S. Samman, S. E. Schlarbaum, and H. B. Kriebel, 28–50. Fort Collins, CO: USDA Forest Service Rocky Mountain Research Station, 2004.

McDowall, R. M. *Gamekeepers for the Nation: The Story of New Zealand's Acclimatisation Societies 1861–1990.* Christchurch: Canterbury University Press, 1994.

McFadyen, A. *Animal Ecology: Aims and Methods.* 2nd ed. London: Pitman, 1963.

McFadyen, R. E. C. "Biological Control of Weeds." *Annual Review of Entomology* 43 (1998): 369–93.

McGaw, B., and G. V. Glass. "Choice of the Metric for Effect Size in Meta-Analysis." *American Educational Research Journal* 17 (1980): 325–37.

McGeary, M. N. *Gifford Pinchot: Forester—Politician.* Princeton, NJ: Princeton University Press, 1960.

McGeoch, M. A., J. D. Shaw, A. Terauds, J. E. Lee, and S. L. Chown. "Monitoring Biological Invasion Across the Broader Antarctic: A Baseline and Indicator Framework." *Global Environmental Change* 32 (2015): 108–25.

McGill, B. J., B. J. Enquist, E. Weiher, and M. Westoby. "Rebuilding Community Ecology from Functional Traits." *Trends in Ecology and Evolution* 21 (2006): 178–85.

McGonigle, B., J. K. Presnail, and N. S. Mutti. *Compositions and Methods for Insecticidal Control of Stinkbugs.* US Patent Application Publication (10) Pub. No. US 2016/0108425A1. Washington, DC: US Patent and Trademark Office, 2016.

McGrath, Z., F. MacDonald, G. Walker, and D. Ward. "A Framework for Predicting Competition Between Native and Exotic Hymenopteran Parasitoids of Lepidopteran Larvae Using Taxonomic Collections and Species Level Traits." *BioControl* 66 (2021): 59–69.

McHarg, I. L. *Design with Nature.* New York: Wiley, 1969.

McIlroy, J. C. "New Techniques for an Old Problem—Recent Advances in Feral Pig Control in Australia." *Journal of Mountain Ecology* 3 (1995): 241–44.

McInerney, C. J., and N. Smith. "The Victorian Weed Alert Program: Weed Spotters: An Integral Surveillance Tool." In *Proceedings of a Symposium Held on Surveillance for Biosecurity: Pre-Border to Pest Management*, edited by K. J. Froud, I. A. Popay, and S. M. Zydenbos, 175–82. Paihia, NZ: New Zealand Plant Protection Society, 2008.

McIntosh, R. P. *The Background of Ecology: Concept and Theory.* Cambridge: Cambridge University Press, 1985.

McKey, D. B., and S. C. Kaufmann. "Naturalization of Exotic *Ficus* Species (Moraceae) in South Florida." In *Proceedings of the Symposium on Exotic Pest Plants*, Technical Report NPS/NREVER/NRTR-91/06, edited by T. D. Center, R. F. Doren, R. L. Hofstetter, R. L. Myers, and L. D. Whiteaker, 221–36. Washington, DC: United States Department of the Interior, National Park Service, 1991.

McKibben, B. "Invasion of the Habitat Snatchers." *Boston Globe*, May 15, 2005. http://archive.boston.com/ae/books/articles/2005/05/15/invasion_of_the_habitat_snatchers/.

McKinney, M. L., and J. L. Lockwood. "Biotic Homogenization: A Few Winners Replacing Many Losers in the Next Mass Extinction." *Trends in Ecology and Evolution* 14 (1999): 450–53.

McKnight, B. N., ed. *Biological Pollution: The Control and Impact of Invasive Exotic Species.* Indianapolis: Indiana Academy of Science, 1993.

McKnight, E., R. Spake, A. Bates, D. A. Smale, and M. Rius. "Non-Native Species Outperform Natives in Coastal Marine Ecosystems Subjected to Warming and Freshening Events." *Global Ecology and Biogeography* 30 (2021): 1698–712.

McLachlan, G. J. *Discriminant Analysis and Statistical Pattern Recognition.* New York: Wiley, 1992.

McLachlan, J. S., J. J. Hellmann, and M. W. Schwartz. "A Framework for Debate of Assisted Migration in an Era of Climate Change." *Conservation Biology* 21 (2007): 297–302.

McLane, S. C., and S. N. Aitken. "Whitebark pine (*Pinus albicaulis*) Assisted Migration Potential: Testing Establishment North of the Species Range." *Ecological Applications* 22 (2012): 142–53.

McLane, W. M. "The Aquatic Plant Business in Relation to Infestations of Exotic Aquatic Plants in Florida Waters." *Journal of Aquatic Plant Management* 8 (1969): 48–49.

McNeely, J. A., H. A. Mooney, L. E. Neville, P. Schei, and J. K. Waage, eds. *A Global Strategy on Invasive Alien Species.* Gland, Switz.: International Union for Conservation of Nature, 2001.

McNeely, J. A., H. A. Mooney, L. E. Neville, P. J. Schei, and J. K. Waage. "A Global Strategy on Invasive Alien Species: Synthesis and Ten Strategic Elements." In Mooney et al., *Invasive Alien Species: A New Synthesis*, 332–45.

McNeill, J. R. "The World According to Jared Diamond." *History Teacher* 34 (2001): 165–74.

McNeill, W. H. "Guns, Germs, and Steel (Reply)." *New York Review of Books* 44, no. 11 (1997): 69–70.

McNeill, W. H. "History Upside Down." *New York Review of Books* 44, no. 8 (1997): 48–50.

McNeill, W. H. *Plagues and Peoples.* New York: Doubleday, 1976.

McNeill, W. H. *The Rise of the West: A History of the Human Community.* Chicago: University of Chicago Press, 1963.

McNeish, R. E., and R. W. McEwan. "A Review on the Invasion Ecology of Amur Honeysuckle (*Lonicera maackii*, Caprifoliaceae), a Case Study of Ecological Impacts at Multiple Scales." *Journal of the Torrey Botanical Society* 143 (2016): 367–85.

Mead, A. R. "Ecological Malacology: With Particular Reference to *Achatina fulica*." In *Pulmonates*, vol. 2b, edited by V. Fretter, V. J. Fretter, and J. Peake. London: Academic Press, 1979.

Medd, R. W., and R. C. G. Smith. "Prediction of the Potential Distribution of *Carduus nutans* (Nodding Thistle) in Australia." *Journal of Applied Ecology* 15 (1978): 603–12.

Medvedev, Z. A. *The Rise and Fall of T. D. Lysenko.* Garden City, NY: Doubleday, 1971.

Meffe, G. K., and C. R. Carroll. *Principles of Conservation Biology.* Sunderland, MA: Sinauer Associates, 1994.

Meffe, G. K., and C. R. Carroll. *Principles of Conservation Biology.* 2nd ed. Sunderland, MA: Sinauer Associates, 1997.

Meinesz, A. *Killer Algae: The True Tale of a Biological Invasion.* Chicago: University of Chicago Press, 1999.

Meinesz, A. *Le roman noir de l'algue "tueuse": Caulerpa taxifolia contre la Méditerranée*. Paris: Éditions Belin, 1997.

Meinig, D. W. *The Great Columbia Plain: A Historical Geography, 1805–1910*. Seattle: University of Washington Press, 1968.

Meinig, D. W. "Prologue: Andrew Hill Clark, Historical Geographer." In *European Settlement and Development in North America: Essays on Geographical Change in Honour and Memory of Andrew Hill Clark*, edited by J. R. Gibson, 3–26. Toronto: University of Toronto Press, 1978.

Memmott, J., S. V. Fowler, and R. L. Hill. "The Effect of Release Size on the Probability of Establishment of Biological Control Agents: Gorse Thrips (*Sericothrips staphylinus*) Released Against Gorse (*Ulex europaeus*) in New Zealand." *Biocontrol Science and Technology* 8 (1998): 103–15.

Memmott, J., N. M. Waser, and M. V. Price. "Tolerance of Pollination Networks to Species Extinctions." *Proceedings of the Royal Society of London, Series B* 271 (2004): 2605–11.

Merrill, E. D. *Plant Life of the Pacific World*. New York: Macmillan, 1945.

Meshaka, W. E., Jr., B. P. Butterfield, and J. B. Hauge. *The Exotic Amphibians and Reptiles of Florida*. Malabar, FL: Krieger, 2004.

Messenger, P. S. "Bioclimatic Studies with Insects." *Annual Review of Entomology* 4 (1959): 183–206.

Messenger, P. S. "Bioclimatic Inputs to Biological Control and Pest Management Programs." In *Concepts of Pest Management*, edited by R. L. Rabb and F. E. Guthrie, 84–99. Raleigh: North Carolina State University, 1970.

Messenger, P. S., and N. E. Flitters. "Bioclimatic Studies of Three Species of Fruit Flies in Hawaii." *Journal of Economic Entomology* 47 (1954): 756–65.

Metcalf, C. L., and W. P. Flint. *Destructive and Useful Insects: Their Habits and Control*. 3rd ed. New York: McGraw-Hill, 1951.

Meudt, H. M., and A. C. Clarke. "Almost Forgotten or Latest Practice? AFLP Applications, Analyses and Advances." *Trends in Plant Science* 12 (2007): 106–17.

Meyer, P. W. "John Bartram: Botanist & Horticulturist." *Journal of Arboriculture* 4 (1978): 165–68.

Meyer, W. M., N. W. Yeung, J. Slapcinsky, and K. A. Hayes. "Two for One: Inadvertent Introduction of *Euglandina* Species During Failed Bio-Control Efforts in Hawaii." *Biological Invasions* 19 (2017): 1399–405.

Meyerson, L. A., and D. Simberloff, eds. "Early Detection and Rapid Response." Special issue, *Biological Invasions* 22, no. 1 (2020).

"Miami Snail Control." *Science* 166 (1969): 355.

Michel, A., R. S. Arias, B. E. Scheffler, S. O. Duke, M. Netherland, and F. E. Dayan. "Somatic Mutation-Mediated Evolution of Herbicide Resistance in the Nonindigenous Invasive Plant Hydrilla (*Hydrilla verticillata*)." *Molecular Ecology* 13 (2004): 3229–37.

Michener, R., and K. Lajtha, eds. *Stable Isotopes in Ecology and Environmental Science*. 2nd ed. Malden, MA: Blackwell, 2007.

Middleton, A. D. "The Distribution of the Grey Squirrel (*Sciurus carolinensis*) in Great Britain in 1935." *Journal of Animal Ecology* 4 (1935): 274–76.

Middleton, A. D. "The Ecology of the American Grey Squirrel (*Sciurus carolinensis* Gmelin) in the British Isles." *Proceedings of the Zoological Society of London* 100 (1930): 809–43.

Middleton, A. D. *The Grey Squirrel: The Introduction and Spread of the American Grey Squirrel in the British Isles, Its Habits, Food, and Relations with the Native Fauna of the Country*. London: Sidgwick & Jackson, 1931.

Middleton, A. D. "The Grey Squirrel (*Sciurus carolinensis*) in the British Isles 1930–1932." *Journal of Animal Ecology* 1 (1932): 166–67.

Migaud, P. "A First Approach to Links Between Animals and Life on Board Sailing Vessels (1500–1800)." *International Journal of Nautical Archaeology* 40 (2011): 283–92.

Mikheyev, A. S., T. Vo, B. Wee, M. C. Singer, and C. Parmesan. "Rapid Microsatellite Isolation from a Butterfly by *de novo* Transcriptome Sequencing: Performance and a Comparison with AFLP-derived Distances." *PLoS One* 5, no. 6 (2010): e11212.

Milgroom, M. G., and P. Cortesi. "Biological Control of Chestnut Blight with Hypovirulence: A Critical Analysis." *Annual Review of Phytopathology* 42 (2004): 311–38.

Millennium Ecosystem Assessment. *Ecosystems and Human Well-Being: Synthesis.* Washington, DC: Island Press, 2005.

Miller, James (Jim) R., and L. J. Gut. "Mating Disruption for the 21st Century: Matching Technology with Mechanism." *Environmental Entomology* 44 (2015): 427–53.

Miller, James Robert, and B. T. Bestelmeyer. "What's Wrong with Novel Ecosystems, Really?" *Restoration Ecology* 24 (2016): 577–82.

Miller, James Robert, and B. T. Bestelmeyer. "What the Novel Ecosystem Concept Provides: A Reply to Kattan et al." *Restoration Ecology* 25 (2017): 488–90.

Miller, M. L., and R. N. Fabian. *Harmful Invasive Species: Legal Responses.* Washington, DC: Environmental Law Institute, 2004.

Miller, R. C. "Ecological Relations of Marine Wood-Boring Organisms in San Francisco Bay." *Ecology* 7 (1926): 247–54.

Miller, R. C. "The Marine Borer Problem on the Pacific Coast." *University of Washington Forest Club Quarterly* 5, no. 3 (1926): 5–11.

Miller, R. R., and J. R. Alcorn. "The Introduced Fishes of Nevada, with a History of Their Introduction." *Transactions of the American Fisheries Society* 73, no. 1943 (1945): 173–93.

Miller-Rushing, A. J., and R. B. Primack. "Global Warming and Flowering Times in Thoreau's Concord: A Community Perspective." *Ecology* 89 (2008): 332–41.

Miller-Rushing, A. J., R. B. Primack, and R. Bonney. "The History of Public Participation in Ecological Research." *Frontiers in Ecology and the Environment* 10 (2012): 285–90.

Millet, C. *Les Merveilles des fleuves et des ruisseaux.* Paris: Hachette, 1871.

Milne, L. J., and M. Milne. *The Balance of Nature.* New York: Knopf, 1960.

Milo, R., S. Shen-Orr, S. Itzkovitz, N. Kashtan, D. Chklovskii, and U. Alon. "Network Motifs: Simple Building Blocks of Complex Networks." *Science* 298 (2002): 824–27.

Ministry of the Environment, Government of Japan. "Declaration of the Eradication of the Small Indian Mongoose (Designated Invasive Alien Species) in Amami Oshima Island." Press release, September 3, 2024. https://www.env.go.jp/en/press/press_03205.html.

Mitchell, C. "The Liberator: How One Man's 15,000 Pest Fish Changed New Zealand's Waterways." Stuff. January 25, 2020. www.stuff.co.nz/national/118845051/the-liberator-how-one-mans-15000-pest-fish-changed-new-zealands-waterways.

Mitchell, R. C., and R. T. Carson. *Using Surveys to Value Public Goods.* New York: RFF Press, 1989.

Mithen, S. J., and J. H. Lawton. "Food-Web Models That Generate Constant Predator-Prey Ratios." *Oecologia* 69 (1986): 542–50.

Mittan, C. S., and K. R. Zamudio. "Rapid Adaptation to Cold in the Invasive Cane Toad *Rhinella marina.*" *Conservation Physiology* 7, no. 1 (2019): coy075. https://doi.org/10.1093/conphys/coy075.

Mlot, C. "Spread of Tropical Livestock Virus Linked to Climate Change." *BioScience* 56 (2006): 1028.

Moffitt, C. M., G. Whelan, and R. Jackson. "Historical Perspectives on Inland Fisheries Management in North America." In *Inland Fisheries Management in North America*. 3rd ed., edited by W. A. Hubert and M. C. Quist, 141. Bethesda, MD: American Fisheries Society, 2010.

Mohr, E. "The Muskrat, *Ondatra zibethica* (Linnaeus), in Europe." *Journal of Mammalogy* 14 (1933): 58–63.

Moles, A. T., H. Flores-Moreno, S. P. Bonser, D. I. Warton, A. Helm, et al. "Invasions: The Trail Behind, the Path Ahead, and a Test of a Disturbing Idea." *Journal of Ecology* 100 (2012): 116–27.

Molinari, N., C. M. D'Antonio, and G. Thomson. "*Carpobrotus* as a Case Study of the Complexities of Species Impacts." In *Ecosystem Engineers: Concepts, Theory and Application in Ecology*, edited by K. Cuddington, J. Byers, A. Hastings, and W. Wilson, 139–62. San Diego, CA: Academic Press, 2007.

Molnar, M. L. *Playing God in the Meadow: How I Learned to Admire My Weeds*. Amherst, MA: Bright Leaf, 2022.

"The Mongoose." *Proceedings of the Agricultural Society of Trinidad and Tobago* 18 (1918): 603–30.

Montague, T. L., ed. *The Brushtail Possum*. Lincoln, New Zealand: Manaaki Whenua Press, 2000.

Montgomery, G. M. "'Never So at Home': Charles Elton and the Woods at Wytham." *Notes and Records* 75 (2021): 223–38.

Moody, M. E., and R. N. Mack. "Controlling the Spread of Plant Invasions: The Importance of Nascent Foci." *Journal of Applied Ecology* 25 (1988): 1009–21.

Mooney, H. A. *The Globalization of Ecological Thought*. Oldendorf (Luhe): Ecology Institute, 1998.

Mooney, H. A. "The Global Invasive Species Program (GISP)." *Biological Invasions* 1 (1999): 97–98.

Mooney, H. A., and G. Bernardi, eds. *Introduction of Genetically Modified Organisms in the Environment*. Chichester: Wiley, 1990.

Mooney, H. A., and J. A. Drake, eds. *Ecology of Biological Invasions of North America and Hawaii*. New York: Springer, 1986.

Mooney, H. A., and J. A. Drake. "The Release of Genetically Designed Organisms in the Environment: Lessons from the Study of the Ecology of Biological Invasions." In Mooney and Bernardi, *Introduction of Genetically Modified Organisms in the Environment*, 117–29.

Mooney, H. A., and P. Ehrlich. "Ecosystem Services: A Fragmentary History." In Daily, *Nature's Services*, 11–19.

Mooney, H. A., P. J. Ferrar, and R. O. Slatyer. "Photosynthetic Capacity and Carbon Allocation Patterns in Diverse Growth forms of *Eucalyptus*." *Oecologia* 36 (1978): 103–11.

Mooney, H. A., S. P. Hamburg, and J. A. Drake. "The Invasion of Plants and Animals into California." In Mooney and Drake, *The Ecology of Biological Invasions of North America and Hawaii*, 250–72.

Mooney, H. A., and R. J. Hobbs, eds. *Invasive Species in a Changing World*. Washington, DC: Island Press, 2000.

Mooney, H. A., R. N. Mack, J. A. McNeely, L. E. Neville, P. J. Schei, and J. K. Waage, eds. *Invasive Alien Species: A New Synthesis*. Washington, DC: Island Press, 2005.

Mooney, J., M. R. Lynch, C. V. Prescott, T. Clegg, M. Loughlin, et al. "VKORC1 Sequence Variants Associated with Resistance to Anticoagulant Rodenticides in Irish Populations of *Rattus norvegicus* and *Mus musculus domesticus*." *Scientific Reports* 8 (2018): 4535.

Moore, D. M. "Human Impact on Island Vegetation." In Holzner et al., *Man's Impact on Vegetation*, 237–46.

Moore, J. P. "Use of Fishes for Control of Mosquitos in Northern Fresh Waters of the United States." Report of the US Commissioner of Fisheries for 1922, Appendix 4, 1–60. Washington, DC: Government Printing Office, 1922.

Moore, N. W. "The Heaths of Dorset and Their Conservation." *Journal of Ecology* 50 (1962): 369–91.

Morais, P., and M. Reichard. "Cryptic Invasions: A Review." *Science of the Total Environment* 613–14 (2018): 1438–48.

Moran, E. V., and J. M. Alexander. "Evolutionary Responses to Global Change: Lessons from Invasive Species." *Ecology Letters* 17 (2014): 637–49.

Moran, V. C., and J. H. Hoffmann. "The Fourteen International Symposia on Biological Control of Weeds, 1969–2014: Delegates, Demographics and Inferences from the Debate on Non-Target Effects." *Biological Control* 87 (2015): 23–31.

Moran, V. C., and H. G. Zimmermann. "The Biological Control of Cactus Weeds: Achievements and Prospects." *Biocontrol News and Information* 5 (1984): 297–320.

Morange, M. *A History of Molecular Biology*. Cambridge, MA: Harvard University Press, 2000.

Moreau, J. "Analysis of Species Changes in Lake Victoria Using ECOPATH, a Multispecies Trophic Model." In *The Impact of Species Changes in African Lakes*, edited by T. J. Pitcher and P. J. V. Hart, 137–61. London: Chapman & Hall, 1995.

Morgan, R., and M. Shorten. "Breeding of the Woodcock in Britain." *Bird Study* 21 (1974): 193–99.

Mori, E., R. Amerini, G. Mazza, S. Bertolino, R. Battiston, et al. "Alien Shades of Grey: New Occurrences and Relevant Spread of *Sciurus carolinensis* in Italy." *European Journal of Ecology* 2 (2016): 13–20.

Morin, R. S., A. M. Liebhold, E. R. Luzader, A. J. Lister, K. W. Gottschalk, and D. B. Twardus. *Mapping Host-Species Abundance of Three Major Exotic Forest Pests*. Forest Service Northeastern Research Station Research Paper NE-726. Washington, DC: US Department of Agriculture, 2004.

Morin, X., and W. Thuiller. "Comparing Niche-and Process-Based Models to Reduce Prediction Uncertainty in Species Range Shifts Under Climate Change." *Ecology* 90 (2009): 1301–13.

Moro, D., M. Byrne, M. Kennedy, S. Campbell, and M. Tizard. "Identifying Knowledge Gaps for Gene Drive Research to Control Invasive Animal Species: The Next CRISPR Step." *Global Ecology and Conservation* 13 (2018): e00363.

Morrison, S. A., A. J. DeNicola, K. Walker, D. Dewey, L. Laughrin, et al. "An Irruption Interrupted: Eradication of Wild Turkeys *Meleagris gallopavo* from Santa Cruz Island, California." *Oryx* 50 (2016): 121–27.

Mortensen, H. S., Y. L. Dupont, and J. M. Olesen. "A Snake in Paradise: Disturbance of Plant Reproduction Following Extirpation of Bird Flower-Visitors on Guam." *Biological Conservation* 141 (2008): 2146–54.

Morton, J. F. "Pestiferous Spread of Many Ornamental and Fruit Species in South Florida." *Proceedings of the Florida State Horticultural Society* 89 (1976): 348–53.

Morton, R. D., and R. Law. "Regional Species Pools and the Assembly of Local Ecological Communities." *Journal of Theoretical Biology* 187 (1997): 321–31.

Moss, V. D. "Antibiotics for Control of Blister Rust on Western White Pine." *Forest Science* 7 (1961): 380–96.

Mott, J. J. "Planned Invasions of Australian Tropical Savannas." In Groves and Burdon, *Ecology of Biological Invasions*, 89–96.

Moulton, M. P., W. P. Cropper Jr., and M. L. Avery. "A Reassessment of the Role of Propagule Pressure in Influencing Fates of Passerine Introductions to New Zealand." *Biodiversity and Conservation* 20 (2011): 607–23.

Moulton, M. P., W. P. Cropper, A. J. Broz, and S. A. Gezan. "Patterns of Success in Game Bird Introductions in the United States." *Biodiversity and Conservation* 27 (2018): 967–79.

Moutinho, S. "A Golden Menace." *Science* 374 (2021): 390–93.

Moyle, P. B. "Fish Introductions into North America: Patterns and Ecological Impact." In Mooney and Drake, *Ecology of Biological Invasions of North America and Hawaii*, 27–43.

Mueller, J. M., and J. J. Hellmann. "An Assessment of Invasion Risk from Assisted Migration." *Conservation Biology* 22 (2008): 562–67.

Mukherjee, S. "The Invasion Equation." *New Yorker* 93, no. 27 (2017): 40–49.

Mullard, A. "Run, Forest, Run: Helping Trees Flee Climate Change." *Walrus*, updated April 18, 2024. https://thewalrus.ca/run-forest-run/.

Mulligan, G. A. "Recent Colonization by Herbaceous Plants in Canada." In Baker and Stebbins, *Genetics of Colonizing Species*, 127–43.

Mullis, K., F. Faloona, S. Scharf, R. Saiki, G. Horn, and H. Erlich. "Specific Enzymatic Amplification of DNA in vitro: The Polymerase Chain Reaction." *Cold Spring Harbor Symposia in Quantitative Biology* 51 (1986): 263–73.

Mulvaney, M. "The Effect of Introduction Pressure on the Naturalization of Ornamental Woody Plants in South-Eastern Australia." In *Weed Risk Assessment*, edited by F. D. Panetta, J. G. Virtue, and R. H. Groves, 186–93. Collingwood Australia: CSIRO, 2001.

Mumford, J. D., O. Booy, R. H. A. Baker, M. Rees, G. H. Copp, et al. "Invasive Non-Native Species Risk Assessment in Great Britain." *Aspects of Applied Biology* 104 (2010): 49–54.

Munakamwe, Z., A. Constantine, and C. McInerney. "The Victorian Weed Spotters: Recruitment and Training of Citizen Scientists." In *Proceedings of the 21st Australasian Weeds Conference*, edited by S. Johnson, L. Weston, H. W. Wu, and B. Auld, 193–96. Muswellbrook: Weed Society of New South Wales, 2018.

Mungall, E. C., and W. J. Sheffield. *Exotics on the Range: The Texas Example*. College Station: Texas A&M University Press, 1994.

Muniappan, R. "Biological Control of the Giant African Snail, *Achatina fulica* Bowdich, in the Maldives." *FAO Plant Protection Bulletin* 35 (1987): 127–33.

Muniappan, R., G. Duhamel, R. M. Santiago, and D. R. Acay. "Giant African Snail Control in Bugsuk Island, Philippines, by *Platydemus manokwari*." *Oléagineux* 41 (1986): 183–86.

Muñoz, E. N. *Caulerpa Conquest: A Biological Eradication on the California Coast*. Bloomington, IN: Open Books Press, 2016.

Muñoz-Fuentes, V., C. Vilà, A. J. Green, J. Negro, and M. D. Sorenson. "Hybridization Between White-Headed Ducks and Introduced Ruddy Ducks in Spain." *Molecular Ecology* 16 (2007): 629–38.

Munro, D., J. Steer, and W. Linklater. "On Allegations of Invasive Species Denialism." *Conservation Biology* 33 (2019): 797–802.

Murcia, C., J. Aronson, G. H. Kattan, D. Moreno-Mateos, K. Dixon, and D. Simberloff. "A Critique of the 'Novel Ecosystem' Concept." *Trends in Ecology and Evolution* 29 (2014): 548–53.

Murdoch, W. W. "Diversity, Complexity, Stability and Pest Control." *Journal of Applied Ecology* 12 (1975): 795–807.

Murdoch, W. W. "The Population Ecology of Certain Carabid Beetles Living in Marshes and Near Fresh Water." DPhil diss., Oxford University, 1963.

Murdoch, W. W., J. Chesson, and P. L. Chesson. "Biological Control in Theory and Practice." *American Naturalist* 125 (1985): 344–66.

Murie, O. J. "Wildlife Introductions in Alaska." *Transactions of the Fifth North American Wildlife Conference* 5 (1941): 432–36.

Murrill, W. A. "A New Chestnut Disease." *Torreya* 6 (1906): 186–89.

Musolin, D. L. "Surviving Winter: Diapause Syndrome in the Southern Green Stink Bug Nezara viridula in the Laboratory, in the Field, and Under Climate Change Conditions." *Physiological Entomology* 37 (2012): 309–22.

Myers, G. S. "An American Cyprinodont Fish, *Jordanella floridae*, Reported from Borneo, with Notes on the Possible Widespread Introduction of Foreign Aquarium Fishes." *Copeia* 1940 (1940): 267–68.

Myers, G. S. "Foreign Introductions of North American Fishes: Inadvisability of Recommending North American Fishes Without Careful Appraisal of Foreign Fishes and Ecology." *Progressive Fish-Culturist* 9 (1947): 177–80.

Myers, G. S. "*Gambusia*, the Fish Destroyer." *Tropical Fish Hobbyist* 13 (1965): 31–32, 53–54.

[Myers, G. S.]. "The Use of Predatory Vertebrate Animals." In *The Biological Control of Insects*, edited by H. L. Sweetman, 318–39. Ithaca, NY: Comstock, 1936. [Although the book does not credit Myers with this contribution, convincing evidence suggests that he wrote it. See E. S. Herald, "Annotated Chronological Bibliography of George Sprague Myers," *Proceedings of the California Academy of Sciences*, 4th ser., 38, no. 2 (1979): 19–52.]

Myers, J. G. "The Mongoose: The Present Position of the Mongoose in the West Indies." *Tropical Agriculture* 8, no. 4 (1931): 94–95.

Myers, J. H., and D. R. Bazely. *Ecology and Control of Introduced Plants*. Cambridge: Cambridge University Press, 2003.

Myers, K. "The Rabbit in Australia." In *Dynamics of Numbers in Populations*, edited by P. J. den Boer and G. R. Bradwell, 478–506. Wageningen, Neth.: Centre for Agricultural Publishing and Documentation, 1971.

Myers, K., I. Parer, D. Wood, and B. D. Cooke. "The Rabbit in Australia." In H. Thompson and King, *European Rabbit*, 108–57.

Myers, N. *The Primary Source: Tropical Forests and Our Future*. New York: W. W. Norton, 1984.

Nadel, H., J. H. Frank, and R. J. Knight. "Escapees and Accomplices—The Naturalization of Exotic *Ficus* and Their Associated Faunas in Florida." *Florida Entomologist* 75 (1992): 29–38.

Nasimovich, A. A. "Some General Problems and Results of Acclimatization of Terrestrial Animals." *Zoologicheskii Zhurnal* 40 (1961): 957–61.

Nasimovich, A. A. "Theoretical Basis of the Acclimatization of Land Mammals." In Yanushevich, *Acclimatization of Animals in the U.S.S.R.*, 15–17.

Nass, M. M. K., S. Nass, and B. A. Afzelius. "The General Occurrence of Mitochondrial DNA." *Experimental Cell Research* 37 (1965): 190–200.

National Academies of Sciences, Engineering, and Medicine. *Gene Drives on the Horizon: Advancing Science, Navigating Uncertainty, and Aligning Research with Public Values*. Washington, DC: National Academies Press, 2016.

National Archives. "Records of the Bureau of Plant Industry, Soils, and Agricultural Engineering (BPISAE): 54.2; General Records of the Bureau of Plant Industry and BPISAE, 1885–1953." Accessed November 11, 2024. https://www.archives.gov/research/guide-fed-records/groups/054.html#54.2.

National Invasive Species Council. *Meeting the Invasive Species Challenge: National Invasive Species Management Plan*. Washington, DC: United States Department of the Interior, 2001.

National Park Service. *Administrative Policies for Natural Areas of the National Park System*. Washington, DC: US Government Printing Office, 1968.

National Park Service. "Mountain Goat Removal Complete for 2020." News release, November 9, 2020. https://www.nps.gov/olym/learn/news/mountain-goat-removal-complete-for-2020.htm.

National Research Council. *Field Testing Genetically Modified Organisms: Framework for Decisions*. Washington, DC: National Academy Press, 1989.

National Research Council. *Introduction of Recombinant DNA-Engineered Organisms into the Environment*. Washington, DC: National Academy Press, 1987.

Naylor, E. "A North American Xanthoid Crab New to Britain." *Nature* 187 (1960): 256–57.

Nazni, W. A., A. A. Hoffmann, A. N. Afizah, Y. L. Cheong, M. V. Mancini, et al. "Establishment of *Wolbachia* Strain wAlbB in Malaysian Populations of *Aedes aegypti* for Dengue Control." *Current Biology* 29 (2019): 4241–48.

Nebeker, F. *Calculating the Weather: Meteorology in the 20th Century*. New York: Academic Press, 1995.

Neidermyer, W. J., and J. J. Hickey. "The Monk Parakeet in the United States, 1970–75." *American Birds* 31 (1977): 273–78.

Neilson, R. P., and L. H. Wullstein. "Biogeography of Two Southwest American Oaks in Relation to Atmospheric Dynamics." *Journal of Biogeography* 10 (1983): 275–97.

Nellis, D. W., and C. O. R. Everard. "The Biology of the Mongoose in the Caribbean." *Studies on the Fauna of Curaçao and Other Caribbean Islands* 64, no. 195 (1983): 1–162.

Nelson, G. "From Candolle to Croizat: Comments on the History of Biogeography." *Journal of the History of Biology* 11 (1978): 269–305.

Nentwig, W., E. Kühnel, and S. Bacher. "A Generic Impact-Scoring System Applied to Alien Mammals in Europe." *Conservation Biology* 24 (2010): 302–11.

Nesler, T. P., and E. P. Bergersen, eds. *Mysids in Fisheries: Hard Lessons from Headlong Introductions*. Bethesda, MD: American Fisheries Society, 1991.

Nettleship, D. N., J. Burger, and M. Gochfeld. *Seabirds on Islands: Threats, Case Studies and Action Plans*. Cambridge: BirdLife International, 1994.

Newhouse, A. E., L. D. Polin-McGuigan, K. A. Baier, K. E. Valletta, W. H. Rottmann, et al. "Transgenic American Chestnuts Show Enhanced Blight Resistance and Transmit the Trait to T1 Progeny." *Plant Science* 228 (2014): 88–97.

Newsom, L. A., and E. S. Wing. *On Land and Sea: Native American Uses of Biological Resources in the West Indies*. Tuscaloosa: University of Alabama Press, 2004.

Newsome, A. E., and I. R. Noble. "Ecological and Physiological Characters of Invading Species." In Groves and Burdon, *Ecology of Biological Invasions*, 1–20.

Newton, I. *Newton's Principia: The Mathematical Principles of Natural Philosophy; First American Edition, Carefully Revised and Corrected*. New York: Daniel Adee, 1848.

Nguyen, T. H., H. L. Nguyen, T. Y. Nguyen, S. N. Vu, N. D. Tran, et al. "Field Evaluation of the Establishment Potential of *w*melpop *Wolbachia* in Australia and Vietnam for Dengue Control." *Parasites and Vectors* 8 (2015): 563.

Nichols, F. H., J. K. Thompson, and L. E. Scheme. "Remarkable Invasion of San Francisco Bay (California, USA) by the Asian Clam *Potamocorbula amurensis*, II: Displacement of a Former Community." *Marine Ecology Progress Series* 66 (1990): 95–101.

Nicholson, A. J. "The Balance of Animal Populations." *Journal of Animal Ecology* 2 (1933): 132–78.

Nicholson, A. J. "Progress in the Control of *Hypericum* by Insects." In *Eighth International Congress of Entomology*, 96–99. Stockholm: Axel B. Elfstroms Boktryckeri AB, 1950.

Nico, L. G., and J. D. Williams. *Risk Assessment on Black Carp (Pisces: Cyprinidae).* Gainesville, FL: United States Department of the Interior, National Biological Service, 1996.

Nico, L. G., J. D. Williams, and H. L. Jelks. *Black Carp: Biological Synopsis and Risk Assessment of an Introduced Fish.* Bethesda, MD: American Fisheries Society, 2005.

Niemelä, P., and W. J. Mattson. "Invasion of North American Forests by European Phytophagous Insects." *BioScience* 46 (1996): 741–53.

Nix, H. A. "A Biogeographic Analysis of Australian Elapid Snakes." In *Atlas of Elapid Snakes of Australia*, Australian Flora and Fauna Series 7, edited by R. Longmore, 4–15. Canberra: Bureau of Flora and Fauna, 1986.

Nix, H. A., J. McMahon, and D. Mackenzie. "Potential Areas of Production and the Future of Pigeon Pea and Other Grain Legumes in Australia." In *The Potential for Pigeon Pea in Australia: Proceedings of Pigeon Pea (Cajanus cajan [L.] Millsp.) Field Day*, edited by E. S. Wallis and P. C. Whiteman, 5/1–5/12. Brisbane: University of Queensland, 1977.

Nix, H. A., and A. J. Wapshere. "Biogeographic Origin of Invading Species." In Groves and Burdon, *Ecology of Biological Invasions*, 155.

Nobel Prize. "The 1973 Nobel Prize in Physiology or Medicine." Nobel Prize Outreach AB 2024. Accessed December 18, 2024. https://www.nobelprize.org/prizes/medicine/1973/summary/.

Noble, C., B. Adlam, G. M. Church, K. M. Esvelt, and M. A. Nowak. "Current CRISPR Gene Drive Systems Are Likely to Be Highly Invasive in Wild Populations." *eLife*, June 19, 2018. https://doi.org/10.7554/eLife.33423.

Noble, I. R. "Attributes of Invaders and the Invading Process: Terrestrial and Vascular Plants." In Drake et al., *Biological Invasions: A Global Perspective*, 301–13.

Noble, I. R., and R. O. Slatyer. "The Effect of Disturbance on Plant Succession." In D. Anderson, *Exotic Species in Australia—Their Establishment and Success*, 135–45.

Norby, R. J., and D. R. Zak. "Ecological Lessons from Free-Air CO_2 Enrichment (FACE) Experiments." *Annual Review of Ecology, Evolution, and Systematics* 42 (2011): 181–203.

Nori, J., J. N. Urbina-Cardona, R. D. Loyola, J. N. Lescano, and G. C. Leynaud. "Climate Change and American Bullfrog Invasion: What Could We Expect in South America?" *PLoS One* 6, no. 10 (2011): e25718.

North-Coombes, A. *The Vindication of François Leguat.* 2nd ed. Stanley Rose-Hill, Mauritius: Editions de l'Océan Indien, 1991.

Norton, Bob. "The Hydrilla Menace." *Water Spectrum* 11, no. 2 (1979): 36–43.

Norton, Bryan G. "Environmental Ethics and Weak Anthropocentrism." *Environmental Ethics* 6 (1984): 131–48.

"Note." *Bulletin of Foreign Plant Introductions*, no. 67 (1911): 468.

Novaczek, I. "The Mermaid's Tresses: Seaplants in the Culture and Economy of Prince Edward Island." In E. MacDonald et al., *Time and Place: An Environmental History of Prince Edward Island*, 140–60.

Novak, M., J. D. Yeakel, A. E. Noble, D. F. Doak, M. Emmerson, et al. "Characterizing Species Interactions to Understand Press Perturbations: What Is the Community Matrix?" *Annual Review of Ecology, Evolution, and Systematics* 47 (2016): 409–32.

Novak, S. J., and R. N. Mack. "Genetic Bottlenecks in Alien Plant Species." In Sax et al., *Species Invasions: Insights into Ecology, Evolution, and Biogeography*, 201–28.

Novoa, A., S. Canavan, K. T. Faulkner, P. Genovesi, D. Lieurance, et al. "Announcing the Invasions Newsletter." *Biological Invasions* 25 (2025): 159. https://doi.org/10.1007/s10530-025-03621-5.

Nowak, D. J., and R. A. Rowntree. "History and Range of Norway Maple." *Journal of Arboriculture* 16 (1990): 291–96.

Nowak, R. S., D. S. Ellsworth, and S. D. Smith. "Functional Responses of Plants to Elevated Atmospheric CO_2—Do Photosynthetic and Productivity Data from FACE Experiments Support Early Predictions?" *New Phytologist* 162 (2004): 253–80.

Noyes, R. *Wordsworth and the Art of Landscape*. Bloomington: Indiana University Press, 1968.

Nuñez, M. A., and T. Amano. "Monolingual Searches Can Limit and Bias Results in Global Literature Reviews." *Nature Ecology and Evolution* 5 (2021): 264.

Nuñez, M. A., J. Barlow, M. Cadotte, K. Lucas, E. Newton, et al. "Assessing the Uneven Global Distribution of Readership, Submissions and Publications in Applied Ecology: Obvious Problems Without Obvious Solutions." *Journal of Applied Ecology* 56 (2019): 4–9.

Nuñez, M. A., M. C. Chiuffo, A. Pauchard, and R. D. Zenni. "Making Ecology Really Global." *Trends in Ecology and Evolution* 36 (2021): 766–69.

Nuñez, M. A., M. C. Chiuffo, H. Seebens, S. Kuebbing, M. A. McCary, et al. "Two Decades of Data Reveal That *Biological Invasions* Needs to Increase Participation Beyond North America, Europe, and Australasia." *Biological Invasions* 24 (2022): 333–40.

Nuñez, M. A., R. D. Dimarco, and D. Simberloff. "Why Some Exotic Species Are Deeply Integrated into Local Cultures While Others Are Reviled." In Rozzi et al., *From Biocultural Homogenization to Biocultural Conservation*, 219–31.

Nuñez, M. A., and A. Pauchard. "Biological Invasions in Developing and Developed Countries: Does One Model Fit All?" *Biological Invasions* 12 (2010): 707–14.

Nuñez, M. A., and D. Simberloff. "Invasive Species and the Cultural Keystone Concept." *Ecology and Society* 10, no. 1 (2005): r4. http://www.ecologyandsociety.org/vol10/iss1/resp4/.

Nuñez, M. A., D. Simberloff, and M. A. Relva. "Seed Predation as a Barrier to Alien Conifer Invasions." *Biological Invasions* 10 (2008): 1389–98.

Nyamukondiwa, C., E. Kleynhans, and J. S. Terblanche. "Phenotypic Plasticity of Thermal Tolerance Contributes to the Invasion Potential of Mediterranean Fruit Flies (*Ceratitis capitata*)." *Ecological Entomology* 35 (2010): 565–75.

O'Brien, W. "Exotic Invasions, Nativism, and Ecological Restoration: On the Persistence of a Contentious Debate." *Ethics, Place, and Environment* 9 (2006): 63–77.

O'Connor, M. P., S. J. Kemp, S. J. Agosta, F. Hansen, A. E. Sieg, et al. "Reconsidering the Mechanistic Basis of the Metabolic Theory of Ecology." *Oikos* 116 (2007): 1058–72.

O'Dowd, D. J., P. T. Green, and P. S. Lake. "Invasional 'Meltdown' on an Oceanic Island." *Ecology Letters* 6 (2003): 812–17.

Odum, E. P. *Basic Ecology*. Philadelphia: Saunders College Publishing, 1983.

Odum, E. P. *Ecology*. New York: Holt, Rinehart and Winston, 1963.

Odum, E. P. *Fundamentals of Ecology*. Philadelphia: W. B. Saunders, 1953.

Odum, E. P. *Fundamentals of Ecology*. 2nd ed. In collaboration with H. Odum. Philadelphia: W. B. Saunders, 1959.

Odum, E. P. *Fundamentals of Ecology*. 3rd ed. Philadelphia: W. B. Saunders, 1971.

Odum, E. P., and G. W. Barrett. *Fundamentals of Ecology*. 5th ed. Belmont, CA: Brooks/Cole, 2005.

Oduor, A. M. O., J. M. Gómez, and S. Y. Strauss. "Exotic Vertebrate and Invertebrate Herbivores Differ in Their Impacts on Native and Exotic Plants: A Meta-Analysis." *Biological Invasions* 12 (2010): 407–19.

Oetting, J. "Controversy Grows over Recipe for Rodent Problem on Farallon Islands: Rat Poison." *Mercury News* (California), June 1, 2021. https://www.mercurynews.com/2021/06/01/controversy-grows-over-recipe-for-rodent-problem-on-farallon-islands-rat-poison/.

Office of Technology Assessment, US Congress. *Harmful Non-Indigenous Species in the United States.* Report no. OTA-F-565. Washington, DC: US Government Printing Office, 1993.

Ogram, A., G. Sayler, and T. Barkay. "The Extraction and Purification of Microbial DNA from Sediments." *Journal of Microbiological Methods* 7 (1987): 57–66.

Okubo, A. *Diffusion and Ecological Problems: Mathematical Models.* Berlin: Springer-Verlag, 1980.

Oldfield, S., C. Lusty, and A. MacKinven. *The World List of Threatened Trees.* Cambridge: World Conservation Press, 1998.

Olesen, J., L. Eskildsen, and S. Venkatasamy. "Invasion of Pollination Networks on Oceanic Islands: Importance of Invader Complexes and Endemic Super Generalists." *Diversity and Distributions* 8 (2002): 181–92.

Oliver, W. R. B. *New Zealand Birds.* Wellington: A. H. and A. D. Reed, 1930.

Olmsted, F. L. "Foreign Plants and American Scenery." Letter to the editor, *Garden and Forest* 1 (1888): 418–19.

Olson, L. J., and S. Roy. "The Economics of Controlling a Stochastic Biological Invasion." *American Journal of Agricultural Economics* 84 (2002): 1311–16.

Olwig, K. R. "Natives and Aliens in the National Landscape." *Landscape Research* 28 (2003): 61–74.

O'Malley, M. A. "'Everything Is Everywhere: But the Environment Selects': Ubiquitous Distribution and Ecological Determinism in Microbial Biogeography." *Studies in History and Philosophy of Science Part C: Studies in History and Philosophy of Biological and Biomedical Sciences* 39 (2008): 314–25.

O'Neil, D. "English as the Lingua Franca of International Publishing." *World Englishes* 37 (2018): 146–65.

O'Neill, G. A., N. K. Ukrainetz, M. Carlson, C. V. Cartwright, B. C. Jaquish, et al. *Assisted Migration to Address Climate Change in British Columbia: Recommendations for Interim Seed Transfer Standards.* Technical Report 048. Victoria: British Columbia Ministry of Forest and Range, Research Branch, 2008.

O'Neill, S. "The Dengue Stopper." *Scientific American* 312, no. 6 (2015): 72–77.

Oommen, M. A., R. Cooney, M. Ramesh, M. Archer, D. Brockington, et al. "The Fatal Flaws of Compassionate Conservation." *Conservation Biology* 33 (2019): 784–87.

Oosting, H. J. *The Study of Plant Communities: An Introduction to Plant Ecology.* San Francisco: W. H. Freeman, 1948.

Oosting, H. J. *The Study of Plant Communities: An Introduction to Plant Ecology.* 2nd ed. San Francisco: W. H. Freeman, 1956.

Oreskes, N., and E. M. Conway. *Merchants of Doubt.* New York: Bloomsbury Press, 2010.

Orians, G. H. "A Diversity of Textbooks: Ecology Comes of Age." *Science* 181 (1973): 1238–39.

Orians, G. H. "An Ecological and Evolutionary Approach to Landscape Aesthetics." In *Landscape Meanings and Values*, edited by C. Penning-Rowsell and D. Lowenthal, 3–25. London: Allen and Unwin, 1986.

Orians, G. H. "Site Characteristics Favoring Invasions." In Mooney and Drake, *Ecology of Biological Invasions of North America and Hawaii*, 133–48.

Orion, T. *Beyond the War on Invasive Species: A Permaculture Approach to Ecosystem Restoration*. White River Junction, VT: Chelsea Green, 2015.

Orledge, G. M., P. A. Smith, and S. E. Reynolds. "The Non-Pest Australasian Fungivore *Cis bilamellatus* Wood (Coleoptera: Ciidae) in Northern Europe: Spread Dynamics and Ecological Impact." *Biological Invasions* 12 (2010): 515–30.

"Ornithology of New Zealand." *Nature* 142 (1872): 218–19.

Orr, R. "Generic Nonindigenous Aquatic Organisms Risk Analysis Review Process." In Ruiz and Carlton, *Invasive Species: Vectors and Management Strategies*, 415–38.

Ortega, Y., D. E. Pearson, and K. S. McKelvey. "Effects of Biological Control Agents and Exotic Plant Invasion on Deer Mouse Populations." *Ecological Applications* 14 (2004): 241–53.

Ortiz-Ospina, E. "The Rise of Social Media." Our World in Data. September 18, 2019. https://ourworldindata.org/rise-of-social-media.

Orwig, D. A., and D. R. Foster. "Forest Response to the Introduced Hemlock Woolly Adelgid in Southern New England, USA." *Journal of the Torrey Botanical Society* 125 (1998): 60–73.

Osborn, H. F. *Men of the Old Stone Age: Their Environment, Life and Art*. New York: Charles Scribner's Sons, 1915.

Ossenblok, T. L. B., F. T. Verleysen, and T. C. E. Engels. "Coauthorship of Journal Articles and Book Chapters in the Social Sciences and Humanities (2000–2010)." *Journal of the Association for Information Science and Technology* 65 (2014): 882–97.

Ottoni, C., W. Van Neer, B. De Cupere, J. Daligault, S. Guimaraes, et al. "The Palaeogenetics of Cat Dispersal in the Ancient World." *Nature Ecology and Evolution* 1 (2017): 0139.

"Our Next World War—Against Insects." *Popular Mechanics* 81, no. 4 (1944): 66–70.

Owre, O. T. "A Consideration of the Exotic Avifauna of Southeastern Florida." *Wilson Bulletin* 85 (1973): 491–500.

Oxitec. "Our Technology." Accessed January 30, 2020. https://www.oxitec.com.en/our-technology.

Pack, D. H. "Meteorology of Air Pollution." In Detwyler, *Man's Impact on Environment*, 98–112.

Paine, R. T. "A Conversation on Refining the Concept of Keystone Species." *Conservation Biology* 9 (1995): 962–64.

Paine, R. T. "Food Web Complexity and Species Diversity." *American Naturalist* 100 (1966): 65–75.

Paine, R. T. "Food Webs: Linkage, Interaction Strength and Community Infrastructure." *Journal of Animal Ecology* 49 (1980): 666–85.

Paine, R. T. "A Note on Trophic Complexity and Community Stability." *American Naturalist* 103 (1969): 91–93.

Paine, R. T., and D. E. Schindler. "Ecological Pork: Novel Resources and the Trophic Reorganization of an Ecosystem." *Proceedings of the National Academy of Sciences of the United States of America* 99 (2002): 554–55.

Paine, R. T., M. J. Tegner, and E. A. Johnson. "Compounded Perturbations Yield Ecological Surprises." *Ecosystems* 1 (1998): 535–45.

Paine, R. T., and T. M. Zaret. "Ecological Gambling: The High Risks and Rewards of Species Introductions." *Journal of the American Medical Association* 231 (1975): 471–73.

Pales, J. C., and C. D. Keeling. "The Concentration of Atmospheric Carbon Dioxide in Hawaii." *Journal of Geophysical Research* 70 (1965): 6053–76.

Palmer, C. "Contested Frameworks in Environmental Ethics." In Rozzi et al., *Linking Ecology and Ethics for a Changing World*, 191–206.

Palmer, C., and B. M. H. Larson. "Should We Move the Whitebark Pine? Assisted Migration, Ethics, and Global Environmental Change." *Environmental Values* 23 (2014): 641–62.

Palmer, T. S. "The Danger of Introducing Noxious Animals and Birds." In US Department of Agriculture, *Yearbook of the Department of Agriculture, 1898*, 87–110. Washington, DC: Government Printing Office, 1899.

Panagiotakopulu, E., and P. C. Buckland. "Early Invaders: Farmers, the Granary Weevil and Other Uninvited Guests in the Neolithic." *Biological Invasions* 20 (2018): 219–33.

Panagiotakopulu, E., and P. C. Buckland. "A Thousand Bites—Insect Introductions and Late Holocene Environments." *Quaternary Science Reviews* 156 (2017): 23–35.

Panetta, F. D., and N. D. Mitchell. "Bioclimatic Prediction of the Potential Distribution of Some Weed Species Prohibited Entry to New Zealand." *New Zealand Journal of Agricultural Research* 34 (1991): 341–50.

Panov, V. E., and S. Gollasch. "*Aquatic Invasions*—The New European Journal of Applied Research on Biological Invasions in Aquatic Ecosystems." *Aquatic Invasions* 1 (2006): 1–3.

Paolucci, E., H. J. MacIsaac, and A. Ricciardi. "Origin Matters: Alien Consumers Inflict Greater Damage on Prey Populations Than Do Native Consumers." *Diversity and Distributions* 19 (2013): 988–95.

Parker, I. M., and P. Kareiva. "Assessing the Risks of Invasion for Genetically Engineered Plants: Acceptable Evidence and Reasonable Doubt." *Biological Conservation* 78 (1996): 193–203.

Parker, I. M., D. Simberloff, W. M. Lonsdale, K. Goodell, M. Wonham, et al. "Impact: Toward a Framework for Understanding the Ecological Effects of Invaders." *Biological Invasions* 1 (1999): 3–19.

Parker, J. D., D. E. Burkepile, and M. E. Hay. "Opposing Effects of Native and Exotic Herbivores on Plant Invasions." *Science* 311 (2006): 1459–61.

Parkin, D. T., and S. R. Cole. "Genetic Differentiation and Rates of Evolution in Some Introduced Populations of the House Sparrow, *Passer domesticus*, in Australia and New Zealand." *Heredity* 54 (1985): 15–23.

Parmesan, C. "Ecological and Evolutionary Responses to Recent Climate Change." *Annual Review of Ecology, Evolution, and Systematics* 37 (2006): 637–69.

Parsons, J. J. "Forest to Pasture: Development or Destruction." Supplement, *Revista Biológica Tropical* 24, no. 1 (1976): 121–38.

Parsons, J. J. "Spread of African Pasture Grasses to the New World Tropics." *Journal of Range Management* 25 (1972): 12–17.

Pascal, M., O. Lorvelec, and J. D. Vigne. *Invasions biologiques et extinctions—11,000 ans d'histoire des vertébrés en France*. Paris: Belin, 2006.

Pasfield-Oliver, S. *The Voyage of François Le Guat of Bresse, to Rodriguez, Mauritius, Java, and the Cape of Good Hope*. London: Hakluyt Society, 1891.

Pattison, R. R., and R. N. Mack. "Potential Distribution of the Invasive Tree *Triadica sebifera* (Euphorbiaceae) in the United States: Evaluating CLIMEX Predictions with Field Trials." *Global Change Biology* 14 (2008): 813–26.

Pauchard, A., C. Kueffer, H. Dietz, C. C. Daehler, J. Alexander, et al. "Ain't No Mountain High Enough: Plant Invasions Reaching High Elevations." *Frontiers in Ecology and the Environment* 7 (2009): 479–86.

Paula, D. P., P. H. B. Togni, V. A. Costa, L. M. Souza, A. A. T. C. Sousa, et al. "Scrutinizing the Enemy Release Hypothesis: Population Effects of Parasitoids on *Harmonia axyridis* and Local Host Coccinellids in Brazil." *BioControl* 66 (2021): 71–82.

Pauly, P. J. "The Beauty and Menace of the Japanese Cherry Trees: Conflicting Visions of American Ecological Independence." *Isis* 87 (1996): 51–73.

Pauly, P. J. *Fruits and Plains: The Horticultural Transformation of America*. Cambridge, MA: Harvard University Press, 2007.

Pauly, P. J. Reply to Simberloff letter to the editor, *Isis* 87 (1996): 677–78.

Paviour-Smith, K. "The Ecology of the Fauna Associated with Macrofungi Growing on Dead or Decaying Trees." DPhil diss., Oxford University, 1959.

Paviour-Smith, K. "How Far Has *Cryptostroma corticale* Spread in Britain?" *Bulletin of the British Mycological Society* 10 (1976): 16–19.

Paviour-Smith, K. "The Invasion of Britain by *Cis bilamellatus* Fowler (Coleoptera: Ciidae)." *Proceedings of the Royal Entomological Society of London A* 35 (1960): 145–55.

Paviour-Smith, K. "A Population Study of *Cis bilamellatus* Wood (Coleoptera, Ciidae)." *Journal of Animal Ecology* 37 (1968): 205–28.

Pavlopoulos, G. A., P. I. Kontou, A. Pavlopoulou, C. Bouyioukos, E. Markou, and P. G. Bagos. "Bipartite Graphs in Systems Biology and Medicine: A Survey of Methods and Applications." *GigaScience* 7, no. 4 (2018): giy014.

Pavlov, M. P. "Results of the Acclimatization of Nutria in the U.S.S.R." In Yanushevich, *Acclimatization of Animals in the U.S.S.R.*, 91–92.

Peaker, M. "Tahiti—Introduced Birds." *Zoology Jottings* (blog), June 24, 2012. https://zoologyweblog.blogspot.com/2012/06/tahiti-introduced-birds.html.

Peaker, M. "Tahiti—Introduced Birds—More Information." *Zoology Jottings* (blog), August 14, 2012. https://zoologyweblog.blogspot.com/2012/08/tahiti-introduced-birds-more-information.html.

Pearce, F. *The New Wild—Why Invasive Species Will Be Nature's Salvation*. Boston: Beacon Press, 2015.

Pearce, F. "On a Remote Island, Lessons in How Ecosystems Function." *Yale Environment 360*, August 26, 2013. http://e360.yale.edu/feature/on_a_remote_island_lessons_in_how_ecosystems_function/2683/.

Pearson, D. E., and R. M. Callaway. "Indirect Effects of Host-Specific Biological Control Agents." *Trends in Ecology and Evolution* 18 (2003): 456–61.

Pearson, D. E., and R. M. Callaway. "Indirect Nontarget Effects of Host-Specific Biological Control Agents: Implications for Biological Control." *Biological Control* 35 (2005): 288–98.

Pearson, D. E., Y. K. Ortega, and S. J. Sears. "Darwin's Naturalization Hypothesis Up-Close: Intermountain Grassland Invaders Differ Morphologically and Phenologically from Native Community Dominants." *Biological Invasions* 14 (2012): 901–13.

Pearson, D. E., Y. K. Ortega, D. Villarreal, Y. Lekberg, M. C. Cock, et al. "The Fluctuating Resource Hypothesis Explains Invasibility, but Not Exotic Advantage Following Disturbance." *Ecology* 99 (2018): 1296–305.

Peay, K. G. "The Mutualistic Niche: Mycorrhizal Symbiosis and Community Dynamics." *Annual Review of Ecology, Evolution, and Systematics* 47 (2016): 143–64.

Pederson, J., ed. *Marine Bioinvasions*. Cambridge, MA: MIT Sea Grant College Program, 1999.

Pedlar, J., D. McKenney, J. Beaulieu, S. Colombo, J. McLachlan, and G. O'Neill. "The Implementation of Assisted Migration in Canadian Forests." *Forestry Chronicle* 87 (2011): 766–77.

Pejchar, L., and H. A. Mooney. "Invasive Species, Ecosystem Services and Human Well-Being." *Trends in Ecology and Evolution* 24 (2009): 497–504.

Pell, J. K., J. Baverstock, H. E. Roy, R. L. Ware, and M. E. N. Majerus. "Intraguild Predation Involving *Harmonia axyridis*: A Review of Current Knowledge and Future Perspectives." *BioControl* 53 (2008): 147–68.

Peltzer, D., P. Bellingham, I. Dickie, G. Houliston, P. Hulme, et al. "Scale and Complexity Implications for Predator-Free New Zealand." *Journal of the Royal Society of New Zealand* 49 (2019): 412–39.

Pemberton, R. W. "*Cactoblastis cactorum* (Lepidoptera: Pyralidae) in the United States: An Immigrant Biological Control Agent or an Introduction of the Nursery Industry?" *American Entomologist* 41 (1995): 230–32.

Pemberton, R. W., and E. Escalona. "Spread and Distribution of the Naturalized Orchid Bee *Euglossa dilemma* in Florida." *Florida Entomologist* 106 (2023): 59–62.

Penev, L., W. J. Kress, S. Knapp, D.-Z. Li, and S. Renner. "Fast, Linked, and Open—The Future of Taxonomic Publishing for Plants: Launching the Journal *PhytoKeys*." *PhytoKeys* 1 (2010): 1–14.

Penfound, W. T., and T. T. Earle. "The Biology of the Water Hyacinth." *Ecological Monographs* 18 (1948): 447–72.

Peng, S., N. L. Kinlock, J. Gurevitch, and S. Peng. "Correlation of Native and Exotic Species Richness: A Global Meta-Analysis Finds No Invasion Paradox Across Scales." *Ecology* 100, no. 1 (2019): e02552.

Peretti, J. H. "Nativism and Nature: Rethinking Biological Invasion." *Environmental Values* 7 (1998): 183–92.

Pereyra, P. J. "Revisiting the Use of the Invasive Species Concept: An Empirical Approach." *Austral Ecology* 4 (2016): 519–28.

Pereyra, P. J. "Rethinking the Native Range Concept." *Conservation Biology* 34 (2020): 373–77.

Perrings, C. "Invasion Economics." In Simberloff and Rejmánek, *Encyclopedia of Biological Invasions*, 375–78.

Perrings, C., H. Mooney, and M. Williamson, eds. *Bioinvasions and Globalization: Ecology, Economics, Management, and Policy*. Oxford: Oxford University Press, 2010.

Perrings, C., M. Williamson, and S. Dalmazzone, eds. *The Economics of Biological Invasions*. Cheltenham: Edward Elgar, 2000.

Perrins, J., M. Williamson, and A. Fitter. "A Survey of Differing Views of Weed Classification: Implications for Regulation of Introductions." *Biological Conservation* 60 (1992): 47–56.

Peters, R. H. *A Critique for Ecology*. Cambridge: Cambridge University Press, 1991.

Peters, R. L., and J. D. S. Darling. "The Greenhouse Effect and Nature Reserves." *BioScience* 35 (1985): 707–17.

Petersen, K. S., K. L. Rasmussen, J. Heinemeier, and N. Rud. "Clams Before Columbus?" *Nature* 359 (1992): 679.

Peterson, A. T. "Ecological Niche Conservatism: A Time-Structured Review of Evidence." *Journal of Biogeography* 38 (2011): 817–27.

Peterson, A. T. "Predicting the Geography of Species' Invasions via Ecological Niche Modeling." *Quarterly Review of Biology* 78 (2003): 419–33.

Peterson, A. T., M. Papeş, and M. Eaton. "Transferability and Model Evaluation in Ecological Niche Modeling: A Comparison of GARP and MaxEnt." *Ecography* 30 (2007): 550–60.

Peterson, A. T., and C. R. Robins. "Using Ecological-Niche Modeling to Predict Barred Owl Invasions with Implications for Spotted Owl Conservation." *Conservation Biology* 17 (2003): 1161–65.

Peterson, A. T., J. Soberón, R. G. Pearson, R. P. Anderson, E. Martínez-Meyer, et al. *Ecological Niches and Geographic Distributions*. Princeton, NJ: Princeton University Press, 2011.

Peterson, A. T., and D. A. Vieglais. "Predicting Species Invasions Using Ecological Niche Modeling: New Approaches from Bioinformatics Attack a Pressing Problem." *BioScience* 51 (2001): 363–71.

Peterson, P. D. "The Barberry Eradication Program in Minnesota for Stem Rust Control: A Case Study." *Annual Review of Phytopathology* 56 (2018): 203–23.

Petitpierre, B., O. Broennimann, C. Kueffer, C. Daehler, and A. Guisan. "Selecting Predictors to Maximize the Transferability of Species Distribution Models: Lessons from Cross-Continental Plant Invasions." *Global Ecology and Biogeography* 26 (2017): 275–87.

Pfitzner, C., M. A. White, S. G. Piltz, M. Scherer, F. Adikusuma, et al. "Progress Toward Zygotic and Germline Gene Drives in Mice." *CRISPR Journal* 3 (2020): 388–97.

Pheloung, P. C. *Determining the Weed Potential of New Plant Introductions to Australia: A Report on the Development of a Weed Risk Assessment System Commissioned by the Australian Weeds Committee and the Plant Industries Committee*. South Perth: Agriculture Protection Board, Western Australia, 1995. https://nre.tas.gov.au/Documents/WA-Weed-Risk-Assessment-REPORT.pdf.

Pheloung, P. C., P. A. Williams, and S. R. Halloy. "A Weed Risk Assessment Model for Use as a Biosecurity Tool Evaluating Plant Introductions." *Journal of Environmental Management* 57 (1999): 239–51.

Philipp, D. P., J. M. Epifanio, and M. J. Jennings. "Point/Counterpoint: Conservation Genetics and Current Stocking Practices—Are They Compatible?" *Fisheries* 18, no. 12 (1993): 14–17.

Phillips, B. L., M. J. Greenlees, G. P. Brown, and R. Shine. "Predator Behaviour and Morphology Mediates the Impact of an Invasive Species: Cane Toads and Death Adders in Australia." *Animal Conservation* 13 (2010): 53–59.

Phillips, B. L., and R. Shine. "Adapting to an Invasive Species: Toxic Cane Toads Induce Morphological Change in Australian Snakes." *Proceedings of the National Academy of Sciences of the United States of America* 101 (2004): 17150–55.

Phillips, B. L., and R. Shine. "An Invasive Species Induces Rapid Adaptive Change in a Native Predator: Cane Toads and Black Snakes in Australia." *Proceedings of the Royal Society B* 273 (2006): 1545–50.

Phillips, E. F. "The Honey Crop from Introduced Plants." *Ecology* 4 (1923): 180–82.

Phillips, J. C. *Wild Birds Introduced or Transplanted in North America*. USDA Technical Bulletin No. 61. Washington, DC: US Department of Agriculture, 1928. Published in cooperation with Bureau of Biological Survey of the USA.

Phillips, J. F. V. "The Biotic Community." *Journal of Ecology* 19 (1931): 1–24.

Phillips, S. J., R. P. Anderson, and R. E. Schapire. "Maximum Entropy Modelling of Species Geographic Distributions." *Ecological Modelling* 190 (2006): 231–59.

Phillips, S. J., and M. Dudík. "Modeling of Species Distributions with Maxent: New Extensions and a Comprehensive Evaluation." *Ecography* 31 (2008): 161–75.

Piaggio, A. J., R. M. Engeman, M. W. Hopken, J. S. Humphrey, K. L. Keacher, et al. "Detecting an Elusive Invasive Species: A Diagnostic PCR to Detect Burmese Python in Florida Waters and an Assessment of Persistence of Environmental DNA." *Molecular Ecology Resources* 14 (2014): 374–80.

Pianka, E. R. *Evolutionary Ecology*. New York: Harper & Row, 1974.

Pianka, E. R. *Evolutionary Ecology*. 2nd ed. New York: Harper & Row, 1978.

Pianka, E. R. *Evolutionary Ecology*. 3rd ed. New York: Harper & Row, 1983.

Pianka, E. R. *Evolutionary Ecology*. 4th ed. New York: HarperCollins, 1987.

Pianka, E. R. *Evolutionary Ecology*. 5th ed. New York: HarperCollins, 1994.

Pianka, E. R. *Evolutionary Ecology*. 6th ed. San Francisco: Addison Wesley Longman, 2000.

Pianka, E. R. *Evolutionary Ecology*. 7th ed. 2011. Google Play Books.

Pickart, A. J. "Dune Restoration over Two Decades at the Lanphere and Ma-le'l Dunes in Northern California." In *Restoration of Coastal Dunes*, edited by M. L. Martinez, J. B. Gallego-Fernandez, and P. A. Hesp, 159–71. Berlin: Springer-Verlag, 2013.

Pickart, A. J. "Mechanical Control." In Simberloff and Rejmánek, *Encyclopedia of Biological Invasions*, 449–52.

Pierce, W. D. "A New Interpretation of the Relationship of Temperature and Humidity to Insect Development." *Journal of Agricultural Research* 5 (1916): 1183–91.

Pigati, J. S., K. B. Springer, J. S. Honke, D. Wahl, M. R. Champagne, et al. "Independent Age Estimates Resolve the Controversy of Ancient Human Footprints at White Sands." *Science* 382 (2023): 73–75.

"Pike Reappear, and a California City Is on Guard." *New York Times*, June 21, 1999, A12.

Pimentel, D. "Animal Population Regulation by the Genetic Feed-Back Mechanism." *American Naturalist* 95 (1961): 65–79.

Pimentel, D., ed. *Biological Invasions: Economic and Environmental Costs of Alien Plant, Animal, and Microbe Species*. Boca Raton, FL: CRC Press, 2002.

Pimentel, D., ed. *Biological Invasions: Economic and Environmental Costs of Alien Plant, Animal, and Microbe Species*. 2nd ed. Boca Raton, FL: CRC Press, 2011.

Pimentel, D. "Biology of the Indian Mongoose in Puerto Rico." *Journal of Mammalogy* 36 (1955): 62–68.

Pimentel, D. "The Control of the Mongoose in Puerto Rico." *American Journal of Tropical Medicine and Hygiene* 41 (1955): 147–51.

Pimentel, D. "Introduction: Non-Native Species in the World." In Pimentel, *Biological Invasions: Economic and Environmental Costs*, 3–8.

Pimentel, D., C. Glenister, S. Fast, and D. Gallahan. "Environmental Risks of Biological Pest Controls." *Oikos* 42 (1984): 283–90.

Pimentel, D., L. Lach, R. Zuniga, and D. Morrison. "Environmental and Economic Costs Associated with Non-Indigenous Species in the United States." In Pimentel, *Biological Invasions: Economic and Environmental Costs*, 285–303.

Pimentel, D., L. Lach, R. Zuniga, and D. Morrison. "Environmental and Economic Costs of Nonindigenous Species in the United States." *BioScience* 50 (2000): 53–65.

Pimm, S. L. *Food Webs*. Chicago: University of Chicago Press, 1982.

Pimm, S. L. "Theories of Predicting Success and Impact of Introduced Species." In Drake et al., *Biological Invasions: A Global Perspective*, 351–67.

Pinchot, G. *Breaking New Ground*. New York: Harcourt, Brace, 1947.

Pinchot, G. *The Fight for Conservation*. New York: Doubleday, Page, 1910.

Pingré, A.-G. "Voyage à Rodrigue." In *Alexandre-Gui Pingré: Voyage à Rodrigue*, edited by S. Hoarau, M.-P. Janiçon, and J.-M. Racault. Paris: SEDES; Saint-Denis: Université de la Réunion; Paris: Le Publieur, 2004.

Piper, I., J. Lee, E. Izadi, and B. Sacks. "Maui's Neglected Grasslands Caused Lahaina Fire to Grow with Deadly Speed." *Washington Post*, September 2, 2023. https://www.washingtonpost.com/investigations/interactive/2023/lahaina-wildfires-invasive-grass-destruction.

Platt, J. R. "Strong Inference." *Science* 146 (1964): 347–53.

Platt, T. "Engaging the Past: Charles M. Goethe, American Eugenics, and Sacramento State University." *Social Justice* 32, no. 2 (2005): 17–33.

Pohl, U. "Physiology Without Borders." *Physiology* 20 (2005): 148.

Poisot, T. "ESM: Ecological Specificity Measures." R Package version 2.0.3-02/r47, 2011.

Poisot, T., A. R. Cirtwill, K. Cazelles, D. Gravel, M.-J. Fortin, and D. B. Stouffer, "The Structure of Probabilistic Networks." *Methods in Ecology and Evolution* 7 (2016): 303–12.

Poletti, M., A. H. N. Maia, and C. Omoto. "Toxicity of Neonicotinoid Insecticides to *Neoseiulus californicus* and *Phytoseiulus macropilis* (Acari: Phytoseiidae) and Their Impact on Functional Response to *Tetranychus urticae* (Acari: Tetranychidae)." *Biological Control* 40 (2007): 30–36.

Polgar, C., A. Gallinat, and R. B. Primack. "Drivers of Leaf-Out Phenology and Their Implications for Species Invasions: Insights from Thoreau's Concord." *New Phytologist* 202 (2014): 106–15.

Polito, M. J., B. Robinson, P. Warzybok, and R. W. Bradley. "Population Dynamics and Resource Availability Drive Seasonal Shifts in the Consumptive and Competitive Impacts of Introduced House Mice (*Mus musculus*) on an Island Ecosystem." *PeerJ* 10 (2022): e13904.

Pollan, M. "Against Nativism." *New York Times Magazine*, May 15, 1994, 52–55.

Pons, A., J. L. de Beaulieu, J. Guiot, and M. Reille. "Vingt cinq ans de recherches en analyse pollinique." *Ecologia mediterranea* 16 (1990): 169–93.

Popper, K. R. *Conjectures and Refutations: The Growth of Scientific Knowledge*. New York: Harper & Row, 1963.

Popper, K. R. *The Logic of Scientific Discovery*. 3rd ed. London: Hutchinson, 1972.

Por, F. D. *Lessepsian Migration: The Influx of Red Sea Biota into the Mediterranean by Way of the Suez Canal*. Berlin: Springer, 1978.

Porter, W. P., and D. M. Gates. "Thermodynamic Equilibria of Animals with Environment." *Ecological Monographs* 39 (1969): 227–44.

Porter, W. P., J. W. Mitchell, W. A. Beckman, and C. B. DeWitt. "Behavioral Implications of Mechanistic Ecology." *Oecologia* 13 (1973): 1–54.

Pouget, R. *Histoire de la lutte contre le phylloxéra de la vigne en France*. Paris: Institut National de la Recherche Agronomique, 1990.

Poulos, H. M., B. Chernoff, P. L. Fuller, and D. Butman. "Ensemble Forecasting of Potential Habitat for Three Invasive Fishes." *Aquatic Invasions* 7 (2012): 59–72.

Poutsma, J., A. J. M. Loomans, B. Aukema, and T. Heijerman. "Predicting the Potential Geographical Distribution of the Harlequin Ladybird, *Harmonia axyridis*, Using the CLIMEX Model." *Biocontrol* 53 (2008): 103–25.

Power, M. E., D. Tilman, J. A. Estes, B. A. Menge, W. J. Bond, et al. "Challenges in the Quest for Keystones." *BioScience* 46 (1996): 609–20.

Pratt, P. D., and T. D. Center. "Biocontrol Without Borders: The Unintended Spread of Introduced Weed Biological Control Agents." *BioControl* 57 (2012): 319–29.

Prebble, M., and J. M. Wilmshurst. "Detecting the Initial Impact of Humans and Introduced Species on Island Environments in Remote Oceania Using Palaeoecology." *Biological Invasions* 11 (2009): 1529–56.

Prendergast, H. D. V., and P. W. Hattersley. "Distribution and Cytology of Australian *Neurachne* and Its Allies (Poaceae), a Group Containing C3, C4 and C3–C4 Intermediate Species." *Australian Journal of Botany* 33 (1985): 317–36.

Preston, D. "The Mystery of Sandia Cave." *New Yorker* 71, no. 16 (1995): 66–83.

Preston, R. E., and M. Wetherwax. "*Verbascum speciosum*" [2012]. In *Jepson eFlora*, edited by Jepson Flora Project, Jepson Herbarium, University of California, Berkeley.

Updated August 22, 2024. https://ucjeps.berkeley.edu/eflora/eflora_display.php?tid=47844.

Price, A. G. *The Explorations of Captain James Cook in the Pacific as Told by Selections of His Own Journals 1768–1779*. New York: Heritage Press, 1958.

Price, A. G. *The Importance of Disease in History*. Adelaide: Libraries Board of South Australia, 1964.

Price, A. G. *The Western Invasions of the Pacific and Its Continents*. Oxford: Oxford University Press, 1963.

Price, A. G. *White Settlers and Native Peoples: An Historical Study of Racial Contacts Between English-Speaking Whites and Aboriginal Peoples in the United States, Canada, Australia and New Zealand*. Melbourne: Cambridge University Press, 1949.

Price, A. G. *White Settlers in the Tropics*. New York: American Geographical Society, 1939.

Price, D. J. S. *Little Science, Big Science . . . and Beyond*. New York: Columbia University Press, 1963.

Prichard, J. C. *Researches into the Physical History of Mankind*. 3rd ed. London: Sherwood, Gilbert, and Piper; J. and A. Arch, 1836.

Primack, R. B. *Essentials of Conservation Biology*. Sunderland, MA: Sinauer Associates, 1993.

Primack, R. B. *Essentials of Conservation Biology*. 2nd ed. Sunderland, MA: Sinauer Associates, 1998.

Primack, R. B. *Essentials of Conservation Biology*. 3rd ed. Sunderland, MA: Sinauer Associates, 2002.

Primack, R. B. *Essentials of Conservation Biology*. 4th ed. Sunderland, MA: Sinauer Associates, 2006.

Primack, R. B. *Essentials of Conservation Biology*. 5th ed. Sunderland, MA: Sinauer Associates, 2010.

Primack, R. B. *Essentials of Conservation Biology*. 6th ed. Sunderland, MA: Sinauer Associates, 2014.

Prince, S. D., R. N. Carter, and K. J. Dancy. "The Geographical Distribution of Prickly Lettuce (*Lactuca serriola*), II: Characteristics of Populations Near Its Distribution Limit in Britain." *Journal of Ecology* 73 (1985): 39–48.

Pringle, R. M. "Nile Perch." In Simberloff and Rejmánek, *Encyclopedia of Biological Invasions*, 484–88.

Proctor, H. S., G. Carder, and A. R. Cornish. "Searching for Animal Sentience: A Systematic Review of the Scientific Literature." *Animals* (Basel) 3 (2013): 882–906.

Proulx, S. R., D. E. L. Promislow, and P. C. Phillips. "Network Thinking in Ecology and Evolution." *Trends in Ecology and Evolution* 20 (2005): 345–53.

Pruett-Jones, S., J. R. Newman, C. M. Newman, M. L. Avery, and J. R. Lindsay. "Population Viability Analysis of Monk Parakeets in the United States and Examination of Alternative Management Strategies." *Human-Wildlife Conflict* 1 (2007): 35–44.

Puri, A., G. E. MacDonald, and W. T. Haller. "Stability of Fluridone-Resistant Hydrilla (*Hydrilla verticillata*) Biotypes over Time." *Weed Science* 55 (2007): 12–15.

Purse, B. V., P. S. Mellor, D. J. Rogers, A. R. Samuel, P. P. Mertens, and M. Baylis. "Climate Change and the Recent Emergence of Bluetongue in Europe." *Nature Reviews Microbiology* 3 (2005): 171–81.

Pyke, G. H. "Plague Minnow or Mosquito Fish? A Review of the Biology and Impacts of Introduced *Gambusia* Species." *Annual Review of Ecology, Evolution, and Systematics* 39 (2008): 171–91.

Pyšek, P., T. M. Blackburn, E. García-Berthou, I. Perglová, and W. Rabitsch. "Displacement and Local Extinction of Native and Endemic Species." In Vilà and Hulme, *Impact of Biological Invasions on Ecosystem Services*, 157–75.

Pyšek, P., and P. Hulme. "Invasion Biology Is a Discipline That's too Young to Die." *Nature* 460 (2009): 324.

Pyšek, P., P. E. Hulme, D. Simberloff, S. Bacher, T. M. Blackburn, et al. "Scientists' Warning on Invasive Alien Species." *Biological Reviews* 95 (2020): 1511–34.

Pyšek, P., and D. Richardson. "Traits Associated with Invasiveness in Alien Plants: Where Do We Stand?" In *Biological Invasions: Ecological Studies*, edited by W. Nentwig, 97–125. Berlin: Springer, 2007.

Pyšek, P., D. M. Richardson, M. Rejmánek, G. L. Webster, M. Williamson, and J. Kirschner. "Alien Plants in Checklists and Floras: Towards Better Communication Between Taxonomists and Ecologists." *Taxon* 53 (2004): 131–43.

Pyšek, P., J. Sádlo, and B. Mandák. "Catalogue of Alien Plants of the Czech Republic." *Preslia* 74 (2002): 97–186.

Quayle, H. J. "Possible Future Distribution and Abundance of the Mediterranean Fruit Fly (*Ceratitis capitata* Wied) in the United States." *Journal of Economic Entomology* 24 (1931): 1064–66.

Raby, P. *Alfred Russel Wallace: A Life*. Princeton, NJ: Princeton University Press, 2001.

Radcliffe, E. B., and K. L. Flanders. "Biological Control of Alfalfa Weevil in North America." *Integrated Pest Management Reviews* 3 (1998): 225–42.

Radtkey, R. R., and M. C. Singer. "Repeated Reversals of Host-Preference Evolution in a Specialist Insect Herbivore." *Evolution* 49 (1995): 351–59.

Raffles, H. "Mother Nature's Melting Pot." *New York Times*, April 3, 2011, WK 12.

Rafoss, T., and M. G. Sæthre. "Spatial and Temporal Distribution of Bioclimatic Potential for the Codling Moth and the Colorado Potato Beetle in Norway: Model Predictions Versus Climate and Field Data from the 1990s." *Agricultural and Forest Entomology* 5 (2003): 75–85.

Raine, A. F., M. Vynne, and S. Driskill. "The Impact of an Introduced Avian Predator, the Barn Owl *Tyto alba*, on Hawaiian Seabirds." *Marine Ornithology* 47 (2019): 3338.

Rainey, J. "Catalina Goats Being Shot to Protect Island's Ecosystem." *Los Angeles Times*, January 26, 1990. https://www.latimes.com/archives/la-xpm-1990-01-26-me-832-story.html.

Rains, M. "Maui's Deadly Blazes Reveal a Fire-Prone Hawaii." *Science* 381 (2023): 718–19.

Rakov, V. A., and D. L. Brodianski. "Ancient Oyster Farming in the Boisman Culture of the Primorye Neolithic." *Archaeology, Ethnology and Anthropology of Eurasia*, no. 3 (2007): 39–43.

Ramakrishnan, P. S. "Biological Invasion in the Tropics: An Overview." In Ramakrishnan, *Ecology of Biological Invasion in the Tropics*, 1–19.

Ramakrishnan, P. S., ed. *Ecology of Biological Invasion in the Tropics*. New Delhi: International Scientific Publications, 1991.

Ramakrishnan, P. S. Preface to Ramakrishnan, *Ecology of Biological Invasion in the Tropics*, vii–viii.

Ramakrishnan, P. S., and P. M. Vitousek. "Ecosystem-Level Processes and the Consequences of Biological Invasions." In Drake et al., *Biological Invasions: A Global Perspective*, 281–300.

Ramp, D., and M. Bekoff. "Compassion as a Practical and Evolved Ethic for Conservation." *BioScience* 65 (2015): 323–27.

Ramsey, D. S., K. J. Campbell, C. Lavoie, N. Macdonald, and S. A. Morrison. "Quantifying the Probability of Detection of Wild Ungulates with the Judas Technique." *Conservation Biology* 36 (2022): e13898.

Rangan, H., K. Bell, D. A. Baum, R. Fowler, P. McConvell, et al. "New Genetic and Linguistic Analyses Show Ancient Human Influence on Baobab Evolution and Distribution in Australia." *PLoS One* 10, no. 4 (2015): e0119758. https://doi.org/10.1371/journal.pone.0119758.

Rangan, H., J. Carney, and T. Denham. "Environmental History of Botanical Exchanges in the Indian Ocean World." *Environment and History* 18 (2012): 311–42.

Rapai, W. *Lake Invaders: Invasive Species and the Battle for the Future of the Great Lakes.* Detroit: Wayne State University Press, 2016.

Rauzon, M. J. Review of *When the Killing's Done*, by T. C. Boyle. *Marine Ornithology* 39 (2011): 280.

Rawtani, D., C. M. Hussain, and N. Khatri, eds. *COVID-19 in the Environment: Impact, Concerns, and Management of Coronavirus.* Amsterdam: Elsevier, 2021.

Read, Q. D., S. M. Hoban, M. B. Eppinga, J. A. Schweitzer, and J. K. Bailey. "Accounting for the Nested Nature of Genetic Variation Across Levels of Organization Improves Our Understanding of Biodiversity and Community Ecology." *Oikos* 125 (2016): 895–904.

Reaser, J. K., S. W. Burgiel, J. Kirkey, K. A. Brantley, S. D. Veatch, and J. Burgos-Rodríguez. "The Early Detection of and Rapid Response (EDRR) to Invasive Species: A Conceptual Framework and Federal Capacities Assessment." *Biological Invasions* 22 (2020): 1–19.

Recart, W., J. D. Ackerman, and A. A. Cuevas. "There Goes the Neighborhood: Apparent Competition Between Invasive and Native Orchids Mediated by a Specialist Florivorous Weevil." *Biological Invasions* 15 (2013): 283–93.

Rédei, D., and A. Torma. "Occurrence of the Southern Green Stink Bug, *Nezara viridula* (Heteroptera: Pentatomidae) in Hungary." *Acta Phytopathologica et Entomologica Hungarica* 38 (2003): 365–67.

Regal, B. *Henry Fairfield Osborn: Race and the Search for the Origins of Man.* Hants, UK: Ashgate Publishing, 2002.

Regal, P. J. "Models of Genetically Engineered Organisms and Their Ecological Impact." In Mooney and Drake, *Ecology of Biological Invasions of North America and Hawaii*, 111–29, 176.

Regal, P. J. "The True Meaning of 'Exotic Species' as a Model for Genetically Engineered Organisms." *Experientia* 49 (1993): 225–34.

Regalado, A. "Multiauthor Papers on the Rise." *Science* 268 (1995): 25.

Regan, T. *The Case for Animal Rights.* Berkeley: University of California Press, 1983.

Reichard, S. H., and P. White. "Horticulture as a Pathway of Invasive Plant Introductions in the United States." *BioScience* 51 (2001): 103–13.

Reid, B. "Survival Status of the Takahe, *Notornis mantelli*, of New Zealand." *Biological Conservation* 1 (1969): 237–40.

Reid, M., L. Morin, P. O. Downey, K. French, and J. G. Virtue. "Does Invasive Plant Management Aid the Restoration of Natural Ecosystems?" *Biological Conservation* 142 (2009): 2342–49.

Rejmánek, M. "Invasibility of Plant Communities." In Drake et al., *Biological Invasions: A Global Perspective*, 369–88.

Rejmánek, M. "Invasive Plant Species and Invadible Ecosystems." In *Invasive Species and Biodiversity Management*, edited by O. T. Sandlund, P. J. Schei, and Å. Viken, 79–102. Dordrecht, Neth.: Kluwer, 1998.

Rejmánek, M. "A Theory of Seed Plant Invasiveness: The First Sketch." *Biological Conservation* 78 (1996): 171–81.

Rejmánek, M., and M. J. Pitcairn. "When Is Eradication of Exotic Pest Plants a Realistic Goal?" In Veitch and Clout, *Turning the Tide: The Eradication of Invasive Species*, 249–53.

Rejmánek, M., and D. M. Richardson. "What Attributes Make Some Plant Species More Invasive?" *Ecology* 77 (1996): 1655–61.

Rejmánek, M., D. M. Richardson, M. G. Barbour, M. Crawley, G. F. Hrusa, et al. "Biological Invasions: Politics and the Discontinuity of Ecological Terminology." *Bulletin of the Ecological Society of America* 83 (2002): 131–33.

Relyea, R. *Ecology: The Economy of Nature*. 9th ed. New York: MacMillan Learning, 2021.

Relyea, R., and R. E. Ricklefs. *Ecology: The Economy of Nature*. 8th ed. New York: W. H. Freeman, 2018.

Renault, M. "Texas Can't Quit the Aoudad." *Atlantic*, January 13, 2020. https://www.theatlantic.com/science/archive/2020/01/how-aoudad-invaded-texas/604834/.

"Reports of 11 SCOPE Working Groups." In SCOPE archives, Paris, n.d.

Rescher, N. *Predicting the Future*. Albany: State University of New York Press, 1998.

Reusch, T., S. Bolte, M. Sparwell, A. Moss, and J. Javidpour. "Microsatellites Reveal Origin and Genetic Diversity of Eurasian Invasions by One of the World's Most Notorious Marine Invaders, *Mnemiopsis leidyi* (Ctenophora)." *Molecular Ecology* 19 (2010): 2690–99.

Reynolds, J. C. "Details of the Geographic Replacement of the Red Squirrel (*Sciurus vulgaris*) by the Grey Squirrel (*Sciurus carolinensis*) in Eastern England." *Journal of Animal Ecology* 54 (1985): 149–62.

Rhymer, J., and D. Simberloff. "Extinction by Hybridization and Introgression." *Annual Review of Ecology and Systematics* 27 (1996): 83–109.

Rhymer, J., M. J. Williams, and M. J. Braun. "Mitochondrial Analysis of Gene Flow between New Zealand Mallards (*Anas platyrhynchos*) and Grey Ducks (*A. superciliosa*)." *Auk* 111 (1994): 970–78.

Ricciardi, A. "Are Modern Biological Invasions an Unprecedented Form of Global Change?" *Conservation Biology* 21 (2007): 329–36.

Ricciardi, A. "Facilitative Interactions Among Aquatic Invaders: Is an 'Invasional Meltdown' Occurring in the Great Lakes?" *Canadian Journal of Fisheries and Aquatic Sciences* 58 (2001): 2513–25.

Ricciardi, A. "Global Range Expansion of the Asian Mussel *Limnoperna fortunei* (Mytilidae): Another Fouling Threat to Freshwater Systems." *Biofouling* 13, no. 2 (1998): 97–106.

Ricciardi, A. "Invasive Species." In *Ecological Systems*, edited by R. Leemans, 161–78. New York: Springer, 2013.

Ricciardi, A. "Predicting the Impacts of an Introduced Species from Its Invasion History: An Empirical Approach Applied to Zebra Mussel Invasions." *Freshwater Biology* 48 (2003): 972–81.

Ricciardi, A., T. M. Blackburn, J. T. Carlton, J. T. A. Dick, P. E. Hulme, et al. "Invasion Science: A Horizon Scan of Emerging Challenges and Opportunities." *Trends in Ecology and Evolution* 32 (2017): 464–74.

Ricciardi, A. and H. J. MacIsaac. "The Book That Began Invasion Biology." *Nature* 452 (2008): 34.

Ricciardi, A., and R. Ryan. "The Exponential Growth of Invasive Species Denialism." *Biological Invasions* 20 (2018): 549–53.

Ricciardi, A., and R. Ryan. "Invasive Species Denialism Revisited: Response to Sagoff." *Biological Invasions* 20 (2018): 2731–38.

Ricciardi, A., and D. Simberloff. "Assisted Colonization: Good Intentions and Dubious Risk Assessment." *Trends in Ecology and Evolution* 24 (2009): 476–77.

Ricciardi, A., and D. Simberloff. "Assisted Colonization Is Not a Viable Conservation Strategy." *Trends in Ecology and Evolution* 24 (2009): 248–53.

Ricciardi, A., and D. Simberloff. "Assisted Colonization Risk Assessment." *Science* 372 (2021): 925.

Rice, W. R., and D. F. Sax. "Testing Fundamental Evolutionary Questions at Large Spatial and Demographic Scales." In Sax et al., *Species Invasions: Insights into Ecology, Evolution, and Biogeography*, 291–308.

Richards, C. L., A. W. Schrey, and M. Pigliucci. "Invasion of Diverse Habitats by Few Japanese Knotweed Genotypes Is Correlated with Epigenetic Differentiation." *Ecology Letters* 15 (2012): 1016–25.

Richardson, D. M. "Editorial." *Diversity and Distributions* 4, no. 1 (1998): 1–2.

Richardson, D. M., ed. *Fifty Years of Invasion Ecology: The Legacy of Charles Elton.* Chichester, UK: Wiley-Blackwell; New York: Wiley, 2011.

Richardson, D. M. Introduction to Richardson, *Fifty Years of Invasion Ecology*, xiii–xix.

Richardson, D. M. "Invasion Science: The Roads Travelled and the Roads Ahead." In Richardson, *Fifty Years of Invasion Ecology*, 397–407.

Richardson, D. M., ed. "Special Section: Economic Costs of Biological Invasions." *Biological Invasions* 24, no. 7 (2022): 1895–2101.

Richardson, D. M., B. Abrahams, N. Boshoff, S. J. Davies, J. Measey, and B. W. van Wilgen. "South Africa's Centre for Invasion Biology: An Experiment in Invasion Science for Society." In van Wilgen et al., *Biological Invasions in South Africa*, 879–914.

Richardson, D. M., N. Allsopp, C. M. D'Antonio, S. J. Milton, and M. Rejmánek. "Plant Invasions—The Role of Mutualisms." *Biological Reviews* 75 (2000): 65–93.

Richardson, D. M., J. J. Hellmann, J. S. McLachlan, D. F. Sax, M. W. Schwartz, et al. "Multidimensional Evaluation of Managed Relocation." *Proceedings of the National Academy of Sciences of the United States of America* 106 (2009): 9721–24.

Richardson, D. M., and P. Pyšek. "Classics in Physical Geography Revisited: Elton, C. S. 1958, *The Ecology of Invasions by Animals and Plants*; Methuen; London." *Progress in Physical Geography* 31 (2007): 659–66.

Richardson, D. M., and P. Pyšek. "Fifty Years of Invasion Ecology—The Legacy of Charles Elton." *Diversity and Distributions* 14 (2008): 161–68.

Richardson, D. M., and P. Pyšek. "Naturalization of Introduced Plants: Ecological Drivers of Biogeographical Patterns." *New Phytologist* 196 (2012): 383–96.

Richardson, D. M., P. Pyšek, M. Rejmánek, M. G. Barbour, F. D. Panetta, and C. J. West. "Naturalization and Invasion of Alien Plants: Concepts and Definitions." *Diversity and Distributions* 6 (2000): 93–107.

Richardson, D. M., and M. Rejmánek. 2004. "Conifers as Invasive Aliens: A Global Survey and Predictive Framework." *Diversity and Distributions* 10 (2004): 321–31.

Richardson, D. M., and A. Ricciardi. "Misleading Criticisms of Invasion Science: A Field Guide." *Diversity and Distributions* 19 (2013): 1461–67.

Ricker, W. E. "Effects of Compensatory Mortality on Populations." *Journal of Wildlife Management* 18 (1954): 45–51.

Ricklefs, R. E. "Disintegration of the Ecological Community." *American Naturalist* 172 (2008): 741–50.

Ricklefs, R. E. *Ecology*. Newton, MA: Chiron, 1973.

Ricklefs, R. E. *Ecology*. 2nd ed. New York: Chiron, 1979.

Ricklefs, R. E. *Ecology*. 3rd ed. New York: W. H. Freeman, 1990.

Ricklefs, R. E., and G. L. Miller. *Ecology*. 4th ed. New York: W. H. Freeman, 2000.

Ricklefs, R. E. *The Economy of Nature*. Portland, OR: Chiron, 1976.

Ricklefs, R. E. *The Economy of Nature*. 6th ed. New York: W. H. Freeman, 2008.

Ricklefs, R. E. "Taxon Cycles." In Sax et al., *Species Invasions: Insights into Ecology, Evolution, and Biogeography*, 165–99.

Ricklefs, R. E., and R. Relyea. *Ecology: The Economy of Nature*. 7th ed. New York: W. H. Freeman, 2014.

Ridley, H. N. *The Dispersal of Plants Throughout the World*. Ashford, UK: Reeve, 1930.

Riechert, S. E., and T. Lockley. "Spiders as Biological Control Agents." *Annual Review of Entomology* 29 (1984): 299–320.

Riis, T., C. Lambertini, B. Olesen, J. S. Clayton, H. Brix, and B. K. Sorrell. "Invasion Strategies in Clonal Aquatic Plants: Are Phenotypic Differences Caused by Phenotypic Plasticity or Local Adaptation?" *Annals of Botany* 106 (2010): 813–22.

Ringel, M., H. Hu, and G. Anderson. "The Stability and Persistence of Mutualisms Embedded in Community Interactions." *Theoretical Population Biology* 50 (1996): 281–97.

Risbey, D. A., M. C. Calver, J. Short, J. S. Bradley, and I. W. Wright. "The Impact of Cats and Foxes on the Small Vertebrate Fauna of Heirisson Prong, Western Australia, II: A Field Experiment." *Wildlife Research* 27 (2000): 223–35.

Ritchie, J. *Birds and Beasts as Farm Pests*. Edinburgh: Oliver and Boyd, 1931.

Ritchie, J. "The Changing Face of Nature, I: Our Prehistoric Ancestors Go A-Hunting." *Listener* 7, no. 171 (1932): 26–27.

Ritchie, J. "The Changing Face of Nature, II: Creatures of Yesterday." *Listener* 7, no. 172 (1932): 34–35.

Ritchie, J. "The Changing Face of Nature III: The Farmer and Wildlife." *Listener* 7, no. 173 (1932): 21–22.

Ritchie, J. "The Changing Face of Nature, IV: War upon Wild Animals." *Listener* 7, no. 174 (1932): 10–11.

Ritchie, J. "The Changing Face of Nature, V: Sport and Commerce Take a Hand in the Game." *Listener* 7, no. 175 (1932): 26–27.

Ritchie, J. "The Changing Face of Nature, VI: Gains and Losses in Animal Life." *Listener* 7, no. 176 (1932): 24–25.

Ritchie, J. *The Influence of Man on Animal Life in Scotland*. Cambridge: Cambridge University Press, 1920.

Rittel, H. W., and M. M. Webber. "Dilemmas in a General Theory of Planning." *Policy Sciences* 4 (1973): 155–69.

Rivero, L. H. "The Introduction of Largemouth Bass, a Predator upon Native Cuban Fishes." *Transactions of the American Fisheries Society* 66 (1937): 367–68.

Robbins, P. "Comparing Invasive Networks: Cultural and Political Biographies of Invasive Species." *Geographical Review* 94 (2004): 139–56.

Röben, P. "Zur Ausbreitung des Waschbären, *Procyon lotor* (Linné, 1758) und des Marderhundes, *Nyctereutes procyonoides* (Gray, 1834) in der Bundesrepublik Deutschland." *Säugetierkundichle Mitteilungen* 23, no. 2 (1975): 93–101.

Robertson, P., T. Adriaens, A. Caizergues, P. Cranswick, K. Devos, et al. "Towards the European Eradication of the North American Ruddy Duck." *Biological Invasions* 17 (2015): 9–12.

Robin, C., and U. Heiniger. "Chestnut Blight in Europe: Diversity of *Cryphonectria parasitica*, Hypovirulence and Biocontrol." *Forest Snow and Landscape Research* 76 (2001): 361–67.

Roche, M. "A. H. Clark's Framing of Geographical Change." *Historical Geography* 41 (2013): 191–206.

Rodewald, A. D., D. P. Shustack, and L. E. Hitchcock. "Exotic Shrubs as Ephemeral Ecological Traps for Nesting Birds." *Biological Invasions* 12 (2010): 33–39.

Rodier, W. *The Rabbit Pest in Australia: Its Cause and Cure.* Sydney: William Brooks, 1902.

Rodman, J. "Restoring Nature: Natives and Exotics." In *The Nature of Things: Language, Politics, and the Environment*, edited by J. Bennett and W. Chaloupka, 139–53. Minneapolis: University of Minnesota Press, 1993.

Rodriguez, L. F. "Can Invasive Species Facilitate Native Species? Evidence of How, When, and Why These Impacts Occur." *Biological Invasions* 8 (2006): 927–39.

Rodriguez, R., J. J. K. Leary, and D. M. Jenkins. "Enhancing Invasive Species Control with Unmanned Aerial Systems and Herbicide Ballistic Technology." Paper no. 1700668, presented at the annual international meeting of the American Society of Agricultural and Biological Engineers, Spokane, WA, July 16–19, 2017.

Rogers, H., J. Hille Ris Lambers, R. Miller, and J. J. Tewksbury. "'Natural Experiment' Demonstrates Top-Down Control of Spiders by Birds on a Landscape Level." *PLoS One* 7, no. 9 (2012): e43446.

Rohwer, Y., and E. Marris. "Clarifying Compassionate Conservation with Hypotheticals: Response to Wallach et al. 2018." *Conservation Biology* 33 (2019): 781–83.

Rolls, E. C. *They All Ran Wild: The Story of Pests on the Land in Australia.* Sydney: Angus & Robertson, 1969.

Rolls, E. C. *They All Ran Wild: The Animals and Plants That Plague Australia.* Sydney: Angus & Robertson, 1984.

Rolston, H., III. *Environmental Ethics: Values in and Duties to the Natural World.* Philadelphia: Temple University Press, 1988.

Rolston, H., III. *Conserving Natural Value.* New York: Columbia University Press, 1994.

Romagosa, C. M., and R. F. Labisky. "Establishment and Dispersal of the Asian Collared-Dove in Florida." *Journal of Field Ornithology* 71 (2000): 159–66.

Roman, J. "Diluting the Founder Effect: Cryptic Invasions Expand a Marine Invader's Range." *Proceedings of the Royal Society B: Biological Sciences* 273 (2006): 2453–59.

Roman, J., and J. A. Darling. "Paradox Lost: Genetic Diversity and the Success of Aquatic Invasions." *Trends in Ecology and Evolution* 22 (2007): 454–64.

Romanuk, T., Y. Zhou, U. Brose, E. Berlow, R. Williams, and N. Martinez. "Predicting Invasion Success in Complex Ecological Networks." *Philosophical Transactions of the Royal Society of London, Series B: Biological Sciences* 364 (2009): 1743–54.

Roose, K. "Bluesky, Smiling at Me." *New York Times*, November 25, 2024, B1. New York Edition.

Roots, C. *Animal Invaders.* New York: Universe Books, 1976.

Rosa, C. A. da, B. R. Ribeiro, V. Bejarano, F. H. Puertas, A. Bocchiglieri, et al. "Neotropical Alien Mammals: A Data Set of Occurrence and Abundance of Alien Mammals in the Neotropics." *Ecology* 101, no. 11 (2020): e03115.

Rosen, D., and P. DeBach. "Diaspididae." In *Introduced Parasites and Predators of Arthropod Pests and Weeds*, Agriculture Handbook 480, edited by C. P. Clausen, 78–128. Washington, DC: United States Department of Agriculture, 1978.

Rosenberg, M. S., H. R. Rothstein, and J. Gurevitch. "Effect Sizes: Conventional Choices and Calculations." In Koricheva et al., *Handbook of Meta-Analysis in Ecology and Evolution*, 61–71.

Rosenfield, A., and R. Mann, eds. *Dispersal of Living Organisms into Aquatic Ecosystems.* College Park: Maryland Sea Grant College, 1992.

Rosenthal, D. M. "How Many Kinds of Consciousness?" *Consciousness and Cognition* 11 (2002): 653–65.

Rosenthal, R. "The File Drawer Problem and Tolerance for Null Results." *Psychological Bulletin* 86 (1979): 638–41.

Rosenthal, R. *Meta-Analytic Procedures for Social Research*. Beverly Hills, CA: Sage, 1984.

Ross, H. A. "Genetic Differentiation of Starling (*Sturnus vulgaris*: Aves) Populations in New Zealand and Great Britain." *Journal of Zoology* 201 (1983): 351–62.

Rotherham, I. D., and R. A. Lambert. "Good Science, Good History, and Pragmatism: Managing the Way Ahead." In Rotherham and Lambert, *Invasive and Introduced Plants and Animals*, 355–66.

Rotherham, I. D., and R. A. Lambert, eds. *Invasive and Introduced Plants and Animals: Human Perceptions, Attitudes and Approaches to Management*. London: Earthscan, 2011.

Rotter, M. C., and L. M. Holeski. "A Meta-Analysis of the Evolution of Increased Competitive Ability Hypothesis: Genetic-Based Trait Variation and Herbivory Resistance Trade-Offs." *Biological Invasions* 20 (2018): 2647–60.

Roughgarden, J. "Competition and Theory in Community Ecology." *American Naturalist* 122 (1983): 583–601.

Roullier, C., L. Benoit, D. B. McKey, and V. Lebot. "Historical Collections Reveal Patterns of Diffusion of Sweet Potato in Oceania Obscured by Modern Plant Movements and Recombination." *Proceedings of the National Academy of Sciences of the United States of America* 110 (2013): 2205–10.

Roura-Pascual, N., L. Brotons, A. T. Peterson, and W. Thuiller. "Consensual Predictions of Potential Distributional Areas for Invasive Species: A Case Study of Argentine Ants in the Iberian Peninsula." *Biological Invasions* 11 (2009): 1017–31.

Rowley, R. "A Graceful Gazelle Becomes a Pest." *High Country News*, October 22, 2001, 33. https://www.hcn.org/issues/213/10797.

Roy, H. E., A. Pauchard, P. Stoett, T. Renard Truong, S. Bacher, et al. *Summary for Policymakers of the Thematic Assessment Report on Invasive Alien Species and Their Control of the Intergovernmental Science-Policy Platform on Biodiversity and Ecosystem Services (IPBES)*. Bonn: IPBES Secretariat, 2023. https://doi.org/10.5281/zenodo.7430692.

Roy, H. E., and E. Wajnberg. "From Biological Control to Invasion: The Ladybird *Harmonia axyridis* as a Model Species." *BioControl* 53 (2008): 1–4.

Roy, J. "In Search of the Characteristics of Plant Invaders." In di Castri et al., *Biological Invasions in Europe and the Mediterranean Basin*, 335–52.

Royal Society for the Protection of Birds. "Saving Species from Extinction: The Gough Island Restoration Programme." Updated January 13, 2022. https://www.goughisland.com/.

Royle, J. A., R. B. Chandler, C. Yackulic, and J. D. Nichols. "Likelihood Analysis of Species Occurrence Probability from Presence-Only Data for Modelling Species Distributions." *Methods in Ecology and Evolution* 3 (2012): 545–54.

Rozzi, R. "Biocultural Ethics: From Biocultural Homogenization Toward Biocultural Conservation." In Rozzi et al., *Linking Ecology and Ethics for a Changing World*, 9–32.

Rozzi, R., R. May, T. Chapin, F. Massardo, M. Gavin, et al., eds. *From Biocultural Homogenization to Biocultural Conservation*. Dordrecht, Neth.: Springer, 2019.

Rozzi, R., S. T. A. Pickett, C. Palmer, J. J. Armesto, and J. B. Callicott, eds. *Linking Ecology and Ethics for a Changing World*. Dordrecht, Neth.: Springer, 2013.

Rudman, S. M., and D. Schluter. "Ecological Impacts of Reverse Speciation in Threespine Stickleback." *Current Biology* 26 (2016): 490–95.

Ruesink, J. L., I. M. Parker, M. J. Groom, and P. Kareiva. "Reducing the Risks of Non-Indigenous Species Introductions: Guilty Until Proven Innocent." *BioScience* 45 (1995): 465–77.

Ruiz, G. M., and J. T. Carlton. *Invasive Species: Vectors and Management Strategies.* Washington, DC: Island Press, 2003.

Ruiz, G. M., and C. L. Hewitt. "Latitudinal Patterns of Biological Invasions in Marine Ecosystems: A Polar Perspective." In *Smithsonian at the Poles: Contributions to International Polar Year Science*, edited by I. Krupnik, M. A. Lang, and S. E. Miller, 347–58. Washington DC: Smithsonian Institution, 2009.

Ruiz-López, M. J., L. Barahona, J. Martínez-de la Puente, M. Pepió, A. Valsecchi, et al. "Widespread Resistance to Anticoagulant Rodenticides in *Mus musculus domesticus* in the City of Barcelona." *Science of the Total Environment* 845 (2022): 157192.

Rukovskii, N. N. "Acclimatization of the Teleut Squirrel in the Habitat of Squirrels of Other Subspecies." In Yanushevich, *Acclimatization of Animals in the U.S.S.R.*, 98.

Ruse, M. *The Philosophy of Biology*. London: Hutchinson University Library, 1973.

Russell, E. P., III. "L. O. Howard Promoted War Metaphors as a Rallying Cry for Economic Entomology." *American Entomologist* 45 (1999): 74–78.

Russell, E. P., III. "'Speaking of Annihilation': Mobilizing for War Against Human and Insect Enemies, 1914–1945." *Journal of American History* 82 (1996): 1505–29.

Russell, E. P., III. *War and Nature: Fighting Humans and Insects with Chemicals from World War I to Silent Spring*. New York: Cambridge University Press, 2001.

Russell, J. C., and T. M. Blackburn. "The Rise of Invasive Species Denialism." *Trends in Ecology and Evolution* 32 (2017): 3–6.

Russell, J. C., H. P. Jones, D. P. Armstrong, F. Courchamp, P. J. Kappes, et al. "Importance of Lethal Control of Invasive Predators for Island Conservation." *Conservation Biology* 30 (2016): 670–72.

Russell, L. M. "Leland Ossian Howard: A Historical Review." *Annual Review of Entomology* 23 (1978): 1–15.

Russell, P. E. "A Century of Fungicide Evolution." *Journal of Agricultural Science* 143 (2005): 11–25.

Russo, L., R. Albert, C. Campbell, and K. Shea. "Experimental Species Introduction Shapes Network Interactions in a Plant-Pollinator Community." *Biological Invasions* 21 (2019): 3505–19.

Russo, L., J. Memmott, D. Montoya, K. Shea, and Y. M. Buckley. "Patterns of Introduced Species Interactions Affect Multiple Aspects of Network Structure in Plant-Pollinator Communities." *Ecology* 95 (2014): 2953–63.

Russo, L., C. Nichol, and K. Shea. "Pollinator Floral Provisioning by a Plant Invader: Quantifying Beneficial Effects of Detrimental Species." *Diversity and Distributions* 22, no. 2 (2016): 189–98.

Rutgers School of Environmental and Biological Sciences, Department of Entomology. "Gypsy Moth." Accessed October 17, 2024. https://entomology.rutgers.edu/history/gypsy-moth.html.

Rutkin, A. "GM Mosquitoes Approved for Field Trial Release in Florida." *New Scientist*, November 22, 2016. https://www.newscientist.com/article/2113627-gm-mosquitoes-approved-for-field-trial-release-in-florida/.

Ryan, J. *A History of the Internet and the Digital Future*. London: Reaktion Books, 2010.

Ryan, P. A., A. P. Turley, G. Wilson, T. P. Hurst, K. Retzki, et al. "Establishment of wMel *Wolbachia* in *Aedes aegypti* Mosquitoes and Reduction of Local Dengue

Transmission in Cairns and Surrounding Locations in Northern Queensland, Australia." *Gates Open Research* 3 (2019): 1547.

Rytwinski, T., J. J. Taylor, L. A. Donaldson, J. R. Britton, and D. R. Browne. "The Effectiveness of Non-Native Fish Removal Techniques in Freshwater Ecosystems: A Systematic Review." *Environmental Reviews* 27 (2019): 71–94.

Saavedra, F., D. W. Inouye, M. V. Price, and J. Harte. "Changes in Flowering and Abundance of *Delphinium nuttallianum* (Ranunculaceae) in Response to a Subalpine Climate Warming Experiment." *Global Change Biology* 9 (2003): 885–94.

Sadowski, J. S., J. A. Gonzalez, S. I. Lonhart, R. Jeppesen, T. M. Grimes, and E. D. Grosholz. "Temperature-Induced Range Expansion of a Subtropical Crab Along the California Coast." *Marine Ecology* 39 (2018): e12528.

Sagoff, M. "Do Non-Native Species Threaten the Natural Environment?" *Journal of Agricultural and Environmental Ethics* 18 (2005): 215–36.

Sagoff, M. "Fact and Value in Invasion Biology." *Conservation Biology* 34 (2020): 581–88.

Sagoff, M. "Invasive Species Denialism: A Reply to Ricciardi and Ryan." *Biological Invasions* 20 (2018): 2723–29.

Sagoff, M. "What's Wrong with Alien Species?" *Report of the Institute for Philosophy and Public Policy* 19, no. 4 (1999): 16–23.

Sahrage, D., and J. Lundbeck. *A History of Fishing*. Berlin: Springer, 1992.

Sailer, R. I. "History of Insect Introductions." In *Exotic Plant Pests and North American Agriculture*, edited by C. Graham and C. Wilson, 15–38. New York: Academic Press, 1983.

Sailer, R. I. "Our Immigrant Insect Fauna." *Bulletin of the Entomological Society of America* 24 (1978): 3–11.

Sale, K. *The Green Revolution: The American Environmental Movement, 1962–1992*. New York: Hill and Wang, 1993.

Salisbury, A., V. M. L. Barclay, S. Reid, and A. Halstead. "The Current Status of the Southern Green Shield Bug, *Nezara viridula* (Hemiptera: Pentatomidae), an Introduced Pest Species Recently Established in South-East England." *British Journal of Entomology and Natural History* 22 (2009): 189–94.

Salo, L. F. "Red Brome (*Bromus rubens* subsp. *madritensis*) in North America: Possible Modes for Early Introductions, Subsequent Spread." *Biological Invasions* 7 (2005): 165–80.

Saltonstall, K. "Cryptic Invasion by Non-Native Genotypes of the Common Reed, *Phragmites australis*, into North America." *Proceedings of the National Academy of Sciences of the United States of America* 99 (2002): 2445–49.

Saltonstall, K., and L. A. Meyerson. "*Phragmites australis*: From Genes to Ecosystems." *Biological Invasions* 18 (2016): 2415–20.

Salvaterra, T., D. S. Green, T. P. Crowe, and E. J. O'Gorman. "Impacts of the Invasive Alga *Sargassum muticum* on Ecosystem Functioning and Food Web Structure." *Biological Invasions* 15 (2013): 2563–76.

Salzman, A. "Eradication of Parakeets Draws Protests." *New York Times*, November 27, 2005, P3.

Samusenko, E. G. "Acclimatization of Fur Animals in Belorussia." In Yanushevich, *Acclimatization of Animals in the U.S.S.R.*, 104.

Sanders, J. G. "Save Us from Invading Pests." *American Forestry* 23 (1917): 147–53.

Sanders, S., C. Castiglione, and M. Hoff. *Risk Assessment Mapping Program: RAMP*. Washington, DC: US Fish and Wildlife Service, 2014.

Sanderson, E. D. "The Influence of Minimum Temperatures in Limiting the Northern Distribution of Insects." *Journal of Economic Entomology* 1 (1908): 245–62.

Sandler, R. "The Value of Species and the Ethical Foundations of Assisted Migration." *Conservation Biology* 24 (2010): 424–31.

Sandler, R., and J. Crane. "On the Moral Considerability of *Homo sapiens* and Other Species." *Environmental Values* 15 (2006): 69–84.

Sandlund, O. T., P. J. Schei, and Å. Viken, eds. *Proceedings of the Norway/UN Conference on Alien Species*. Trondheim: Directorate for Nature Management and Norwegian Institute for Nature Research, 1996.

San Miguel, K., and J. G. Scott. "The Next Generation of Insecticides: dsRNA Is Stable as a Foliar-Applied Insecticide." *Pest Management Science* 4 (2016): 801–9.

Sargent, C. S. "Editorial." *Garden and Forest* 1 (1888): 266.

Sauer, C. O. "The Agency of Man on the Earth." In Thomas, *Man's Role in Changing the Face of the Earth*, 49–69.

Sauer, C. O. "Early Relations of Man to Plants." *Geographical Review* 37 (1947): 1–25.

Sauer, C. O. "A Geographic Sketch of Early Man in America." *Geographical Review* 34 (1944): 529–73.

Sauer, C. O. "Man in the Ecology of Tropical America." In *Proceedings of the Ninth Pacific Science Congress of the Pacific Science Association (1957)*, vol. 20, *Special Symposium on Climate, Vegetation, and Rational Land Utilization in the Humid Tropics*, 104–10. Bangkok: Secretariat, Ninth Pacific Science Congress, Department of Science, 1958.

Sauer, C. O. "The Morphology of Landscape." *University of California Publications in Geography* 2, no. 2 (1925): 19–53.

Sauer, J. D. "Biogeographical Theory and Cultural Analogies." *World Archaeology* 8 (1977): 320–31.

Sauer, J. D. "Oceanic Islands and Biogeographical Theory." *Geographical Review* 59 (1969): 582–93.

Sauer, J. D. *Plant Migration: The Dynamics of Geographic Patterning in Seed Plant Species*. Berkeley: University of California Press, 1988.

Saunders, G. R., M. N. Gentle, and C. R. Dickman. "The Impacts and Management of Foxes *Vulpes vulpes* in Australia." *Mammal Review* 40 (2010): 181–211.

Sauper, H., dir. *Darwin's Nightmare*. Mille et Une Productions, Coop99 Filmproduktion, and Saga Film, 2004. 107 min.

Sauve, A., C. Fontaine, and E. Thébault. "Structure–Stability Relationships in Networks Combining Mutualistic and Antagonistic Interactions." *Oikos* 123 (2014): 378–84.

Savaya-Alkalay, A., O. Ovadia, A. Barki, and A. Sagi. "Size-Selective Predation by All-Male Prawns: Implications for Sustainable Biocontrol of Snail Invasions." *Biological Invasions* 20 (2018): 137–49.

Savidge, J. "Extinction of an Island Forest Avifauna by an Introduced Snake." *Ecology* 68 (1987): 660–68.

Savona-Ventura, C. N. D. "The Herpetofauna of the Maltese Islands." *Civilization* 25 (1985): 693–95.

Sax, D. F., and J. H. Brown. "The Paradox of Invasion." *Global Ecology and Biogeography* 9 (2000): 363–71.

Sax, D. F., J. J. Stachowicz, and S. D. Gaines, eds. *Species Invasions: Insights into Ecology, Evolution, and Biogeography*. Sunderland, MA: Sinauer, 2005.

SCEP [Study of Critical Ecosystem Problems]. *Man's Impact on the Global Environment*. Cambridge, MA: MIT Press, 1970.

Schardt, J. D. "Maintenance Control." In Simberloff, Schmitz, and Brown, *Strangers in Paradise: Impact and Management of Nonindigenous Species in Florida*, 229–43.

Schatz, G., E. Haslbrunner, and H. Tuppy. "Deoxyribonucleic Acid Associated with Yeast Mitochondria." *Biochemical and Biophysical Research Communications* 15 (1964): 127–35.

Scheiner, S. M. "Additional Thoughts on 'A Critique for Ecology.'" *Bulletin of the Ecological Society of America* 74 (1993): 179–80.

Schill, D. J., J. A. Heindel, M. R. Campbell, K. A. Meyer, and E. R. Mamer. "Production of a YY Male Brook Trout Broodstock for Potential Eradication of Undesired Brook Trout Populations." *North American Journal of Aquaculture* 78 (2016): 72–83.

Schill, D. J., K. A. Meyer, and M. J. Hansen. "Simulated Effects of YY-Male Stocking and Manual Suppression for Eradicating Nonnative Brook Trout Populations." *North American Journal of Fisheries Management* 37 (2017): 1054–66.

Schlaepfer, M. A., D. F. Sax, and J. D. Olden. "The Potential Conservation Value of Non-Native Species." *Conservation Biology* 25 (2011): 428–37.

Schluter, D., and J. D. McPhail. "Ecological Character Displacement and Speciation in Sticklebacks." *American Naturalist* 140 (1992): 85–108.

Schmitz, D. C. "The Historical Introductions of Water Hyacinth and Hydrilla into the United States." *Aquatics* 38, no. 1 (2016): 25–29.

Schmitz, D. C., and T. C. Brown, eds. *An Assessment of Invasive Non-Indigenous Species in Florida's Public Lands.* Tallahassee: Florida Department of Environmental Protection, 1994.

Schmitz, D. C., B. V. Nelson, L. E. Nall, and J. D. Schardt. "Exotic Aquatic Plants in Florida: A Historical Perspective and Review of the Present Aquatic Plant Regulation Program." In *Proceedings of the Symposium on Exotic Pest Plants,* US National Park Service Technical Report NPS/NREVER/NRTR-91/06, edited by T. D. Center, R. F. Doren, R. L. Hofstetter, R. L. Myers, and L. D. Whiteaker, 303–23. Denver: US National Park Service, 1991.

Schneider, D. W., C. D. Ellis, and K. S. Cummings. "A Transportation Model Assessment of Risk to Native Mussel Communities of Zebra Mussel Spread." *Conservation Biology* 12 (1998): 788–800.

Schooler, S. S. "*Alternanthera philoxeroides* (Martius) Grisebach (Alligator Weed)." In *A Handbook of Global Freshwater Invasive Species,* edited by R. A. Francis, 25–35. London: Earthscan, 2012.

Schrader, L., J. W. Kim, D. Ence, A. Zimin, A. Klein, et al. "Transposable Element Islands Facilitate Adaptation to Novel Environments in an Invasive Species." *Nature Communications* 5, no. 1 (2014): 5495.

Schreiner, C., III. "Uses of Exotic Animals in a Commercial Hunting Program." In *Introduction of Exotic Animals: Ecological and Socioeconomic Considerations,* 13–16. College Station: Texas A&M University, Caesar Kleberg Research Program in Wildlife Ecology, 1968.

Schrey, A. W., M. T. Grispo, M. Awad, M. Cook, E. D. McCoy, et al. "Broad-Scale Latitudinal Patterns of Genetic Diversity Among Native European and Introduced House Sparrow (*Passer domesticus*) Populations." *Molecular Ecology* 20 (2011): 133–43.

Schrey, A. W., C. A. C. Coon, M. T. Grispo, M. Awad, T. Imboma, et al. "Epigenetic Variation May Compensate for Decreased Genetic Variation with Introductions: A Case Study Using House Sparrows (*Passer domesticus*) on Two Continents." *Genetics Research International,* February 9, 2012, 7. https://doi.org/10.1155/2012/979751.

Schüttler, E., R. Rozzi, and K. Jax. "Towards a Societal Discourse on Invasive Species Management: A Case Study of Public Perceptions of Mink and Beavers in Cape Horn." *Journal of Nature Conservation* 19 (2011): 175–84.

Schuyler, P. T., D. Garcelon, and S. Escover. "Control of Feral Goats (*Capra hircus*) on Santa Catalina Island, California, USA." In Veitch and Clout, *Turning the Tide: The Eradication of Invasive Species*, 412–13.

Schweinitz, L. D. de. "Remarks on the Plants of Europe Which Have Become Naturalized in a More or Less Degree, in the United States." *Annals of the Lyceum of Natural History of New-York* 3 (1836): 148–55.

Schwindt, E., T. A. August, S. Vanderhoeven, M. A. McGeoch, S. Bacher, et al. "Overwhelming Evidence Galvanizes a Global Consensus on the Need for Action Against Invasive Alien Species." *Biological Invasions* 26 (2023): 621–26.

Sclater, P. L. "On the General Geographical Distribution of the Members of the Class Aves." *Journal of the Proceedings of the Linnean Society of London* 2 (1858): 130–45.

Sclater, P. L. "The Rabbit Pest." *Nature* 39 (1889): 493–94.

Scriver, M., A. Marinich, C. Wilson, and J. Freeland. "Development of Species-Specific Environmental DNA (eDNA) Markers for Invasive Aquatic Plants." *Aquatic Botany* 122 (2015): 27–31.

Secord, A. W. "Studies in the Narrative Method of Defoe." *University of Illinois Studies in Language and Literature* 9 (1924): 1–248.

Secretariat of the International Plant Protection Convention. *International Plant Protection Convention (1997)*. 2nd ed. Rome: Food and Agriculture Organization of the United Nations, 2024. https://doi.org/10.4060/cd0175en.

Seftel, H. "Government Regulation and the Rise of the California Fruit Industry: The Entrepreneurial Attack on Fruit Pests, 1880–1920." *Business History Review* 59 (1985): 369–402.

Seibert, P. *Toad Overload*. Brookfield, CT: Millbrook Press, 1996.

Selander, R. K., and R. F. Johnston. "Evolution in the House Sparrow: Intrapopulation Variation in North America." *Condor* 69 (1967): 217–58.

Selander, R. K., and D. W. Kaufman. "Self-Fertilization and Genetic Population Structure in a Colonizing Land Snail." *Proceedings of the National Academy of Sciences of the United States of America* 70 (1973): 1186–90.

Sellars, R. W. "The Rise and Decline of Ecological Attitudes in National Park System Management, 1929–1940, Part II." *George Wright Society Forum* 10, no. 2 (1993): 79–108.

Sellars, R. W. *Preserving Nature in the National Parks*. New Haven, CT: Yale University Press, 1997.

Semmens, B. X., E. R. Buhle, A. K. Salomon, and C. V. Pattengill-Semmens. "A Hotspot of Non-Native Aquarium Fishes: Evidence for the Aquarium Trade as an Invasion Pathway." *Marine Ecology Progress Series* 266 (2004): 239–44.

Senner, J. W. "Inbreeding Depression and the Survival of Zoo Populations." In Soulé and Wilcox, *Conservation Biology: An Evolutionary-Ecological Perspective*, 209–24.

Serbesoff-King, K. "Melaleuca in Florida: A Literature Review on the Taxonomy, Distribution, Biology, Ecology, Economic Importance and Control Measures." *Journal of Aquatic Plant Management* 41 (2003): 98–112.

Serniak, L. T., S. S. Chan, and K. Lajtha. "Predicting Habitat Suitability for *Amynthas* spp. in the United States: A Retrospective Analysis Using Citizen Science Data from iNaturalist." *Biological Invasions* 25 (2023): 817–25.

Servick, K. "Winged Warriors." *Science* 354 (2016): 164–67.

Sexton, J. P., J. K. McKay, and A. Sala. "Plasticity and Genetic Diversity May Allow Saltcedar to Invade Cold Climates in North America." *Ecological Applications* 12 (2002): 1652–60.

Shabani, F., M. Ahmadi, L. Kumar, S. Solhjouy-fard, M. S. Tehrany, et al. "Invasive Weed Species' Threats to Global Biodiversity: Future Scenarios of Changes in the Number of Invasive Species in a Changing Climate." *Ecological Indicators* 116 (2020): 106436.

Shackleton, R. T., G. Vimercati, A. F. Probert, S. Bacher, C. A. Kull, and A. Novoa. "Consensus and Controversy in the Discipline of Invasion Science." *Conservation Biology* 36, no. 5 (2022): e13931. https://doi.org/10.1111/cobi.13931.

Shaffer, M. L. "Minimum Population Sizes for Species Conservation." *BioScience* 31 (1981): 131–34.

Shaffer, M. L. "Minimum Viable Populations, Coping with Uncertainty." In *Viable Populations for Conservation*, edited by M. E. Soulé, 69–86. Cambridge: Cambridge University Press, 1987.

Shafland, P. L. "The Continuing Problem of Non-Native Fishes in Florida." *Fisheries* 1, no. 6 (1976): 25.

Shafland, P. L. "Exotic Fish Assessments: An Alternative View." *Reviews in Fisheries Science* 4 (1996): 123–32.

Shafland, P. L. "The Introduced Butterfly Peacock (*Cichla ocellaris*) in Florida, I: Fish Community Analyses." *Reviews in Fisheries Science* 7, no. 2 (1999): 71–94.

Shafland, P. L. "The Introduced Butterfly Peacock (*Cichla ocellaris*) in Florida, II: Food and Reproductive Biology." *Reviews in Fisheries Science* 7, no. 2 (1999): 95–113.

Shafland, P. L. "The Introduced Butterfly Peacock (*Cichla ocellaris*) in Florida, III: Length Distribution Analyses." *Reviews in Fisheries Science* 7, no. 2 (1999): 115–26.

Shafland, P. L. "Non-Native Fish Introductions with Special Reference to Florida." *Fisheries* 4, no. 3 (1979): 18–24.

Shafland, P. L. *A Proposal for Introducing Peacock Bass (Cichla spp.) in Southeast Florida Canals*. Contribution 40. Boca Raton: Non-Native Fish Research Laboratory, Florida Game and Fresh Water Fish Commission, 1984.

Shafland, P. L. "A Review of Florida's Efforts to Regulate, Assess and Manage Exotic Fishes." *Fisheries* 11, no. 2 (1986): 20–25.

Shafland, P. L., and K. J. Foote. "A Reproducing Population of *Serrasalmus humeralis* Valenciennes in Southern Florida." *Florida Scientist* 42 (1979): 206–14.

Shafland, P. L., K. B. Gestring, and M. S. Stanford. "Florida's Exotic Freshwater Fishes–2007." *Florida Scientist* 71 (2008): 220–45.

Shafland, P. L., and J. P. Pestrak. "Lower Lethal Temperatures for Fourteen Non-Native Fishes in Florida." *Environmental Biology of Fishes* 7 (1982): 149–56.

Shafland, P. L., and M. S. Stanford. "The Introduced Butterfly Peacock (*Cichla ocellaris*) in Florida, IV: Socioeconomic Analyses." *Reviews in Fisheries Science* 7, no. 2 (1999): 127–35.

Shah, S. *The Next Great Migration: The Beauty and Terror of Life on the Move*. New York: Bloomsbury, 2020.

Shantz, H. L. "Frederic Edward Clements." *Ecology* 26 (1945): 317–19.

Shardlow, M. E., and R. Taylor. "Is the Southern Green Shield Bug, *Nezara viridula* (L.) (Hemiptera: Pentatomidae) Another Species Colonising Britain Due to Climate Change?" *British Journal of Entomology and Natural History* 17 (2004): 143–46.

Sharov, A. A., D. Leonard, A. M. Liebhold, E. A. Roberts, and W. Dickerson. "'Slow the Spread': A National Program to Contain the Gypsy Moth." *Journal of Forestry* 100 (2002): 30–36.

Sharpe, D. M. T., L. F. De León, R. González, and M. E. Torchin. "Tropical Fish Community Does Not Recover 45 Years After Predator Introduction." *Ecology* 98 (2017): 412–24.

Sharples, F. E. "Ecological Aspects of Hazard Identification for Environmental Uses of Genetically Engineered Organisms." In M. Levin and Strauss, *Risk Assessment in Genetic Engineering*, 18–31.

Sharples, F. E. "Spread of Organisms with Novel Genotypes: Thoughts from an Ecological Perspective." *Recombinant DNA Technical Bulletin* 6, no. 2 (1983): 43–56.

Sheldon, K. S. "Climate Change in the Tropics: Ecological and Evolutionary Responses at Low Latitudes." *Annual Review of Ecology, Evolution, and Systematics* 50 (2019): 303–33.

Shelford, V. E. *Animal Communities in Temperate America as Illustrated in the Chicago Region: A Study in Animal Ecology*. Bulletin 5. Chicago: Geographic Society of Chicago, 1913.

Shelford, V. E. "An Experimental Approach to the Study of Bird Populations." *Wilson Bulletin* 66 (1954): 253–58.

Shelford, V. E. *Laboratory and Field Ecology*. Baltimore: Williams and Wilkins, 1929.

Shelford, V. E. "Physiological Animal Geography." *Journal of Morphology* 22 (1911): 551–618.

Shelford, V. E. "Some Concepts of Bioecology." *Ecology* 12 (1931): 455–67.

Shelly, T. E., J. Edu, and J. Nishimoto. "Chilling and Flight Ability and Mating Competitiveness of Sterile Males of the Mediterranean Fruit Fly." *Journal of Applied Entomology* 137, no. S1 (2013): 11–18.

Shelton, J. M., M. J. Samways, and J. A. Day. "Predatory Impact of Non-Native Rainbow Trout on Endemic Fish Populations in Headwater Streams in the Cape Floristic Region of South Africa." *Biological Invasions* 17 (2015): 365–79.

Shi, M. Z., J. Y. Li, B. Ding, J. W. Fu, L. Z. Zheng, and H. Chi. "Indirect Effect of Elevated CO_2 on Population Parameters and Growth of *Agasicles hygrophila* (Coleoptera: Chrysomelidae), a Biocontrol Agent of Alligatorweed (Amaranthaceae)." *Journal of Economic Entomology* 112 (2019): 1120–29.

Shields, M. W., A. C. Johnson, S. Pandey, R. Cullen, M. González-Chang, et al. "History, Current Situation and Challenges for Conservation Biological Control." *Biological Control* 131 (2019): 25–35.

Shiels, A. B., C. A. Flores, A. Khamsing, P. D. Krushelnycky, S. M. Mosher, and D. R. Drake. "Dietary Niche Differentiation Among Three Species of Invasive Rodents (*Rattus rattus, R. exulans, Mus musculus*)." *Biological Invasions* 15 (2013): 1037–48.

Shigesada, N., and K. Kawasaki. *Biological Invasions: Theory and Practice*. Oxford: Oxford University Press, 1997.

Shine, R. "Invasive Species as Drivers of Evolutionary Change: Cane Toads in Tropical Australia." *Evolutionary Applications* 5 (2012): 107–16.

Shōji, T. "The Man Who Trained Japan's First World Boxing Champion: Alvin Robert Cahn." Nippon.com. August 15, 2019. https://www.nippon.com/en/japan-topics/g00711/the-man-who-trained-japan's-first-world-boxing-champion-alvin-robert-cahn.html.

Shorten, M. "Damage Caused by Squirrels in Forestry Commission Areas, 1954–6." *Forestry* 30 (1957): 151–71.

Shorten, M. "Inheritance of Melanism in Grey Squirrels." *Nature* 156 (1945): 46–47.

Shorten, M. *Squirrels*. London: Collins, 1954.

Shorten, M. "Squirrels in England, Wales and Scotland, 1955." *Journal of Animal Ecology* 26 (1957): 287–94.

Shorten, M. "A Survey of the Distribution of the American Grey Squirrel (*Sciurus carolinensis*) and the British Red Squirrel (*S. vulgaris leucourus*) in England and Wales in 1944–45." *Journal of Animal Ecology* 15 (1946): 82–92.

Shorten, M., and F. A. Courtier. "A Population Study of the Grey Squirrel (*Sciurus carolinensis*) in May 1954." *Annals of Applied Biology* 43 (1955): 494–510.

Shorten, M., and C. S. Elton. "Some Aspects of the Biology of the Grey Squirrel (*Sciurus carolinensis*) in Great Britain." *Proceedings of the Zoological Society of London* 121 (1951): 427–59.

Shurin, J. B. "Top-Down and Bottom-Up Regulation of Communities." Oxford Bibliographies. May 23, 2012. https://doi.org/10.1093/obo/9780199830060-0029.

Shuttleworth, C. M., P. W. W. Lurz, and J. Gurnell, eds. *The Grey Squirrel: Ecology and Management of an Invasive Species in Europe.* Stoneleigh Park, UK: European Squirrel Initiative, 2016.

Shuttleworth, S. J. *Cognition, Evolution, and Behavior.* New York: Oxford University Press, 1998.

Signorile, A. L., D. Paoloni, and D. C. Reuman. "Grey Squirrels in Central Italy: A New Threat for Endemic Red Squirrel Subspecies." *Biological Invasions* 16 (2014): 2339–50.

Sih, A., A. M. Bell, and J. C. Johnson. "Behavioral Syndromes: An Ecological and Evolutionary Overview." *Trends in Ecology and Evolution* 19 (2004): 372–78.

Sih, A., A. M. Bell, J. C. Johnson, and R. E. Ziemba. "Behavioral Syndromes: An Integrative Overview." *Quarterly Review of Biology* 79 (2004): 241–77.

Sikka, V., V. K. Chattu, R. K. Popli, S. C. Galwankar, D. Kelkar, et al. "The Emergence of Zika Virus as a Global Health Security Threat: A Review and a Consensus Statement of the INDUSEM Joint Working Group (JWG)." *Journal of Global Infectious Diseases* 8 (2016): 3–15.

Sikora, E. J., T. W. Allen, K. A. Wise, G. Bergstrom, C. A. Bradley, et al. "A Coordinated Effort to Manage Soybean Rust in North America: A Success Story in Soybean Disease Monitoring." *Plant Disease* 98 (2014): 864–75.

Silliman, B. R., B. B. Hughes, Y. S. Zhang, and Q. He. "Business as Usual Leads to Underperformance in Coastal Restoration." In Kareiva et al., *Effective Conservation Science*, 173–78.

Simberloff, D. "Biogeographic Models, Species' Distributions and Community Organization." In *Evolution, Time and Space: The Emergence of the Biosphere*, edited by R. W. Sims, J. H. Price, and P. E. S. Whaley, 57–83. London: Academic Press, 1983.

Simberloff, D. "Biogeography: The Unification and Maturation of a Science." In *Perspectives in Ornithology*, edited by A. H. Brush and G. C. Clark Jr., 411–55. Cambridge: Cambridge University Press, 1983.

Simberloff, D. "Biological Invasions in the National Parks and in *Park Science*." In *Science, Conservation, and National Parks*, edited by S. R. Beissinger, D. D. Ackerly, H. Doremus, and G. E. Machlis, 161–88. Chicago: University of Chicago Press, 2017.

Simberloff, D. "Biological Invasions: What's Worth Fighting and What Can Be Won?" *Ecological Engineering* 65 (2014): 112–21.

Simberloff, D. "Charles Elton." Oxford Bibliographies. Modified May 28, 2013. https://doi.org/10.1093/OBO/9780199830060-0090.

Simberloff, D. "Charles Elton—Neither Founder nor Siren, but Prophet." In Richardson, *Fifty Years of Invasion Ecology*, 11–24.

Simberloff, D. "Charles Elton: Pioneer Conservation Biologist." *Environment and History* 18 (2012): 183–202.

Simberloff, D. "Community Ecology: Is It Time to Move On?" *American Naturalist* 163 (2004): 787–99.

Simberloff, D. "Community Effects of Introduced Species." In *Biotic Crises in Ecological and Evolutionary Time*, edited by M. H. Nitecki, 53–81. New York: Academic Press, 1981.

Simberloff, D. "Competition Theory, Hypothesis-Testing, and Other Community Ecological Buzzwords." *American Naturalist* 122 (1983): 626–35.

Simberloff, D. "Confronting Introduced Species: A Form of Xenophobia?" *Biological Invasions* 5 (2003): 179–92.

Simberloff, D. "The Contribution of Population and Community Ecology to Conservation Science." *Annual Review of Ecology and Systematics* 19 (1988): 473–511.

Simberloff, D. "Elton, Charles S." In Simberloff and Rejmánek, *Encyclopedia of Biological Invasions*, 187–89.

Simberloff, D. "Equilibrium Theory of Island Biogeography and Ecology." *Annual Review of Ecology and Systematics* 5 (1974): 161–82.

Simberloff, D. "Eradication and Maintenance Management of Established Aquatic Invaders." *Hydrobiologia* 848 (2021): 2399–420.

Simberloff, D. Foreword to *The Ecology of Invasion by Animals and Plants*, edited by C. S. Elton, vii–xiv. Chicago: University of Chicago Press, 2000.

Simberloff, D. "How Common Are Invasion-Induced Ecosystem Impacts?" *Biological Invasions* 13 (2011): 1255–68.

Simberloff, D. "Integrity, Stability, and Beauty: Aldo Leopold's Evolving View of Non-Native Species." *Environmental History* 17 (2012): 487–511.

Simberloff, D. "Introduced Insects: A Biogeographic and Systematic Perspective." In Mooney and Drake, *Ecology of Biological Invasions of North America and Hawaii*, 3–26.

Simberloff, D. "Invasional Meltdown 6 Years Later: Important Phenomenon, Unfortunate Metaphor, or Both?" *Ecology Letters* 9 (2006): 912–19.

Simberloff, D. "Invasions of Plant Communities—More of the Same, Something Very Different, or Both?" *American Midland Naturalist* 163 (2010): 220–33.

Simberloff, D. "Keystone Species and Community Effects of Biological Introductions." In *Assessing Ecological Risks of Biotechnology*, edited by L. R. Ginzburg, 1–19. Boston: Butterworth-Heinemann, 1991.

Simberloff, D. Letter to the editor, *Isis* 87 (1996): 676–77.

Simberloff, D. "Native Invaders." In Simberloff and Rejmánek, *Encyclopedia of Biological Invasions*, 472–75.

Simberloff, D. "Nature, Culture, and Natureculture: The Role of Nonnative Species in Biocultures." In Rozzi et al., *From Biocultural Homogenization to Biocultural Conservation*, 207–18.

Simberloff, D. "Nature, Natives, Nativism, and Management: Worldviews Underlying Controversies in Invasion Biology." *Environmental Ethics* 34 (2012): 5–25.

Simberloff, D. "New Zealand as a Leader in Conservation Practice and Invasion Management." *Journal of the Royal Society of New Zealand* 49 (2019): 259–80.

Simberloff, D. "Non-Native Species Do Threaten the Natural Environment!" *Journal of Agricultural and Environmental Ethics* 18 (2005): 595–607.

Simberloff, D. "The Politics of Assessing Risk for Biological Invasions: The USA as a Case Study." *Trends in Ecology and Evolution* 20 (2005): 216–22.

Simberloff, D. Preface to *Invasions biologiques et extinctions: 11,000 ans d'histoire des vertébrés en France*, edited by M. Pascal, O. Lorvelec, and J.-D. Vigne, 5–7. Paris: Éditions Belin, 2006.

Simberloff, D. "Rats Are Not the Only Introduced Rodents Producing Ecosystem Impacts on Islands." *Biological Invasions* 11 (2009): 1735–42.

Simberloff, D. Review of *Invasion Biology*, by M. A. Davis. *Ecological Restoration* 28 (2010): 101–3.

Simberloff, D. "Risk Assessments, Blacklists, and White Lists for Introduced Species: Are Predictions Good Enough to Be Useful?" *Agricultural and Resource Economics Review* 35 (2006): 1–10.

Simberloff, D. "Risks of Biological Control for Conservation Purposes." *BioControl* 57 (2012): 263–76.

Simberloff, D. "The Role of Propagule Pressure in Biological Invasions." *Annual Review of Ecology, Evolution and Systematics* 40 (2009): 81–102.

Simberloff, D. "SCOPE Project." In Simberloff and Rejmánek, *Encyclopedia of Biological Invasions*, 617–19.

Simberloff, D. Review of *When the Killing's Done*, by T. C. Boyle. *Biological Invasions* 13 (2011): 2163–66.

Simberloff, D. "Which Insect Introductions Succeed and Which Fail?" In Drake et al., *Biological Invasions: A Global Perspective*, 61–75.

Simberloff, D., and M. Alexander. "Assessing Risks to Ecological Systems from Biological Introductions (Excluding Genetically Modified Organisms)." In *Handbook of Environmental Risk Assessment and Management*, edited by P. Calow, 147–76. Oxford: Blackwell Science, 1998.

Simberloff, D., and M. Alexander. "Issue Paper on Biological Stressors." In *Environmental Protection Agency, Ecological Risk Assessment Issue Papers*. EPA/630/R-94/009, 6–1 to 6–59. Washington, DC: US Environmental Protection Agency Risk Assessment Forum, 1994.

Simberloff, D., and W. Boecklen. "Patterns of Extinction in the Introduced Hawaiian Avifauna: A Reexamination of the Role of Competition." *American Naturalist* 138 (1991): 300–327.

Simberloff, D., T. Dayan, C. Jones, and G. Ogura. "Character Displacement and Release in the Small Indian Mongoose, *Herpestes javanicus*." *Ecology* 81 (2000): 2086–99.

Simberloff, D., and L. Gibbons. "Now You See Them, Now You Don't! Population Crashes of Established Introduced Species." *Biological Invasions* 6 (2004): 161–72.

Simberloff, D., B. Keitt, D. Will, N. Holmes, E. Pickett, and P. Genovesi. "Yes We Can! Exciting Progress and Prospects for Controlling Invasives on Islands and Beyond." *Western North American Naturalist* 78 (2018): 942–58.

Simberloff, D., J.-L. Martin, P. Genovesi, V. Maris, D. A. Wardle, et al. "Impacts of Biological Invasions—What's What and the Way Forward." *Trends in Ecology and Evolution* 28 (2013): 58–66.

Simberloff, D., L. A. Meyerson, P. Pyšek, and D. M. Richardson. "Honoring Harold A. Mooney: Citizen of the World and Catalyst for Invasion Science." *Biological Invasions* 19 (2017): 2219–24.

Simberloff, D., and M. Rejmánek, eds. *Encyclopedia of Biological Invasions*. Berkeley: University of California Press, 2011.

Simberloff, D., D. C. Schmitz, and T. C. Brown, eds. *Strangers in Paradise: Impact and Management of Nonindigenous Species in Florida*. Washington, DC: Island Press, 1997.

Simberloff, D., L. Souza, M. A. Nuñez, M. N. Barrios-Garcia, and W. Bunn. "The Natives Are Restless, but Not Often and Mostly When Disturbed." *Ecology* 93 (2012): 598–607.

Simberloff, D., and P. Stiling. "How Risky Is Biological Control?" *Ecology* 77 (1996): 1965–74.

Simberloff, D., and D. R. Strong. "Counterpoint: Scientists Offer a Dissenting View on Ascension Island." *Yale Environment 360*, September 11, 2013. https://e360.yale.edu/digest/counterpoint_scientists_offer_dissenting_view_on_ascension_island.

Simberloff, D., and J. R. S. Vitule. "A Call for an End to Calls for the End of Invasion Biology." *Oikos* 123 (2014): 408–13.

Simberloff, D., and B. Von Holle. "Positive Interactions of Nonindigenous Species: Invasional Meltdown?" *Biological Invasions* 1 (1999): 21–32.

Simmonds, F. J., and F. D. Bennett. "Biological Control of *Opuntia* spp. by *Cactoblastis cactorum* in the Leeward Islands (West Indies)." *Entomophaga* 11 (1966): 183–89.

Simmonds, F. J., J. M. Franz, and R. I. Sailer. "History of Biological Control." In *Theory and Practice of Biological Control*, edited by C. B. Huffaker and P. S. Messenger, 17–39. New York: Academic Press, 1976.

Simon, K. S., and C. R. Townsend. "Impacts of Freshwater Invaders at Different Levels of Ecological Organisation, with Emphasis on Salmonids and Ecosystem Consequences." *Freshwater Biology* 48 (2003): 982–94.

Sinatra, G. M., and B. K. Hofer. *Science Denial: Why It Happens and What to Do About It.* Oxford: Oxford University Press, 2021.

Sinclair, B. J., K. E. Marshall, M. A. Sewell, D. L. Levesque, C. S. Willett, et al. "Can We Predict Ectotherm Responses to Climate Change Using Thermal Performance Curves and Body Temperatures?" *Ecology Letters* 19 (2016): 1372–85.

Sindel, B. M., and P. W. Michael. "Spread and Potential Distribution of *Senecio madagascariensis* Poir. (Fireweed) in Australia." *Australian Journal of Ecology* 17 (1992): 21–26.

Singer, M. C. "Evolution of Food-Plant Preferences in the Butterfly, *Euphydryas editha*." *Evolution* 25 (1971): 383–89.

Singer, M. C., and C. S. McBride. "Multitrait, Host-Associated Divergence Among Sets of Butterfly Populations: Implications for Reproductive Isolation and Ecological Speciation." *Evolution* 64 (2010): 921–33.

Singer, M. C., C. D. Thomas, and C. Parmesan. "Rapid Human-Induced Evolution of Insect-Host Associations." *Nature* 366 (1993): 681–83.

Singer, P. *Animal Liberation.* New York: Random House, 1975.

Singer, P. *Practical Ethics.* 2nd ed. Cambridge: Cambridge University Press, 1993.

Skellam, J. G. "Random Dispersal in Theoretical Populations." *Biometrika* 38 (1951): 196–218.

Skelly, D. K., L. N. Joseph, H. P. Possingham, L. K. Freidenburg, T. J. Farrugia, et al. "Evolutionary Responses to Climate Change." *Conservation Biology* 21 (2007): 1353–55.

Slatis, H. M. "An Analysis of Inbreeding in the European Bison." *Genetics* 45 (1960): 275–87.

Slatyer, R. O. "Ecological Reserves: Size, Structure and Management." In *A National System of Ecological Reserves in Australia: Proceedings of a Symposium Held in Canberra, 31 October 1974 Under the Auspices of the Australian Academy of Science Through Its International Biological Programme*, edited by F. Fenner, 22–38. Netley, Australia: Griffin Press, 1975.

Slobodkin, L. B. "The Good, the Bad and the Reified." *Evolutionary Ecology Research* 3 (2001): 1–13.

Sloggett, J., A. Magro, F. Verheggen, J.-L. Hemptinne, W. D. Hutchison, and E. W. Riddick. "The Chemical Ecology of *Harmonia axyridis*." *BioControl* 56 (2011): 643–61.

Sludskii, A. A. "Theory and Practice of Acclimatization of Game Animals." In Yanushevich, *Acclimatization of Animals in the U.S.S.R.*, 112–16.

Smith, C. M. "Changed and Changing Vegetation." In *Science in New Zealand*, edited by F. R. Callaghan, 123–36. Wellington: A. H. & A. W. Reed, 1957.

Smith, C. S., W. M. Lonsdale, and J. Fortune. "When to Ignore Advice: Invasion Predictions and Decision Theory." *Biological Invasions* 1 (1999): 89–96.

Smith, E. H. "The Entomological Society of America: The First Hundred Years, 1889–1989." *Bulletin of the Entomological Society of America* 35, no. 3 (1989): 10–32.

Smith, H. M. "A Review of the History and Results of the Attempts to Acclimatize Fish and Other Water Animals in the Pacific States." *Bulletin of the United States Fish Commission* 15 (1896): 379–472.

Smith, H. S. "Racial Segregation in Insect Populations and Its Significance in Applied Entomology." *Journal of Economic Entomology* 34 (1941): 1–18.

Smolik, M. G., S. Dullinger, F. Essl. I. Kleinbauer, M. Leitner, et al. "Integrating Species Distribution Models and Interacting Particle Systems to Predict the Spread of an Invasive Alien Plant." *Journal of Biogeography* 37 (2010): 411–22.

Smout, T. C. "The Alien Species in 20th-Century Britain: Constructing a New Vermin." *Landscape Research* 28 (2003): 11–20.

"Snails as Big as Your Fist." *Guardian* (UK edition), September 29, 1969.

Soberón, J. "Grinnellian and Eltonian Niches and Geographic Distributions of Species." *Ecology Letters* 10 (2007): 1115–23.

Sobey, W. R., and D. Conolly. "Myxomatosis: The Introduction of the European Rabbit Flea *Spilopsyllus cuniculi* (Dale) into Wild Rabbit Populations in Australia." *Journal of Hygiene* (Cambridge) 69 (1971): 331–46.

Soboleski, H. "Island History for Friday, November 13, 2009." *Garden Island*, November 13, 2009. https://www.thegardenisland.com/2009/11/13/hawaii-news/island-history-for-friday-november-13-2009/.

Soboleski, H. "Bird Lover, Amateur Ornithologist Alexander H. Isenberg." *Garden Island*, November 11, 2018. https://www.thegardenisland.com/2018/11/11/lifestyles/bird-lover-amateur-ornithologist-alexander-h-isenberg/.

Solomon, M. E. "The Natural Control of Animal Populations." *Journal of Animal Ecology* 18 (1949): 1–35.

Solomon, R. "Kudzu Is So Much More Than the 'Vine That Ate the South.'" *Slate*, August 28, 2021. https://slate.com/news-and-politics/2021/08/kudzu-south-japan-metaphor.html.

Soper, F. L., and D. B. Wilson. *Anopheles gambiae in Brazil 1930–1940*. New York: Rockefeller Foundation, 1943.

Sorte, C. J., I. Ibáñez, D. M. Blumenthal, N. A. Molinari, L. P. Miller, et al. "Poised to Prosper? A Cross-System Comparison of Climate Change Effects on Native and Non-Native Species Performance." *Ecology Letters* 16 (2013): 261–70.

Soulé, M. E. "Thresholds for Survival: Maintaining Fitness and Evolutionary Potential." In Soulé and Wilcox, *Conservation Biology: An Evolutionary-Ecological Perspective*, 111–24.

Soulé, M. E., and B. A. Wilcox, eds. *Conservation Biology: An Evolutionary-Ecological Perspective*. Sunderland, MA: Sinauer, 1980.

Sousa, W. P. "The Role of Disturbance in Natural Communities." *Annual Review of Ecology and Systematics* 15 (1984): 353–91.

South, J., J. T. A. Dick, M. McCard, D. Barrios-O'Neill, and A. Anton. "Predicting Predatory Impact of Juvenile Invasive Lionfish (*Pterois volitans*) on a Crustacean Prey Using Functional Response Analysis: Effects of Temperature, Habitat Complexity and Light Regimes." *Environmental Biology of Fishes* 100 (2017): 1155–65.

Southward, A. J., R. S. Burton, S. L. Coles, P. R. Dando, R. DeFelice, et al. "Invasion of Hawaiian Shores by an Atlantic Barnacle." *Marine Ecology Progress Series* 165 (1998): 119–26.

Southwood, R., and J. R. Clarke. "Charles Sutherland Elton." *Biographical Memoirs of Fellows of the Royal Society of London* 45 (1999): 129–40.

Spear, R. J. *The Great Gypsy Moth War: A History of the First Campaign in Massachusetts to Eradicate the Gypsy Moth, 1890–1901*. Amherst: University of Massachusetts Press, 2005.

Spencer, C. N., B. R. McClelland, and J. A. Stanford. "Shrimp Stocking, Salmon Collapse, and Eagle Displacement." *BioScience* 41 (1991): 14–21.

Spencer, P. B., J. O. Hampton, C. Pacioni, M. S. Kennedy, K. Saalfeld, et al. "Genetic Relationships Within Social Groups Influence the Application of the Judas Technique: A Case Study with Wild Dromedary Camels." *Journal of Wildlife Management* 79 (2015): 102–11.

Spiro, J. P. *Defending the Master Race: Conservation, Eugenics, and the Legacy of Madison Grant*. Burlington: University of Vermont Press, 2009.

Spiro, J. P. "Puritan Racist: The Evolution of Madison Grant." PhD diss., University of California, Berkeley, 2000.

Spitzy, R. "The Prehistoric Man as a Possible Crayfish Transplantor." *Freshwater Crayfish* 4 (1978): 221–26.

Spotswood, E. N., J.-Y. Meyer, and J. W. Bartolome. "An Invasive Tree Alters the Structure of Seed Dispersal Networks Between Birds and Plants in French Polynesia." *Journal of Biogeography* 39 (2012): 2007–20.

Springer, K. "Eradication of Invasive Species on Macquarie Island to Restore the Natural Ecosystem." In *Recovering Australian Threatened Species: A Book of Hope*, edited by S. Garnett, P. Latch, D. Lindenmayer, and J. Woinarski, 13–22. Clayton, Australia: CSIRO Publishing, 2018.

Srivastava, D. S., and M. Vellend. "Biodiversity-Ecosystem Function Research: Is It Relevant to Conservation?" *Annual Review of Ecology, Evolution, and Systematics* 36 (2005): 267–94.

Stace, C. A., and M. J. Crawley. *Alien Plants*. London: HarperCollins, 2015.

Stachowicz, J. J., J. R. Terwin, R. B. Whitlatch, and R. W. Osman. "Linking Climate Change and Biological Invasions: Ocean Warming Facilitates Nonindigenous Species Invasions." *Proceedings of the National Academy of Sciences of the United States of America* 99 (2002): 15497–500.

Stadler, J., A. Trefflich, S. Klotz, and R. Brandl. "Exotic Plant Species Invade Diversity Hot Spots: The Alien Flora of Northwestern Kenya." *Ecography* 23 (2000): 169–76.

Standen, A. *Insect Invaders*. Boston: Houghton Mifflin, 1943.

Stanković, D., A. J. Crivelli, and A. Snoj. "Rainbow Trout in Europe: Introduction, Naturalization, and Impacts." *Reviews in Fisheries Science & Aquaculture* 23 (2015): 39–71.

Stanley, D. L. "Local Perception of Public Goods: Recent Assessments of Willingness-to-Pay for Endangered Species." *Contemporary Economic Policy* 23 (2005): 165–79.

Stansly, P. A., and M. E. Rogers. "Soft-Bodied Insects Attacking Foliage and Fruit." In *2016 Florida Citrus Pest Management Guide*. Gainesville: Institute of Food and Agriculture Sciences, University of Florida, 2016. https://nueces.agrilife.org/files/2020/05/Citrus-Management-soft-bodied-insects-and-pesticide-applications.pdf.

Stanton, D. W. G., J. A. Mulville, and M. W. Bruford. "Colonization of the Scottish Islands via Long-Distance Neolithic Transport of Red Deer (*Cervus elaphus*)." *Proceedings of the Royal Society B: Biological Sciences* 283 (2016): 20160095.

Stapley, J., A. W. Santure, and S. R. Dennis. "Transposable Elements as Agents of Rapid Adaptation May Explain the Genetic Paradox of Invasive Species." *Molecular Ecology* 24 (2015): 2241–52.

State of Victoria, Department of Primary Industries. *Invasive Plants and Animals Policy Framework*. Melbourne: Department of Primary Industries, 2010. https://

agriculture.vic.gov.au/__data/assets/pdf_file/0009/582255/Invasive-Plants-and-Animals-Framework-Sep-22.pdf.

Steadman, D. W. *Extinction and Biogeography of Tropical Pacific Birds*. Chicago: University of Chicago Press, 2006.

Stearns, S., J. Meyer, and J. Shykoff. "COGENE 88 at Basel." *Trends in Ecology and Evolution* 3 (1988): S2–S3.

Steinitz, C., P. Parker, and L. Jordan. "Hand-Drawn Overlays: Their History and Prospective Uses." *Landscape Architecture* 66 (1976): 444–55.

Stejneger, L. "The Land Reptiles of the Hawaiian Islands." *Proceedings of the United States National Museum* 21 (1899): 783–813.

Stephenson, W. "The Validity of the Community Concept in Marine Biology." *Proceedings of the Royal Society of Queensland* 84 (1973): 73–86.

Sternberg, H. O. "Man and Environmental Change in South America." In *Biogeography and Ecology in South America*, edited by E. J. Fittkau, J. Illies, H. Klinge, G. H. Schwabe, and H. Sioli, 413–45. The Hague: Dr. W. Junk, 1969.

Stevenson, I. "The Diffusion of Disaster: The Phylloxera Outbreak in the Département of the Hérault, 1862–1880." *Journal of Historical Geography* 6 (1980): 47–63.

Stewart, B. "The Invasion of the Soil Snatchers: Newcomers Push Out Native Plants." *New York Times*, April 24, 2001, B1.

Stewart, G. "Meta-Analysis in Applied Ecology." *Biology Letters* 6 (2010): 78–81.

Stewart, P. L. C. F., L. R. Brown, G. Richards, and A. Bernard. "An Alternative Baiting Method of Yellow Crazy Ants (*Anoplolepis gracilipes*) on Christmas Island, Indian Ocean." *International Journal of Ecology and Development* 23, no. 3 (2012): 40–59.

Stiling, P. "A Worm That Turned." *Natural History* 109, no. 5 (2000): 40–43.

Stöck, M., F. Reisch, M. Elmeros, D. Gabriel, W. Kloas, et al. "The Potential of VKORC1 Polymorphisms in Mustelidae for Evolving Anticoagulant Resistance Through Selection Along the Food Chain." *PLoS One* 14 (2019): e0221706.

Stockwell, D. R. B., and I. R. Noble. "Induction of Sets of Rules from Animal Distribution Data: A Robust and Informative Method of Analysis." *Mathematics and Computers in Simulation* 33 (1992): 385–90.

Stockwell, D. R. B., and D. P. Peters. "The GARP Modelling System: Problems and Solutions to Automated Spatial Prediction." *International Journal of Geographic Information Systems* 13 (1999): 143–58.

Stohlgren, T. J., D. Binkley, G. W. Chong, M. A. Kalkhan, L. D. Schell, et al. "Exotic Plant Species Invade Hot Spots of Native Plant Diversity." *Ecological Monographs* 69 (1999): 25–46.

Stohlgren, T. J., P. Ma, S. Kumar, M. Rocca, J. T. Morisette, et al. "Ensemble Habitat Mapping of Invasive Plant Species." *Risk Analysis* 30 (2010): 224–35.

Stokstad, E. "DNA Plucked from Air Identifies Nearby Animals." *Science* 373 (2021): 376.

Stolzenburg, W. *Rat Island: Predators in Paradise and the World's Greatest Wildlife Rescue*. New York: Bloomsbury, 2011.

Stone, C. P. "Toward Ethical Treatment of Animals in Hawai'i's Natural Areas." *Pacific Science* 49 (1995): 98–108.

Stone, L., D. Simberloff, and Y. Artzy-Randrup. "Network Motifs and Their Origins." *PLoS Computational Biology* 15, no. 4 (2019): e1006749.

Strauss, S. Y. "Ecological and Evolutionary Responses in Complex Communities: Implications for Invasions and Eco-Evolutionary Feedbacks." *Oikos* 123 (2014): 257–66.

Strauss, S. Y., and A. A. Agrawal. "The Ecology and Evolution of Plant Tolerance to Herbivory." *Trends in Ecology and Evolution* 14 (1999): 179–85.

Strauss, S. Y., J. A. Lau, and S. P. Carroll. "Evolutionary Responses of Natives to Introduced Species: What Do Introductions Tell Us About Natural Communities?" *Ecology Letters* 9 (2006): 354–71.

Strauss, S. Y., C. O. Webb, and N. Salamin. "Exotic Taxa Less Related to Native Species Are More Invasive." *Proceedings of the National Academy of Sciences of the United States of America* 103 (2006): 5841–45.

Stromberg, J. C., M. K. Chew, P. L. Nagler, and E. P. Glenn. "Changing Perceptions of Change: The Role of Scientists in *Tamarix* and River Management." *Restoration Ecology* 17 (2009): 177–86.

Stromberg, J. C., S. J. Lite, R. Marler, C. Paradzick, P. B. Shafroth, et al. "Altered Stream-Flow Regimes and Invasive Plant Species: The *Tamarix* Case." *Global Ecology and Biogeography* 16 (2007): 381–93.

Strong, D. R. "Density-Vague Population Change." *Trends in Ecology and Evolution* 1 (1986): 39–42.

Strong, D. R., and D. R. Ayres. "Ecological and Evolutionary Misadventures of *Spartina*." *Annual Review of Ecology, Evolution, and Systematics* 44 (2013): 389–410.

Strong, D. R., J. H. Lawton, and R. Southwood. *Insects on Plants.* Cambridge, MA: Harvard University Press, 1984.

Strong, J. S., and S. J. Leroux. "Impact of Non-Native Terrestrial Mammals on the Structure of the Terrestrial Mammal Food Web of Newfoundland, Canada." *PLoS One* 9, no. 8 (2014): e106264.

Stuart, A. I. "Insularity and Quaternary Vertebrate Faunas in Britain and Ireland." In *Island Britain: A Quaternary Perspective*, Special Publication 96, edited by R. C. Preece, 111–25. London: Geology Society of London, 1995.

Stuhrmann, C. "'It Felt More Like a Revolution': How Behavioral Ecology Succeeded Ethology, 1970–1990." *Berichte zur Wissenschaftsgeschichte* 45 (2022): 135–63.

Sturgeon, R. K. "*Achatina fulica* Infestation in North Miami, Florida." *Biologist* 53 (1971): 93–103.

Subalusky, A. L., E. P. Anderson, G. Jiménez, D. M. Post, D. E. Lopez, et al. "Potential Ecological and Socio-Economic Effects of a Novel Megaherbivore Introduction: The Hippopotamus in Colombia." *Oryx* 55 (2021): 105–13.

Subramaniam, B. "The Aliens Have Landed! Reflections on the Rhetoric of Biological Invasions." *Meridians: Feminism, Race, Transnationalism* 2, no. 1 (2001): 26–40.

Subramaniam, B. "The Aliens in Our Midst: Managing Our Ecosystems." In *Controversies in Science and Technology, from Sustainability to Surveillance*, vol. 4, edited by D. L. Kleinman, K. A. Cloud-Hansen, and J. Handelsman, 228–40. New York: Oxford University Press, 2014.

Subramaniam, B. *Botany of Empire.* Seattle: University of Washington Press, 2024.

Suckling, D. M., and R. F. H. Sforza. "What Magnitude Are Observed Non-Target Impacts from Weed Biocontrol?" *PLoS One* 9 (2014): e84847.

Suckling, D. M., L. D. Stringer, and J. M. Kean. "Trends in Arthropod Eradication Programmes from the Global Eradication Database, GERDA." In *Area-Wide Integrated Pest Management: Development and Field Application*, edited by J. Hendrichs, R. Pereira, and M. J. B. Vreysen, 505–18. Boca Raton, FL: CRC Press, 2020.

Sugiura, S. "Species Interactions—Area Relationships: Biological Invasions and Network Structure in Relation to Island Area." *Proceedings of the Royal Society B: Biological Sciences* 277 (2010): 1807–15.

Suhr, S., and M.-C. Senut. *Eradication of Invasive Quagga and Zebra Mussels Using Engineered Disseminated Neoplasia.* Lansing, MI: BiomiLab, 2019. https://

invasivemusselcollaborative.net/wp-content/uploads/2019/03/Suhr-Biomilab-Eradication-of-Invasive-Mussels.pdf.

Sukopp, H. "Neophyten in natürlichen Pflanzengesellschaften Mitteleuropas." *Berichte der Deutschen Botanischen Gesellschaft* 75 (1962): 193–205.

Sultan, S. E., and S. Matesanz. "An Ideal Weed: Plasticity and Invasiveness in *Polygonum cespitosum*." *Annals of the New York Academy of Sciences* 1360 (2015): 101–19.

Summerhayes, V. S., and C. S. Elton. "Bear Island." *Journal of Ecology* 11 (1923): 216–33.

Sutherland, S. "What Makes a Weed a Weed: Life History Traits of Native and Exotic Plants in the USA." *Oecologia* 141 (2004): 24–39.

Sutherland, W. J. "The Importance of Behavioural Studies in Conservation Biology." *Animal Behaviour* 56 (1998): 801–9.

Sutherst, R. W., and G. F. Maywald. "A Computerised System for Matching Climates in Ecology." *Agriculture, Ecosystems and Environment* 13 (1985): 281–300.

Sutherst, R. W., G. F. Maywald, and W. Bottomley. "From CLIMEX to PESKY, a Generic Expert System for Pest Risk Assessment." *EPPO Bulletin* 21 (1991): 595–608.

Sutherst, R. W., G. F. Maywald, and D. B. Skarratt. "Predicting Insect Distributions in a Changed Climate." In *Insects in a Changing Environment*, edited by R. Harrington and N. E. Stork, 59–82. London: Academic Press, 1995.

Swartzmann, G. L., and T. M. Zaret. "Modeling Fish Species Introduction and Prey Extermination: The Invasion of *Cichla ocellaris* to Gatun Lake, Panama." *Developments in Environmental Monitoring* 5 (1983): 361–73.

Sweetman, H. L. *The Biological Control of Insects*. Ithaca, NY: Comstock, 1936.

Sweetman, H. L. "Precipitation and Irrigation as Factors in the Distribution of the Mexican Bean Beetle *Epilachna corrupta* Muls." *Ecology* 10 (1929): 228–44.

Sweetman, H. L. *The Principles of Biological Control*. Dubuque, IA: Wm. C. Brown, 1958.

Swincer, D. E. "Physical Characteristics of Sites in Relation to Invasions." In Groves and Burdon, *Ecology of Biological Invasions*, 66–76.

Syrett, P., D. T. Briese, and J. H. Hoffmann. "Success in Biological Control of Terrestrial Weeds by Arthropods." In *Biological Control: Measures of Success*, edited by G. Gurr and S. Wratten, 189–230. Dordrecht, Neth.: Kluwer, 2000.

Takacs, D. *The Idea of Biodiversity*. Baltimore: Johns Hopkins University Press, 1996.

Talbott, H. "The Invasion of the Potomac." *Transactions of the American Fisheries Society* 39 (1910): 168–73.

Talekar, N. S., and A. M. Shelton. "Biology, Ecology, and Management of the Diamondback Moth." *Annual Review of Entomology* 38 (1993): 275–301.

Tansley, A. G. "The Use and Abuse of Vegetational Concepts and Terms." *Ecology* 16 (1935): 284–307.

Target Malaria. "Who We Are." Accessed February 15, 2025. https://targetmalaria.org/about-us/.

Tassin, J. *La grande invasion: Qui a peur des espèces invasives?* Paris: Odile Jacob, 2014.

Tassin, J., K. Thompson, S. P. Carroll, and C. D. Thomas. "Determining Whether the Impacts of Introduced Species Are Negative Cannot Be Based Solely on Science: A Response to Russell and Blackburn." *Trends in Ecology and Evolution* 32 (2017): 230–31.

Tatarinov, K. A. "Problems of Enrichment of the Carpathian Fauna." In Yanushevich, *Acclimatization of Animals in the U.S.S.R.*, 120–21.

Taylor, D., and L. Katahira. "Radio Telemetry as an Aid in Eradicating Remnant Feral Goats." *Wildlife Society Bulletin* 16 (1988): 297–99.

Taylor, K. D., M. Shorten, H. G. Lloyd, and F. A. Courtier. "Movements of the Grey Squirrel as Revealed by Trapping." *Journal of Applied Ecology* 8 (1971): 123–46.

Taylor, P. W. "The Ethics of Respect for Nature." *Environmental Ethics* 3 (1981): 197–218.

Taylor, P. W. *Respect for Nature*. Princeton, NJ: Princeton University Press, 1986.

Taylor, P. W. *Respect for Nature: A Theory of Environmental Ethics*. Princeton, NJ: Princeton University Press, 2011.

Taylor, R. S., and V. L. Friesen. "The Role of Allochrony in Speciation." *Molecular Ecology* 26 (2017): 3330–42.

Taylor, T. G. *The Control of Settlement by Humidity and Temperature (with Special Reference to Australia and the Empire): An Introduction to Comparative Climatology*. Bulletin No. 14. Melbourne: Commonwealth Bureau of Meteorology, 1916.

Teem, J. L., and J. B. Gutierrez. "A Theoretical Strategy for Eradication of Asian Carps Using a Trojan Y Chromosome to Shift the Sex Ratio of the Population." *American Fisheries Society Symposium* 74 (2010): 1–12.

Terblanche, J. S., A. A. Hoffmann, K. A. Mitchell, L. Rako, P. C. le Roux, and S. L. Chown. "Ecologically Relevant Measures of Tolerance to Potentially Lethal Temperatures." *Journal of Experimental Biology* 214 (2011): 3713–25.

Terborgh, J. "Faunal Equilibria and the Design of Wildlife Preserves." In *Tropical Ecological Systems: Trends in Terrestrial and Aquatic Research*, edited by F. Golley and E. Medina, 369–80. New York: Springer, 1975.

Terborgh, J. "Preservation of Natural Diversity: The Problem of Extinction-Prone Species." *BioScience* 24 (1974): 715–22.

Thakur, M. P., W. H. Van der Putten, M. M. P. Cobben, M. Van Kleunen, and S. Geisen. "Microbial Invasions in Terrestrial Ecosystems: From Processes to Impacts and Implications." *Nature Reviews Microbiology* 17 (2019): 621–31.

Thebaud, C., and D. Simberloff. "Are Plants Really Bigger in Their Introduced Ranges?" *American Naturalist* 157 (2001): 231–36.

Thellung, A. "Pflanzenwanderungen unter dem Einfluss des Menschen." *Botanische Jahrbücher für Systematik, Pflanzengeschichte und Pflanzengeographie* 53, no. 116 (1915): 37–66.

Theodoropoulos, D. I. *Invasion Biology: Critique of a Pseudoscience*. Blythe, CA: Avvar Books, 2003.

Theoharides, K. A., and J. S. Dukes. "Plant Invasion Across Space and Time: Factors Affecting Nonindigenous Species Success During Four Stages of Invasion." *New Phytologist* 176 (2007): 256–73.

Therriault, T. W., and J. Kolasa. "Explicit Links Among Physical Stress, Habitat Heterogeneity and Biodiversity." *Oikos* 89 (2000): 387–91.

Thiele, J., J. Kollmann, B. Markussen, and A. Otte. "Impact Assessment Revisited: Improving the Theoretical Basis for Management of Invasive Alien Species." *Biological Invasions* 12 (2010): 2025–35.

Thomas, C. D. "Climate, Climate Change and Range Boundaries." *Diversity and Distributions* 16 (2010): 488–95.

Thomas, C. D., D. Ng, M. C. Singer, L. B. Mallet, C. Parmesan, and H. L. Billington. "Incorporation of a European Weed into the Diet of a North American Herbivore." *Evolution* 41 (1987): 892–901.

Thomas, M. B., and A. M. Reid. "Are Exotic Natural Enemies an Effective Way of Controlling Invasive Plants?" *Trends in Ecology and Evolution* 22 (2007): 447–53.

Thomas, M. D. "A Radio Telemetry System for Animal Tracking in New Zealand." *New Zealand Journal of Science* 25 (1982): 245–52.

Thomas, W. L., Jr., ed. *Man's Role in Changing the Face of the Earth.* Chicago: University of Chicago Press, 1956.

Thomas, Y., S. Pouvreau, M. Alunno-Bruscia, L. Barillé, F. Gohin, et al. "Global Change and Climate-Driven Invasion of the Pacific Oyster (*Crassostrea gigas*) Along European Coasts: A Bioenergetics Modelling Approach." *Journal of Biogeography* 43 (2016): 568–79.

Thompson, H. V. "The Wild European Rabbit and Possible Dangers of Its Introduction into the U.S.A." *Journal of Wildlife Management* 19 (1955): 8–13.

Thompson, H. V., and C. M. King, eds. *The European Rabbit: The History and Biology of a Successful Colonizer*. Oxford: Oxford University Press, 1994.

Thompson, K. *Where Do Camels Belong? The Story and Science of Invasive Species*. London: Profile Books, 2014.

Thompson, K., J. G. Hodgson, and T. C. G. Rich. "Native and Alien Invasive Plants: More of the Same?" *Ecography* 18 (1995): 390–402.

Thompson, W. R. "Notes on the Pupation and Hibernation of Tachinid Parasites." *Journal of Economic Entomology* 3 (1910): 283–95.

Thomsen, M. S., and P. M. South. "Communities and Attachment Networks Associated with Primary, Secondary and Alternative Foundation Species: A Case Study of Stressed and Disturbed Stands of Southern Bull Kelp." *Diversity* 11 (2019): 56.

Thomson, G. M. *The Naturalisation of Animals and Plants in New Zealand*. Cambridge: Cambridge University Press, 1922.

Thuiller, W. "Patterns and Uncertainties of Species' Range Shifts Under Climate Change." *Global Change Biology* 10 (2004): 2020–27.

Thuiller, W., B. Lafourcade, R. Engler, and M. B. Araújo. "BIOMOD—A Platform for Ensemble Forecasting of Species Distributions." *Ecography* 32 (2009): 369–73.

Tiedje, J. M., R. K. Colwell, Y. L. Grossman, R. E. Hodson, R. E. Lenski, et al. "The Planned Introduction of Genetically Engineered Organisms: Ecological Considerations and Recommendations." *Ecology* 70 (1989): 298–315.

Tilman, D. "Resource Competition Between Plankton Algae: An Experimental and Theoretical Approach." *Ecology* 58 (1977): 338–48.

Tilman, D., F. Isbell, and J. M. Cowles. "Biodiversity and Ecosystem Functioning." *Annual Review of Ecology, Evolution, and Systematics* 45 (2014): 471–93.

Timmons, F. L. "A History of Weed Control in the United States and Canada." *Weed Science* 53 (2005): 748–61.

Tinghitella, R. M. "Rapid Evolutionary Change in a Sexual Signal: Genetic Control of the Mutation "Flatwing" That Renders Male Field Crickets (*Teleogryllus oceanicus*) Mute." *Heredity* 100 (2008): 261–67.

Tinghitella, R. M., J. W. Wang, and M. Zuk. "Preexisting Behavior Renders a Mutation Adaptive: Flexibility in Male Phonotaxis Behavior and the Loss of Singing Ability in the Field Cricket *Teleogryllus oceanicus*." *Behavioral Ecology* 20 (2009): 722–28.

Tjossem, S. "Victor Shelford." Oxford Bibliographies. March 31, 2016. https://doi.org/10.1093/obo/9780199830060-0145.

Todd, J. H., B. M. Pearce, and B. I. P. Barratt. "Using Qualitative Food Webs to Predict Species at Risk of Indirect Effects from a Proposed Biological Control Agent." *BioControl* 66 (2021): 45–58.

Todd, K. *Tinkering with Eden: A Natural History of Exotics in America*. New York: W. W. Norton, 2001.

Tomes, N. *The Gospel of Germs: Men, Women, and the Microbe in American Life*. Cambridge, MA: Harvard University Press, 1998.

Toochin, R., and M. Meredith. *Status and Occurrence of Eurasian Skylark (Alauda arvensis) in British Columbia*. April 15, 2018. https://ibis.geog.ubc.ca/biodiversity/efauna/documents/EUSK-article-RT-MM.pdf.

Torreya Guardians. "Assisted Migration." Accessed February 15, 2025. http://www.torreyaguardians.org.

"Total Insect War Urged." *Science News Letter* 47, no. 1 (1945): 5.

Townsend, C. R. "Individual, Population, Community, and Ecosystem Consequences of a Fish Invader in New Zealand Streams." *Conservation Biology* 17 (2003): 38–47.

Townsend, C. R. "Invasion Biology and Ecological Impacts of Brown Trout *Salmo trutta* in New Zealand." *Biological Conservation* 78 (1996): 13–22.

Townsend, M. "Is the Social Construction of Native Species a Threat to Biodiversity?" *ECOS* 26, no. 3/4 (2005): 1–9.

Tracy, C. R. "Biophysical Modelling in Reptilian Physiology and Ecology." In *Physiology C: Physiological Ecology*, edited by C. Gans and F. H. Pough, 275–321. London: Academic Press, 1982.

Traill, L. W., C. J. Bradshaw, and B. W. Brook. "Minimum Viable Population Size: A Meta-Analysis of 30 Years of Published Estimates." *Biological Conservation* 139 (2007): 159–66.

Transeau, E. N. "Vegetation Types and Insect Devastations: Distribution of the Mexican Bean Beetle and European Corn Borer in Ohio." *Ecology* 8 (1927): 285–88.

Tremblay, É., M. O. Duceppe, G. B. Thurston, M. C. Gagnon, M. J. Côté, and G. J. Bilodeau. "High-Resolution Biomonitoring of Plant Pathogens and Plant Species Using Metabarcoding of Pollen Pellet Contents Collected from a Honey Bee Hive." *Environmental DNA* 1 (2019): 155–75.

Tremblay, N., E. J. Guerra-Castro, F. Díaz, G. Rodríguez-Fuentes, N. Simões, et al. "Cold Temperature Tolerance of the Alien Indo-Pacific Damselfish *Neopomacentrus cyanomos* from the Southern Gulf of Mexico." *Journal of Experimental Marine Biology and Ecology* 524 (2020): 151308.

Trexler, J. C., W. F. Loftus, F. Jordan, J. J. Lorenz, J. H. Chick, and R. M. Kobza. "Empirical Assessment of Fish Introductions in a Subtropical Wetland: An Evaluation of Contrasting Views." *Biological Invasions* 2 (2000): 265–77.

Trigger, D. S., and L. Head. "Restored Nature, Familiar Culture: Contesting Visions for Preferred Environments in Australian Cities." *Nature and Culture* 5 (2010): 231–50.

Trinidad-Domínguez, C. D., C. Mota-Vargas, and O. Rojas-Soto. "Geographic and Ecological Invasion of the Cattle Egret (*Bubulcus i. ibis*) in the Americas and Its Current Seasonal Patterns." Research Square. March 15, 2022. https://doi.org/10.21203/rs.3.rs-1340663/v1.

Tromp, M., and J. V. Dudgeon. "Differentiating Dietary and Non-Dietary Microfossils Extracted from Human Dental Calculus: The Importance of Sweet Potato to Ancient Diet on Rapa Nui." *Journal of Archaeological Science* 54 (2015): 54–63.

Tsing, A. L. *Friction: An Ethnography of Global Connection*. Princeton, NJ: Princeton University Press, 2005.

Tsutsui, N. D., A. V. Suarez, D. A. Holway, and T. J. Case. "Reduced Genetic Variation and the Success of an Invasive Species." *Proceedings of the National Academy of Sciences of the United States of America* 97 (2000): 5948–53.

Tucker, K. C., and D. M. Richardson. "An Expert System for Screening Potentially Invasive Alien Plants in South African Fynbos." *Journal of Environmental Management* 44 (1995): 309–38.

Tulian, E. A. "Acclimatization of American Fishes in Argentina." *Bulletin of the Bureau of Fisheries* 28 (1908): 955–65.

Turbelin, A. J., B. D. Malamud, and R. A. Francis. "Mapping the Global State of Invasive Alien Species: Patterns of Invasion and Policy Responses." *Global Ecology and Biogeography* 26 (2017): 78–92.

Turner, B. L., II, W. C. Clark, R. W. Kates, J. F. Richards, J. T. Matthews, and W. B. Meyer, eds. *The Earth as Transformed by Human Action.* Cambridge: Cambridge University Press, 1990.

Twigg, L. E., G. R. Martin, and T. J. Lowe. "Evidence of Pesticide Resistance in Medium-Sized Mammalian Pests: A Case Study with 1080 Poison and Australian Rabbits." *Journal of Applied Ecology* 39 (2002): 549–60.

Twomey, A. C. "Climographic Studies of Certain Introduced and Migratory Birds." *Ecology* 17 (1936): 122–32.

Tyndall, J. "The Bakerian Lecture: On the Absorption and Radiation of Heat by Gases and Vapours, and on the Physical Connexion of Radiation, Absorption, and Conduction." *Philosophical Transactions of the Royal Society* 151 (1861): 1–36.

Tyrwhitt, J. "Surveys for Planning." In *Town and Country Planning Textbook*, edited by APRR, 146–96. London: Architectural Press, 1950.

Uekötter, F. *The Green and the Brown: A History of Conservation in Nazi Germany.* Cambridge: Cambridge University Press, 2006.

Unckless, R. L., A. G. Clark, and P. W. Messer. "Evolution of Resistance Against CRISPR/Cas9 Gene Drive." *Genetics* 205 (2017): 827–41.

UNESCO. "UNESCO Universal Declaration on Cultural Diversity." November 2, 2001. https://www.unesco.org/en/legal-affairs/unesco-universal-declaration-cultural-diversity.

United States Congress, Office of Technology Assessment. *Harmful Non-Indigenous Species in the United States.* Report no. OTA-F-565. Washington, DC: US Government Printing Office, 1993.

United States Department of Agriculture Forest Service. *Pest Risk Assessment of the Importation of Larch from Siberia and the Soviet Far East.* Miscellaneous Publication No. 1495. Washington, DC: USDA, 1991.

United States Department of Agriculture Forest Service. *Pest Risk Assessment of the Importation of Pinus radiata and Douglas-fir Logs from New Zealand.* Miscellaneous Publication No. 1508. Washington, DC: USDA, 1992.

United States Department of Agriculture Forest Service. *Pest Risk Assessment of the Importation of Pinus radiata, Nothofagus dombeyi, and Laurelia philippiana Logs from Chile.* Miscellaneous Publication No. 1517. Washington, DC: USDA, 1993.

United States Department of the Interior. "Department of the Interior Proposes Expanding Conservation Technique as Climate Change Threatens Greater Species Extinction." Press release, June 6, 2022. https://www.doi.gov/pressreleases/department-interior-proposes-expanding-conservation-technique-climate-change-threatens.

United States Fish and Wildlife Service. *Butterfly Peacock Bass (*Cichla ocellaris*) Ecological Risk Screening Summary.* Revised February 2017. https://www.fws.gov/sites/default/files/documents/Ecological-Risk-Screening-Summary-Butterfly-Peacock-Bass.pdf.

United States Fish and Wildlife Service. "Dr. Gardiner Bump Joins Research Staff of FWS." Press release PN 55680, April 29, 1949.

United States Fish and Wildlife Service. "Endangered and Threatened Wildlife and Plants; Designation of Experimental Populations." Proposed Rule, FR Doc. 2022-12061. *Federal Register.* June 6, 2022. https://www.federalregister.gov/documents/2022/06/07/2022-12061/

endangered-and-threatened-wildlife-and-plants-designation-of-experimental-populations.

United States General Accounting Office (GAO). *Invasive Species: Obstacles Hinder Federal Rapid Response to Growing Threat.* Report no. GAO-01-724. Washington, DC: GAO, 2001.

United States National Research Council. *Carbon Dioxide and Climate: A Scientific Assessment.* Washington, DC: National Academy of Sciences, 1979.

United States National Research Council. *Energy and Climate: Studies in Geophysics.* Washington, DC: National Academy of Sciences, 1977.

United States National Research Council. *Global Change: Ecosystems Research.* Washington, DC: National Academies Press, 2000.

Urban, M. C., G. Bocedi, A. P. Hendry, J. B. Mihoub, G. Pe'er, et al. "Improving the Forecast for Biodiversity Under Climate Change." *Science* 353, no. 6304 (2016): aad8466. https://doi.org/10.1126/science.aad8466.

Urban, M. C., B. L. Phillips, D. K. Skelly, and R. Shine. "The Cane Toad's (*Chaunus* [*Bufo*] *marinus*) Increasing Ability to Invade Australia Is Revealed by a Dynamically Updated Range Model." *Proceedings of the Royal Society B: Biological Sciences* 274 (2007): 1413–19.

Urich, F. W. "The Mongoose in Trinidad." *Tropical Agriculture* 8, no. 4 (1931): 95–97.

Urich, F. W. "The Mongoose in Trinidad and Methods of Destroying It." *Board of Agriculture of Trinidad and Tobago Circular* 12 (1914): 5–12.

US Army Corps of Engineers, Jacksonville District. "The Invasion Curve." Accessed December 7, 2024. https://www.saj.usace.army.mil/Media/Images/igphoto/2000809451/.

US Bureau of Reclamation. "Bureau of Reclamation Selects Three Submissions to Receive Prizes for the Eradication of Invasive Mussels in Open Water Prize Competition." News release, October 29, 2018. https://www.usbr.gov/newsroom/news release/detail.cfm?RecordID=63426 (content no longer available). Image of web page viewable via Internet Archive Wayback Machine, https://www.usbr.gov/newsroom/newsroomold/newsrelease/detail.cfm?RecordID=63426.

Usher, M. B., F. J. Kruger, I. A. W. Macdonald, L. L. Loope, and R. E. Brockie. "The Ecology of Biological Invasions into Nature Reserves: An Introduction." *Biological Conservation* 44 (1988): 1–8.

Valdovinos, F. S., E. L. Berlow, P. M. de Espanes, R. Ramos-Jiliberto, D. P. Vazquez, and N. D. Martinez. "Species Traits and Network Structure Predict the Success and Impacts of Pollinator Invasions." *Nature Communications* 9 (2018): 2153.

Valdovinos, F. S., R. Ramos-Jiliberto, J. D. Flores, C. Espinoza, and G. López. "Structure and Dynamics of Pollination Networks: The Role of Alien Plants." *Oikos* 118 (2009): 1190–200.

Valentin, R. E., D. M. Fonseca, A. L. Nielsen, T. C. Leskey, and J. L. Lockwood. "Early Detection of Invasive Exotic Insect Infestations Using eDNA from Crop Surfaces." *Frontiers in Ecology and the Environment* 16 (2018): 265–70.

Valéry, L., H. Fritz, and J.-C. Lefeuvre. "Another Call for the End of Invasion Biology." *Oikos* 122 (2013): 1143–46.

Valverde, M. P., D. M. T. Sharpe, M. E. Torchin, D. G. Buck, and L. J. Chapman. "Trophic Shifts in a Native Predator Following the Introduction of a Top Predator in a Tropical Lake." *Biological Invasions* 22 (2020): 643–61.

Vandepitte, K., T. De Meyer, K. Helsen, K. Van Acker, I. Roldán-Ruiz, et al. "Rapid Genetic Adaptation Precedes the Spread of an Exotic Plant Species." *Molecular Ecology* 23 (2014): 2157–64.

Vanderhoeven, S., C. S. Brown, C. K. Tepolt, N. D. Tsutsui, V. Vanparys, et al. "Linking Concepts in the Ecology and Evolution of Invasive Plants: Network Analysis Shows What Has Been Most Studied and Identifies Knowledge Gaps." *Evolutionary Applications* 3 (2010): 193–202.

Van der Putten, W. H., L. E. M. Vet, J. A. Harvey, and F. L. Wackers. "Linking Above- and Belowground Multitrophic Interactions of Plants, Herbivores, Pathogens and Their Antagonists." *Trends in Ecology and Evolution* 16 (2001): 547–54.

van der Weijden, W., R. Leewis, and P. Bol. *Biological Globalisation: Bio-Invasions and Their Impacts on Nature, the Economy and Public Health.* Utrecht, Neth.: KNNV Publishing, 2017.

Van Driesche, J., and R. Van Driesche. *Nature Out of Place: Biological Invasions in the Global Age.* Washington, DC: Island Press, 2000.

Van Driesche, R. D., P. D. Pratt, T. D. Center, M. B. Rayamajhi, P. Tipping, et al. "Cases of Biological Control Restoring Natural Systems." In R. D. Van Driesche et al., *Integrating Biological Control into Conservation Practice,* 208–46.

Van Driesche, R. D., D. Simberloff, B. Blossey, C. Causton, M. Hoddle, et al., eds. *Integrating Biological Control into Conservation Practice.* Chichester: Wiley-Blackwell, 2016.

Van Driesche, R. G., R. I. Carruthers, T. Center, M. S. Hoddle, J. Hough-Goldstein, L. Morin, et al. "Classical Biological Control for the Protection of Natural Ecosystems." *Biological Control Supplement* 1 (2010): S2–S33.

van Eeden, L. M., C. R. Dickman, E. G. Ritchie, and T. M. Newsome. "Shifting Public Values and What They Mean for Increasing Democracy in Wildlife Management Decisions." *Biodiversity and Conservation* 26 (2017): 2759–63.

Van Kleunen, M., W. Dawson, and P. Dostal. "Research on Invasive-Plant Traits Tells Us a Lot." *Trends in Ecology and Evolution* 26 (2011): 317.

Van Kleunen, M., W. Dawson, F. Essl, J. Pergl, M. Winter, et al. "Global Exchange and Accumulation of Non-Native Plants." *Nature* 525 (2015): 100–103.

Van Kleunen, M., P. Pyšek, W. Dawson, H. Kreft, J. Pergl, et al. "The Global Naturalized Alien Flora (GloNAF) Database." *Ecology* 100 (2019): e02542.

Van Kleunen, M., E. Weber, and M. Fischer. "A Meta-Analysis of Trait Differences Between Invasive and Non-Invasive Plant Species." *Ecology Letters* 13 (2010): 235–45.

Van Lenteren, J. C., D. Babendreier, F. Bigler, G. Burgio, H. M. T. Hokkanen, et al. "Environmental Risk Assessment of Exotic Natural Enemies Used in Inundative Biological Control." *BioControl* 48 (2003): 3–38.

Van Oosten, J. "The Dispersal of Smelt, *Osmerus mordax* (Mitchill), in the Great Lakes Region." *Transactions of the American Fisheries Society* 66 (1937):160–71.

van Wilgen, B. W. "A Brief, Selective History of Research Related to Alien Species and Invasion Science in South Africa." In van Wilgen et al., *Biological Invasions in South Africa,* 33–64.

van Wilgen, B. W., J. Measey, D. M. Richardson, J. R. Wilson, and T. A. Zengeya, eds. *Biological Invasions in South Africa.* Cham, Switz.: Springer, 2020.

Varner, G. *In Nature's Interests? Interests, Animal Rights and Environmental Ethics.* Oxford: Oxford University Press, 1998.

Vats, S. "Herbicides: History, Classification and Genetic Manipulation of Plants for Herbicide Resistance." In *Sustainable Agriculture Reviews,* vol. 15, edited by E. Lichtfouse, 153–92. Cham, Switz.: Springer, 2015.

Vaz, A. S., C. Kueffer, C. A. Kull, D. M. Richardson, S. Schindler, et al. "The Progress of Interdisciplinarity in Invasion Science." *Ambio* 46 (2017): 428–42.

Veitch, C. R., and M. N. Clout, eds. *Turning the Tide: The Eradication of Invasive Species.* Gland, Switz.: International Union for Conservation of Nature, 2002.

Veitch, C. R., M. N Clout, A. R. Martin, J. C. Russell, and C. J. West, eds. *Island Invasives: Scaling Up to Meet the Challenge.* Gland, Switz.: International Union for Conservation of Nature, 2019.

Veitch, C. R., M. N. Clout, and D. R. Towns, eds. *Island Invasives: Eradication and Management; Proceedings of the International Conference on Island Invasives.* Gland, Switz.: International Union for Conservation of Nature; Auckland: Centre for Biodiversity and Biosecurity, 2011.

Velema, G. J., J. S. Rosenfeld, and E. B. Taylor. "Effects of Invasive American Signal Crayfish (*Pacifastacus leniusculus*) on the Reproductive Behaviour of Threespine Stickleback (*Gasterosteus aculeatus*) Sympatric Species Pairs." *Canadian Journal of Zoology* 90 (2012): 1328–38.

Vellend, M. "Conceptual Synthesis in Community Ecology." *Quarterly Review of Biology* 85 (2010): 183–206.

Veloz, S. D., J. W. Williams, J. L. Blois, F. He, B. Otto-Bliesner, and Z. Liu. "No-Analog Climates and Shifting Realized Niches During the Late Quaternary: Implications for 21st-Century Predictions by Species Distribution Models." *Global Change Biology* 18 (2012): 1698–713.

Veltman, C. J., S. Nee, and M. J. Crawley. "Correlates of Introduction Success in Exotic New Zealand Birds." *American Naturalist* 147 (1996): 542–57.

Verberk, W. C., J. Overgaard, R. Ern, M. Bayley, T. Wang, et al. "Does Oxygen Limit Thermal Tolerance in Arthropods? A Critical Review of Current Evidence." *Comparative Biochemistry and Physiology Part A: Molecular and Integrative Physiology* 192 (2016): 64–78.

Verkaik, A. J. "The Muskrat in the Netherlands." *Proceedings of the Koninklijke Nederlandse Akademie van Wetenschappen, Series C: Biological and Medical Sciences* 90 (1987): 67–72.

Vermeij, G. J. "An Agenda for Invasion Biology." *Biological Conservation* 78 (1996): 3–9.

Vetter, D., G. Rücker, and I. Storch. "Meta-Analysis: A Need for Well-Defined Usage in Ecology and Conservation Biology." *Ecosphere* 4 (2013): 74.

Vidal, O., N. Sanz, R. M. Araguas, R. Fernández-Cebrian, D. Diez-del-Molino, and J. L. García-Marín. "SNP Diversity in Introduced Populations of the Invasive *Gambusia holbrooki.*" *Ecology of Freshwater Fish* 21 (2012): 100–108.

Vigne, J.-D. "The Large 'True' Mediterranean Islands as a Model for the Holocene Human Impact on the European Vertebrate Fauna? Recent Data and New Reflections." In *The Holocene History of the European Vertebrate Fauna*, edited by N. Benecke, 295–322. Rahden, Ger.: Marie Leidorf, 1999.

Vigne, J-D. "The Origins of Mammals on the Mediterranean Islands as an Indicator of Early Voyaging." *Eurasian Prehistory* 10, no. 1–2 (2014): 45–56.

Vigne, J.-D., F. Briois, A. Zazzo, G. Willcox, T. Cucchi, et al. "First Wave of Cultivators Spread to Cyprus at Least 10,600 y Ago." *Proceedings of the National Academy of Sciences of the United States of America* 109 (2012): 8445–49.

Vigne, J. D., A. Zazzo, J.-F. Saliege, F. Poplin, J. Guilaine, and A. Simmons. "Pre-Neolithic Wild Boar Management and Introduction to Cyprus more than 11,400 Years Ago." *Proceedings of the National Academy of Sciences of the United States of America* 106 (2009): 16135–38.

Vilà, M., C. Basnou, P. Pyšek, M. Josefsson, P. Genovesi, et al. "How Well Do We Understand the Impacts of Alien Species on Ecosystem Services? A Pan-European

Cross-Taxa Assessment." *Frontiers of Ecology and the Environment* 8 (2010): 135–44.

Vilà, M., J. L. Espinar, M. Hejda, P. E. Hulme, V. Jarošik, et al. "Ecological Impacts of Invasive Alien Plants: A Meta-Analysis of Their Effects on Species, Communities and Ecosystems." *Ecology Letters* 14 (2011): 702–8.

Vilà, M., and P. E. Hulme, eds. *Impact of Biological Invasions on Ecosystem Services.* Cham, Switz.: Springer, 2017.

Vilà, M., and P. E. Hulme. "Non-Native Species, Ecosystem Services, and Human Well-Being." In Vilà and Hulme, *Impact of Biological Invasions on Ecosystem Services,* 1–14.

Vilà, M., M. Williamson, and M. Lonsdale. "Competition Experiments on Alien Weeds with Crops: Lessons for Measuring Plant Invasion Impact?" *Biological Invasions* 6 (2004): 59–69.

Vimercati, G., S. Kumschick, A. F. Probert, L. Volery, and S. Bacher. "The Importance of Assessing Positive and Beneficial Impacts of Alien Species." *NeoBiota* 62 (2020): 525–45.

Vimercati, G., A. F. Probert, L. Volery, R. Bernardo-Madrid, S. Bertolino, et al. "The EICAT+ Framework Enables Classification of Positive Impacts of Alien Taxa on Native Biodiversity." *PLoS Biology* 20 (2022): e3001729.

Vinal, S. C. "The European Corn Borer, *Pyrausta nubilalis* Hübner, a Recently Established Pest in Massachusetts." *Massachusetts Agricultural Experiment Station Bulletin* 178 (1917): 147–52.

Vince, G. "Embracing Invasives." *Science* 331 (2011): 1383–84.

Violle, C., P. B. Reich, S. W. Pacala, B. J. Enquist, and J. Kattge. "The Emergence and Promise of Functional Biogeography." *Proceedings of the National Academy of Sciences of the United States of America* 111 (2014): 13690–96.

Vitousek, P. M. "Nutrient Cycling and Nutrient Use Efficiency." *American Naturalist* 119 (1982): 553–72.

Vitousek, P. M. "Biological Invasions and Ecosystem Properties: Can Species Make a Difference?" In Mooney and Drake, *Ecology of Biological Invasions of North America and Hawaii,* 163–76.

Vitousek, P. M. "Biological Invasions and Ecosystem Processes: Towards an Integration of Population Biology and Ecosystem Studies." *Oikos* 57 (1990): 7–13.

Vitousek, P. M. "The Regulation of Element Concentrations in Mountain Streams in the Northeastern United States." PhD diss., Dartmouth College, 1975.

Vitousek, P. M., C. D'Antonio, and G. P. Asner. "Invasions and Ecosystems: Vulnerabilities and the Contribution of New Technologies." In Richardson, *Fifty Years of Invasion Ecology,* 277–88.

Vitousek, P. M., C. M. D'Antonio, L. L. Loope, M. Rejmánek, and R. Westbrooks. "Introduced Species: A Significant Component of Human-Caused Global Change." *New Zealand Journal of Ecology* 21 (1997): 1–16.

Vitousek, P. M., L. L. Loope, and C. P. Stone. "Introduced Species in Hawaii: Biological Effects and Opportunities for Ecological Research." *Trends in Ecology and Evolution* 2 (1987): 224–27.

Vitousek, P. M., K. Van Cleve, N. Balakrishnan, and D. Mueller-Dombois. "Soil Development and Nitrogen Turnover in Montane Rainforest Soils in Hawai'i." *Biotropica* 15 (1983): 268–74.

Vitousek, P. M., and L. R. Walker. "Biological Invasion by *Myrica faya* in Hawai'i: Plant Demography, Nitrogen Fixation, Ecosystem Effects." *Ecological Monographs* 59 (1989): 247–65.

Vitule, J. R. S. "Introdução de peixes em ecossistemas continentais brasileiros: Revisão, comentários e sugestões de ações contra o inimigo quase invisível." *Neotropical Biology and Conservation* 4 (2009): 111–22.

Vitule, J. R. S., H. Bornatowski, C. A. Freire, and V. Abilhoa. "Extralimital Introductions of *Salminus brasiliensis* (Cuvier, 1816) (Teleostei, Characidae) for Sport Fishing Purposes: A Growing Challenge for the Conservation of Biodiversity in Neotropical Aquatic Ecosystems." *BioInvasions Records* 3 (2014): 291–96.

Vitule, J. R. S., C. A. Freire, D. P. Vazquez, M. A. Nuñez, and D. Simberloff. "Revisiting the Potential Conservation Value of Non-Native Species." *Conservation Biology* 26 (2012): 1153–55.

Vizentin-Bugoni, J., C. E. Tarwater, J. T. Foster, D. R. Drake, J. M. Gleditsch, et al. "Structure, Spatial Dynamics, and Stability of Novel Seed Dispersal Mutualistic Networks in Hawai'i." *Science* 364 (2019): 78–82.

Vizoso, A. D., M. R. Vizoso, and R. Hay. "Isolation of a Virus Resembling Encephalomyocarditis from a Red Squirrel." *Nature* 201 (1964): 849–50.

Volery, L., T. M. Blackburn, S. Bertolino, T. Evans, P. Genovesi, et al. "Improving the Environmental Impact Classification for Alien Taxa (EICAT): A Summary of Revisions to the Framework and Guidelines." *NeoBiota* 62 (2020): 547–67.

Volpe, E. P. *Understanding Evolution*. 4th ed. Dubuque, IA: Wm. C. Brown, 1981.

Vonesh, J., M. McCoy, R. Altwegg, P. Landi, and J. Measey. "Functional Responses Can't Unify Invasion Ecology." *Biological Invasions* 19 (2017): 1673–76.

von Humboldt, A. *Ideen zu einer Physiognomik der Gewächse*. Tübingen: Cotta, 1806.

von Humboldt, A. "Sur les lois que l'on observe dans la distribution des formes végétales: Extrait d'un mémoire lu à l'Institut, dans la séance du 5 février 1816." *Annales de chimie et de physique* 1 (1816): 225–39.

von Humboldt, A., and A. Bonpland. *Essai sur la géographie des plantes*. Paris: Levrault, Schoell, 1805.

von Mandelslo, J. A. *The Voyages and Travels of Ambassadors Sent by Frederick Duke of Holstein to the Grand Duke of Muscovy and the King of Persia Begun in 1633 and Finished in 1634 Wherein Are Added the Travels of J. A. de Mandelslo*. 2nd ed. London: John Starkey and Thomas Basset, 1669. Originally published in 1639 in German.

Voosen, P. "Myth-Busting Scientist Pushes Greens Past Reliance on 'Horror Stories.'" *Greenwire*, April 3, 2012. http://www.eenews.net/stories/1059962401.

Vos, P., R. Hogers, M. Bleeker, M. Reijans, T. V. D. Lee, et al. "AFLP: A New Technique for DNA Fingerprinting." *Nucleic Acids Research* 23 (1995): 4407–14.

Vredenburg, V. T. "Reversing Introduced Species Effects: Experimental Removal of Introduced Fish Leads to Rapid Recovery of a Declining Frog." *Proceedings of the National Academy of Sciences of the United States of America* 101 (2004): 7646–50.

Vucetich, J. A., and M. P. Nelson. "The Infirm Ethical Foundations of Conservation." In Bekoff, *Ignoring Nature No More*, 9–25.

Vucetich, J. A., and M. P. Nelson. "What Are 60 Warblers Worth? Killing in the Name of Conservation." *Oikos* 116 (2007): 1267–78.

Vuilleumier, F. "Dean of American Ornithologists: The Multiple Legacies of Frank M. Chapman of the American Museum of Natural History." *Auk* 122 (2005): 389–402.

Wahba, G., and J. Wendelberger. "Some New Mathematical Methods for Variational Objective Analysis Using Splines and Cross Validation." *Monthly Weather Review* 108 (1980): 1122–43.

Walker, S., J. Bastow Wilson, and W. G. Lee. "Does Fluctuating Resource Availability Increase Invasibility? Evidence from Field Experiments in New Zealand Short Tussock Grassland." *Biological Invasions* 7 (2005): 195–211.

Wallace, A. R. *Darwinism: An Exposition of the Theory of Natural Selection with Some of Its Applications*. London: Macmillan, 1890.

Wallace, A. R. *The Geographical Distribution of Animals*. New York: Harper, 1876.

Wallace, A. R. *Island Life: or, the Phenomena and Causes of Insular Faunas and Floras*. London: Macmillan, 1880.

Wallace, R. D., C. T. Bargeron, D. J. Moorhead, and J. H. LaForest. "IveGot1, Reporting and Tracking Invasive Species in Florida." *Southeastern Naturalist* 15 (2016): 51–62.

Wallach, A. D., M. Bekoff, C. Batavia, M. P. Nelson, and D. Ramp. "Summoning Compassion to Address the Challenges of Conservation." *Conservation Biology* 32 (2018): 1255–65.

Wallach, A. D., C. Batavia, M. Bekoff, S. Alexander, L. Baker, et al. "Recognizing Animal Personhood in Compassionate Conservation." *Conservation Biology* 34 (2020): 1097–1106.

Wallach, A. D., M. Bekoff, M. P. Nelson, and D. Ramp. "Promoting Predators and Compassionate Conservation." *Conservation Biology* 29 (2015): 1481–84.

Walters, M. L., R. Sequeira, R. Staten, O. El-Lissy, and N. Moses-Gonzales. "Eradication: Strategies and Tactics." In *Integrated Pest Management: Concepts, Tactics, Strategies and Case Studies*, edited by E. Radcliffe, W. Hutchison, and R. Cancelado, 298–308. Cambridge: Cambridge University Press, 2009.

Wang, Y., T. E. Cerling, and W. R. Effland. "Stable Isotope Ratios of Soil Carbonate and Soil Organic Matter as Indicators of Forest Invasion of Prairie Near Ames, Iowa." *Oecologia* 95 (1993): 365–69.

"War on Insects." *Time* 46, no. 9 (1945): 65.

Ward, D. "Modelling the Potential Geographic Distribution of Invasive Ant Species in New Zealand." *Biological Invasions* 9 (2007): 723–35.

Ward, S. M., R. D. Cousens, M. V. Bagavathiannan, J. N. Barney, H. J. Beckie, et al. "Agricultural Weed Research: A Critique and Two Proposals." *Weed Science* 62 (2014): 672–78.

Wardle, D. A. *Communities and Ecosystems: Linking the Aboveground and Belowground Components*. Princeton, NJ: Princeton University Press, New Jersey, 2002.

Wardle, D. A., R. D. Bardgett, J. N. Klironomos, H. Setälä, W. H. Van der Putten, and D. H. Wall. "Ecological Linkages Between Aboveground and Belowground Biota." *Science* 304 (2004): 1625–33.

Wardle, D. A., G. M. Barker, G. W. Yeates, K. I. Bonner, and A. Ghani. "Introduced Browsing Mammals in New Zealand Natural Forests: Aboveground and Belowground Consequences." *Ecological Monographs* 71 (2001): 587–614.

Wardle, D. A., H. A. Verhoef, and M. Clarholm. "Trophic Relationships in the Soil Microfood-Web: Predicting the Responses to a Changing Global Environment." *Global Change Biology* 4 (1998): 713–28.

Wares, J. P., A. R. Hughes, and R. K. Grosberg. "Mechanisms That Drive Evolutionary Change." In Sax et al., *Species Invasions: Insights into Ecology, Evolution, and Biogeography*, 229–57.

Warren, C. R. "Nativeness and Nationhood: What Species 'Belong' in Post-Devolution Scotland?" In Rotherham and Lambert, *Invasive and Introduced Plants and Animals*, 67–79.

Warren, C. R. "Perspectives on the 'Alien' Versus 'Native' Species Debate: A Critique of Concepts, Language and Practice." *Progress in Human Geography* 31 (2007): 427–46.

Warren, J. L. *Aldo Leopold's Odyssey*. 2nd ed. Washington, DC: Island Press, 2016.

Warren, M. S., J. K. Hill, J. A. Thomas, J. Asher, R. Fox, et al. "Rapid Responses of British Butterflies to Opposing Forces of Climate and Habitat Change." *Nature* 414 (2001): 65–69.

Warwick, T. "A Contribution to the Ecology of the Musk-Rat (*Ondatra zibethica*) in the British Isles." *Proceedings of the Zoological Society of London A* 110 (1940): 165–201.

Warwick, T. "The Distribution of the Muskrat (*Fiber zibethicus*) in the British Isles." *Journal of Animal Ecology* 3 (1934): 250–67.

Warwick, T. "Some Escapes of Coypus (*Myopotamus coypu*) from Nutria Farms in Great Britain." *Journal of Animal Ecology* 4 (1935): 146–47.

Wasserman, E. A., and T. R. Zentall. *Comparative Cognition: Experimental Explorations of Animal Intelligence*. New York: Oxford University Press, 2006.

Watanabe, S., R. E. Scheibling, and A. Metaxas. "Contrasting Patterns of Spread in Interacting Invasive Species: *Membranipora membranacea* and *Codium fragile* off Nova Scotia." *Biological Invasions* 12 (2010): 2329–42.

Watari, Y., K. Fukasawa, T. Hashimoto, T. Morosawa, H. Komine, et al. "Final Stage of the Mongoose-Eradication Project on Amami-Oshima Island, Japan." Paper presented at Island Biology conference, Angra do Heroismo, Azores, Portugal, July 18–22, 2016.

Watson, H. C. *Cybele Brittanica; British Plants and Their Geographical Relations*. Vol. 1. London: Longman, 1847.

Watson, H. C. "On the Altitude of the Habitats of Plants in Cumberland, with Localities of the Rarer Mountain Species." *Magazine of Natural History* 7 (1834): 20–24.

Watson, H. C. "Outlines of the Geographical Distribution of British Plants, Belonging to the Division Vascularès, or Cotylédones." Unpublished manuscript, quoted in an anonymous review in *Magazine of Natural History* 6 (1833): 265–67.

Watson, J., C. Warman, D. Todd, and V. Laboudallon. "The Seychelles Magpie Robin *Copsychus sechellarum*: Ecology and Conservation of an Endangered Species." *Biological Conservation* 61 (1992): 93–106.

Watson, W. "An Account of a Treatise, (Intitled, *D. Alberti Halleri* Archiatri Regii & Elect. Medicin. Anatomiae, Botan. Praelect. &c. Enumeratio Methodica *Stirpium Helvetiae* Indigenarum. Qua ominium brevis Descriptio & Synonymia, Compendium Virium Medicarum, dubiarum Declaratio, Novarum & Rariorum Uberior Historia & Icones Continentur. *Gottingiae, 1742*. in Folio)." *Philosophical Transactions of the Royal Society* 42 (1743): 369–80.

Weart, S. R. *The Discovery of Global Warming*. Cambridge, MA: Harvard University Press, 2008.

Weathersbee, A. A., R. C. Bullock, T. D. Panchal, and P. M. Dang. "Differentiation of *Diaprepes abbreviatus* and *Pachnaeus litus* (Coleoptera: Curculionidae) Egg Masses: PCR–Restriction Fragment-Length Polymorphism and Species-Specific PCR Amplification of 18S rDNA Products." *Annals of the Entomological Society of America* 96 (2003): 637–42.

Webber, B. L., and J. K. Scott. "Rapid Global Change: Implications for Defining Natives and Aliens." *Global Ecology and Biogeography* 21 (2012): 305–11.

Weber, D. "Pinta, Galápagos: Une île à sauver." *Biological Conservation* 4 (1971): 8–12.

Wee, [B.] P.-S. "Effects of Geographic Distance, Landscape Features and Host Association on Genetic Differentiation of Checkerspot Butterflies." PhD diss., University of Texas at Austin, 2004.

Weigl, G. L., R. G. Domey, and W. R. Courtenay Jr. "Survival and Range Expansion of the Curly-Tailed Lizard, *Leiocephalus carinatus armouri*, in Florida." *Copeia* 1969 (1969): 841–42.

Weiner, D. R. "Community Ecology in Stalin's Russia." *Isis* 75 (1984): 684–96.

Weiner, D. R. "The Historical Origins of Soviet Environmentalism." *Environmental Review* 6, no. 2 (1982): 42–62.

Weiner, D. R. *A Little Corner of Freedom: Russian Nature Protection from Stalin to Gorbachëv*. Berkeley: University of California Press, 1999.

Weiner, H. "Congress Threatens Wild Immigrants." *Earth Island Journal* 11, no. 4 (1996): 19.

Weiner, J. "On the Practice of Ecology." *Journal of Ecology* 83 (1995): 153–58.

Weinhold, B. "Do Noxious Neighbors Spread Disease?" *Environmental Health Perspectives* 118 (2010): A524.

Weis, J. S., and D. Sol, eds. *Biological Invasions and Animal Behaviour*. Cambridge: Cambridge University Press, 2016.

Welch, P. G. "Save the Parakeets." *New York Times*, April 26, 1973, 42.

Welcomme, R. L. *International Introductions of Inland Aquatic Species*. FAO Fisheries Technical Paper 294. Rome: Food and Agriculture Organization, 1988.

Welcomme, R. L. *Register of International Transfers of Inland Fish Species*. FAO Fisheries Technical Paper 213. Rome: Food and Agriculture Organization, 1981.

Weller, P., A. J. Jeffreys, V. Wilson, and A. Blanchetot. "Organization of the Human Myoglobin Gene." *EMBO Journal* 3 (1984): 439–46.

Wells, M. J., R. J. Poynton, A. A. Balsinhas, C. F. Musil, H. Joffe, et al. "The History of Introduction of Invasive Alien Plants to Southern Africa." In Macdonald, Kruger, and Ferrar, *Ecology and Management of Biological Invasions in Southern Africa*, 21–35.

Wells, P. K. "'An Enemy of the Rabbit': The Social Context of Acclimatisation of an Immigrant Killer." *Environment and History* 12 (2006): 297–324.

Weltzin, J. F., R. T. Belote, and N. J. Sanders. "Biological Invaders in a Greenhouse World: Will Elevated CO_2 Fuel Plant Invasions?" *Frontiers in Ecology and the Environment* 1 (2003): 146–53.

Weltzin, J. F., R. T. Belote, L. T. Williams, J. K. Keller, and E. C. Engel. "Authorship in Ecology: Attribution, Accountability, and Responsibility." *Frontiers in Ecology and the Environment* 4 (2006): 435–41.

Werenkraut, V., F. Baudino, and H. E. Roy. "Citizen Science Reveals the Distribution of the Invasive Harlequin Ladybird (*Harmonia axyridis* Pallas) in Argentina." *Biological Invasions* 22 (2020): 2915–21.

Werner, D. J. "*Nezara viridula* (Linnaeus, 1758) in Köln und in Deutschland (Heteroptera, Pentatomidae)." *Heteropteron* 21 (2005): 29–31.

West, G. B., J. H. Brown, and B. J. Enquist. "A General Model for the Origin of Allometric Scaling Laws in Biology." *Science* 276 (1997): 122–26.

West, K., C. Collins, O. Kardailsky, J. Kahn, T. L. Hunt, et al. "The Pacific Rat Race to Easter Island: Tracking the Prehistoric Dispersal of *Rattus exulans* Using Ancient Mitochondrial Genomes." *Frontiers in Ecology and Evolution* 5 (2017): 52.

Westbrooks, R. G. "New Approaches for Early Detection and Rapid Response to Invasive Plants in the United States." *Weed Technology* 18 (2004): 1468–71.

Westbrooks, R. G., and R. E. Eplee. "Early Detection and Rapid Response." In Simberloff and Rejmánek, *Encyclopedia of Biological Invasions*, 169–77.

Whalen, J. "In Britain, a Game of Cat and Mouse Breaks Out over a Parakeet." *Wall Street Journal*, November 23, 2013, A1, A14.

Whelan, W. J. "The Purpose of the Meeting." In *Recombinant DNA and Genetic Experimentation*, edited by J. Morgan and W. J. Whelan, xxi–xxiv. Oxford: Pergamon, 1979.

Whisson, D. A. "Pesticides (Mammal)." In Simberloff and Rejmánek, *Encyclopedia of Biological Invasions*, 535–38.

White, P. S., and S. T. A. Pickett. "Natural Disturbance and Patch Dynamics: An Introduction." In *Natural Disturbance and Patch Dynamics*, edited by S. T. A. Pickett and P. S. White, 3–13. Orlando: Academic Press, 1985.

White, R. P. *A Century of Service: A History of the Nursery Industry Associations of the United States*. Washington, DC: American Association of Nurserymen, 1975.

Whitfield, P. E., T. Gardner, S. P. Vives, M. R. Gilligan, W. R. Courtenay Jr., et al. "Biological Invasion of the Indo-Pacific Lionfish *Pterois volitans* Along the Atlantic Coast of North America." *Marine Ecology Progress Series* 235 (2002): 289–97.

Wikelski, M., and S. J. Cooke. "Conservation Physiology." *Trends in Ecology and Evolution* 21 (2006): 38–46.

Wilkinson, D. M. "The Disturbing History of Intermediate Disturbance." *Oikos* 84 (1999): 145–47.

Wilkinson, D. M. "Ecology Before Ecology: Biogeography and Ecology in Lyell's *Principles*." *Journal of Biogeography* 29 (2002): 1109–15.

Willdenow, K. L. *The Principles of Botany and of Vegetable Physiology*. Edinburgh: William Blackwood, 1805.

Williams, A. E., H. C. Duthie, and R. E. Hecky. "Water Hyacinth in Lake Victoria: Why Did It Vanish So Quickly and Will It Return?" *Aquatic Botany* 81 (2005): 300–314.

Williams, C. K., and R. J. Moore. "Phenotypic Adaptation and Natural Selection in the Wild Rabbit, *Oryctolagus cuniculus*, in Australia." *Journal of Animal Ecology* 58 (1989): 495–507.

Williams, C. L., S. L. Goldson, D. B. Baird, and D. W. Bullock. "Geographical Origin of an Introduced Insect Pest, *Listronotus bonariensis* (Kuschel), Determined by RAPD Analysis." *Heredity* 72 (1994): 412–19.

Williams, D. F., ed. *Exotic Ants*. Boulder, CO: Westview Press, 1994.

Williams, G. R. "The Dispersal from New Zealand and Australia of Some Introduced European Passerines." *Ibis* 95 (1953): 676–92.

Williams, I. C. "The New Chestnut Bark Disease." *Science* 34 (1911): 397–400.

Williams, J. G. K., A. R. Kubelik, K. J. Livak, J. A. Rafalski, and S. V. Tingey. "DNA Polymorphisms Amplified by Arbitrary Primers Are Useful as Genetic Markers." *Nucleic Acids Research* 18 (1990): 6531–35.

Williams, J. W., and S. T. Jackson. "Novel Climates, No-Analog Communities, and Ecological Surprises." *Frontiers in Ecology and the Environment* 5 (2007): 475–82.

Williams, K. E., K. P. Huyvaert, K. C. Vercauteren, A. J. Davis, and A. J. Piaggio. "Detection and Persistence of Environmental DNA from an Invasive, Terrestrial Mammal." *Ecology and Evolution* 8 (2018): 688–95.

Williams, T. "Killer Weeds." *Audubon* 99, no. 2 (1997): 24–31.

Williamson, M. H. *Biological Invasions*. London: Chapman and Hall, 1996.

Williamson, M. H. "Environmental Risks from the Release of Genetically Modified Organisms (GMOs)—The Need for Molecular Ecology." *Molecular Ecology* 1 (1992): 3–8.

Williamson, M. H. "Explaining and Predicting the Success of Invading Species at Different Stages of Invasion." *Biological Invasions* 8 (2006): 1561–68.

Williamson, M. H. "Invaders, Weeds and the Risk from Genetically Modified Organisms." *Experientia* 49 (1993): 219–24.

Williamson, M. H. "Measuring the Impact of Plant Invaders in Britain." In *Plant Invasions: Ecological Mechanisms and Human Responses*, edited by U. Starfinger, K. Edwards, I. Kowarik, and M. Williamson, 57–68. Leiden, Neth.: Backhuys, 1998.

Williamson, M. H. "Sir Joseph Hooker's Lecture on Insular Floras." *Biological Journal of the Linnean Society* 22 (1984): 55–77. Originally published by J. D. Hooker as "Insular Floras" in *Gardeners' Chronicle*, 1867. Page references are to the Williamson edition.

Williamson, M. H., and K. C. Brown. "The Analysis and Modelling of British Invasions." *Philosophical Transactions of the Royal Society of London B* 314 (1986): 505–22.

Williamson, M. H., and A. Fitter. "The Characters of Successful Invaders." *Biological Conservation* 78 (1996): 163–70.

Williamson, M. H., and A. Fitter. "The Varying Success of Invaders." *Ecology* 77 (1996): 1661–66.

Willis, C. G., B. R. Ruhfel, R. B. Primack, A. J. Miller-Rushing, J. B. Losos, and C. C. Davis. "Favorable Climate Change Response Explains Non-Native Species' Success in Thoreau's Woods." *PLoS One* 5, no. 1 (2010): e8878.

Willis, S. C., M. S. Nunes, C. G. Montaña, I. P. Farias, and N. R. Lovejoy. "Systematics, Biogeography, and Evolution of the Neotropical Peacock Basses *Cichla* (Perciformes: Cichlidae)." *Molecular Phylogenetics and Evolution* 44 (2007): 291–307.

Wilmé, L., P. O. Waeber, and J. U. Ganzhorn. "Human Translocation as an Alternative Hypothesis to Explain the Presence of Giant Tortoises on Remote Islands in the South-Western Indian Ocean." *Journal of Biogeography* 44 (2017): 1–7.

Wilmshurst, J. M., T. L. Hunt, C. P. Lipo, and A. J. Anderson. "High Precision Radiocarbon Dating Shows Recent and Rapid Initial Human Colonization of East Polynesia." *Proceedings of the National Academy of Sciences of the United States of America* 108 (2011): 1815–20.

Wilson, E. O., ed. *Biodiversity*. Washington, DC: National Academy Press, 1986.

Wilson, E. O. "The Fire Ant." *Scientific American* 198, no. 3 (1958): 36–41.

Wilson, E. O. "The Nature of the Taxon Cycle in the Melanesian Ant Fauna." *American Naturalist* 95 (1961): 169–93.

Wilson, E. O., and W. L. Brown. "Recent Changes in the Introduced Population of the Fire Ant *Solenopsis saevissima* (Fr. Smith)." *Evolution* 12 (1958): 211–18.

Wilson, E. O., and E. O. Willis. "Applied Biogeography." In Cody and Diamond, *Ecology and Evolution of Communities*, 523–34.

Wilson, J. R. U. "Definitions Can Confuse: Why the 'Neonative' Neologism Is Bad for Conservation." *BioScience* 70 (2020): 110–11.

Wilson, J. R. U., O. Ajuonu, T. D. Center, M. P. Hill, M. H. Julien, et al. "The Decline of Water Hyacinth on Lake Victoria Was Due to Biological Control by *Neochetina spp*." *Aquatic Botany* 87 (2007): 90–93.

Wilson, J. R. U., E. E. Dormontt, P. J. Prentis, A. J. Lowe, and D. M. Richardson. "Something in the Way You Move: Dispersal Pathways Affect Invasion Success." *Trends in Ecology and Evolution* 24 (2009): 136–44.

Wilson, J. R. U., P. García-Díaz, P. Cassey, D. M. Richardson, P. Pyšek, and T. M. Blackburn. "Biological Invasions and Natural Colonisations Are Different—The Need for Invasion Science." *NeoBiota* 31 (2016): 87–98.

Wilson, L. D., and L. Porras. *The Ecological Impact of Man on the South Florida Herpetofauna*. Special Publication No. 9. Lawrence: Museum of Natural History, University of Kansas, 1983.

Winberry, J. J., and D. M. Jones. "Rise and Decline of the 'Miracle Vine': Kudzu in the Southern Landscape." *Southeastern Geographer* 13 (1973): 61–70.

Winch, N. J. *An Essay on the Geographical Distribution of Plants, Through the Counties of Northumberland, Cumberland, and Durham*. Newcastle: E. Charney, 1819.

Windelband, W. "History and Natural Science," trans. G. Oakes. *History and Theory* 19 (1980): 169–85. First published in 1894 in German.

Wing, E. S., and S. R. Wing. "The Introduction of Animals as an Adaptation to Colonization of Islands: An Example from the West Indies." *Anthropozoologica* 1997, no. 25–26 (1997): 269–78.

Wing, M. G., and P. Bettinger. *Geographic Information Systems: Applications in Natural Resource Management.* 2nd ed. Don Mills, ON: Oxford University Press, 2008.

Winter, W. H., A. G. Cameron, R. Reid, T. G. Stockwell, and M. C. Page. "Improved Pasture Plants." In *Agro-Research for the Semi-arid Tropics: North-west Australia*, edited by R. C. Muchow, 227–69. Brisbane: University of Queensland Press, 1985.

Wisz, M. S., J. Pottier, W. D. Kissling, L. Pellissier, J. Lenoir, et al. "The Role of Biotic Interactions in Shaping Distributions and Realised Assemblages of Species: Implications for Species Distribution Modelling." *Biological Reviews* 88 (2013): 15–30.

Wittenberg, R., and M. J. W. Cock, eds. *Invasive Alien Species: A Toolkit of Best Prevention and Management Practices.* Wallingford, UK: CABI, 2001.

Wodzicki, K. "Ecology and Management of Introduced Ungulates in New Zealand." *Terre et vie* 1 (1961): 130–57.

Wodzicki, K. Foreword to C. King, *Handbook of New Zealand Mammals*, vii.

Wodzicki, K. "Introduced Mammals in New Zealand Forests." *Acta Biologica Cracoviensia Series Zoologia* 6 (1963): 111–34.

Wodzicki, K. *Introduced Mammals of New Zealand.* Wellington: Department of Scientific and Industrial Research, 1950.

Wodzicki, K. "The Status of Some Exotic Vertebrates in the Ecology of New Zealand." In Baker and Stebbins, *Genetics of Colonizing Species*, 425–58.

Woldendorp, G., and M. Bomford. *Weed Eradication Strategies, Timeframes and Costs.* Canberra: Bureau of Resource Sciences, 2004.

Wolf, M., and F. J. Weissing. "Animal Personalities: Consequences for Ecology and Evolution." *Trends in Ecology and Evolution* 27 (2012): 452–61.

Wolkovich, E. M., and E. E. Cleland. "The Phenology of Plant Invasions: A Community Ecology Perspective." *Frontiers in Ecology and the Environment* 9 (2011): 287–94.

Wolkovich, E. M., T. J. Davies, H. Schaefer, E. E. Cleland, B. I. Cook, et al. "Temperature-Dependent Shifts in Phenology Contribute to the Success of Exotic Species with Climate Change." *American Journal of Botany* 100 (2013): 1407–21.

Wolschke-Bulmahn, J. Introduction to Wolschke-Bulmahn, *Nature and Ideology*, 1–9.

Wolschke-Bulmahn, J. "The Nationalization of Nature and the Naturalization of the German Nation: 'Teutonic' Trends in Early Twentieth-Century Landscape Design." In Wolschke-Bulmahn, *Nature and Ideology*, 187–219.

Wolschke-Bulmahn, J. *Nature and Ideology: Natural Garden Design in the Twentieth Century.* Washington, DC: Dumbarton Oaks, 1997.

Wolschke-Bulmahn, J. " 'The Peculiar Garden'—The Advent and the Destruction of Modernism in German Garden Design." In *Proceedings of the Garden Conservancy Symposium*, vol. 3. edited by R. Karson, 17–30. Cold Spring Harbor, NY: Garden Conservancy, 1994.

Wolschke-Bulmahn, J. Review of *Jens Jensen: Maker of Natural Parks and Gardens*, by R. E. Grese. *Journal of Garden History* 15 (1995): 54–55.

Wolschke-Bulmahn, J. "The 'Wild Garden' and the 'Nature Garden'—Aspects of the Garden Ideology of William Robinson and Willy Lange." *Journal of Garden History* 12 (1992): 183–206.

Wolschke-Bulmahn, J., and G. Gröning. "The Ideology of the Nature Garden: Nationalistic Trends in Garden Design in Germany During the Early Twentieth Century." *Journal of Garden History* 12 (1992): 73–80.

Wolters, V., W. L. Silver, D. E. Bignell, D. C. Coleman, P. Lavelle, et al. "Effects of Global Changes on Above- and Belowground Biodiversity in Terrestrial Ecosystems: Implications for Ecosystem Functioning." *BioScience* 50 (2000): 1089–98.

Wonham, M. J., K. Goodell, B. Von Holle, and I. M. Parker. "Ecological and Geographic Patterns of Impact in Invaded Communities: A Meta-Analysis." *Bulletin of the Ecological Society of America* 81 (2000): 236.

Wood, W. *New England's Prospect.* Edited by A. T. Vaughan. Amherst: University of Massachusetts Press, 1977. Originally published in 1634.

Woodwell, G. M. "The Carbon Dioxide Question." *Scientific American* 238 (1978): 34–43.

Woodwell, G. M., and H. H. Smith, eds. *Diversity and Stability in Ecological Systems.* Brookhaven Symposia in Biology 22. Upton, NY: Brookhaven National Laboratory, 1969.

Woodwell, G. M., R. Whittaker, W. A. Reiners, G. E. Likens, C. C. Delwiche, and D. B. Botkin. "The Biota and the World Carbon Budget: The Terrestrial Biomass Appears to Be a Net Source of Carbon Dioxide for the Atmosphere." *Science* 199 (1978): 141–46.

Wordsworth, W. A. *A Guide Through the District of the Lakes.* Malvern, UK: Tantivy Press, 1948. Facsimile of the Definitive 5th Edition published in 1835.

"Workshop: Role Model Interventions—US." In *Best Management Practices for Preventing and Controlling Invasive Alien Species,* edited by G. Preston, G. Brown, and E. van Wyk, 271–72. Cape Town: Working for Water Programme, 2000.

WorldClim. "Maps, Graphs, Tables, and Data of the Global Climate." Accessed December 4, 2024. https://www.worldclim.org/.

Worner, S. P. "Ecoclimatic Assessment of Potential Establishment of Exotic Pests." *Journal of Economic Entomology* 81 (1988): 973–83.

Worster, D. A. *Nature's Economy: A History of Ecological Ideas.* Cambridge: Cambridge University Press, 1994.

Worthington, E. B. "The Ecology of Introductions." *Environmental Conservation* 1 (1974): 284.

Wright, M. G., M. P. Hoffmann, T. P. Kuhar, J. Gardner, and S. A. Pitcher. "Evaluating Risks of Biological Control Introductions: A Probabilistic Risk-Assessment Approach." *Biological Control* 35 (2005): 338–47.

Wright, T. F., J. R. Eberhard, E. A. Hobson, M. L. Avery, and M. A. Russello. "Behavioral Flexibility and Species Invasions: The Adaptive Flexibility Hypothesis." *Ethology, Ecology, and Evolution* 22 (2010): 393–404.

Wu, H., M. Ismail, and J. Ding. "Global Warming Increases the Interspecific Competitiveness of the Invasive Plant Alligator Weed, *Alternanthera philoxeroides.*" *Science of the Total Environment* 575 (2017): 1415–22.

Wulf, A. *The Brother Gardeners: Botany, Empire and the Birth of an Obsession.* New York: Knopf, 2009.

Wulff, E. V. *Введение в историческую географию растений* (Vvedenie v Istoricheskuiu Geofgrafiiu Rastenii). Moscow: Sel'khozgiz, 1932.

Wulff, E. V. *An Introduction to Historical Plant Geography.* Translated by E. Brissenden. Waltham, MA: Chronica Botanica, 1943.

Wynn, G. "Museums, Laboratories, Showcases: Prince Edward and Other Islands in Environmental History." In E. MacDonald et al., *Time and Place: An Environmental History of Prince Edward Island,* 36–58.

Xie, X., A. P. Michel, D. Schwarz, J. Rull, S. Velez, et al. "Radiation and Divergence in the *Rhagoletis pomonella* Species Complex: Inferences from DNA Sequence Data." *Journal of Evolutionary Biology* 21 (2008): 900–913.

Xu, T., and M. F. Hutchinson. *ANUCLIM Version 6.1 User Guide.* Canberra: Fenner School of the Environment, Australian National University, 2011.

Yackulic, C. B., L. L. Bailey, K. M. Dugger, R. J. Davis, A. B. Franklin, et al. "The Past and Future Roles of Competition and Habitat in the Rangewide Occupancy Dynamics of Northern Spotted Owls." *Ecological Applications* 29, no. 3 (2019): e01861.

Yactayo-Chang, J. P., H. V. Tang, J. Mendoza, S. A. Christensen, and A. K. Block. "Plant Defense Chemicals Against Insect Pests." *Agronomy* 10, no. 8 (2020): 1156.

Yadav, G., and S. Babu. "NEXCADE: Perturbation Analysis for Complex Networks." *PLoS One* 7 (2012): e41827.

Yan, H., L. Feng, Y. Zhao, L. Feng, D. Wu, and C. Zhu. "Prediction of the Spatial Distribution of *Alternanthera philoxeroides* in China Based on ArcGIS and MaxEnt." *Global Ecology and Conservation* 21 (2020): e00856.

Yang, L. H., J. L. Bastow, K. O. Spence, and A. N. Wright. "What Can We Learn from Resource Pulses?" *Ecology* 89 (2008): 621–34.

Yanushevich, A. I., ed. *Acclimatization of Animals in the U.S.S.R.* Jerusalem: Israel Program for Scientific Translations, 1966. Translation of proceedings of a Russian conference held and published in 1963.

Yeh, L.-L. "U.S. Can't Handle Today's Tide of Immigrants." *Christian Science Monitor*, March 23, 1995, 19.

Young, D. *Our Islands, Our Selves: A History of Conservation in New Zealand.* Dunedin: University of Otago Press, 2004.

Young, N., L. Carter, and P. Evangelista. *A MaxEnt Model v3. 3.3 e tutorial* (ArcGIS v10). Fort Collins: Natural Resource Ecology Laboratory, Colorado State University and the National Institute of Invasive Species Science, 2011.

Yu, H., F. H. Wan, and J. Y. Guo. "Different Thermal Tolerance and HSP Gene Expression in Invasive and Indigenous Sibling Species of *Bemisia tabaci*." *Biological Invasions* 14 (2012): 1587–95.

Zaitzevsky, C. *Frederick Law Olmsted and the Boston Park System.* Cambridge, MA: Harvard University Press, 1982.

Zaret, T. M. "The Ecology of Introductions—A Case-Study from a Central American Lake." *Environmental Conservation* 1 (1974): 308–9.

Zaret, T. M. "Life History and Growth Relationships of *Cichla ocellaris*, a Predatory South American Cichlid." *Biotropica* 12 (1980): 144–57.

Zaret, T. M. *Predation and Freshwater Communities.* New Haven, CT: Yale University Press, 1980.

Zaret, T. M. "Predation in Freshwater Fish Communities." In *Predator-Prey Systems in Fisheries Management*, edited by H. Clepper and R. H. Stroud, 135–50. Washington, DC: Sport Fishing Institute, 1979.

Zaret, T. M. "Predator-Prey Interaction in a Tropical Lacustrine Ecosystem." *Ecology* 53 (1972): 248–57.

Zaret, T. M. "The Stability/Diversity Controversy: A Test of Hypotheses." *Ecology* 63 (1982): 721–31.

Zaret, T. M. "Studies on Predator-Prey Interaction in a Tropical Lacustrine Ecosystem." Doctoral diss., Yale University, 1971.

Zaret, T. M., and R. T. Paine. "Species Introduction in a Tropical Lake." *Science* 182 (1973): 449–55.

Zavaleta, E. S. "It's Often Better to Eradicate, but Can We Eradicate Better?" In Veitch and Clout, *Turning the Tide: The Eradication of Invasive Species*, 393–403.

Zavaleta, E. S., R. J. Hobbs, and H. A. Mooney. "Viewing Invasive Species Removal in a Whole-Ecosystem Context." *Trends in Ecology and Evolution* 16 (2001): 454–59.

Zenni, R. D., F. Essl, E. García-Berthou, and S. M. McDermott, eds. "The Economic Costs of Biological Invasions Around the World." Special issue, *NeoBiota*, no. 67 (2021): 1–9.

Zenni, R. D., and M. A. Nuñez. "The Elephant in the Room: The Role of Failed Invasions in Understanding Invasion Biology." *Oikos* 122 (2013): 801–15.

Zenni, R. D., S. R. Ziller, A. Pauchard, M. Rodriguez-Cabal, and M. A. Nuñez. "Invasion Science in the Developing World: A Response to Ricciardi et al." *Trends in Ecology & Evolution* 32 (2017): 807–8.

Zettlemoyer, M. A., S. L. Ellis, C. W. Hale, E. C. Horne, R. D. Thoen, and M. L. DeMarche. "Limited Evidence for Phenological Differences Between Non-Native and Native Species." *Frontiers in Ecology and the Environment* 10 (2022): 983172.

Zhang, D. X., and D. L. Nuss. "Engineering Super Mycovirus Donor Strains of Chestnut Blight Fungus by Systematic Disruption of Multilocus *vic* Genes." *Proceedings of the National Academy of Sciences of the United States of America* 113 (2016): 2062–67.

Zhang, S. "The EPA Quietly Approved Monsanto's New Genetic-Engineering Technology." *Atlantic*, June 23, 2017. https://www.theatlantic.com/science/archive/2017/06/monsanto-rna-interference/531288/.

Zhang, Z., S. Mammola, C. L. McLay, C. Capinha, and M. Yokota. "To Invade or Not to Invade? Exploring the Niche-Based Processes Underlying the Failure of a Biological Invasion Using the Invasive Chinese Mitten Crab." *Science of the Total Environment* 728 (2020): 138815.

Zhitkov, B. M. *Akklimatizatsiya zhivotnykh* [Acclimatization of animals]. Moscow: Biomedgiz, 1934.

Zierer, C. M. "Brisbane—River Metropolis of Queensland." *Economic Geography* 17 (1941): 325–44.

Zimdahl, R. L. *A History of Weed Science in the United States*. London: Elsevier, 2010.

Zimmerman, E. C. *Insects of Hawaii*. Vol. 8, *Lepidoptera: Pyraloidea*. Honolulu: University of Hawaii Press, 1958.

Zimmerman, E. C. "Possible Evidence of Rapid Evolution in Hawaiian Moths." *Evolution* 14 (1960): 137–38.

Zimmermann, H. G., V. C. Moran, and J. J. Hoffmann. "The Renowned Cactus Moth, *Cactoblastis cactorum*: Its Natural History and Threat to Native *Opuntia* Floras in Mexico and the United States of America." *Diversity and Distributions* 6 (2000): 259–69.

Zimmermann, H. G., V. C. Moran, and J. H. Hoffmann. "The Renowned Cactus Moth, *Cactoblastis cactorum* (Lepidoptera: Pyralidae): Its Natural History and Threat to Native *Opuntia* Floras in Mexico and the United States of America." *Florida Entomologist* 84 (2001): 543–51.

Zinn, K. E., M. Tunc-Ozdemir, and J. F. Harper. "Temperature Stress and Plant Sexual Reproduction: Uncovering the Weakest Links." *Journal of Experimental Botany* 61 (2010): 1959–68.

Zotti, M., E. A. dos Santos, D. Cagliari, O. Christiaens, C. N. T. Taning, and G. Smagghe. "RNAi Technology in Crop Protection Against Arthropod Pests, Pathogens and Nematodes." *Pest Management Science* 74 (2018): 1239–50.

Zuk, M., J. T. Rotenberry, and L. W. Simmons. "Calling Songs of Field Crickets (*Teleogryllus oceanicus*) With and Without Phonotactic Parasitoid Infection." *Evolution* 52 (1998): 166–71.

Zuk, M., J. T. Rotenberry, and R. M. Tinghitella. "Silent Night: Adaptive Disappearance of a Sexual Signal in a Parasitized Population of Field Crickets." *Biology Letters* 2 (2006): 521–24.

Zuk, M., L. W. Simmons, and L. Cupp. "Calling Characteristics of Parasitized and Unparasitized Populations of the Field Cricket *Teleogryllus oceanicus*." *Behavioral Ecology and Sociobiology* 33 (1993): 339–43.

Zwölfer, H., and P. Harris. "Biology and Host Specificity of *Rhinocyllus conicus* (Froel.) (Col., Curculionidae), a Successful Agent for Biocontrol of the Thistle, *Carduus nutans*." *Zeitschrift für angewandte Entomologie* 97 (1984): 36–62.

Species Index

Subject Index